ANATOMY

ANATOMY

Joel Zimmerman, Ph.D.

Adjunct Assistant Professor of Dermatology,
New York University School of Medicine, New York;
Formerly Assistant Professor, Department of Anatomy,
University of Alabama at Birmingham, School of Medicine, Birmingham

Stanley Jacobson, Ph.D.

Professor, Department of Anatomy and Cell Biology,
Tufts University School of Medicine, Boston

Illustrated primarily by

Ronald A. Brandon

LITTLE, BROWN AND COMPANY Boston/Toronto

Figures 2-5B, 2-14, 2-21, 2-22, 3-14B, 3-22B, 3-28B, 3-29, 4-2, and 7-35 are taken from the Diagnostic Radiological Health Sciences Learning Laboratory, as developed by the Radiological Health Sciences Education Project, University of California, San Francisco, under contract with the Bureau of Radiological Health, Food and Drug Administration, and in cooperation with the American College of Radiology. (Additional information is available from the American College of Radiology, 560 Lennon Lane, Walnut Creek, CA 94598.) With permission.

First Edition

Library of Congress Catalog Card No. 88-83927

ISBN 0-316-98838-3

Printed in the United States of America

MV-NY

Contents

Preface

When we first began writing *Anatomy*, our idea was to create a book for students of medicine and related fields that lay somewhere between a traditional exhaustive text and a traditional review book, incorporating the best features of both. Since that initial work, *Anatomy* has developed into the inaugural volume of the Essentials of Basic Science Series, but its basic thrust and philosophy have remained intact. *Anatomy* focuses on basic facts without undue elaboration, makes extensive use of tables, and is heavily illustrated with original anatomical drawings, a point we feel is worth emphasizing given the visual nature of the subject matter. In addition, clear chapter objectives, National Board type questions, and annotated answers round out the material. We think that this combination is appropriate for medical students in basic science courses as well as those with an interest in anatomy looking either to develop a foundation in the subject or to review and refresh already existing knowledge.

We are grateful to all of our friends and colleagues who have encouraged and provided support for us during the long development and execution time for this book. We would also like to thank the people of Little, Brown (both past and present) whose patience and prodding helped to bring the book to fruition. Finally, we owe a great deal of thanks to Ron Brandon, our illustrator, who was able to translate our ideas into the reality of his excellent illustrations.

J. Z.
S. J.

ANATOMY

1 Introduction

Objectives

After reading this chapter, you should know the following:

Terms for orientations, planes, and movements commonly used in anatomical discussions

Composition and development of cancellous and compact bone

Composition and function of hyaline and elastic cartilage and white fibrocartilage

Structure and movement capabilities of the different types of joints

Organization, components, and functions of the basic circulatory system and lymphatic system

Composition and functions of the skin and its glands

Basic organization, functional units, and structure of the peripheral and central nervous systems

Position, Organization, and Movement

The anatomical position is erect and forward facing, with the hands by the side and the palms forward, and with the heels and toes together and the feet turned toward the front.

The **median plane** bisects the body into left and right halves. A **sagittal plane** is any plane parallel to the median (Fig. 1-1). A **coronal (frontal) plane** is also a vertical plane, but it is at right angles to the median plane. A **transverse plane** is a horizontal plane and thus perpendicular to the other two.

The following terms are commonly used in anatomical discussions:

Term	*Definition*
Ventral, anterior	Front
Dorsal, posterior	Back
Cranial	Head
Craniad	Headward
Caudal	Tail
Caudad	Tailward
Medial	Toward the median plane
Lateral	Away from the median plane
Proximal	Toward the center
Distal	Away from the center
Superficial	Toward the surface
Deep	Farther in

Many of these are terms of comparison; thus, the ribs are superficial to the heart, or the elbow is distal to the shoulder.

The body may be considered to be organized on five levels. The first level is the cellular level, cells being the smallest anatomical unit. On the second level, similar cells are organized into tissues; on the third are organs composed of several types of cells or tissues. On the fourth level, several organs are linked together to form the skeletal,

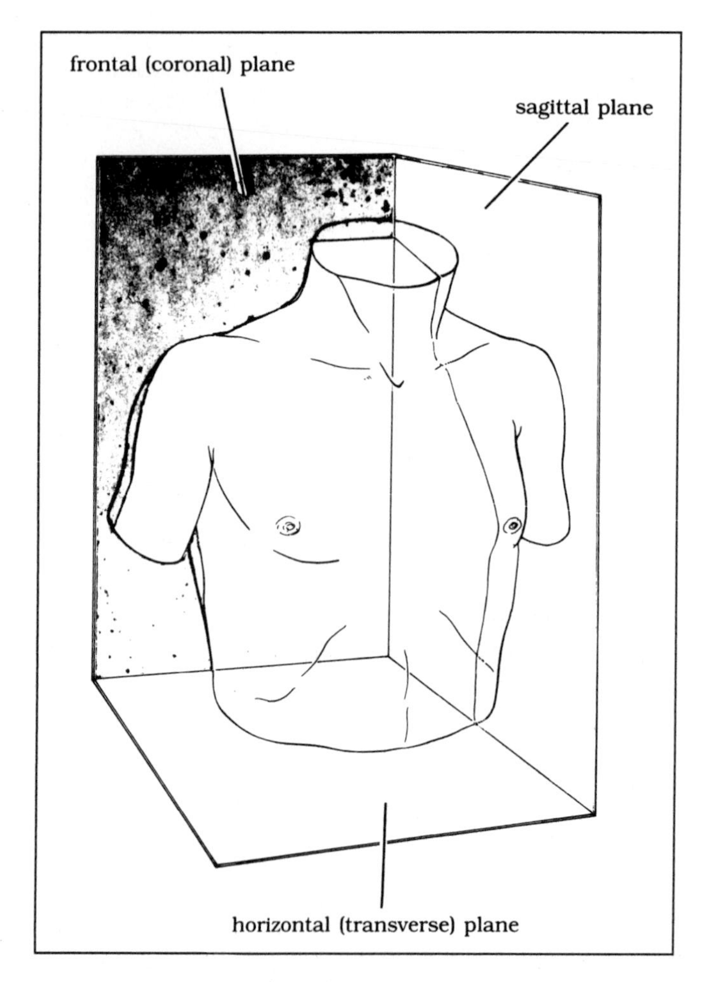

Fig. 1-1. Torso with planes of section.

articular, muscular, nervous, circulatory, lymphatic, skin, and visceral systems. Finally, regions consist of parts of several systems. The body can be divided into the head and neck, upper limb, thorax, abdomen, back, pelvis, perineum, and lower limb regions.

Movement can be described in terms of the motion of muscles across a joint. **Flexion** (Fig. 1-2) decreases the angle between bones, and **extension** increases the angle, except at the foot, where bending of the ankle produces either dorsiflexion (toward the foot's upper surface) or plantar flexion (toward the sole), and at the shoulder, where forward movement of the arm is flexion and backward movement is extension. **Abduction** is movement away from the midline, whereas **adduction** is movement toward the midline.

Circumduction is a combination of the four movements flexion, extension, adduction, and abduction. **Rotation** is the turning of a bone around its long axis. In medial rotation the anterior surface moves toward the midline. In lateral rotation the anterior surface moves toward the side.

In addition there are two movements peculiar to the forearm and two movements peculiar to

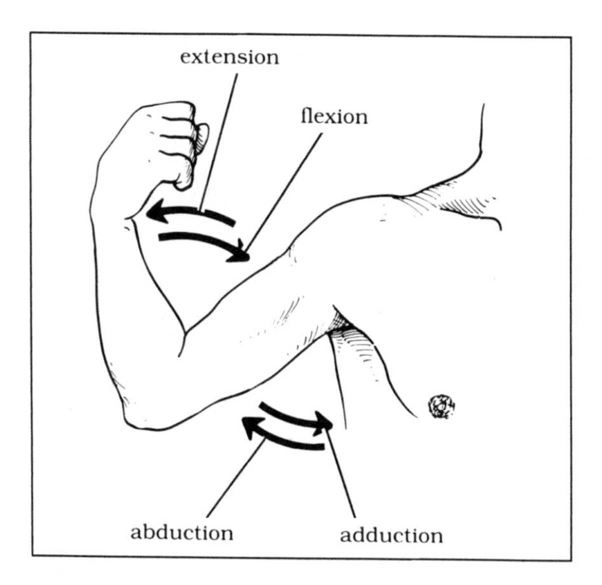

Fig. 1-2. Arm indicating movements of flexion, extension, adduction, and abduction.

the foot. **Pronation** is movement of the forearm so that the palm faces backward; **supination** is movement so that the palm faces forward. **Inversion** moves the sole of the foot to the median plane, whereas **eversion** moves the sole away from the median plane.

Bone

Bone, composed of a mixture of organic fibrous tissue and inorganic salt (calcium phosphate), serves many functions in the body. It provides a supporting framework for the other tissues of the body and, forming the skull and rib cage, serves to protect important organs from trauma. Its locomotive functions include serving as the attachment site for muscles and as a lever at a joint. Finally, its marrow is involved in the formation of blood cells, and it serves as a storehouse for calcium and phosphate.

Bones are classified by their shape: long, short, flat, irregular, and sesamoid. The length of long bones is greater than their diameter. A typical long bone contains a shaft (diaphysis) and articular ends (epiphyses). These ends, having a covering of articular cartilage, are usually called facets, heads, or condyles. Short bones are cuboidal in shape. Flat bones are flat plates separated by a thin narrow space. Irregular bones, such as the vertebrae, have a mixed shape. Sesamoid bones are found in some tendons at joints; they protect the tendon from wear and alter the angle at which the tendon approaches its insertion.

At the point where tendon attaches to bone there is usually a raised elevation known as a line, ridge, crest, tubercle, tuberosity, spine, or trochanter. A depression in bone is known as a pit, fovea, or fossa, in order of increasing size. A line may be called a groove or sulcus. An opening, or hole, is known as a foramen. If it has depth, it is called a canal.

There are two histological types of bone: compact and cancellous (Fig. 1-3). **Compact bone** is

Fig. 1-3. Longitudinal section of a long bone.

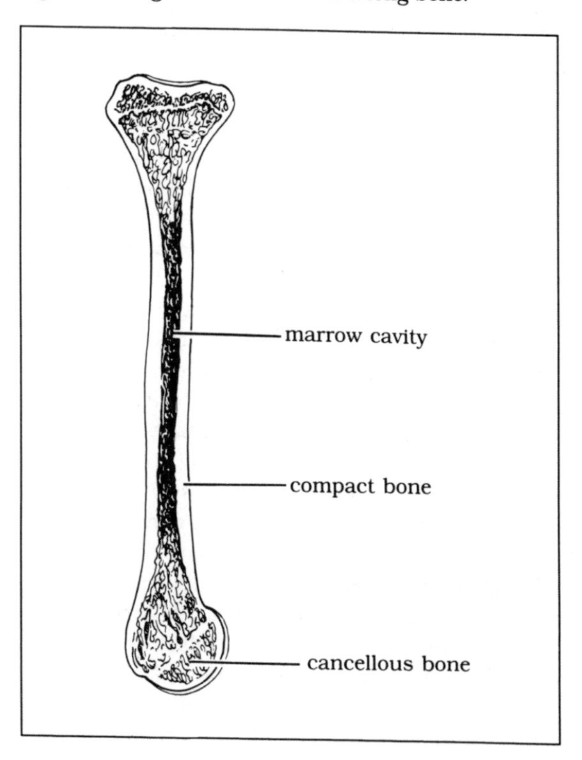

dense and small spaces can be found within it. It is laid down in lamellae and is found on the surface of bone. **Cancellous bone** is trabecular or spongy. It is found in the interior of bone and in the articular ends. Its meshwork of easily visible, interconnected spaces contains marrow and communicates with the central marrow cavity of the bone. In flat bones the trabeculae are known as diploë.

Bones are formed by two processes. In the first, cartilaginous models are replaced by bone. This process is known as endochondral bone development and results in cancellous bone. In the second, intramembranous development, the deepest layer of the periosteum (connective tissue covering) lays down layers of bone, resulting in compact bone. Both processes are accompanied by bone destruction, resulting in the formation of a marrow cavity and an increase in length. Length growth is also achieved by growth of the epiphyseal cartilage and its replacement by bone through endochondral development. Most bones usually have more than one center for ossification (e.g., one in the diaphysis and one at each epiphysis). The fusion of these centers is called synostosis.

Bone is intimately associated with bone marrow and cartilage. At birth red bone marrow generates all of the red blood cells in the body, as well as some of the white cells. By puberty this marrow is found only in the cancellous trabeculae. Gradually it is replaced, except in the sternum, by yellow bone marrow, which consists of fat.

Cartilage has a solid resilient matrix lacking the rigidity of bone. There are three types of cartilage: hyaline cartilage, white fibrocartilage, and elastic cartilage. **Hyaline cartilage** is bluish white in appearance. In developing bone it serves as the cartilage model. In the adult it persists as articular cartilage, the respiratory cartilages, and costal cartilage. The latter two may ossify in old age. **White fibrocartilage** is a mixture of white fibrous connective tissue and cartilage matrix. It is both tougher and more flexible than hyaline cartilage and is found in intervertebral and articular disks. **Elastic cartilage** is a mixture of matrix and elastic fibers. It is resilient and is found only in the external ear, auditory tube, epiglottis, and some laryngeal structures.

Cartilage receives no vascular supply or innervation, but bone receives both. Bones are supplied by both articular and nutrient arteries, each of which gives rise to many branches in the bone marrow. Because the two ends of a bone do not stop growing at the same time, the nutrient artery tends to enter the bone close to the end that ceases growth earlier. Venous drainage of bone follows the arterial pathways. Nerves supply the periosteum of the bone, and vasomotor branches accompany the arteries into the bone.

Joints

A joint is a region of contact between two or more bones and may be movable **(synovial or fibrous)**, immovable **(cartilaginous or fibrous)**, or in between. **Fibrous joints** (Fig. 1-4 A,B) are either sutures or syndesmoses. A suture joins the flat bones of the skull. The fibrous tissue separating the bones is continuous with the periosteum of the bones, and there is little or no movement across the joint. Fusion of the bones across the suture line, synostosis, occurs late in life. In a syndesmosis the opposed bones are held together by intervening fibrous tissue in the form of an interosseous membrane or ligament. The

Fig. 1-4. Fibrous joints. A. A sutural joint. B. A syndesmosis.

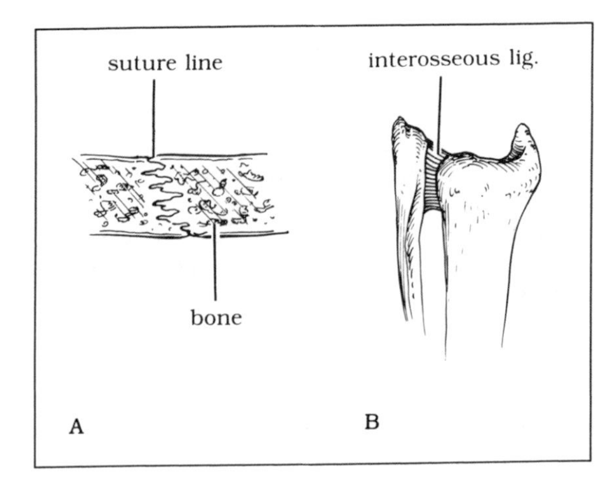

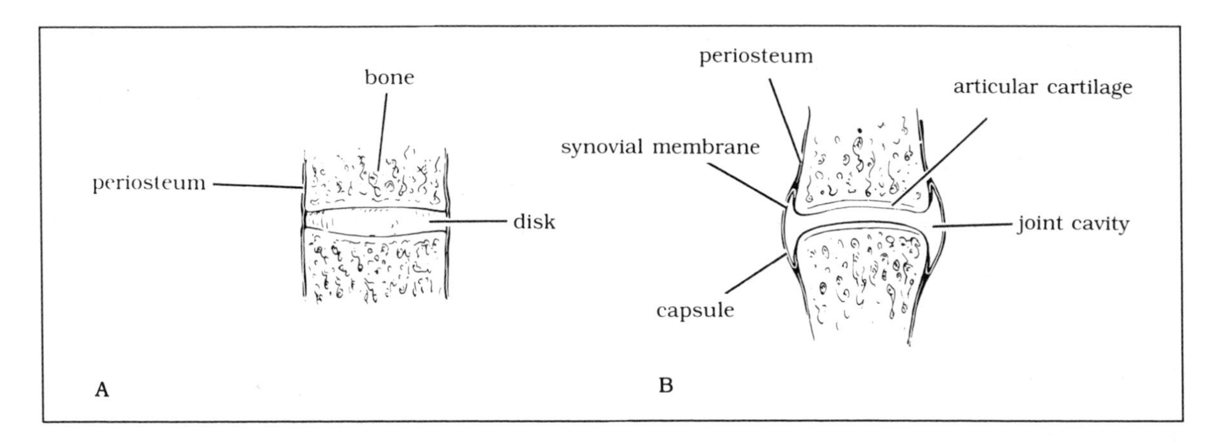

Fig. 1-5. Cartilaginous joints. A. A symphysis. B. A synovial joint.

distance between the bones and the flexibility of the fibrous tissue determine the degree of movement.

Cartilaginous joints (Fig. 1-5A,B) are also of two types: synchondroses and symphyses. A synchondrosis is a temporary joint. It is a transitional stage in bone formation, separating two centers of ossification. A symphysis is composed of hyaline cartilage on the bones connected by fibrocartilage. It is capable of compression and displacement.

Synovial joints are the most common joints, as well as the most important functionally. The ends of the bones are covered by articular cartilage, which imparts smoothness. The joint cavity, except for the hyaline-covered bearing surfaces and any articular disks or menisci, is covered by a synovial membrane. The synovial membrane produces synovial fluid, which decreases the friction in the cavity and nourishes the cartilage. Within the joint cavity, articular disks of fibrocartilage or subsynovial fat deposits may exist. These joints are surrounded by a fibrous articular capsule that is often strengthened by accessory ligaments that limit motion in certain directions.

There are seven types of synovial joints, each providing a different type of movement. In a **plane joint,** such as the acromioclavicular joint, the articulating surfaces are flat, allowing sliding or gliding movement. A **hinge joint,** such as the elbow joint, allows movement around a single axis that is perpendicular to the long axis of the bone (i.e., flexion and extension). In this type of joint the capsule is thin where bending occurs and thick on the sides. A **pivot joint,** such as the atlantoaxial joint (between the first and second cervical vertebrae), allows movement around a single axis that is parallel to the long axis of the bone (i.e., rotation). In this case the rounded bone process rotates on its long axis.

Condylar, ellipsoid, and saddle joints allow movement in two directions perpendicular to each other. **Condylar joints** have two convex and two opposing concave surfaces that allow flexion and extension, and a small amount of rotation; knuckle joints are good examples of this type of joint. **Ellipsoid joints,** such as the radiocarpal joint, have an elliptical convex and an elliptical concave surface, allowing flexion, extension, abduction, and adduction, but no rotation. **Saddle joints,** such as the carpometacarpal joint of the thumb, have reciprocal convex and concave surfaces and allow flexion, extension, abduction, adduction, and rotation.

The joint allowing the greatest freedom of movement is the **ball-and-socket joint** (shoulder and hip joints). Here movement is allowed through many axes passing through the center of the joint.

Muscles, tendons, and accessory ligaments may act on all these joints to strengthen them and to hold the articular surfaces in contact.

Joints are well supplied with nerves derived from those supplying the overlying skin and the surrounding muscles. Joint nerves transmit important information concerning position and degree and direction of movement. Vasomotor and vasosensory fibers and pain fibers respond to twisting and stretching. The numerous articular arteries supply the capsule, ligaments, and synovial membrane of the joint. There are many anastomotic connections between the arteries. Venous drainage accompanies the arterial supply.

Muscle

There are three types of muscles in the body: smooth (nonstriated, involuntary, visceral); cardiac (striated, involuntary); and skeletal (striated, voluntary, somatic). **Smooth muscle** forms the muscular wall of the viscera and blood vessels. Its cell shape is elongate and fusiform, with each cell having only one nucleus. The cells are slower and less powerful in their contraction than the cells in skeletal muscle, but are able to hold the contraction longer. **Cardiac muscle** is also involuntary, but it is striated. Its cell fibers branch and anastomose with each other. Cardiac muscle has its own automatic contractile rhythm that is modified only by nerve cells.

The rest of this discussion applies solely to **skeletal muscle.** This muscle type, which comprises 40 percent of body mass, is made up of multinucleated striated fibers under voluntary control. Skeletal muscle acts by pulling, usually across a joint, and is capable of contracting to 57 percent of its relaxed length. Muscle fibers undergo all-or-none contraction, with the strength of contraction determined by the number of fibers contracting.

Skeletal muscle is composed of bundles of fibers (Fig. 1-6). Each fiber is surrounded by a connective tissue layer, the endomysium. The bundles are surrounded by a layer called the perimysium. The entire muscle is in turn surrounded by a layer called the epimysium. All of these layers blend gradually into tendon at the tendon-muscle junction. Most skeletal muscles end in a connective tissue tendon that connects the muscle to bone, skin, joint capsule, or other connective tissue structure. **Tendons,** which have low metabolic requirements, are actually arrays of closely packed collagen fibers that are flexible but strong. Aponeuroses, such as are found in the abdomen, are simply flat-sheet tendons.

Muscles have at least two attachments, at the origin and at the insertion. The **origin** is usually the proximal end of lesser movement. The **inser-**

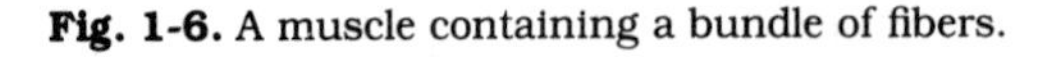

Fig. 1-6. A muscle containing a bundle of fibers.

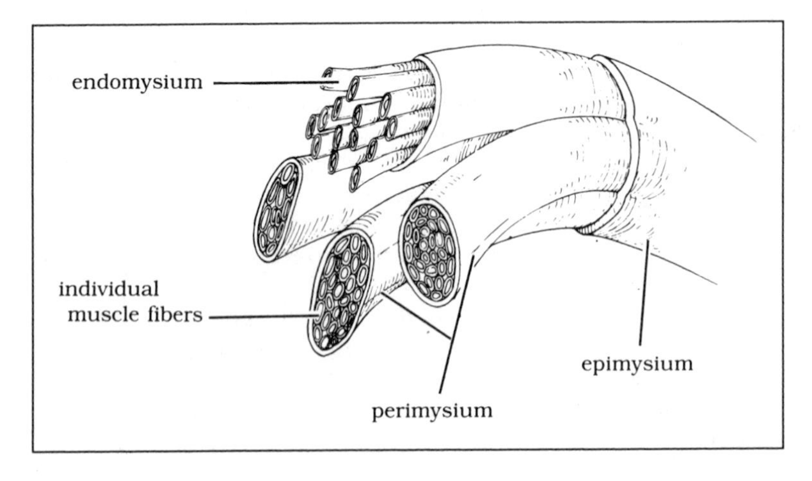

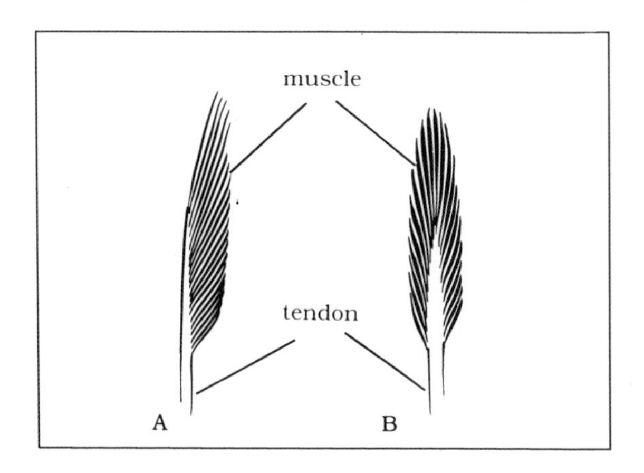

Fig. 1-7. Muscle forms. A. Unipennate. B. Bipennate.

tion is usually the distal end, where greater movement occurs.

Skeletal muscles exist in four forms: quadrilateral (straplike), fusiform (spindle-shaped), spiralized, and pennate. Quadrilateral and fusiform muscles contain parallel bundles of fibers with the tendon at the end of the muscle. In **pennate muscles** (Fig. 1-7A,B) the muscle fiber bundles insert into the tendon obliquely along its length. Pennate muscles may be either unipennate, bipennate, multipennate, or circumpennate, and exert more power than fusiform muscles. Spiralized muscles vary in the degree of spiralization within the muscle. Contraction tends to involve some form of either despiralization or rotation. For any given muscle action there is usually one prime mover, which may be aided by synergists or opposed by antagonists.

Muscles often have several structures associated with them. Subtendinous bursae are serous fluid–containing pouches often underlying a tendon as it crosses over a joint. They act to reduce the wear and tear on a tendon by reducing friction. A **synovial sheath** (Fig. 1-8) is a double-walled tube that surrounds a tendon. Between its walls it contains synovial fluid, which also reduces friction across a joint.

Each muscle receives its innervation from at least one nerve. If a muscle is innervated by more than one nerve, the nerve fibers arise from adjacent spinal cord segments. Pain fibers and proprioception fibers, which transmit information on the degree of contraction and the tension in the tendon, are among the nerve fibers innervating muscle.

The **motor unit** (Fig. 1-9) consists of the nerve cell body, axon, motor end plate (effector ending), and all the muscle fibers innervated by that nerve cell. Fewer fibers are innervated per nerve cell in muscles involved in fine movement than in muscles involved in coarse movement.

Because of their high metabolic requirements, muscles have a rich, highly anastomosed blood supply. Blood vessels accompany nerves into muscles, and in large muscles, may enter alone.

Circulatory System

The components of the circulatory system are the blood, heart, arteries, capillaries, and veins. In addition lymph and its associated vessels are part of this system. (They are covered in the next section.) The functions of the circulatory system are (1) to provide cells with a fluid environment, (2) to furnish cells with oxygen, (3) to transport food, hormones, and other cellular products between cells, and (4) to remove wastes from cells.

Fig. 1-8. Tendon in synovial sheath.

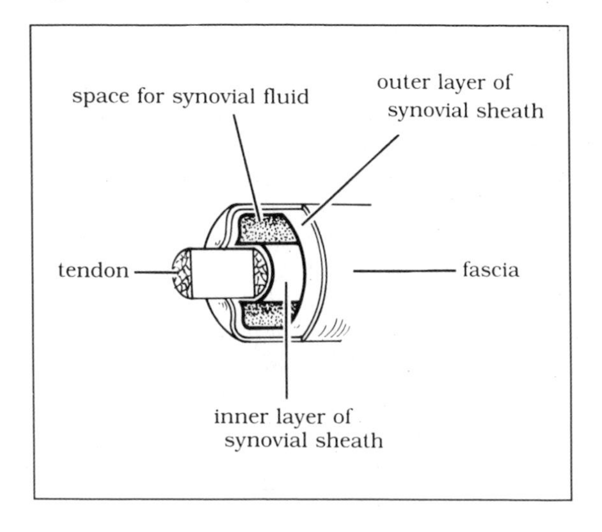

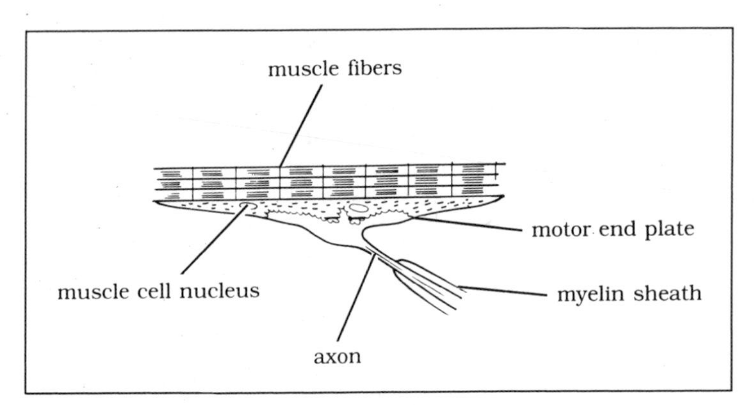

Fig. 1-9. Motor unit.

The **blood** is the prime medium in which all the above activities are carried out. Blood consists of plasma—which is primarily water with dissolved salts, proteins, and other macromolecules—and blood cells. There are red blood cells, for the transport of oxygen, and six types of white blood cells involved in the immune system and the transport of toxic wastes.

The **heart** (Fig. 1-10) is the pump of the circulatory system. It is composed of muscular tissue lying within a serous sac, the pericardial sac. The right atrium receives the venous blood from all over the body except from the lungs. The right ventricle then pumps this blood to the lungs. The oxygenated blood is returned to the left atrium via the pulmonary veins. The left ventricle then pumps this blood out to the rest of the body. (See Chapter 2 for a more detailed discussion.)

The **great vessels** of the heart are (1) the pulmonary trunk, which carries blood to the lungs; (2) the pulmonary veins, which bring blood back from the lungs; (3) the aorta, which sends blood to the rest of the body; and (4) the caval veins, which return blood to the heart from the body.

Arteries, the aorta being the greatest, conduct blood from the heart to a capillary bed by dividing and branching into smaller arteries, and finally ending as arterioles. An arterial wall has three layers. The tunica intima is the inner endothelial lining containing connective and elastic tissues. The tunica media is the middle smooth muscle and elastic tissue coat. Finally the tunica adventitia forms the outer covering, composed primarily of collagen fibers. As arteries decrease in size, the amount of smooth muscle in the tunica media increases. This increased musculature allows the small arteries and arterioles to regulate the flow of blood to the capillary bed.

Capillaries (Fig. 1-11) provide the intimate fluid environment for the body's cells. Capillaries have walls that consist only of an endothelial cell lining and are just wide enough for a red blood cell to pass through. Plasma is thus allowed to move out of the capillaries carrying soluble food and oxgyen with it. At the venous end of the capillary bed plasma moves back into the capillaries, carrying wastes. Some of the plasma, however, returns not via the venous system, but rather via the lymphatic system.

In addition to plasma, white blood cells may pass through the capillary walls. Furthermore in the intestines, liver, and lungs, the capillary bed adds soluble food, detoxifies wastes, and oxygenates red blood cells, respectively.

The function of **veins** is to return blood to the heart. These vessels are thin walled and large in diameter, with a tunica media that is poor in smooth muscle and elastic tissue. Veins usually accompany arteries, although there may be more than one vein for a given artery. In addition, in certain regions of the body, they may form highly interconnected venous plexuses.

Since there is little pressure in veins, they have

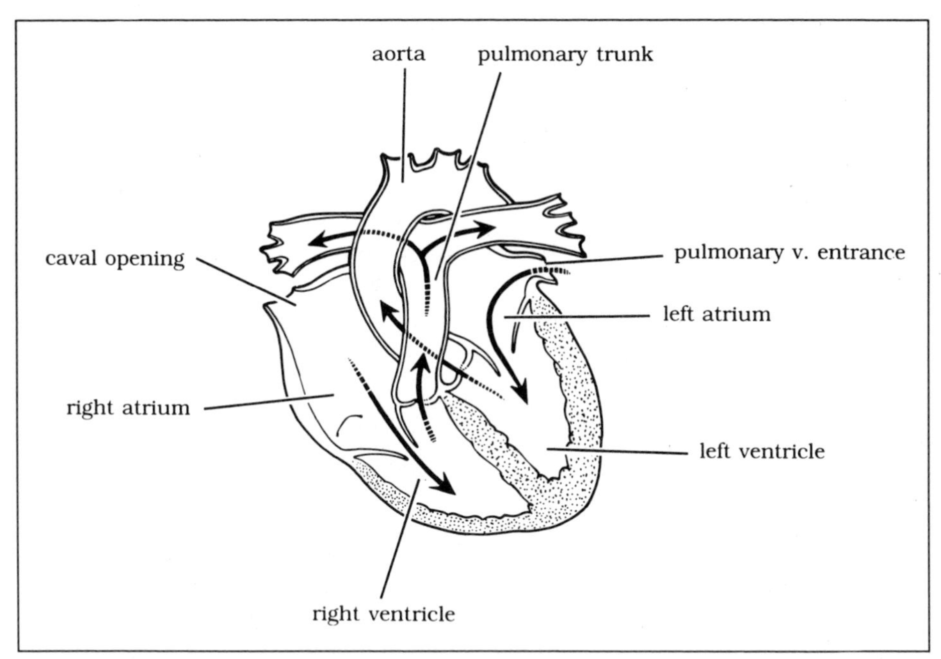

Fig. 1-10. Path of blood through the heart and great vessels.

Fig. 1-11. Capillary bed and arteriovenous anastomosis.

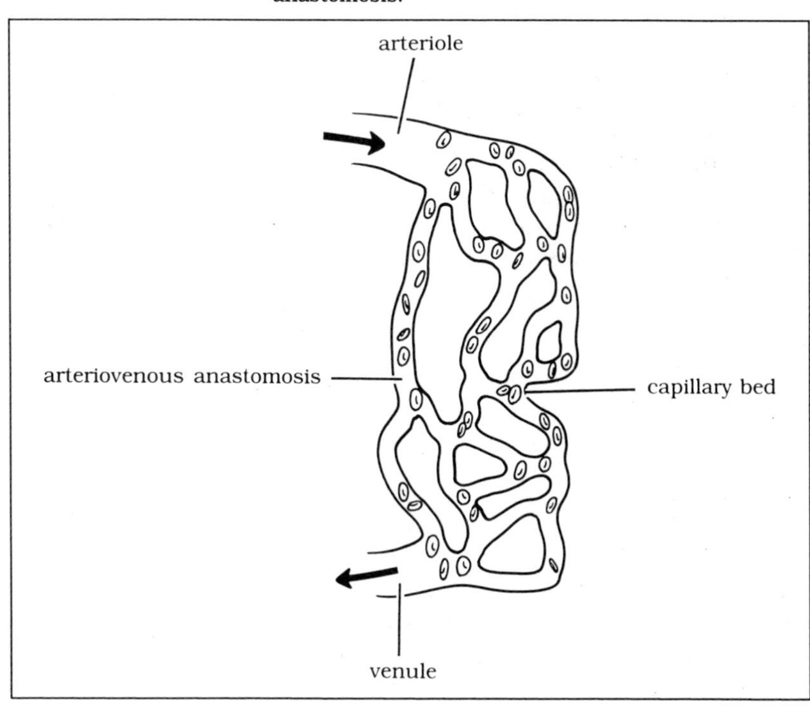

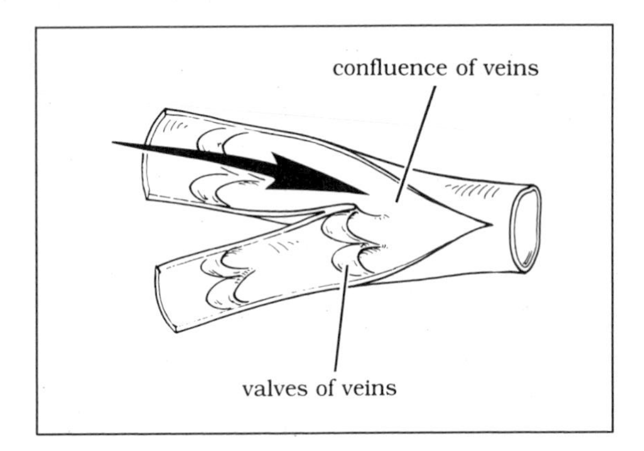

Fig. 1-12. Open vein with internal valves.

valves (Fig. 1-12) that ensure that the blood flows in only one direction, toward the heart. The valves also help to support the weight of the blood column. To assist in the return of blood, the muscles of the limbs help to provide a kneading action that serves to pump blood toward the heart.

Since all body systems do not need the maximum flow of blood through their capillary bed at all times (e.g., muscles at rest) and since it is impossible to pass optimal amounts of blood through all capillary beds at the same time, the flow of blood into the capillary beds is regulated by **arteriovenous anastomoses** (see Fig. 1-11). These direct shunts between arterioles and venules have good muscular walls so that blood flow can be shut down, directing blood into the capillary bed, or opened up, partially bypassing the capillaries.

The nerves innervating the circulatory system are almost entirely automatic and involuntary. The nervous system coordinates the heartbeat rate and the blood flow through arteries, arterioles, and arteriovenous anastomoses. Nerves innervate the smooth muscle components of blood vessel walls, causing vasoconstriction and sometimes vasodilatation. In addition, the nervous system monitors and regulates blood pressure through the carotid sinus and the chemical composition of blood at the carotid body.

Lymphatic system

The lymphatic system is actually a subdivision of the circulatory system. **Lymph,** the fluid of the system, is similar to plasma in that it carries proteins, salts, and wastes. The other components of the system are lymphocytes, lymphatic vessels, and lymph nodes. Additional lymphatic organs are the tonsils, spleen, and thymus gland.

The **lymphocytes** arise in the nodes and pass through the lymphatic vessels to reach the bloodstream. Lymphocytes play an active role in the immune system.

The capillaries of the lymph system are simple endothelial tubes. They are distributed unevenly throughout the body and are particularly abundant in the dermis and mucosa of the digestive tract. Capillaries flow into collecting vessels that are like veins, except that they lack well-defined layers in their walls. The main collecting vessel of the lymphatic system is the **thoracic duct,** which collects lymph from the lower half and the upper left side of the body. Other lymph trunks drain other regions.

The **lymph nodes** (Fig. 1-13) are variable in size and lie along the course of the lymphatic vessels. They have a fibroelastic capsule. Internally they are filled with lymphocytes, which are formed there, and serve as a deposit site for particulate matter and debris collected in the lymphatic capillaries. During infection the affected nodes may enlarge as they become engaged in phagocytizing bacteria.

The lymphatic system is the primary transport system for soluble fats from the intestines to the bloodstream. Ninety-five percent of its contents come from the intestines and liver, and emulsified fats may be found in the thoracic duct after meals.

Skin

The skin (Fig. 1-14), the outermost covering of the body, serves many functions. It is a sensory organ for touch, pressure, temperature, and pain sensation. It is a strong, flexible protective covering that shields the body from ultraviolet radia-

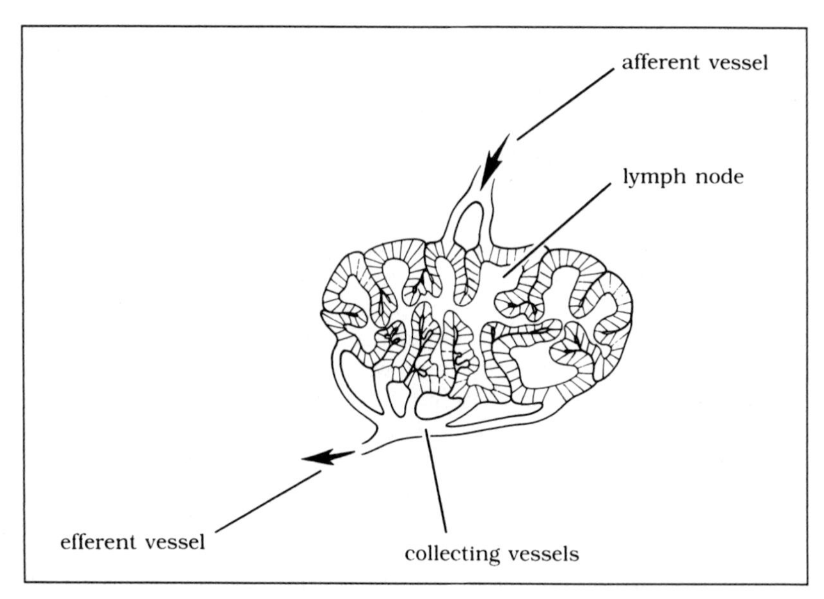

Fig. 1-13. Lymph node and collecting vessels.

Fig. 1-14. Skin with hair.

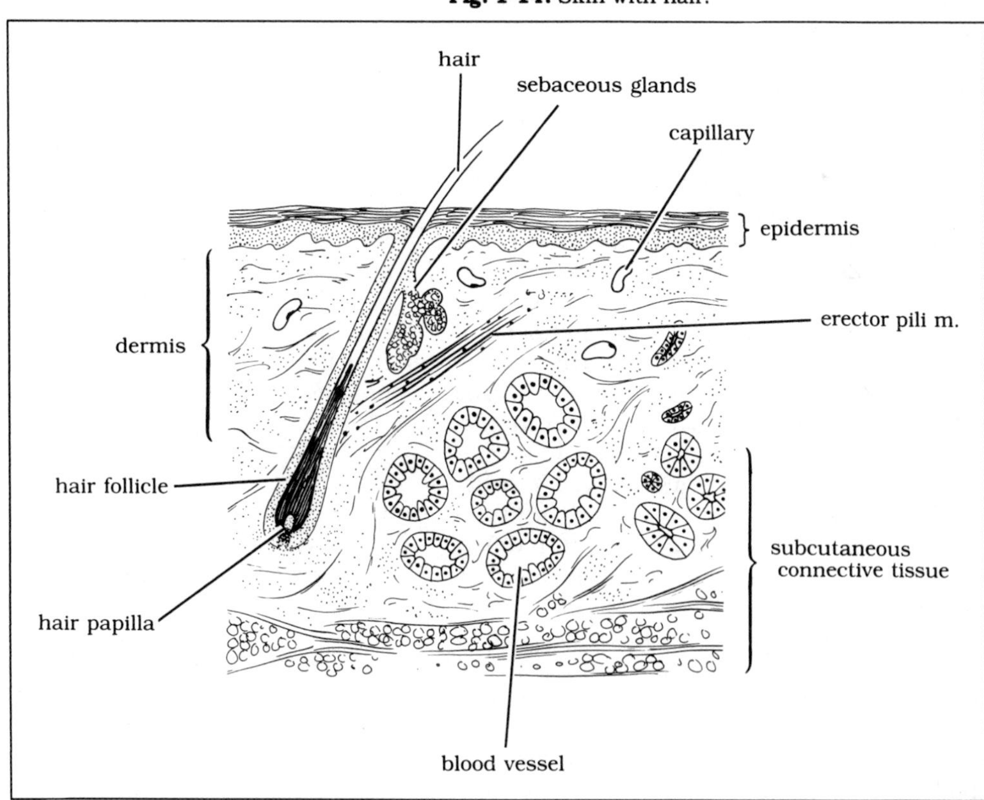

tion and prevents dehydration. The skin is also involved in the regulation of body temperature and the production of vitamin D. Furthermore it serves both an excretory and absorptive function, and at the fingers and toes, it is involved in gripping.

The skin tends to be thicker on dorsal and extensor surfaces than on ventral and flexor surfaces. It may be either loosely applied and mobile or firmly attached to the underlying connective tissue.

Skin consists of three layers: the epidermis, the dermis, and the hypodermis. The **epidermis,** or surface layer, is a nonvascular layer of cornified epithelium. This epithelium is continually turned over—the most superficial layers are flaked or rubbed off. Pigmentation of the skin is due to the production of melanin by some cells of the deepest layer of the epidermis. The epidermis also contains some sensory nerve terminals.

Underlying and interlocking with the epidermis is the **dermis.** This layer, a connective tissue network, contains capillaries, fat, lymph channels, tactile nerve endings, hair follicles, sweat glands, sebaceous glands, and smooth muscle. The deepest layer of the dermis contains both collagenous fibers and elastic fibers that give the skin toughness and flexibility.

The dermis interdigitates with the underlying subcutaneous connective tissue. This layer contains fat, elastic fibers, white fibrous connective tissue, nerves, pressure receptors, and parts of sweat glands. In the head and neck region there is also subcutaneous voluntary muscle in this layer. In some regions (e.g., nipple), subcutaneous involuntary smooth muscle is found here.

Hair, nails, and glands are accessory structures of the skin. **Hair** covers almost the entire body, varying in thickness and length in different regions. The hair shaft projects from the skin, its root being in the skin. The hair root is surrounded and protected by a hair follicle into which sebaceous glands open. These glands are squeezed and the hair is erected by a smooth muscle, the erector pili muscle.

The **nail plate** consists of lamellae of cornified epithelial cells that are closely welded together. The plate is firmly bound to the underlying periosteum through the nail bed.

There are two kinds of glands found in skin: sebaceous glands and sweat glands. **Sebaceous glands,** found around hairs, lips, glans penis, internal layer of the prepuce, clitoris, labia minora, areola, and nipple, secrete sebum. Sebum is an oily lubricant that protects the hair and skin from drying. Acne results when the gland gets plugged and inflamed and sebum accumulates.

Sweat glands are found widely, except at the outer ear and wherever sebaceous glands are found without hair. Sweat is a clear cellfree fluid involved in temperature regulation. However, around the anus and axilla, cell breakdown products may be found in sweat.

The nerves to the skin provide afferent somatic fibers for pain, touch, pressure, and temperature and efferent fibers for the smooth muscles found in blood vessels and glands.

Nervous System

The nervous system is composed of the central nervous system and the peripheral nervous system. The nervous system functions in the coordination of the other systems of the body. It receives information in the form of stimuli, processes it, and coordinates a response to the stimuli. These stimuli may be of either external or internal origin and the response may take several different forms (e.g., motor, glandular). In humans the nervous system also performs the function of abstract thought.

NEURON

The basic functional unit of the nervous system is the neuron. Essentially the neuron (Fig. 1-15) is composed of a cell body and cell processes. The cell body of the most common type of neuron tends to be multipolar, with a large, pale nucleus and cytoplasm containing granular Nissl bodies.

The many cell processes give this class of cells its multipolar appearance. There are two types of processes: dendrites and axons. The **dendrites,** usually several, extend from the cell body in a

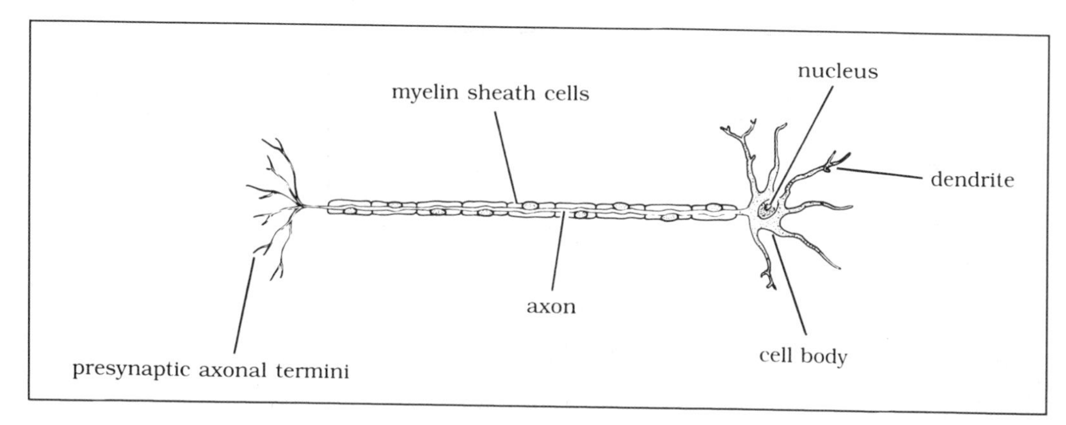

Fig. 1-15. Neuron.

branching array. They receive input from other nerve cells. The single **axon** carries the impulse outward from the cell body and transmits it across a synapse to the next cell. It branches only at its termination. The axon and its sheath constitute a nerve fiber. The axon sheath is composed of myelin, and it is formed by glial cells, which wrap around the axon. The sheath, when present, is interrupted approximately every 0.5 mm at nodes. Nerves are composed of bundles of nerve fibers bound together by connective tissue sheaths.

PERIPHERAL NERVOUS SYSTEM

The peripheral nervous system is involved in the transmission of impulses between the brain and the rest of the body. It is composed of 31 pairs of spinal nerves and 12 pairs of cranial nerves. Each pair of nerves has a specific distribution, which may be "visualized" on the surface by the dermatome distribution (Fig. 1-16A,B).

Within each nerve, some fibers receive stimuli and transmit impulses to the central nervous system, while other fibers carry the response back to the body. The fibers carrying impulses elicited by stimuli are known as **sensory,** or **afferent,** fibers. The fibers carrying impulses resulting in a response are known as **motor,** or **efferent,** fibers. Each spinal nerve is composed of both sensory and motor fibers. Furthermore, nerves can be differentiated into those dealing with the body (somatic) and those dealing with organs (visceral). The result is that each spinal nerve contains four types of fibers: general somatic afferent, general somatic efferent, general visceral afferent, and general visceral efferent.

Cranial nerves are slightly more complicated. Vision and hearing are considered special somatic senses; therefore, their fibers are special somatic afferents. Taste and smell are considered special visceral senses; therefore, their fibers are special visceral afferents. Finally skeletal muscles related to the feeding mechanism (e.g., muscles of mastication) are innervated by special visceral efferent fibers. Cranial nerves may also have the same components as spinal nerves.

Although peripheral nerves contain both somatic and visceral fibers, these fibers do not reach the target organs by the same pathway. General somatic efferents are uninterrupted in their course from the spinal cord to the effector organ, with the neuron cell body lying in the spinal cord. General somatic afferents are uninterrupted between the receptor organ and the spinal cord, with the neuron cell body lying in the **dorsal root ganglion.** General visceral efferents synapse once outside the spinal cord, either in a chain of ganglia known as the **sympathetic chain** or near the organ that they innervate. General visceral afferents are uninterrupted between the receptor organ and the spinal cord. However, the neuron cell body may lie either in the dorsal

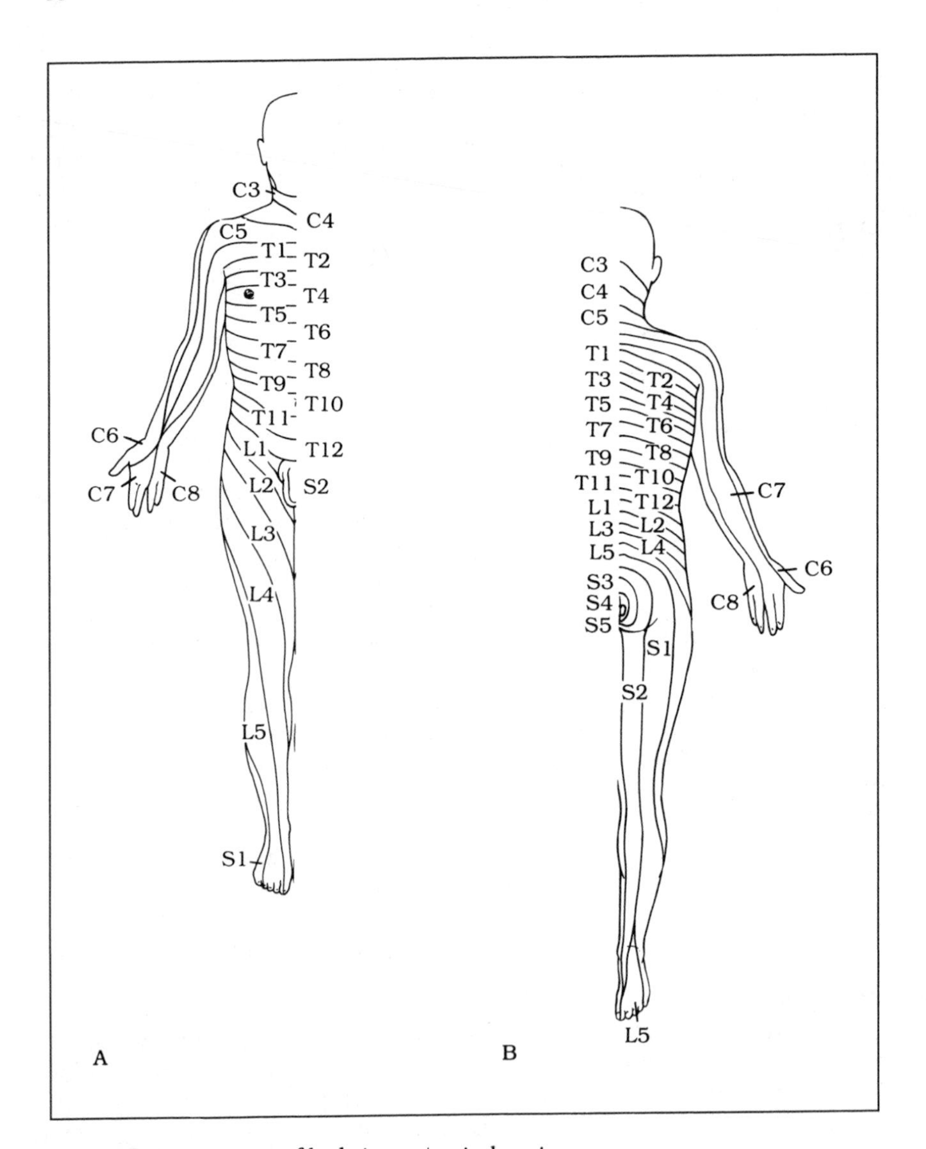

Fig. 1-16. Dermatomes of body in anatomical position. A. From front. B. From back.

root ganglion or in special ganglia on the cranial nerves.

This system of visceral innervation is known as the **autonomic system** (Fig. 1-17). It is a system by which the body automatically regulates visceral function. Regulation is achieved by balancing two antagonistic impulses (e.g., one to speed up, the other to slow down). These antagonistic impulses are known as the sympathetic and parasympathetic portions of the autonomic system. The **sympathetic division** is a defensive system, organizing the body on a widespread level for danger. The **parasympathetic division** acts to conserve and preserve the body in its normal state. It acts on a local and discrete level. The effects of the sympathetic and parasympathetic divisions on a given organ depend on the organ involved. For example, sympathetic nerve fibers

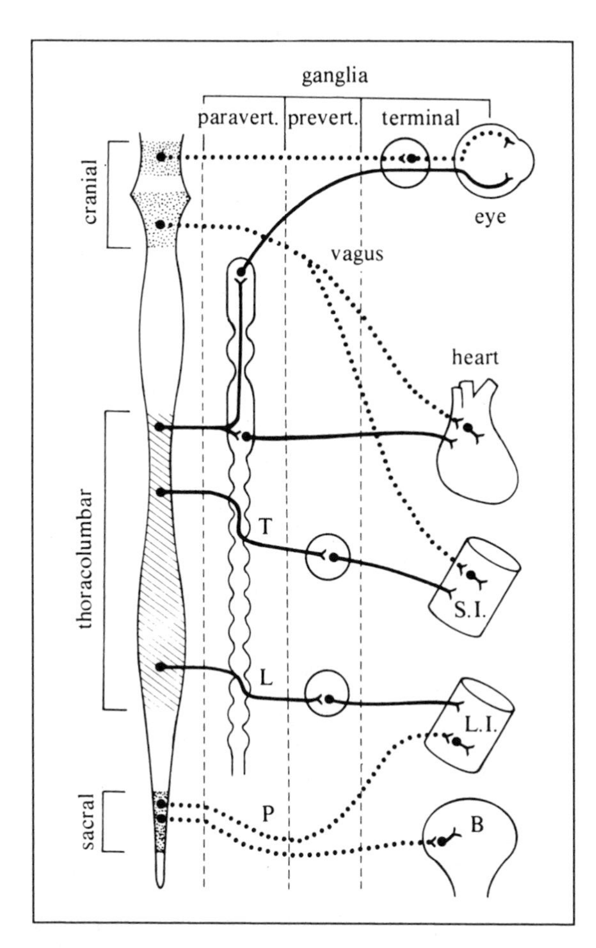

Fig. 1-17. General plan of the autonomic nervous system. The thoracolumbar outflow comprises the sympathetic system, and some of these fibers form thoracic (T) and lumbar (L) splanchnic nerves. The cranial and sacral outflows comprise the parasympathetic system, and some of these fibers form pelvic (P) splanchnic nerves. Sympathetic fibers are shown by continuous lines, parasympathetic fibers by dots. (B = bladder; L.I. = large intestine; S.I. = small intestine.) (From R. O'Rahilly, *Gardner, Gray and O'Rahilly Anatomy: A Regional Study of Human Structure* (5th ed.). Philadelphia: Saunders, 1986. P. 35. With permission.)

speed up the heart but slow down digestion.

Sympathetic fibers originate within the thoracic and lumbar regions of the spinal cord, exit the spinal cord through its ventral root, and pass to the paravertebral sympathetic chain. Fibers may exit the sympathetic chain either above, below, or at the level at which they entered.

The fibers leaving the ventral root to enter the sympathetic chain are presynaptic (preganglionic) fibers. These fibers synapse either in one of the chain ganglia or in a collateral ganglion. Thus the sympathetic fibers that reach the end organ are postsynaptic (postganglionic) fibers.

Parasympathetic fibers arise in either cranial or sacral regions. Generally they synapse near or in the organ of innervation. Thus the parasympathetic presynaptic fibers are long, whereas the postsynaptic fibers are quite short.

General visceral afferent fibers travel with both sympathetic and parasympathetic nerves. Those that travel with the sympathetic nerves tend to have cell bodies in the dorsal root ganglia at the same level as that at which the corresponding efferent nerve leaves the spinal cord. Parasympathetic visceral afferents arising from cranial levels have cell bodies in special ganglia, and those arising from sacral levels have cell bodies in dorsal root ganglia.

GANGLIA

Ganglia are foci of nerve cell bodies outside the central nervous system. There are essentially four types of ganglia. The dorsal root ganglia (Fig. 1-18) contain the cell bodies of all the general somatic afferent spinal nerve fibers. In addition they contain the cell bodies of some general visceral afferent fibers of spinal nerves. There are no synapses in the dorsal root ganglia.

The **sympathetic chain ganglia** and their collateral ganglia (lying along the major abdominal blood vessels) contain the cell bodies of the sympathetic general visceral efferent postsynaptic fibers. The myelinated, preganglionic fibers reach the chain via the short white rami communicantes. These fibers synapse in the ganglia, and the unmyelinated postganglionic fibers leave

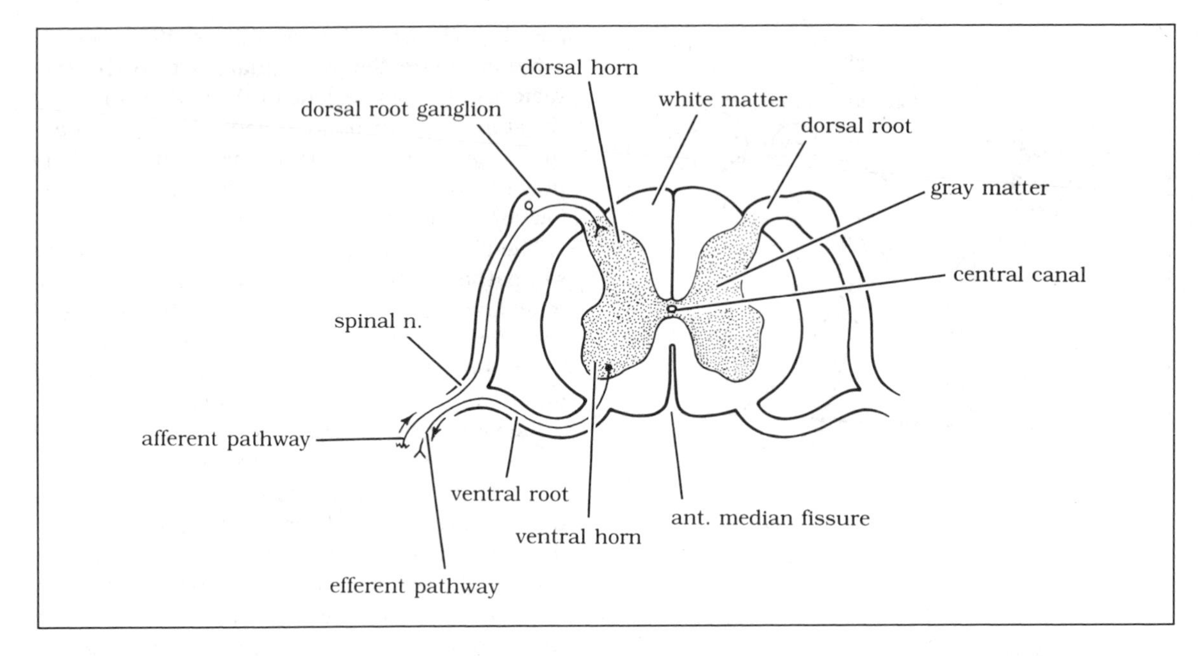

Fig. 1-18. Schematic cross section of spinal cord.

the chain via the long gray rami communicantes.

The **parasympathetic ganglia** contain the cell bodies of the postsynaptic general visceral efferent fibers of the parasympathetic system. These ganglia are located near or in the organs they innervate and are reached by long presynaptic fibers. The postsynaptic fibers tend to be very short.

Finally there is a special set of ganglia associated with some cranial nerves that contain the cell bodies of special visceral afferent fibers, some general visceral afferent fibers, and general somatic afferent fibers.

CENTRAL NERVOUS SYSTEM

The central nervous system is composed of the spinal cord and brain and is the central control and processing region of the body. The **spinal cord,** the caudal continuation of the central nervous system, lies in the vertebral canal. It contains both cell bodies and nerve fibers. In cross section (see Fig. 1-18) the spinal cord appears as a butterfly-shaped gray area surrounded by a white zone. The gray area contains cell bodies; the white area contains myelinated ascending or descending nerve fibers.

The **dorsal root** of the spinal cord contains afferent fibers entering the cord. The cell bodies of these fibers are found in the dorsal root ganglion. The **ventral root** of the spinal cord contains efferent fibers. The cell bodies of these fibers lie in the ventral horn if they are somatic and in the lateral horn if they are visceral (at thoracic and lumbar levels only).

The spinal peripheral nerves are formed by the joining of the dorsal and ventral root fibers lateral to the spinal cord. Also existing within the gray area of the spinal cord are **interneurons.** These short neurons connect sensory (afferent) and motor (efferent) neurons of the same level of the cord, thus completing a sensorimotor pathway. This pathway is the basis for the reflex arc. In the **reflex arc** sensory input entering the spinal cord is transmitted to the interneuron. From here it passes to the motor neuron, thus bringing about a response.

Interneurons may receive impulses from more than one sensory neuron. They may also receive impulses from neurons whose cell bodies are located at higher or lower levels of the central nervous system. Sensory and motor fibers do not have to synapse with interneurons; instead they may ascend to, or descend from, higher levels of the central nervous system.

National Board Type Questions

Select the one best response for each of the following.

1. Which one of the following statements is false?
   A. An adducted limb has been moved closer to the midline.
   B. Circumduction involves flexion, extension, adduction, and abduction.
   C. Rotation is a combination of circumduction and pronation.
   D. When the foot is inverted, the sole is turned toward the midline.
   E. To touch your nose with the palm of your hand, the hand must be supinated.
2. All of the following statements are true **except**
   A. There are three types of cartilage.
   B. Moveable joints are of the synovial or fibrous type.
   C. Synovial joints contain articular cartilage.
   D. Every surface in a synovial joint must be covered with synovial membrane.
   E. A syndesmosis is a type of fibrous joint.

Select the response most closely associated with each numbered item. (The headings may be used once, more than once, or not at all.)

A. Hinge joint
B. Saddle joint
C. Ball-and-socket joint
D. Plane joint
E. Pivot joint

3. Allows greatest freedom of movement
4. Rotation along the long axis of the bone
5. Allows movement in two directions perpendicular to each other

---

Select the response most closely associated with each numbered item.

A. Central nervous system
B. Peripheral nervous system
C. Both
D. Neither

6. Spinal cord
7. Spinal nerves and cranial nerves
8. Contains axons
9. Contains cell bodies and nerve fibers

For the following, select

A. if only *1, 2, and 3* are correct
B. if only *1 and 3* are correct
C. if only *2 and 4* are correct
D. if only *4* is correct
E. if *all* are correct

10. With regard to the muscles of the body,
    1. All striated muscle is voluntary.
    2. Smooth muscle lines the viscera and blood vessels.
    3. An aponeurosis is found at the end of each bundle of smooth muscle.
    4. Skeletal muscle is striated and voluntary.
11. The great vessels of the heart include the
    1. Aorta
    2. Caval veins
    3. Pulmonary trunk
    4. Pulmonary veins
12. The lymphatic system
    1. Drains to the venous system
    2. Is composed of nodes that give rise to lymphocytes
    3. Includes the tonsils and spleen
    4. Acts only as part of the immune system of the body

Annotated Answers

1. C. Circumduction is a combination of the movements listed. Rotation occurs along the central axis of the bone (see Question 4).
2. D. Articular cartilage is an essential component of a synovial joint, and it is not covered by synovial membrane.

3. C. A ball-and-socket joint, by definition, allows the greatest freedom of movement, although joint movement can be greatly restricted by accessory ligaments.
4. E. The pivot joint allows rotation along the long axis. (See the description of the proximal radio-ulnar joint in Chap. 5.)
5. B.
6. A.
7. B. and 8. C. The peripheral nervous system is made up of the spinal and cranial nerves. Both systems must contain axons for impulses to travel.
9. C.
10. C. This question is difficult. Remember that by definition cardiac muscle is striated and is under involuntary control.
11. E. The word *caval* refers to the two great veins that return blood to the heart and are two of the great vessels.
12. A. Certainly playing an extremely important role in the immune system, the lymphatics also transport fats from the intestine to the bloodstream.

2 Thorax

Objectives

After reading this chapter, you should know the following:

Surface landmarks of the thorax and surface projections of the respiratory organs and the heart and valves

Skeletal structure, including articulations and movements of the different ribs, and musculature of the thoracic wall

General organization of the thoracic cavity

Organization, components, and function of the respiratory system (trachea, bronchi, pleurae, and lungs)

Mechanism of respiration

Composition, structure, and function of the heart and pericardium, including the great vessels

Circulation of blood through the heart

Contents of the mediastinum

Blood supply, lymphatics, and innervations of the thoracic wall, cavity, and organs

Surface Anatomy

The thorax contains the organs of respiration and the heart and great vessels. Because these structures are of great importance in the examination of a patient, a knowledge of their surface projections and relation to the surface anatomy of the region is necessary.

SURFACE LANDMARKS

The musculature visible on the surface of the thorax is primarily concerned with the movement of the upper limbs and includes the pectoralis major, serratus anterior, and lattissimus dorsi muscles (Fig. 2-1A,B,C,D). The sternocleidomastoid muscle, arising from the manubrium and clavicle, is also visible.

The **jugular** (suprasternal) **notch** (Fig. 2-2A,B; see also Fig. 2-6), the superior margin of the manubrium, is easily palpated. It marks the insertion of the sternocleidomastoid muscle onto the sternum and is also the region of attachment of the clavicle to the sternum. The notch lies at the level of the lower border of the second thoracic vertebra.

The **sternal angle** is the angle formed by the junction of the manubrium and the body of the sternum. This landmark is especially important because it is where the second costal cartilage inserts; the sternal angle thus serves as a landmark for counting the ribs. The angle is at the level of the intervertebral disk between the fourth and fifth thoracic vertebrae.

The **xiphosternal joint** connects the body of the sternum with the xiphoid process and lies at the level of the ninth thoracic vertebra.

The **costal margin** is formed by the lower borders of the costal cartilages of the seventh through the twelfth ribs. The costal margins of the right and left sides converge in the anterior midline to form the subcostal angle.

The **clavicle** (see Fig. 2-2) is palpable under the

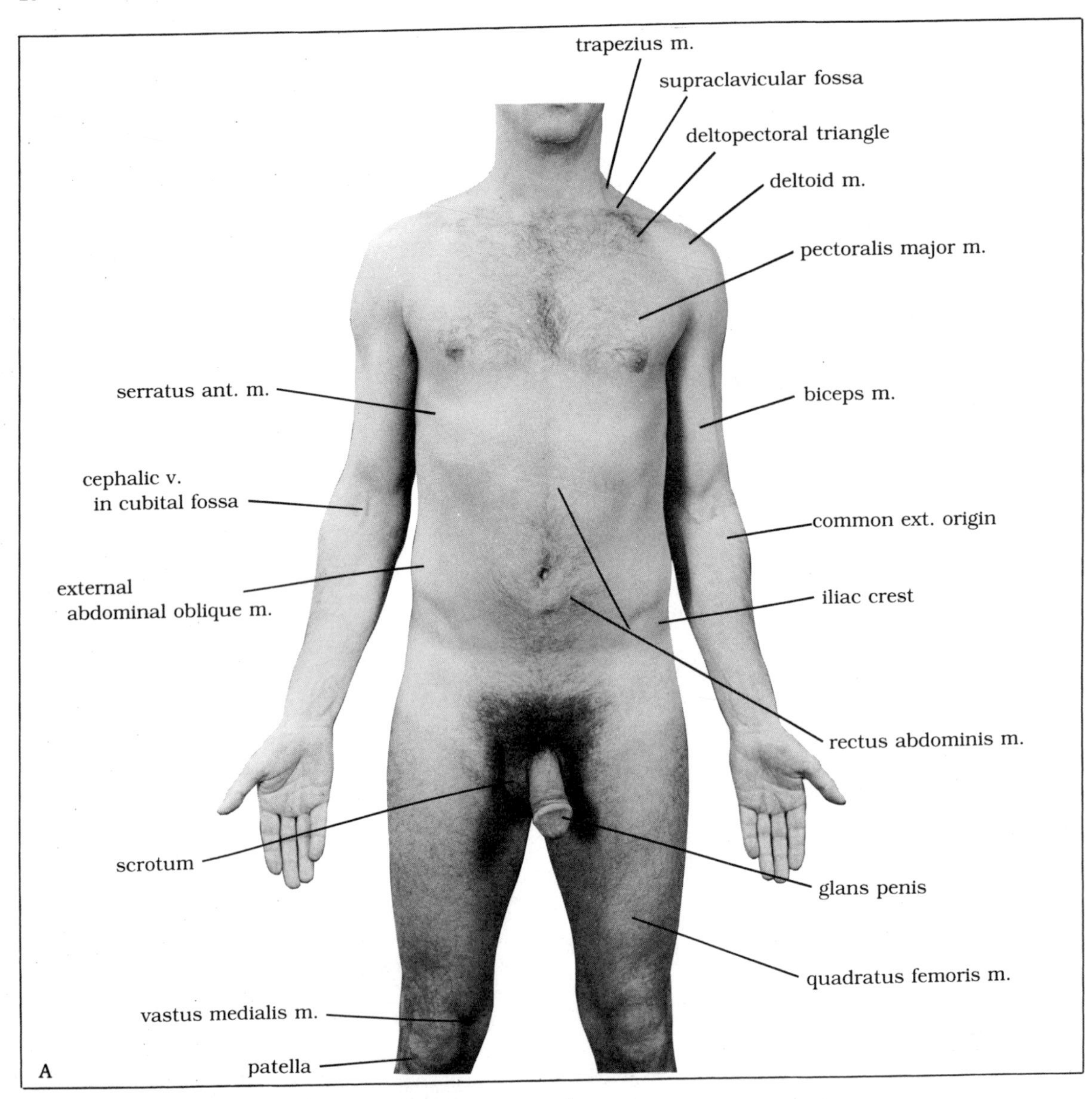

Fig. 2-1. Surface anatomy of the thorax. A. Male, anterior view. B. Female, anterior view. C. Male, lateral view. D. Female, lateral view.

skin from the suprasternal notch of the manubrium to the acromion of the scapula.

The anterior and posterior **axillary folds** are formed by the pectoralis major and latissimus dorsi muscles, respectively.

In the posterior thorax the spinous processes of the vertebrae (see Fig. 2-2) may be counted using the seventh cervical vertebra as a starting point. This process may be palpated as the prominence at the root of the neck when the head is bent forward. It is important to note that the spinous process descends downward from the vertebral body so that the tip is actually posterior to the vertebral body of the next lower vertebra.

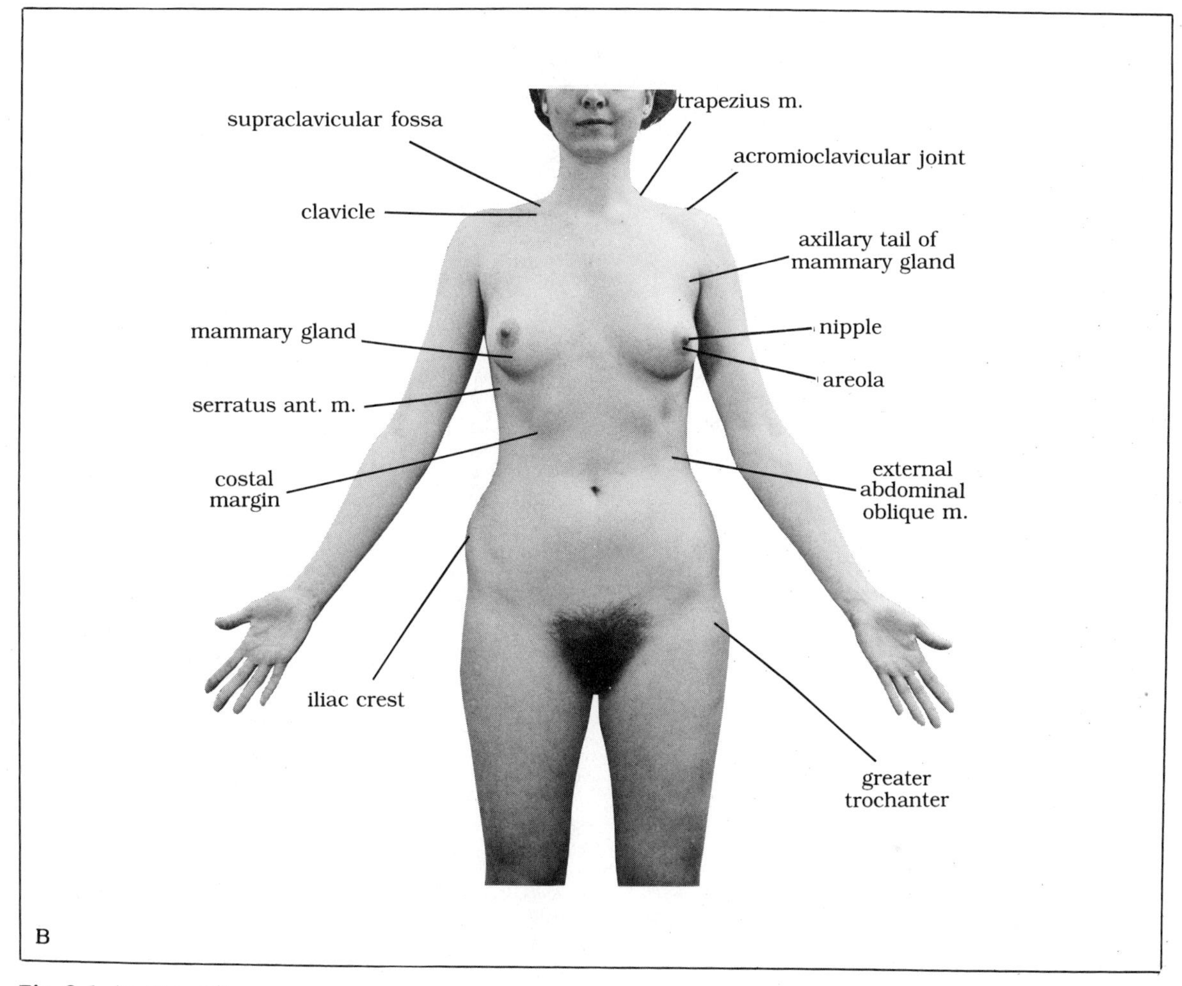

Fig. 2-1. (continued).

For orientation several vertical lines may be drawn on the body (Fig. 2-3A,B). The **midsternal line** marks the midline of the body, lying over the middle of the sternum. The **midclavicular** (or mammary) **line** descends vertically from the middle of the clavicle. In males and young females the nipple is situated in the fourth intercostal space along this line. The **anterior** and **posterior axillary lines** descend from their respective folds. The **midaxillary line** descends from the apex of the axilla midway between the anterior and posterior axillary lines. The **scapular line** descends on the posterior thorax through the inferior angle of the scapula.

SURFACE PROJECTIONS

Organs of Respiration

The superior (apical) surface projection of the lung is a dome-shaped region extending approximately 2.5 cm into the neck whose peak is at the

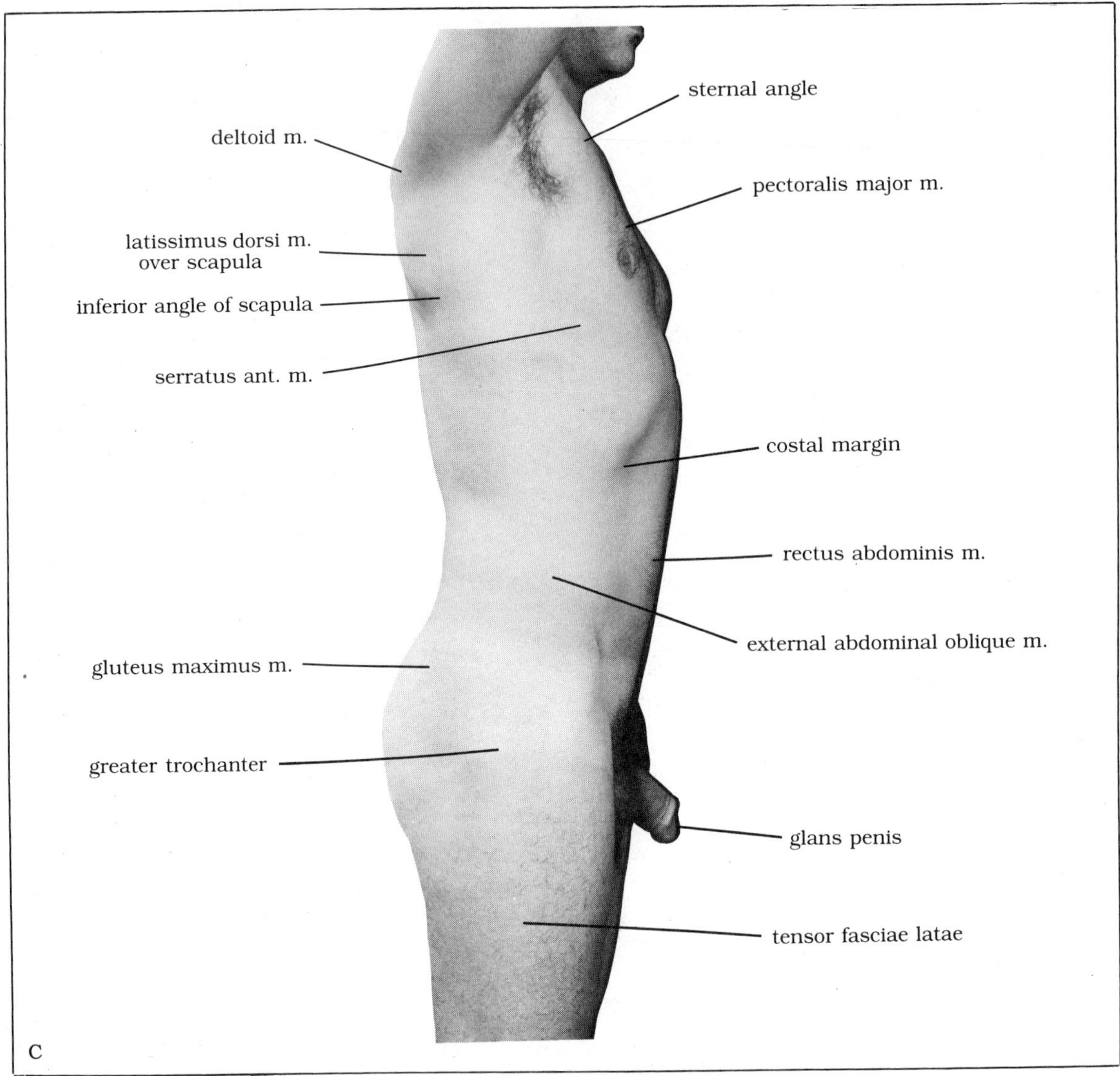

Fig. 2-1. (continued).

junction of the medial and intermediate third of the clavicle (Fig. 2-4). On the right side the lung descends from the sternoclavicular joint toward the midsternal line, reaching that line at the sternal angle and descending along it until approximately the fourth costal cartilage. From there it gradually diverges, exiting from behind the sternum at the level of the sixth cartilage. The bottom of the lung then crosses the midclavicular line behind the sixth rib, the midaxillary line behind the eighth rib, and the scapular line behind the tenth rib. From here the inferior margin continues at this level to a point approximately 4 cm from the posterior midline, where the border of the lung ascends to the cervical dome.

The left lung deviates from this course, allowing space for the heart (cardiac notch) by extending laterally at the level of the fourth costal cartilage to a point approximately halfway to the midclavicular line. The border of the lung then descends, crossing the sixth costal cartilage just medial to the costochondral junction.

The surface projection of the parietal pleura

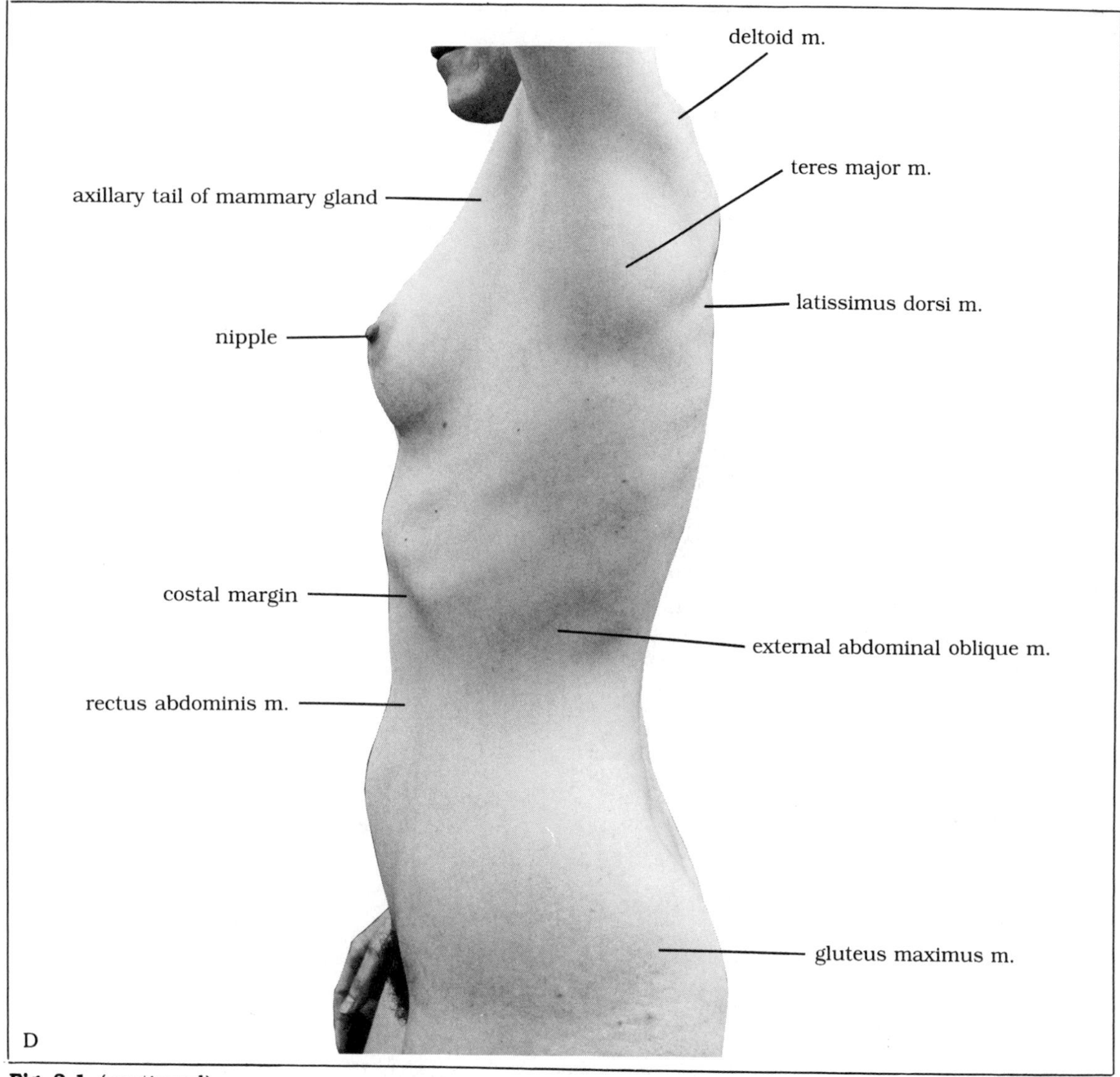

Fig. 2-1. (continued).

parallels that of the lungs in the apical region. On the right, however, the pleura does not exit from behind the sternum until the seventh costal cartilage. It then crosses the midclavicular line at the eighth costochondral junction, the midaxillary line at the tenth rib, and the scapular line at the eleventh rib. The pleura then descends below the neck of the twelfth rib to the body of the twelfth thoracic vertebra, from which it ascends to the cervical dome.

On the left side the pleura deviates laterally, leaving the sternum at the fifth intercostal space. From here it crosses the sixth costal cartilage 1.5 cm lateral to the sternum and then follows the same pattern as that followed on the right side beginning with the seventh rib. The inferior region between the lungs and pleura is the **costodiaphragmatic recess.**

Heart and Valves

In the upright position the apex of the heart, where the heartbeat can be most clearly heard, is

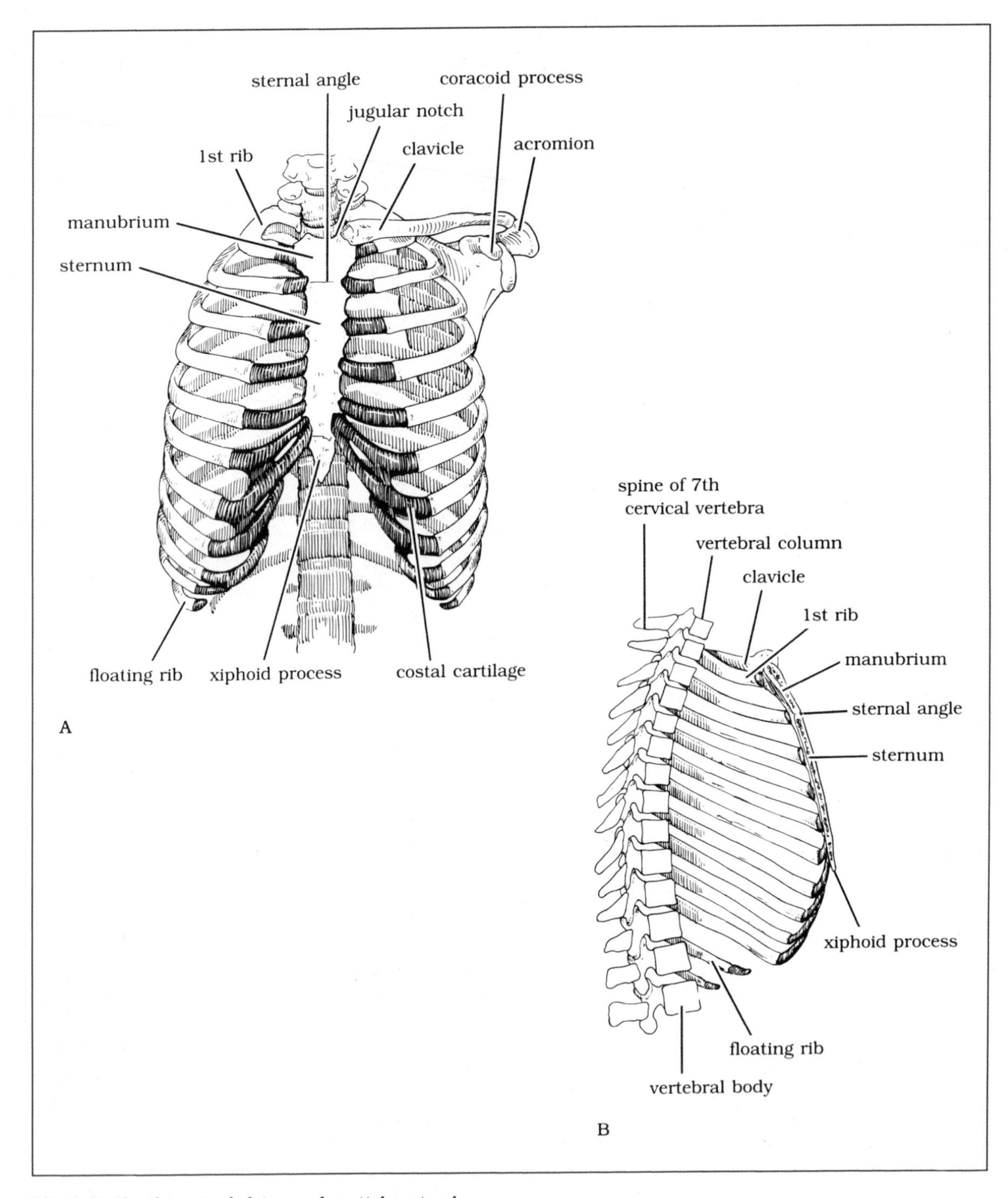

Fig. 2-2. The thoracic skeleton and partial pectoral girdle. A. Anterior view. B. View from the midline looking laterally.

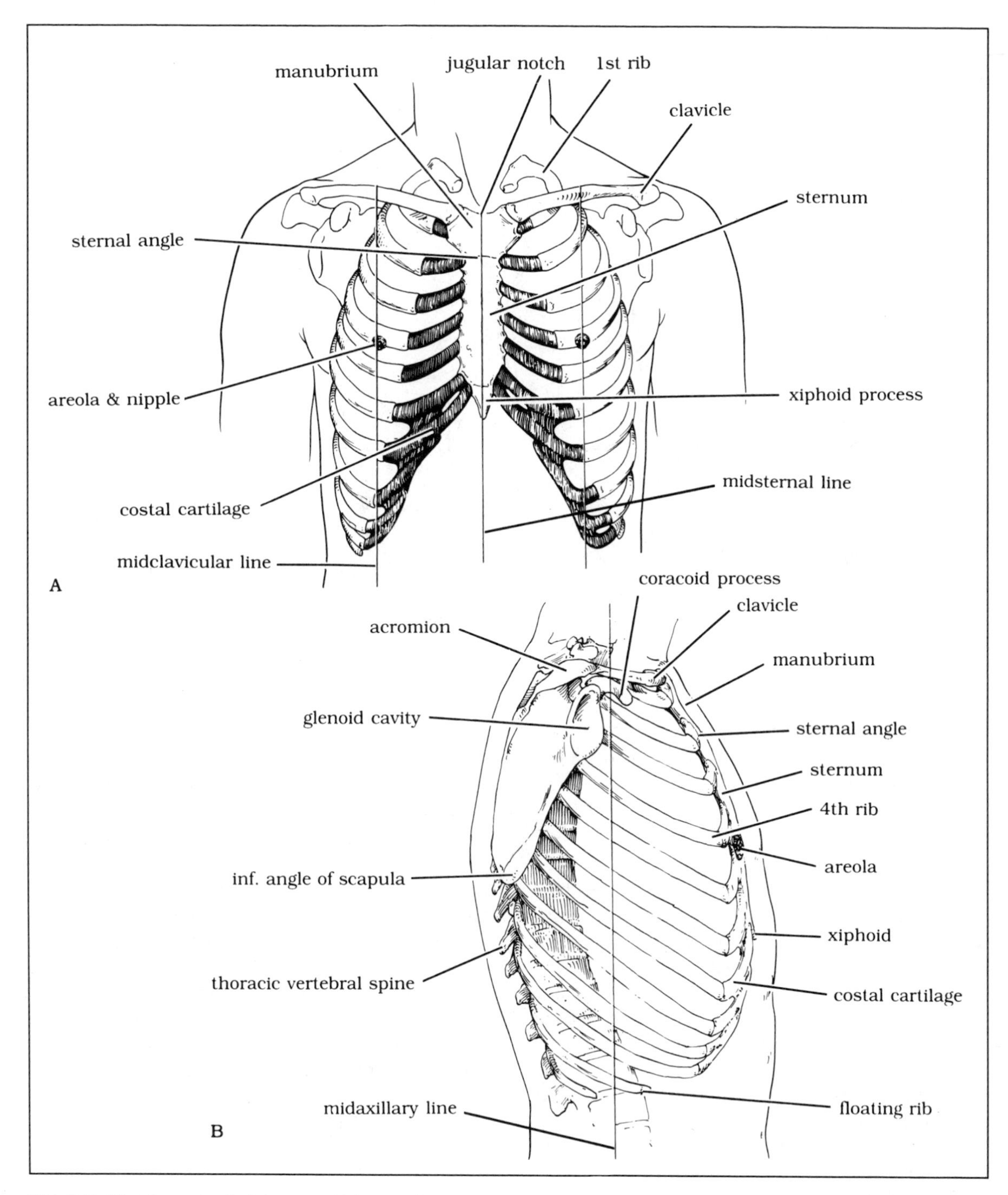

Fig. 2-3. The thoracic skeleton in relation to the surface. A. Anterior view. B. Lateral view.

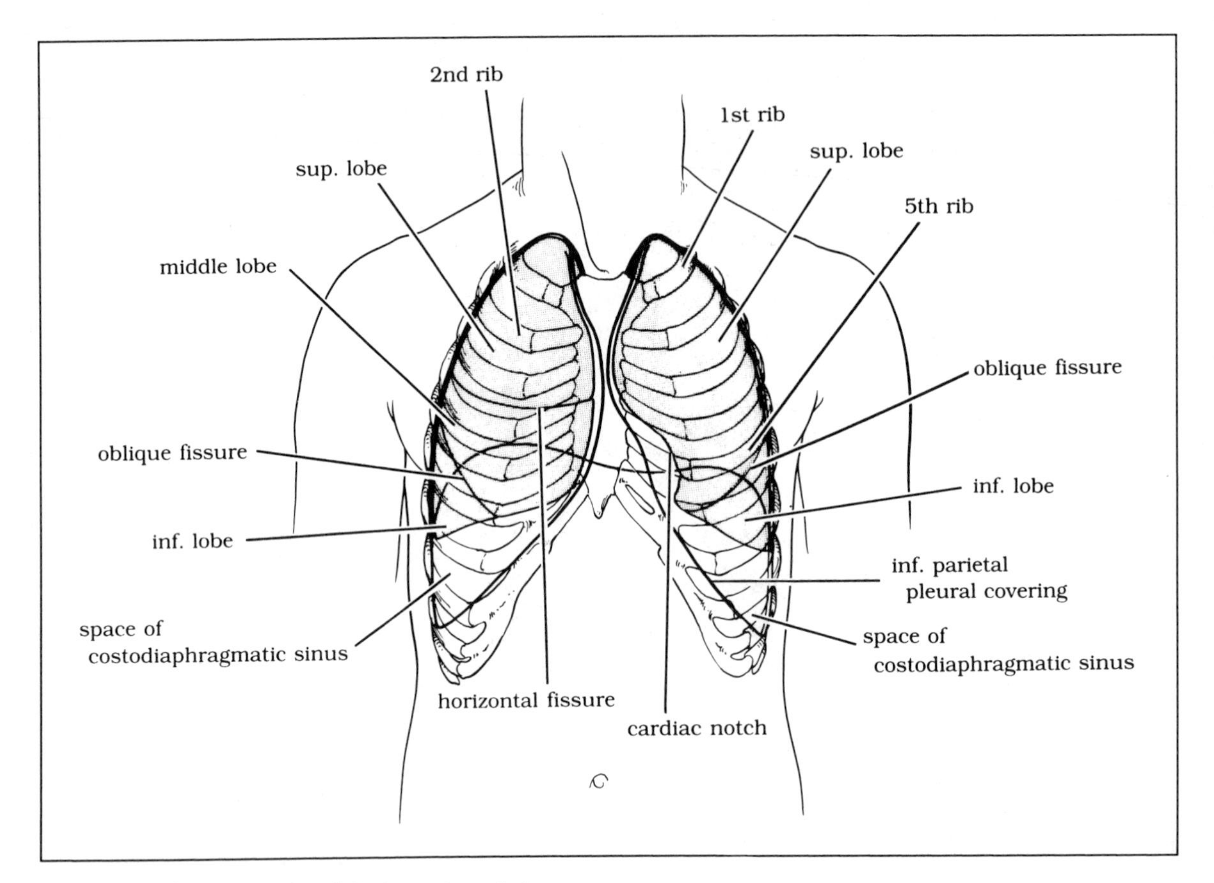

Fig. 2-4. Surface projection of the lungs (stippled region) and pleura on the anterior body wall.

found in the fifth intercostal space approximately 9 cm from the midline (Fig. 2-5A,B). The left border of the heart ascends from there to a point approximately 2 cm from the sternal margin in the second intercostal space. The base of the heart, from which the great vessels emerge, is a line from this point to a point 1 cm to the right of the sternal margin at the level of the third costal cartilage. The right border then descends to the level of the sixth or seventh costal cartilage. Finally the diaphragmatic border continues from this point behind the xiphosternal junction to the apex.

The projection of the pericardium differs only in that it extends superiorly to the sternal angle. This is due to the fact that the pericardium covers the root of the great vessels.

The valves of the heart are found at costal levels as follows: the pulmonary valve at the third costal cartilage, left side of the sternum; the aortic valve at the third intercostal space, left side of the sternum; the mitral (bicuspid or left atrioventricular) valve at the fourth costal cartilage, left side of the sternum; and the tricuspid (right atrioventricular) valve at the fourth intercostal space, right side of the sternum. This arrangement is most easily remembered as the series 3, 3½, 4, 4½.

Thoracic Wall

SKELETON

Unlike in the extremities, where the bony tissue lies at the core of the limb, in the thorax the bony

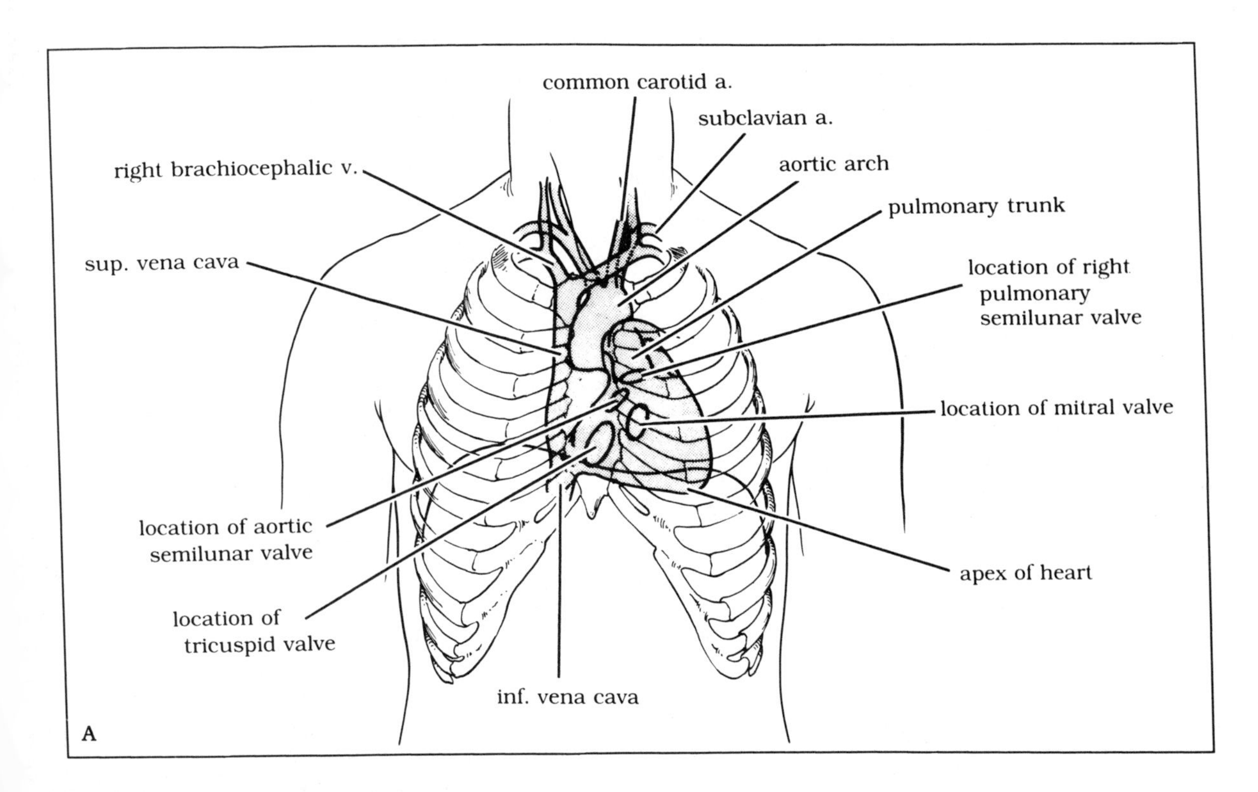

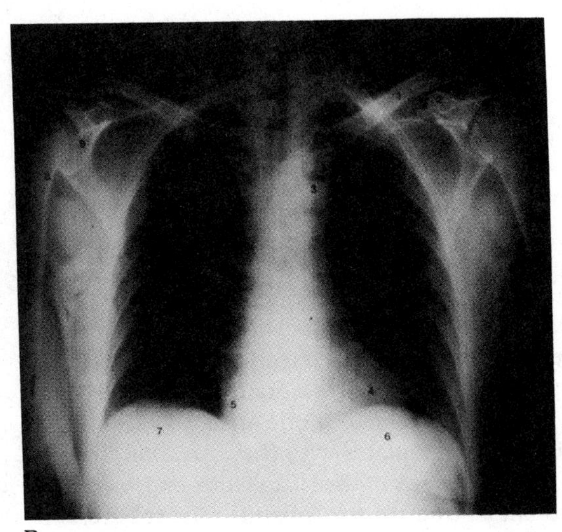

B

Fig. 2-5. A. Surface projection of the heart and great vessels on the anterior body wall. B. X-ray film of the chest, posteroanterior view. (1 = clavicle; 2 = coracoid process; 3 = aortic arch; 4 = left ventricle; 5 = right atrium; 6 = left dome of diaphragm; 7 = right dome of diaphragm; 8 = humeral body; 9 = glenoid cavity.)

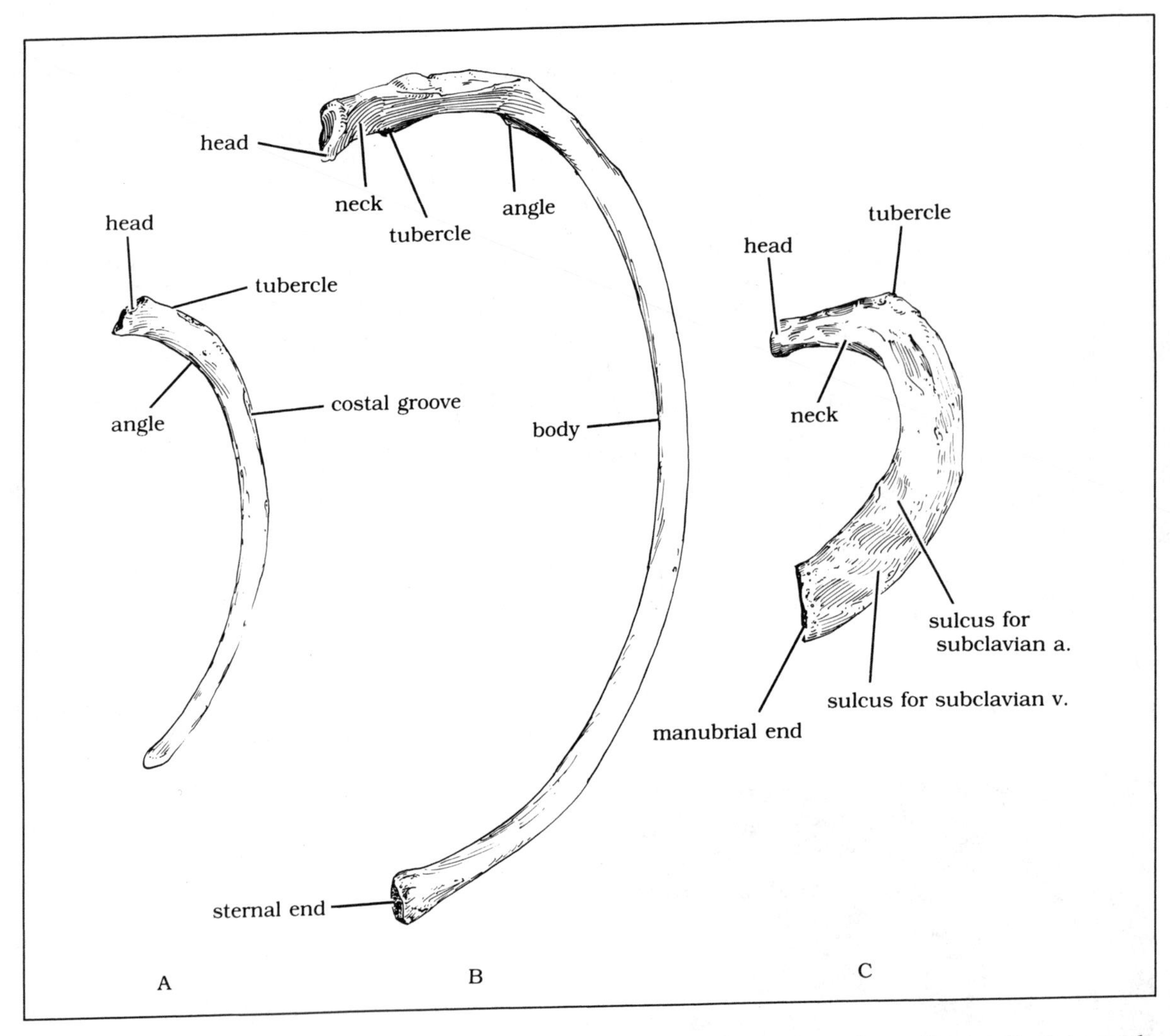

Fig. 2-6. Ribs. A. Floating rib (from above). B. True rib (from below). C. First rib (from below).

skeleton surrounds the major thoracic structures. The thoracic skeleton (see Fig. 2-2) consists of twelve ribs (seven true and five false, two of which are floating), twelve vertebrae, and the sternum (manubrium, body of the sternum, and xiphoid process).

The **ribs** (Fig. 2-6A,B,C) each have a head, neck, tubercle, and body (Table 2-1). The **head** has two articular facets, separated by a crest, which articulate by gliding joints with the adjacent vertebrae. The inferior facet articulates with the superior articular facet of the corresponding vertebra; the superior facet articulates with the inferior facet of the vertebra above the rib. (Ribs 1 and 10–12 have only one articular facet on the head that articulates with the corresponding vertebra.) The **neck** of the rib extends approximately 2.5 cm laterally to the tubercle. The **tubercle,** lying at the junction of the neck and shaft, articulates with the transverse process of its vertebra; the eleventh and twelfth ribs lack tubercular articular facets.

The **body** of the rib is both twisted and curved.

Table 2-1. Ribs

Part	*Rib Number*	*Special Features*	*Articulation*
Head	2–9	Inferior articular facet	With own vertebra
		Superior articular facet	With vertebra above
	1, 10–12	Articular facet	With own vertebra
Neck	1–12	Approximately 2.5 cm long	
Tubercle	1–10	Articular facet	With transverse process of own vertebra
	11, 12	No articular facet	
Shaft	2–11	Angle—change in curvature, point of greatest twisting	
Costal cartilage	1–7	True rib	Inserts into sternum
	8–10	False rib	Inserts into superior cartilage
	11, 12	False rib	Floats

The greatest twist and change in curvature occur at the angle of the rib. These turns in the body serve to make the anterior end of the rib lower than the posterior end. The superior border of the body is smooth and rounded; the inferior border is thin and sharp, having a groove, the costal groove, for passage of the vessels and nerves.

The first through seventh ribs end in costal cartilages that insert independently into the sternum and are thus **true ribs.** The eighth through tenth ribs end in costal cartilages that articulate with the cartilage of the rib above each and are thus **false ribs.** Finally the costal cartilages of the eleventh and twelfth ribs fail to articulate anteriorly and so the ribs **float** in the body wall.

Each rib articulates with the vertebral column at two places: between the head of the rib and the vertebral body, and between the neck and tubercle of the rib and the transverse process on the corresponding vertebra. The articulation between head and body is a plane joint that exhibits gliding motion that is restricted by its ligamentous binding. The articulation between neck, tubercles, and transverse process provides the major locus for motion (although this motion too is tightly constrained) guided by the shape of the articular surfaces.

On the upper six ribs the articular surfaces of the tubercles (convex) and the transverse processes (concave) allow rotation of the rib along the long axis of the neck, consequently moving the distal (anterior) tip upward or downward. On the seventh through tenth ribs the articular surface is flat and on the upper part of the transverse process rather than its anterior surface, where it is for the first six ribs. The ribs are therefore carried backward (upward and medially) or forward (downward and laterally).

The **sternum** (Fig. 2-7A,B and Table 2-2) consists of the manubrium, body of the sternum, and xiphoid process. Lying anteriorly in the midline it serves as the anterior site of articulation

Table 2-2. Sternum

Part	*Articulation*
Manubrium	With rib 1, clavicle, rib 2 (superior demifacet), body of sternum
Body	With rib 2 (inferior demifacet), ribs 3–6, manubrium, xiphoid, rib 7 (superior demifacet)
Xiphoid	With rib 7 (inferior demifacet), body of sternum

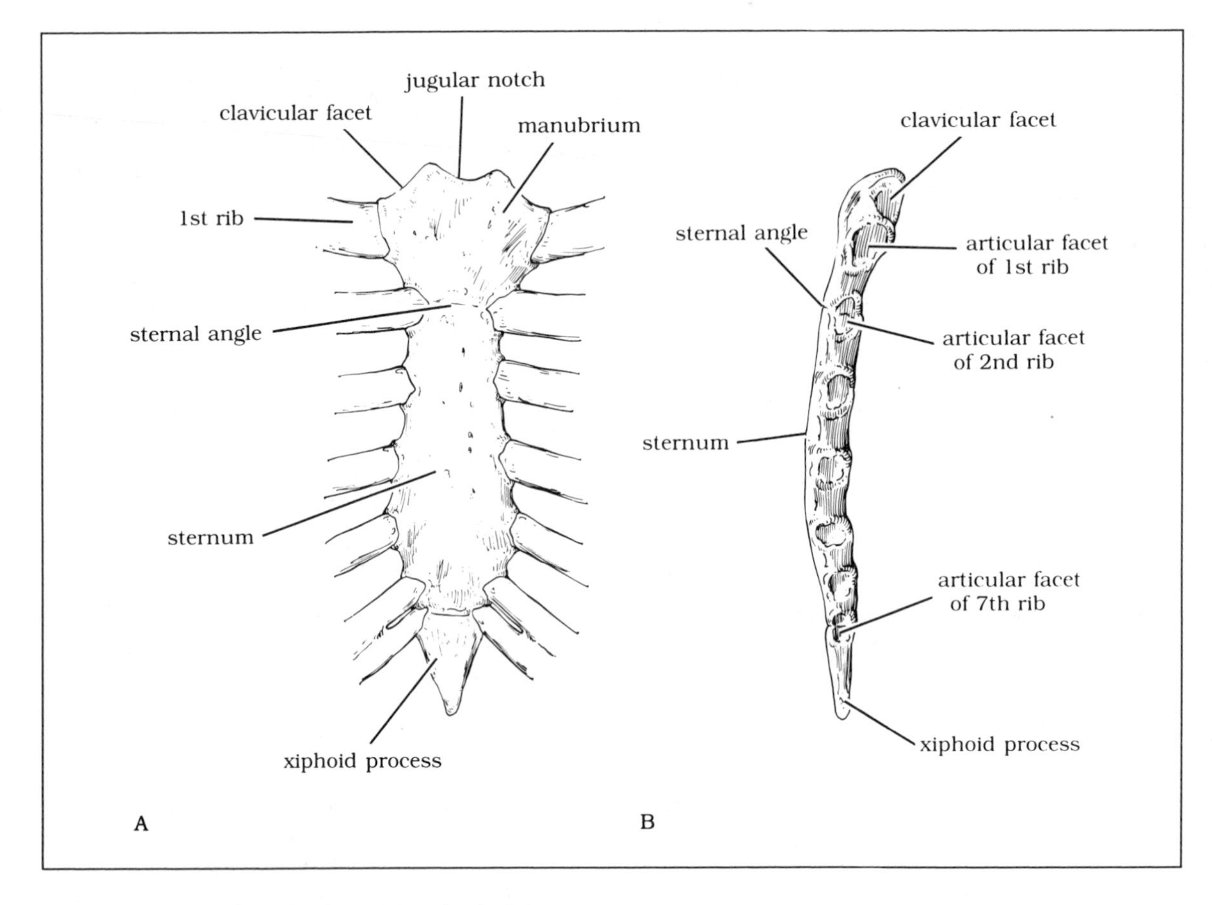

Fig. 2-7. Sternum from the front (A) and side (B).

for the true ribs. The **manubrium** articulates with the body of the sternum at the sternal angle. Above the angle the manubrium articulates with the first rib laterally, and the clavicle superiorly. Between the insertions of the clavicles lies the jugular notch. On the inferiolateral borders of the manubrium, at the level of the sternal angle, two demifacets for the insertion of the second costal cartilages exist.

Similarly, at the level of the sternal angle on the superolateral border of the **body of the sternum** two other demifacets exist for the insertion of the inferior portion of the second costal cartilages. The third through seventh ribs also insert into the body of the sternum. The seventh rib ends in a demifacet at the superolateral border of the **xiphoid process.**

The thoracic vertebrae are discussed in Chap. 9.

MUSCULATURE

The musculature of the thoracic wall consists of the external and internal intercostal, subcostal, and tranverse thoracic muscles (Table 2-3 and Fig. 2-8). (The levatores costarum and serratus posterior superior and inferior muscles are discussed in Chap. 9, and the diaphragm in Chap. 3.)

The **external intercostal muscles** occupy the 11 intercostal spaces from the tubercle of the ribs to the costochondral junction. Anterior to the junction, the muscle continues as a thin mem-

Table 2-3. Musculature of the Thoracic Wall

Muscle	*Origin*	*Insertion*	*Innervation*
External intercostal	Lower border, upper rib	Upper border, lower rib	Intercostal
Internal intercostal	Upper border, lower rib	Lower border, upper rib	Intercostal
Subcostal	Subset of internal intercostal covering two or more intercostal spaces		
Transverse thoracic	Lower sternum and xiphoid	Costal cartilages 2–6	Intercostal

Fig. 2-8. Schematic cross section of the thoracic wall.

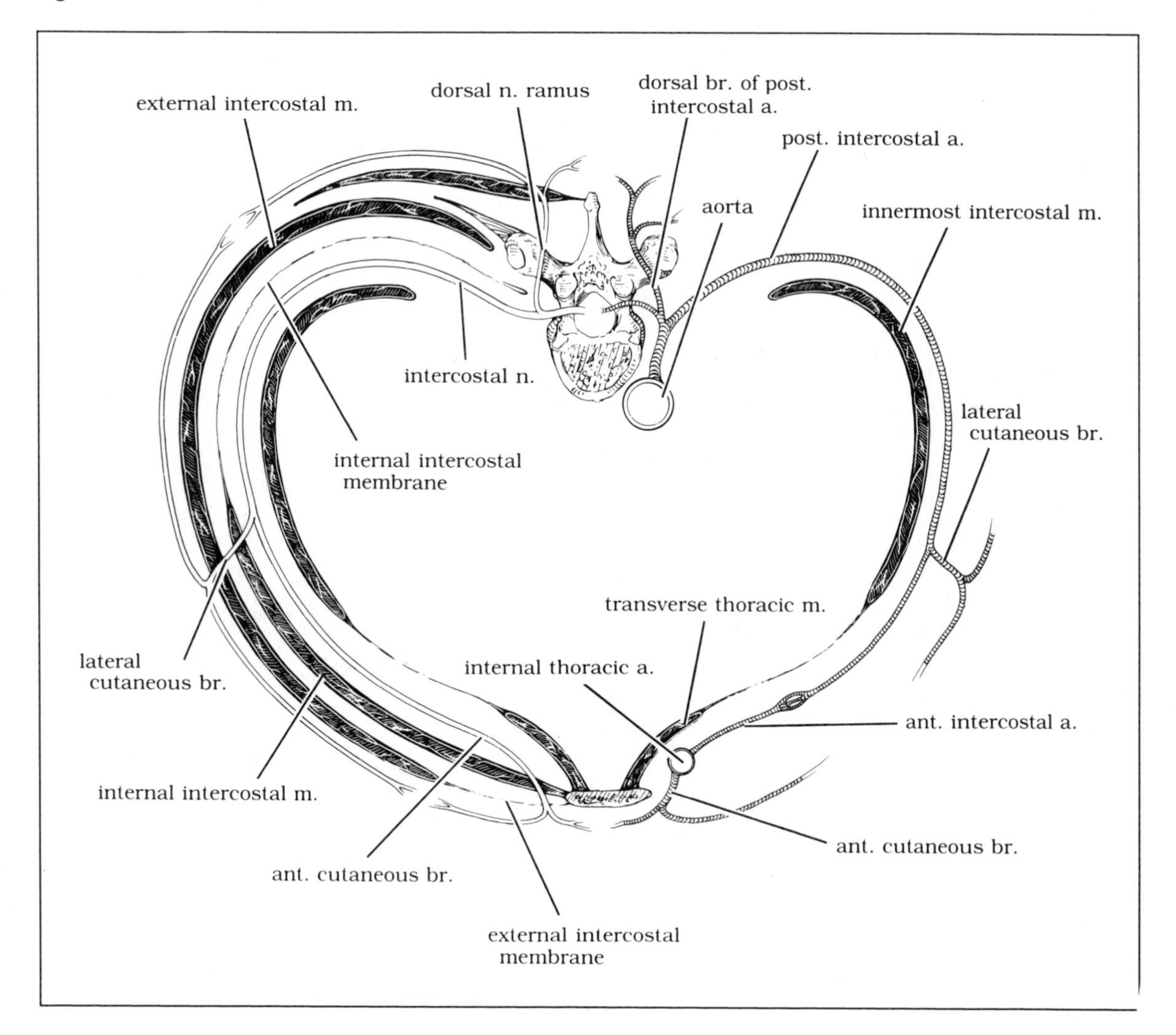

brane, the external intercostal membrane (see Fig. 2-8). The muscle arises from the lower border of the upper rib and inserts into the upper border of the lower rib. On the back the fibers run downward and laterally, and in the front they run downward and medially. They are innervated by the intercostal nerves.

The **internal intercostal muscles** (Fig. 2-9) occupy the 11 intercostal spaces from the border of the sternum to the angle of the ribs. Posteriorly, the muscle continues as the internal intercostal membrane. The muscle arises from the upper border of the lower rib and inserts into the lower border of the upper rib. The fibers, running perpendicular to those of the external intercostal, run upward and forward. A deep lamina of this muscle is known as the innermost intercostal muscle. It is innervated by branches of the intercostal nerves.

The **subcostal muscle** is a subset of the internal intercostal muscle, comprising the muscular fascicles that extend over two or more intercostal spaces.

The **transverse thoracic muscle** (see Figs. 2-8 and 2-9) is the deepest muscular layer lying on the sternochondral portion of the thoracic wall.

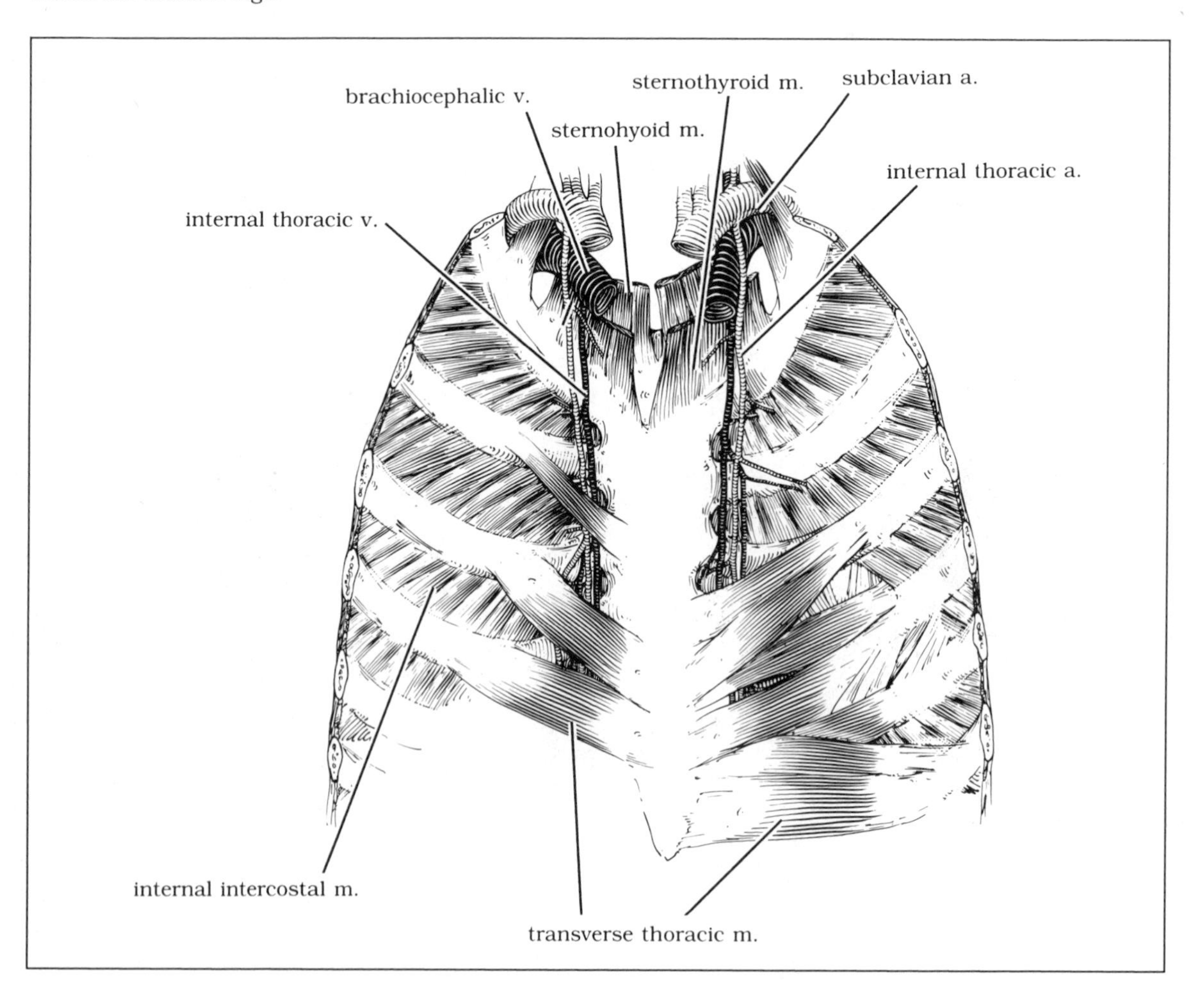

Fig. 2-9. The anterior thoracic wall viewed from within the thoracic cage.

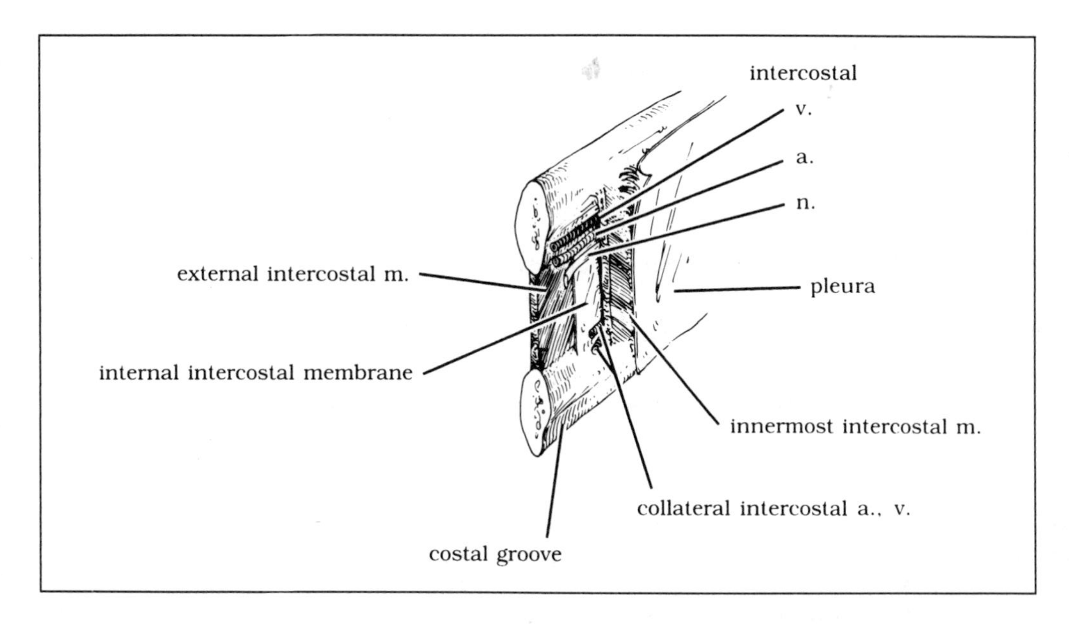

Fig. 2-10. Section through the posterior portion of an intercostal space showing muscles, vessels, and nerve.

The fibers arise from the lower body of the sternum and the xiphoid process and run upward and laterally to insert into the second to the sixth costal cartilages.

Deep to all the muscular layers lies a layer of fascia, the **endothoracic fascia,** which lines the entire thoracic cavity.

The veins, arteries, and nerves (V.A.N.) lie in the intercostal space in the costal groove (Fig. 2-10) in that order from superior to inferior.

BLOOD SUPPLY

The first two intercostal spaces are supplied by branches of the costocervical trunk of the subclavian artery and the superior thoracic branch of the axillary artery. The remaining intercostal spaces are supplied by nine pairs of **posterior intercostal arteries** and one pair of **subcostal arteries,** all of which arise from the thoracic aorta (see Fig. 2-8) and divide into anterior and posterior branches. The posterior branches run dorsally with the dorsal ramus of the spinal nerve. The anterior branches run with, and superior to, the intercostal nerve. The anterior artery gives off a lateral cutaneous branch that runs with a nerve of the same name, and a collateral branch that runs on the upper border of the lower rib (see Fig. 2-10). The artery ends by anastomosing with the corresponding anterior intercostal artery, which arises from the internal thoracic artery (see Figs. 2-8 and 2-9).

There are eleven pairs of posterior intercostal veins and one pair of subcostal veins that accompany the corresponding arteries, lying most superiorly in the intercostal space. The first posterior intercostal vein empties into the brachiocephalic vein. The second, third, and fourth join to form the superior intercostal vein (see Fig. 2-24), which empties into the brachiocephalic vein on the left side and the azygos on the right side. The remaining veins empty into the azygos on the right and the hemiazygos or accessory hemiazygos on the left.

Internal Thoracic Artery and Vein

The **internal thoracic artery** is a branch of the subclavian artery running 1 cm from the margin of the sternum ventral to the transverse thoracic

Table 2-4. Internal Thoracic Artery

Branch	*Supplies*	*Anastomoses*
Anterior intercostal (ribs 1–6)	Intercostal space	With posterior intercostal
Anterior perforating	Cutaneous	
Pericardiacophrenic	Pericardium, pleura, diaphragm	
Musculophrenic	Diaphragm, anterior intercostal (ribs 7–11)	
Superior epigastric	Abdominal wall	With inferior epigastric (occasionally)

muscle (see Figs. 2-8 and 2-9; Table 2-4). It gives off (1) **anterior intercostal arteries,** which run laterally to anastomose with the posterior intercostal arteries; (2) **anterior perforating arteries,** which accompany the anterior cutaneous nerves; (3) the **pericardiacophrenic artery,** which supplies blood to the pericardium, pleura, and diaphragm. At the level of the sixth intercostal space the internal thoracic artery splits into two terminal branches; (4) the **musculophrenic artery,** which gives off the seventh to eleventh anterior intercostal arteries and supplies the diaphragm; and (5) the **superior epigastric artery,** which supplies the abdominal wall by running in the rectus sheath and sometimes anastomosing with the inferior epigastric artery.

The internal thoracic veins are paired venae comitantes of the internal thoracic artery (see Fig. 2-9). They arise by fusion of the superior epigastric and musculophrenic veins between the first and third intercostal spaces. The veins end in the brachiocephalic vein.

INNERVATION

The **intercostal nerves** are the ventral rami of the first 11 thoracic spinal nerves (Table 2-5). The twelfth forms the **subcostal nerve.** These nerves contribute preganglionic fibers (white rami) to the sympathetic ganglia and receive postganglionic fibers (gray rami) from them.

The **thoracic intercostal nerves** (thoracic spinal nerves 1–6) pass posteriorly between the internal and external intercostal muscles (see Figs. 2-8 and 2-10). As they progress anteriorly they come to lie between the internal intercostal and transverse thoracic muscles. At the approximate level of the midaxillary line they give off **lateral cutaneous branches,** which in turn divide into anterior and posterior branches. These branches supply much of the ventral, lateral, and dorsal cutaneous innervation. The thoracic intercostal nerves terminate by piercing the intercostal and pectoral musculature to become **anterior cutaneous nerves,** which supply ventral cutaneous regions. The muscles innervated by the thoracic intercostal nerves are the external and internal intercostal, subcostal, and transverse thoracic and serratus posterior superior muscles.

The seventh to eleventh intercostal nerves constitute the **thoracoabdominal intercostal nerves.** They follow the same course as the thoracic intercostal nerves, except at their anterior ends, where they pass behind the costal cartilages to penetrate the abdominal wall by passing through the transverse abdominal and internal abdominal oblique muscles and end as anterior cutaneous branches. They supply the internal and external intercostals, the transverse abdominal and rectus abdominis, the internal and external abdominal oblique and the serratus posterior inferior.

The **subcostal** (twelfth thoracic) **nerve** has no

Table 2-5. Thoracic Spinal Nerves

Nerve	*Thoracic Spinal Nerves*	*Branches*	*Muscular Innervation*
Thoracic intercostal	1–6	Lateral and anterior cutaneous	External and internal intercostal, subcostal, transverse thoracic, serratus posterior superior
Thoracoabdominal intercostal	7–11	Lateral and anterior cutaneous	External and internal intercostal, external and internal abdominal oblique, transverse abdominal and rectus abdominis, serratus posterior inferior
Subcostal	12	Lateral and anterior cutaneous	External and internal abdominal oblique, transverse abdominal and rectus abdominis, serratus posterior inferior, pyramidalis

thoracic distribution (see Table 2-5). It crosses ventral to the quadratus lumborum, penetrates the transverse abdominal muscle, and distributes like a thoracoabdominal intercostal nerve. It also innervates the pyramidalis, when this muscle is present.

LYMPHATICS

The lymph nodes of the thoracic wall can be classified according to where they are located. The **sternal nodes** lie along the intercostal arteries on the upper four intercostal spaces. The **intercostal nodes** lie either singly or in pairs in the intercostal space near the head of each rib. The **phrenic nodes** are located on the thoracic surface of the diaphragm. The anterior phrenic nodes are found behind and beside the xiphoid process, the middle phrenic nodes are found near the termination of the phrenic nerve, and the posterior phrenic nodes are located next to the aorta.

Thoracic Cavity

GENERAL THORACIC ORGANIZATION

The thoracic cavity, contained within the thoracic wall, may be divided into three parts: two lateral pleural compartments containing the lungs and a central compartment, the mediastinum, which completely separates the pulmonary compartments and contains the heart and great vessels.

The **mediastinum** contains all the thoracic viscera except the lungs. Its lateral limits are the parietal pleura of the right and left pleural sacs. In a dorsoventral direction it extends from the vertebral column to the sternum, and craniocaudally its limits are the superior thoracic aperture and the diaphragm. Thus the mediastinum serves to divide the thoracic cavity completely.

RESPIRATORY SYSTEM

Trachea and Bronchi

The trachea, composed of C-shaped cartilaginous rings, begins in the neck at the cricoid cartilage (see Chap. 8). In the thorax the trachea lies in the posterior portion of the superior mediastinum. It ends at the level of the sternal angle (fifth thoracic vertebra) just to the right of the midline by dividing into right and left bronchi. The last tracheal ring marks the beginning of the bronchi and has a central ridge inside, the **carina,** which divides the entrance to the bronchi (Fig. 2-11).

The **right bronchus** is shorter (2.5 cm long), straighter, and wider than the left (5 cm long). It is the more-direct continuation of the trachea and is therefore more likely to receive foreign objects. The right bronchus enters the lung at the level of the fifth thoracic vertebra, the azygos vein

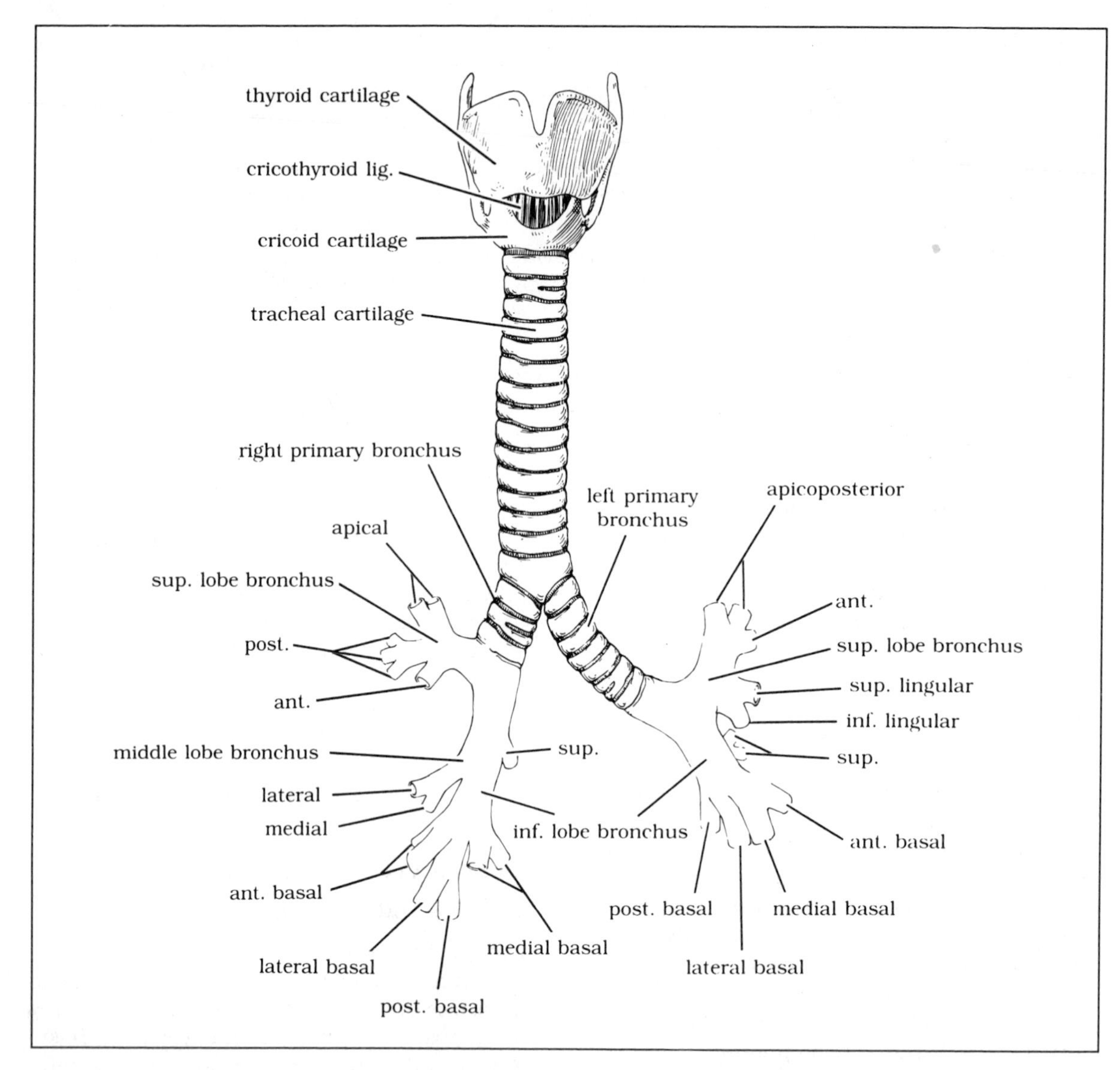

Fig. 2-11. The trachea and associated respiratory structures (segmental bronchi).

arching over it and the pulmonary artery lying ventral to it (see Fig. 2-25A). Just superior to the artery the bronchus gives off its first branch, the **superior lobe bronchus.** The remaining two branches, the **middle and inferior lobe bronchi,** are given off below the pulmonary artery.

The **left bronchus** passes under the aortic arch anterior to the esophagus and thoracic aorta. At the root of the lung it is inferior to the pulmonary artery. Its branches are the **superior lobe bronchus** (also supplying the lingula) and the **inferior lobe bronchus.** Further divisions of the lobar bronchi follow those of the lung (Table 2-6 and pp. 37–38).

Pleurae

The limits of the pleural cavities and the lines of reflection of the pleura are discussed at the beginning of this chapter. There are both parietal and visceral pleurae, and the pleural cavity is the potential space between them. This space is normally filled with a serous fluid that facilitates the movement of the lungs by lubricating the pleurae.

The **visceral pleura** completely and intimately invests the lungs, dipping down into the fissures separating the lobes. It is continuous with the parietal pleura at the root of the lungs. The **parietal pleura** (see Fig. 2-4) lines the surface of the thoracic wall, diaphragm, superior thoracic aperture, and mediastinum. It is considered to have four parts, which are named after the surfaces they cover: costal, diaphragmatic, cervical, and mediastinal. The mediastinal pleura is continuous with the visceral pleura at the root of the lungs.

During quiet inspiration the parietal pleura comes in contact with itself at two places: the **costodiaphragmatic recess** and the **costomediastinal recess.** In the former, costal and diaphragmatic pleurae come in contact with each other; in the latter, costal and mediastinal pleurae make contact in the anterior mediastinal region.

Lungs

The surface projection of the lungs is discussed at the beginning of this chapter. The lungs are the essential organs of respiration, allowing for an exchange of gases with the blood at the level of the pulmonary capillaries. The lung tissue is smooth and shiny because of its covering of visceral pleura. The lung itself is light, spongy, and elastic. Each lung has an apex and a base, costal and mediastinal surfaces, and anterior, posterior, and inferior borders.

The **apex** of the lungs (Fig. 2-12A,B) is round and occupies the cervical dome of the pleura. The **base** is broad and concave, resting on the diaphragm.

The **anterior** and **inferior borders** are thin and sharp. The anterior border of the left lung is indented by the cardiac notch (see Fig. 2-12A). The **posterior border** is rounded and smooth.

The **costal surface** is large and convex, and in fixed tissue it is indented by the ribs. The **mediastinal surface** (see Fig. 2-12B) is concave, containing the root of the lung and the cardiac impression (which is deeper and larger on the left side). At the root of the lung, mediastinal pleura reflects onto the lung to become visceral pleura; a downward extension of this reflection is the **pulmonary ligament.**

In fixed tissue the mediastinal surface of the lung is marked by the structures that it comes in contact with in the mediastinum. On the left lung above the hilum there are grooves for the arch of the aorta, subclavian artery, and brachiocephalic vein. Behind the hilum there is a groove for the thoracic aorta. At the base of the lung there is a depression in which lies the esophagus. On the right lung above the root there are grooves for the arch of the azygos vein, superior vena cava, brachiocephalic vein, and subclavian artery. Behind the hilum are depressions for the esophagus and inferior vena cava.

Each lung has an **oblique fissure** separating it into superior and inferior lobes. The superior lobe of the right lung is further divided by the **horizontal fissure** into superior and middle lobes. The lingula of the left lung is homologous to the middle lobe of the right lung.

Each lung is further divided into **bronchopulmonary segments** (Fig. 2-13A,B and Table 2-6),

Table 2-6. Bronchopulmonary Segments

Lobe	*Right Lung*	*Left Lung*
Inferior	Posterior, lateral, apical, medial, anterior	Posterior, apical, lateral, anteromedial
Middle (lingula)	Lateral, medial	Superior, inferior
Superior	Posterior, apical, anterior	Apicoposterior, anterior

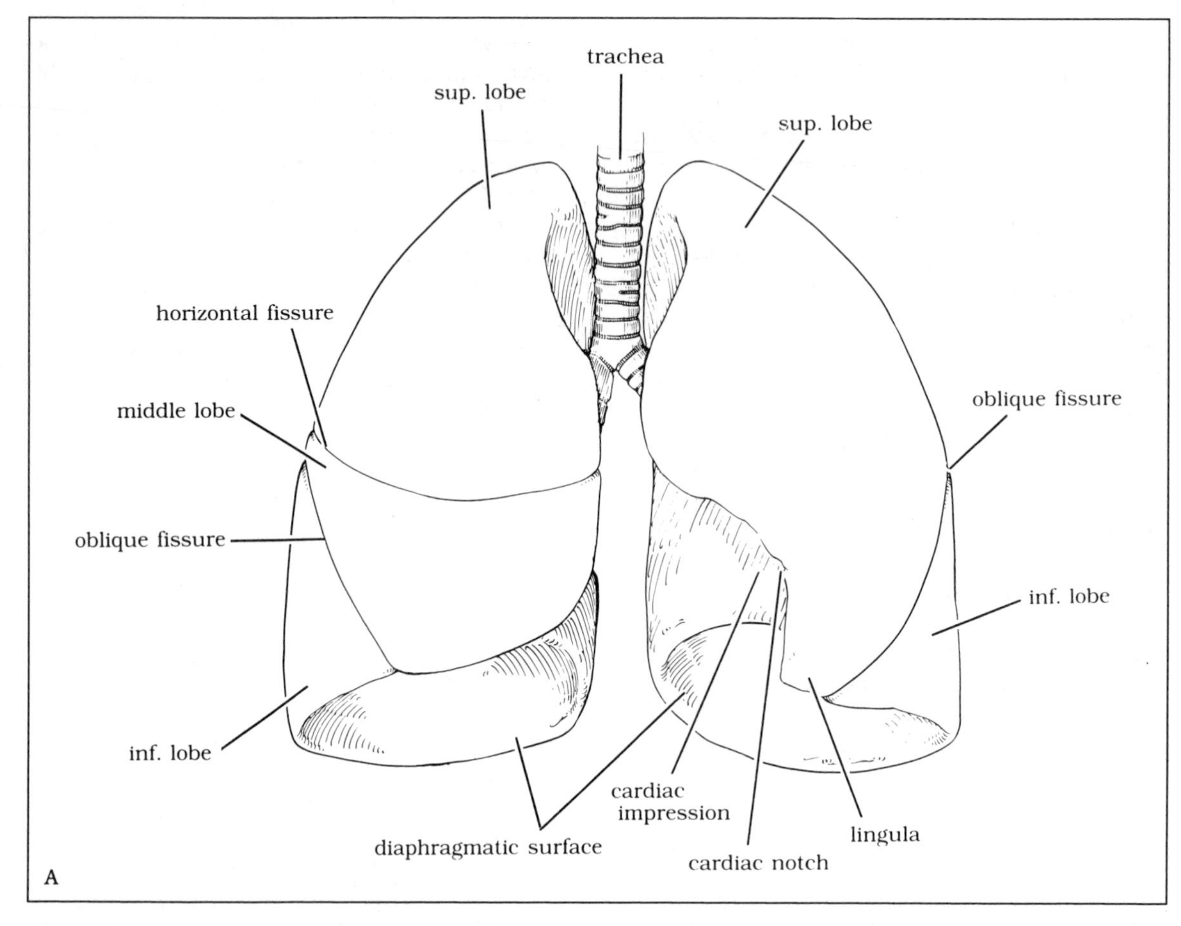

Fig. 2-12. The lungs. A. Anterior view. B. Medial view.

ten on the right and eight on the left. The bronchopulmonary segments of the right lung may be easily learned as follows. The inferior lobe is like a pyramid having anterior, posterior, medial, lateral, and apical segments. The middle and superior lobes form a second pyramid. The middle lobe has medial and lateral segments, whereas the superior lobe has anterior, posterior, and apical segments.

The left lung's bronchopulmonary segments are similar. The inferior lobe is the same as the right except that because of the position of the heart the anterior and medial segments are fused into an anteromedial segment. The lingula has superior and inferior segments, and the apical and posterior segments of the superior lobe are fused into an apicoposterior segment.

The **root** of the lung (see Fig. 2-12B) is that portion of the mediastinal surface of the lung through which nerves and vessels enter and leave. The order of the bronchi, pulmonary arteries, and pulmonary veins is similar on both sides. In a dorsoventral direction the bronchi are dorsal, the veins ventral, and the arteries in between. Moving from superior to inferior, the arteries are superior, the veins inferior, and the bronchi in between. The only exception to this arrangement is the superior lobe bronchus of the right lung, which lies above the artery (eparterial).

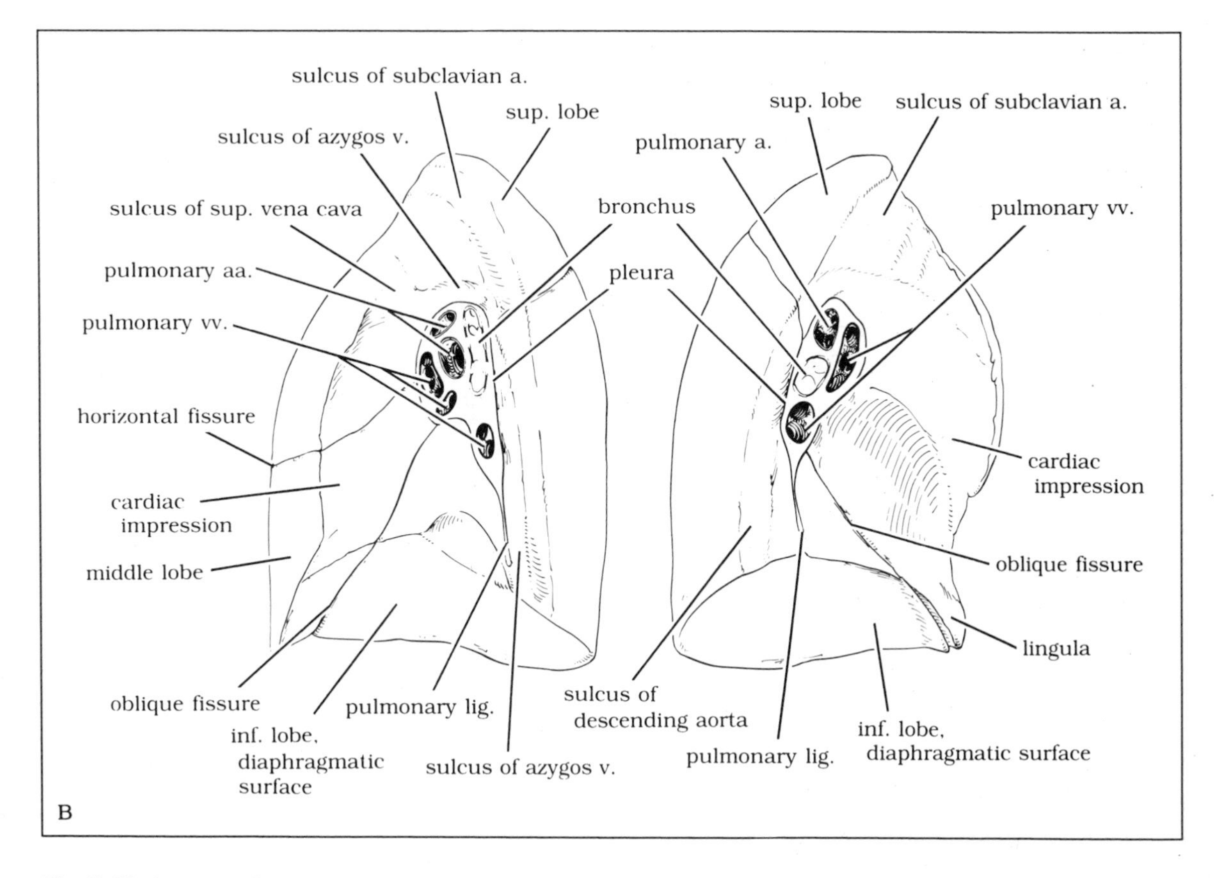

Fig. 2-12. (continued).

Blood Supply

Blood supply to and from the lung tissue is provided by the **bronchial arteries** and **veins.** All arteries in the lungs travel with the bronchi, branching into smaller and smaller units as they travel from lobes to bronchopulmonary segments and finally to alveoli. Blood is brought to the lungs by the pulmonary arteries for aeration (Fig. 2-14). The **pulmonary trunk** divides under the arch of the aorta (see Figs. 2-15A and 2-17B) into right and left pulmonary arteries.

The **right pulmonary artery,** longer and slightly larger than the left, passes dorsal to the ascending aorta and superior vena cava to come to the root of the lung, where it divides into an anterior trunk (to the superior lobe) and an interlobar trunk (to the middle and inferior lobes). The trunks then subdivide to supply the bronchopulmonary segments.

The **left pulmonary artery,** connected to the arch of the aorta by the ligamentum arteriosum, arches over the left bronchus as it enters the root of the lung. Dorsal to the bronchi it divides into apical segmental, posterior segmental, anterior segmental, and interlobar arteries. These arteries then subdivide to supply the bronchopulmonary segments.

The **pulmonary veins** (two per lung) (see Fig. 2-17A) return the aerated blood to the left atrium of the heart. The left pulmonary veins come from the superior and inferior lobes separately. On the right side the middle and superior lobe veins fuse to leave the lung as one, in company with the vein from the inferior lobe.

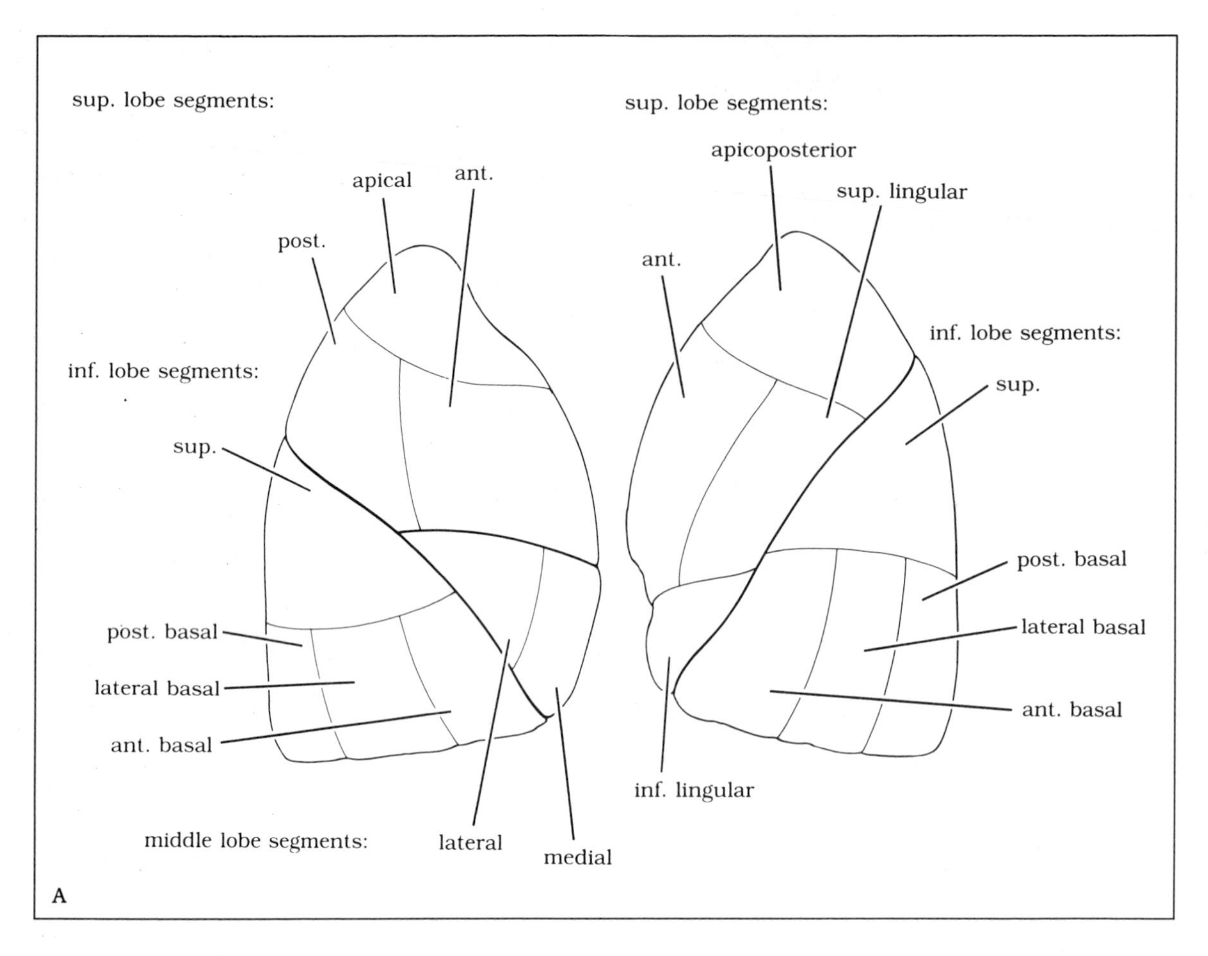

Fig. 2-13. Bronchopulmonary segments. A. Lateral view. B. Medial view.

Mechanism of Respiration

Respiration is the exchange of external oxygen for internally generated carbon dioxide. This exchange is carried out in the lungs, in particular in the alveoli, and thus the lungs must be continuously filled and purged of gas. The primary mechanism of normal respiration is the contraction and relaxation of the diaphragm. Contraction of this muscle pushes it down into the abdomen to approximately the level of the eleventh vertebra, increasing intraabdominal pressure and decreasing thoracic pressure by changing the volumes of the two compartments. Under decreased pressure, air is passively pulled into the lungs, where gas exchange may proceed. Conversely, relaxation of the diaphragm increases thoracic pressure and forces air from the lungs.

The lower intercostal muscles are thought to assist the diaphragm in increasing the transverse diameter, and so the volume, of the thoracic cavity. The remaining muscles of the thoracic wall appear to be involved in shape maintenance and organ constraint. However, in deep and forced respiration the diaphragm contracts more forcefully, creating a larger pressure drop, and higher levels of intercostal muscles become involved in cavity enlargement. The scalene and the sternocleidomastoid muscles of the neck may act to pull up the rib cage, and the erector

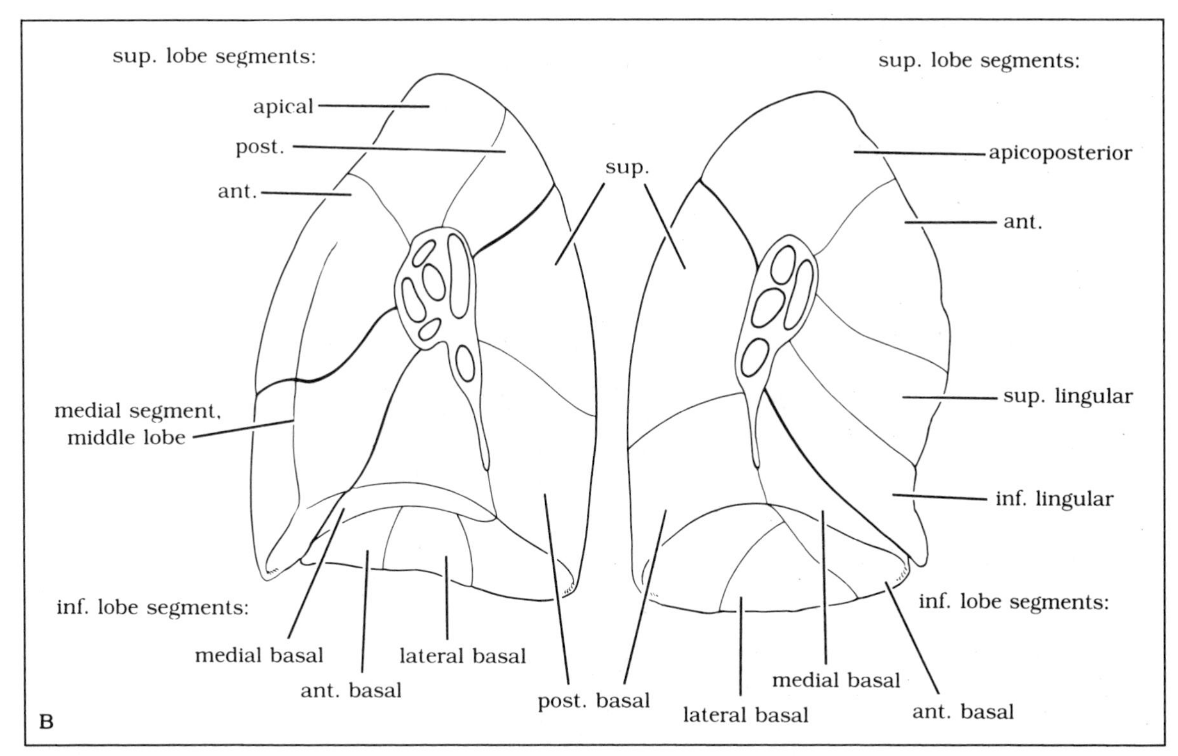

Fig. 2-13. (continued).

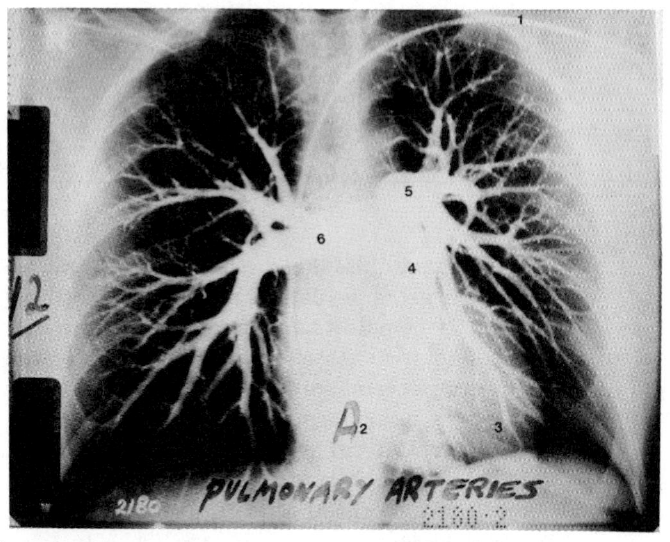

Fig. 2-14. X-ray film of the pulmonary arteries. (1 = catheter to right ventricle; 2 = right ventricle; 3 = left ventricle; 4 = pulmonary trunk; 5 = left pulmonary artery; 6 = right pulmonary artery.)

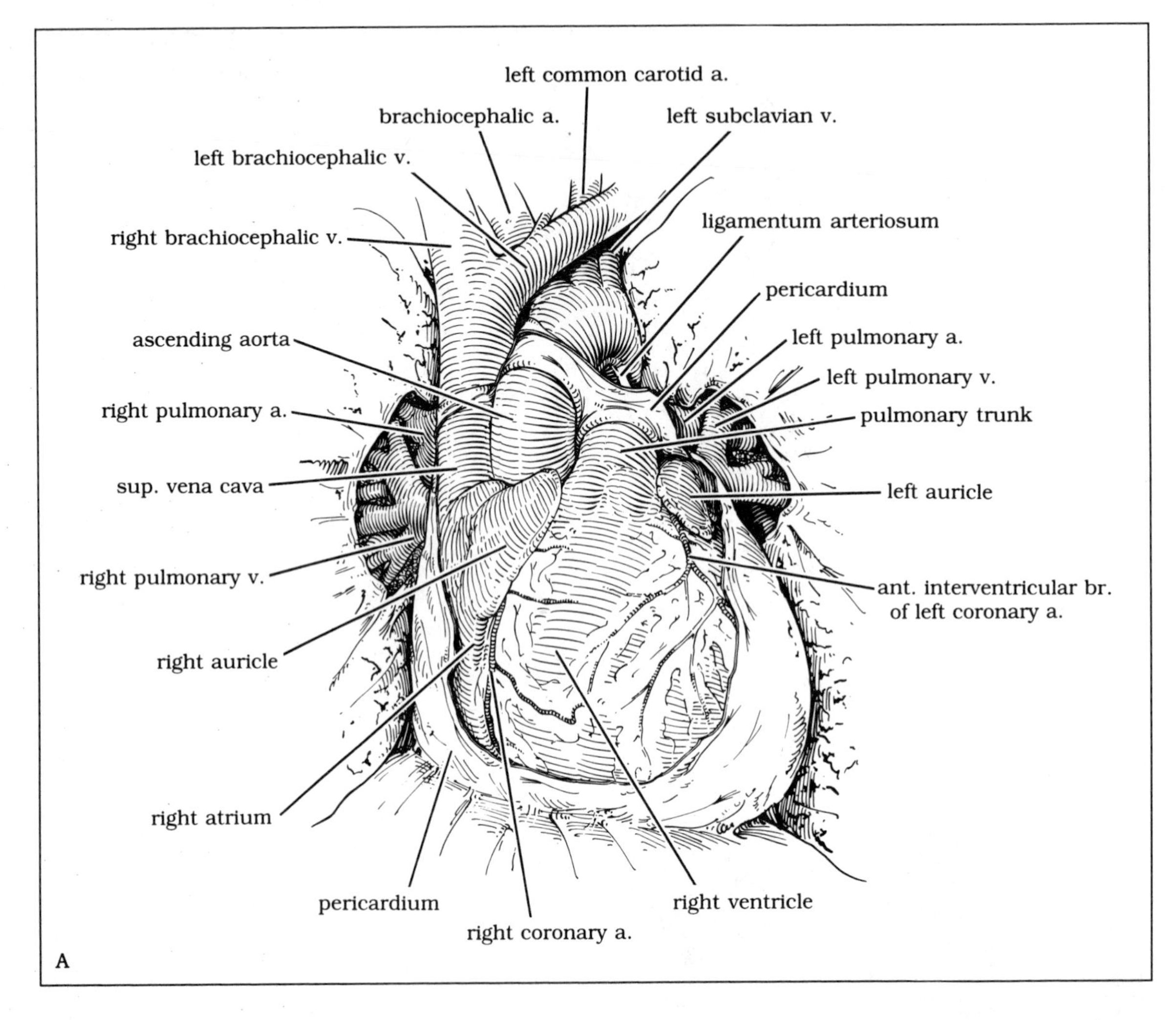

Fig. 2-15. A. The pericardium, heart, and great vessels. B. The pericardial sac with the heart removed, showing the sinuses.

spinae muscles of the back may act to decrease the concavity of the thoracic vertebral column, allowing greater motion of the ribs and expansion of the rib cage.

HEART AND PERICARDIUM (MIDDLE MEDIASTINUM)

The surface projection of the pericardium and the heart and its valves is discussed at the beginning of this chapter.

The pericardium (Fig. 2-15A,B) is a sac enclosing the heart musculature, and it and everything within it constitute the middle mediastinum: heart and great vessels, roots of the lungs, arch of the azygos vein, and phrenic nerves. The top of the sac is closed around the great vessels (approximately 2.5 cm from their roots). The sac extends from the diaphragm, where it is firmly attached to the central tendon, to the level of the sternal angle.

The outer wall of the pericardial sac is a dense fibrous connective tissue; the inner wall is a serous membrane. This serous pericardium lines both the inner wall of the sac and the outside of

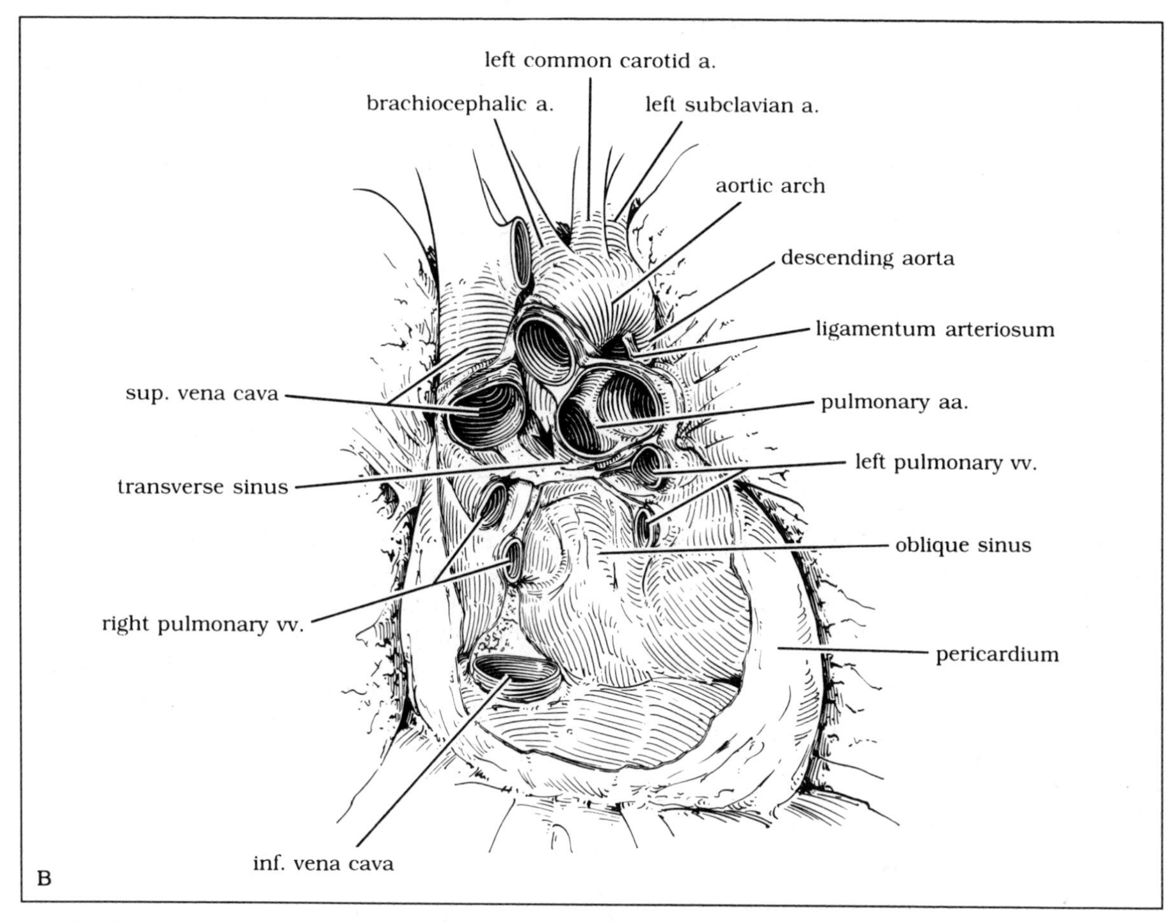

Fig. 2-15. (continued).

the heart itself and so may be designated either parietal or visceral pericardium. The latter is also known as epicardium.

The points at which the visceral pericardium leaves the heart to become parietal pericardium constitute the **pericardial reflections.** These reflections occur at the roots of all vessels entering or leaving the heart (see Fig. 2-15B). In the dorsal wall of the pericardium the reflection for the veins creates a cul-de-sac known as the oblique sinus. The space between the venous and arterial reflections also constitutes a sinus known as the transverse sinus.

The **pericardial cavity** is the potential space between the visceral and parietal pericardia. In the healthy individual this is filled with a thin layer of serous fluid that lubricates the movements of the heart within the pericardium.

Chambers of and Circulation Through the Heart

The human heart is composed of four chambers: two atria and two ventricles. (Table 2-7 summarizes the blood circulation through these chambers.) The apex of the heart is made up by the ventricles; the base by the atria and beginnings of the great vessels. The sternocostal surface of the heart (see Fig. 2-15A) is made up by the right atrium and ventricle with a small contribution by the left ventricle; the diaphragmatic surface is formed by the two ventricles.

Table 2-7. Circulation of Blood Through the Heart

Chamber	*Blood from*	*Blood to*	*Valves*	*Internal Structures*
Right atrium	Superior and inferior venae cavae and coronary sinus	Right ventricle	Tricuspid (right atrioventricular)	Pectinate muscles, crista terminalis, fossa ovalis
Right ventricle	Right atrium	Pulmonary artery	Tricuspid (right atrioventricular), pulmonary	Trabeculae carneae, papillary muscles, chordae tendineae
Left atrium	Pulmonary vein	Left ventricle	Bicuspid (left atrioventricular)	Pectinate muscles, valve of foramen ovale
Left ventricle	Left atrium	Aorta	Bicuspid (left atrioventricular), aortic	Trabeculae carneae, papillary muscles, chordae tendineae

The surface of the heart has two major sulci. The **coronary sulcus** (atrioventricular sulcus) encircles the heart, separating the atria and the ventricles on both the anterior and posterior surfaces of the heart. The **interventricular sulcus** lies between the ventricles on both the anterior and the posterior surfaces of the heart.

The **right atrium** (Fig. 2-16A) receives venous blood from the entire body, except the pulmonary veins. It is larger than the left atrium, although its walls are thinner. The atrium has two cavities, the auricle and the principal cavity. The

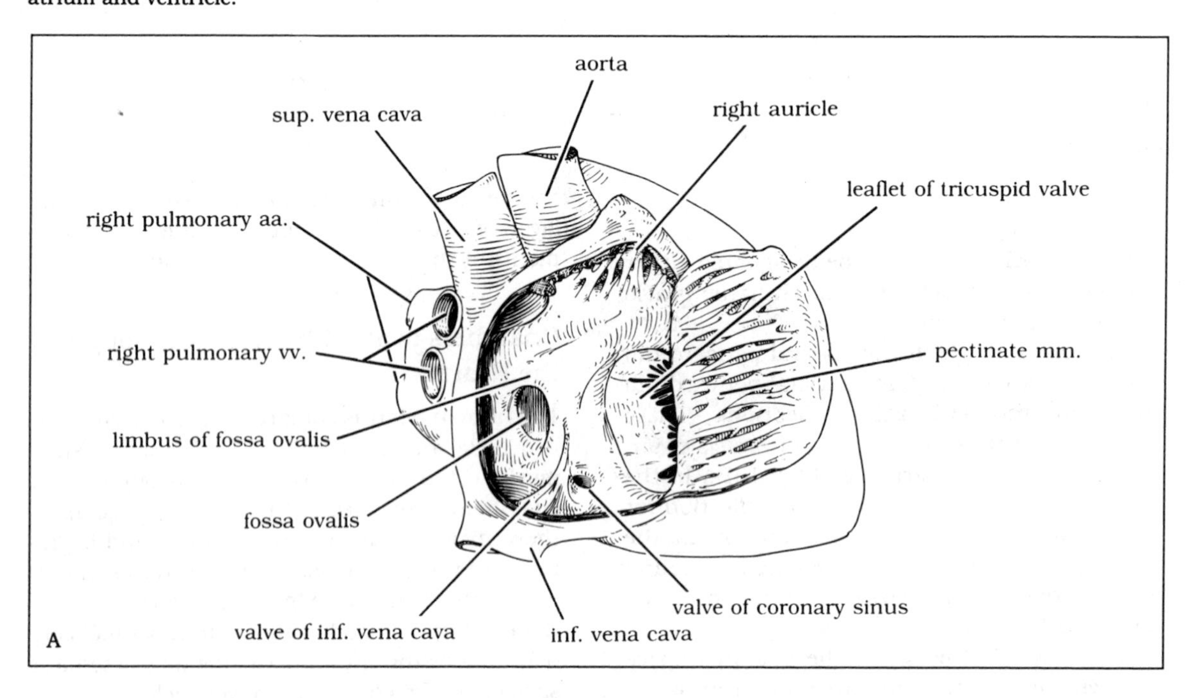

Fig. 2-16. Interior views of the chambers of the heart. A. Right atrium. B. Right ventricle. C. Left atrium and ventricle.

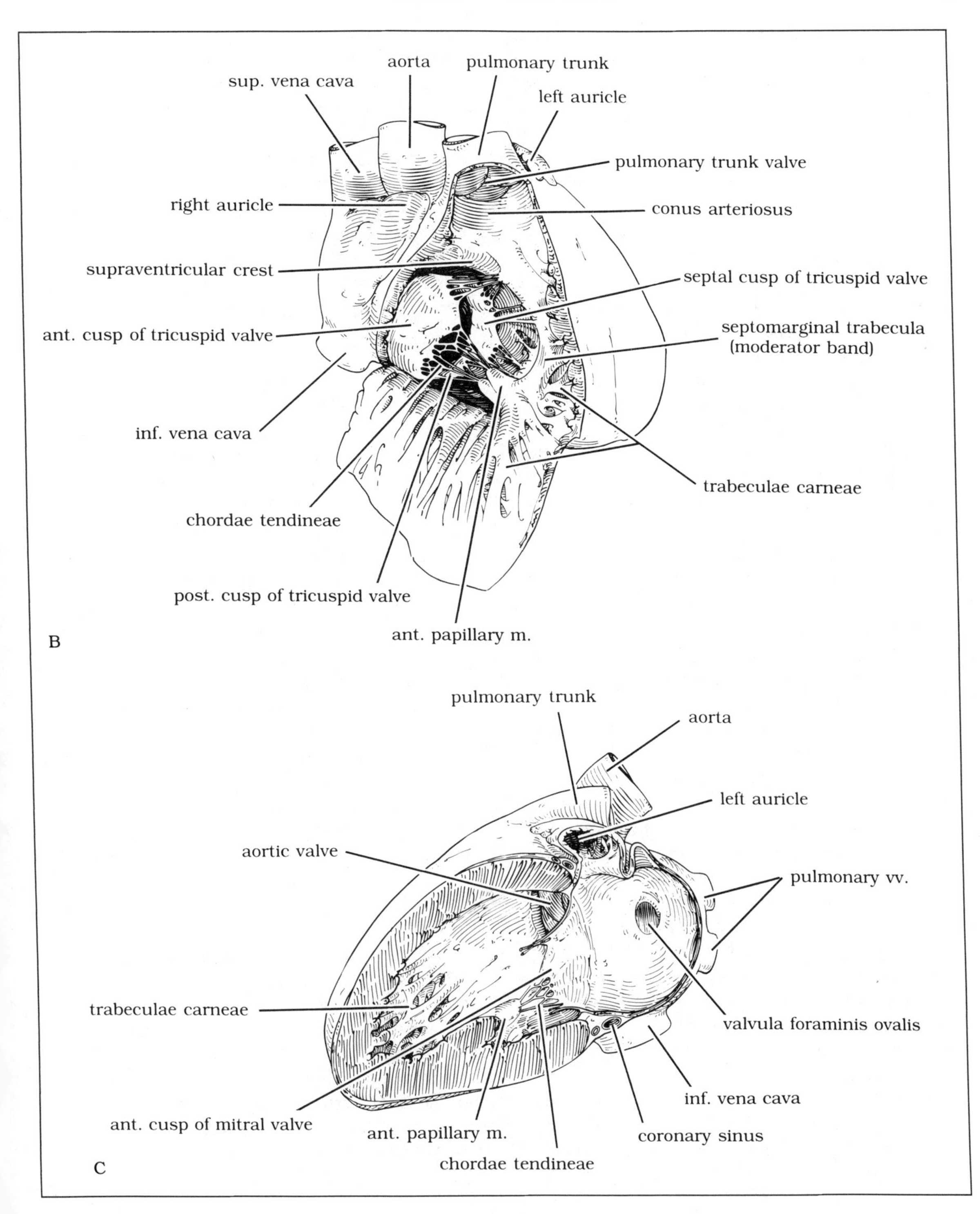
aorta
pulmonary trunk
sup. vena cava
left auricle
pulmonary trunk valve
right auricle
conus arteriosus
supraventricular crest
septal cusp of tricuspid valve
septomarginal trabecula
(moderator band)
ant. cusp of tricuspid valve
inf. vena cava
trabeculae carneae
chordae tendineae
post. cusp of tricuspid valve
ant. papillary m.
B
pulmonary trunk
aorta
left auricle
aortic valve
pulmonary vv.
trabeculae carneae
valvula foraminis ovalis
inf. vena cava
ant. cusp of mitral valve
ant. papillary m.
coronary sinus
chordae tendineae
C

auricle is a small, rough-walled, blind pouch. The **principal cavity** receives blood from the superior and inferior venae cavae and the coronary sinus. The openings of the inferior vena cava and coronary sinus have valves that are thin crescentic folds.

The wall of the atrium has both smooth and rough portions. The ridge between them is the **crista terminalis** (terminal crest). The rough part of the wall, primarily in the auricle, is formed by the **pectinate muscles** (musculi pectinati). On the right side in the interatrial septum is an oval depression, the **fossa ovalis,** which corresponds to the foramen ovale in the embryo. The margin of the fossa is limited by a limbus.

The **atrioventricular opening** on the right side is closed by the **tricuspid** (right atrioventricular) valve. Blood flows through this into the **right ventricle** (Fig. 2-16B). The cusps of the valve are attached to **chordae tendineae,** strong fibrous cords that are involved in keeping the valve shut during ventricular contraction and so preventing backflow of blood into the atrium. The chordae run from the cusps to **papillary muscles,** which are projections of the ventricular wall involved in maintaining the position of the cusps. The anterior papillary muscle is the largest muscle on the right side.

The ventricular wall is also ridged by the **trabeculae carneae,** an irregular muscular network covering almost the entire inner ventricular surface.

Blood exits the right ventricle to the pulmonary artery through the **pulmonary valve,** which is a semilunar valve with three cusps.

After traveling through the pulmonary system, blood returns to the heart via the four pulmonary veins, entering the left atrium (Fig. 2-16C). The **left atrium** is primarily smooth walled, the pectinate muscles being limited almost entirely to the auricle. On the left side of the interatrial septum is a ridge, the remains of the embryonic valve of the foramen ovale.

Blood exits the left atrium through the **mitral,** or left atrioventricular, valve into the **left ventricle.** The structure of this valve is essentially the same as that of the tricuspid, except it has only two cusps. These cusps are also connected to chordae tendineae that end in papillary muscles, as on the right side. The walls of the left ventricle are trabeculated and thicker than those of the right ventricle.

Blood leaves the left ventricle through the **aortic semilunar valve,** which has three cusps. Behind two of the cusps are the openings for the **coronary arteries.**

The **interventricular septum** may be divided into two parts: a thicker muscular part below, and a thinner membranous part above.

Great Vessels

The great vessels (Fig. 2-17A,B) entering and leaving the heart are the superior vena cava, inferior vena cava, aorta, pulmonary trunk, and pulmonary veins. The **superior vena cava** returns blood to the heart from the upper limbs, head, and neck. It is formed by the fusion of the two brachiocephalic veins at the level of the first costal cartilage, behind the manubrium. The vessel runs caudally to enter the right atrium at the level of the third costal cartilage. For approximately half its course it is covered with pericardium.

The **inferior vena cava** brings blood to the heart from the lower body. It has a short course in the thorax; piercing the diaphragm at the level of the eighth thoracic vertebra, it runs only 1 to 2 cm before entering the right atrium.

The **aorta** arises from the left ventricle at the level of the third costal cartilage (the level of the aortic semilunar valve). It then ascends, covered by pericardium, to the level of the middle of the manubrium, where it arches dorsally and leftward, prior to its descent. The **arch of the aorta** is connected to the left pulmonary artery by the ligamentum arteriosum.

The **pulmonary trunk** arises from the right ventricle and runs for approximately 5 cm before bifurcating. The division into right and left pulmonary arteries takes place at the level of the fifth thoracic intervertebral space, under the arch of the aorta. The pulmonary trunk is covered by pericardium for its entire length.

The four **pulmonary veins** return blood from

the lungs to the left atrium. There is one vein each from the superior and inferior lobes of the left lung and one from the inferior lobe of the right lung. The fourth vein drains the middle and superior lobes of the right lung.

Blood Supply and Innervation

Arising from the ascending aorta in the right aortic sinus (the cusp of the right aortic semilunar valve) is the **right coronary artery** (Fig. 2-18A). This artery, along with the left coronary artery, supplies blood to the musculature of the heart. The right coronary artery runs in the coronary sulcus to the right margin of the heart. There it gives rise to the **marginal artery** (which runs along the right margin toward the apex) and then continues in the sulcus onto the posterior surface of the heart. At the posterior interventricular sulcus it gives rise to the **posterior interventricular artery,** which runs toward the apex.

The **left coronary artery** arises from the left aortic sinus. The artery runs between the left atrium and the pulmonary trunk to the coronary sulcus in the region of the anterior interventricular sulcus. There it gives rise to the **anterior interventricular artery** and continues in the coronary sulcus as the **circumflex artery.** The anterior interventricular artery passes toward the apex, continuing to the posterior surface to anastomose with the posterior interventricular artery. The circumflex artery ends by anastomosing with the right coronary artery on the posterior surface of the heart.

The **coronary veins** (Fig. 2-18B) return the blood to the right atrium of the heart. The **great cardiac vein** runs in the anterior interventricular sulcus to the coronary sulcus, where it turns on to the posterior surface of the heart to become the coronary sinus. The **middle cardiac vein** runs in the posterior interventricular sulcus to the coronary sinus. The **small cardiac vein** runs in the right coronary sulcus to the coronary sinus. The **posterior vein** of the left ventricle runs along the diaphragmatic surface of the heart to the coronary sinus. The **coronary sinus,** receiving the major veins of the heart, runs in the coronary sulcus and enters the right atrium between the inferior vena cava and the atrioventricular opening.

In addition to these vessels many small arteries and veins run in the heart musculature. These vessels communicate directly with the chambers of the heart.

The heart is innervated by way of the **cardiac plexus.** This plexus, formed by parasympathetic (vagal) preganglionic fibers and sympathetic postganglionic fibers, is located at the base of the heart within the arch of the aorta between the bifurcation of the trachea and the bifurcation of the pulmonary trunk. The parasympathetic fibers, involved in slowing down the heart, synapse either within the plexus or within the heart wall. The sympathetic fibers are involved in increasing the speed and strength of the heartbeat.

Internally the heartbeat is coordinated by the **sinoatrial node,** which lies between the superior vena cava and the right atrium and receives fibers from the cardiac plexus. Impulses are transmitted through the atrial wall to the **atrioventricular** (AV) **node,** which also receives fibers from the cardiac plexus. This node lies just above the aperture of the coronary sinus. From the AV node stimuli pass down the interventricular septum via the AV bundle and are distributed to the ventricular walls, in particular to the papillary muscles.

Mediastinum

The mediastinum has previously been described as a partition separating the two pleural cavities. In addition it serves as a corridor through which structures from the head and neck region connect with lower regions of the body and vice versa. The mediastinum is divided into four anatomical regions (Fig. 2-19): superior, anterior, middle, and posterior. The superior mediastinum lies above a line drawn from the sternal angle to the lower border of the fourth thoracic vertebra. Its upper limit is the plane of the first rib. The middle mediastinum consists of the con-

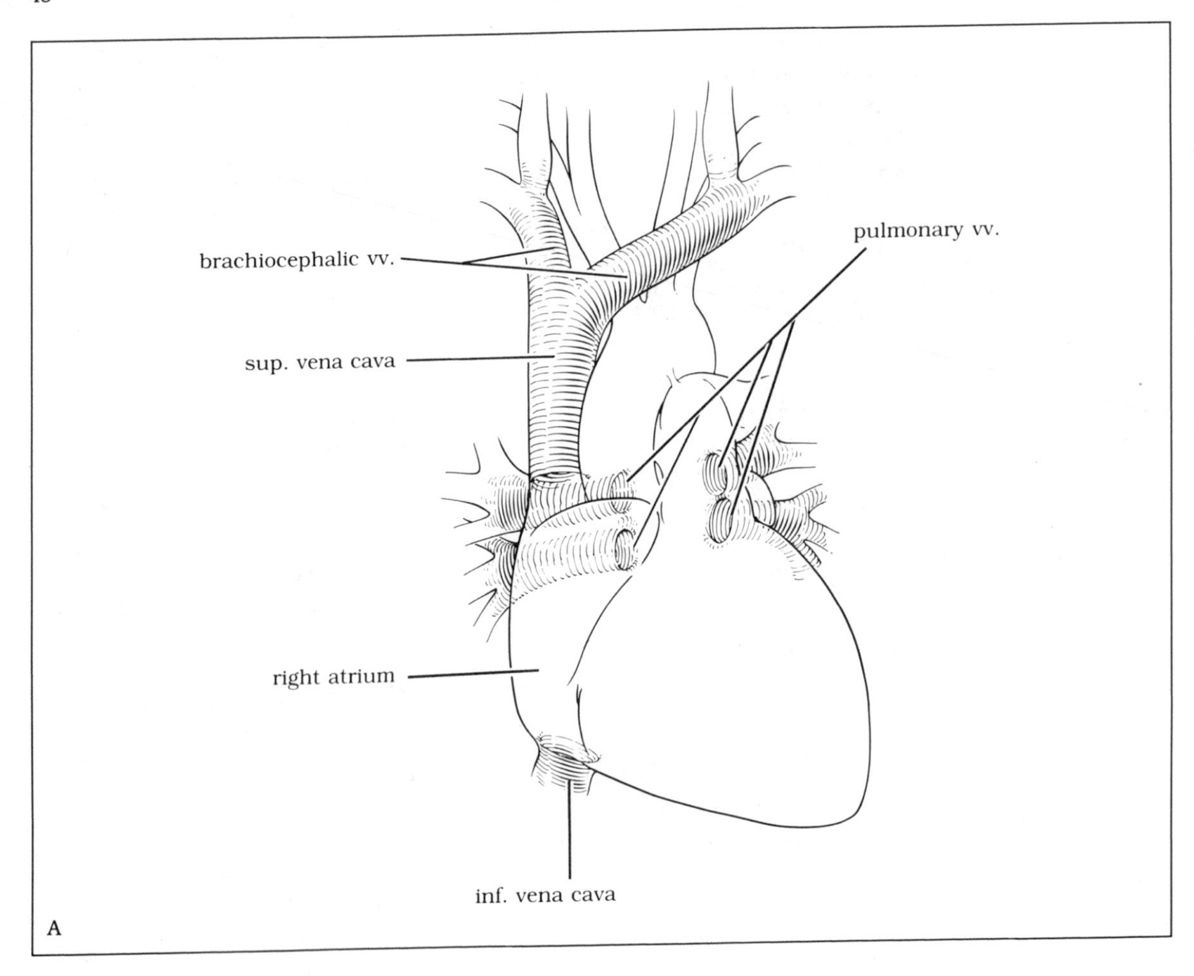

Fig. 2-17. Major vessels entering and leaving the heart. A. Venous drainage to the heart. B. Arterial flow from the heart.

tents of the pericardium (heart and great vessels), the roots of the lungs, the arch of the azygos vein, and the phrenic nerves (see previous discussion). The anterior and posterior mediastina are defined in relation to the middle mediastinum: The anterior mediastinum is anterior to the pericardium and posterior to the thoracic wall; the posterior mediastinum is posterior to the pericardium and anterior to the vertebral column.

The **anterior mediastinum** consists primarily of loose areolar connective tissue. In addition lymph nodes and lymphatic vessels may be found in the region. The inferior sternopericardial ligament, which holds the pericardium to the xiphoid process, passes through the anterior mediastinum.

In the **middle mediastinum,** the phrenic nerves (which innervate the diaphragm) enter the thorax by passing between the subclavian artery and vein (Fig. 2-20). As they pass the origin of the internal thoracic artery they are joined by the pericardiacophrenic vessels. The right nerve passes directly to the diaphragm, lying against the superior vena cava as it does so. The left nerve descends between the common carotid and subclavian arteries to cross the arch of the aorta. It then descends, between the pleura and the pericardium, anterior to the root of the lungs to reach the diaphragm.

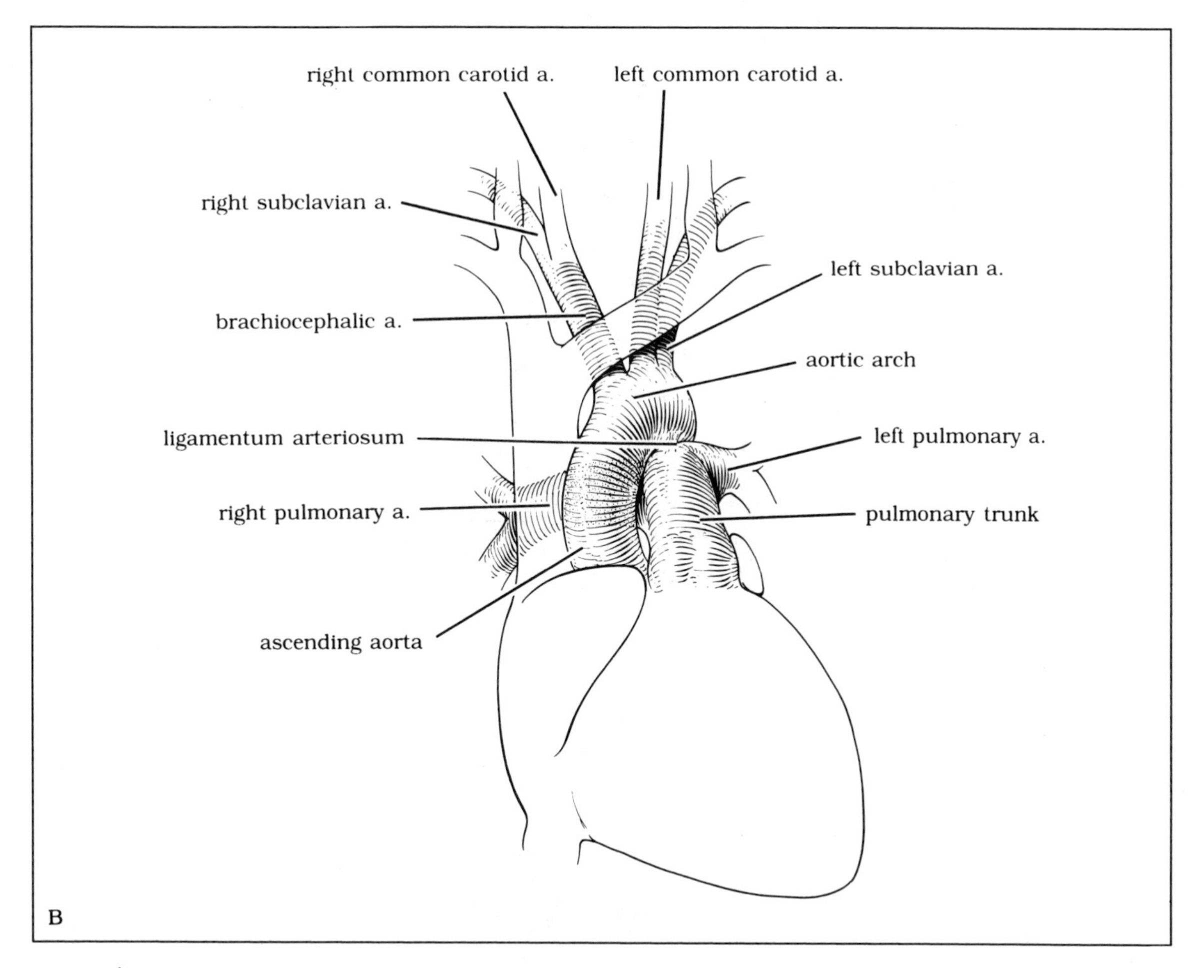

Fig. 2-17. (continued).

SUPERIOR MEDIASTINUM

The **thymus gland,** a bilobed organ, lies in the superior mediastinum (Table 2-8 and Fig. 2-20) posterior to the manubrium and anterior to the great vessels. In the adult the majority of glandular tissue has been replaced by fat.

The arch of the aorta (see Figs. 2-17B and 2-24B) lies immediately posterior to the thymus. The arch carries the aorta dorsally and leftward, turning downward at the fourth thoracic vertebra to become the descending thoracic aorta at the level of the fifth thoracic vertebra. The branches of the arch of the aorta are the brachiocephalic, left common carotid, and left subclavian arteries (Fig. 2-21). The brachiocephalic artery ascends backward and to the right for 5 cm. As it passes behind the sternoclavicular junction it divides into the right subclavian and common carotid arteries.

The left common carotid artery, the second branch of the aortic arch, arises from the top of the arch and ascends behind the sternoclavicular joint to enter the neck. The left subclavian artery arises 1 cm distal to the left common carotid. Lying against the lung it ascends in a leftward direction to pass over the first rib and leave the thorax.

The trachea lies in the superior mediastinum posterior to and slightly to the right of the arch of

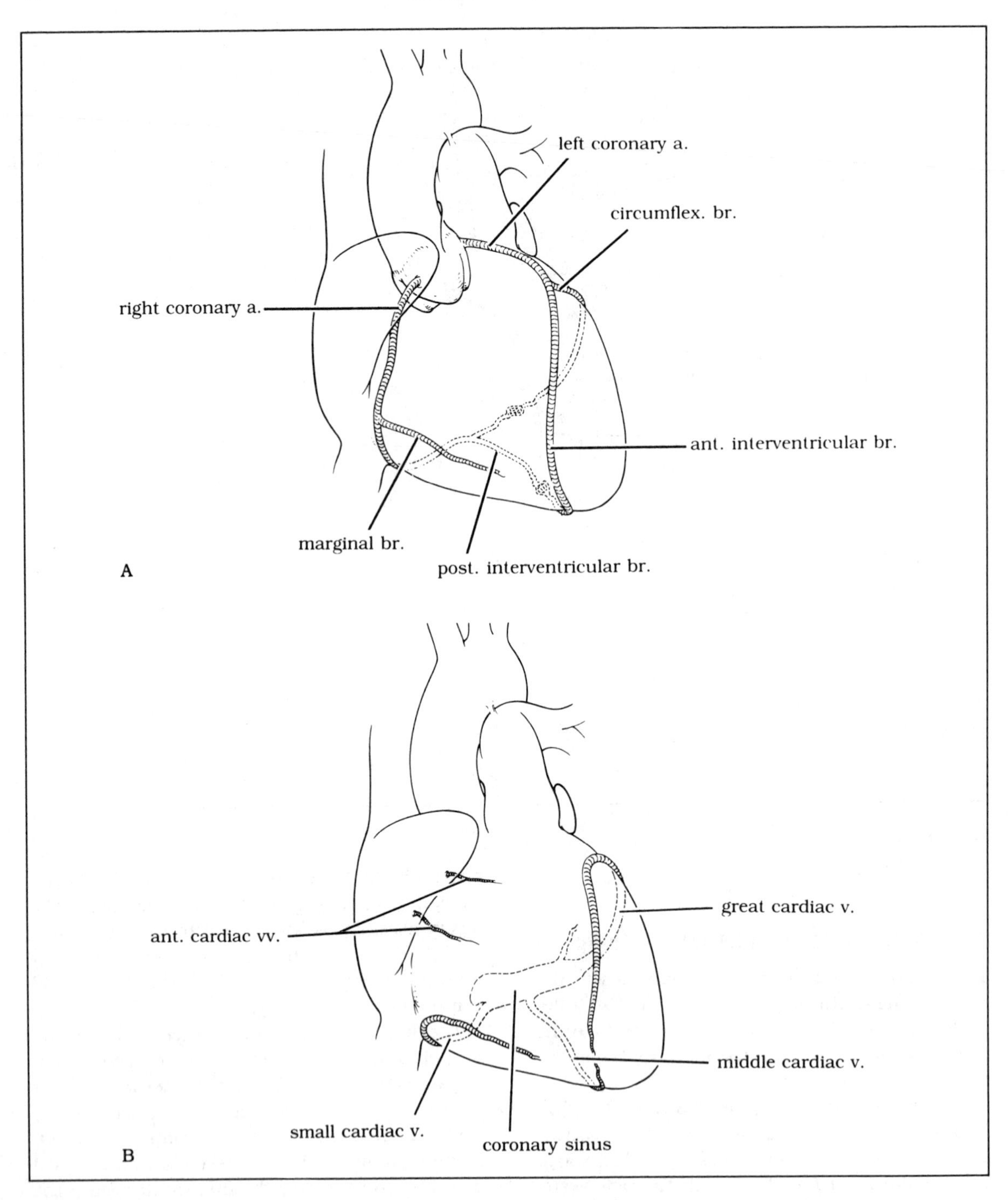

Fig. 2-18. A. Coronary arterial supply. B. Coronary venous drainage.

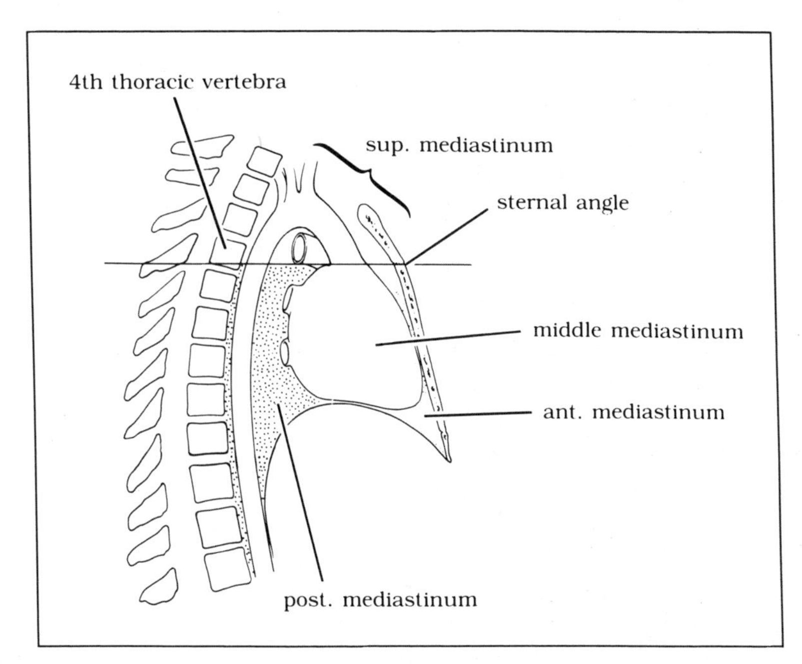

Fig. 2-19. Schematic view of the mediastinum.

Fig. 2-20. The superior mediastinum and root of the neck.

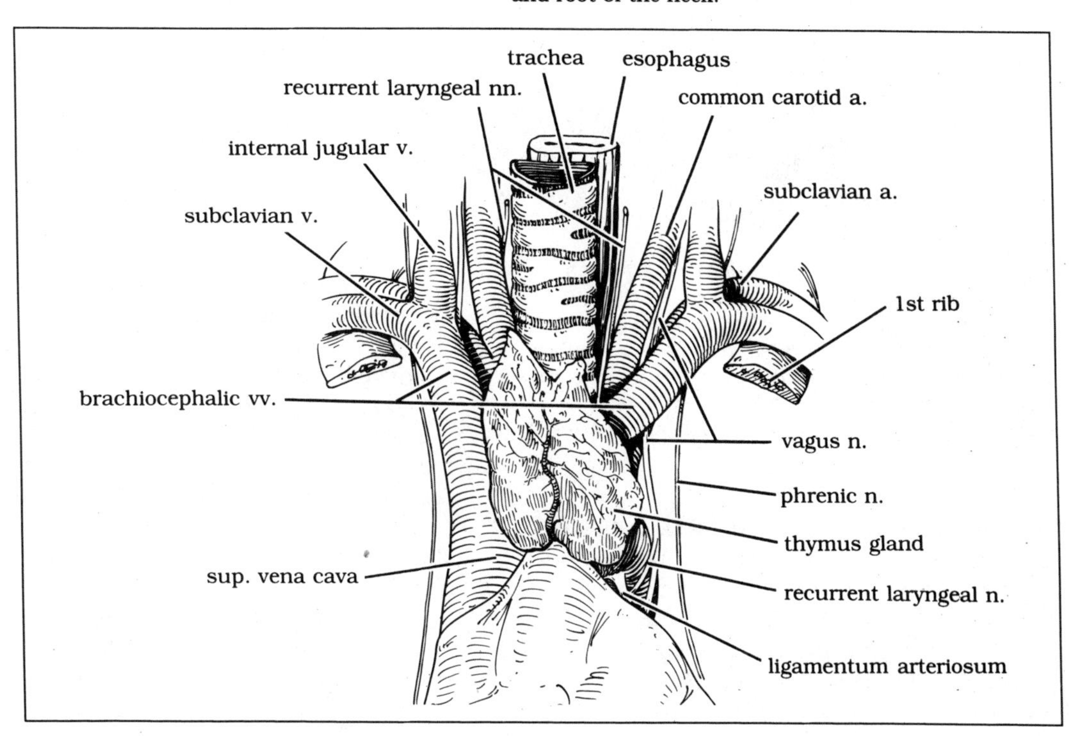

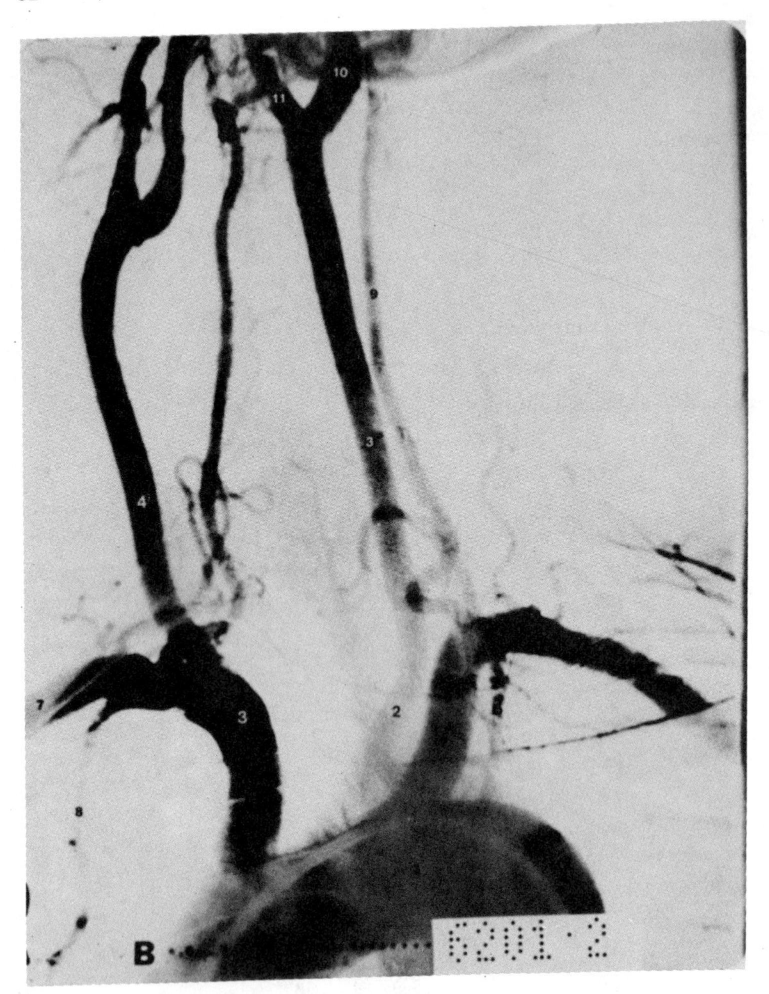

Fig. 2-21. X-ray film of vessels off the aortic arch. (1 = aortic arch; 2 = left common carotid artery; 3 = brachiocephalic artery; 4 = right common carotid artery; 5 = left internal thoracic artery; 6 = left subclavian artery; 7 = right subclavian artery; 8 = right internal thoracic artery; 9 = vertebral arteries; 10 = external carotid artery; 11 = internal carotid artery.)

the aorta (see Figs. 2-20 and 2-23). It divides at the level of the sternal angle into right and left bronchi. Deep to the trachea is the **esophagus** (Fig. 2-22; see also Fig. 2-25A), which lies just anterior to the thoracic vertebrae. In the superior mediastinum the esophagus is primarily a midline structure. As it descends through the posterior mediastinum it shifts leftward and comes to lie on the thoracic aorta at the level of the eighth thoracic vertebra.

Lying to the right of the ascending aorta and the aortic arch is the superior vena cava (see Figs. 2-17A, 2-20, and 2-25A). It is formed by the fusion of the two brachiocephalic veins at the level of the first costal cartilage. Lengthwise, about half of it is within the superior mediastinum.

The brachiocephalic veins are formed behind the sternoclavicular joints by the junction of the internal jugular and subclavian veins. The right brachiocephalic vein descends vertically, to the

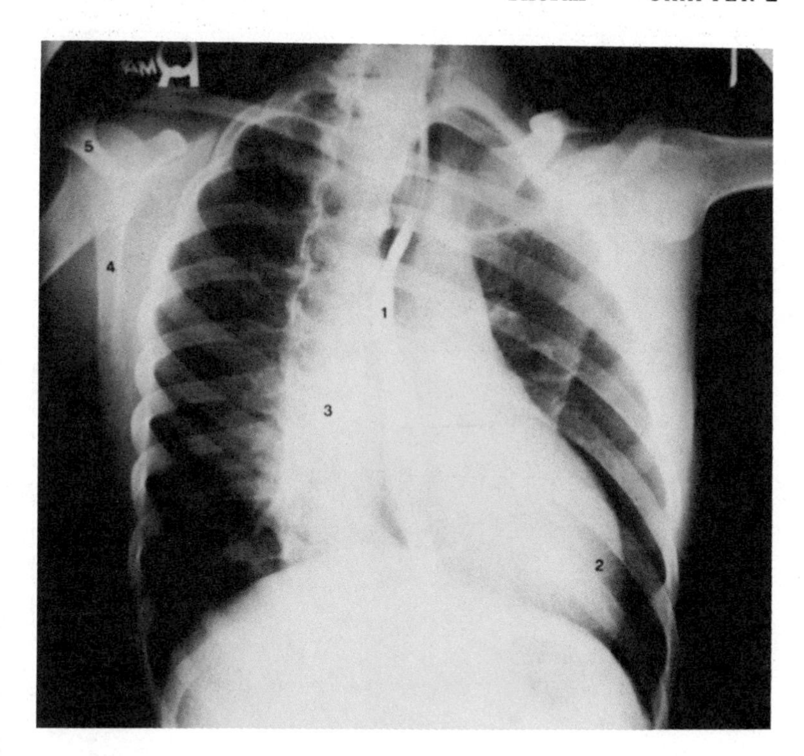

Fig. 2-22. Barium image of the esophagus. (1 = esophagus; 2 = left ventricle; 3 = vertebral column; 4 = blade of scapula; 5 = acromion.)

right of the brachiocephalic artery, for 2.5 cm to join the left brachiocephalic vein and form the superior vena cava. The left brachiocephalic vein passes rightward and ventrally, above the arch of the aorta, to the right brachiocephalic vein. It is separated from the manubrium by the thymus and connective tissue.

The **right vagus nerve** (see Figs. 2-20, 2-23, and 2-25A) enters the thorax by crossing the subclavian artery. It descends between the brachiocephalic artery and vein, finally coming to lie next to the trachea. It then passes behind the root of the lung to form the posterior pulmonary plexus. The **left vagus nerve** (see Fig. 2-25B) descends between the left common carotid and left subclavian arteries and crosses over the aortic arch. As it passes the ligamentum arteriosum it gives off the left recurrent laryngeal nerve. The nerve then passes posterior to the left pulmonary artery to

Table 2-8. Contents of the Superior Mediastinum

Structure	*Branches Within the Superior Mediastinum*
Thymus	
Aorta (arch)	Brachiocephalic trunk, left common carotid artery, left subclavian artery
Superior vena cava	Left and right brachiocephalic veins
Trachea	Left and right bronchi
Esophagus	
Right vagus nerve	
Left vagus nerve	Left recurrent laryngeal nerve
Phrenic nerves	
Thoracic duct	

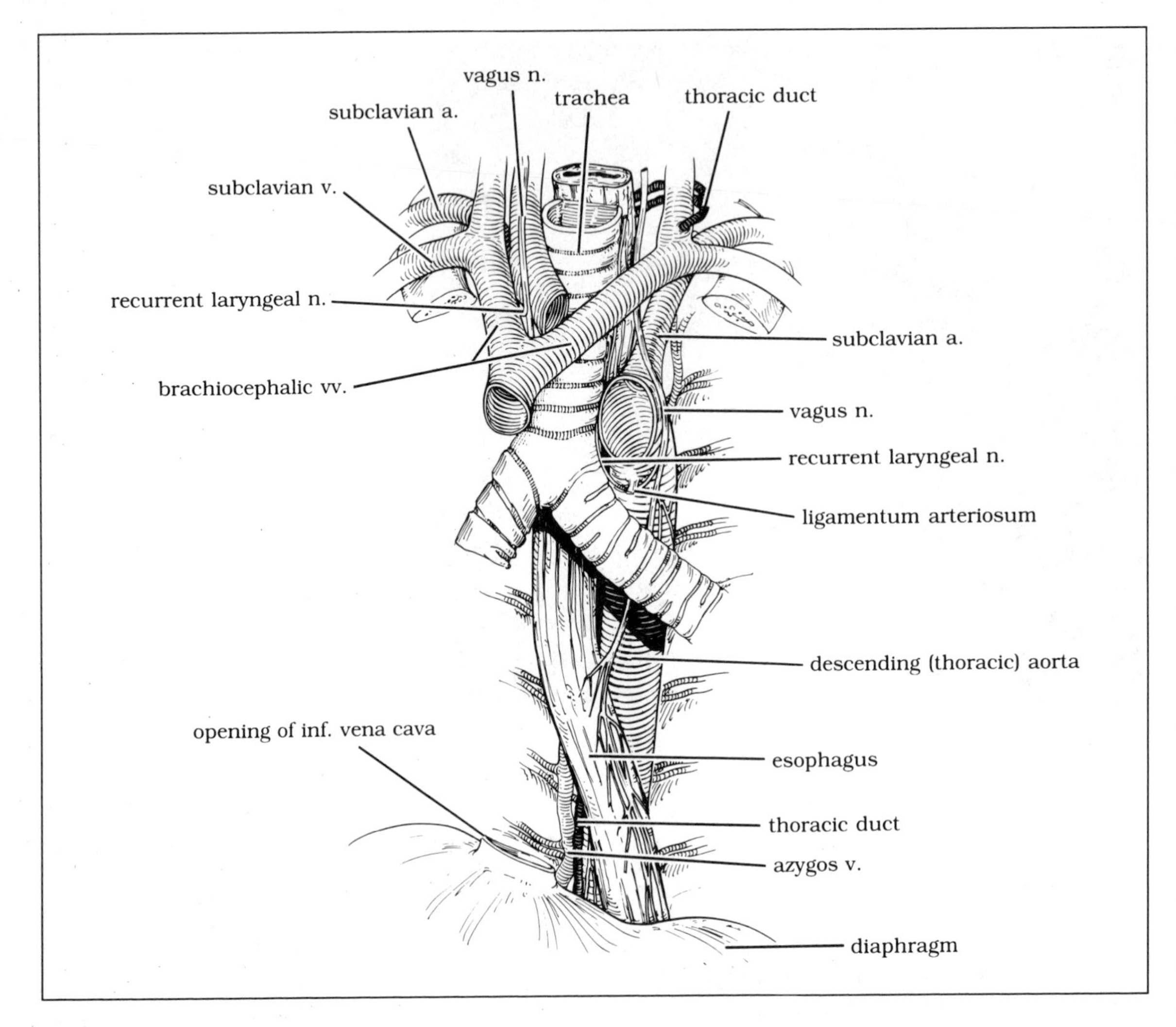

Fig. 2-23. The posterior mediastinum and root of the neck.

end in the posterior pulmonary plexus behind the root of the left lung. The left recurrent laryngeal nerve ascends to the larynx along the left border of the trachea.

POSTERIOR MEDIASTINUM

The contents of the posterior mediastinum (Fig. 2-23) are the esophagus, thoracic duct, azygos and hemiazygos veins, aorta, vagi, and thoracic splanchnic nerves. The **thoracic duct** (Fig. 2-24) provides the primary lymphatic drainage for the entire left side of the body and for the right side of the body below the diaphragm. Arising just below the diaphragm in the abdomen, the thoracic duct accompanies the aorta through the diaphragm. It passes superiorly, lying against the right side of the vertebral column, surrounded by the aorta and azygos vein, to the level of the fifth thoracic vertebra. There, as it enters the superior mediastinum, it deviates to the left side, ascending on the left side of the esophagus to the base of the neck, where it arches first laterally and then

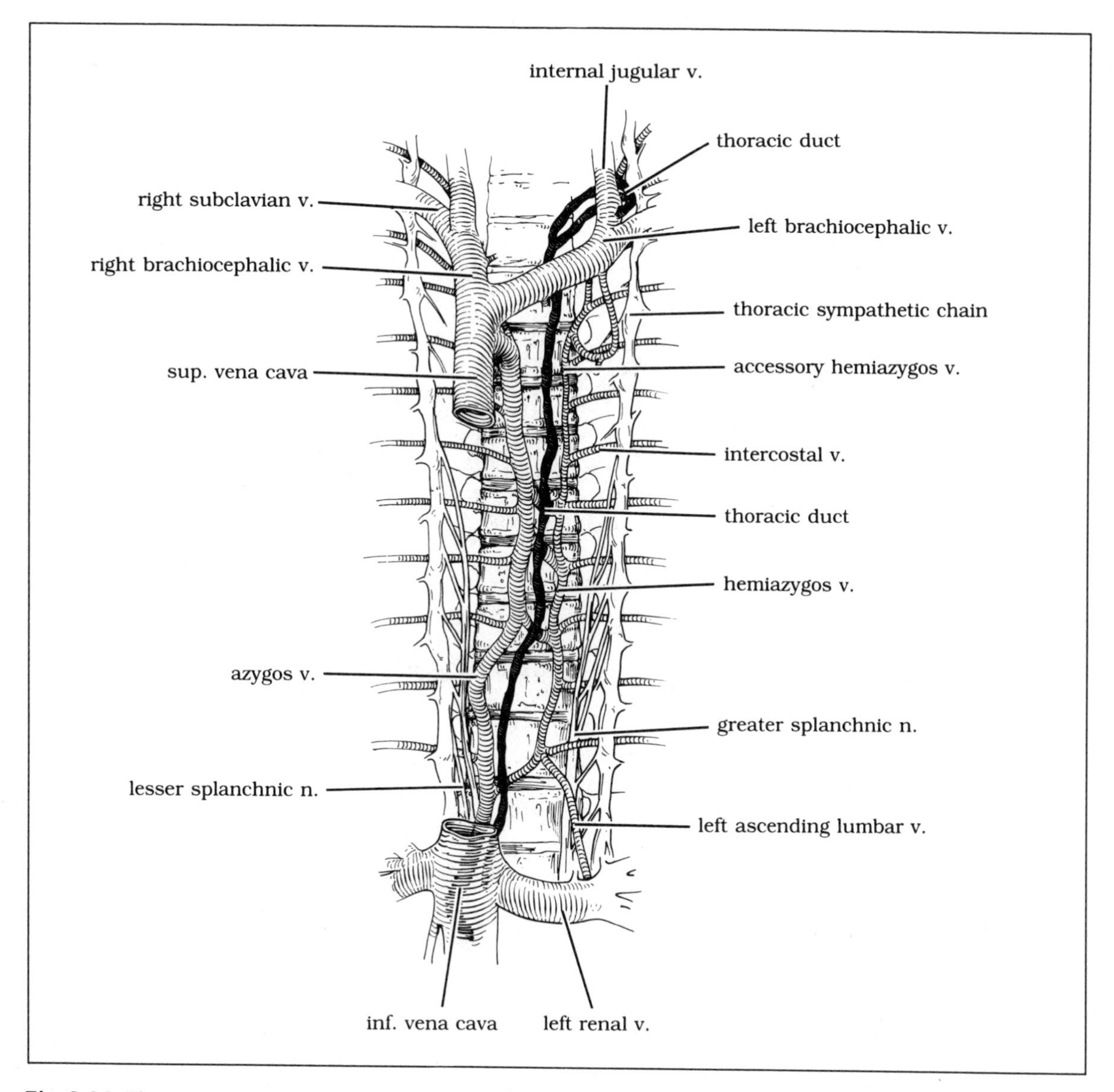

Fig. 2-24. The azygos venous system.

downward and anteriorly to end at the junction of the subclavian and internal jugular veins.

Accompanying the thoracic duct along part of its length is the **azygos vein** (Figs. 2-24 and 2-25A). This vein is formed by the junction of the ascending lumbar vein and the right subcostal vein at the level of the diaphragm. The azygos vein ascends through the thorax, lying against the vertebral column, to the right of the aorta. At the level of the fourth thoracic vertebra the azygos vein arches ventrally over the root of the right lung to enter the superior vena cava. The azygos vein drains the right chest wall by receiving the right posterior intercostal veins and right

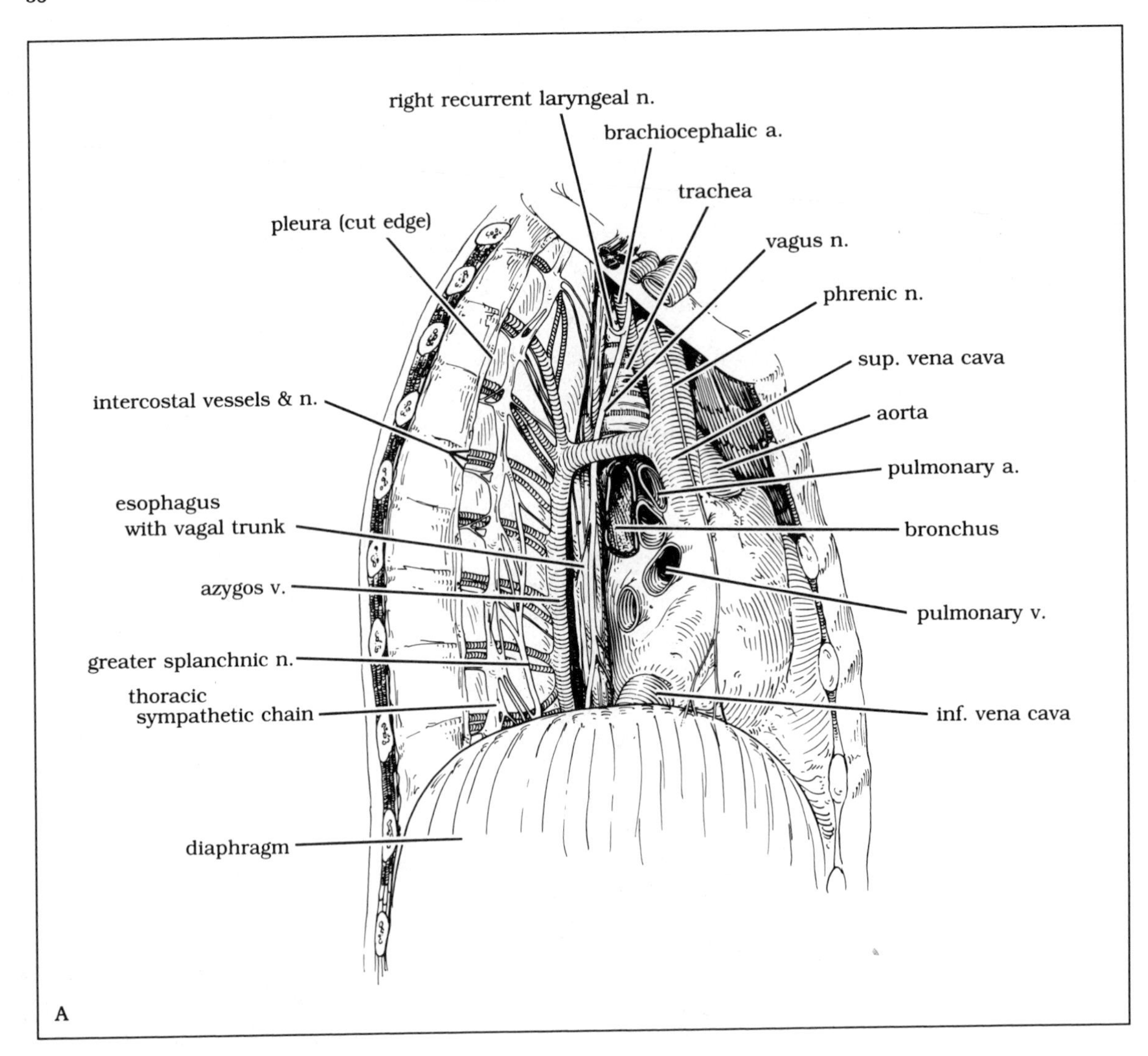

Fig. 2-25. The mediastinum. A. Right side view. B. Left side view.

superior intercostal vein. The left side is drained indirectly through the hemiazygos vein and accessory hemiazygos vein. The azygos vein also receives the venous drainage from the esophagus, pericardium, mediastinum, and right bronchus.

The **hemiazygos vein** (see Figs. 2-24 and 2-25B) begins parallel to the azygos vein on the left side and rises to the level of the ninth thoracic vertebra, where it crosses the vertebral column to enter the azygos vein. The hemiazygos vein drains the lower left posterior intercostal veins. The **accessory hemiazygos vein** is formed by the junction of the left posterior intercostal veins of the third to eighth intercostal spaces. The vein descends to the eighth intercostal space, where it crosses over the vertebral column to enter the azygos vein. Both hemiazygos and accessory hemiazygos veins may be absent, in which case the left posterior intercostal veins cross over the

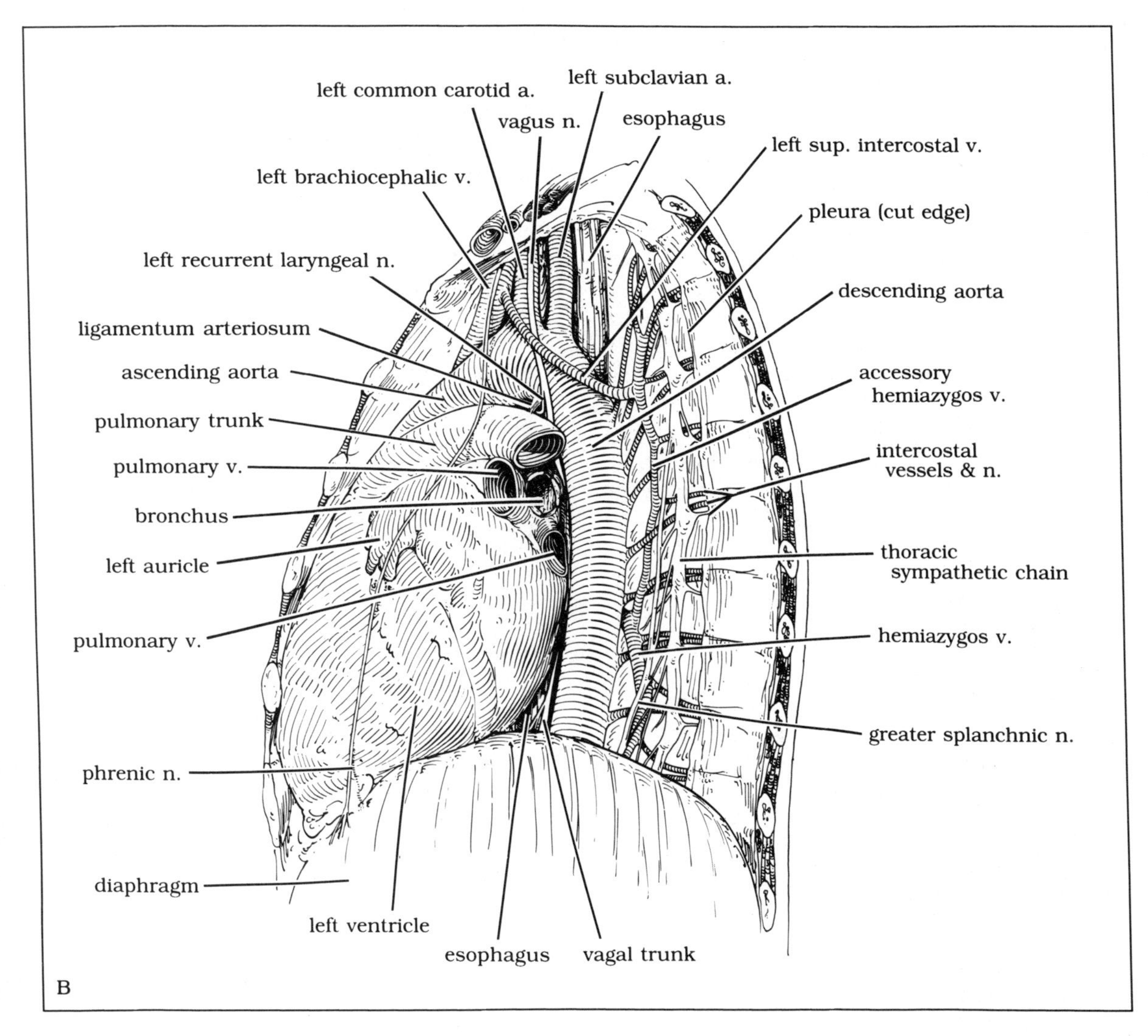

Fig. 2-25. (continued).

vertebral column to enter the azygos vein directly.

The thoracic aorta (see Figs. 2-23 and 2-25B) is the direct continuation of the aortic arch. It initially lies to the left of the vertebral column, but, as it descends, it gradually passes to a midline position directly over the vertebrae. In the thorax it gives off bronchial arteries and posterior intercostal and subcostal arteries. In addition it sends arteries to the pericardium, esophagus, mediastinum, and superior surface of the diaphragm. The thoracic aorta exits from the thorax through the aortic hiatus of the diaphragm.

The vagal nerves continue in the distal one-third of the posterior mediastinum as anterior and posterior vagal trunks (see Figs. 2-23 and 2-25). After passing through the posterior pulmonary plexus the right vagus continues inferiorly on the posterior surface of the esophagus as the posterior vagal trunk. The left vagus continues as the anterior vagal trunk on the anterior surface of the esophagus. Both trunks contribute to the esophageal plexus of nerves, which is a fine meshwork of nerves on the surface of the esophagus. The trunks may or may not lose their identity in this plexus, but in either case they leave the thorax as distinct trunks on the surface of the esophagus.

The thoracic sympathetic chain (see Figs. 2-24 and 2-25) is discussed on page 88. The thoracic splanchnic nerves are preganglionic sympathetic nerves that exit the sympathetic chain to arch medially over the vertebral column. These nerves pass through the diaphragm to synapse in abdominal ganglia and provide sympathetic innervation to many of the abdominal viscera. The greater splanchnic nerve arises from the fifth to the ninth thoracic ganglia, the lesser splanchnic nerve from the tenth and eleventh thoracic ganglia, and the lowest splanchnic nerve from the twelfth thoracic ganglion.

National Board Type Questions

Select the one best response for each of the following.

1. Which of the following arteries is **not** a branch of the internal thoracic artery?
   A. Pericardiacophrenic
   B. Anterior intercostal
   C. Lateral thoracic
   D. Musculophrenic
   E. Superior epigastric
2. The triangle of auscultation is bordered laterally by the
   A. Trapezius
   B. Scapula
   C. Latissimus dorsi
   D. Deltoid
   E. Rhomboids
3. In an infarction (lack of blood caused by a blocked artery) of the right atrium the occluded vessel would be the
   A. Coronary sinus
   B. Anterior interventricular artery
   C. Circumflex artery
   D. Left coronary artery
   E. Right coronary artery
4. The sternocostal surface of the heart is primarily formed by the
   A. Left ventricle
   B. Right ventricle
   C. Base
   D. Left atrium
   E. Right atrium
5. What is a common site of the thoracic duct in the posterior mediastinum?
   A. Between the azygos vein and thoracic aorta
   B. Through the vena caval opening of diaphragm
   C. Anterior to the esophagus
   D. At the right of the superior vena cava
   E. Anterior to the aortic arch

Select the response most closely associated with each numbered item. (The headings may be used once, more than once, or not at all.)

Which part of the heart is found within which lettered area?
   A. Part of the right atrium
   B. Part of the left atrium
   C. Part of the right ventricle
   D. Part of the left ventricle
   E. None of the above

6. Crista terminalis
7. Septomarginal (moderator) band
8. Coronary ostia
9. Sinoatrial node

Each part of the mediastinum contains which lettered item?
   A. Subclavian artery
   B. Left ventricle
   C. Apical bronchopulmonary segment
   D. Arch of aorta
   E. Hemiazygos vein

10. Posterior mediastinum
11. Middle mediastinum
12. Superior mediastinum

Which landmark is found at which lettered level?
   A. Third intercostal space
   B. Sixth rib on the right

C. Fifth intercostal space on the left
D. Second rib
E. Fourth intercostal space
13. Sternal angle (angle of Louis)
14. Horizontal fissure of right lung in anterior axillary line
15. Nipple (in the male)
16. Cardiac apex

Where do the following numbered vessels terminate?
A. In the azygos vein
B. In the superior vena cava
C. In the brachiocephalic vein
D. In the junction of left subclavian and left jugular veins
E. In the internal thoracic vein
17. Hemiazygos vein
18. Thoracic duct
19. Azygos vein
20. Anterior intercostal veins
21. Left highest intercostal vein

For the following, select
A. if only *1, 2, and 3* are correct
B. if only *1 and 3* are correct
C. if only *2 and 4* are correct
D. if only *4* is correct
E. if *all* are correct
22. The aortic arch
1. Encompasses both the bifurcations of the trachea and the pulmonary arteries
2. Begins at the level of the sternal angle
3. May be partially seen in posteroanterior films of the chest
4. Is crossed on its left aspect by both left phrenic and left vagus nerves
23. The fifth rib articulates with the
1. Vertebral body of T5
2. Vertebral body of T4
3. Transverse process of T5
4. Transverse process of T4
24. The chordae tendineae are attached to
1. Semilunar valve cusps (valvules)
2. Leaflets (cusps) of atrioventricular valves
3. Trabeculae carneae
4. Papillary muscles
25. The heart receives autonomic innervation from the
1. Preganglionic axons of the vagus nerves
2. Deep cardiac plexus
3. Cervical chain ganglion
4. Sympathetics via the phrenic nerves
26. During inspiration the volume of the thorax can be increased in a number of ways. Which of the following statements is (are) true?
1. The transverse diameter is mainly increased by the "pump-handle" movement.
2. The vertical diameter is increased mainly by the "pump-handle" movement.
3. The anteroposterior diameter is increased mainly by the "bucket-handle" movement.
4. The anteroposterior diameter is increased mainly by the "pump-handle" movement.
27. The lymphatic channels of the breast drain via the
1. Axillary lymph vessels
2. Intercostal lymph vessels
3. Infraclavicular lymph vessels
4. Internal thoracic lymph vessels
28. The left phrenic nerve
1. Lies posterior to the left vagus nerve as they enter the thorax
2. Originates partly from the left fourth cervical nerve
3. Accompanies the left internal thoracic artery through the thorax
4. Passes anterior to the left subclavian artery
29. On the right thoracic wall,
1. Pleura reaches to sixth intercostal space just lateral to sternum.
2. Pleura reaches to eighth intercostal space just lateral to sternum.
3. Lung can be found in the sixth intercostal space at the midaxillary line.
4. Lung can be found in the eighth intercostal space at the midclavicular line.

Annotated Answers

1. C. The internal thoracic artery is an important artery supplying the anterior thoracic region and also has the potential for making important anastomotic connections to other areas, most notably via the superior epigastric and anterior intercostal branches. The lateral thoracic is a branch of the axillary artery.
2. B.
3. E. The right coronary artery supplies the right atrium. The anterior interventricular and circumflex arteries are branches of the left coronary artery.
4. B. During development the heart rotates so that the right ventricle becomes closely associated with the sternum and costal cartilages. Remember this when examining a chest x-ray film.
5. A.
6. A. The crista terminalis separates the smooth portion of the right atrium (primitive sinus venosum) from the roughened (pectinate) portion (atrium proper).
7. C. The septomarginal band is a portion of the trabecular muscle of the right ventricle that contains conducting fibers of the atrioventricular bundle.
8. E.
9. A. The sinoatrial node, the pacemaker of the heart, is located at the junction of the superior vena cava and the right atrium. (How does this relate to Question 6?)
10. E.
11. B. The mediastinum is the space between the two pleural sacs and contains the thoracic viscera, except the lungs. Thus the left ventricle is in the middle mediastinum, whereas the apical bronchopulmonary segment cannot be in any portion of the mediastinum.
12. D.
13. D. The ribs and their interspaces are extremely useful landmarks for locating underlying structures (see Question 29). The sternal angle forms the boundary between the manubrium and body of the sternum and is the site for attachment of the second rib.
14. E.
15. E.
16. C. The apex of the heart, in the upright position, is close to the fifth left intercostal space and the heartbeat can be heard best here.
17. A.
18. D. The thoracic duct, the main lymphatic channel of the body, which traverses the entire thorax, terminates at the base of the neck at the junction of the internal jugular and subclavian veins.
19. B.
20. E.
21. C. The majority of the intercostal veins drain to the azygos system. The exceptions are the highest and superior intercostals, which drain to the brachiocephalic vein.
22. E. As it courses deep to the manubrium (in the superior mediastinum) the arch of the aorta has important relations to the trachea and pulmonary artery bifurcations. Remember also that it can be seen on a normal posteroanterior view of the chest as a "knuckle" protruding just lateral to the manubrium.
23. A.
24. C.
25. A. The phrenic nerve supplies the diaphragm, but though in close association with the pericardium, does not supply the heart muscle.
26. D. During quiet respiration diaphragmatic movement is responsible for the major changes in thoracic volume. However, the movement of the ribs upward and outward (pump-handle) adds to the anteroposterior dimension.
27. E.
28. C.
29. B. Pleural reflections and extent of the lung fields are important landmarks for where, for example, fluid could collect, air-tissue shadows should be seen on x-ray films, and approaches could be made to reach underlying structures without damaging lung tissue. Remember that pleura descends farther than lung in many areas and that the associated pleural reflections are of clinical significance.

3 Abdomen

Objectives

After reading this chapter, you should know the following:

Surface landmarks of the abdomen and the surface projections of the abdominal viscera

Organization, musculature, and the 11 specializations of the abdominal wall

General organization of the abdominal cavity and viscera, including mesenteries

Organization, components, and function of the digestive system (stomach, omenta, small and large intestines, pancreas, spleen, liver, gallbladder)

Organization, components, and function of the urinary system and its associated structures (kidneys, ureters, suprarenal glands)

Musculature, tendons, and function of the diaphragm, including its involvement in breathing

Relationship of some structures, including the lumbar plexus, to the lower limb

Blood supply, lymphatics, and innervation of the abdominal wall, cavity, and organs

Surface Anatomy

SURFACE LANDMARKS

Because the walls of the abdomen are primarily composed of muscles, the bony landmarks of the region are limited to its boundaries (Fig. 3-1; see also Fig. 2-1). The costal margin of the ribs (see Chap. 2) marks the superior border of the abdomen. The ilium forms the lateral inferior boundary, and the pubis marks the anterior inferior margin.

The **iliac crest,** the upper curved edge of the ilium, is easily palpated on the body. The crest ends in the anterior superior iliac spine, which is palpable as the most anterior point of the crest. The **pubis** may be felt at the lower border of the abdomen. Its lateral border is the pubic tubercle, and its medial border forms the pubic symphysis with the medial border of the pubis of the other side. In between the pubic tubercle and symphysis is the pubic crest. The **inguinal ligament,** marking the lower anterolateral margin of the abdomen, stretches from the pubic tubercle laterally to the anterior superior iliac spine.

The **linea alba,** the midabdominal line, lies between the two rectus abdominis muscles and marks the junction of the abdominal wall aponeuroses of the two sides. The umbilicus lies along this line. Lying lateral to the rectus abdominis muscles are the two **lineae semilunaris,** marking the line of fusion of the muscular aponeuroses of each side to form the rectus sheath.

Three tendinous intersections cross the rectus muscles. The first is just below the xiphoid process, the second halfway to the umbilicus, and the third at the level of the umbilicus.

SURFACE PROJECTIONS

Several of the abdominal viscera may be palpated

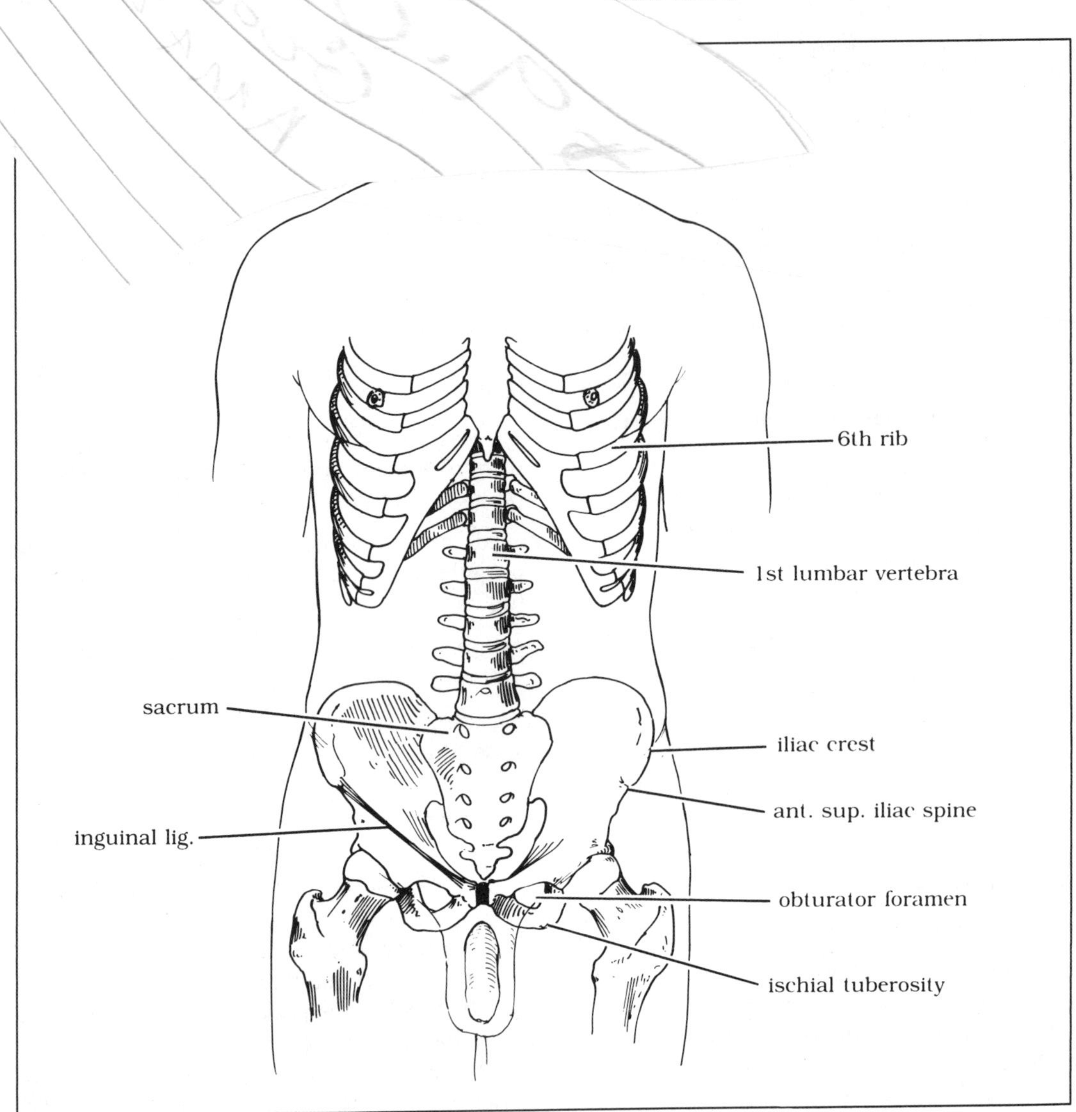

CHAPTER 3

Fig. 3-1. Surface projection of abdominal skeleton.

through the abdominal wall (Fig. 3-2). The liver may be found in the upper right quadrant lying predominantly under the lower margin of the ribs. The stomach lies primarily below the ribs on the left side, descending to the midline at approximately the level of the second lumbar vertebra.

The small bowel occupies most of the central abdominal region, descending into the pelvis. The ascending colon lies at the right lateral border of the pelvis, the descending colon at the left, with the transverse colon connecting the two just above the umbilicus. The root of the appendix may usually be found in the lower right quadrant at the intersection of two lines, one running upward from the middle of the inguinal ligament and the second connecting the two anterior superior iliac spines.

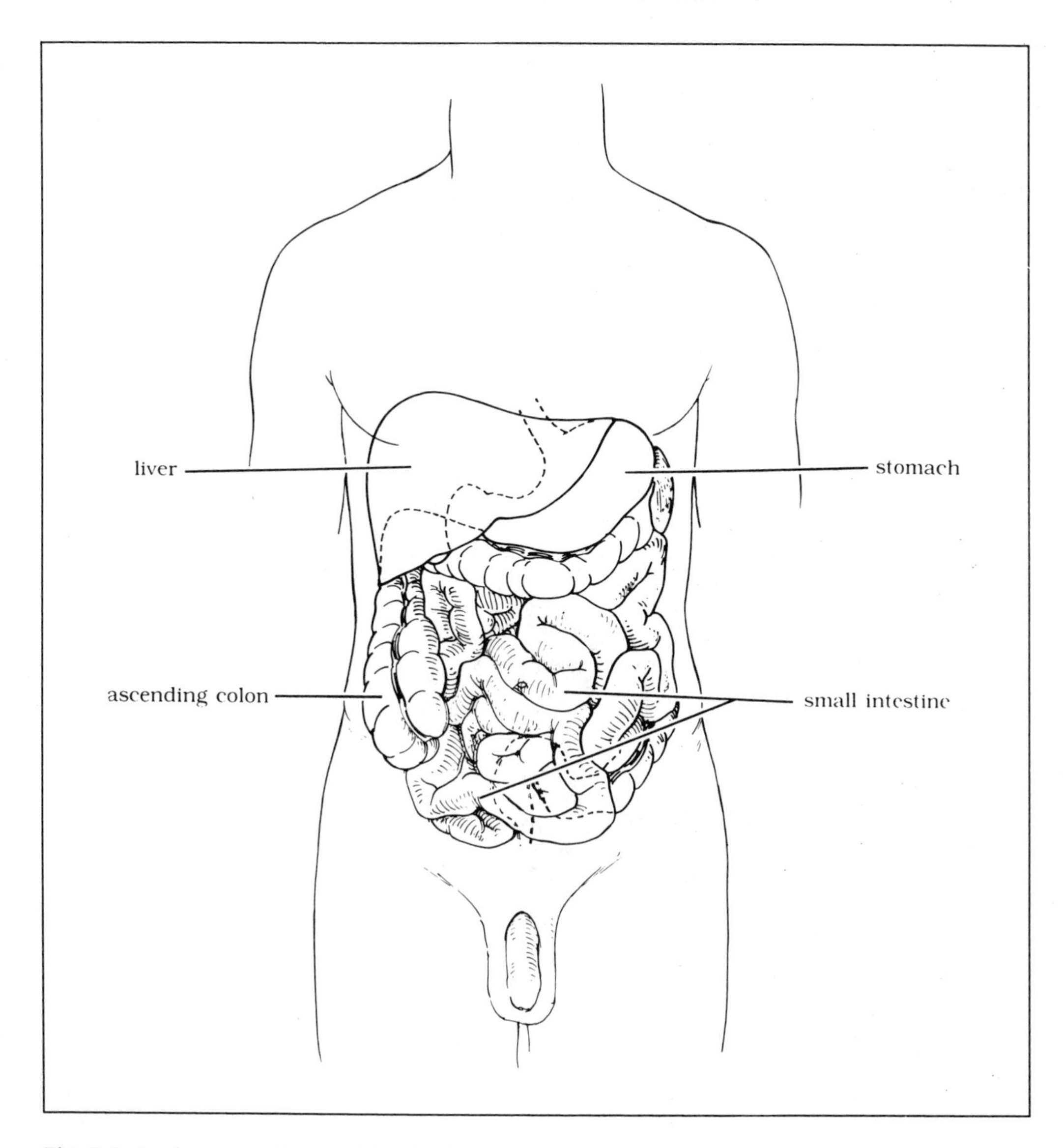

Fig. 3-2. Surface projection of abdominal viscera.

The spleen lies between the midaxillary line of the left side and a line 4 cm from the midback, between the ninth and eleventh ribs. The kidneys project onto the posterior abdominal wall from the level of the eleventh rib to the third lumbar vertebra, between 2.5 and 9.5 cm from the midline of the back (Fig. 3-3).

Abdominal Wall

The structure and organization of the abdominal wall is similar to that of the thoracic wall, the major differences being the presence of the ribs in the thorax and the rectus abdominis muscle in the abdomen. Otherwise, the direction of the muscle fibers and the innervation of the wall are similar in both regions, a fact that should be kept

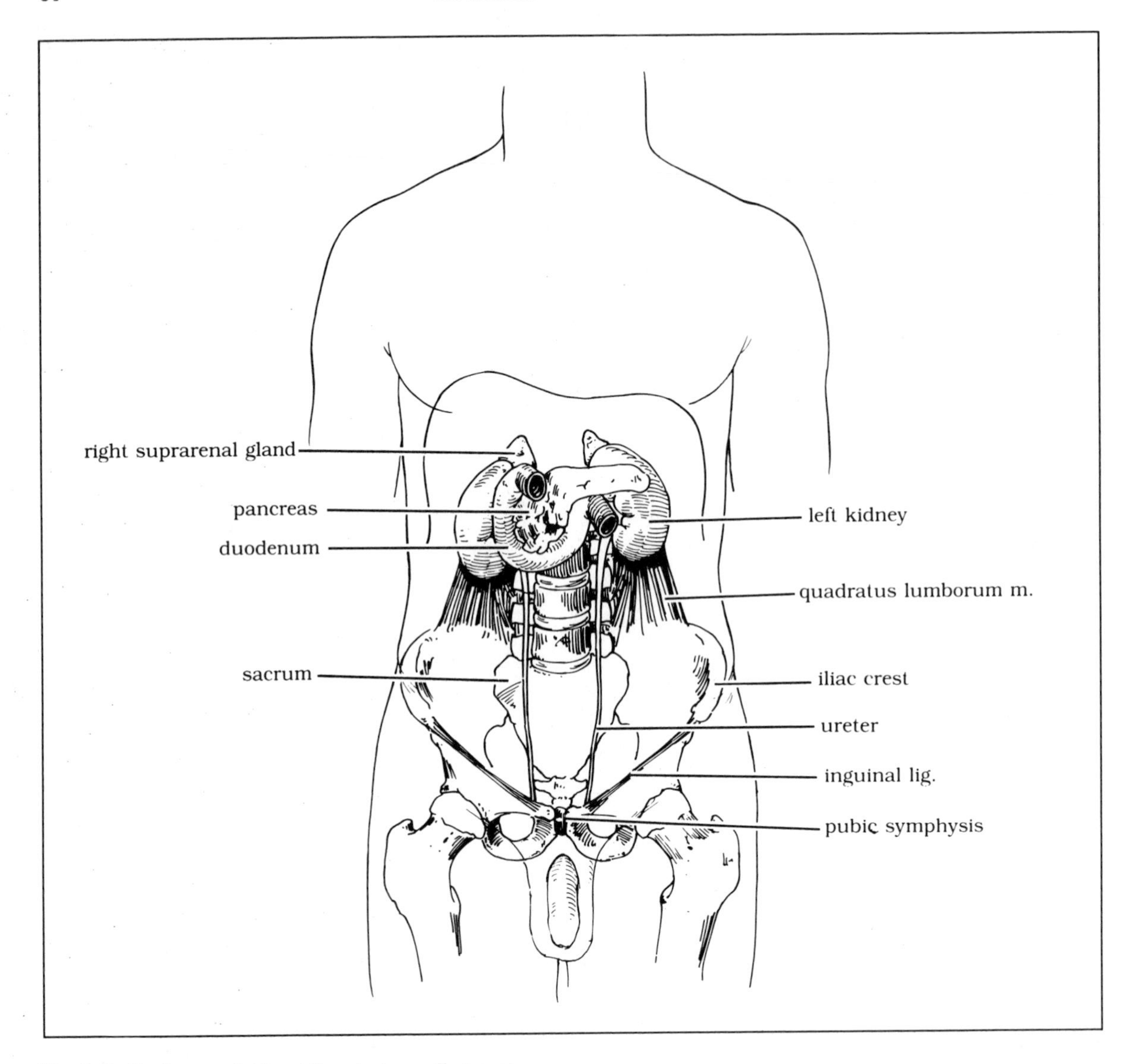

Fig. 3-3. Surface projection of posterior wall viscera.

in mind by the student as an aid to understanding the organization of both regions.

The abdominal wall is almost entirely muscular in nature; the only bony regions of the wall are the spine (see Chap. 9) and floating ribs (see Chap. 2). The bony limits of the abdomen are the rib cage superiorly (see Chap. 2) and the ileum and pubis inferiorly (see Chap. 4). The muscles of the wall are the external abdominal oblique, internal abdominal oblique, transverse abdominal, rectus abdominis, pyramidalis, and quadratus lumborum. All but the last belong to the anterolateral group of muscles; the quadratus lumborum is the muscle of the posterior abdominal wall.

The abdominal wall has several regions of specialization that result in alterations of the basic structure, the inguinal region being the primary one. This region provides support for the abdominal wall and viscera through the inguinal ligament, a tough tendinous band stretching along the inferior border of the abdomen from the anterior superior iliac spine to the pubic tubercle. This region is also modified in the male by the passage of the testes into the scrotum and the presence of the spermatic cord; in the female the spermatic cord is replaced by the round ligament. (These specializations will be discussed after a basic understanding of the structure of the region has been established.)

MUSCULATURE

The three muscles of the lateral abdominal wall are essentially fan shaped and overlapping. They can be remembered thus: The external abdominal oblique may be visualized by placing the heel of the hand on the lower lateral border of the rib cage and pointing the hand toward the pubis and spreading the fingers. The internal abdominal oblique may be visualized by placing the hand on the top of the iliac crest and pointing the fourth finger toward the umbilicus and spreading the others. Finally, the transverse abdominal muscle may be visualized by placing the hand on the side midway between the rib cage and the iliac crest, pointing it toward the midline, and spreading the fingers.

The **external abdominal oblique muscle** (Fig. 3-4A) arises from the lower eight ribs, interdigitating at its origin with the latissimus dorsi (lower three ribs) and serratus anterior (upper five slips) muscles. The muscle inserts into the iliac crest, pubic symphysis, pubic crest, and rectus sheath, forming a broad aponeurosis that contributes to the structure of the rectus sheath.

The fibers of the **internal abdominal oblique muscle** (Fig. 3-4B) run perpendicular to the fibers of the external abdominal oblique. This muscle lies just under the external abdominal oblique, arising from the thoracolumbar fascia, the anterior two-thirds of the iliac crest, the lateral two-thirds of the inguinal ligament, and a portion (6 cm) of the iliacus fascia medial to the anterior superior iliac spine. The muscle inserts into the costal cartilages of the lower four ribs and, via its aponeurosis, into the rectus sheath and conjoint tendon.

The third and deepest muscle of the abdominal wall is the **transverse abdominal muscle** (Fig. 3-4C). This muscle originates from the deep surface of the costal cartilages of the lower six ribs, the thoracolumbar fascia, the anterior three-fourths of the iliac crest, and the iliacus fascia behind the lateral part of the inguinal ligament. The muscle inserts via its aponeurosis into the rectus sheath or the conjoint tendon to the pubis. Deep to the transverse abdominal muscle lies the **transversalis fascia** (Fig. 3-5). The fascia lines both the abdominal and the pelvic cavities, separating the peritoneum from the wall. The fascia also contributes to the covering of the spermatic cord.

The **rectus abdominis** muscle (Fig. 3-6) forms the anterior abdominal wall. It lies in a sheath formed by the aponeuroses of the lateral wall muscles. The rectus abdominis muscle arises from the crest and symphysis of the pubis and inserts on the fifth to seventh costal cartilages. The muscle is crossed by three tendinous intersections that attach the muscle anteriorly to the rectus sheath. The first inscription is just below the xiphoid process, the second halfway to the umbilicus, and the third at the level of the umbilicus.

The **pyramidalis** (see Fig. 3-6) is a small triangular-shaped muscle with no significant function. It is often absent. The muscle arises from the pubis and inserts into the linea alba.

The **quadratus lumborum** muscle lies in the posterior abdominal wall (see Fig. 3-6). It arises from the iliolumbar ligament and the posterior 5 cm of the iliac crest. Its fibers run upward to insert into the lower border of the twelfth rib and the upper four lumbar transverse processes. Characteristics of these muscles are summarized in Table 3-1.

SPECIALIZATIONS

1. The **rectus sheath** (Fig. 3-7A,B) is a tendinous sheath enclosing the rectus abdominis muscle and is formed by the aponeurosis of the

Table 3-1. Musculature of the Abdominal Wall

Muscle	*Origin*	*Insertion*	*Innervation*
External abdominal oblique	Lower borders of lower 8 ribs	Iliac crest, rectus sheath, pubic symphysis, pubic crest	Intercostals 8–12, ilioinguinal, iliohypogastric
Internal abdominal oblique	Thoracolumbar fascia, iliac crest, inguinal ligament, iliacus fascia	Costal cartilages of lower 4 ribs, rectus sheath, conjoint tendon	Intercostals 8–12, ilioinguinal, iliohypogastric
Transverse abdominal	Costal cartilages of lower 6 ribs, thoracolumbar fascia, iliac crest, iliacus fascia	Rectus sheath, conjoint tendon	Intercostals 7–12, ilioinguinal, iliohypogastric
Rectus abdominis	Pubic crest, pubic symphysis	Costal cartilages 5–7	Intercostals 7–12
Pyramidalis	Pubis	Linea alba	Subcostal (intercostal 12)
Quadratus lumborum	Iliac crest, iliolumbar ligament	Lower border of rib 12, transverse processes of vertebrae L1–L4	Thoracic 12 to lumbar 3

lateral abdominal wall muscles. The lateral edge of the sheath, where the aponeuroses fuse, forms the linea semilunaris. The midline of the sheath, where the fibers from both sides decussate to form a tendinous line, is the linea alba.

The posterior wall of the rectus sheath changes in quality about midway between the pubis and the umbilicus. Below this line, the **arcuate line,** the posterior sheath consists only of transversalis fascia; above this line the posterior sheath is composed of transversalis fascia plus the aponeurosis of the transverse abdominal muscle and part of (usually one-half of) the aponeurosis of the internal abdominal oblique muscle.

The remaining fibers of the internal abdominal oblique aponeurosis fuse with the aponeurosis of the external abdominal oblique muscle to form the anterior sheath of the rectus muscle above the arcuate line. Below the level of the arcuate line all the fibers of the internal abdominal oblique and transverse abdominal aponeuroses fuse with the aponeurosis of the external abdominal oblique muscle to form the anterior rectus sheath.

2. The **inguinal ligament** (Fig. 3-8) stretches from the anterior superior iliac spine to the pubic tubercle. It is the thickened, rolled-under lower edge of the aponeurosis of the external abdominal oblique muscle and is a major supporter of the anterior abdominal wall and the contents of the abdominal cavity.

3. The **lacunar ligament** (see Fig. 3-8) is formed by the medial fibers of the inguinal ligament. These fibers roll under the medial end of the inguinal ligament, flattening out and attaching along the medial 2 cm of the iliopectineal line.

4. The **pectineal ligament** (see Fig. 3-8) is the lateral continuation of the lacunar ligament along the pectineal line.

5. The **conjoint tendon** (falx inguinalis) (see Fig. 3-11) is formed by the fusion of fibers of the internal abdominal oblique aponeurosis and the transverse abdominal aponeurosis. These fibers, arching medially to the superficial inguinal ring, fuse and continue on to insert into the pubic crest and the medial 2 cm of the iliopectineal line.

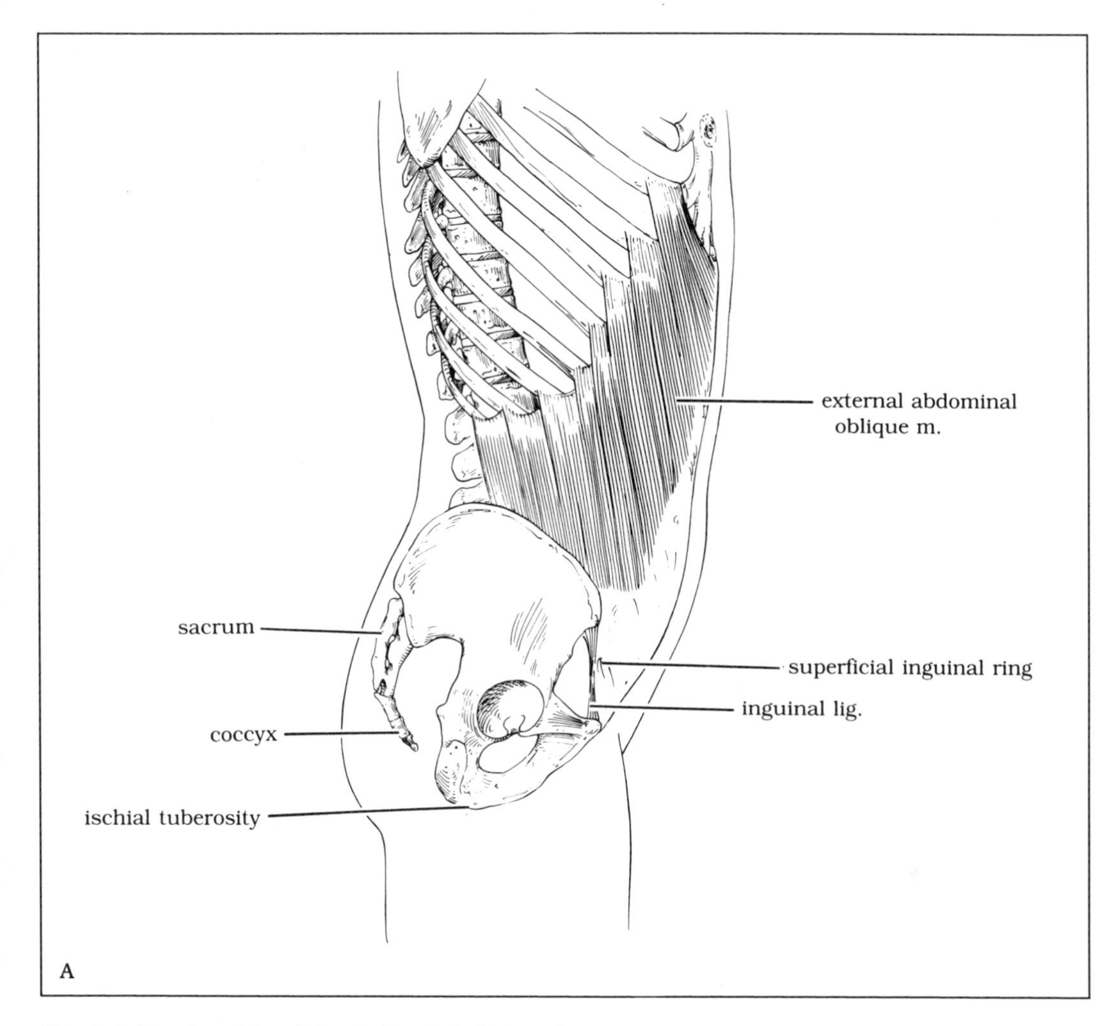

Fig. 3-4. Muscles of the abdominal wall. A. External abdominal oblique muscle. B. Internal abdominal oblique muscle. C. Transverse abdominal muscle.

6. The **superficial inguinal ring** (Fig. 3-9) is a region of the external abdominal oblique aponeurosis stretched for the passage of the spermatic cord. The ring is located just above and lateral to the pubic tubercle. Its medial and lateral boundaries are strengthened by the medial and lateral crura, respectively. Its superior border is somewhat strengthened by intercrural fibers.

7. The **deep inguinal ring** (Fig. 3-10) is the outpocketing in the transversalis fascia for the passage of the spermatic cord.

8. Since the **spermatic cord** (Figs. 3-9, 3-11) traverses the abdominal wall obliquely—its entrance (the deep inguinal ring) being lateral to its exit (the superficial inguinal ring)—for a distance of 4 cm, a passageway or canal, the **inguinal canal,** is formed in the abdominal wall. Herniation of the abdominal contents down this canal is known as an indirect inguinal hernia. A direct inguinal hernia involves extrusion of the abdominal viscera directly through the superficial inguinal ring without first being passed down the inguinal canal.

In addition to the spermatic cord, the inguinal canal provides passage through the abdominal

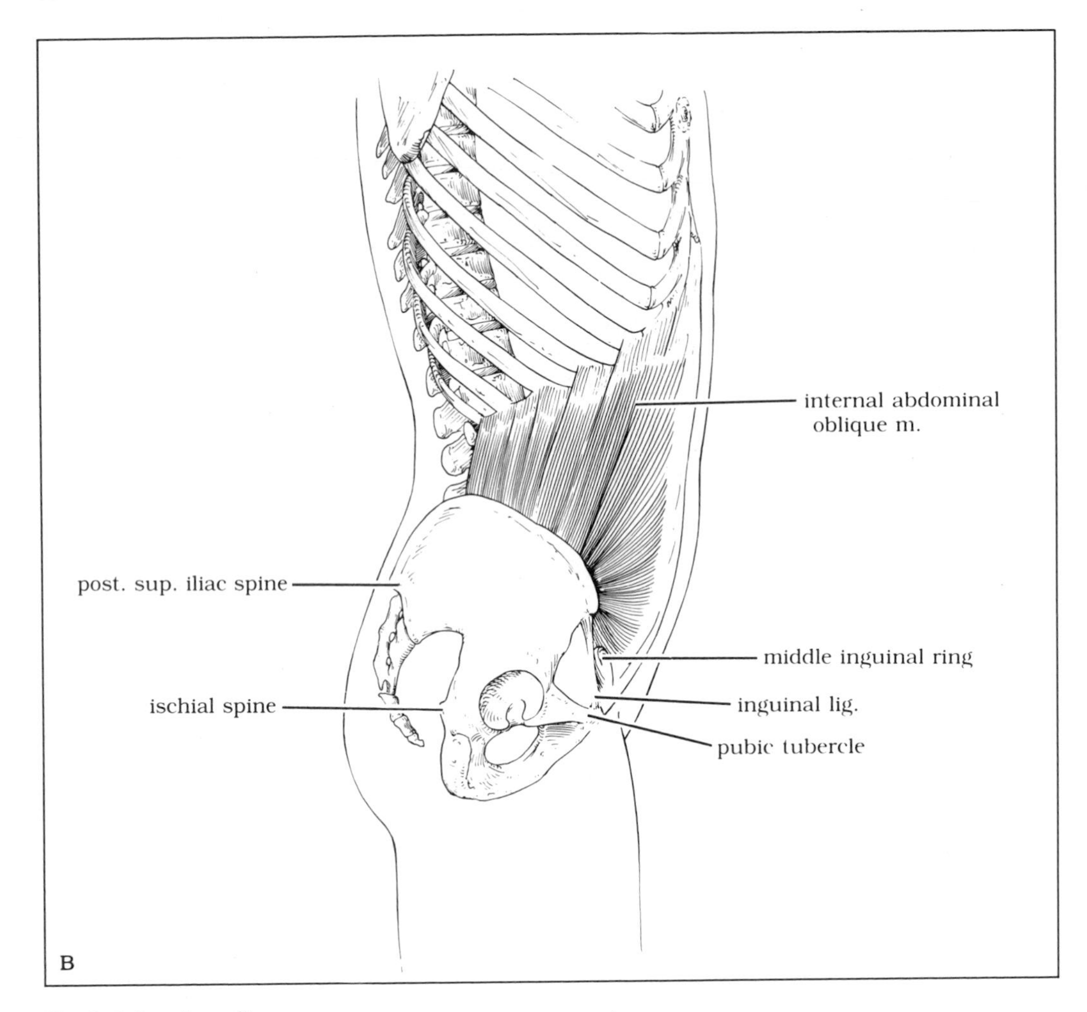

Fig. 3-4. (continued).

wall for the **ilioinguinal nerve,** the genital branch of the **genitofemoral nerve,** and the **cremasteric artery.**

9. As the spermatic cord (see Fig. 3-11) passes through each of the inguinal rings, it receives from each layer a fascial contribution that becomes one of the covering layers of the cord and its contents. The transversalis fascia contributes the **internal spermatic fascia;** the internal abdominal oblique contributes the **cremaster muscle** and **cremasteric fascia;** and the external abdominal oblique contributes the **external spermatic fascia.** The contents of the spermatic cord (see Fig. 3-11) covered by these fascia are the **ductus deferens,** the **deferential artery** and **vein,** the **testicular artery,** and the **pampiniform plexus** of veins, lymphatics, and nerves.

10. The **scrotum,** a saclike appendage in the groin, contains the testes and part of the spermatic cord. It is divided into two compartments by a septum. The scrotal layers are the skin and the tunica dartos. The **tunica dartos** is a layer of smooth muscle in the scrotal wall continuous with the fatty and membranous layers of the abdominal wall.

11. The **testis** (see Fig. 3-11) is covered by the same layers as cover the spermatic cord (see 9 above) (Table 3-2). In addition the testis is cov-

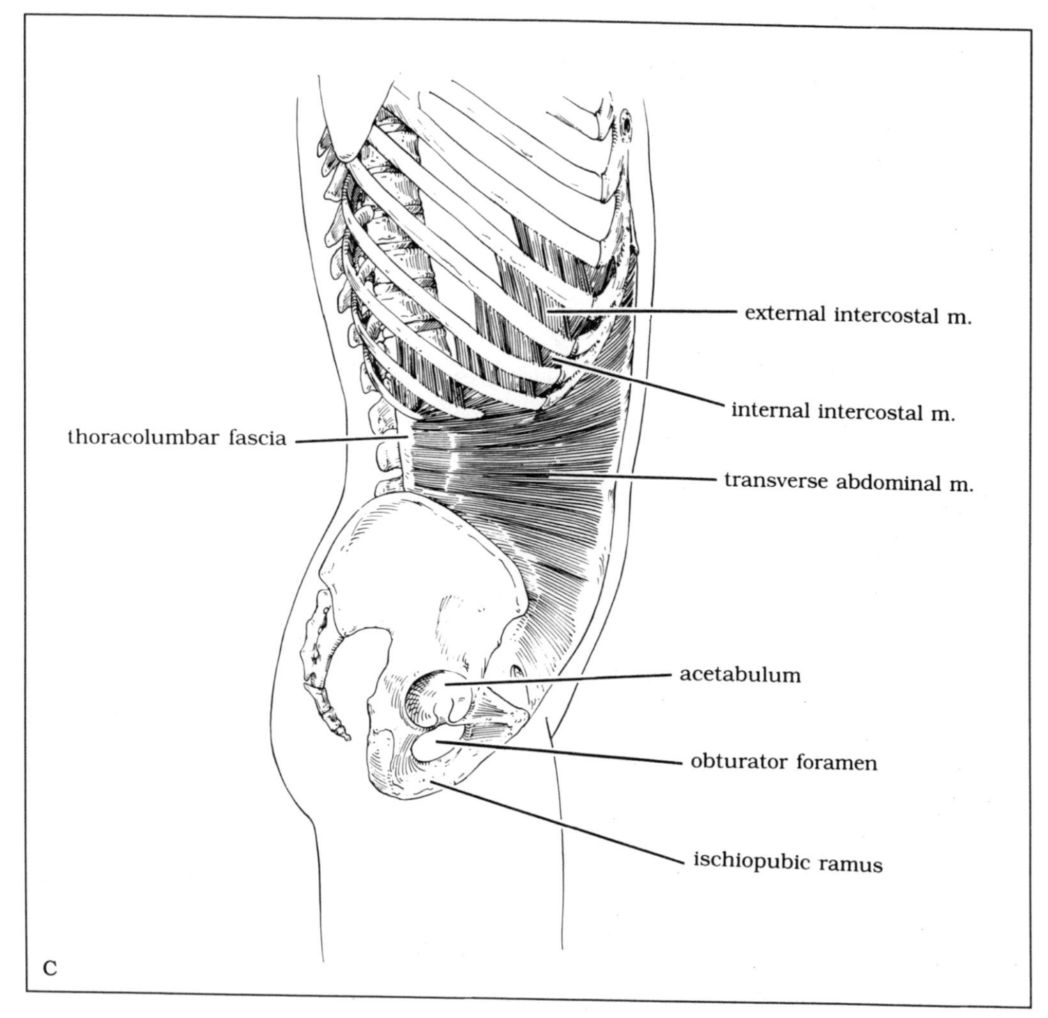

Fig. 3-4. (continued).

ered by the **tunica vaginalis testis,** a serous sac with visceral and parietal layers. The tunica vaginalis testis is the remnant of the peritoneal contribution to the fascial coverings of the testis.

Table 3-3 summarizes these specializations of the abdominal wall.

BLOOD SUPPLY, LYMPHATICS, AND INNERVATION

Blood is supplied to the skin and the subcutaneous regions of the inferior abdominal wall by

Table 3-2. Fascial Layers of the Spermatic Cord and Testis

Fascia	*Contributed by*
External spermatic	External abdominal oblique muscle
Cremaster muscle and fascia	Internal abdominal oblique muscle
Internal spermatic	Transversalis fascia
Tunica vaginalis testis	Peritoneum

Table 3-3. Specializations of the Abdominal Wall

Structure	*Contributing Layers*	*Points of Attachment*	*Comments*
Anterior rectus sheath	Above the arcuate line—aponeuroses of external abdominal oblique and internal abdominal oblique Below the arcuate line—aponeuroses as above plus transverse abdominal aponeurosis	Linea semilunaris to linea alba	
Posterior rectus sheath	Above the arcuate line—aponeuroses of internal abdominal oblique and transverse abdominal and transversalis fascia Below the arcuate line—transversalis fascia	Linea semilunaris to linea alba	
Inguinal ligament	Aponeurosis of external abdominal oblique	Anterior superior iliac spine to pubic tubercle	
Lacunar ligament	Inguinal ligament	Iliopectineal line	
Pectineal ligament	Lacunar ligament	Pectineal line	
Conjoint tendon	Aponeuroses of internal abdominal oblique and transverse abdominal	Pubic crest and iliopectineal line	
Superficial inguinal ring	Aponeurosis of external abdominal oblique		Surrounds spermatic cord, surrounded by medial and lateral crura
Deep inguinal ring	Transversalis fascia		Surrounds spermatic cord
Inguinal canal	Abdominal wall	Deep to superficial inguinal ring	Passageway for spermatic cord, ilioinguinal nerve, genital branch of genitofemoral nerve, cremasteric artery

branches of the femoral artery (Fig. 3-12) that arise just below the inguinal ligament: the superficial epigastric, superficial circumflex iliac, and superficial external pudendal arteries. The **superficial epigastric artery** runs medially from the inguinal ligament toward the umbilicus. The **superficial circumflex iliac artery** runs laterally along the line of the inguinal ligament and the iliac crest. The **superficial external pudendal artery** crosses upward over the spermatic cord, toward the pubis.

The vessels of the abdominal wall travel in the fascial plane between the transverse abdominal muscle and the internal abdominal oblique muscle. The lateral and posterior abdominal wall is supplied by the last two posterior intercostal ar-

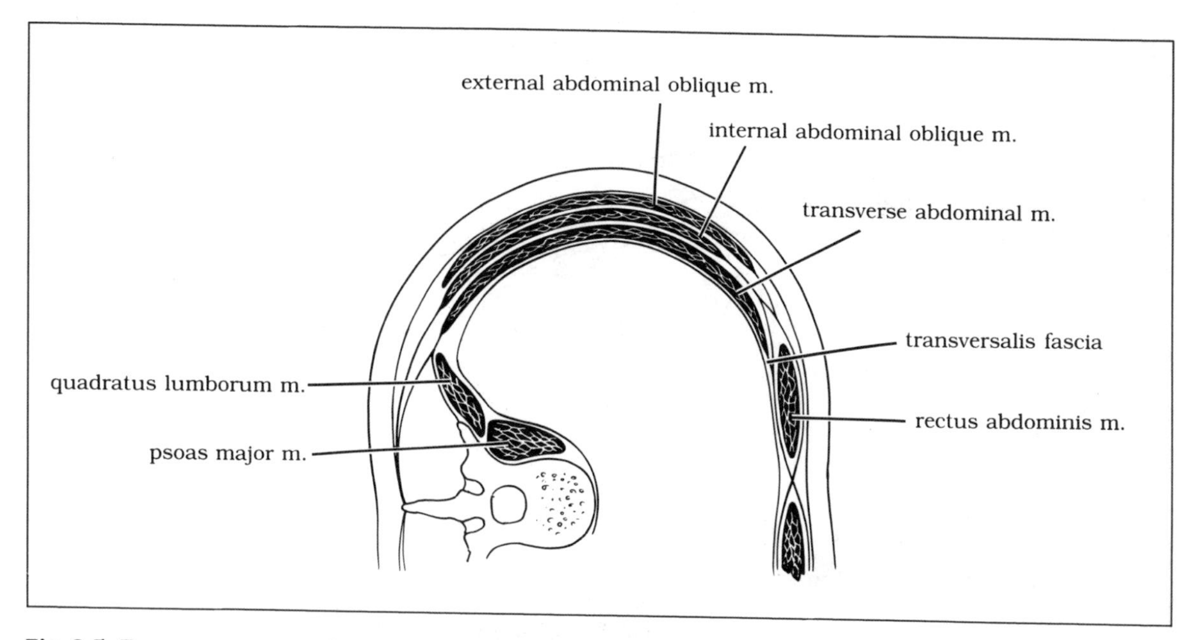

Fig. 3-5. Transverse section of abdominal wall.

Fig. 3-6. Rectus abdominis muscle.

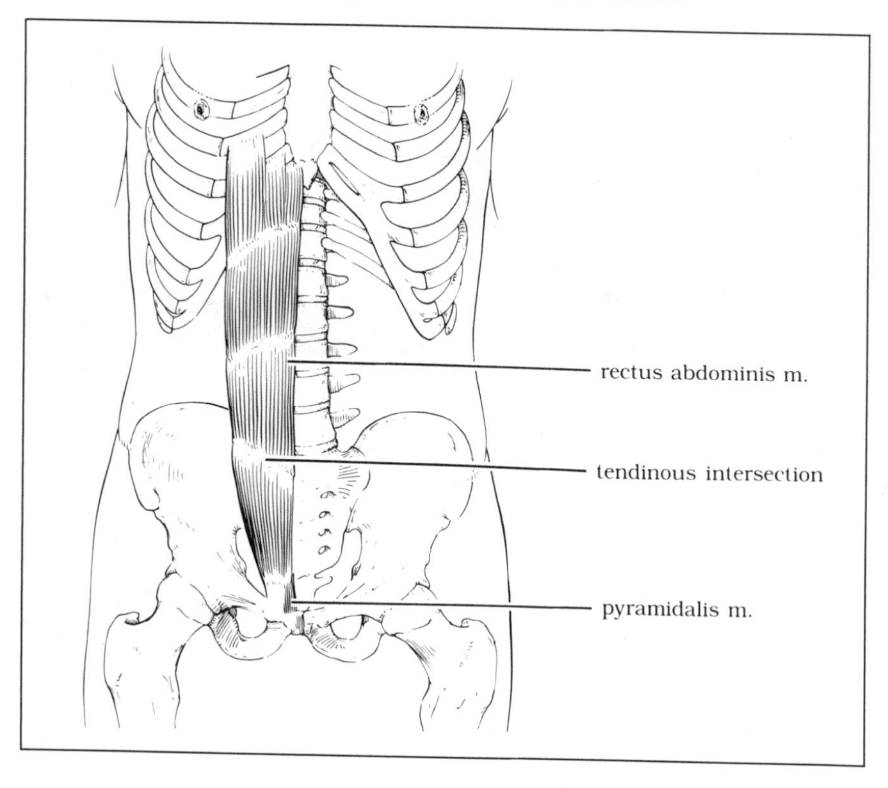

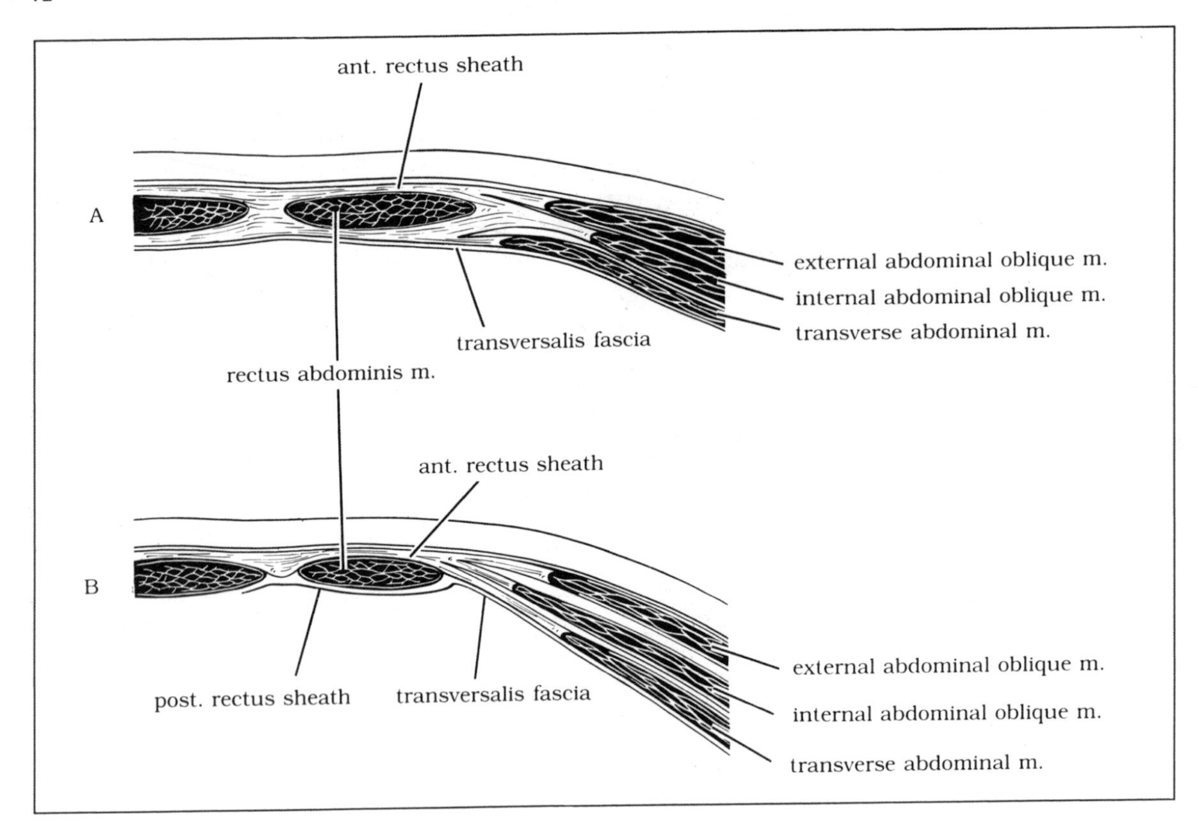

Fig. 3-7. Composition of the rectus sheath. A. Above the arcuate line. B. Below the arcuate line.

Fig. 3-8. Ligaments of the inguinal region.

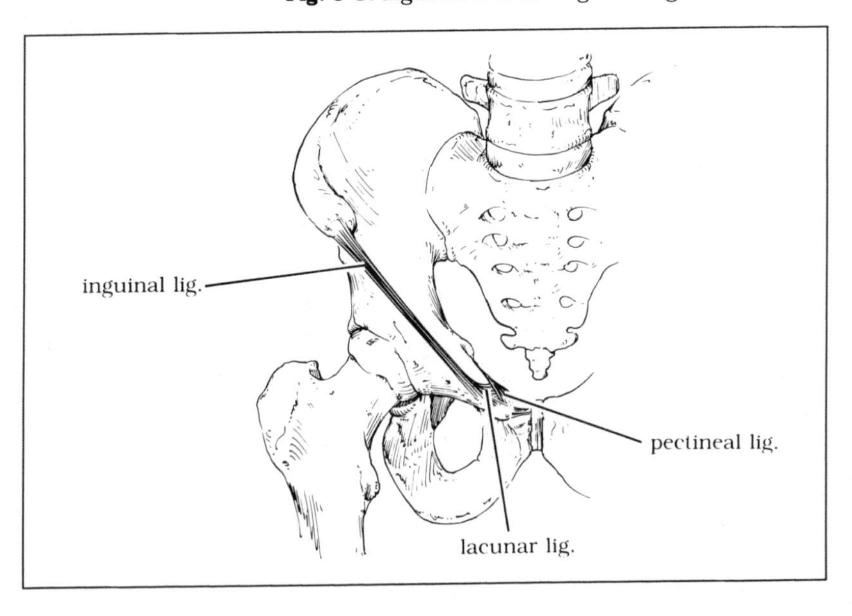

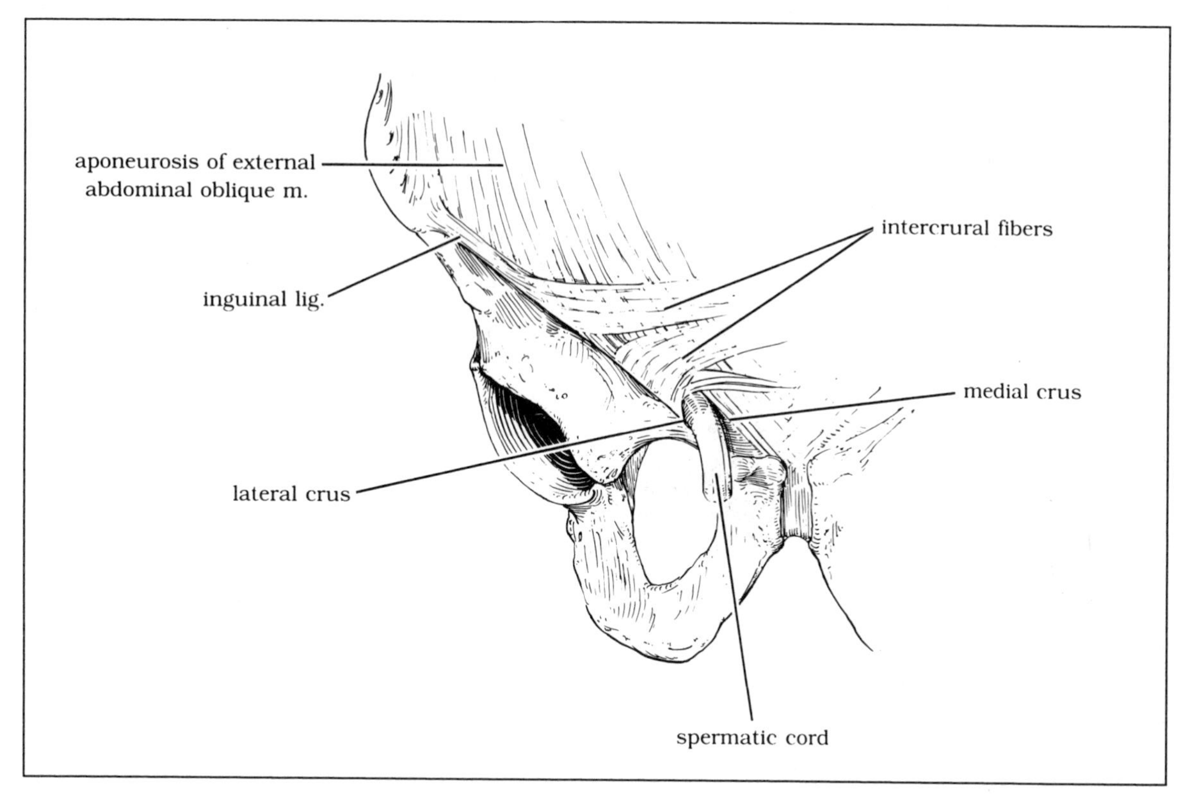

Fig. 3-9. Superficial inguinal ring and spermatic cord.

Fig. 3-10. Pathways of direct (right arrow) and indirect (left arrow) inguinal hernias.

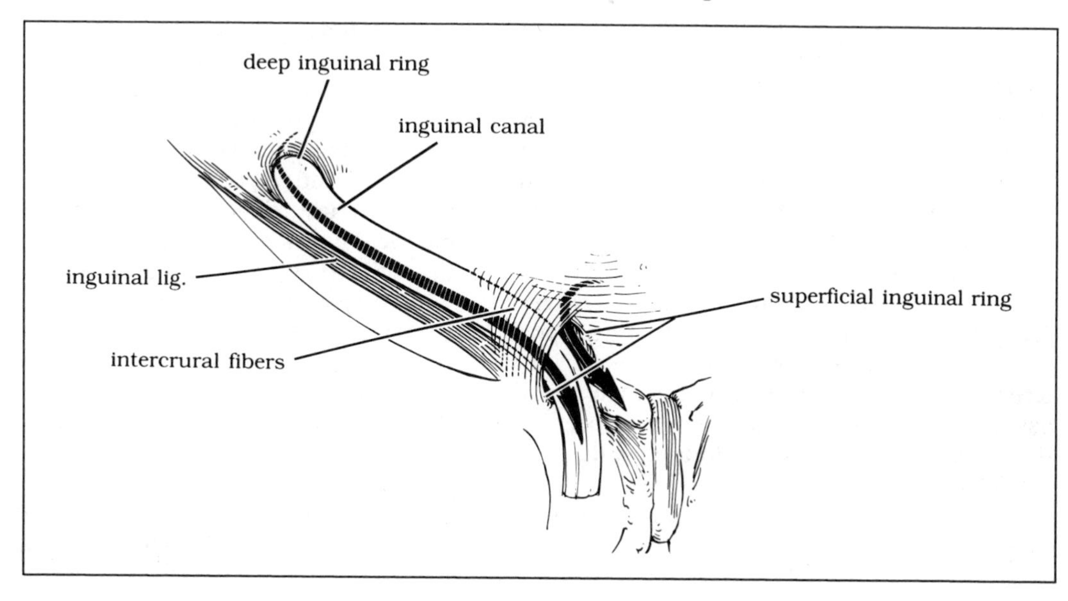

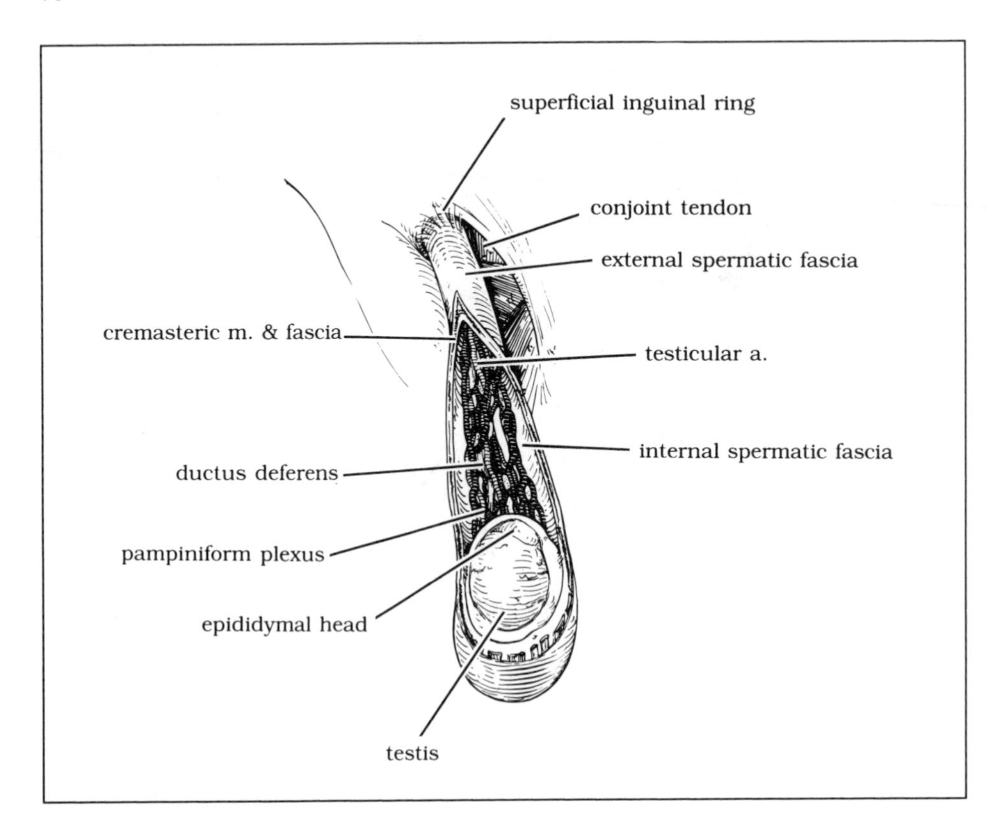

Fig. 3-11. Spermatic cord and its contents.

teries and the subcostal artery (see Chap. 2) and the four lumbar segmental arteries off the aorta, which follow the course of the posterior intercostal arteries. The **iliolumbar artery** substitutes for a fifth lumbar artery. It arises from the posterior trunk of the internal iliac artery in the pelvis and passes behind the medial border of the psoas major muscle to divide into lumbar and iliac branches to the posterior wall.

The branches of the external iliac artery (see Fig. 3-12) that supply the abdominal wall, the inferior epigastric and deep circumflex iliac arteries, arise just before the external iliac artery passes beneath the inguinal ligament to become the femoral artery. The **deep circumflex iliac artery** runs, in a course parallel to the superficial circumflex iliac artery, first between the transversalis and iliacus fasciae and then between the transverse abdominal and internal abdominal oblique muscles.

The **inferior epigastric artery** arises just medial to the deep inguinal ring; it therefore lies medial to an indirect inguinal hernia and lateral to a direct one. Thus it may be used as a landmark in diagnosis. The artery crosses toward the umbilicus and pierces the rectus sheath to lie on the posterior surface of the rectus abdominis muscle. There it ascends, sometimes anastomosing with the superior epigastric artery.

The **superior epigastric artery** is a terminal branch of the internal thoracic artery (see Chap. 2). It descends on the posterior surface of the rectus abdominis muscle, within the rectus sheath, to anastomose with the inferior epigastric artery.

The veins of the abdominal wall follow the same course as the arteries they accompany.

The lymphatics of the abdominal wall may be

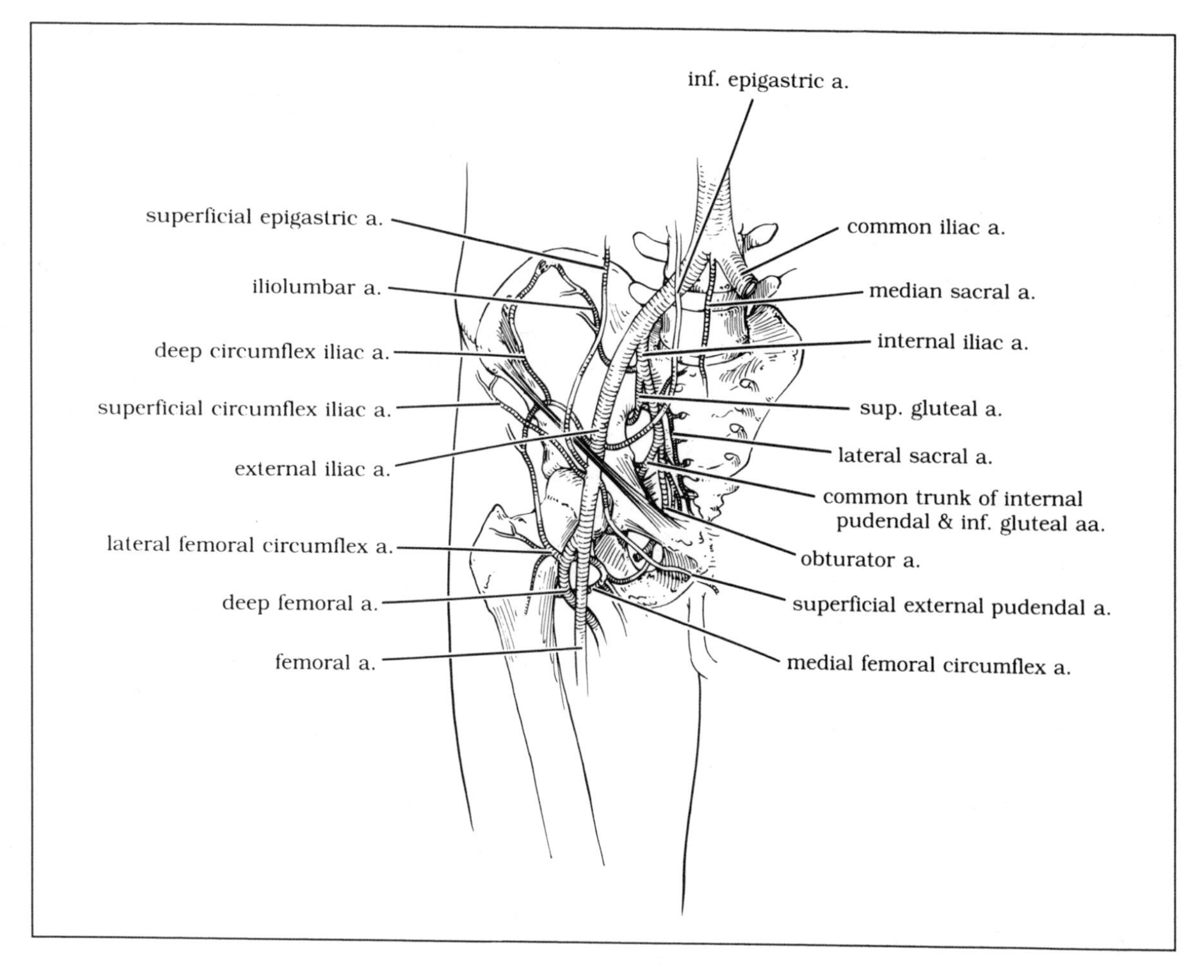

Fig. 3-12. Region of arterial distribution of the iliac arteries and some of their branches.

divided into four groups: (1) those that accompany the inferior epigastric artery and end in nodes lying along the external iliac artery **(external iliac nodes)**; (2) those that accompany the deep circumflex iliac artery and also end in the external iliac nodes; (3) those that accompany the lumbar vessels and end in **lumbar nodes,** which lie along the aorta; and (4) those that accompany the superior epigastric artery and end in **sternal nodes** (see p. 35).

The thoracoabdominal intercostal nerves are discussed in Chapter 2. The **iliohypogastric nerve** is formed by contributions from spinal cord segments T12 and L1; its course is the same as that of the thoracoabdominal intercostals. The **ilioinguinal nerve** (L1) has in part the same course as the above nerve. It also sends fibers through the inguinal canal to innervate the skin of the upper and medial thigh. An additional branch, the anterior scrotal nerve, innervates the skin of the anterior scrotum and the root of the penis. The nerves of the abdominal wall can be found between the transverse abdominal muscle and the internal abdominal oblique muscle in the fascial plane.

The innervation of the muscles of the wall is detailed in Table 3-1.

Abdominal Cavity and Viscera

GENERAL ORGANIZATION

The abdominal cavity (Fig. 3-13) contains most of the viscera involved in the digestion of food and the excretion of body wastes. The cavity is lined by a serous membrane, the **peritoneum,** that is similar in structure and function to the pleura and serous pericardium of the thorax. The peritoneum may be divided into two subunits: the parietal peritoneum and the visceral peritoneum.

The **parietal peritoneum** covers the inner surface of the abdominal body wall, being separated from the transversalis fascia by a layer of extraperitoneal connective tissue containing varying amounts of fatty deposits. The **visceral peritoneum** is a layer of peritoneum that closely adheres to the organs it covers. The **peritoneal cavity** is the potential space between these layers of peritoneum. It is usually filled by a thin layer of lubricating serous fluid.

Fig. 3-13. The abdominal cavity with viscera in place.

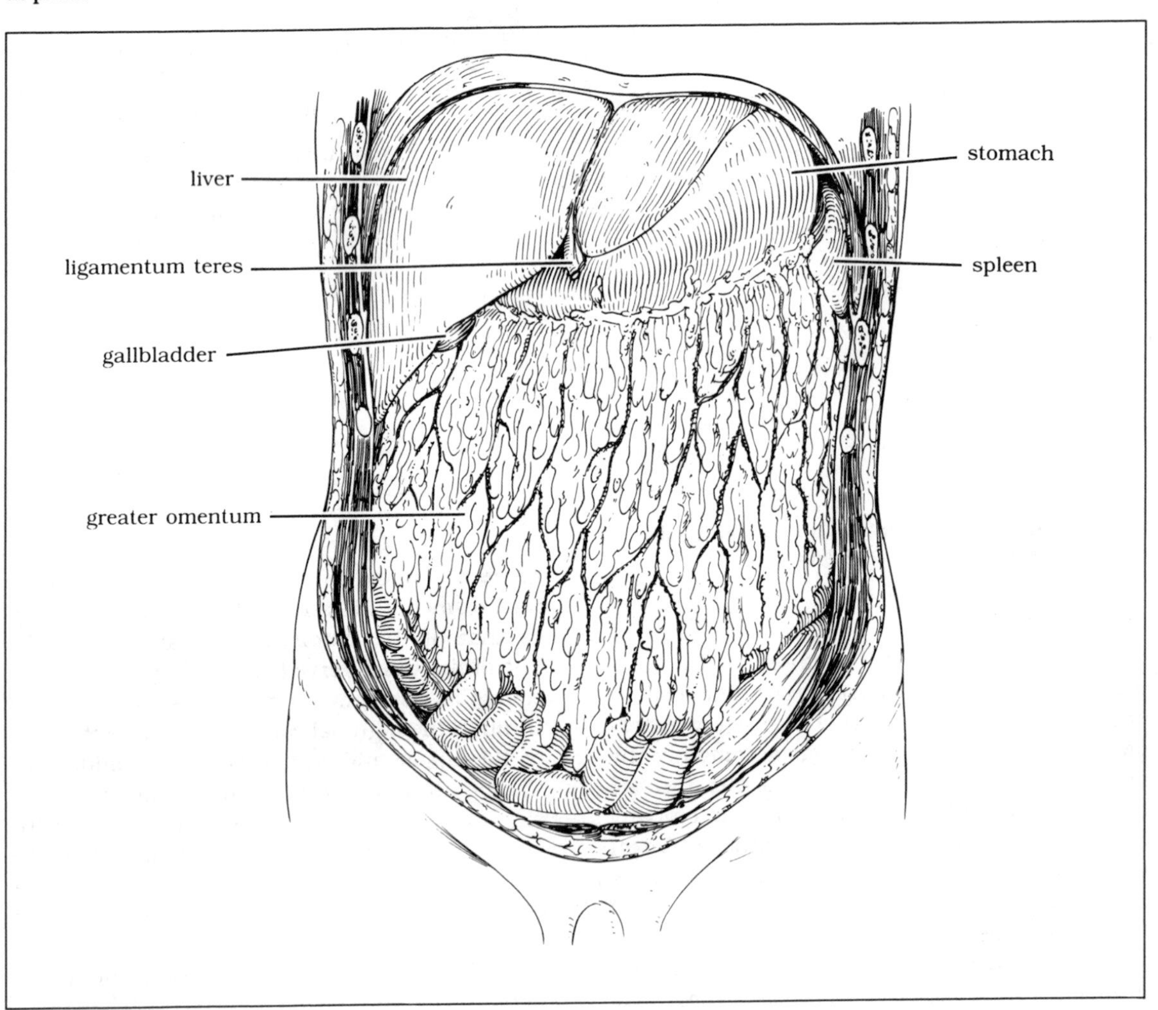

Table 3-4. Mesenteries

Embryonic Mesentery	*Adult Mesentery*	*Subdivisions of Adult Mesentery*
Dorsal mesogastrium	Greater omentum	Gastrophrenic ligament, gastrolienal ligament, gastrocolic ligament, lienorenal ligament
	Phrenicocolic ligament	
Dorsal common mesentery	Mesentery proper, transverse mesocolon, sigmoid mesocolon, mesoappendix	
Ventral mesogastrium	Lesser omentum	Hepatogastric ligament, hepatoduodenal ligament
	Falciform ligament, coronary ligament, right and left triangular ligaments	

The organs of the abdominal cavity either hang from the dorsal body wall or are implanted in the body wall. Those organs that hang into the abdominal cavity are attached to the body wall by a double layer of peritoneum known as a mesentery, through which vessels and nerves pass (Table 3-4). Those organs that are implanted in the body wall, and thus lack a mesentery, are known as retroperitoneal organs.

Embryonically all the digestive organs are attached by a mesentery to the dorsal body wall. This mesentery is defined by the organ to which it attaches: the mesogastrium attaches to the stomach and the mesocolon attaches to the colon. The adult derivatives of the embryonic dorsal mesogastrium are the greater omentum and the phrenicocolic ligament. The derivatives of the embryonic dorsal common mesentery are the mesentery (the mesentery to the small intestine), the transverse mesocolon, the sigmoid mesocolon, and the mesoappendix. In addition there is an embryonic ventral mesogastrium whose derivatives are the lesser omentum, the falciform ligament, the coronary ligament, and the right and left triangular ligaments.

It should be noted that the ligaments within the abdominal cavity are not true ligaments in that they are not composed of tendinous connective tissue. Rather they are folds of peritoneum stretching between two organs, or an organ and the body wall, and are usually named after their points of attachment; for example, the lienorenal ligament attaches from the spleen to the kidney. In addition, folds of peritoneum may also surround pouches of peritoneum. These pouches (or sacs or fossae) are known as recesses and may be of clinical importance should a portion of the intestine become trapped within one.

Structures in the abdomen that are retroperitoneal in the adult are the ascending colon, the descending colon, the pancreas, the duodenum, the kidneys and suprarenal glands, and the major blood vessels.

The following section on the organs of digestion is organized around the blood supply to these viscera. The digestive tract in the abdomen receives blood through three arteries: the celiac, the superior mesenteric, and the inferior mesenteric. The celiac artery supplies the organs of the upper abdominal tract, from the lower esophagus to the first part of the duodenum. The superior mesenteric artery furnishes blood from the duodenum through the first two-thirds of the transverse colon. The inferior mesenteric artery supplies the remainder of the abdominal digestive tract.

ORGANS OF DIGESTION

Organs Supplied by the Celiac Artery

The esophagus, which carries food from the mouth to the stomach, is discussed in Chapter 2.

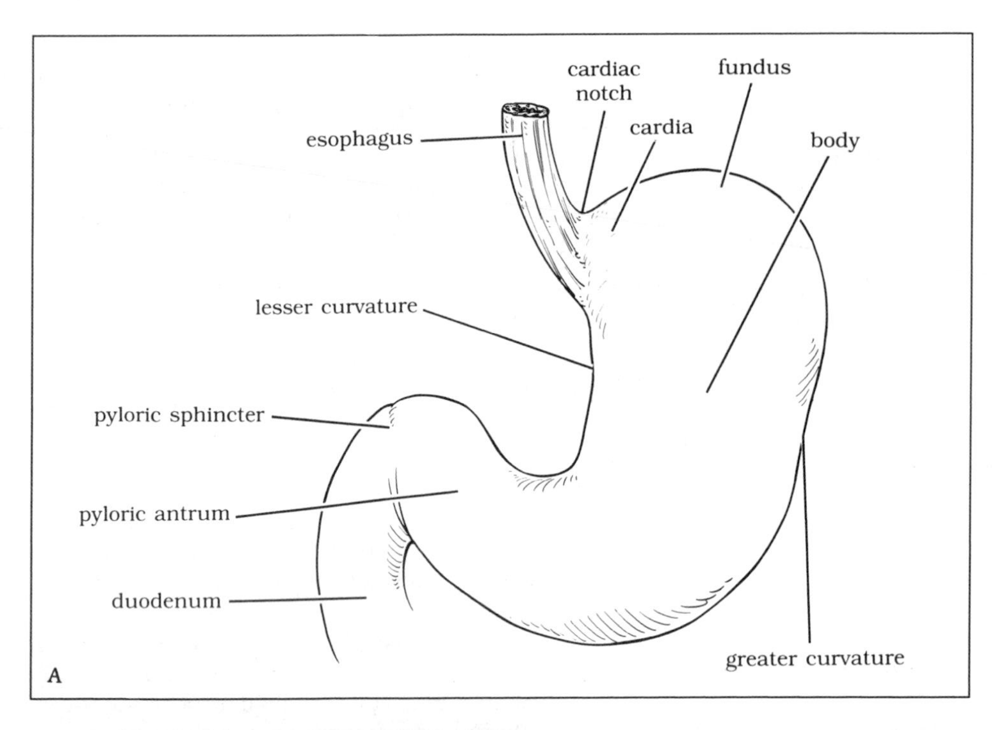

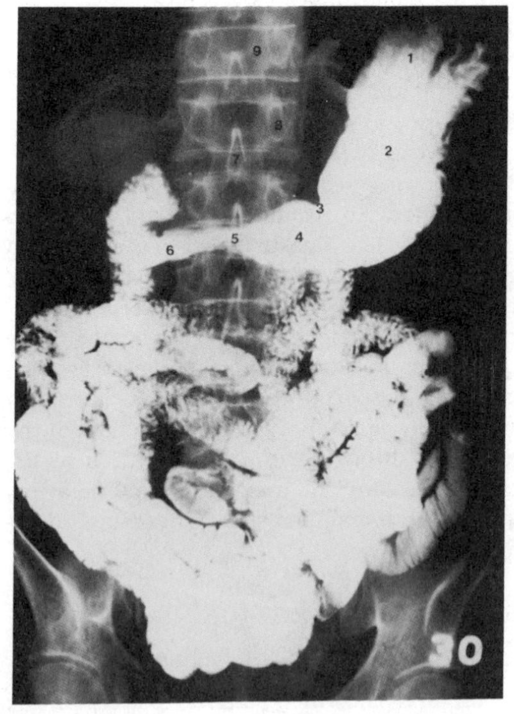

Fig. 3-14. A. The stomach. B. Barium image of the stomach and small intestine. (1 = fundus of stomach; 2 = body of stomach; 3 = angular notch; 4 = pyloric antrum; 5 = pylorus; 6 = first part of duodenum; 7 = vertebral spine; 8 = pedicle of vertebra; 9 = body and lamina of vertebra.)

STOMACH AND OMENTA

The stomach (gaster) (Fig. 3-14), lying in the upper left quadrant of the abdominal cavity, is a large organ of flexible shape. Its shape may vary depending on (1) the general stature of the person, (2) the person's position (i.e., standing versus reclining), and (3) the amount of material within the stomach and neighboring intestines. The stomach has several regions—the cardia, fundus, body, pyloric antrum, and pylorus—and two curvatures—the greater and lesser curvatures. It begins at the esophagus and ends in the pylorus (Table 3-5).

The **cardia** is the region of the stomach where the esophagus enters. The **fundus** is the superior portion of the stomach lying entirely under the

Table 3-5. Stomach

Divisions	*Mesenteries*	*Arterial Supply*	*Venous Drainage*	*Innervation*
Cardia, fundus, body, pyloric antrum, pylorus	Lesser omentum, greater omentum	Left and right gastrics, left and right gastroepiploics, short gastrics	Left and right gastrics, left and right gastroepiploics, short gastrics	Vagal trunks, splanchnics via celiac ganglion

greater curvature. The **body,** or major portion, of the stomach lies below the fundus and cardia between the greater and lesser curvatures. The body ends at the level of a line descending vertically from the angular notch of the lesser curvature. The **pyloric antrum** lies between the body and the pylorus and serves as a vestibule to the pylorus. The **pylorus** is a valvular region at the distal end of the stomach that regulates the flow of material into the duodenum. The walls of the pylorus are highly muscular and thickened.

The medial curvature of the stomach is the **lesser curvature.** Toward its pyloric end it is grooved by the angular notch, marking the line between the body and the pyloric antrum (see Fig. 3-14B). The lesser curvature is continuous with the medial right wall of the esophagus. The lateral curvature of the stomach is the **greater curvature.** It begins at the cardiac notch, where the esophagus and fundus come together, forms the top of the fundus and lateral wall of the body of the stomach, and ends in the pylorus.

The outer surface of the stomach is entirely covered with peritoneum. The inner mucosal surface of the stomach is covered with raised folds, the rugae, which are transient in position.

Hanging off the greater curvature of the stomach is the **greater omentum** (Fig. 3-15; see also Fig. 3-13). This structure is a derivative of the embryonic dorsal mesogastrium. It is formed by the stretching and folding of the mesentery as the stomach is rotated into the final adult position. This process results in a four-layered structure, with an actual or potential space separating the two original mesenteries of two layers each. The greater omentum acts as both a site of fat storage and a lymphatic organ. To perform this latter function, the omentum often migrates to sites of inflammation within the abdomen. In the adult the greater omentum is said to have several subsections: the gastrophrenic, gastrocolic, gastrolienal, and lienorenal ligaments. These subsections connect the various organs to which their name alludes.

The **lesser omentum** (see Fig. 3-15) stretches from the lesser curvature of the stomach and the adjacent portion of the duodenum to the liver. It roofs the omental bursa, or sac, and its free medial border forms the **hepatoduodenal ligament,** through which run the hepatic artery, proper portal vein, and common bile duct. Deep to the hepatoduodenal ligament is the **epiploic foramen** (Fig. 3-16), which is the entrance into the omental bursa. The **omental bursa** is the potential space under the stomach between the liver and the points at which the ligaments of the greater omentum attach to the abdominal wall. The posterior boundary of the lesser omental sac is the peritoneum of the body wall, its anterior boundary is the stomach and lesser omentum, and it is entered through the epiploic foramen.

The stomach lies in front of the diaphragm, left kidney, left suprarenal gland, pancreas, spleen, left colic flexure, and transverse mesocolon. These structures and their peritoneal coverings form the floor of the lesser omental sac.

A description of the vessels, nerves, and lymphatics of the stomach and other organs of this region may be found at the end of this section (see pp. 83–88).

DUODENUM

The **duodenum** (Fig. 3-17; see also Figs. 3-3 and 3-15) is the continuation of the digestive tract past the pylorus of the stomach. It may be di-

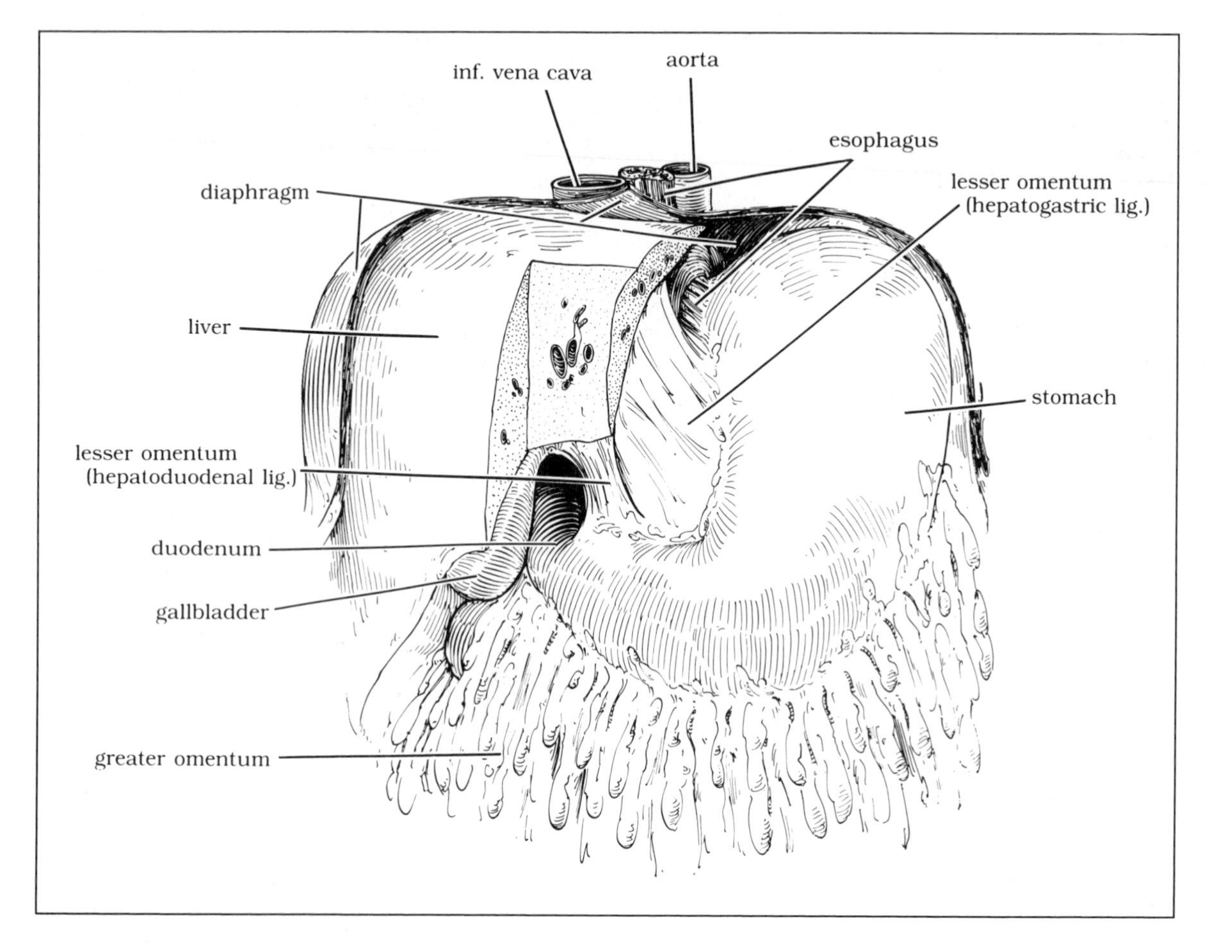

Fig. 3-15. The stomach in relation to surrounding viscera.

vided into four parts: superior, descending, horizontal, and ascending (Table 3-6). The last three parts of the duodenum are retroperitoneal; the first, or superior, portion has its own mesentery. This section, approximately 5 cm in length, is connected to the liver by the hepatoduodenal ligament (free edge of the lesser omentum), through which the hepatic artery, proper portal vein, and common bile duct run. The bile duct and the pancreatic duct together enter the descending portion (10 cm in length) of the duodenum about midway in its length, whereupon they form a chamber (ampulla) that is marked on the interior wall of the duodenum by a papilla. The horizontal portion of the duodenum is 8 cm long, and the ascending portion is 2 cm long. The ascending portion ends by becoming the jejunum. The last three sections of the duodenum form a C-shaped structure in which the head of the pancreas is lodged.

PANCREAS

The **pancreas** (see Figs. 3-3 and 3-17 and Table 3-6) stretches across the posterior abdominal wall from the C-shaped cup of the duodenum to the hilum of the spleen. It serves a double role in the body, having both digestive and endocrine functions. The pancreas may be divided into four regions: head, neck, body, and tail.

Table 3-6. Duodenum, Pancreas, and Spleen

Organ	*Divisions*	*Arterial Supply*	*Venous Drainage*	*Innervation*
Duodenum	Superior, descending, horizontal, ascending	Right gastric, superior and inferior pancreaticoduodenal	Splenic, superior mesenteric	Celiac plexus
Pancreas	Head, uncinate process, neck, body, tail	Superior and inferior pancreaticoduodenal, pancreatics of splenic artery	Splenic, superior mesenteric	Celiac plexus
Spleen		Splenic	Splenic	Celiac plexus

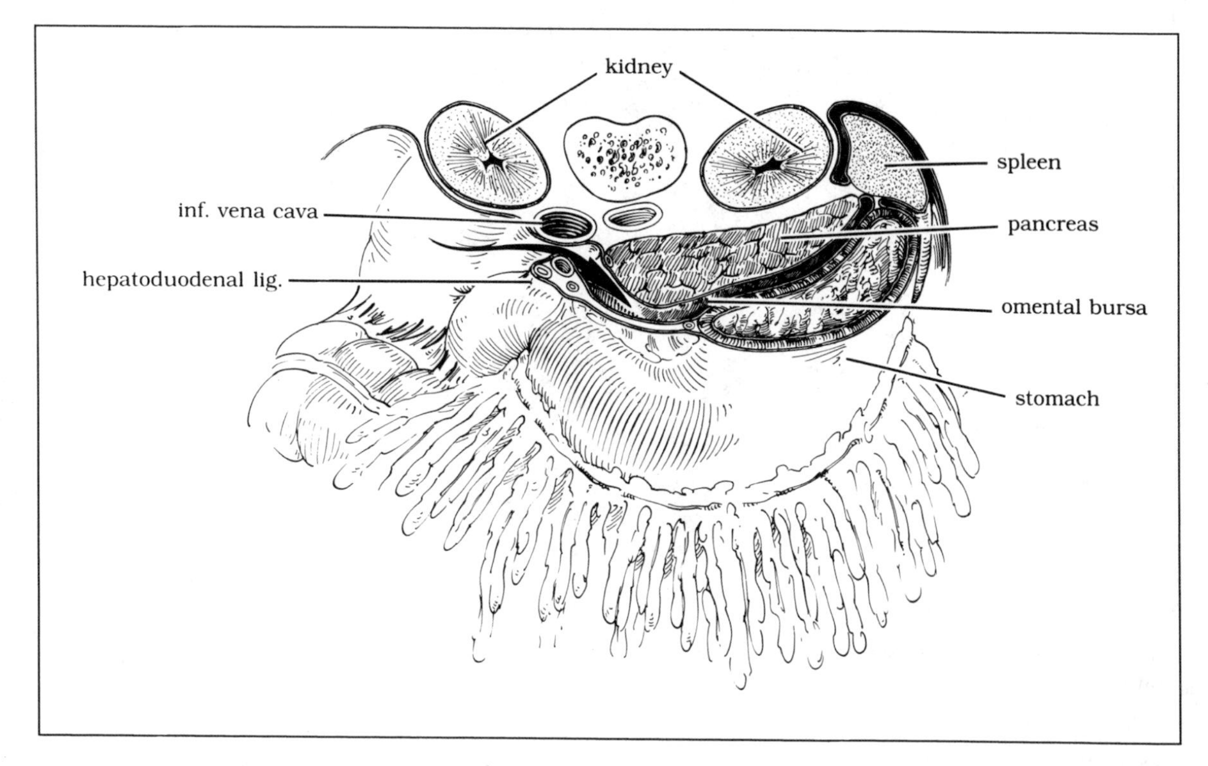

Fig. 3-16. The lesser omental bursa and surrounding viscera.

The **head** lies within the duodenal cup and has a continuation, the uncinate process, that runs toward the lower left and lies behind the superior mesenteric artery and vein. The **neck** of the pancreas is a narrowed region lying in front of the superior mesenteric vessels and connecting the head and body of the pancreas. The **body** is the major horizontal portion of the pancreas, running leftward in front of the aorta and left kidney. The **tail** of the pancreas ends in or near the hilum of the spleen.

The **pancreatic duct** is a Y-shaped structure. The major arm of the Y runs horizontally from tail to head, and the side arm ascends rightward within the uncinate process to join the duct in

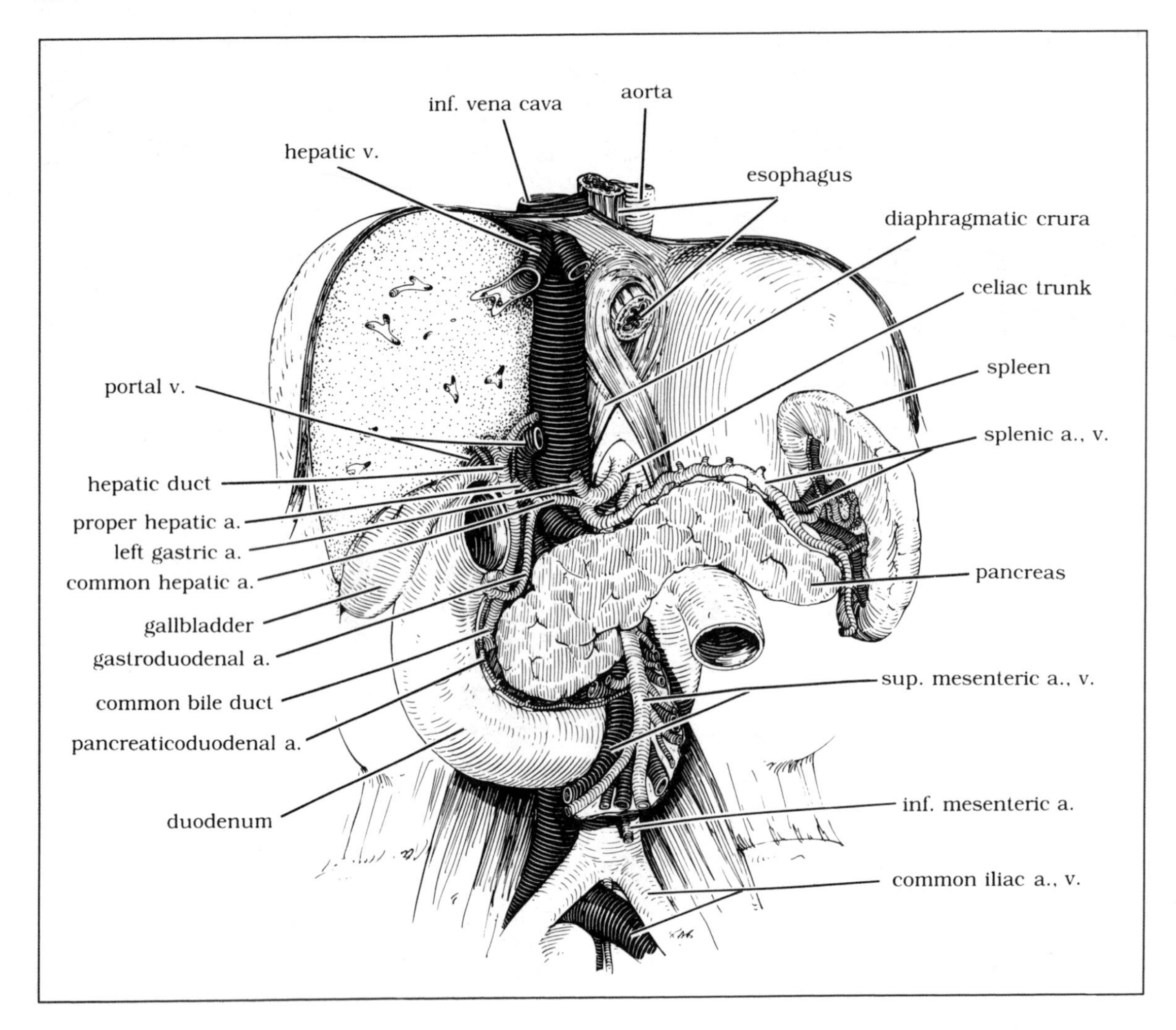

Fig. 3-17. The duodenum, pancreas, and spleen in situ.

Table 3-7. Liver and Gallbladder

Organ	*Divisions*	*Arterial Supply*	*Bile Drainage*	*Portal Venous Supply*
Liver	Right lobe	Right hepatic	Right hepatic	Right branch portal vein
	Left lobe	Left hepatic	Left hepatic	Left branch portal vein
	Caudate lobe	Left hepatic	Left hepatic	Left branch portal vein
	Quadrate lobe	Left hepatic	Left hepatic	Left branch portal vein
Gallbladder	Fundus, body, neck	Cystic	Cystic	

the head of the pancreas. The pancreatic duct then fuses with the bile duct before emptying into the duodenum. There is often an accessory pancreatic duct whose terminal portion lies superior to the main duct and enters the duodenum independently.

SPLEEN

The spleen (lien) (see Figs. 3-16 and 3-17 and Table 3-6) lies lateral and posterior to the greater curvature of the stomach and is normally covered by the rib cage. It develops in the embryo within the dorsal mesogastrium. In the adult it is connected to the stomach by the gastrolienal ligament and to the body wall by the lienorenal ligament. The spleen is pyramidal in shape, the base of the pyramid being the hilum and the apex directed upward toward the rib cage.

The spleen is involved in the processing of blood and lymphocytes. It is therefore well supplied with blood from the splenic artery, which breaks into several smaller splenic arteries before it enters the hilum of the spleen.

LIVER AND GALLBLADDER

The liver and gallbladder (see Figs. 3-15 and 3-17) are also supplied by branches of the celiac artery. The liver (hepar), lying in the upper right quadrant of the abdomen, is shaped like a large mushroom cap and has only two surfaces—diaphragmatic and visceral—that are separated by the liver's inferior border. It is covered by peritoneum, except in the area bounded by the coronary and triangular ligaments on the diaphragmatic surface (bare area) and at the hilum (Fig. 3-18). The coronary and triangular ligaments carry the peritoneum of the liver onto the diaphragm and are continuous with each other. They are connected ventrally to the **falciform ligament.** The anterior thick, fibrous portion of the falciform ligament is the **ligamentum teres;** the posterior inferior portion contains the **ligamentum venosum.**

The lesser omentum is continuous with the falciform ligament at the hilum of the liver (on its visceral surface). It is through the hepatoduodenal portion of the lesser omentum (see Fig. 3-16) that the portal vein and proper hepatic artery enter the liver at the hilum (porta hepatis) and the bile duct leaves.

The liver is divided into four lobes: right, left, caudate, and quadrate (Table 3-7; see also Fig. 3-18C). Functionally the liver is divided into right and left halves, with the caudate and quadrate lobes belonging to the left half. Each half has its own blood supply, blood drainage, and bile drainage. The right and left hepatic arteries are branches of the hepatic artery proper; the right and left hepatic ducts fuse externally to the liver to form the common hepatic duct (see Fig. 3-17). The common hepatic duct fuses with the cystic duct of the gallbladder to form the **common bile duct,** which, after passing behind the first part of the duodenum, fuses with the pancreatic duct before entering the duodenal wall.

The hepatic veins, which drain the liver substance, empty directly into the inferior vena cava. The inferior vena cava is closely applied to, if not embedded in, the posterior surface of the liver (see Fig. 3-18B,C). The portal venous system is discussed on pages 101 to 103.

The **gallbladder** (see Figs. 3-15 and 3-17 and Table 3-7), supplied by the cystic branch of the right hepatic artery, is a blind sac lying on the visceral surface of the liver. It consists of a fundus, body, and neck, and its main function is the storage and concentration of bile. The **fundus,** the most inferior portion, is the roof of the pouch. The **body** is the main portion of the gallbladder and leads into the neck. The **neck** is the narrowed portion of the gallbladder connecting it to the **cystic duct.** This duct continues on to fuse with the common hepatic duct to form the common bile duct. Within both the neck and cystic duct the walls are ridged, forming a spiral fold of mucosa.

CELIAC ARTERY AND ITS BRANCHES

The celiac artery (Fig. 3-19) is the first visceral branch of the aorta below the diaphragm. Approximately 1.5 cm long, it passes ventrally from

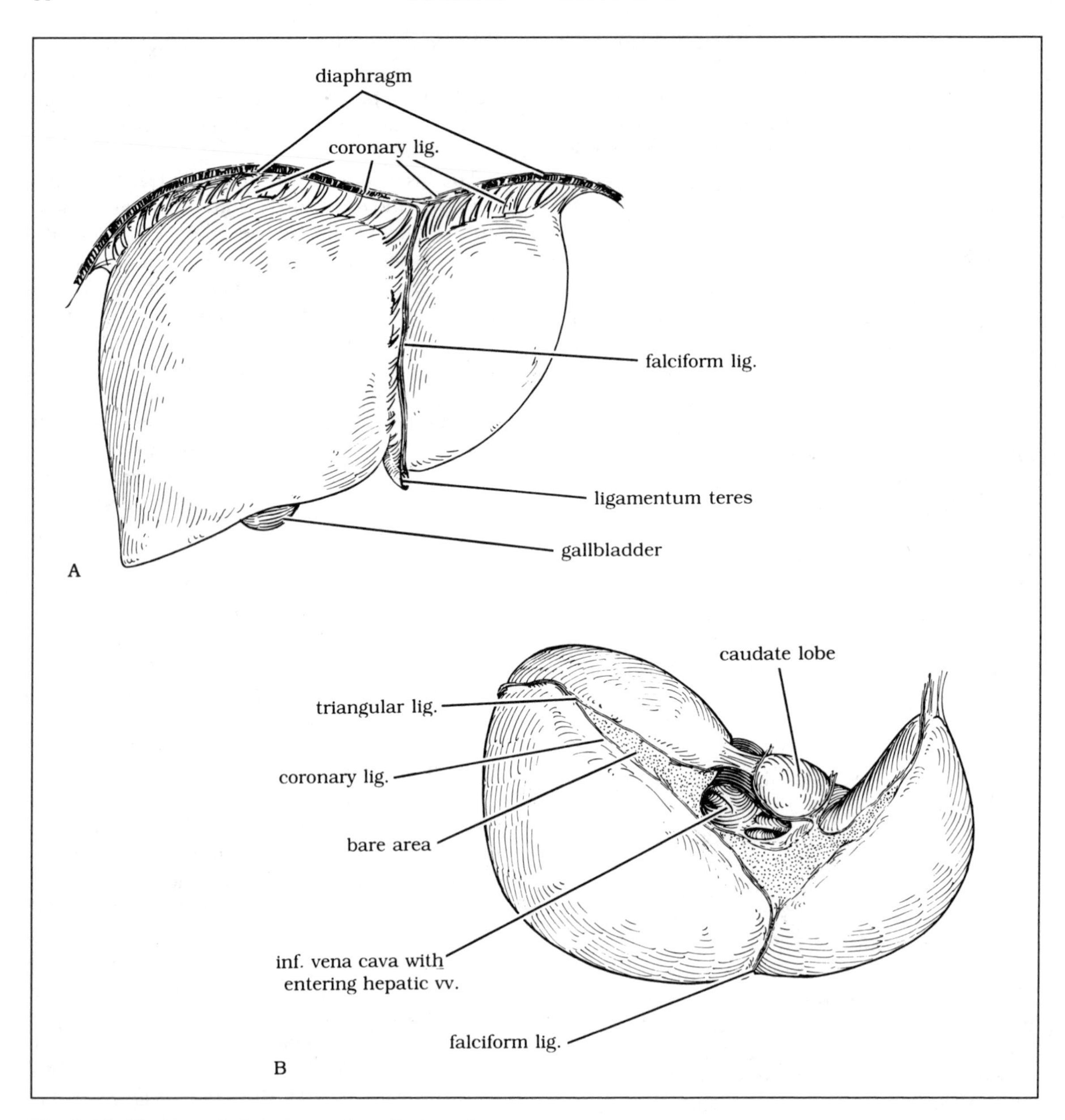

Fig. 3-18. The liver. A. Anterior surface. B. Superior surface. C. From below.

the aorta to split into its three major branches: the left gastric, splenic, and common hepatic arteries (Table 3-8).

The **left gastric artery** (see Fig. 3-19) arches upward and to the left to reach the cardia of the stomach. From here the artery turns to run downward along the lesser curvature of the stomach, splitting as it does so into dorsal and ventral branches. Prior to its bifurcation the ar-

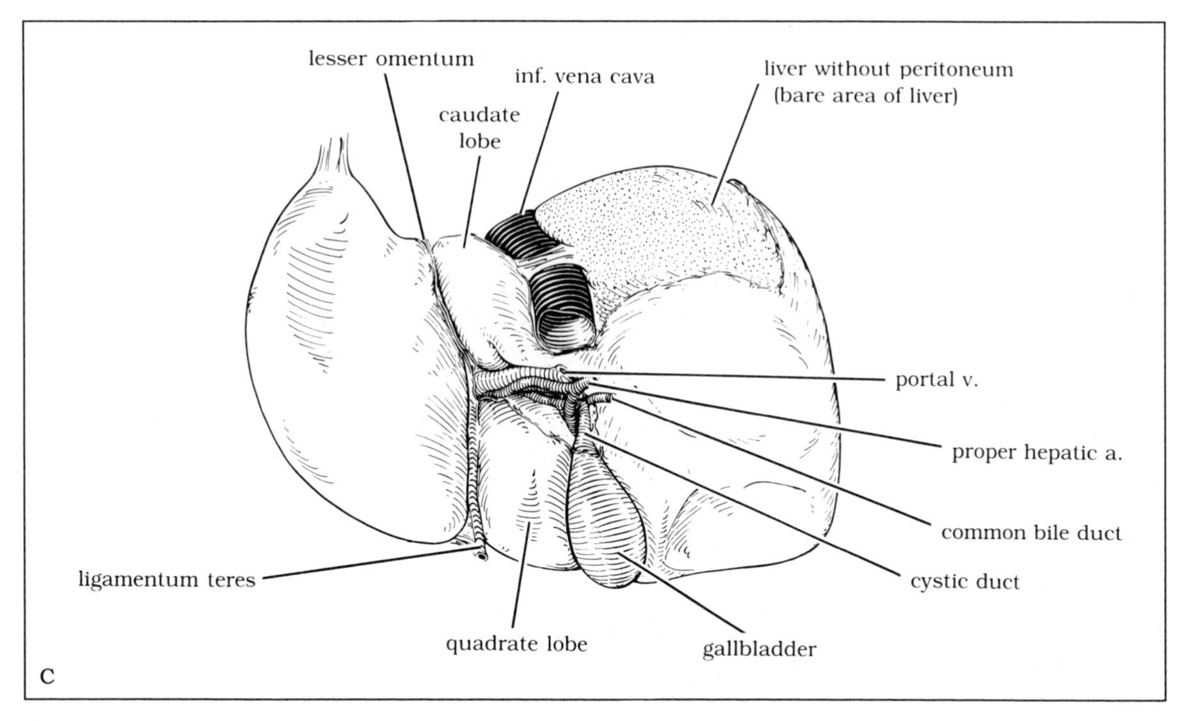

Fig. 3-18. (continued).

Table 3-8. Celiac Arterial Branches and the Organs They Supply

Artery	*Branches*	*Organs Supplied*
Left gastric		Lesser curvature of the stomach, lower esophagus
Common hepatic	Right gastric	Pylorus and lesser curvature of the stomach
	Right hepatic	Right lobe of the liver (and usually the gallbladder through its cystic branch)
	Left hepatic	Left lobe of the liver
	Gastroduodenal	
	Right gastroepiploic	Pylorus, greater curvature of the stomach, greater omentum
	Anterior and posterior superior pancreaticoduodenal	Pancreas, duodenum, bile duct
Splenic	Pancreatic	Pancreas
	Left gastroepiploic	Greater curvature of the stomach, greater omentum
	Short gastrics	Greater curvature of stomach
	Splenics	Spleen

tery gives off a branch, the **esophageal ramus,** which travels upward to supply blood to the lower esophagus.

The **splenic artery** (see Fig. 3-19) runs leftward, toward the spleen, in the dorsal body wall. It is an extremely tortuous artery, lying for much of its length on or within the pancreas. The artery gives rise to several major branches supplying the pancreas, stomach, and omentum before reaching the spleen. The pancreatic branches—three major ones and several minor ones—supply the superior aspect of the pancreas and anastomose with pancreatic branches from the superior mesenteric and gastroduodenal arteries. The major branches are the **dorsal pancreatic artery,** the **pancreatica magna artery,** and the **caudal pancreatic artery.**

The splenic branches to the stomach and greater omentum are the left gastroepiploic artery and the short gastric arteries. The **left gastroepiploic artery** runs along the greater curvature of the stomach toward the pylorus, supplying both the body of the stomach and the greater omentum. It ends by anastomosing with the right gastroepiploic artery. The **short gastric arteries** arise near the end of the splenic artery. They pass through the gastrolienal ligament to reach the body of the stomach, where they distribute along the upper portions of the body of the stomach. Shortly before reaching the spleen the splenic artery divides into several terminal branches that penetrate the spleen at its hilum, often anastomosing between themselves there.

The **common hepatic artery** (Fig. 3-19) has the most extensive distribution in the region, providing blood to the liver, gallbladder, stomach, duodenum, and pancreas. The common hepatic artery proceeds rightward from the common bifurcation point of the celiac artery. It gives off the gastroduodenal, right gastric, right hepatic, and left hepatic arteries. The **gastroduodenal artery** is usually the first branch of the common hepatic artery. It descends behind the duodenum, giving off the anterior and posterior superior pancreaticoduodenal arteries and terminating in the right gastroepiploic artery. The **right gastroepiploic artery** passes away from the pylorus along the greater curvature of the stomach, supplying both the stomach and the greater omentum. It ends by anastomosing with the left gastroepiploic artery. The **superior pancreaticoduodenal arteries** supply the pancreas and duodenum with blood. In addition, the posterior superior pancreaticoduodenal artery provides blood to the bile duct. These arteries descend in the ventral (anterior superior pancreaticoduodenal) and dorsal (posterior superior pancreaticoduodenal) substance of the pancreas. They end by forming, respectively, the anterior and posterior pancreaticoduodenal arcades in conjunction with the anterior and posterior inferior pancreaticoduodenal arterial branches of the superior mesenteric artery.

The **right gastric artery** is the next branch of the common hepatic artery. After branching off the common hepatic artery, the right gastric artery crosses to the pylorus and then continues along the lesser curvature of the stomach, supplying these structures with blood. The artery ends by anastomosing with the left gastric artery.

After giving rise to the gastroduodenal and right gastric arteries, the common hepatic artery becomes the **proper hepatic artery.** This artery crosses toward the liver through the hepatoduodenal ligament in company with the bile duct and the portal vein. Before reaching the porta hepatis, or hilum of the liver, the proper hepatic artery terminates in right and left hepatic arteries. These arteries enter the liver, branching out within its substance to supply tissue. The **left hepatic artery** supplies the left and caudate lobes of the liver. The **right hepatic artery** supplies the right lobe of the liver. Either or both may be involved in delivering blood to the quadrate lobe. In addition, the right hepatic artery usually gives rise to the **cystic artery.** This artery travels up the cystic duct, supplying it and the gallbladder.

The venous blood from these abdominal organs, except the liver, returns to the heart via the portal venous system (see pp. 101–103). In the liver the arterial blood brought in by the hepatic

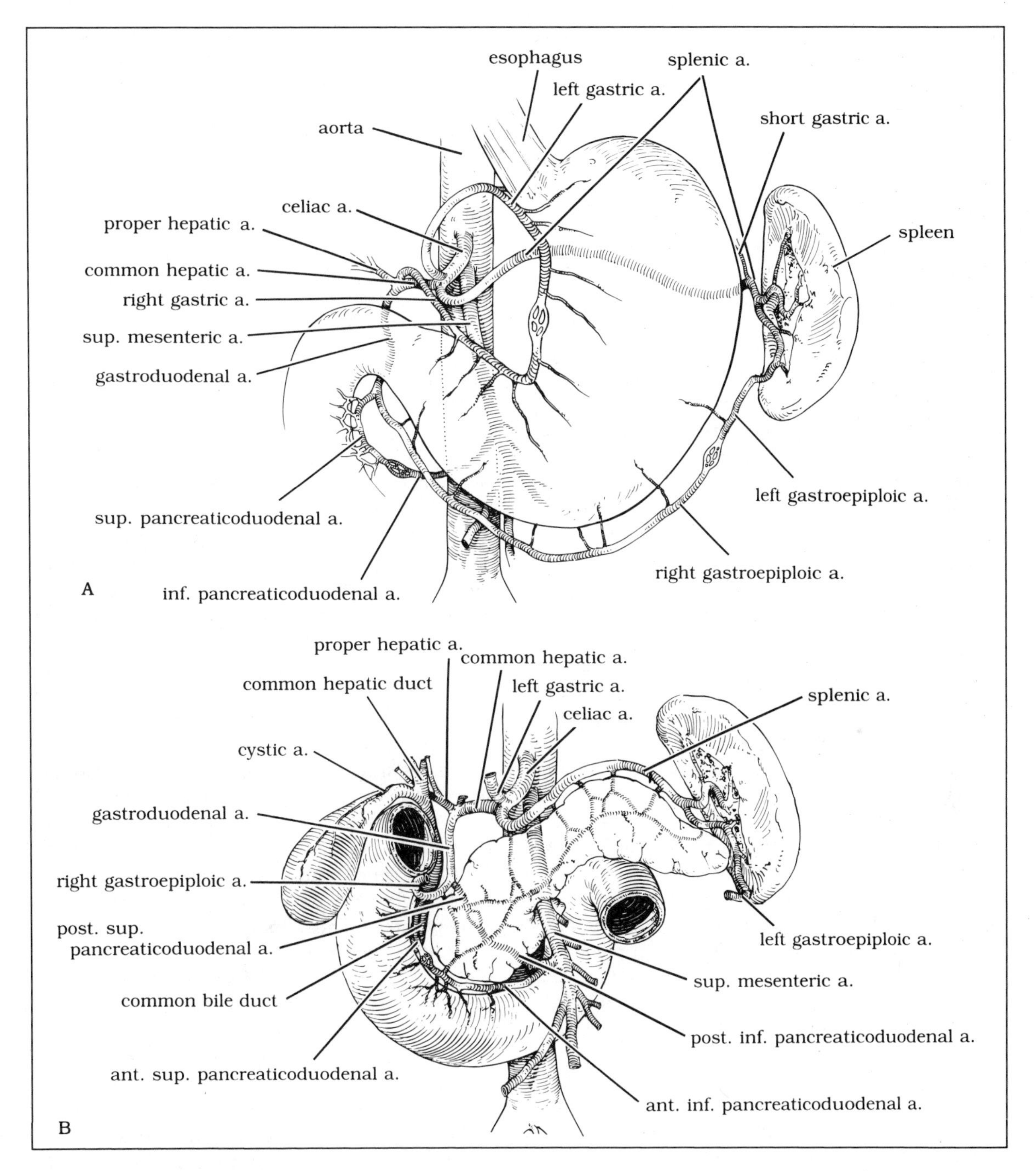

Fig. 3-19. Branches of the celiac artery. A. To the stomach and spleen. B. To the pancreas and duodenum.

arteries passes through a capillary network into either the right or the left hepatic veins. These veins empty directly into the inferior vena cava.

INNERVATION AND LYMPHATICS

Innervation of these abdominal viscera is supplied by the autonomic nervous system (see Chap. 1). The **sympathetic splanchnic nerves** arise in the thorax (see Chap. 2) and travel along the vertebral column to enter the abdomen. In the abdomen these nerves enter the **celiac ganglia,** where they synapse before distributing themselves via the celiac plexus and blood vessels. The parasympathetic nerves (see Chap. 2) enter the upper abdomen as **vagal trunks** on the esophagus. The nerves distribute to their organs via the **celiac plexus** and along the available blood vessels. The celiac ganglia and plexus are found around the celiac artery in the posterior abdominal wall.

The lymphatic drainage of these abdominal viscera follows the blood vasculature already described. For the stomach, four lines of drainage are present. The first two follow the lesser curvature of the stomach either upward to the **left gastric nodes** or downward to the **right gastric nodes.** The left gastric nodes in turn drain to the **celiac nodes,** while the right gastric nodes end in the **hepatic nodes.**

Drainage along the greater curvature of the stomach is into either the **right gastroepiploic nodes** or, following the short gastric and left gastroepiploic arteries, the **pancreaticolienal nodes.** The right gastroepiploic nodes, along with the pyloric nodes, end in the hepatic nodes, which in turn drain into the celiac nodes. The pancreaticolienal nodes drain the stomach, spleen, and pancreas. They also in turn empty into the celiac nodes.

Thus all lymphatic drainage from the stomach ends in the celiac nodes. In addition, the celiac nodes receive lymph from the spleen and pancreas (pancreaticolienal nodes), liver (hepatic nodes), and duodenum (pyloric nodes). The celiac nodes empty into the **cisterna chyli** (when it is present), which is where the thoracic duct originates. The pancreas and duodenum also drain into the superior mesenteric nodes, which also empty into the cisterna chyli.

Organs Supplied by the Mesenteric Arteries

The superior mesenteric artery supplies the inferior portions of the pancreas and duodenum; the whole jejunum, ileum, cecum (and appendix), and ascending colon; and the first two-thirds of the transverse colon. The pancreas and duodenum are discussed earlier in this chapter.

The **jejunum** and **ileum**—which, with the duodenum, constitute the small intestine—may be considered together as the jejunoileum. The jejunoileum (see Figs. 3-22 and 3-24) is approximately 6 meters long. It is suspended from a mesentery that extends for approximately 17.5 cm on the dorsal body wall. The jejunum constitutes about 40 percent of the jejunoileum, the ileum the remaining 60 percent. The bowel occupies most of the abdominal cavity, taking up part of the upper cavity, most of the middle cavity, and a large portion of the pelvic cavity. The jejunoileum receives its blood supply and innervation through its mesentery.

The ileum ends by opening into the **cecum,** a large blind sac at the begining of the large intestine (Figs. 3-20 and 3-21). The large intestine (Fig. 3-22) consists of the cecum, appendix, ascending colon, transverse colon, descending colon, sigmoid colon, and rectum. It is characterized by three longitudinal bands of muscle fibers, the teniae coli, running its length. The **teniae coli** arise at the root of the **appendix** (see Figs. 3-20 and 3-24), which is the tapered tip of the cecum. Because the muscle bundles of the teniae coli are shorter than the length of colon they cover, they cause the colon to bunch up into a series of pouches, the **haustra.** These are not found past the sigmoid colon. The appendix, cecum, and transverse and sigmoid colons (see Fig. 3-23) are all mobile in the abdominal cavity, as they have mesenteries. The ascending and de-

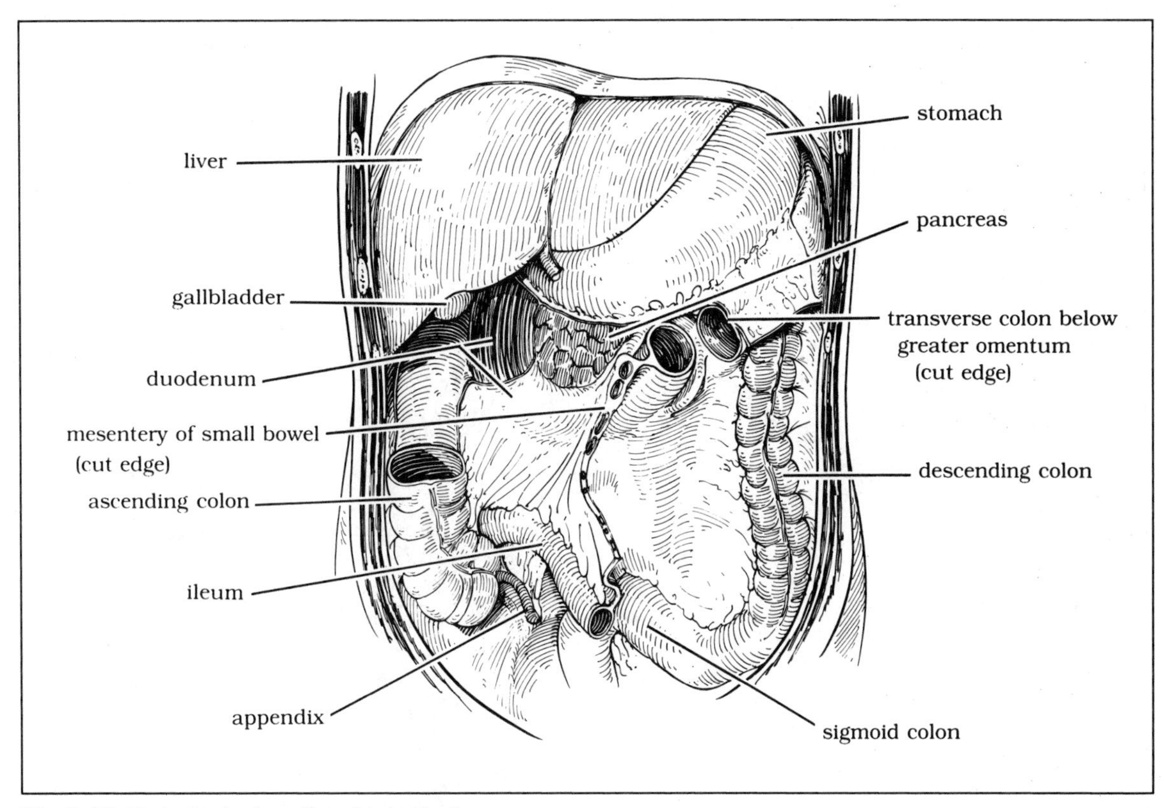

Fig. 3-20. Posterior body wall and intestinal mesentery.

Fig. 3-21. Ileocecal junction.

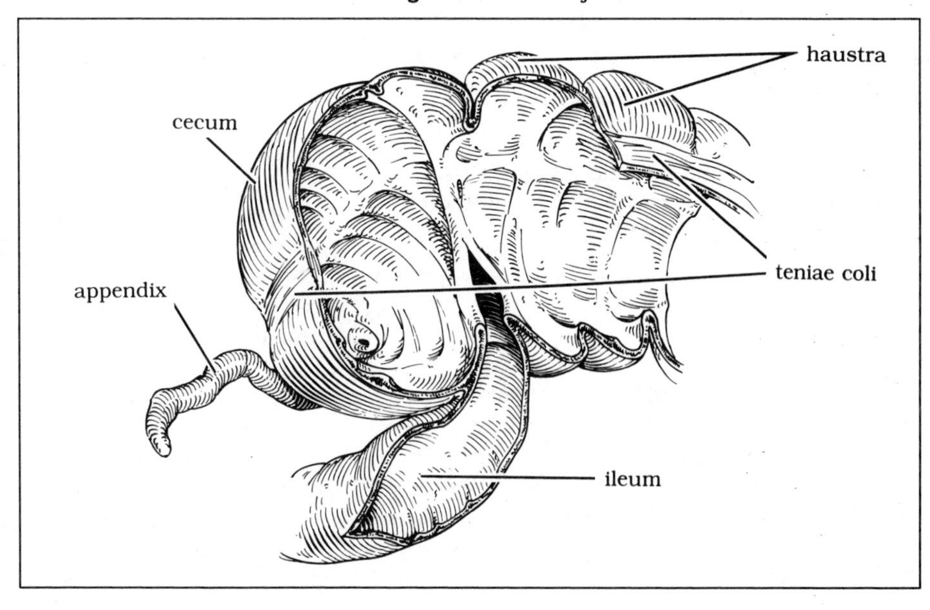

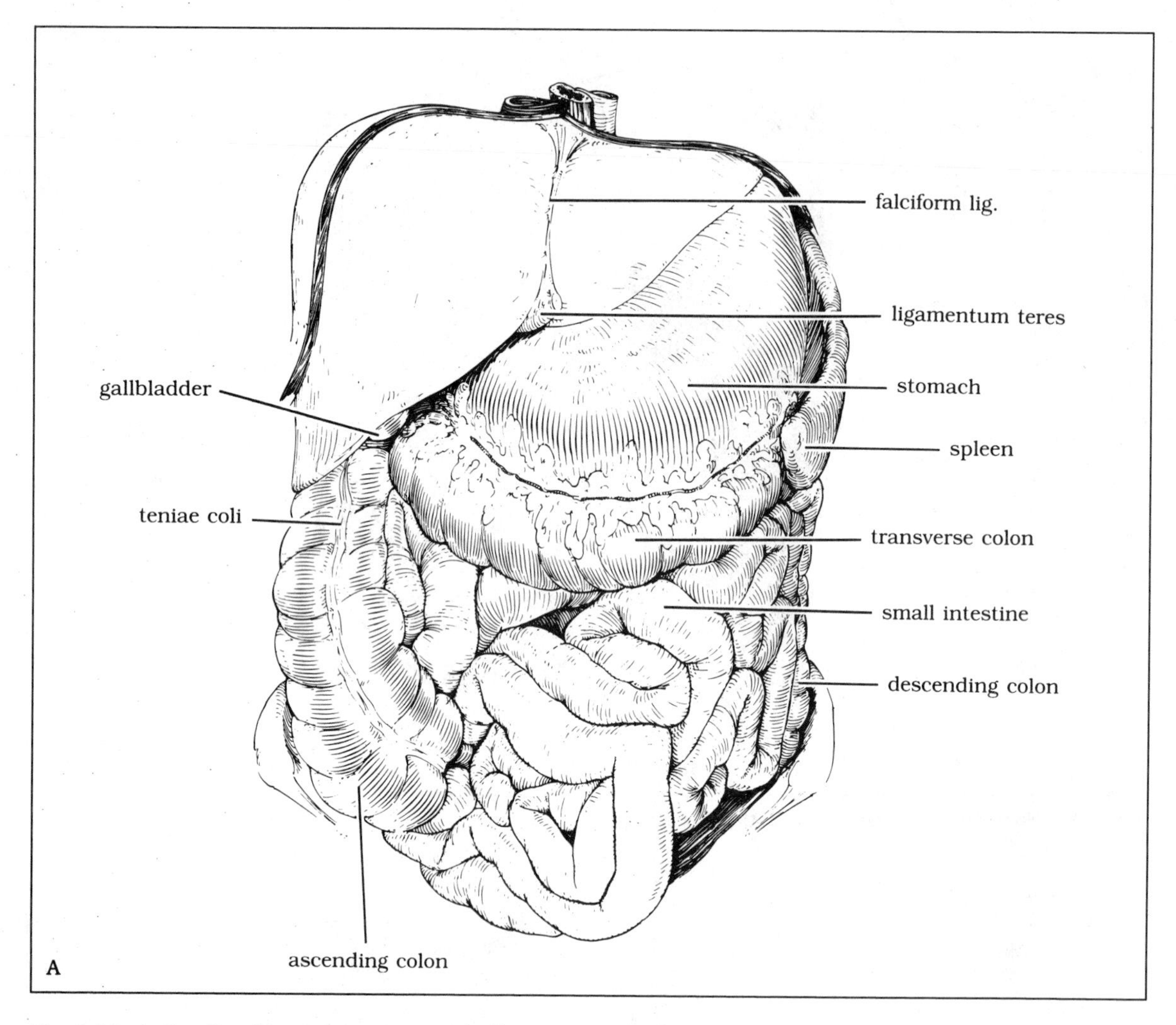

Fig. 3-22. A. Small and large intestine in situ. B. Barium image of the large intestine. (1 = ileum; 2 = ascending colon; 3 = right colic flexure; 4 = transverse colon; 5 = left colic flexure; 6 = descending colon; 7 = sigmoid colon; 8 = rectum; 9 = anal canal.)

scending colons are fixed in position to the posterior abdominal wall.

The **ascending colon** (see Figs. 3-22 and 3-25), beginning at the cecum, rises on the right side of the body to end at the right colic flexure. The **transverse colon** extends across the body from the right to left colic flexures. It is almost entirely covered by the greater omentum, and portions of its mesentery may be fused with the omentum. The first two-thirds of the transverse colon receive blood via the superior mesenteric artery and innervation from thoracic splanchnic and vagal nerves. The last third of the transverse colon and the descending and sigmoid colons are supplied by the inferior mesenteric artery and the lumbar sympathetic and sacral parasympathetic nerves.

The **descending colon** descends from the left colic (splenic) flexure to the pelvic brim and is fixed in the posterior abdominal wall. The **sig-**

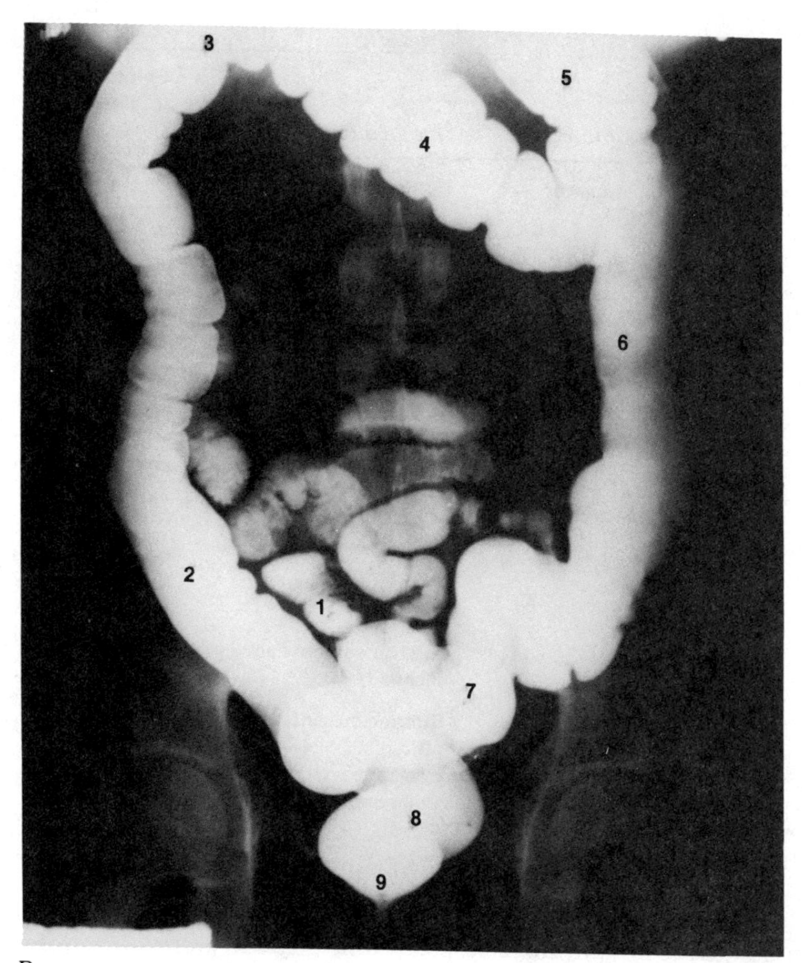

B
Fig. 3-22. (continued).

moid colon (Fig. 3-23), an S-shaped section of the colon, begins at the pelvic brim and ends at a point along the median line of the sacrum. Here it loses its mesentery and becomes the rectum, which is covered in Chapter 4.

Table 3-9 provides a summary of the small and large intestines.

BLOOD SUPPLY, LYMPHATICS, AND INNERVATION

Superior Mesenteric Structures

The **superior mesenteric artery** (Fig. 3-24, Table 3-10) arises from the aorta just below the celiac artery. It then passes under the pancreas, where it gives rise to the **inferior pancreaticoduodenal artery.** The superior mesenteric artery distributes to the intestines through the mesentery. The intestinal branches to the jejunoileum arise from its left side. The right side gives rise to the ileocolic, right colic, and middle colic arteries. The **ileocolic artery** supplies blood to the region of the ileocecal junction; the **right colic artery,** to the ascending colon; the **middle colic artery,** to the transverse colon. The ileocolic, right colic, and middle colic arteries are anastomotically connected by the **marginal artery,** which forms a continuous arterial ring on the inner side of the large bowel. This artery also connects the supe-

Table 3-9. Small and Large Intestines

Region	*Arterial Supply*	*Venous Drainage*	*Innervation**	*Connected to Wall by Mesentery*
Duodenum	Right gastric, superior and inferior pancreaticoduodenal	Splenic, superior mesenteric	Celiac plexus	First part only
Jejunum	Intestinal	Intestinal	Superior mesenteric plexus	Yes
Ileum	Intestinal, ileocecal	Intestinal, ileocecal	Superior mesenteric plexus	Yes
Cecum and appendix	Ileocecal	Ileocecal	Superior mesenteric plexus	Mesocecum Mesoappendix
Ascending colon	Right colic	Right colic	Superior mesenteric plexus	
Transverse colon	Middle colic	Middle colic	Superior mesenteric plexus	Transverse mesocolon
Descending colon	Left colic	Left colic	Inferior mesenteric plexus	
Sigmoid colon	Sigmoid	Sigmoid	Inferior mesenteric plexus	Sigmoid mesocolon

*The sympathetic supply to the celiac and superior mesenteric regions consists of thoracic splanchnic nerves; the parasympathetic supply is vagal. The sympathetic supply to the inferior mesenteric region consists of lumbar sympathetics; the parasympathetic supply is via the pelvic splanchnic nerve.

rior mesenteric blood supply with that of the inferior mesenteric artery through anastomoses between the middle colic and left colic arteries. The venous supply of the system follows the arterial flow. It is considered with the portal vein, on pages 101 to 102.

The **superior mesenteric lymph nodes** are found near the root of the superior mesenteric artery. They are supplied by four smaller groups of nodes: (1) the **mesenteric nodes,** lying within the mesentery; (2) the **ileocolic nodes,** along the ileocolic artery; (3) the **right colic nodes,** along the right colic artery; and (4) the **middle colic nodes,** along the middle colic artery.

Innervation to the organs of this region is via the **superior mesenteric ganglion** and **plexus.** Presynaptic sympathetics are supplied by the thoracic splanchnic nerves; presynaptic parasympathetic fibers are supplied by the vagal trunks. The nerves distribute to the organs via the blood vessels.

Inferior Mesenteric Structures

The **inferior mesenteric artery** (Fig. 3-25; see also Table 3-10) arises approximately 3.5 cm above the bifurcation of the aorta. The artery supplies the last third of the transverse colon and the descending colon, sigmoid colon, and superior rectum. Its branches are the **left colic, sigmoid,** and **superior rectal arteries.** The left colic and sigmoid arteries are connected to the marginal artery, which anastomoses with branches of the superior mesenteric artery. The branches of the inferior mesenteric vein follow

Table 3-10. Superior and Inferior Mesenteric Arteries

Artery	*Branches*	*Organ Supplied*
Superior mesenteric	Inferior pancreaticoduodenal, intestinal, ileocolic, right colic, middle colic	Jejunum, pancreas, duodenum, ileum, ascending colon, transverse colon
Inferior mesenteric	Left colic, sigmoid, superior rectal	Descending colon, sigmoid colon, rectum

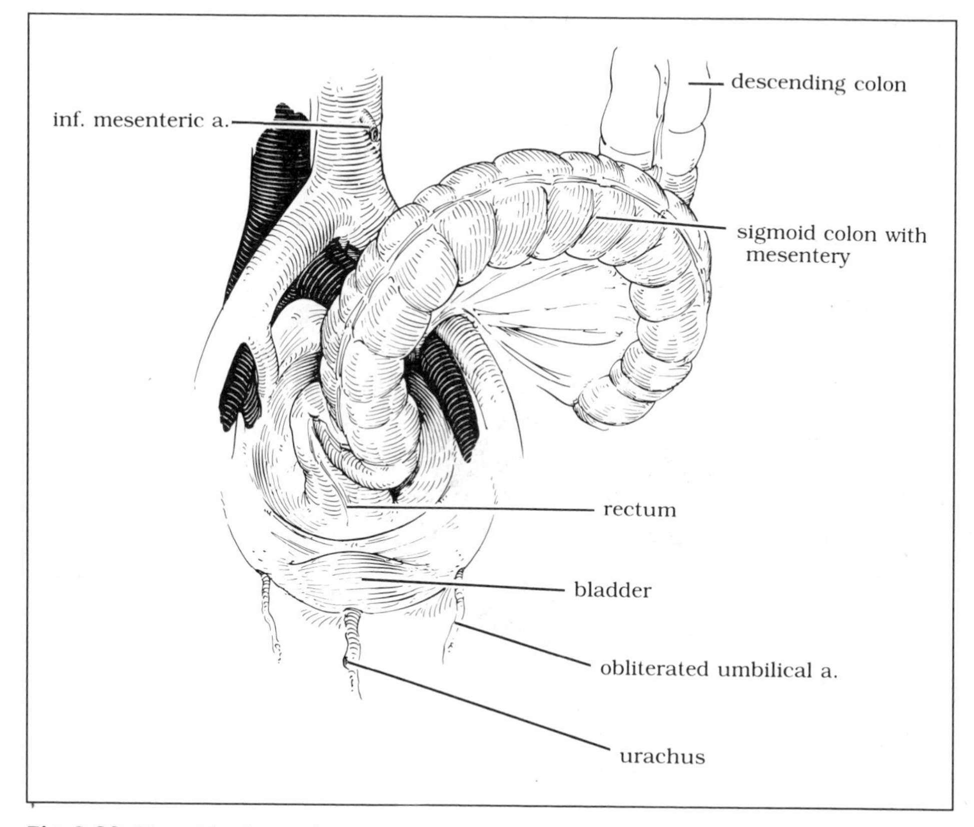

Fig. 3-23. Sigmoid colon and rectum.

the branches of the inferior mesenteric artery. This vein is further discussed later in the chapter, with the abdominal vasculature.

The **inferior mesenteric lymph nodes** may be found near the roots of the vessels. These nodes receive lymph from nodes distributed along the arterial supply to the large bowel and send lymph to the superior mesenteric nodes.

The sympathetic nerves that distribute to the organs supplied by the inferior mesenteric artery come from the first three lumbar segments. The ganglion for these nerves is the **inferior mesenteric ganglion** and is located at the root of the vessel. The sympathetic nerves distribute along with the vessels to the region.

The parasympathetic nerves are supplied by sacral segments S2, S3, and S4 and are known

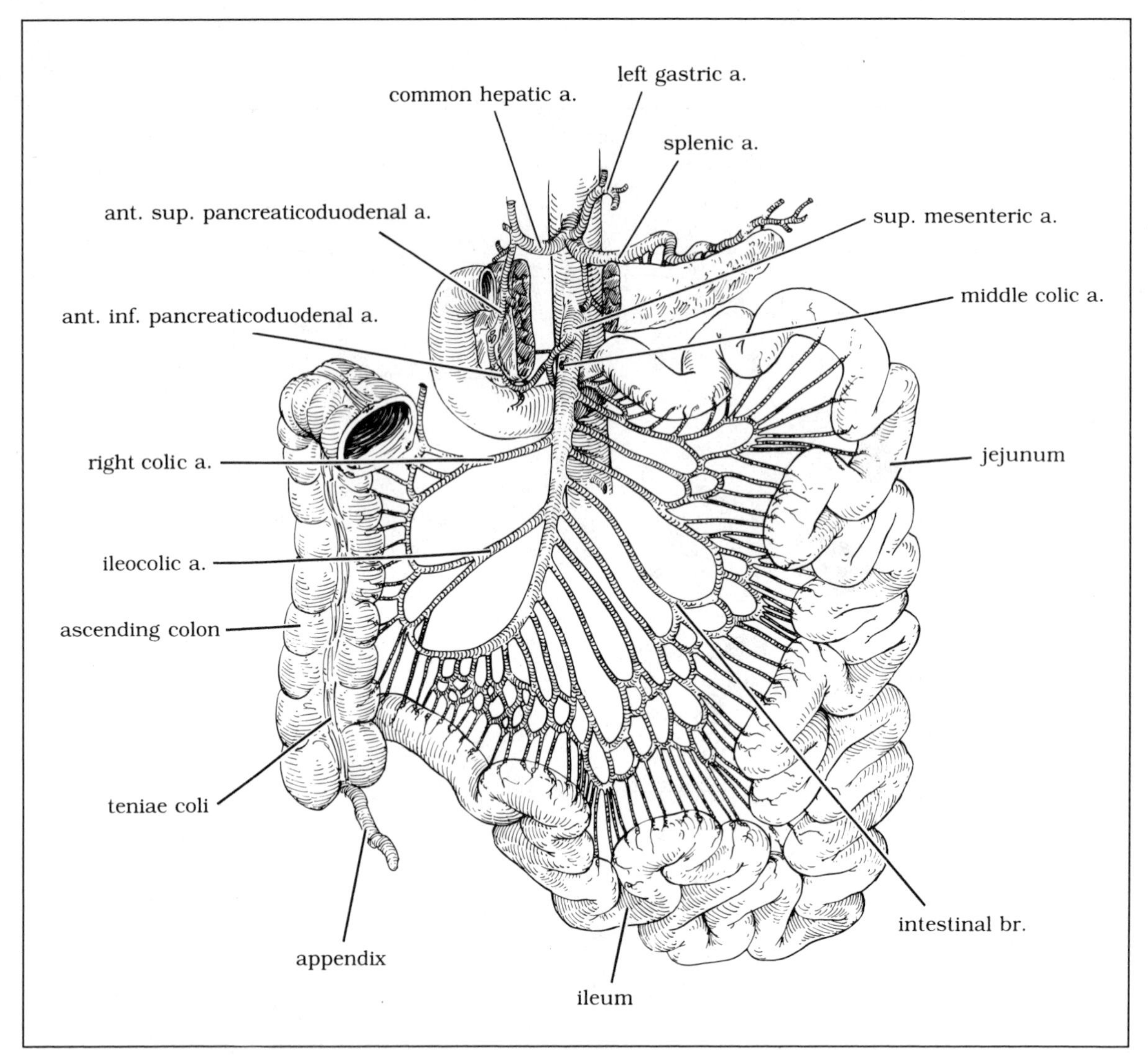

Fig. 3-24. Distribution of branches of the superior mesenteric artery.

as the **pelvic splanchnics.** These nerves reach the organs they innervate by running within the mesentery or the posterior body wall.

URINARY SYSTEM AND ASSOCIATED STRUCTURES

The urinary system consists of the kidneys and ureters in the abdomen (Fig. 3-26), and the bladder and urethra in the pelvis. The pelvic structures, including the pelvic portion of the ureter, are discussed in Chapter 4. In the abdomen the kidneys are in close association with the suprarenal glands, and these, along with the renal vessels, are discussed below.

Kidneys and Ureters

The kidneys are situated on the posterior abdominal wall (see Fig. 3-3), flanking the vertebral col-

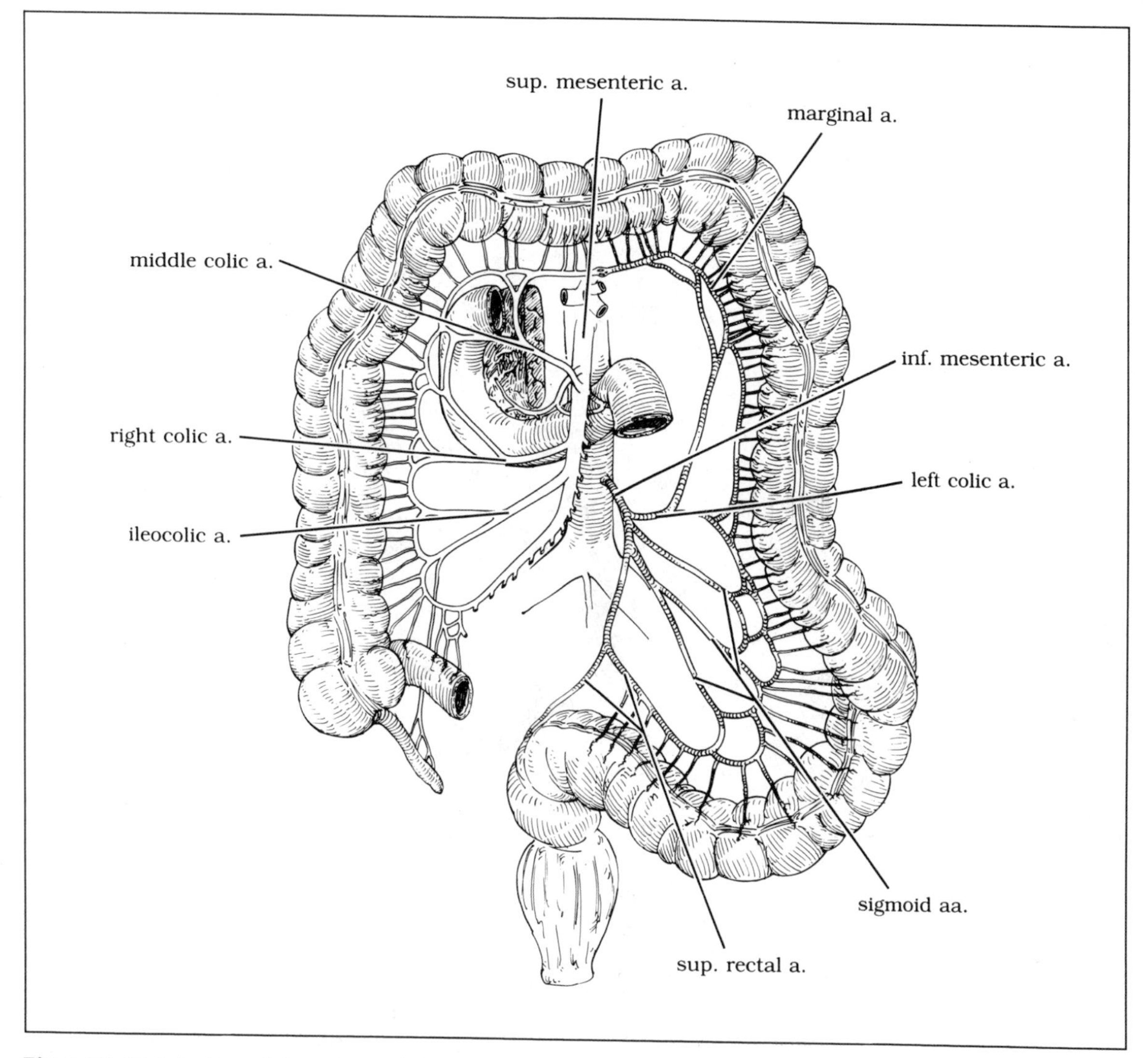

Fig. 3-25. Distribution of the inferior mesenteric artery and its relationship to the superior mesenteric artery and its branches.

umn. They are approximately 10 cm long, 5 cm wide, and 2.5 cm deep, extending from the eleventh or twelfth thoracic vertebra to the third lumbar vertebra (the right kidney being slightly lower than the left). Each kidney has medial and lateral borders, anterior and posterior surfaces, and superior and inferior poles.

The kidneys are embedded in a fat pad **(perirenal fat)** whose layers are separated by layers of renal fascia. The posterior surface and its associated fatty layers are adjacent to the psoas major muscle, the quadratus lumborum muscle and the diaphragm. The suprarenal glands sit on the superior poles, also embedded in fat but separated from the kidney by a plane of renal fascia.

The medial border of the kidney contains the **sinus,** a depression at the hilum of the kidney where the artery, vein, and ureter penetrate the

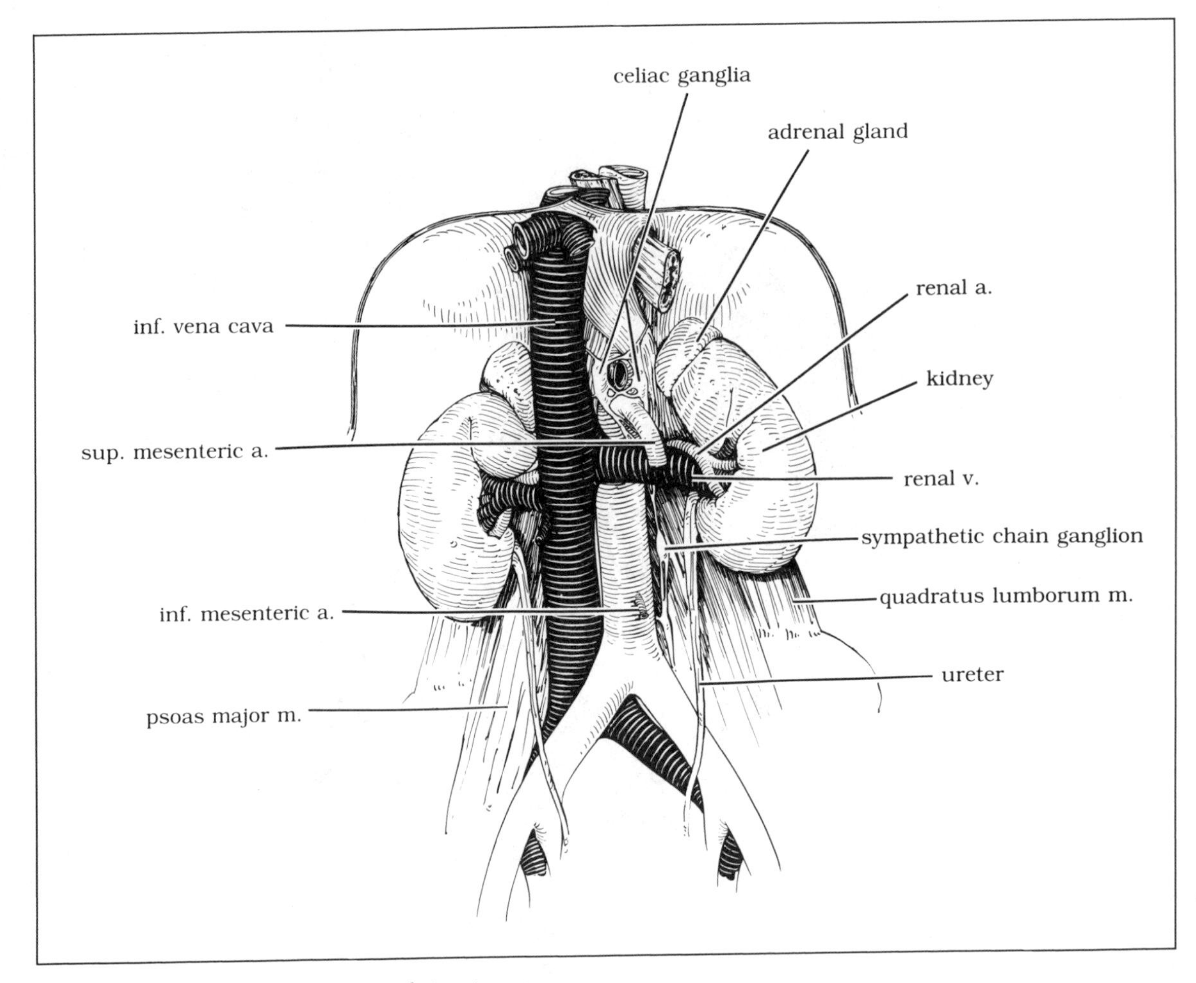

Fig. 3-26. The kidneys and associated structures.

tough fibrous capsule of the kidney. Within the sinus the ureter widens to become the **renal pelvis** (Fig. 3-27), which is formed by the junction of two **major calyces,** which in turn are formed by the junction of several **minor calyces.** The calyces and pelvis are simply a system of converging sacs for the collection of urine. Projecting into the minor calyces are the renal papillae, which are the tips of the medullary pyramids. Through them urine enters the calyces.

Internally the kidney is organized into two regions: the cortex and medulla. The **medulla** consists of the **medullary rays,** which originate in the cortex, and the **pyramids.** The medullary system is primarily a system of collecting tubules that converge, within the medullary rays, to form the pyramids, and end as the **renal papillae.**

The cortical tissue of the kidney—the cortex and renal columns—holds the glomeruli, the blood filtering unit of the kidney. The **cortex** forms the peripheral layer of the kidney. The **renal columns** are extensions of the cortex that pass between the medullary pyramids. They also contain the **interlobar arteries,** which run through the columns to the base of the pyramids. Here they branch to form **arcuate arteries** over the bases. The arcuate arteries give off **interlobular branches** that penetrate the cortical sub-

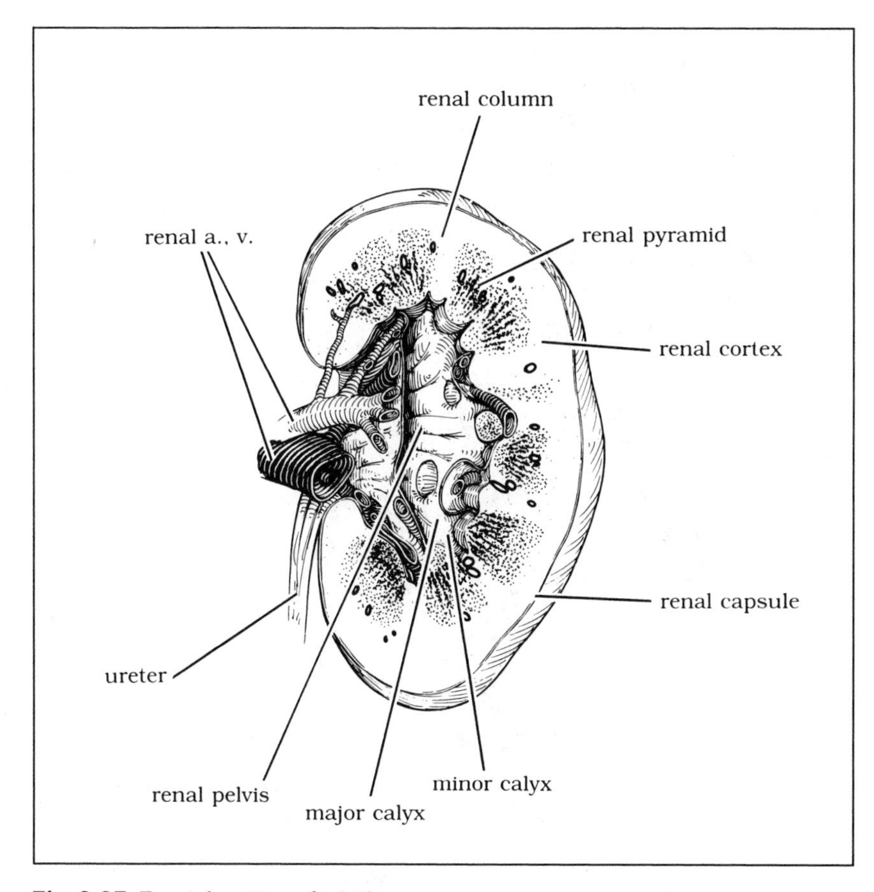

Fig. 3-27. Frontal section of a kidney.

stance and end by dividing into arterioles that are associated with the nephron, the functional unit of the kidney.

The **renal arteries** (see Fig. 3-26) branch from the aorta at the level of the second lumbar vertebra, approximately 1 cm below the superior mesenteric artery. The right (longer) artery runs behind the inferior vena cava to reach the right kidney. In the sinus of the kidneys the renal arteries give off **suprarenal** and **ureteric branches** before dividing into anterior and posterior branches. These anterior and posterior arteries divide into the interlobar arteries described above.

The **renal veins** run anterior to the arteries. They are formed by tributaries that correspond to the branches of the renal arteries. In addition the left renal vein receives the left suprarenal, testicular, and inferior phrenic veins.

The lymphatic vessels of the kidneys lie beneath the capsule of the kidney within the perirenal fat pad. The vessels follow the renal veins to lumbar lymph nodes, which in turn drain into the cisterna chyli when it is present.

The kidney is innervated by the **renal plexus,** a derivative of the celiac plexus. Thoracic splanchnic sympathetics synapse in the **aorticorenal ganglion.**

The **ureter** (see Figs. 3-27 and 3-28) begins at the renal pelvis and runs for approximately 25 cm through the abdomen and pelvis. Approximately half its length is abdominal, and half pelvic. The pelvic portion of the ureter is discussed in Chapter 4. In the abdomen the ureter

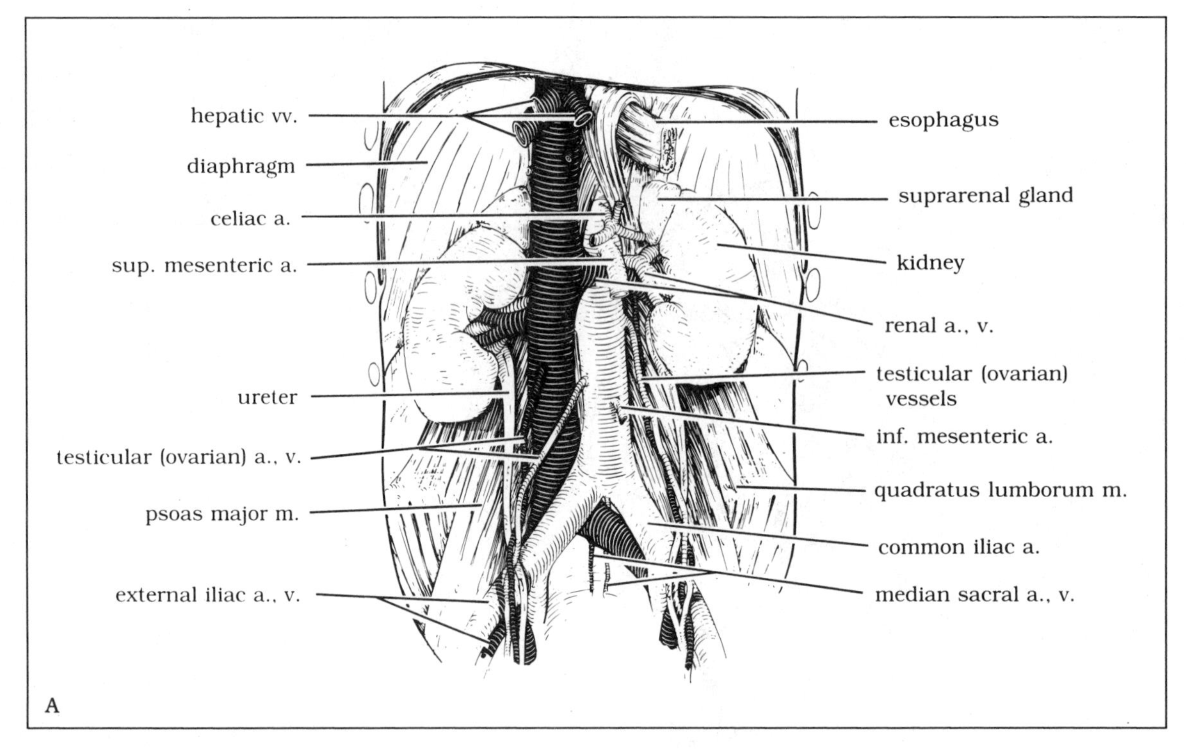

Fig. 3-28. A. Structures of the posterior abdominal wall (duodenum and pancreas removed). B. X-ray film of kidneys, ureter, and bladder. (1 = kidney; 2 = minor calyx; 3 = major calyx; 4 = ureter; 5 = bladder; 6 = transverse process of lumbar vertebra; 7 = body of lumbar vertebra; 8 = spinous processes of lumbar vertebra; 9 = twelfth rib; 10 = sacrum, lateral mass; 11 = psoas major.)

runs along the psoas major muscle, following the line formed by the tips of the transverse vertebral processes. It passes anterior to the external iliac artery at or near the bifurcation of the common iliac artery. From here the ureter passes over the pelvic brim and enters the pelvis. The wall of the ureter is muscular and pushes, through peristaltic contractions, the urine into the bladder.

In the abdomen each ureter is crossed by the testicular (ovarian) vessels. Blood is supplied to the abdominal ureter and drained by the renal or testicular (ovarian) vessels or both. The abdominal ureter is innervated through the renal plexus.

Suprarenal (Adrenal) Glands

The suprarenal glands (see Figs. 3-26 and 3-28) sit on top of the superior pole of the kidneys. The right suprarenal gland is pyramidal in shape; the left is semilunar. Each gland has its own hilum, from which its vein leaves. The glands are enclosed in the renal fat and fascia but are separated from the kidney by a fascial plane. Both suprarenal glands rest on the diaphragm on their posterior surfaces.

The suprarenal gland is actually composed of two separate endocrine glands: the suprarenal cortex and medulla. The external **suprarenal cortex** excretes steroids and is involved in fluid and electrolyte balance, carbohydrate metabolism, and the regulation of the reaction to stress. The internally located **suprarenal medulla** is actually an extension of the sympathetic nervous system

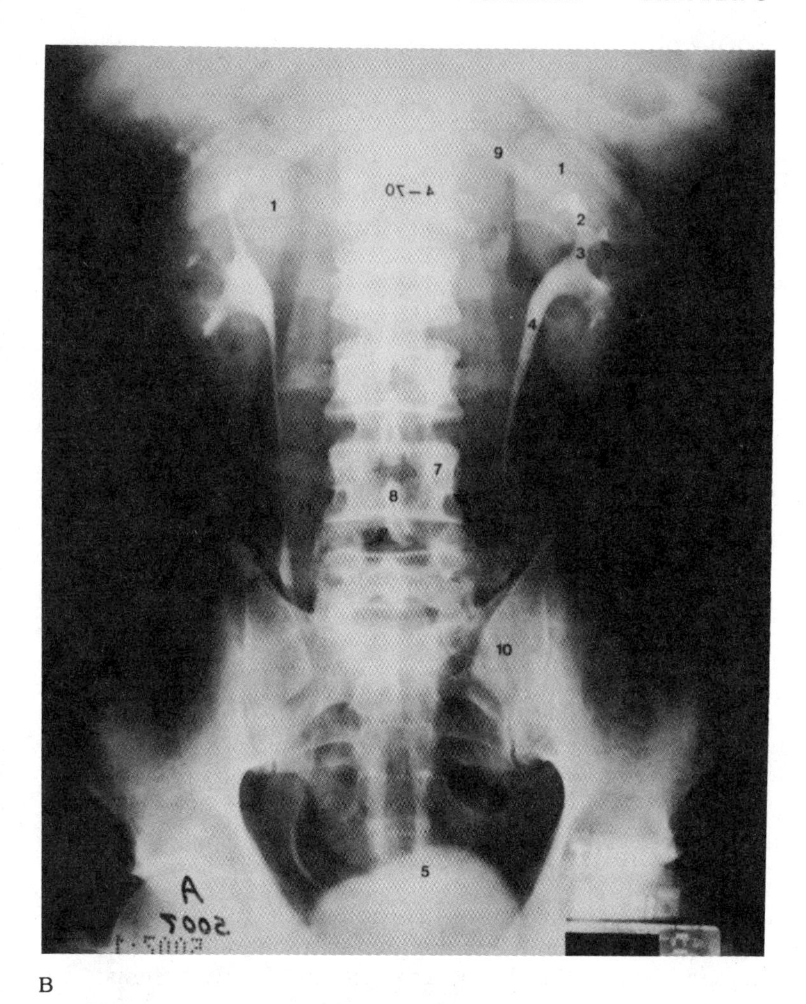

B

Fig. 3-28. (continued).

and is most closely related to the sympathetic chain ganglia. As such it is involved in fight-or-flight reactions associated with fear, anger, and stress. The medulla acts by secreting adrenergic-type neurotransmitters into the blood system, thus inducing broad-spectrum effects on structures such as the heart, digestive tract, and vasculature.

Branches of the renal arteries, aorta (middle suprarenals), and inferior phrenic arteries constitute the arterial supply to the suprarenals. The **right suprarenal vein** drains directly into the inferior vena cava. The **left suprarenal vein** joins the left renal vein. Lymphatic vessels follow the suprarenal veins to lumbar lymph nodes.

The suprarenal glands are innervated via the celiac plexus and thoracic splanchnic sympathetic nerves.

BLOOD SUPPLY, LYMPHATICS, AND INNERVATION OF THE ABDOMINAL CAVITY

Arteries

The **abdominal aorta** (Fig. 3-28 and Table 3-11) descends slightly to the left of the midline in the abdominal cavity for approximately 13 cm. Lying

Table 3-11. Abdominal Aortic Branches and the Structures They Supply

Artery	*Branches*	*Structures Supplied*
Digestive		
Celiac	Left gastric, common hepatic, splenic	Stomach, liver, gallbladder, spleen, duodenum, pancreas
Superior mesenteric	Inferior pancreaticoduodenal, intestinal, ileocolic, right colic, middle colic	Pancreas, duodenum, jejunum, ileum, ascending colon, transverse colon, cecum, appendix
Inferior mesenteric	Left colic, sigmoid, superior rectal	Descending colon, sigmoid colon, rectum
Renal and suprarenal		
Middle suprarenal		Suprarenal gland
Renal		Kidney, suprarenal gland, ureter
Muscular		
Inferior phrenic		Diaphragm, suprarenal gland
Lumbar		Muscles of the abdominal wall
Genital		
Testicular (male only)		Epididymis, testis, ductus deferens, ureter
Ovarian (female only)		Ovary, ureter, uterine tube, uterus
Terminal		
Median sacral		Sacrum, coccyx
Common iliac	External iliac, internal iliac	Lower limb, pelvis

against the vertebral column, it ends in front of the fourth lumbar vertebra by dividing into two common iliac arteries. Its branches are the celiac, superior mesenteric, inferior mesenteric, middle suprarenal, renal, inferior phrenic, lumbar (four branches), testicular or ovarian, median sacral, and common iliac arteries. The anastomoses of the first three are summarized in Table 3-12. The testicular or ovarian, median sacral, and common iliac arteries are discussed here; all other branches are covered elsewhere in the chapter.

The **testicular arteries,** in the male, arise just below the renal arteries. They descend obliquely across the psoas major muscle to the deep inguinal ring. In doing so they cross over the ureters, to which they extend some branches, and the external iliac arteries. Upon reaching the deep inguinal ring, they enter and pass through the inguinal canal to reach the testis. In addition, they supply the epididymis, and some of their branches supply the ductus deferens.

The **ovarian arteries,** in the female, follow the same course as the testicular arteries to the brim of the pelvis. There they turn into the pelvis, running along the suspensory ligament of the ovary to the broad ligament of the uterus. The arteries then run medially and posteriorly to reach the mesovarium and hilum of the ovary. They supply blood to the ovary, ureter, uterine tube, and uterus.

The **median sacral artery** is the continuation of the dorsal aorta past the bifurcation of the common iliacs. The artery descends in the midline over the last two lumbar vertebrae, the sacrum and coccyx, supplying these structures.

The **common iliac arteries** are the two terminal branches of the abdominal aorta. They travel

Table 3-12. Anastomoses of Digestive Visceral Arteries

Celiac

- Left gastric to right gastric
- Splenic
  - Left gastroepiploic to right gastroepiploic
- Common hepatic
  - Gastroduodenal
    - Right gastroepiploic to left gastroepiploic
    - Superior pancreaticoduodenal to inferior pancreaticoduodenal
  - Right gastric to left gastric

Superior Mesenteric

- Inferior pancreaticoduodenal to superior pancreaticoduodenal
- Ileocolic to right and middle colics via marginal
- Middle colic to ileocolic and right and left colics via marginal

Inferior Mesenteric

- Left colic to middle colic and branches, and sigmoids via marginal
- Superior rectal to inferior rectal

a short distance, 5 cm, before dividing into the internal and external iliac arteries. The internal iliac artery supplies the pelvis. The external iliac artery (the direct continuation of the common iliac) passes under the midinguinal point to become the femoral artery of the lower limb. Before leaving the abdomen, the external iliac artery gives off the deep circumflex iliac artery and the inferior epigastric artery.

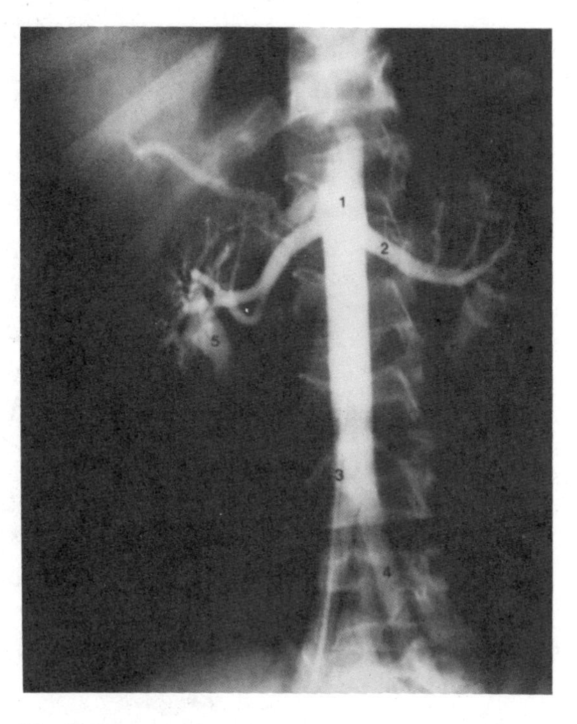

Fig. 3-29. X-ray film of the lower inferior vena cava. (1 = inferior vena cava; 2 = left renal vein; 3 = right genital vein; 4 = left common iliac vein; 5 = renal pelvis.)

Veins

The **inferior vena cava** (Fig. 3-29) is formed by the junction of the two common iliac veins. It brings the nonportal blood of the abdomen and pelvis and the blood of the lower limbs to the heart. Lying to the right of the aorta it ascends for approximately 20 cm just anterior to the vertebral column. The inferior vena cava leaves the abdomen through the caval hiatus in the diaphragm.

The inferior vena cava receives, in addition to the common iliac veins, the lumbar veins, the right testicular or ovarian vein, the renal veins, the right suprarenal vein, the right inferior phrenic vein, and the hepatic vein (see Fig. 3-29). The testicular vein arises from the pampiniform plexus of veins (see Fig. 3-11), a series of anastomotic connections surrounding the ductus deferens. All of the other veins, except the hepatic vein, follow their corresponding artery.

The **hepatic vein** carries blood from the liver directly into the inferior vena cava as the vena cava passes in, or next to, the liver. This vein then remains within the substance of the liver. The blood it carries enters the liver through either the hepatic artery or the portal vein.

The **portal vein** (Fig. 3-30) carries blood to the liver from the gastrointestinal system, the

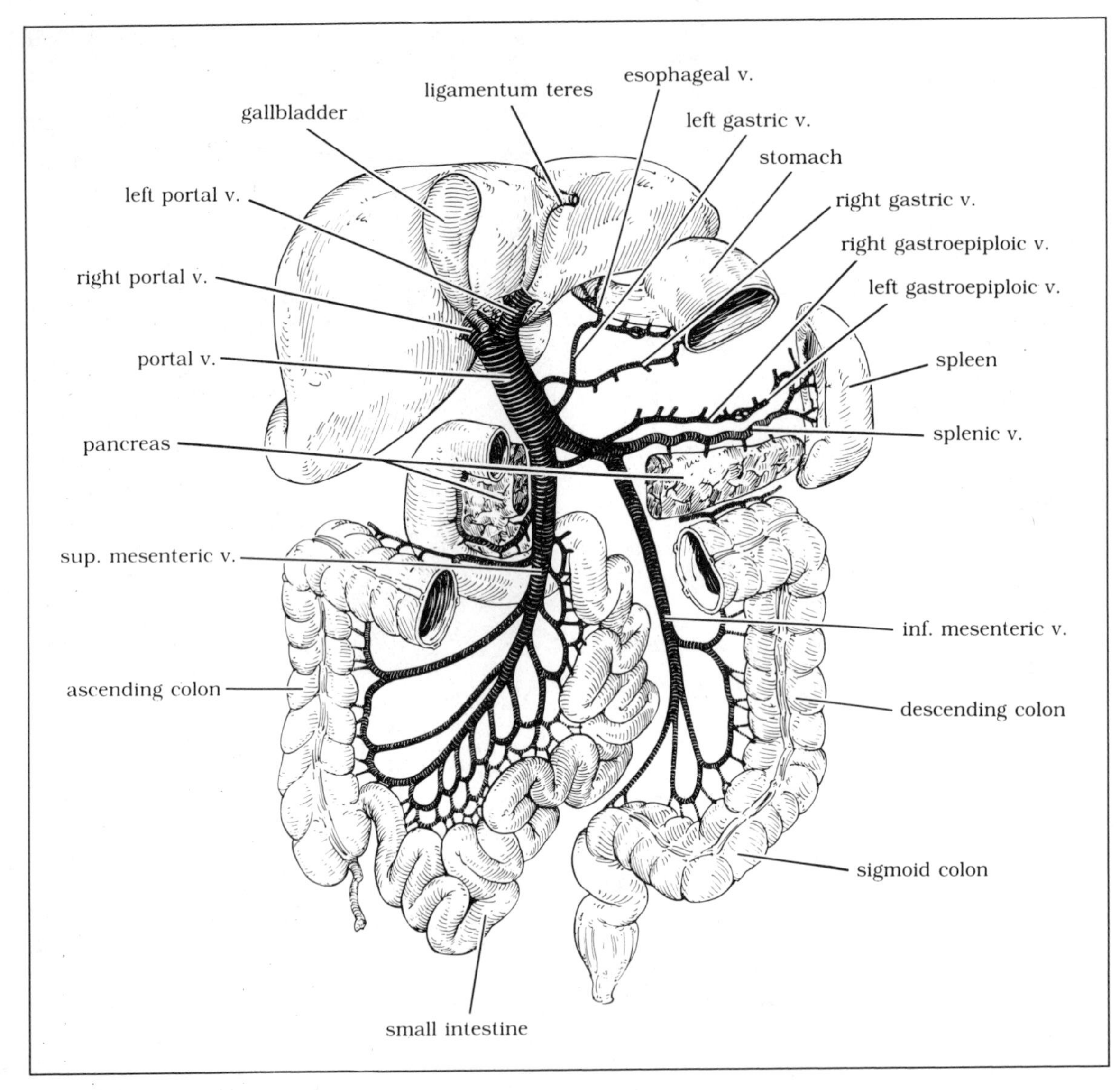

Fig. 3-30. The portal venous system.

spleen, and the pancreas. Within the liver the vein branches like an artery, finally ending in a capillary network—the liver sinusoids. From the sinusoids the blood is recollected by tributaries of the hepatic vein.

The portal vein is formed by the union of the superior mesenteric and splenic veins behind the neck of the pancreas. It then ascends behind the duodenum to the hepatoduodenal ligament. The portal vein traverses this ligament posterior to the bile duct and proper hepatic artery. The vein receives the blood of organs supplied by the celiac, superior mesenteric, and inferior mesenteric arteries. The inferior mesenteric vein joins the system by emptying into either the superior mesenteric or splenic veins.

In cases in which portal or systemic venous blood flow is blocked or slowed down, four sets of anastomoses between the two systems allow for the shunting of blood from one system to the other and thus the return of all the blood from the abdomen, pelvis, and lower limbs to the heart. The first, the esophageal anastomoses, occur on the inferior portion of the esophagus between tributaries of the left gastric vein and the esophageal veins. The second, the rectal plexus, is a link between the superior rectal vein (which normally returns blood via the portal system) and the middle and inferior rectal veins (which are systemic). The third and fourth connections are not of as great importance. The third is the paraumbilical network of veins. These veins connect those of the liver, via the falciform ligament, with the abdominal wall vasculature. Finally the retroperitoneal veins connect tributaries of the veins of the large intestine with the inferior vena cava via renal and testicular or ovarian branches.

Lymphatics

The parietal lymph nodes of the abdomen lie along the aorta and its branches. They fall into three groups: external iliac, common iliac, and lumbar nodes.

The **external iliac nodes** lie along the external iliac vessels. They are continuous with the nodes of the femoral canal and receive lymph from the lower leg through the deep and superficial inguinal nodes. In addition lymph from the anterior abdominal wall below the umbilicus and from the pelvis drains into the external iliac nodes.

The **common iliac nodes** receive lymph from the internal and external iliac nodes. Therefore, they drain the lower limb, pelvic wall and viscera, perineum, and lower abdominal wall. The common iliac nodes drain into the lumbar nodes.

The **lumbar nodes** lie along both sides of the aorta. They receive lymph from the iliac nodes, testis (or ovary), uterus, kidneys, adrenals, abdominal wall, and abdominal surface of the diaphragm. The nodes on each side of the aorta are interconnected to form a single trunk. These trunks converge superiorly and end in the thoracic duct. The left lumbar trunk usually receives the lymph from the intestinal tract. The intestinal trunk is formed by tributaries from the celiac and superior mesenteric nodes. The superior mesenteric nodes receive lymph from the inferior mesenteric nodes.

Innervation

The thoracic splanchnic and vagal nerves supply visceral innervation to the abdomen as far as the first two-thirds of the transverse colon. The rest of the abdominal viscera are innervated by the lumbar sympathetic and pelvic splanchnic nerves (Fig. 3-31).

The **thoracic splanchnic nerves** synapse in either the celiac (see Fig. 3-26), superior mesenteric, aorticorenal, or renal ganglion. They then distribute to the viscera via the plexuses that surround the blood vessels. The **vagal nerve fibers** distribute via the same plexuses to the viscera. These nerves usually synapse within the walls of the organs they innervate.

The **lumbar sympathetic chain** is a continuation of the thoracic sympathetic chain in the abdomen. Nerves from these ganglia distribute either by lumbar spinal nerves to the abdominal wall or by the intermesenteric or superior hypogastric plexuses to the remaining abdominal and pelvic viscera.

The **pelvic splanchnic nerves** are parasympathetic nerves of sacral origin. They distribute by the inferior hypogastric plexus to pelvic viscera and by discrete nerve bundles to the intestine.

Diaphragm

The diaphragm (Fig. 3-32), a domelike muscular structure, constitutes the superior surface of the abdomen, separating it from the thorax. It has three types of fibers that arise from three regions of the thoracic wall that give the fibers their name. The **sternal fibers** arise from the posterior surface of the xiphoid process. The **costal fibers** arise from the costal cartilages and adjacent bone of the sixth to twelfth ribs, interdigitating with fibers of the transverse abdominal muscle. The **lumbar fibers** arise from four lumbocostal arcuate ligaments (two per side) laterally and two crura medially. All of the fibers insert into a central tendon.

The **medial lumbocostal arcuate ligament** of each side arches over the psoas major muscle. It connects the body of the first or second lumbar vertebra to the transverse process of the first lumbar vertebra. The **lateral lumbocostal arcuate ligament** of each side crosses over the quadratus lumborum muscle. It arises from the tips of the transverse process of the first lumbar vertebra and inserts into the tip of the caudal margin of the twelfth rib.

The two **crura** have tendinous origins blending with the ventral vertebral ligaments. The longer and larger right crus arises from the bodies and intervertebral fibrocartilage of the first three lumbar vertebrae. The left crus arises from the same structures only of the first two lumbar vertebrae. The two crura arch medially over the aorta to form the aortic hiatus. The fibers of the right crus continue to ascend leftward to form the left side of the esophageal hiatus. The lateral borders of both crura blend with their adjacent medial lumbocostal arches.

The **central tendon of the diaphragm** is tripartite, having right, left, and central portions. The right portion is the largest, the left the smallest. The superior surface of the central tendon is partially blended with the fibrous pericardium.

Several structures pass from the thoracic cavity into the abdomen through openings in the diaphragm. The three largest openings are the aortic, esophageal, and caval hiatuses. The **aortic hiatus,** the most dorsal, lies at the level of the twelfth thoracic vertebra, slightly to the left of the midline. It is bounded by the crura and the body of the first lumbar vertebra. Through the aortic hiatus pass the aorta, azygos vein, and thoracic duct. The **esophageal hiatus** lies ventral to and left of the aortic hiatus at the level of the tenth

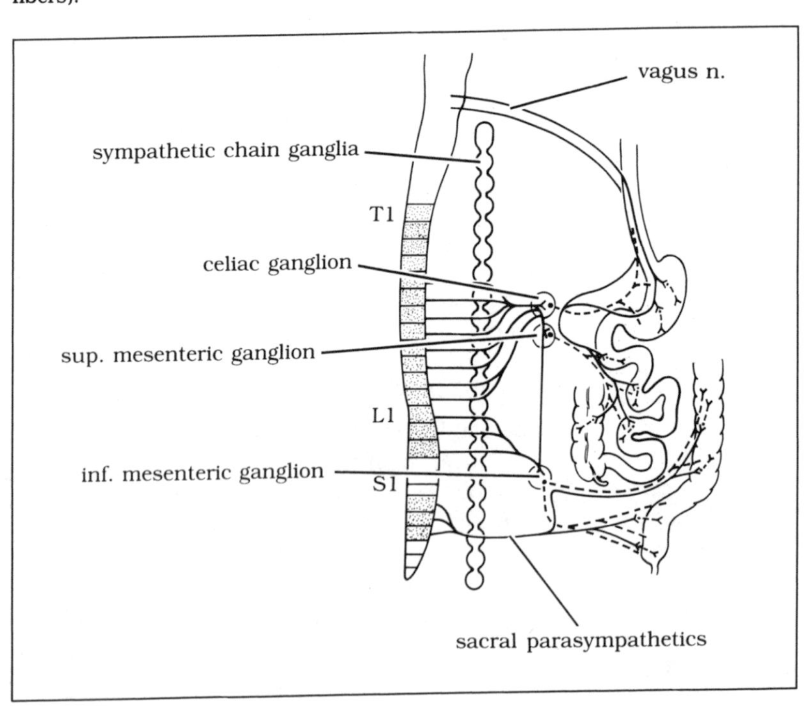

Fig. 3-31. Autonomic abdominal innervation (solid line = presynaptic fibers; dotted line = postsynaptic fibers).

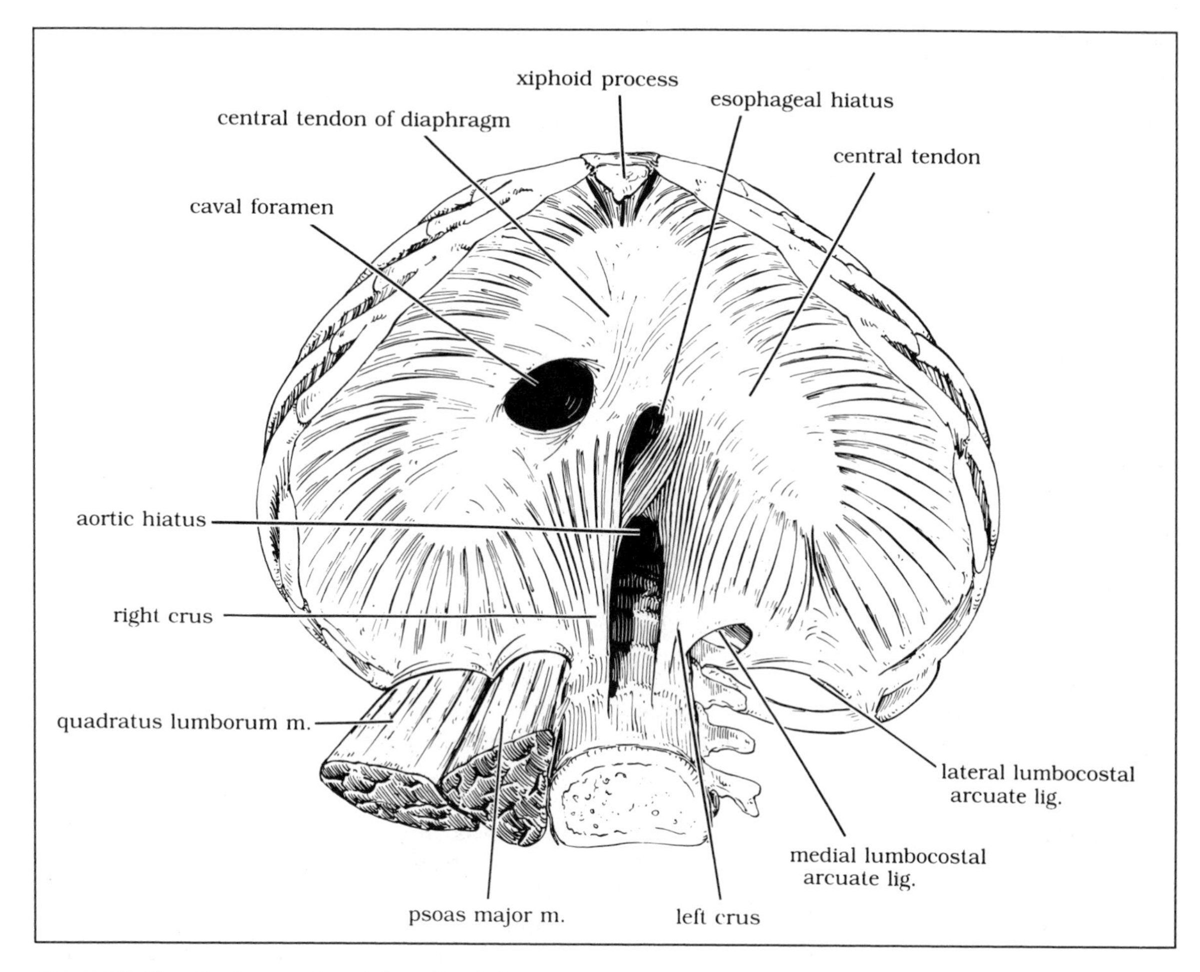

Fig. 3-32. The diaphragm viewed from the abdomen.

thoracic vertebra. The esophagus, vagal trunks, and some small blood vessels pass through it. The **hiatus for the inferior vena cava** is the most ventral opening and lies at the level of the eighth to ninth thoracic vertebrae. Besides the inferior vena cava, it transmits branches of the right phrenic nerve. In addition there are small openings in the right crus for the splanchnic nerves and in the left crus for the splanchnic nerves and hemiazygos vein.

The superior surface of the diaphragm receives blood from three sources: the pericardiacophrenic artery, the musculophrenic artery, and the superior phrenic arteries. The **pericardiacophrenic artery** arises from the internal thoracic artery and accompanies the phrenic nerve to the diaphragm (see Chap. 2). The **musculophrenic artery** is one of the terminal branches of the internal thoracic artery and is described in Chapter 2. The **superior phrenic arteries** arise from the distal thoracic aorta and distribute to the surface of the diaphragm.

The inferior surface of the diaphragm receives blood from the **inferior phrenic arteries.** These arteries may arise separately or together from the aorta, the celiac artery, or occasionally from a renal artery. As they pass to the diaphragm they give off superior suprarenal branches. The arteries on the superior and inferior surfaces of the diaphragm anastomose freely with each other.

The veins follow the arteries in their course, with only one exception: The **right inferior phrenic vein** flows to the inferior vena cava, while the **left inferior phrenic vein** empties into the inferior vena cava or the left renal vein or both.

The lymphatic drainage is through plexuses, one on the thoracic side, the other on the abdominal. Both plexuses communicate freely. The vessels on the abdominal side converge on the **lumbar lymph nodes.** The vessels of the thoracic side empty into anterior, middle, or posterior **phrenic lymph nodes.** The anterior nodes lie at the base of the xiphoid process and drain into sternal nodes. The middle nodes lie at the point of entrance of the phrenic nerve into the diaphragm and drain into the posterior mediastinal nodes. The posterior nodes lie on the back of the crura. They are connected to both the posterior mediastinal and lumbar nodes.

The diaphragm is innervated by the **phrenic nerve.** This nerve is part of the cervical plexus arising from the third to fifth cervical nerves (see Chap. 8).

The diaphragm is involved in respiration. Contraction of its muscular fibers draws the central tendon downward, which increases thoracic volume and decreases thoracic pressure. It is this decrease in pressure that draws air into the lungs. In addition the downward movement of the diaphragm serves to increase abdominal pressure and decrease abdominal volume. When this is accompanied by a contraction of the muscles of the abdominal wall, sufficient pressure is generated to aid in the discharge of wastes from the abdominal and pelvic viscera.

Structures Related to the Lower Limb

Several structures that are abdominal in origin are functional in the lower limb. These are the psoas major and minor muscles, the iliacus muscle, and the lumbar plexus of nerves.

The **psoas major muscle** (Fig. 3-33) arises from the bodies and intervertebral disks of the twelfth thoracic to fifth lumbar vertebrae. The muscle tapers as it approaches the thigh. Crossing under the inguinal ligament lateral to the iliopubic eminence, it ends by inserting into the lesser trochanter of the femur through a tendon (iliopsoas tendon) common to the iliacus muscle. The psoas muscle flexes the thigh when the trunk is held straight; when the leg is kept straight, it flexes and bends the lumbar vertebral column. It receives innervation from the second to fourth lumbar nerves via the lumbar plexus.

Table 3-13. Lumbar Plexus

Segment	*Branch*	*Nerves Formed*
L1	Superior	Ilioinguinal, iliohypogastric
L1	Inferior	Genitofemoral
L2	Superior	
L2	Inferoanterior	
L3	Anterior	Obturator
L4	Anterior	
L2	Inferoposterior	
L3	Posterior	Femoral, lateral femoral cutaneous
L4	Posterior	
L4	To sacral plexus	

The **psoas minor muscle** is an inconstant muscle. Arising from the bodies and intervertebral disks of the twelfth thoracic and first lumbar vertebrae, it inserts into the iliopectineal eminence. The psoas minor muscle flexes the pelvis on the trunk. It is innervated by the first and second lumbar nerves.

The **iliacus muscle** (see Fig. 3-33) lies in the iliac fossa. It is covered by the iliac fascia (a heavy aponeurotic sheath). The iliacus arises from the iliac crest, iliolumbar ligament, ventral sacral ligament, and the ala of the sacrum. The muscle fibers converge to join those of the psoas major muscle in forming the iliopsoas tendon. It is innervated by the femoral nerve (second to fourth lumbar nerves).

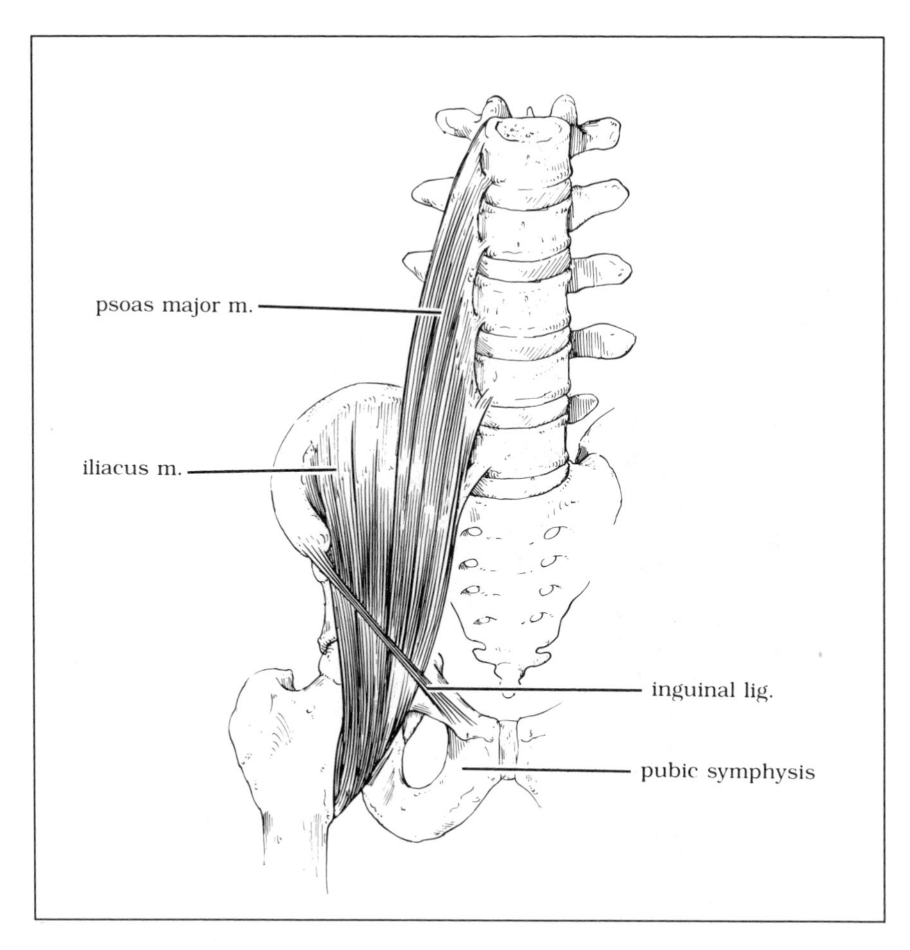

Fig. 3-33. Iliacus and psoas muscles.

The **lumbar plexus** (Fig. 3-34 and Table 3-13) is located within or dorsal to the psoas major muscle. It consists of the ventral divisions of the first through fourth lumbar nerves, with a contribution from the twelfth thoracic nerve. The first lumbar nerve divides into superor and inferior branches. The superior branch gives rise to the **ilioinguinal** and **iliohypogastric nerves.** The inferior branch joins with the superior branch of L2 to form the genitofemoral nerve.

The **genitofemoral nerve** gives rise to a genital branch that passes through the inguinal canal to innervate the cremaster muscle and the skin and fascia of the scrotum and adjacent thigh. The femoral branch of the genitofemoral nerve passes under the inguinal ligament to innervate the skin over the femoral triangle.

The inferior division of the second lumbar nerve joins with fibers from L3 and L4. The nerves divide into anterior and posterior portions. The anterior portions unite to form the **obturator** (and when present, the **accessory obturator**) **nerve.** This nerve passes through the obturator canal, along with the obturator artery and vein, to the thigh. There it divides into anterior and posterior branches that innervate the adductor muscles, hip and knee joints, and skin of the medial thigh (see Chap. 6).

The posterior divisions form the femoral and lateral femoral cutaneous nerves (see Chap. 6).

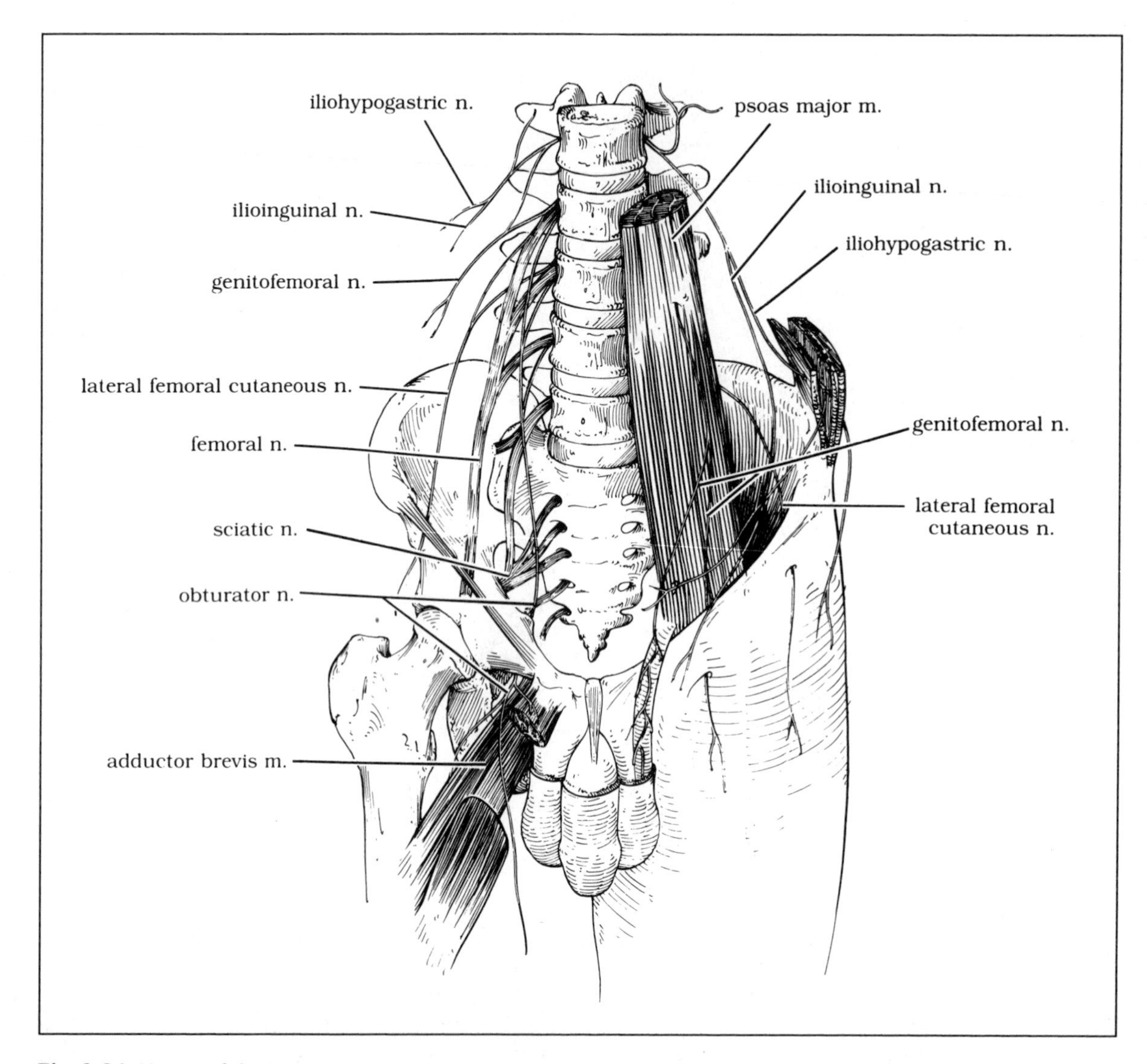

Fig. 3-34. Nerves of the lumbar and sacral plexuses.

The **femoral nerve** descends between the tendons of the psoas and iliacus muscles under the inguinal ligament. On entering the femoral triangle it divides into many branches. The **lateral femoral cutaneous nerve** passes under the lateral edge of the inguinal ligament to the skin of the lateral thigh.

Some fibers of the fourth lumbar nerve also contribute to the sacral plexus.

National Board Type Questions

Select the one best response for each of the following.

1. The division between fore- and midgut is at the
   - A. Cardiac orifice of the stomach
   - B. Pyloric orifice of the stomach
   - C. Entrance of the common bile duct
   - D. Duodenojejunal junction

E. Ileocecal valve

2. The internal spermatic fascia is a continuation of the
   A. Transversalis fascia
   B. Intercrural fascia
   C. Membranous layer of superficial fascia
   D. Fatty layer of superficial fascia
   E. Extraperitoneal connective tissue
3. Choose the statement that is **not** true. An indirect inguinal hernia
   A. Enters the inguinal canal at the deep inguinal ring
   B. Originates lateral to the inferior epigastric vessels
   C. Protrudes through a weakness in the abdominal muscle
   D. Communicates with the peritoneal cavity
   E. Exits the inguinal canal at the superficial inguinal ring

---

Select the response most closely associated with each numbered item. (The headings may be used once, more than once, or not at all.)

The following structures serve as passage for which artery?

A. Transverse mesocolon
B. Mesojejunum
C. Lienorenal ligament
D. Lesser omentum
E. Gastrocolic ligament

4. Common hepatic artery
5. Superior mesenteric artery
6. Middle colic artery

Relate the following vertebral levels to where the structures listed below pass through the diaphragm.

A. T6
B. T8
C. T10
D. T12
E. L2

7. Aorta
8. Esophagus
9. Inferior vena cava

The following organs have which structural characteristics?

A. Stomach
B. Duodenum
C. Jejunum
D. Ileum
E. Colon

10. Haustra
11. Fundus
12. Teniae

The following types of fibers have which anatomical relationship?

A. Preganglionic parasympathetic fibers
B. Preganglionic sympathetic fibers
C. Postganglionic parasympathetic fibers
D. Postganglionic sympathetic fibers

13. Terminate in the celiac ganglion
14. Distribution ends near left colic flexure
15. Begin in the celiac ganglion

For the following, select

A. if only *1, 2, and 3* are correct
B. if only *1 and 3* are correct
C. if only *2 and 4* are correct
D. if only *4* is correct
E. if *all* are correct

16. After occlusion of the inferior mesenteric artery, the blood supply of the left portion of the colon is most likely maintained by the
    1. Left colic artery
    2. Marginal artery (of Drummond)
    3. Left lumbar artery
    4. Anastomoses between the superior, middle, and inferior rectal arteries
17. Which of the following pairs of arteries represents reasonable arterial anastomoses?
    1. Superior pancreaticoduodenal arteries—inferior pancreaticoduodenal arteries
    2. Left gastric artery—splenic artery
    3. Middle colic artery—left colic artery
    4. Jejunal artery—right gastroepiploic artery

18. The abdominal aorta
    1. Bifurcates at the level of the promontory of the sacrum
    2. Has lumbar branches that supply the posterior abdominal wall musculature
    3. Enters the abdomen through a hiatus in the diaphragm that is just anterior to the esophageal hiatus
    4. Supplies the entire abdominal gastrointestinal tract through just three unpaired branches
19. The following statement(s) is (are) true for the duodenum:
    1. The ascending (fourth) portion is directly continuous with the jejunum.
    2. The common bile duct passes behind the first (superior) portion.
    3. The third (horizontal) portion is retroperitoneal.
    4. The ampulla of the common bile duct opens into the second (descending) portion.
20. The lacunar ligament
    1. Is a derivative of the external oblique muscle
    2. Continues as the pectineal ligament
    3. Is medial to the femoral canal
    4. Lies at the lateral end of the inguinal canal
21. The right kidney
    1. Lies more caudally than the left
    2. Has a shorter renal vein than the left
    3. Contacts the duodenum, hepatic flexure of colon, and right lobe of liver
    4. Projects below the twelfth rib posteriorly
22. In the liver
    1. The coronary and triangular ligaments surround the bare area.
    2. The portal vein, hepatic artery, and bile duct reach and leave the liver via the hepatoduodenal ligament.
    3. The left half is functionally equivalent to the caudate, quadrate, and left lobes.
    4. The hepatoduodenal ligament forms the ventral wall of the epiploic foramen.
23. Indicate which of the following statements about the left vagus nerve is (are) true
    1. Passes dorsal to the root of the lung
    2. Contains preganglionic efferent nerve fibers
    3. Passes through the diaphragm at T10 level
    4. Supplies the anterior part of the stomach
24. The pancreas
    1. Drains into the third portion of the duodenum
    2. Lies immediately posterior to the dorsal attachment of the transverse mesocolon
    3. Has an uncinate process that lies ventral to the superior mesenteric artery
    4. Has a tail that ends near the hilum of the spleen

Annotated Answers

1. C. Since the duodenum is part of the fore- and midgut, the division between the two must exist at some point along its length and not at its beginning or end.
2. A. The layers of the spermatic cord are derived from the layers of the abdominal wall, "through" which the testis descended to reach the adult position in the scrotum. The innermost layer is derived from the innermost part of the abdomen—the transversalis fascia.
3. C. Indirect inguinal hernias pass down the inguinal canal (thus following an indirect route out of the abdomen) and enter the deep inguinal ring, which lies lateral to the inferior epigastric vessels.
4. D. The common hepatic artery travels in the edge of the lesser omentum (could you call it ventral mesentery?) to approach the liver.
5. B.
6. A. The middle colic artery reaches the transverse colon by traveling in the substance of the mesentery of this section—transverse mesocolon.
7. D.
8. C. The diaphragm arches cranially. Thus the vertebral level (where it could be related in an

x-ray film, for example) for the esophageal hiatus is higher than the aorta but lower than the inferior vena cava (which pierces the central tendon).

9. B.
10. E., 11. A., and 12. E. Each portion of the gastrointestinal tract has characteristics that help identify it in x-ray films or dissection; you should be aware of these. Thus the fundus of the stomach may be detected because gas may collect there, creating contrast, while the haustra (recess) of the colon may stand out in an x-ray study of the abdomen when a contrast agent has been used. The teniae coli also help identify the colon during dissection because they are three distinct bands of longitudinal smooth muscle.
13. B. The celiac ganglion is a peripheral ganglion of the sympathetic system and thus receives preganglionic fibers from the sympathetic chain (thoracic) via the splanchnic nerves.
14. A.
15. D.
16. C.
17. A.
18. C. The abdominal aorta, besides supplying the gastrointestinal tract through unpaired branches, also sends out, bilaterally, branches to the musculature enclosing the abdomen.
19. E.
20. A.
21. E.
22. E. The triangular and coronary (coronary meaning "corona" or "crown," not the heart vessels) ligaments, named for shape and position, are reflections of the peritoneal membrane that covers all the organs in the body.
23. E. Remember that the preganglionic fibers in the left vagus nerve course very close to the arch of the aorta, pass dorsal to the root of the lung, and distribute onto the anterior surface of the stomach (predominantly due to rotation of the stomach) after emerging into the abdomen with the esophagus at the level of T10.
24. C. The "root" of the transverse mesocolon runs across the body of the pancreas, while the superior mesenteric artery emerges between the two portions of the head with the uncinate process hooked under and thus lying dorsally.

4 Pelvis and Perineum

Objectives

After reading this chapter, you should know the following:

Bones, joints, and important measurements of the bony pelvis

Tissues, structure, and function of the male perineum (penis, scrotum)

Tissues, structure, and function of the female perineum (labia, clitoris, vestibule)

Structure, musculature, and function of the urogenital and anal triangles

Musculature of the pelvis

Organization, structure, and function of the pelvic viscera (bladder, ureters, and urethra; rectum and anal canal)

Gender differences of the bony and muscular pelvis, perineum, and pelvic viscera

Blood supply, lymphatics, and innervation of the pelvis and perineum

The pelvis, or basin, represents the bottom of the abdomen and as such contains the continuation of many abdominal viscera. The false pelvis is bounded by the alae of the ilia. The true pelvis is the area below the pelvic brim (terminal line) ending inferiorly at the muscular pelvic diaphragm.

The perineum represents the region of the inferior pelvic aperture. It is a diamond-shaped region lying between the pelvic diaphragm and the skin, and, as such, contains superficial structures, such as the external genitalia.

Bones

The bones of the pelvis are the hip bones, sacrum, and coccyx (Figs. 4-1 and 4-2). The hip bone is composed of three fused bones: the ilium, pubis, and ischium (Fig. 4-3). The **ilium** is a fan- or wing-shaped bone, the alae being the wings. Each ala is topped by a crest, or rim, that ends in front at the anterior superior iliac spine and in back at the posterior superior iliac spine. On the lateral side of each crest, approximately 6 cm behind the anterior superior iliac spine, there is a tubercle (the iliac tubercle). Inferior to both the anterior and posterior superior iliac spines are anterior and posterior inferior iliac spines, respectively. Below the posterior inferior iliac spine the bone is curved to form the greater sciatic notch. Posterior to the iliac fossa is the articular surface for the sacroiliac joint.

The **false pelvis** is a bowl formed by the two wings of the ilia. It ends below in a ridge, the pelvic brim, composed of both the arcuate line and the pecten pubis.

Each **pubis** is composed of two rami and a body. The **superior pubic ramus** extends from the acetabulum medially to the body. The body is in the region of the **pubic symphysis.** The pubic

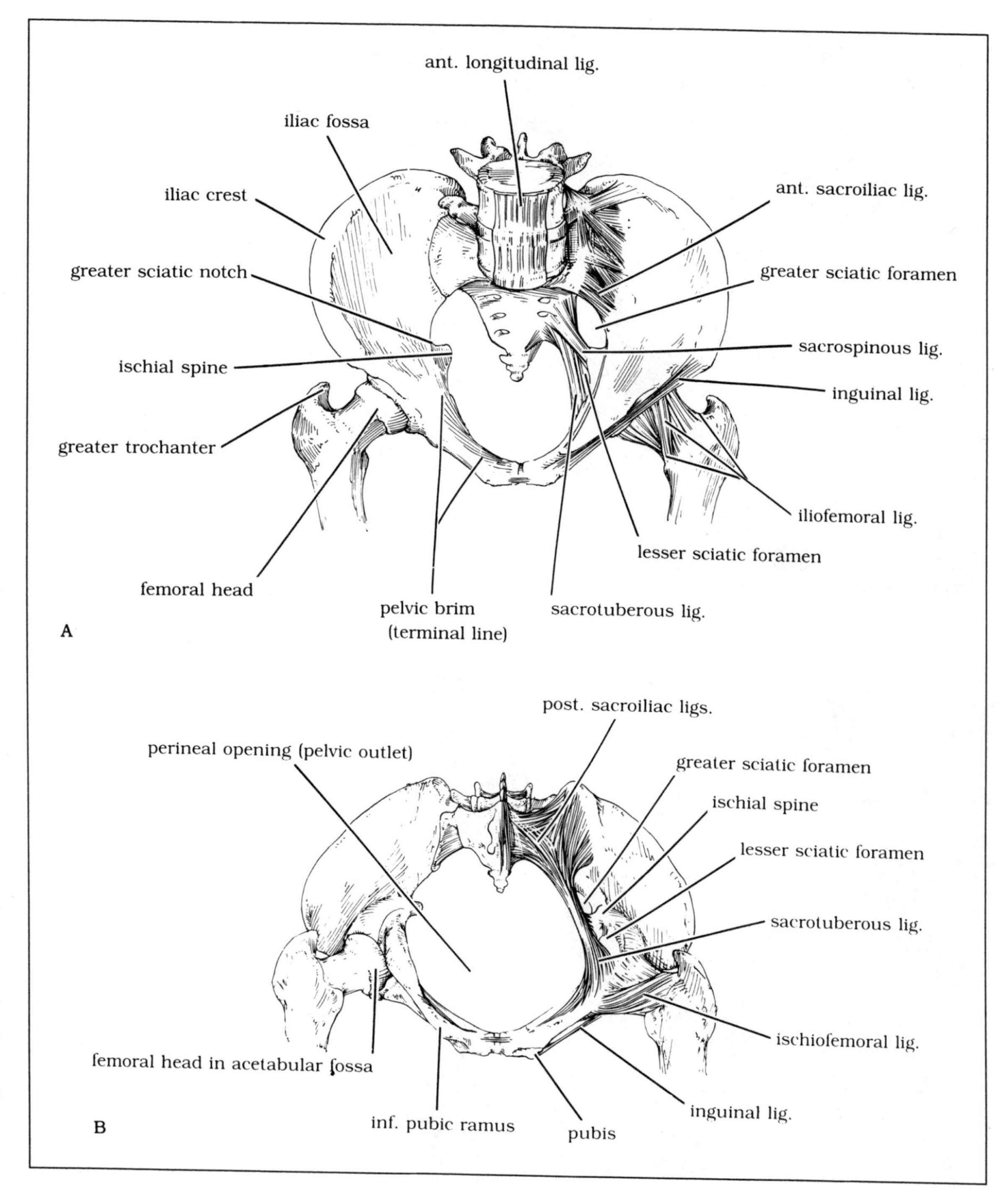

Fig. 4-1. The bony pelvis. A. Anterior view. B. Inferior view.

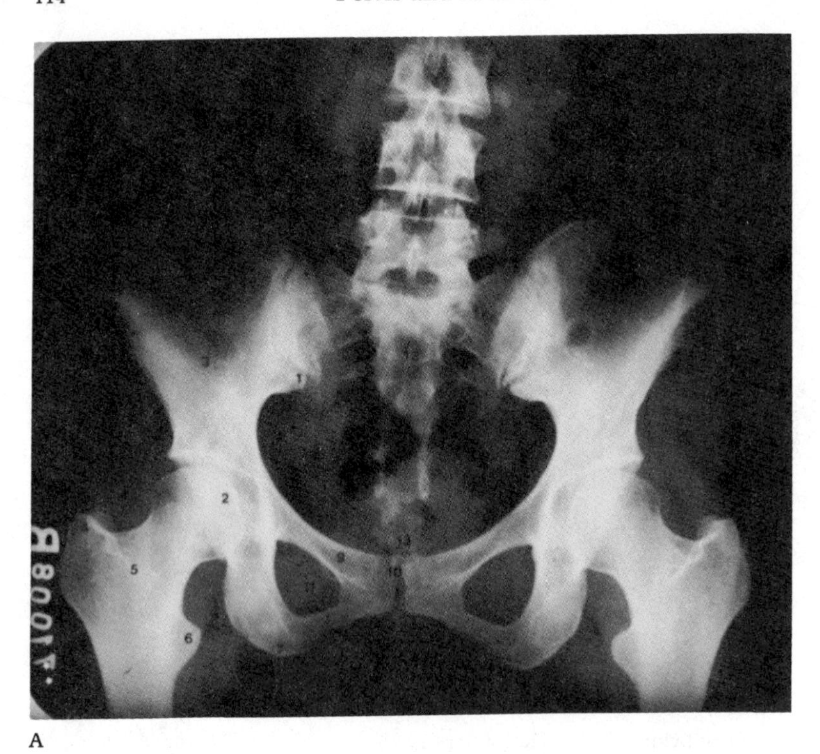

A

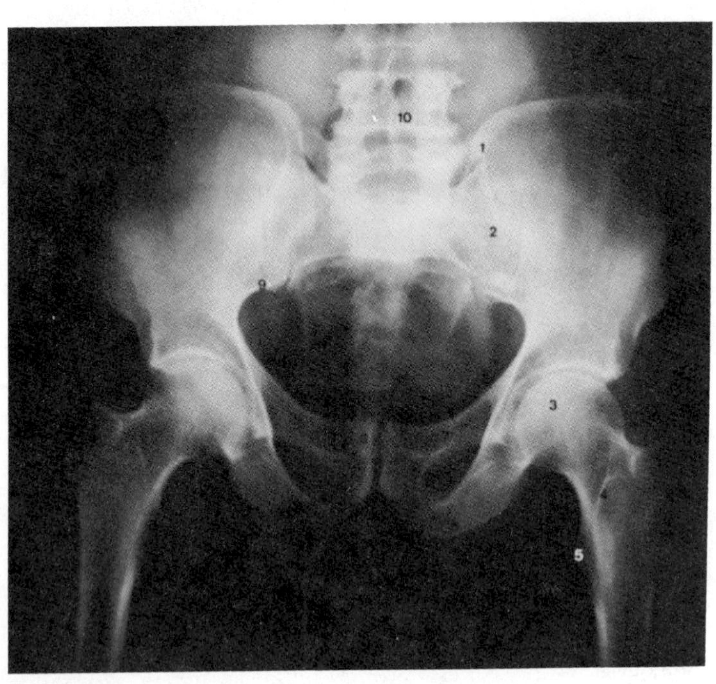

B

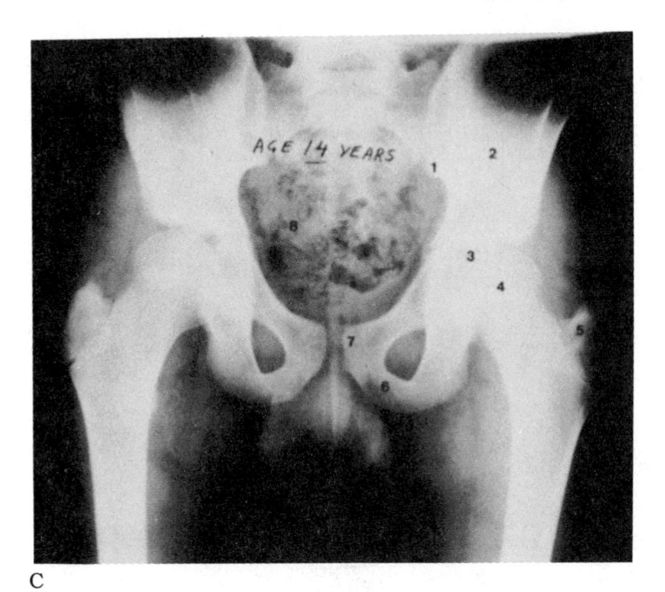

C

Fig. 4-2. X-ray films of the pelvis. A. Male adult, posteroanterior view. (1 = posterior inferior iliac spine; 2 = femoral head in acetabular fossa; 3 = iliac ala; 4 = greater trochanter; 5 = intertrochanteric line; 6 = lesser trochanter; 7 = ischial tuberosity; 8 = inferior pubic ramus; 9 = superior pubic ramus; 10 = pubic symphysis; 11 = obturator foramen; 12 = sacral body; 13 = coccyx.) B. Female adult, anteroposterior view. (1 = posterior superior iliac spine; 2 = lateral sacral mass; 3 = femoral head in acetabular fossa; 4 = intertrochanteric crest; 5 = lesser trochanter; 6 = ischial tuberosity; 7 = superior pubic ramus; 8 = pubic symphysis; 9 = posterior inferior iliac spine; 10 = spine of lumbar vertebra.) C. Male child, posteroanterior view. (1 = posterior inferior iliac spine; 2 = iliac ala; 3 = line of growth cartilage of fusion of pelvic bones; 4 = femoral head; 5 = ossifying greater trochanter; 6 = inferior pubic ramus; 7 = pubic symphysis; 8 = small bowel in pelvic bowl.)

crest is found at the symphysis, and the pubic tubercle lies lateral to the crest. On the upper surface of the superior pubic ramus there is a ridge, the pecten pubis, which is part of the pelvic brim. The **inferior pubic ramus** extends downward from the symphysis to join with the ischial ramus. It forms part of the bony margin of the obturator foramen, along with the superior pubic ramus and the ischium.

The **ischium** is a V-shaped bone with a body and a ramus. The body forms part of the acetabulum. Posteriorly there is a small bony projection, the ischial spine, where the sacrospinous ligament attaches, converting the greater sciatic notch into a foramen. The apex of the V forms the ischial tuberosity. Between the ischial spine and the ischial tuberosity is the lesser sciatic notch, which is turned into a foramen by the attachments of the sacrospinous and sacrotuberous ligaments. The ischial ramus extends from the body of the ischium to the inferior pubic ramus.

The **acetabulum** is a deep fossa that articulates with the femur (Fig. 4-4). It is formed from parts of all three hip bones (the ischium and the ilium each make up 40 percent and the pubis makes up 20 percent). The acetabulum is open below as the acetabular notch.

The joints of the pelvis are (1) the lumbosacral joint, composed of the vertebral joint between the fifth lumbar and first sacral vertebrae and the iliolumbar ligament; (2) the sacrococcygeal joint, composed of the vertebral joint between the sacrum and coccyx and the sacrococcygeal ligaments; (3) the sacroiliac joint; and (4) the pubic symphysis. The **sacroiliac joint,** a synovial joint with a limited amount of movement, adjoins the articular surfaces of the sacrum and ilium. It is

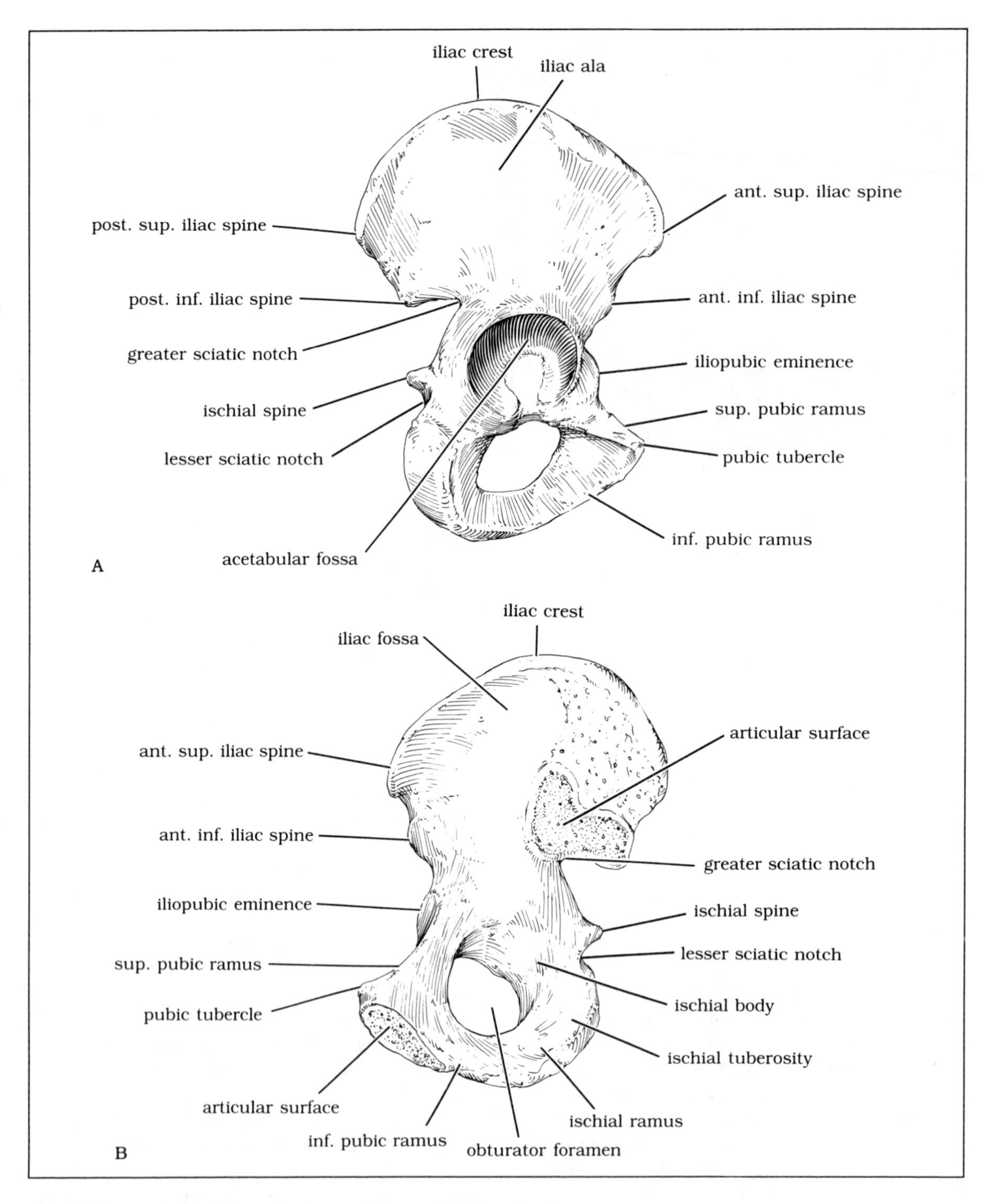

Fig. 4-3. The hip (coxal) bone. A. External aspect. B. Internal aspect.

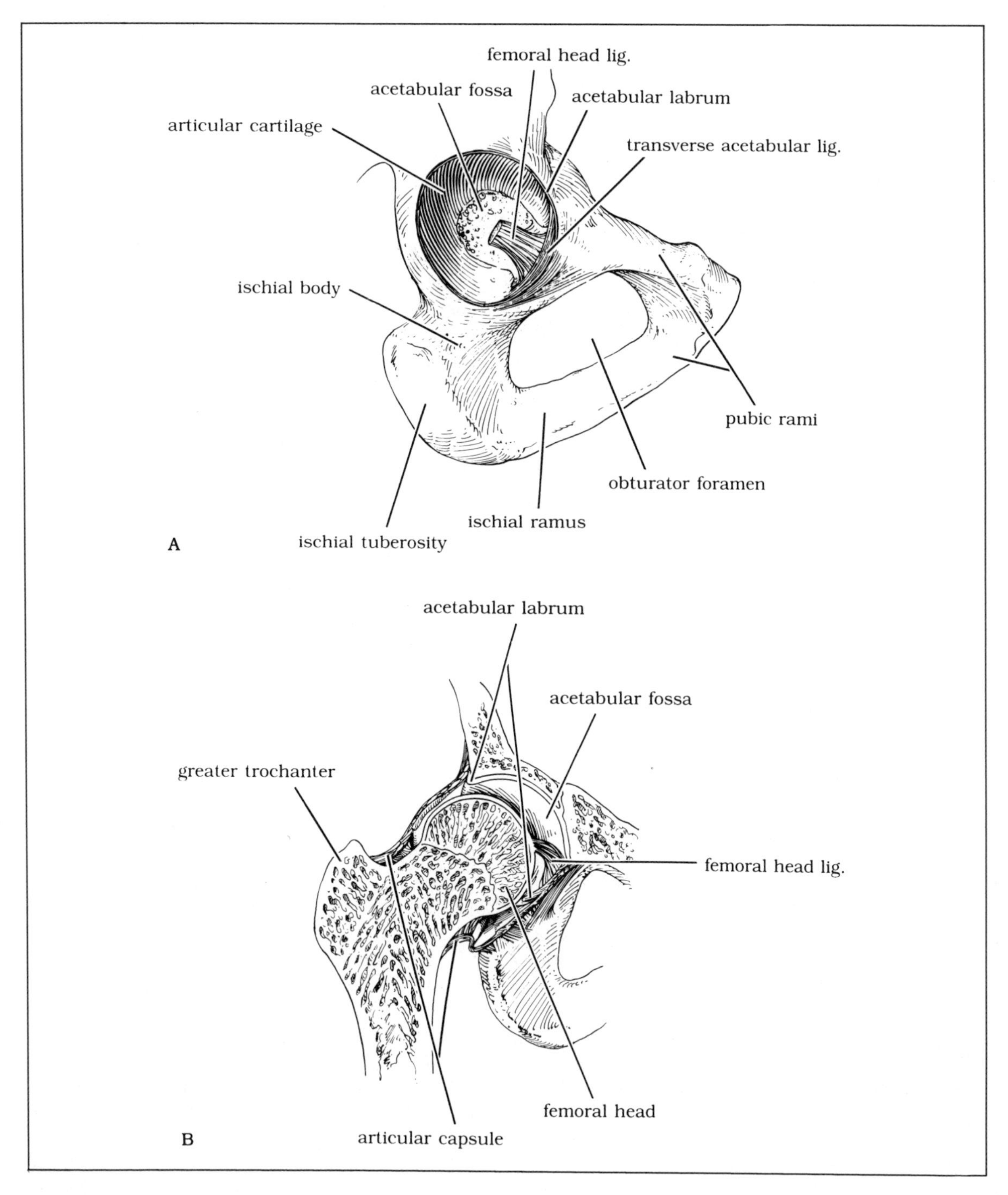

Fig. 4-4. A. The acetabulum with the femur removed. B. Frontal section of the hip joint.

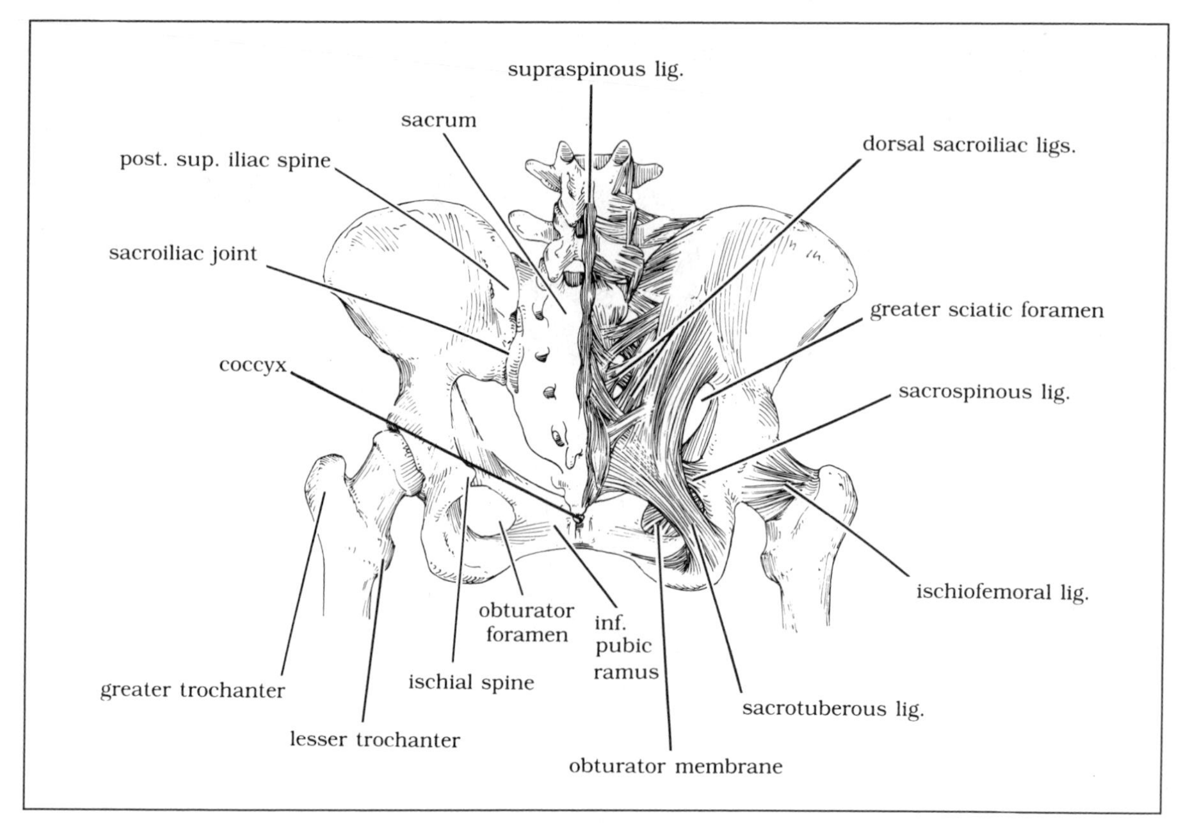

Fig. 4-5. Posterior view of the bony pelvis with ligaments.

strengthened by the strong sacroiliac ligaments (dorsal, ventral, and interosseous), with help from the accessory sacrospinous, iliolumbar, and sacrotuberous ligaments (Fig. 4-5). The **pubic symphysis** is a fibrocartilaginous symphyseal joint between the two pubic bodies. It is strengthened by the superior pubic and arcuate pubic ligaments.

Pelvic Measurements

The **true pelvis** has an inlet bounded by the pubic symphysis, sacral promontory, and iliopectineal lines; an outlet bounded by the pubic arch, coccyx, and ischial tuberosities; and a cavity that lies between the inlet and the outlet. Measurements of the female pelvic inlet and outlet are important in obstetrics. Three conjugate diameters are measured in the female pelvic inlet: the **true conjugate** (about 10 cm), which is the diameter from the sacral promontory to the superior margin of the pubic symphysis; the **diagonal conjugate** (about 10.5 cm), which is the measurement from the sacral promontory to the inferior margin of the pubic symphysis; and the **obstetric conjugate,** which is a measurement of the smallest anteroposterior diameter from the sacral promontory to a point a few millimeters below the superior margin of the pubic symphysis. Pelvic inlet measurements also include the **transverse and oblique diameters,** which are the widest space across the pelvic brim (about 13.5 cm) and the distance from the sacroiliac joint to the contralateral iliopectineal eminence (about 12.5 cm), respectively. Measurements of the female pelvic outlet include the **transverse diameter,** the distance between the ischial tuberosities; the **transverse midplane diameter,** the distance between the ischial spines; and the

anteroposterior (or **sagittal) diameter,** the distance from the lower margin of the pubic symphysis to the sacrococcygeal joint.

The male pelvis and the female pelvis differ in several ways. The male pelvis is thicker and heavier, and the female pelvis is wider and shallower, with a larger pelvic inlet and outlet. In the female, the hip bones are farther apart because of the broader sacrum, and the ischial tuberosities are farther apart. Some of these differences are evident in Figure 4-2.

Fig. 4-6. Surface anatomy of the male perineum, with scrotum raised. A. Schematic drawing. B. Photograph.

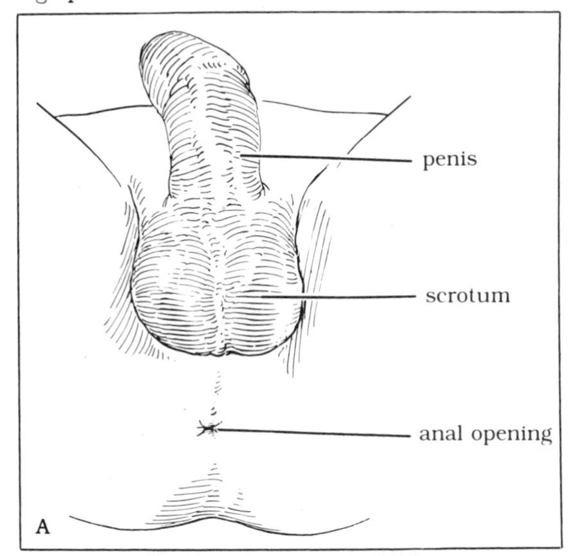

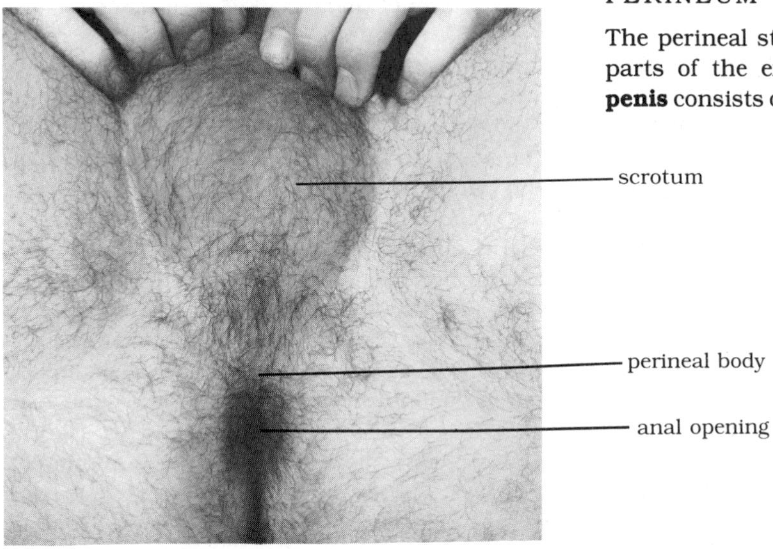

Perineum

The perineum is a diamond-shaped region with the pubic symphysis as its anterior limit. Moving posterolaterally it is delimited by the inferior pubic ramus and ischial ramus down to the ischial tuberosity. Extending posteromedially from the ischial tuberosity along the sacrotuberous ligaments, the limits converge on the coccyx posteriorly. The soft tissue within this cavity is therefore within the perineum.

For study the perineum is usually divided, by a line connecting the two ischial tuberosities, into two equal triangles, an anterior urogenital triangle and a posterior anal triangle. The **urogenital triangle** contains the external genitalia and urethra (except the prostatic portion) and associated structures. The **anal triangle** contains the anus and ischiorectal fossae and associated structures.

STRUCTURES OF THE MALE PERINEUM

The perineal structures unique to the male are parts of the external genitalia (Fig. 4-6). The **penis** consists of three bodies (corpora) of erectile

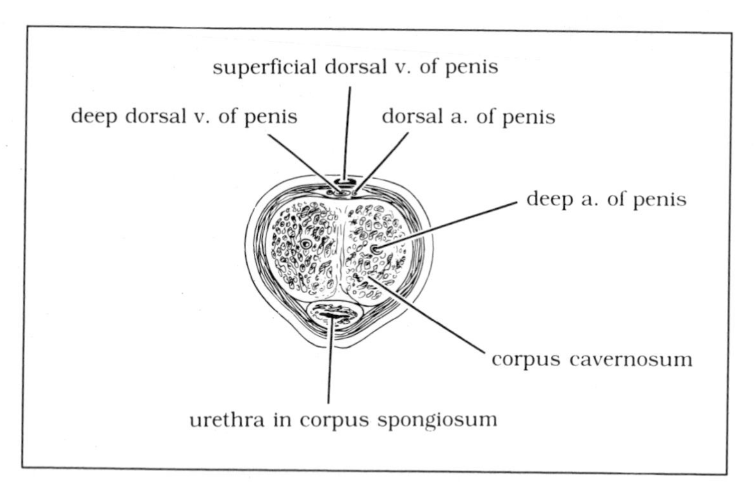

Fig. 4-7. Cross section of the penis.

tissue held together by dense fibrous connective tissue with a covering of skin. The two dorsal bodies are the corpora cavernosa penis and the ventral, midline body is the corpus spongiosum penis (Fig. 4-7).

The **corpus spongiosum penis** contains the urethra, which ends at its distal end at the external urethral orifice. The corpus spongiosum is expanded at its distal end into the **glans penis,** a cap that extends over the two corpora cavernosa. Covering the glans is a fold of skin, the prepuce or foreskin.

The superficial dorsal vein lies in the subcutaneous connective tissue in the midline of the dorsum of the penis. A deep fascial layer underlies this vein, followed by an area containing the midline deep dorsal vein and the dorsal arteries and dorsal nerves. Finally a deep artery of the penis runs through each corpus cavernosum.

At the base of the penis the two **corpora cavernosa penis** diverge laterally so that each lies next to an ischiopubic ramus and attaches to it. Here they are known as the crura of the penis. They extend posteriorly almost to the ischial tuberosity. The corpus spongiosum continues its midline position posteriorly, forming at its root the bulb of the penis. All three erectile bodies are attached proximally in the perineum to the perineal membrane.

Lying at the base of the penis is the **scrotum** (see Fig. 4-6), a pouch containing the testes and part of the spermatic cord. The scrotal raphe is a midline ridge of the skin, indicating where the epidermal contributions from the two sides fused during development. Lying deep to the raphe is a septum dividing the scrotum into two compartments.

STRUCTURES OF THE FEMALE PERINEUM

The **labia majora** of the female are homologous to the scrotum of the male (Fig. 4-8). They are two swellings surrounding the urogenital vestibule. Anteriorly they form the mound of the mons pubis and are connected by the anterior labial commissure. The **labia minora** are two smaller cutaneous folds within the labia majora. They run posteriorly from the clitoris, contributing to the prepuce of the clitoris.

The **clitoris** also consists of three erectile bodies, but it does not contain the female urethra. The two corpora cavernosa form the clitoris and are rooted along the ischiopubic rami as the crura of the clitoris. The corpus spongiosum lies in the midline, joining the corpora cavernosa at the base of the clitoris. Posterior to the clitoris the corpus spongiosum splits in the midline to surround the vestibule. Its two roots are known as the bulbs of the vestibule. The **vestibule** is the

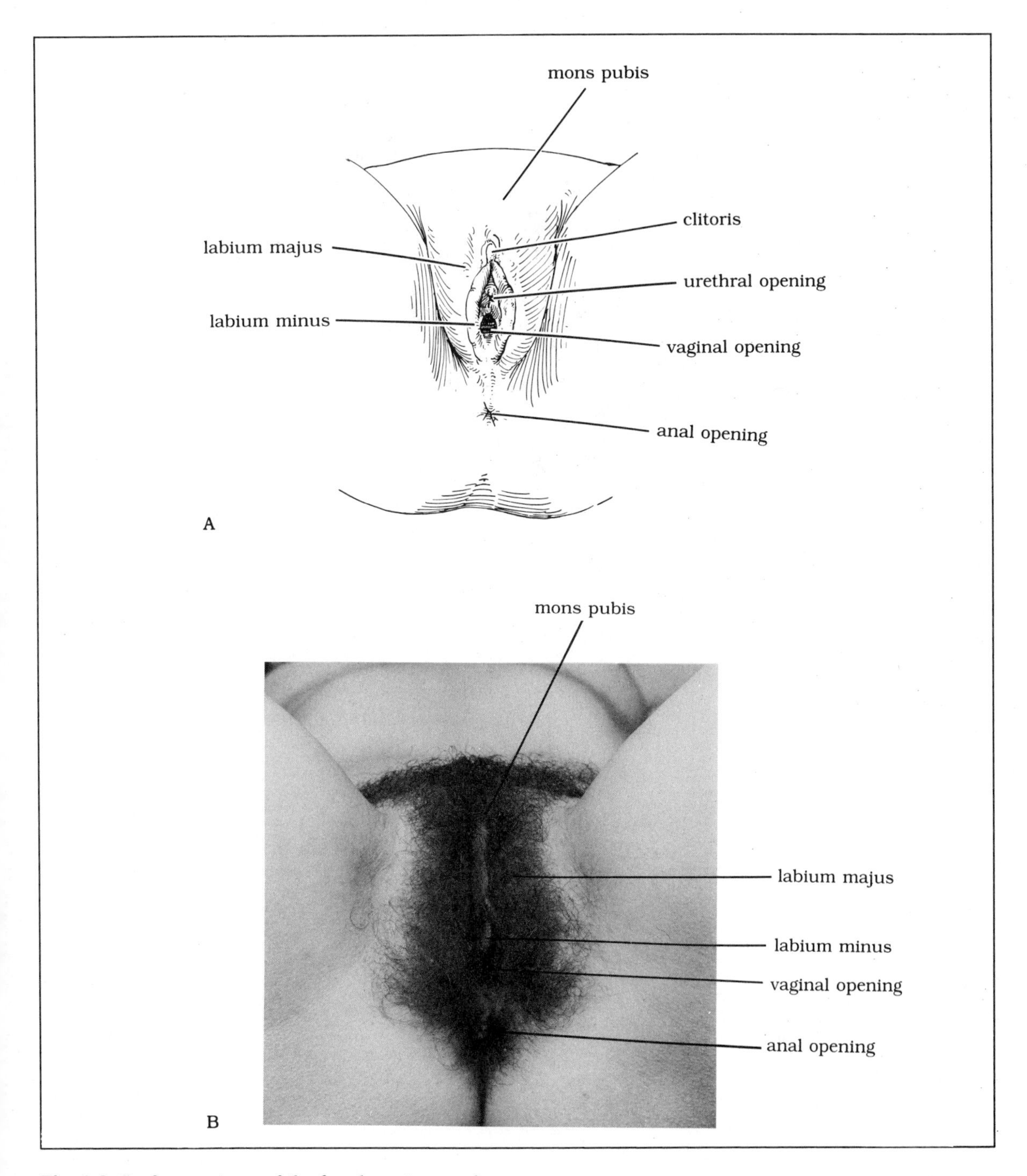

Fig. 4-8. Surface anatomy of the female perineum. A. Schematic drawing. B. Photograph. C. Photograph with opened labia majora.

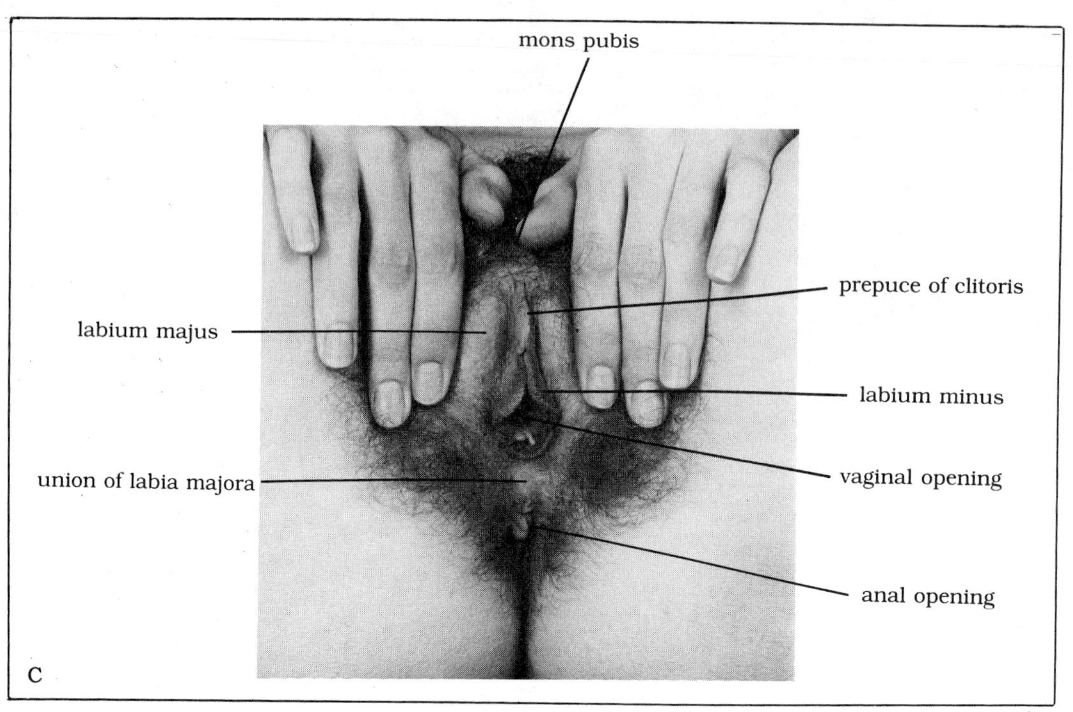

Fig. 4-8. (continued).

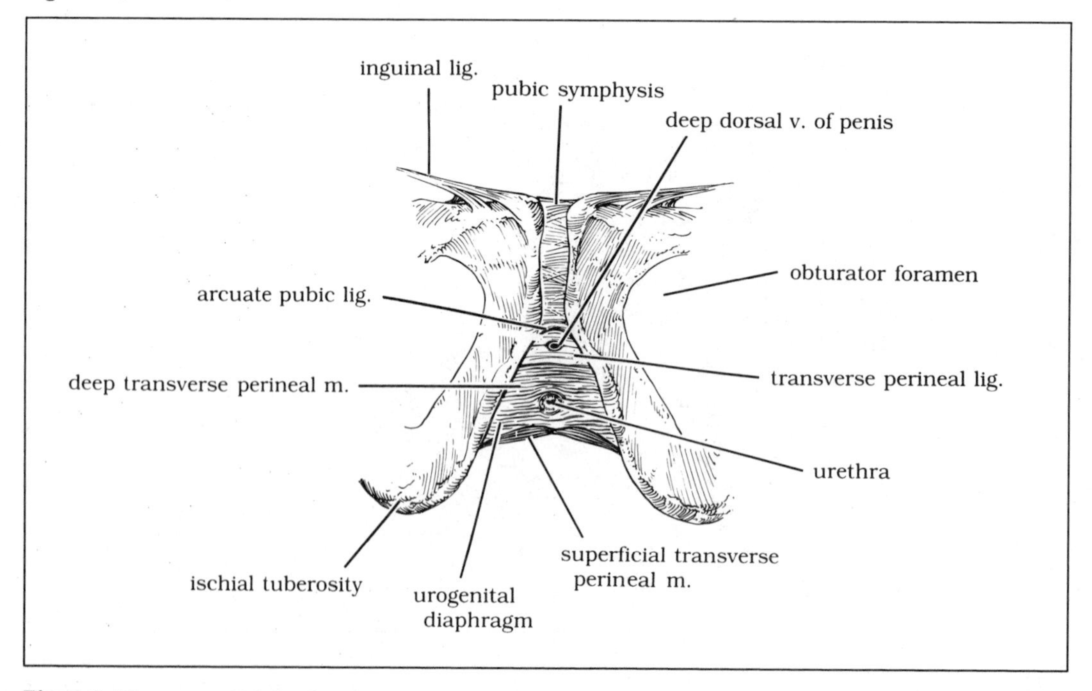

Fig. 4-9. The urogenital diaphragm.

opening between the labia minora that contains the external urethral orifice, the vaginal opening, and the openings for the greater vestibular glands.

UROGENITAL TRIANGLE

The muscles of the urogenital region of the perineum are common to males and females. The **urogenital diaphragm** (Fig. 4-9) consists of the deep transverse perineal muscle stretching between the two ischiopubic rami and the sphincter urethrae, a band of circular fibers in the deep transverse perineal muscle.

Lying superficial to the urogenital diaphragm (Fig. 4-10) are the two superficial transverse perineal muscles. These thin muscles converge on the central tendinous point (the perineal body). This point, in the center of the perineum, acts as an anchor for the two superficial transverse perineal muscles, the external anal sphincter, and the bulbospongiosus muscle. The bulbospongiosus muscle overlies the bulb of the penis in the male, uniting in a midline raphe. In the female the muscles overlie the bulbs of the vestibule. In addition the two crura of the penis or the clitoris are covered by ischiocavernosus muscles.

The urogenital region at the perineum contains two pouches or areas of fascial containment. The urogenital diaphragm is enclosed by inferior and superior fascial layers that are said to compose the deep perineal pouch. Between the inferior fascial layer of the urogenital diaphragm, otherwise known as the perineal membrane, and the membranous layer of subcutaneous connective tissue, is the superficial pouch. This pouch contains the superficial transverse perineal muscles and the erectile bodies and their surrounding structures.

ANAL TRIANGLE

The anal triangle contains the anus and the ischiorectal fossae (Fig. 4-11). The anal canal, terminating in the anus, is the final portion of the large bowel. The external anal sphincter attaches anteriorly to the central tendinous point. This voluntary sphincter muscle has both superficial and deep portions.

The **ischiorectal fossae** are two communicating wedge-shaped spaces between the anal canal and the obturator internus muscle overlying the ischium. Thus, the medial walls are formed by the inferior fascia of the pelvic diaphragm, and the lateral walls by the obturator internus fascia. The fossae contain primarily fat and loose areolar connective tissue. Anteriorly the two fossae extend superiorly to the urogenital diaphragm, forming the anterior recess.

BLOOD SUPPLY, LYMPHATICS, AND INNERVATION

The **internal pudendal artery** supplies the structures of the perineum. One of the terminal branches of the internal iliac artery, the internal pudendal artery exits the pelvis through the greater sciatic foramen. It then crosses the spine of the ischium to enter the perineum through the lesser sciatic foramen. On entering the perineum it lies in the fascia of the obturator internus muscle on the lateral wall of the ischiorectal fossa. The fascial split through which it passes is known as the **pudendal canal.**

The inferior rectal branches of the internal pudendal artery cross the ischiorectal fossa to reach the anus. They supply the muscles and skin of this region and anastomose with the superior and middle rectal arteries.

The perineal artery branch of the internal pudendal artery supplies the muscles of the superficial pouch and the skin of the scrotum. Other branches supply the bulb of the penis (vestibule) and the urethra.

The terminal branches of the internal pudendal artery are the dorsal artery of the penis or clitoris, which runs under the deep fascia, and the deep artery of the penis or clitoris, which supplies the corpora cavernosa.

The venous drainage of the region follows the course of the arterial supply with two exceptions. The superficial dorsal vein of the penis or clitoris (there is no corresponding artery) drains into the superficial external pudendal veins. The deep dorsal vein of the penis or clitoris passes proxi-

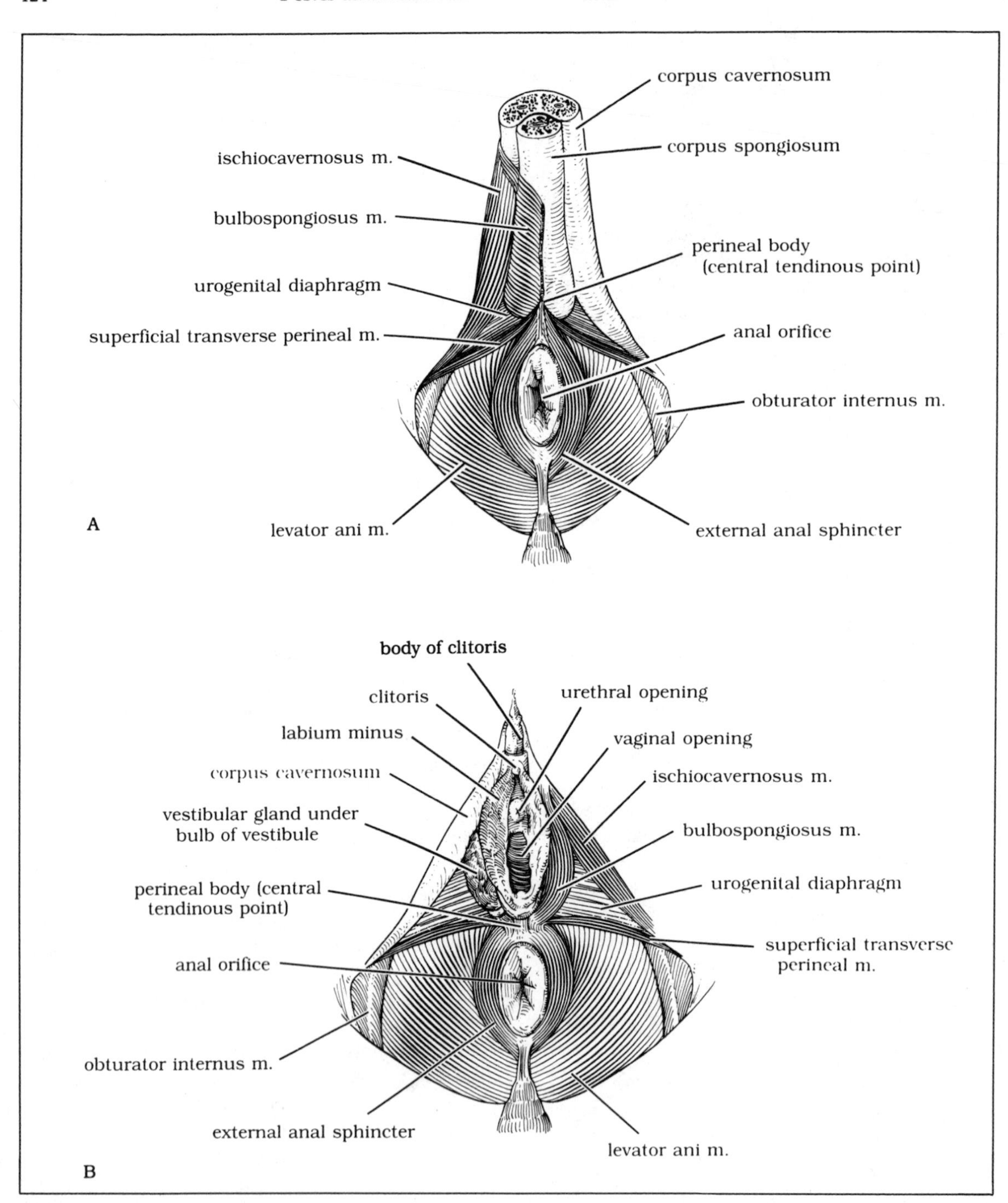

Fig. 4-10. The superficial perineal space. A. Male.
B. Female.

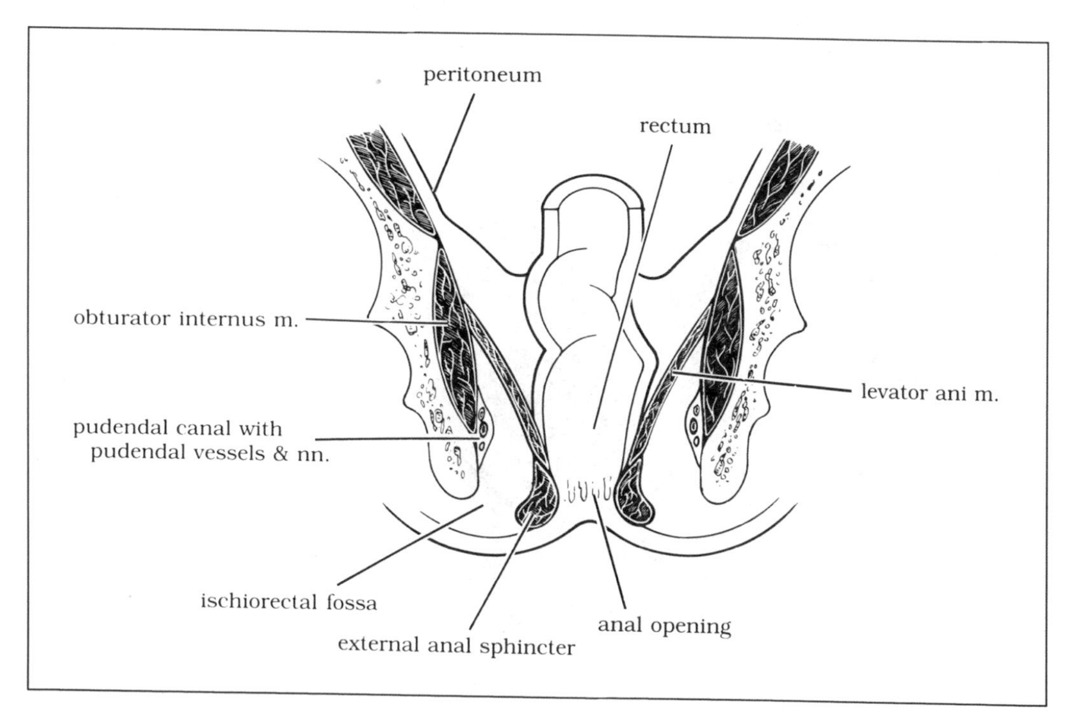

Fig. 4-11. Schematic frontal section of the rectum and ischiorectal fossa.

mally between the urogenital diaphragm and the pubic arch to drain into either the prostatic plexus in the male or the vesical plexus in the female.

All lymphatic drainage of the perineal region is subcutaneous, except for the testicular drainage to the lumbar nodes. The superficial lymphatic vessels drain to the superficial inguinal lymph nodes.

The **pudendal nerve** arises from sacral roots S2 to S4. It accompanies the internal pudendal artery through the two sciatic foramina and into the pudendal canal. Its branches are the inferior rectal nerve to the external anal sphincter; the perineal nerve to the muscles of the deep and superficial pouches and the skin of the scrotum; and the dorsal nerve of the penis or clitoris, which, running deep to the perineal membrane, supplies the crura and then travels with the dorsal artery to supply the penis or clitoris.

The pudendal nerve may be reached by local anesthetic (pudendal nerve block) when performing vaginal procedures without general anesthesia. To reach the nerve a needle is guided through the vaginal wall to the ischial spine and sacrospinous ligament using a finger as a guide.

Pelvis

MUSCULATURE

The bones of the pelvis are described at the beginning of this chapter. Held within the bony pelvis, and constituting the walls and floor of the pelvis, are four muscles: the piriformis, levator ani, coccygeus, and the obturator internus (Fig. 4-12 and Table 4-1). The levator ani and coccygeus constitute what is called the pelvic diaphragm.

The **piriformis** muscle arises from the second to the fourth sacral vertebrae and inserts into the upper and medial side of the greater trochanter, exiting the pelvis through the greater sciatic foramen. Because of this configuration, the muscle serves as a marker in the gluteal region for the identification of nerves and vessels. The pirifor-

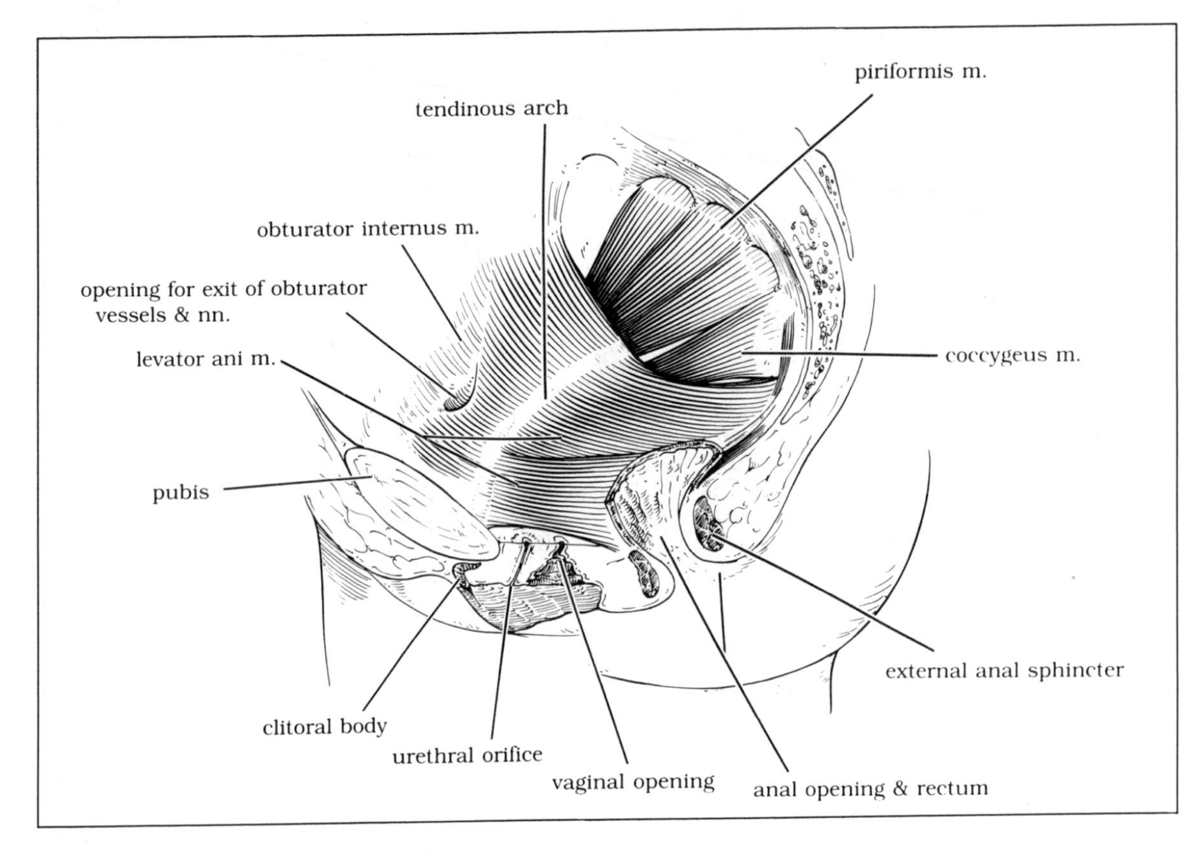

Fig. 4-12. Medial view of the muscles of the female pelvis.

mis rotates the thigh laterally and is innervated by S1 and S2 spinal segments.

The **obturator internus** arises from the inside margin of the obturator foramen, the obturator membrane, the obturator fascia, and the bone above and behind the obturator foramen. The muscle passes through the lesser sciatic foramen to insert onto the medial surface of the greater trochanter. The obturator internus acts as a lateral rotator and abductor of the thigh and is innervated by spinal nerves L5 to S1.

The **obturator membrane** is a dense connective tissue layer that closes off the obturator foramen, except at the **obturator canal,** where the obturator vein, artery, and nerve leave the pelvis. The **obturator fascia** is the fascial covering of the obturator internus muscle. It provides origin for some muscle fibers from both the obturator internus and levator ani muscles.

The **pelvic diaphragm** closes off the pelvic outlet and supports the pelvic viscera. It is composed of the levator ani and coccygeus muscles. The two halves of the diaphragm separate at the **genital hiatus** to allow the rectum, urethra, and, in the female, vagina to pass through to the outside.

The **levator ani** muscle has three parts: the puborectalis, pubococcygeus, and iliococcygeus. The **puborectalis** is the most medial division. It arises from the pubic symphysis and arches back behind the rectum to join its equivalent from the opposite side. Its most medial fibers constitute the levator prostatae in the male and the pubovaginalis in the female. The **pubococcygeus** runs from the superior pubic ramus to the front of the coccyx and the anococcygeal raphe. This raphe represents the line of fusion of the muscles from

Table 4-1. Musculature of the Pelvis

Muscle	*Origin*	*Insertion*	*Innervation*	*Action*
Piriformis	Vertebrae S2–S4	Greater trochanter	S1, S2	Rotation of thigh laterally
Obturator internus	Margin of the obturator foramen, obturator membrane and fascia, adjacent bone	Greater trochanter	L5–S1	Rotation of thigh laterally, abduction
Levator ani			S3, S4	Support pelvic viscera, regulate abdominal pressure
Puborectalis	Pubic symphysis	Muscle of opposite side, median raphe		
Pubococcygeus	Superior pubic ramus	Muscle of opposite side, coccyx		
Iliococcygeus	Obturator fascia, spine of ischium	Coccyx, anococcygeal raphe		
Coccygeus	Ischial spine, sacrospinous ligament	Coccyx, sacrum	S3, S4	Raise pelvic floor, pull coccyx forward

the two sides. The **iliococcygeus** arises from a thickened portion of the obturator fascia, the tendinous arch of the levator ani (arcus tendineus musculi levatoris ani), and the spine of the ischium. It inserts into the coccyx and the anococcygeal raphe.

The levator ani supports and raises the pelvic viscera. It is also involved in the regulation of internal abdominal pressure during breathing, micturation, and defecation. The muscle is innervated by spinal segments S3 and S4.

The **coccygeus** muscle arises from the ischial spine and the sacrospinous ligament. It inserts into the coccyx and the lower portion of the sacrum. The coccygeus raises the pelvic floor and pulls the coccyx forward. It is innervated by spinal nerves S3 and S4.

The muscles of the pelvic diaphragm are covered by fascia. The superior fascial layer is continuous with the transversalis fascia of the abdominal wall.

VISCERA

The peritoneal covering of the pelvis does not reach down to the floor of the pelvis. Thus most of the pelvic viscera lie below the peritoneum and are only partially covered by it. Instead they are embedded in the extraperitoneal connective tissue known as the **endopelvic fascia.**

Several fossae are formed in the pelvic peritoneum. The rectovesical fossa, or pouch, in the male is formed by the peritoneum from the bladder to the rectum (Fig. 4-13A). In the female this space is blocked by the uterus and two pouches are thus formed: the vesicouterine and rectouterine pouches (Fig. 4-13B). Leaving the uterus laterally the peritoneum forms a double-walled sheet, the broad ligament (see Fig. 4-18), that extends laterally to the side walls of the pelvis. In addition there are small fossae on the sides of both the bladder and the rectum: the paravesical fossae and the pararectal fossae, respectively.

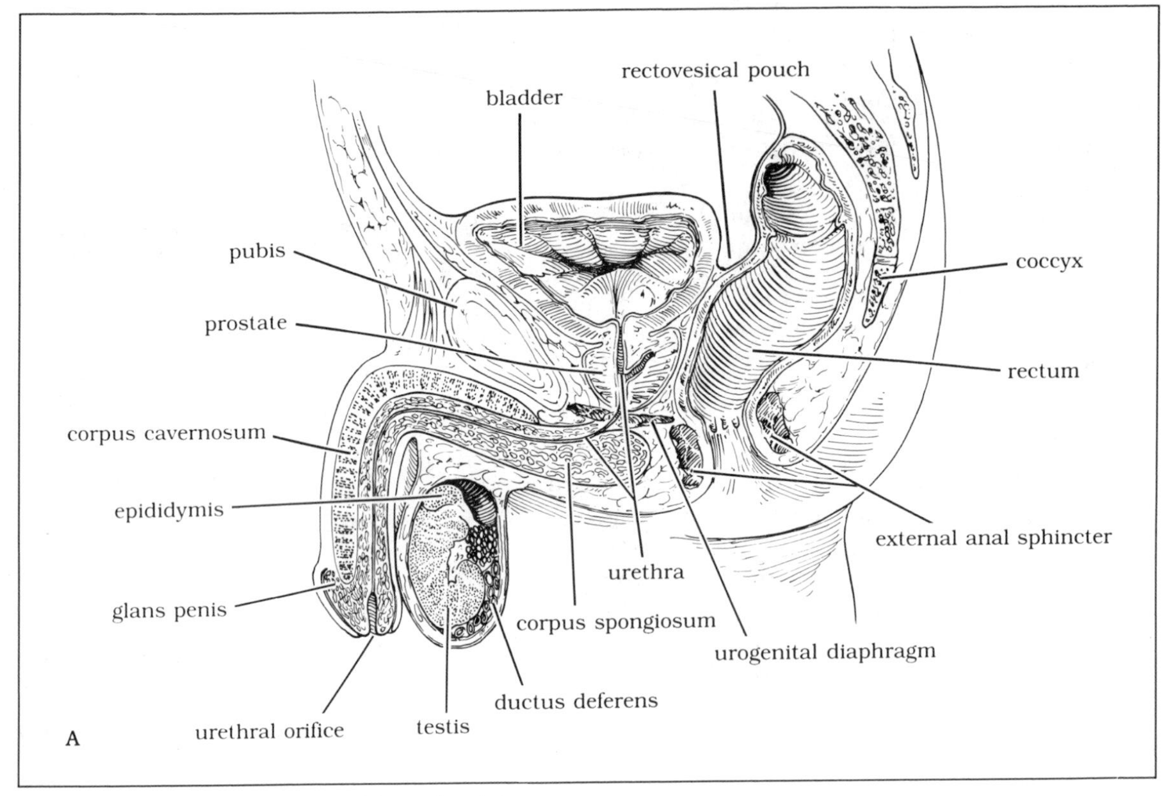

Fig. 4-13. A. Midsagittal section of the male pelvis. B. Midsagittal section of the female pelvis.

Bladder, Ureters, and Urethra

The bladder is the most-anterior pelvic organ, lying against the pubic symphysis (Fig. 4-14). Being hollow, it serves as a reservoir for urine, its muscular walls aiding in expulsion. When full the bladder is globular, when empty it is flat.

Inferiorly, the bladder rests on the prostate (see Fig. 4-13A) in the male and the pelvic diaphragm (see Fig. 4-13B) in the female. Superiorly and anteriorly a fibrous strand, the **median umbilical ligament,** extends to the umbilicus from the bladder's apex. The strand is the remnant of the embryonic urachus. The fundus of the bladder constitutes its posteroinferior portion.

The bladder is lined with mucous membrane that is usually wrinkled and folded, except in one triangular area, the **vesical trigone** (Fig. 4-15). In this region between the two ureteric openings and the inferior urethral aperture the membrane is smooth. An interureteric fold separates the two ureteric openings at the base of the triangle. The walls of the bladder are composed of intrinsic involuntary muscle, except in the region of the trigone, where in addition to the normal muscles of the bladder wall there is an extension of the muscle of the ureter. At the neck of the bladder, the urethral aperture, there is sphincter action but no identifiable sphincter muscle.

The bladder receives blood through the superior and inferior vesical arteries, which branch from the internal iliac artery. Blood drains via the vesical plexus. Lymph drains to both the internal and external iliac nodes. Preganglionic parasympathetic pelvic splanchnics and postganglionic lumbar sympathetics innervate the bladder via the inferior hypogastric plexus.

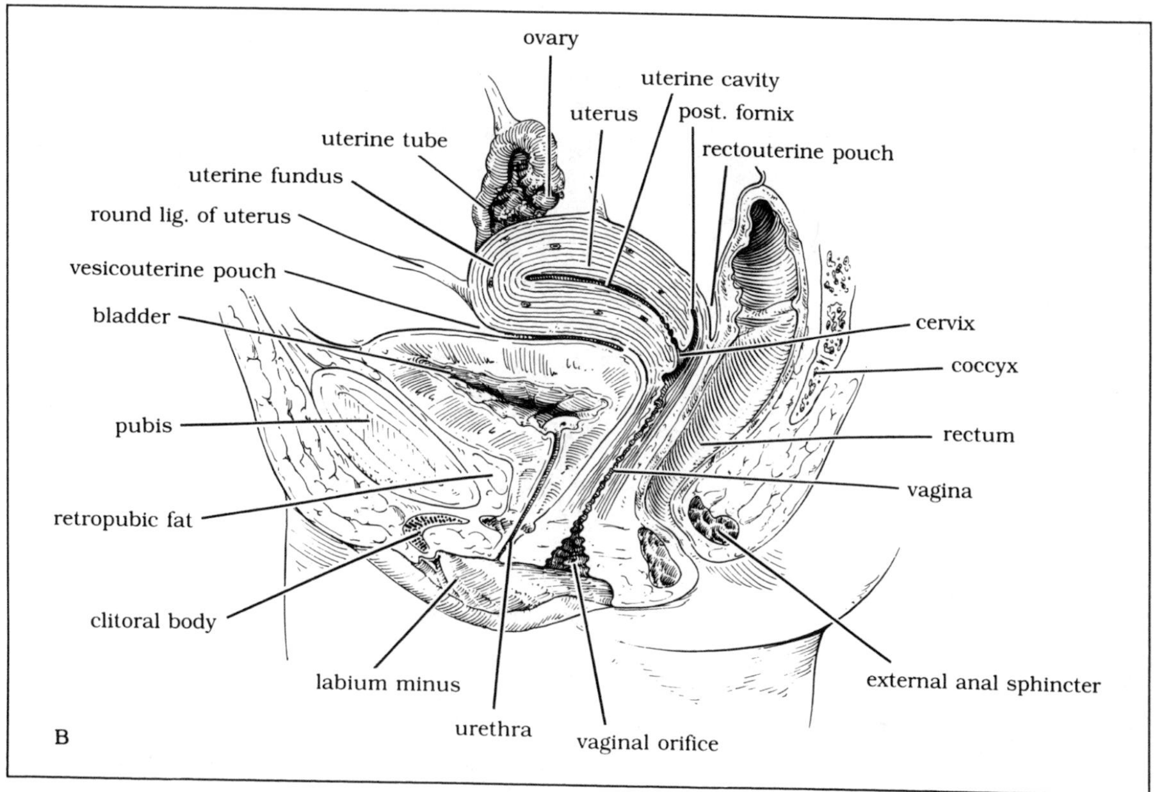

Fig. 4-13. (continued).

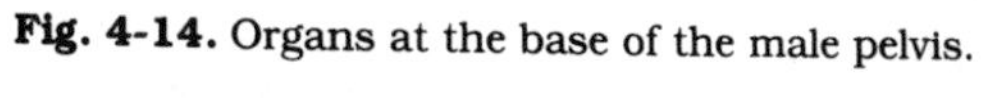

Fig. 4-14. Organs at the base of the male pelvis.

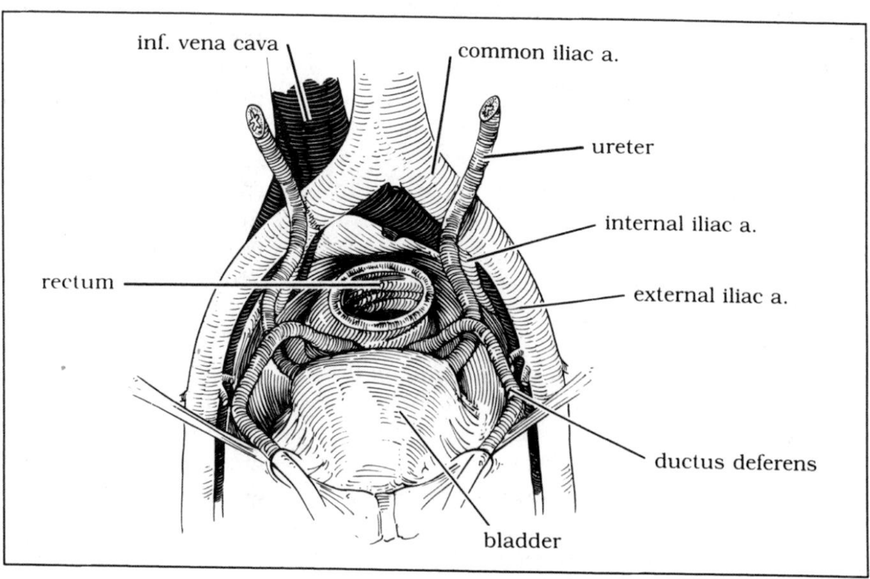

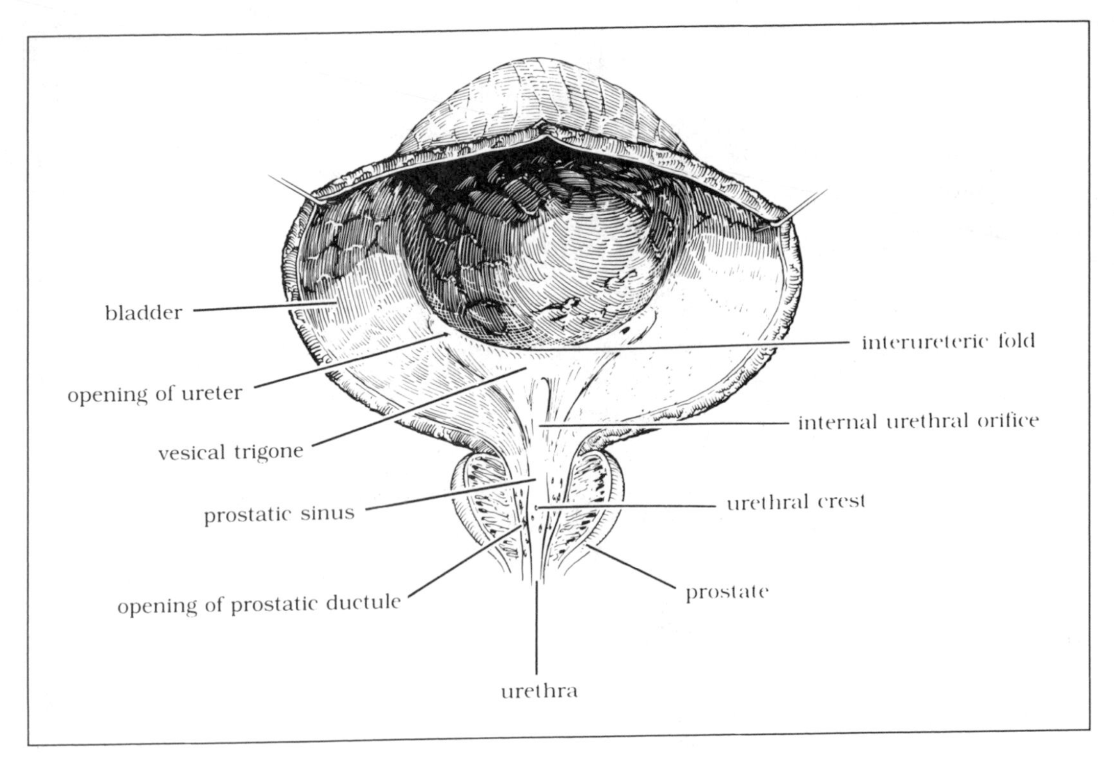

Fig. 4-15. Internal aspect of the bladder and prostate.

The pelvic portions of the **ureters** begin where they cross the pelvic brim and the common, or external, iliac arteries. The ureters descend retroperitoneally, anteriorly, and medially, crossing the obturator nerve and the umbilical artery to reach the posterolateral aspect of the bladder. Here they lie near the seminal vesicles and are crossed by the ductus deferens in the male (Fig. 4-16; see also Fig. 4-14). The ureters then run obliquely through the bladder wall to their openings in the trigone. In the female the ureter lies both posterior and inferior to the ovary and is crossed by the uterine artery. The ureters receive blood from the superior vesical artery and, in the female, the uterine artery.

The **urethra** extends from the neck of the bladder to the external urethral orifice. In the male it is composed of prostatic, membranous, and spongy parts (see Figs. 4-13A and 4-15). The prostatic portion, within the prostate gland, is 3 to 4 cm in length. Here the urethra receives the ejaculatory and prostatic ducts. The short, 1 cm, membranous portion passes through the sphincter urethrae muscle. The spongy portion, the longest portion of the male urethra, extends through the length of the penis to reach the external urethral orifice. The ducts of the bulbourethral glands enter the urethra in the spongy portion.

In the female the urethra is approximately 4 cm long. It passes from the neck of the bladder through the sphincter urethrae muscle to reach the external urethral orifice (see Fig. 4-13B), which lies anterior to the vaginal opening.

Rectum and Anal Canal

The rectum is the continuation of the sigmoid colon in the pelvis. It starts where the colon begins to become retroperitoneal, and it curves downward and forward on the sacrum and coccyx. It then pierces the pelvic diaphragm to become the anal canal (Fig. 4-17). Just distal to this

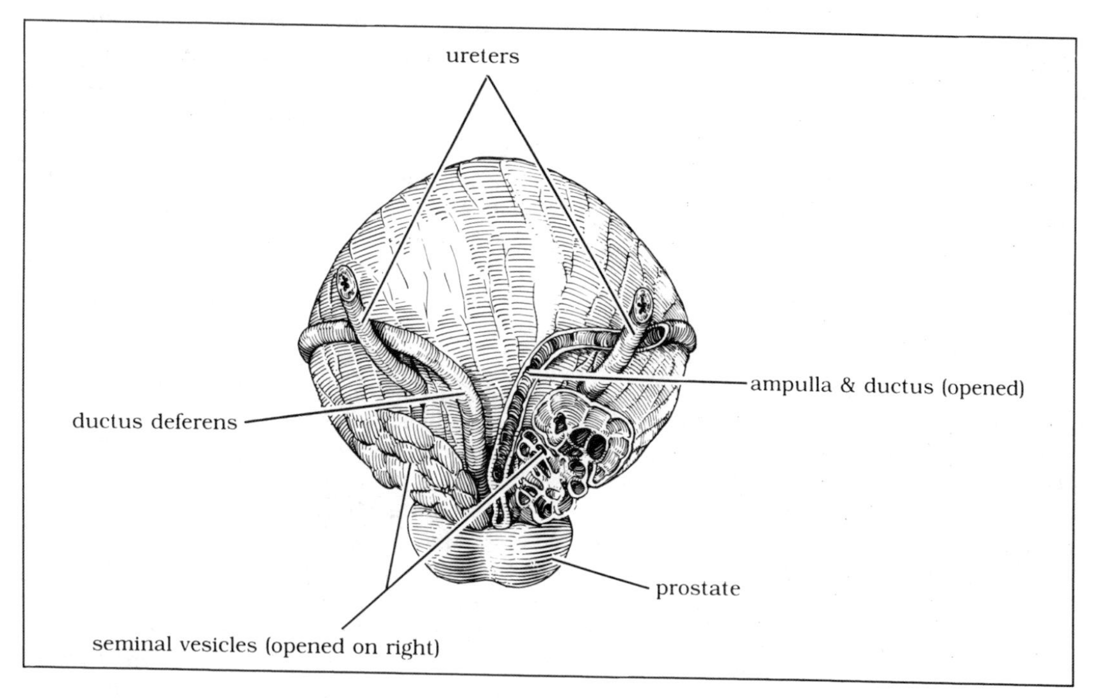

Fig. 4-16. Posterior view of the bladder and prostate.

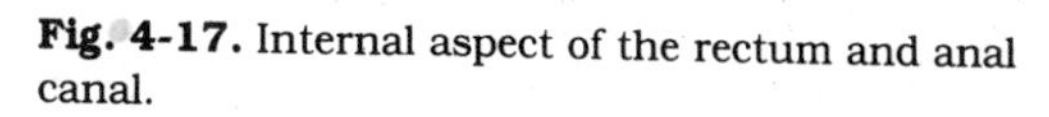
Fig. 4-17. Internal aspect of the rectum and anal canal.

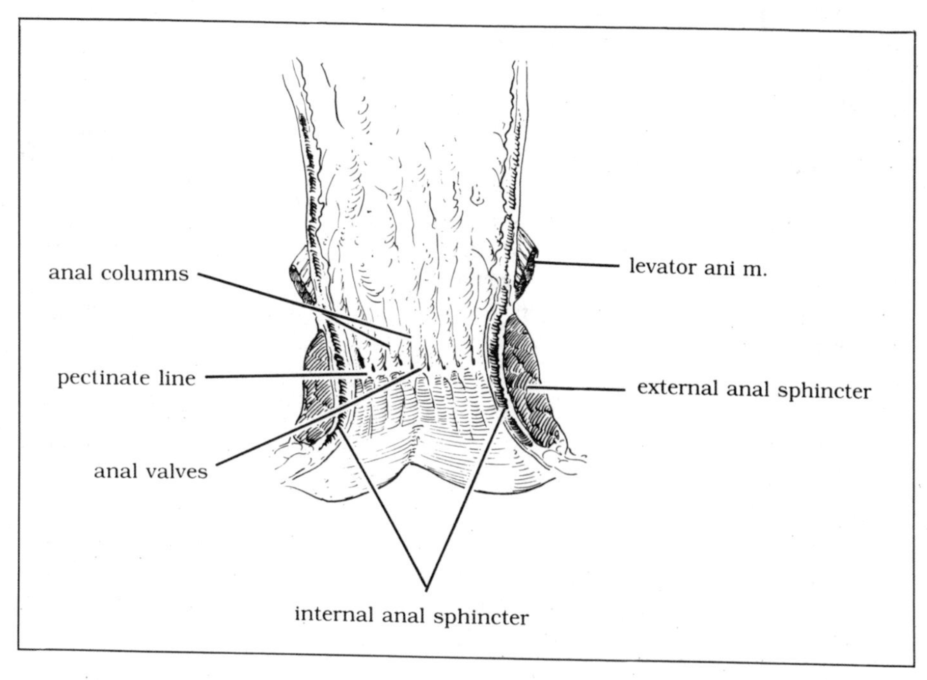

point the rectum dilates into what is known as the rectal ampulla.

The upper third of the rectum is covered with peritoneum on the sides and the front, the middle third only on the front. The last third has no peritoneal covering at all.

The inner circular muscular layer of the rectum is complete. However, the three longitudinal muscles of the colon are present on the rectum as only two wide muscular bands, one anterior and one posterior.

The rectum receives blood from the superior, middle, and inferior rectal arteries. Veins drain through vessels of the same name. The veins lying below the anal valves and anal columns connect the superior rectal vein of the portal system to the middle and inferior rectal veins of the systemic circulation. Given the origins, terminations, and anastomotic connections of the vessels of the region, the rectal blood supply constitutes one of the major shunts from the portal visceral system to the somatic system and vice versa.

The anal canal continues the rectum below the pelvic diaphragm for 2.5 to 3.5 cm. It ends in the anus, which lies in the perineum. The circular muscular layer of the wall of the anal canal, the internal anal sphincter, is continuous with that of the rectum. The external anal sphincter is a voluntary muscle located between the anal canal and the skin. Both sphincters aid in keeping the lumen of the anus closed.

The anal columns are folds of mucous membrane overlying the veins in the anal canal. They are connected inferiorly by folds of membrane known as the anal valves. Below these valves the mucous membrane is in transition to the skin. This line of transition is known as the pectinate line.

Male Pelvic Viscera

The pelvic viscera peculiar to the male are the ductus deferens, seminal vesicles and ejaculatory ducts, and prostate.

The **ductus deferens** serves as the passageway for sperm from the testis to the ejaculatory duct. It begins at the tail of the epididymis and travels upward in the spermatic cord to the inguinal ring. From here it passes into the abdomen and the pelvic brim, where it descends medial to the obliterated umbilical artery. It then descends along the back of the bladder to the fundus, passing medially to both the ureter and the seminal vesicles and dilating to form the ampulla (see Fig. 4-16). As it enters the tissue of the prostate it joins with the duct of the seminal vesicle to form the ejaculatory duct, which empties into the prostatic urethra.

The **seminal vesicles** (see Fig. 4-16) are lobulated pouches that secrete a basic component of semen. Lying on the posterior surface of the bladder, they form a V. Their ducts join the ductus deferens to form the ejaculatory ducts.

The **prostate gland** secretes additional fluid components into semen. It lies underneath the bladder (see Figs. 4-13A and 4-16), and its inferior surface rests on the superior fascia of the urogenital diaphragm. Posteriorly the prostate is in contact with the rectal ampulla. The urethra enters the prostate (see Fig. 4-15) at the middle of its base and leaves at its apex. The ejaculatory ducts run through it to reach the prostatic urethra.

The encapsulated prostate consists of four lobes: two lateral, one posterior, and one middle. The posterior lobe lies below the ejaculatory ducts, the middle lobe above. Muscular tissue in the middle of the prostate connects the lobes. A sheath of endopelvic fascia covers the entire gland.

The inferior vesical and middle rectal arteries supply the prostate. Veins drain via the prostatic plexus to the internal iliac vein. The hypogastric plexus innervates the prostate via the prostatic plexus.

Female Pelvic Viscera

The female pelvic viscera are the ovaries, uterus, uterine tubes, and vagina and their associated ligaments (Fig. 4-18).

The **ovaries** are the gonads responsible for the generation of eggs. They lie in the ovarian fossae against the lateral pelvic wall and are covered by a layer of germinal epithelium. The superior pole of

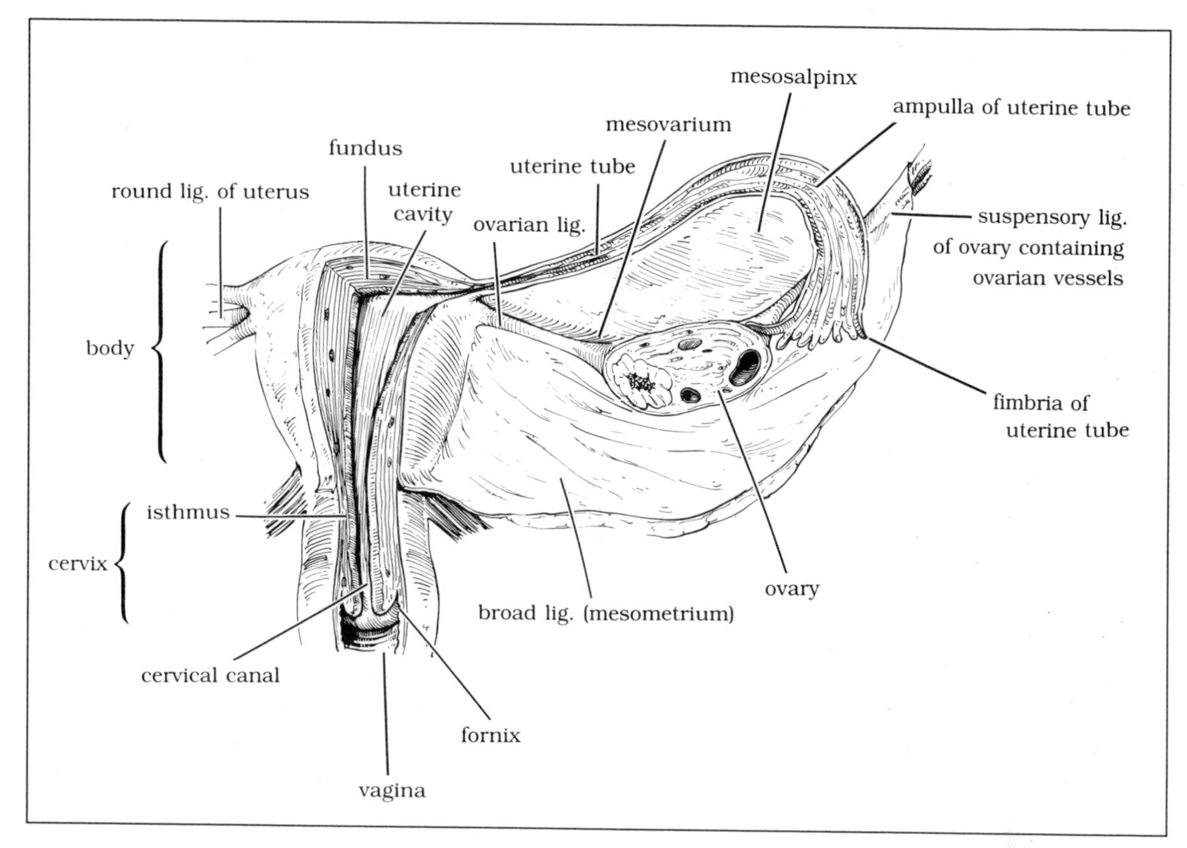

Fig. 4-18. The uterus and ovaries and their mesenteries in relation to each other.

the ovary is covered by the fimbriated portion of the uterine tube. Vessels and nerves reach the ovary through the suspensory ligament of the ovary that passes from the pelvic wall to its superior pole. The ovarian ligament, a fold of the peritoneum, arises from the inferior pole of the ovary and attaches the ovary to the uterus.

The ovarian artery is a branch of the abdominal aorta. The right vein of the ovary drains directly into the inferior vena cava, the left vein into the left renal vein.

The **uterine tube** serves as the passageway for eggs from the ovary to the uterus. Extending from the superior angle of the cavity of the uterus, the tube wraps around the ovary along its mesovarian border to reach the superior region of the ovary. Here it turns backward to cover the superior pole. The ovarian end of the uterine tube is open to the abdominopelvic cavity.

The uterine tube has three parts: a narrow isthmus at the uterus, a long and wide ampulla, and the fimbriated infundibulum at the ovary (see Fig. 4-18). The fimbriae are fingerlike projections covering part of the ovary. They catch the egg as it is released by the ovary and sweep it up into the uterine tube for passage to the uterus.

The **uterus** is a thick-walled muscular organ for the development of the fertilized egg. It normally rests horizontally on top of the bladder, thereby creating an angle between it and the vagina of 100 to 110 degrees.

The uterus is composed of four regions: the fundus, body, isthmus, and cervix. The fundus is the top of the uterus, lying above the entrance of the uterine tubes. The body is the main area of

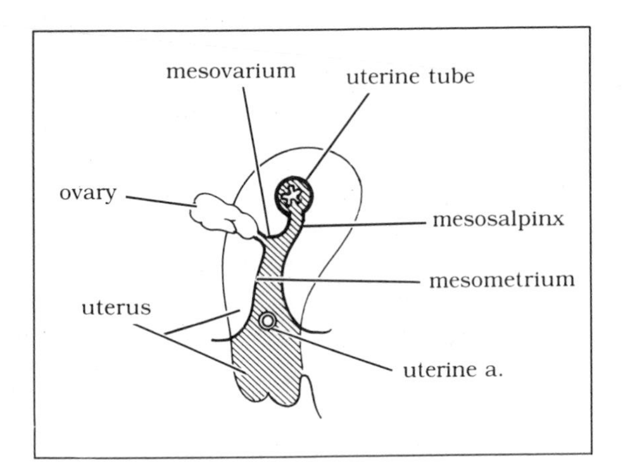

Fig. 4-19. Schematic parasagittal section of the broad ligament and its related structures.

the uterus. The various tubes and ligaments of the female pelvis attach along its lateral walls. The isthmus is a short, constricted region connecting the body with the cervix. Finally the cervix is the inferior end of the uterus; past it lies the vagina. The canal of the cervix is the narrow opening from the interior of the uterus to the vagina.

The uterus receives blood from both the uterine and ovarian arteries. The uterine artery, a branch of the internal iliac artery, also gives off a vaginal branch to the vagina. The veins follow the arteries, draining into the internal iliac vein. The uterus is innervated via fibers from the inferior hypogastric plexus.

The **vagina,** posterior and inferior to the bladder and urethra, is a corridor from the vestibule to the uterus. Superiorly it encloses the bottom of the cervix, attaching above the inferior margin of the cervix. This space between the inferior margin of the cervix and the line of vaginal attachment is the fornix. There are two lateral fornices and anterior and posterior fornices. The vaginal artery is a branch of the internal iliac artery. The vagina also receives blood from the uterine and middle rectal arteries.

The **broad ligament of the uterus** (Fig. 4-19; see also Fig. 4-18) is a double-layered fold of peritoneum. It extends from the uterus and pelvic floor to the lateral walls of the pelvis, covering and containing the uterus, uterine tubes, ovaries, ovarian ligament, and round ligament of the uterus. The uterine tube lies in its free upper edge, in a subdivision known as the mesosalpinx. In addition there is an extension out of the plane of the broad ligament that attaches to the ovaries. This extension is known as the mesovarium. The remainder of the broad ligament is known as the mesometrium.

BLOOD SUPPLY

The internal iliac artery arises at the bifurcation of the common iliac artery (Fig. 4-20 and Table 4-2). It supplies blood to the pelvis, gluteal region, medial thigh, and perineum. Although the internal iliac artery is usually considered to have an anterior and a posterior division, its order of branching is highly irregular.

The posterior division is entirely somatic in distribution, supplying the pelvic wall and gluteal region. The branches of the posterior division are the iliolumbar, lateral sacral, and superior gluteal arteries. The somatic branches of the anterior division are the obturator, internal pudendal, and inferior gluteal arteries. The visceral branches are the umbilical, inferior vesical, middle rectal, uterine, and vaginal arteries.

Arteries supplying the pelvis include the following.

The **iliolumbar** artery supplies the iliac fossa.

Table 4-2. Anastomoses of the Internal Iliac Artery in the Pelvis

Anastomoses
Posterior division
Lateral sacral to medial sacral
Internal pudendal to vaginal
Inferior rectal branches to superior and middle rectal
Anterior division
Middle rectal to superior and inferior rectals and inferior vesical
Vaginal to internal pudendal and uterine
Uterine to vaginal

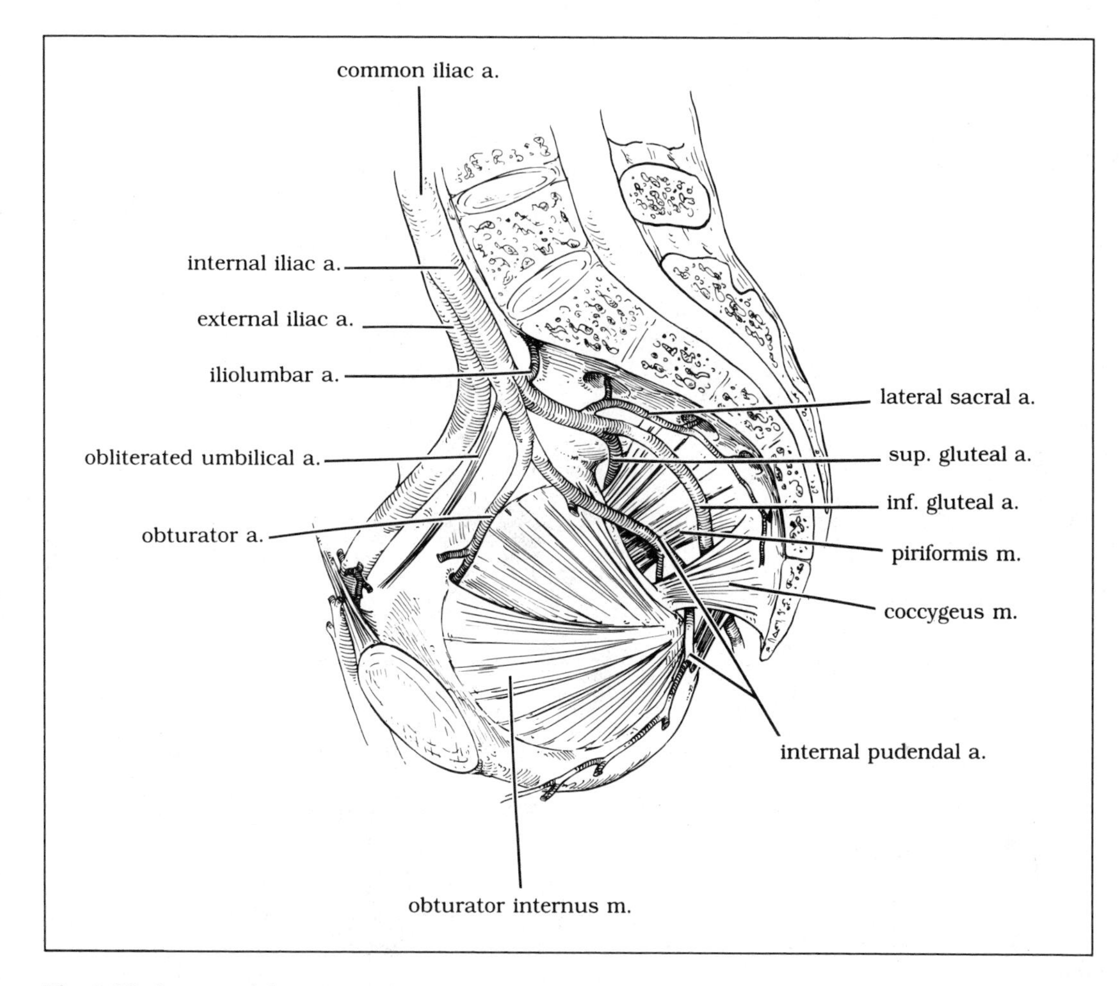

Fig. 4-20. Arteries of the pelvic wall.

The **lateral sacral artery** splits into superior and inferior branches. It supplies the region of the sacrum and anastomoses with the medial sacral artery.

The **superior gluteal artery** exits the pelvis through the greater sciatic foramen above the piriformis muscle. It supplies blood to all three gluteal muscles.

The **obturator artery** exits the pelvis through the obturator canal. It divides into anterior and posterior divisions on entering the medial thigh, which it supplies.

The **internal pudendal artery** often arises along with the inferior gluteal artery. It leaves the pelvis by passing below the piriformis muscle and through the greater sciatic foramen. It then passes over the ischial spine to reach the perineum, which it supplies, by passing through the lesser sciatic foramen.

The **inferior gluteal artery** leaves the pelvis along with the internal pudendal artery. In the gluteal region it supplies the gluteus maximus muscle and the structures around the ischial tuberosity.

The **umbilical artery** serves in the fetus to provide blood to the placenta. In the adult it forms the fibrotic medial umbilical ligament on the abdominal wall. Prior to reaching the wall,

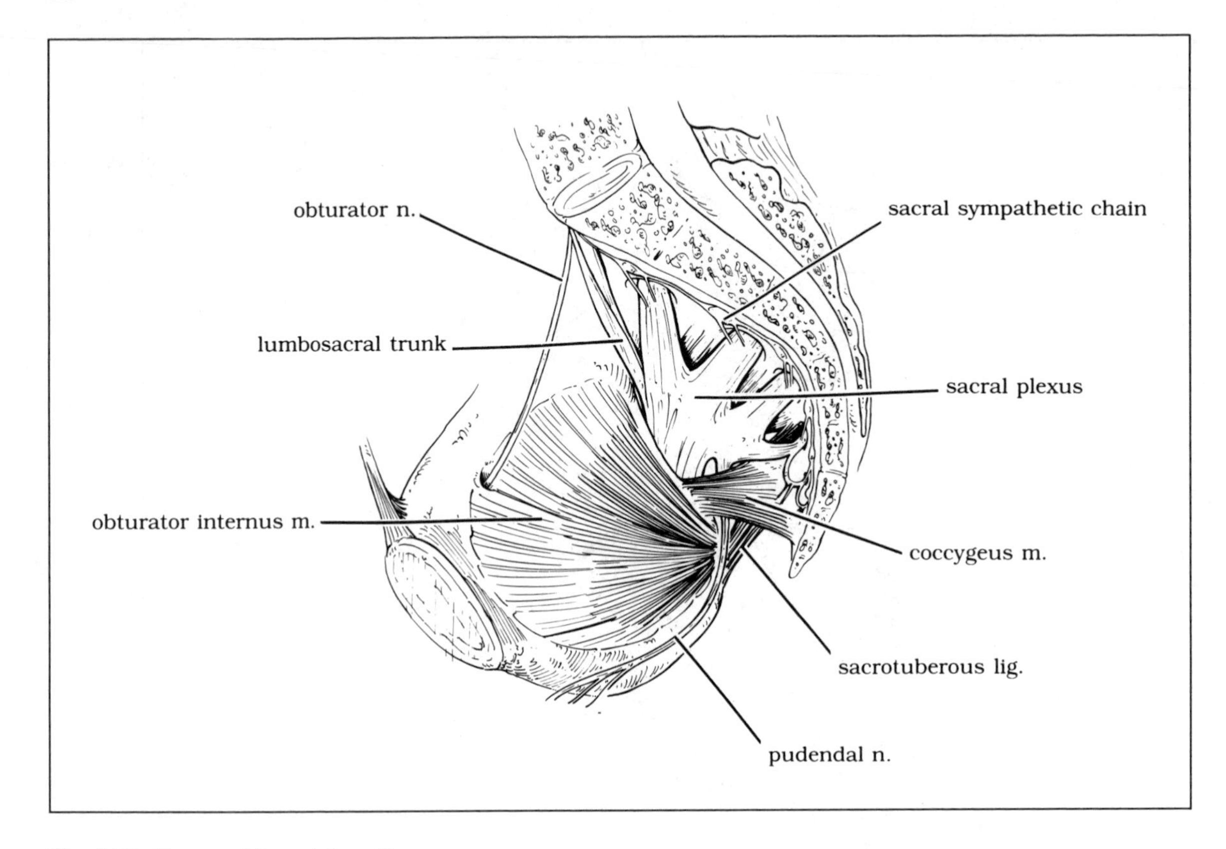

Fig. 4-21. Nerves of the pelvic wall.

however, it gives off the **superior vesical artery,** which supplies the bladder.

The **deferential artery** arises from either the superior or the inferior vesical artery.

The **inferior vesical artery** supplies the fundus of the bladder. It may arise along with the middle rectal artery.

The **middle rectal artery** supplies the rectum and the prostate and seminal vesicles in the male and the vagina in the female. It anastomoses with both the superior and the inferior rectal arteries and the inferior vesical artery.

The **uterine artery** runs through the broad ligament alongside the uterus. It also gives a branch to the vagina.

The **vaginal artery** supplies the vagina via anterior and posterior divisions and also supplies the bulb of the vestibule, bladder, and rectum. It anastomoses with the uterine and internal pudendal arteries.

The **pelvic veins** usually follow the course of their corresponding arteries. In addition, surrounding each of the pelvic viscera are the plexuses: the rectal, vesical, prostatic, uterine, and vaginal plexuses.

INNERVATION

The **sacral sympathetic nerves** supplying the pelvis are a continuation of the lumbar chain that descends on the sacrum and coccyx (Fig. 4-21). They distribute via the superior hypogastric plexus to the inferior hypogastric plexus. These two plexuses are connected by the hypogastric nerves, which also supply the colon, ureter, testis or ovary, and vessels. The inferior hypogastric plexus distributes, via secondary plexuses, to the rectum, bladder, and ductus deferens and prostate or uterus and vagina.

The parasympathetic supply to the pelvis is via the pelvic splanchnic nerves. These distribute through the inferior hypogastric plexus to the pelvis, perineum, and those abdominal viscera that receive their blood from the inferior mesenteric artery.

The following **somatic nerves** of the region constitute the sacral plexus, ventral rami of L4 to S4. These nerves are formed on the dorsal wall of the pelvis and converge toward the greater sciatic foramen. L4 and L5 constitute the lumbosacral trunk.

The primary nerve of the sacral plexus is the sciatic nerve (L4–S3). This nerve, really the tibial and common peroneal nerves combined, leaves the pelvis through the greater sciatic foramen below the piriformis muscle. It supplies much of the lower limb.

The nerve to the piriformis (S1 and S2) supplies the muscle of the same name.

The nerves to the coccygeus and levator ani (S3 and S4) supply the pelvic diaphragm.

The superior gluteal nerve (L4–S1) exits the pelvis through the greater sciatic foramen above the piriformis muscle. It supplies the gluteus medius, gluteus minimus, and tensor fasciae latae muscles.

The inferior gluteal nerve (L5–S2) leaves the pelvis through the greater sciatic foramen below the piriformis muscle. It supplies the gluteus maximus muscle.

The nerve to the quadratus femoris (L4–S1) exits through the lower part of the greater sciatic foramen. It supplies the muscle of the same name and the inferior gemellus muscle. It also supplies articular innervation to the hip.

The nerve to the obturator internus (L5–S2) exits through the lower part of the greater sciatic foramen to supply this muscle and the superior gemellus muscle.

The pudendal nerve (S2–S4) accompanies the internal pudendal artery to the perineum.

LYMPHATICS

Lymph from the anterior and lateral pelvic wall drains to the external iliac lymph nodes. The pelvic viscera and the remaining pelvic wall drain to either the internal iliac or the sacral lymph nodes. The internal iliac nodes, lying at branch points of the artery, also receive lymph from the gluteal region and thigh. These nodes pass lymph to either the external iliac nodes or the common iliac nodes. The sacral lymph nodes lie along the lateral sacral arteries. They receive drainage from the posterior pelvic wall, rectum, and prostate or uterus and vagina. They drain to the common iliac nodes.

National Board Type Questions

Select the one best response for each of the following.

1. Since carcinoma of the cervix invades nearby structures, a patient with early invasion would be most likely to show signs due to obstruction of the
   A. Ovarian artery
   B. Rectum
   C. Urethra
   D. Ureter
   E. Inferior vena cava
2. All of the following nerves supply the skin of the penis except the
   A. Ilioinguinal
   B. Genitofemoral
   C. Pudendal
   D. Perineal
   E. Dorsal nerve of the penis
3. The prostate gland
   A. Surrounds the urethra
   B. Releases secretions via the ejaculatory duct
   C. Contains the voluntary sphincter of the urethra
   D. Is connected to the urinary bladder by a mesentery
   E. Stores sperm

Select the response most closely associated with each numbered item. (The headings may be used once, more than once, or not at all.)

A. Round ligament of the uterus
B. Broad ligament
C. Mesosalpinx
D. Suspensory ligament of the ovary
E. Mesovarium

4. Transmits the ovarian artery and vein
5. Passes through the deep (abdominal) inguinal ring
6. Covers the uterus and ovarian ligament

Select the response most closely associated with each numbered item.

A. Levator ani
B. External anal sphincter
C. Both
D. Neither

7. Composed of smooth muscle
8. Acts on the anal canal
9. Innervated by the pudendal nerve

For the following, select

A. if only *1, 2, and 3* are correct
B. if only *1 and 3* are correct
C. if only *2 and 4* are correct
D. if only *4* is correct
E. if *all* are correct

10. Muscles meeting at the central tendon of the perineum (perineal body) include the
    1. Superficial transverse perineal muscle
    2. External anal sphincter
    3. Bulbospongiosus muscle
    4. Pubococcygeus
11. Contents of the deep perineal space include the
    1. Sphincter urethrae muscle
    2. Bulbocavernosus
    3. Deep transverse perineal muscle
    4. Corpus spongiosum
12. The ischiorectal fossa is bounded by the
    1. Obturator internus muscle
    2. Levator ani muscle
    3. Anal canal
    4. Gluteus maximus muscle
13. The pelvic brim (terminal line) includes the
    1. Pubic crest
    2. Arcuate line of ilium
    3. Pecten pubis
    4. Iliac crest of ilium
14. Which of the following structures, because of their close relationship to the anal canal, can be palpated on rectal examination?
    1. Prostate gland
    2. Puborectal sling
    3. External anal sphincter
    4. Cervix uteri
15. The anal canal
    1. Is surrounded by the internal anal sphincter
    2. Has folds of mucous membrane that cover veins
    3. Is marked by the pectinate line, which designates the transition from gut (visceral nerves) to skin (somatic nerves)
    4. Has venous drainage to both the portal and systemic circulations

Annotated Answers

1. D. A very important relationship exists at the caudal end of the uterus. Just lateral to the cervix, in the base of the broad ligament, the ureter passes from dorsal to ventral, inferior to the uterine artery.
2. D. This question is difficult but helps point out that the pudendal nerve gives rise to the dorsal nerve of the penis, which innervates the skin of the penis, and the perineal nerve, which innervates the scrotum but not the penis.
3. A. Remember that one part of the urethra is termed the *prostatic urethra.*
4. D. Remember that the ovary is suspended off the dorsal side of the broad ligament, and that for vessels to reach structures suspended in the abdominal cavity they must use folds of peritoneum, in this case the suspensory ligament, as roadbeds.
5. A. The round ligament in the female traces

the path of the male spermatic cord and therefore must pass through the deep inguinal ring.

6. B.
7. D. Both muscles are composed of voluntary (striated) muscle.
8. C.
9. B. The pudendal nerve supplies the external anal sphincter, while the levator ani receives fibers directly from pelvic ventral rami.
10. E. The central tendon serves as an anchoring point for all of these muscles.
11. B.
12. E. The ischiorectal fossa, a fat-filled space in the pelvis, is bounded by all of the structures listed and therefore has important relationships to them.
13. A.
14. E. Not all of these structures, of course, could be palpated in one individual, but all of them are in close relationship to the anal canal and therefore can be palpated.
15. E. The venous drainage (as well as arterial supply) follows two important pathways. These pathways are of clinical significance because of potential anastomotic connections and presentations of signs (hemorrhoids) that can indicate disease processes elsewhere in the abdominal cavity.

5 Upper Limb

Objectives

After reading this chapter, you should know the following:

Surface anatomy of the upper limb

Bones, joints, and articulations of the upper limb and shoulder

Musculature and movements of the back and the scapular, pectoral, and axillary regions

Structure and function of the mammary glands

Organization, components, and functions of the brachial plexus

Pathway and importance of axillary lymphatics

Collateral circulation throughout the upper limb

Musculature and movements of the arm, forearm, hand, and fingers.

Effects of injuries to the nerves of the upper limb

Blood supply, lymphatics, and innervation of the upper limb

The upper limb includes the clavicle, scapular region, pectoral muscles, axilla, arm, forearm, and hand. In addition many of the superficial muscles of the neck and back are involved in movements of the upper limb and are therefore considered in this chapter.

Surface Anatomy

The clavicle is palpable and its projection is visible anteriorly at the root of the neck (see Fig. 2-1). It begins at the medial jugular notch and extends laterally to the shoulder, where it meets the acromion of the scapula. Lying 2 cm below the clavicle in the deltopectoral triangle, the coracoid process is palpable.

The acromion represents the termination of the spine of the scapula where it becomes free of the blade. The spine extends from the acromion inferomedially across the back, ending at the medial border of the scapula. The inferior angle of the body of the scapula lies at the level of T7 with the arms at the side (Fig. 5-1). It is covered by the latissimus dorsi muscle, which forms the posterior axillary fold.

Overlying portions of the scapula and latissimus dorsi muscle is the trapezius muscle, which forms the posteriorly viewed curve of the neck. The curve of the shoulder is formed by the deltoid muscle (Fig. 5-2).

Moving anteriorly the upper portion of the thoracic wall is covered by the pectoralis major muscle, which constitutes the anterior axillary fold. This muscle is often visible in the male (see Fig. 2-1A), but in the female it is obscured by breast tissue (see Fig. 2-1B). On the arm the biceps muscles form the anterior mass, with the triceps muscle forming the posterior mass (see Fig. 5-1). Often visible on the arm are the basilic vein (medial) and the cephalic vein (lateral).

At the elbow the palpable bony structures are the humeral epicondyles, the head of the radius

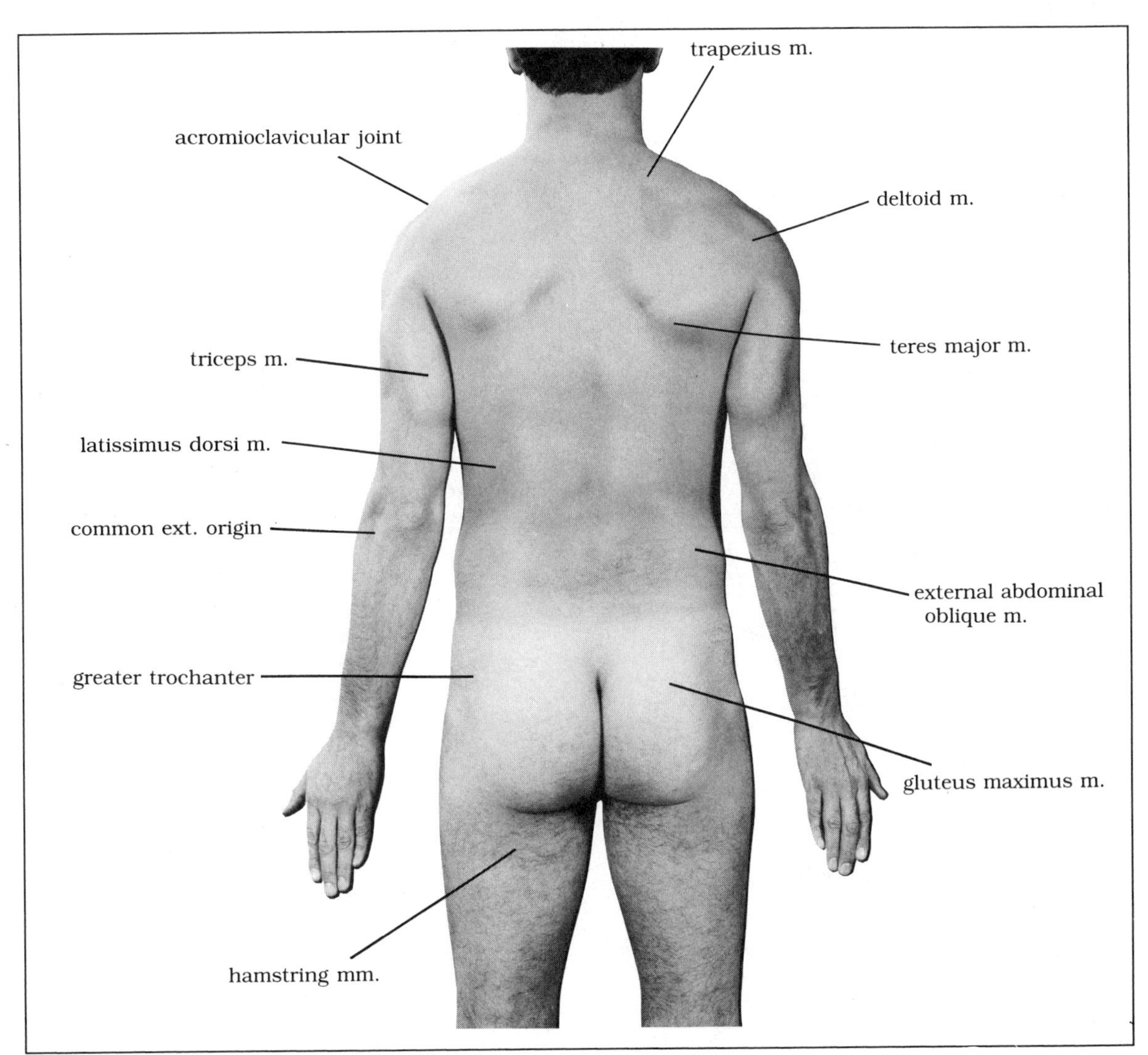

Fig. 5-1. Surface anatomy of the back and upper limb.

(below the lateral epicondyle), and the olecranon. The cubital fossa is the anterior pit at the elbow. It is crossed by the median cubital vein.

In the forearm the flexor muscle mass is visible medially, with the extensor muscle mass visible laterally. Posteromedially the ulna becomes subcutaneous, ending at the wrist in the palpable styloid process. The radius is also subcutaneous for its lower half and ends at the wrist in its styloid process.

The wrist begins at a line connecting the two styloid processes. The distal radius forms the predominant articulation with the carpal bones, the disk of the distal radioulnar articulation having a smaller contribution. Extension of the thumb causes the tendons of the extensor pollicis longus and brevis muscles to become visible. Lying between these tendons is the anatomical snuff box, at the base of which the pulse of the radial artery may be palpated.

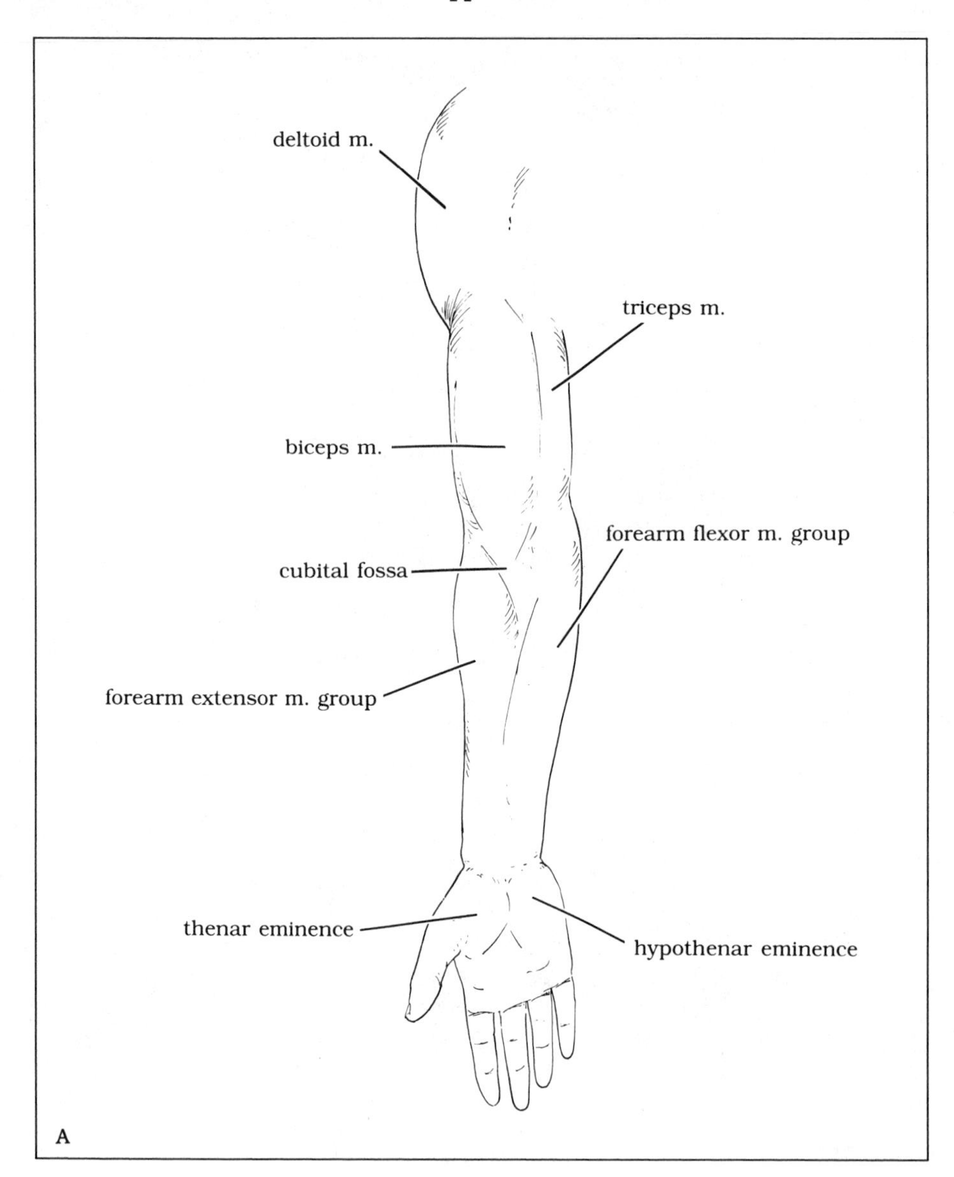

Fig. 5-2. Surface anatomy of the upper limb.
A. Schematic drawing of anterior upper limb.
B. Arm and hand in flexion.

On the anterior surface of the wrist, flexion of the hand causes the flexor carpi ulnaris tendon to become visible. This tendon ends at the pisiform bone. Flexion of the hand combined with opposition of the thumb and fifth digit brings out the palmaris longus tendon. The scaphoid bone lies just lateral to this tendon.

The dorsum of the hand is covered by loose skin, under which the metacarpal and phalangeal bones are palpable. The visible knuckles represent the distal ends of the proximal bones.

The palm is covered by thick skin that has no hair or sebaceous glands and little fat. The trans-

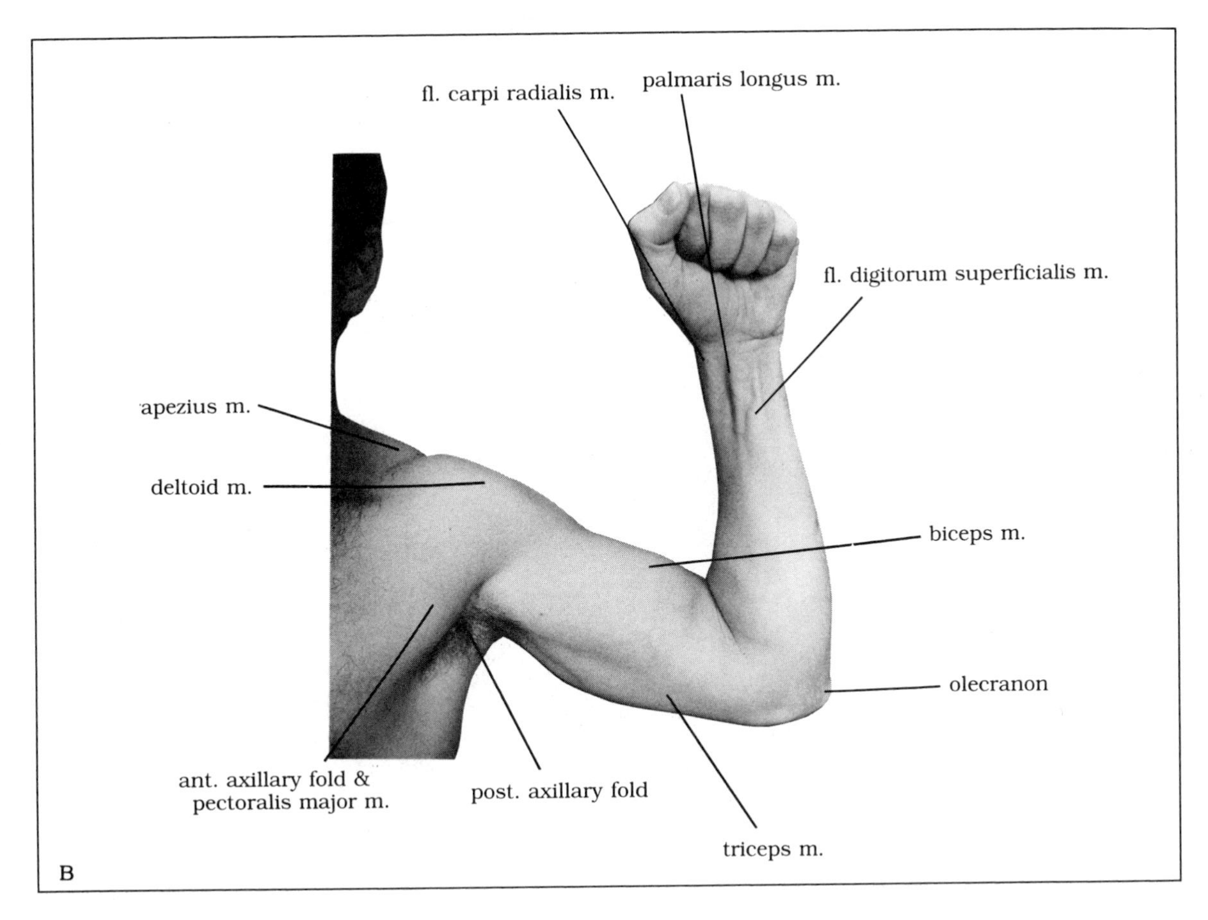

Fig. 5-2 (continued).

verse creases in the middle of the palm mark the metacarpophalangeal joints.

On the digits the proximal crease is just distal to the metacarpophalangeal joint. The middle crease is at the level of the joint between proximal and middle phalanges. The distal crease is proximal to the joint between the middle and distal phalanges.

Bones

CLAVICLE

The clavicle connects the upper limb to the body. Its S shape increases its resiliency and makes the shoulder stand out from the chest (Fig. 5-3). The proximal articulation of the clavicle is the sternoclavicular joint, which has the capabilities of a ball-and-socket joint in allowing the upper limb to have great freedom of movement (see discussion of joints at end of this chapter). The distal articulation is with the scapula at the acromioclavicular joint (Fig. 5-4). The sternoclavicular joint always includes an articular disk between the sternum and clavicle. The acromioclavicular joint usually includes an incomplete disk between the clavicle and acromion.

The medial third of the clavicle lies in close proximity to the first rib. It is attached to this rib by the costoclavicular ligament, which arises from the costal tuberosity on its inferior surface. The sternocleidomastoid muscle ends on this portion of the clavicle and one of the origins of the pectoralis major muscle is here.

The middle third of the clavicle lies superior

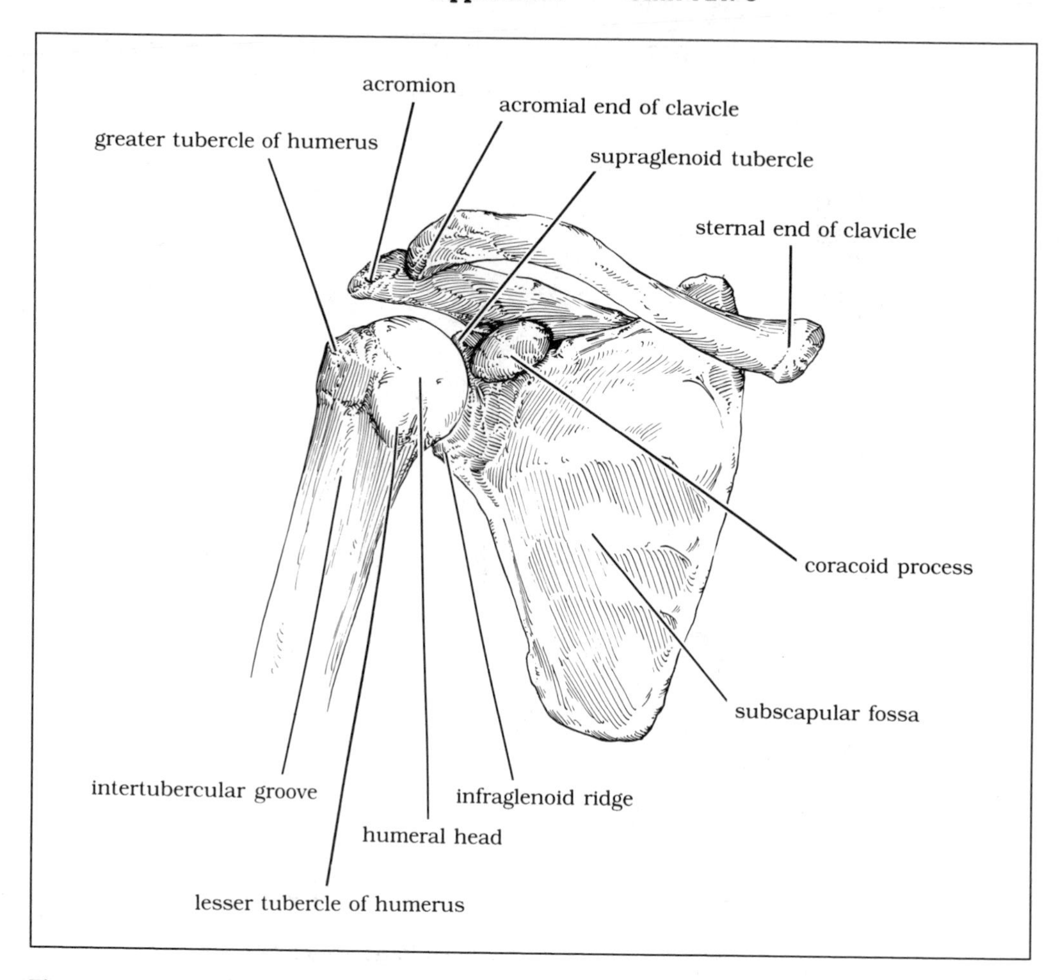

Fig. 5-3. The scapula, clavicle, and humerus. Anterior view.

and lateral to the axilla and its related vessels and nerves. The subclavius muscle inserts on this portion of the clavicle.

The distal third of the clavicle is in proximity to the coracoid process and the acromion of the scapula. The conoid tubercle serves as an attachment for the conoid part of the coracoclavicular ligament. The trapezoid portion of this ligament attaches to the trapezoid line. Fibers of the deltoid muscle arise from the deltoid tubercle on this portion of the clavicle.

SCAPULA

The scapula is a flat, triangular bone overlying the second to the seventh ribs on the dorsum. The body of the scapula is a thin plate having a costal surface and a dorsal surface. The costal surface has a concavity, the subscapular fossa (Fig. 5-5A), in which the subscapularis muscle sits (Fig. 5-6A). In addition the serratus anterior muscle attaches to the costal surface. This surface has no joints or bony contacts with the underlying rib cage.

The dorsum of the scapula is convex and is divided by the scapular spine into supraspinous and infraspinous fossae (Fig. 5-5B). The supraspinous fossa occupies about one-fourth of the dorsal surface and is the point of attachment for the supraspinatus muscle (Fig. 5-6B). The infraspinous fossa occupies the remainder of the dorsum and serves as the origin for the infraspinatus muscle.

As a triangular bone, the scapula has three borders (lateral, medial, and superior) and three angles (lateral, inferior, and superior). The lat-

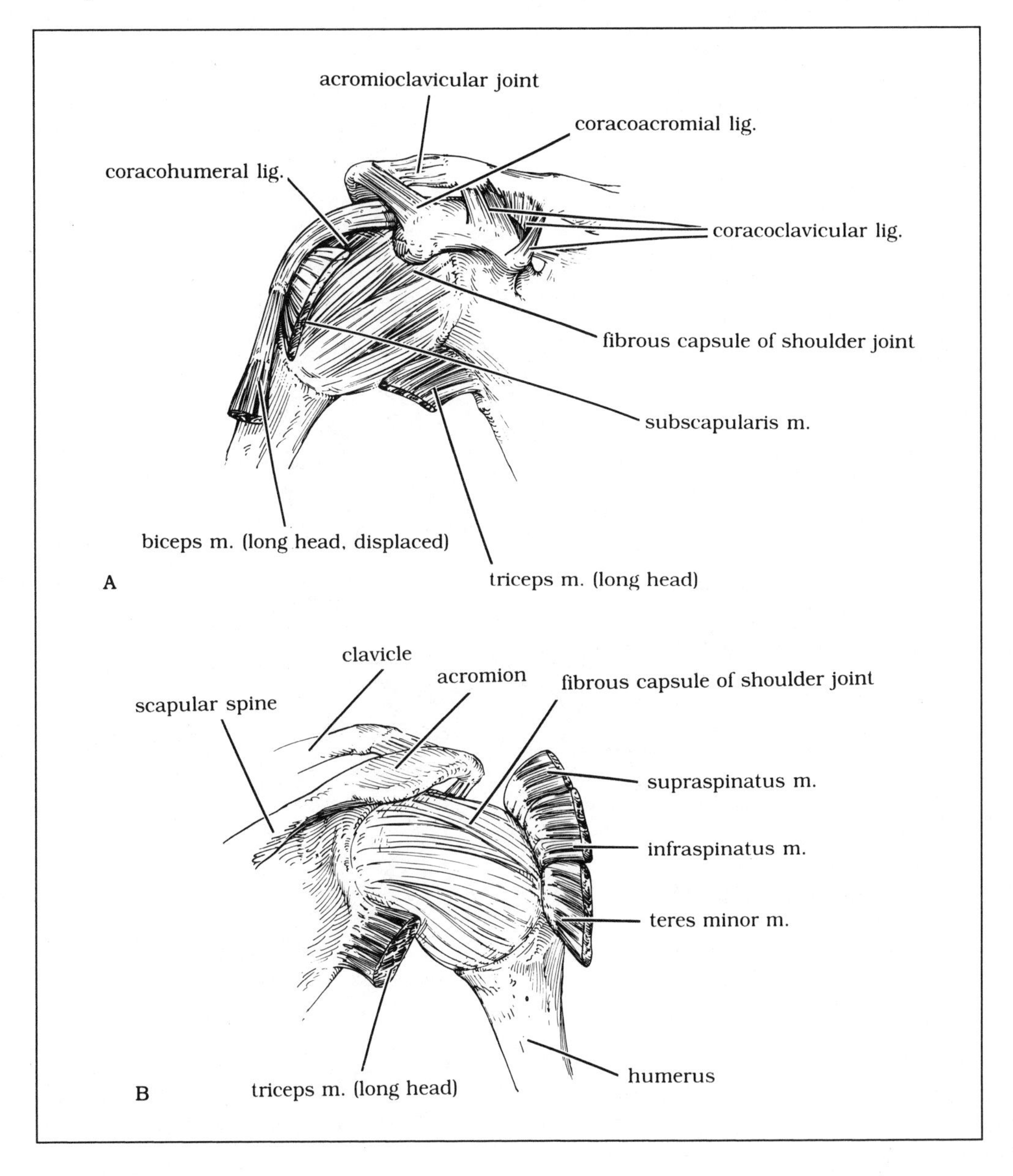

Fig. 5-4. Ligaments and tendons of the scapulohumeral joint. A. Anterior view. B. Posterior view.

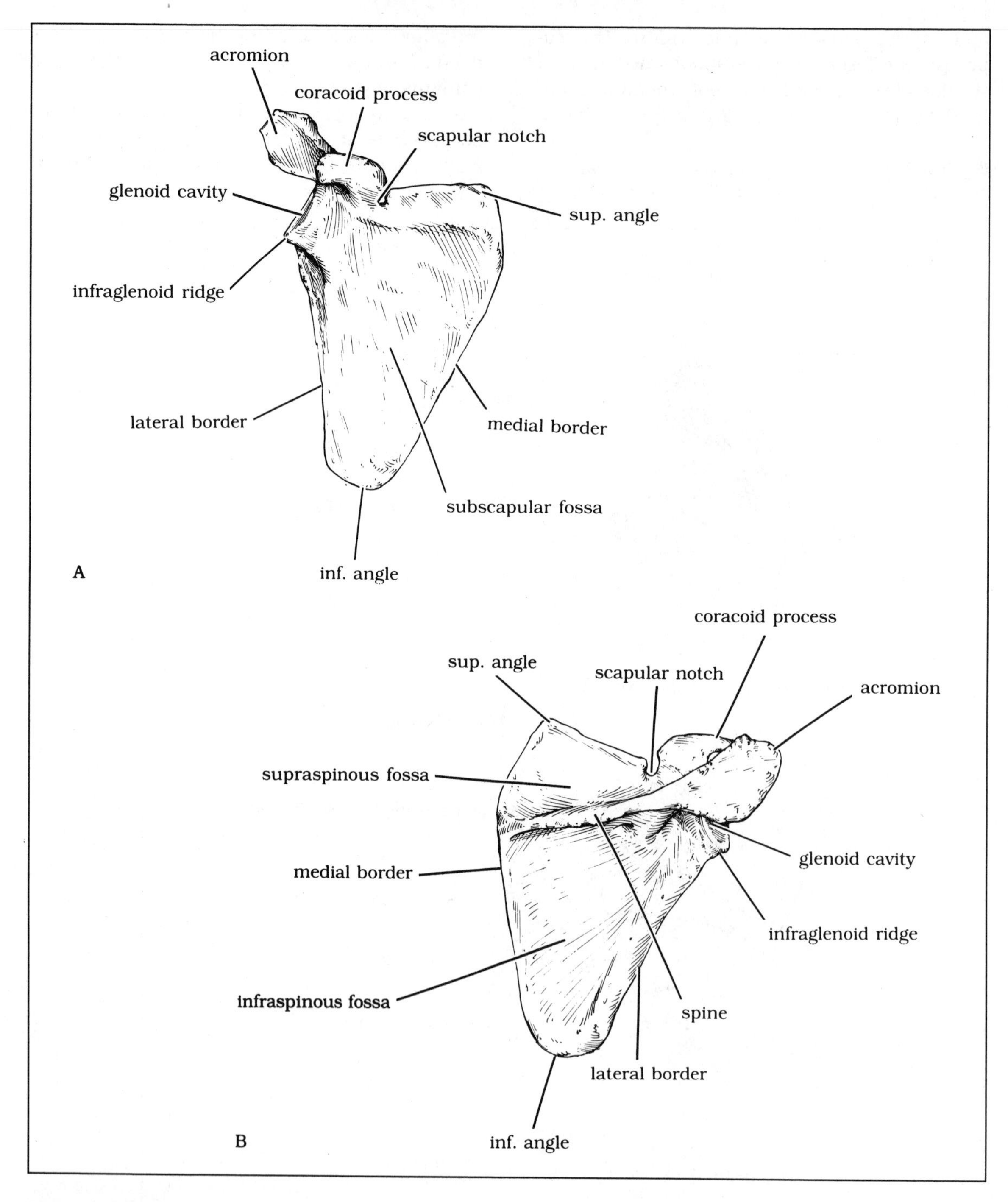

Fig. 5-5. The scapula. A. Anterior costal view.
B. Posterior view.

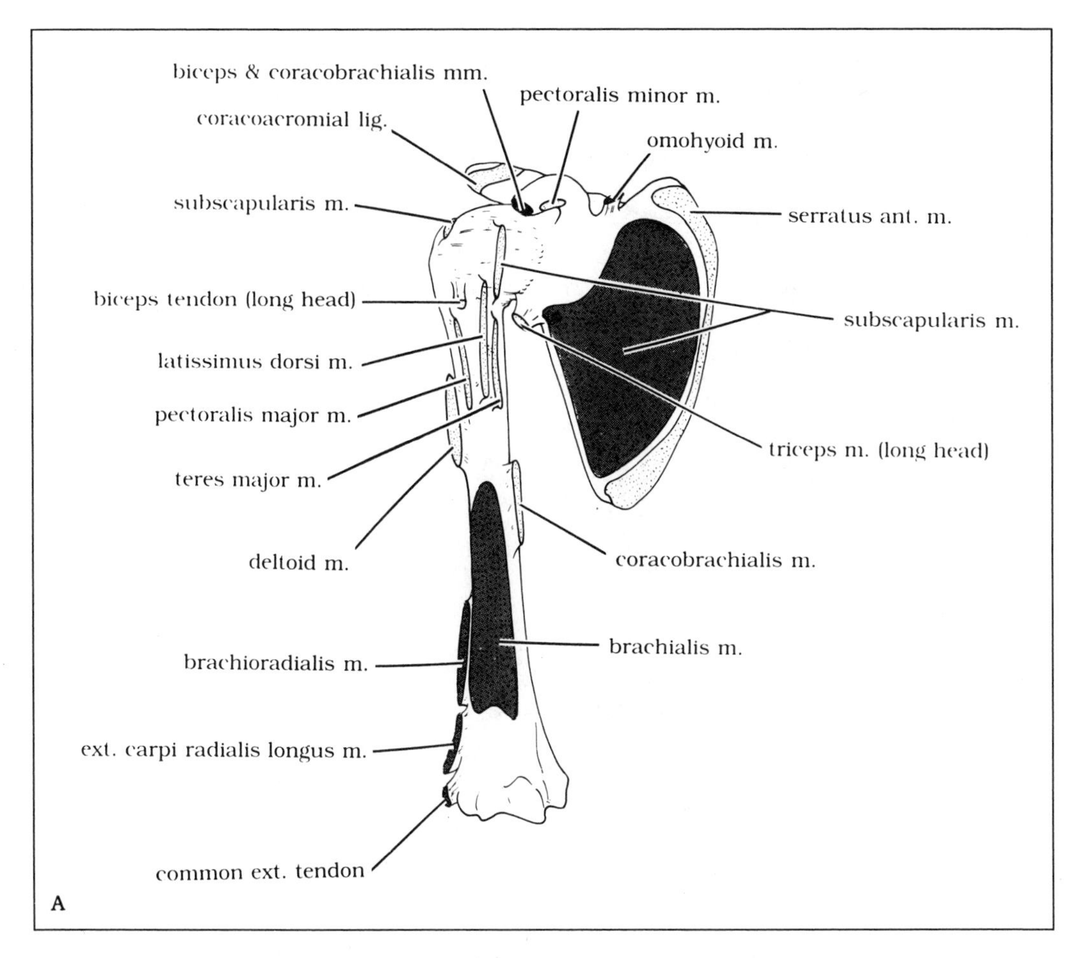

Fig. 5-6. Muscle origins (shaded areas) and insertions on the scapula and humerus. A. Anterior view. B. Posterior view.

eral border of the scapula is its thickest edge (Fig. 5-7). It has a groove in which run the circumflex scapular artery and vein and serves as the origin for the teres major and minor muscles. The superior portion of the lateral border forms the infraglenoid ridge, which serves as the origin of the long head of the triceps muscle.

The medial border of the scapula is its longest border. The spine of the scapula arises from this border approximately one-third of the way down. Above the origin of the spine is the insertion of the levator scapulae muscle; below is the insertion of the rhomboid major muscle. At the base of the spine along the medial border is the insertion of the rhomboid minor muscle (see Fig. 5-6).

The superior border of the scapula is its thinnest and shortest edge. Near its lateral end is a notch, the scapular notch (see Fig. 5-5B), through which the suprascapular nerve passes to reach the supraspinous fossa. This notch is bridged by a transverse ligament, the superior transverse scapular ligament, over which the suprascapular artery passes. Medial to the notch is the origin of the omohyoid muscle (see Fig. 5-6B).

The lateral angle of the scapula is at the junction of the superior and lateral borders. The glenoid cavity is located at the end of the incon-

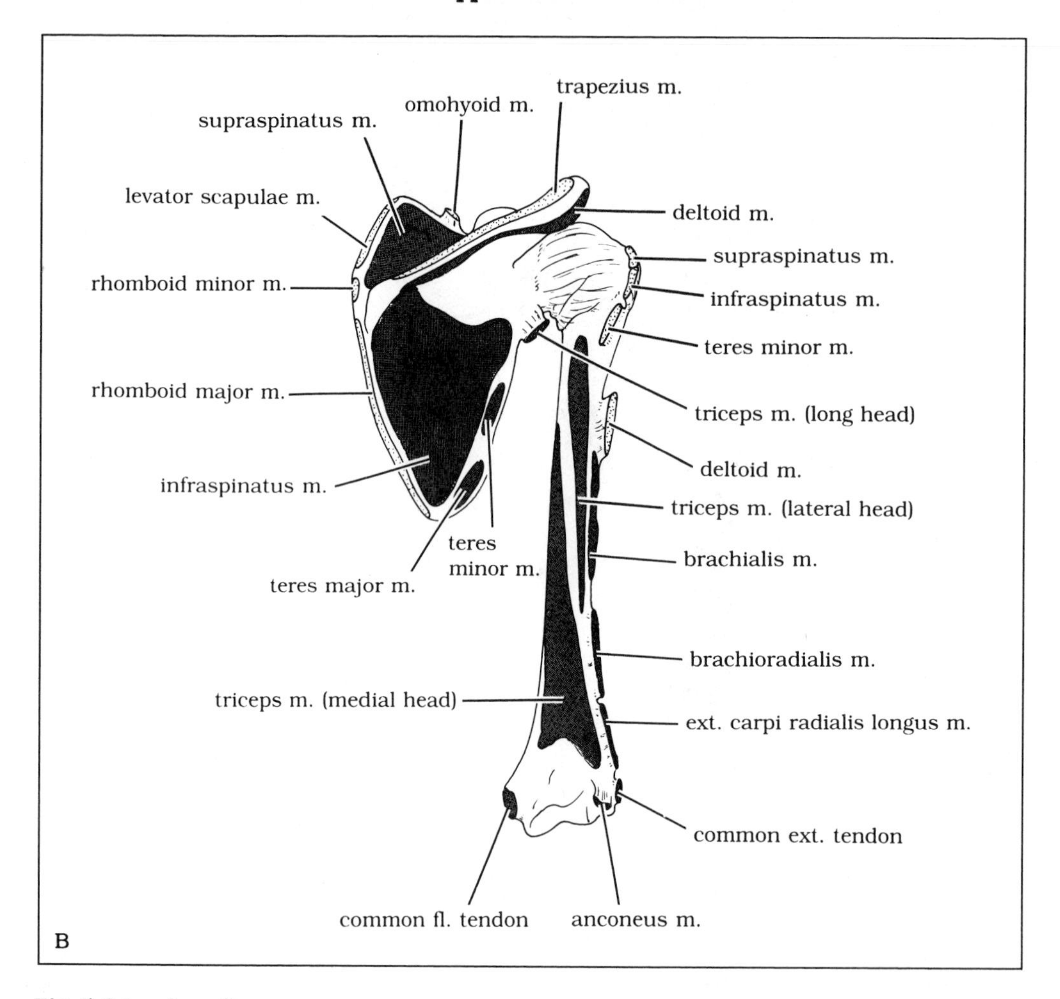

Fig. 5-6 (continued).

spicuous neck of the scapula. The superior angle is at the junction of the superior and medial borders, while the inferior angle is at the junction of the medial and lateral borders. The inferior angle lies at the level of T7.

The spine of the scapula begins at the medial border and extends superolaterally (see Fig. 5-5B). It ends in a notch at the neck of the scapula, just short of the glenoid cavity. The suprascapular vessels and nerves pass under the notch to reach the infraspinous fossa. The posterior edge of the spine, the crest, continues past the notch to become the acromion. The trapezius muscle inserts on this crest, while part of the deltoid muscle arises here (see Fig. 5-6B).

The **acromion** of the scapula forms the point of the shoulder (see Figs. 5-3 and 5-5A). It and the glenoid cavity are the only two articular points for the scapula. The acromion articulates with the clavicle and forms the superior limit of the glenoid articular region (Fig. 5-8). Part of the deltoid muscle originates from the acromion, and the coracoacromial ligament ends here.

The **coracoid process** is an upward, forward, and lateral projection from the scapular neck. The pectoralis minor muscle inserts here, while the short head of the biceps and coracobrachialis

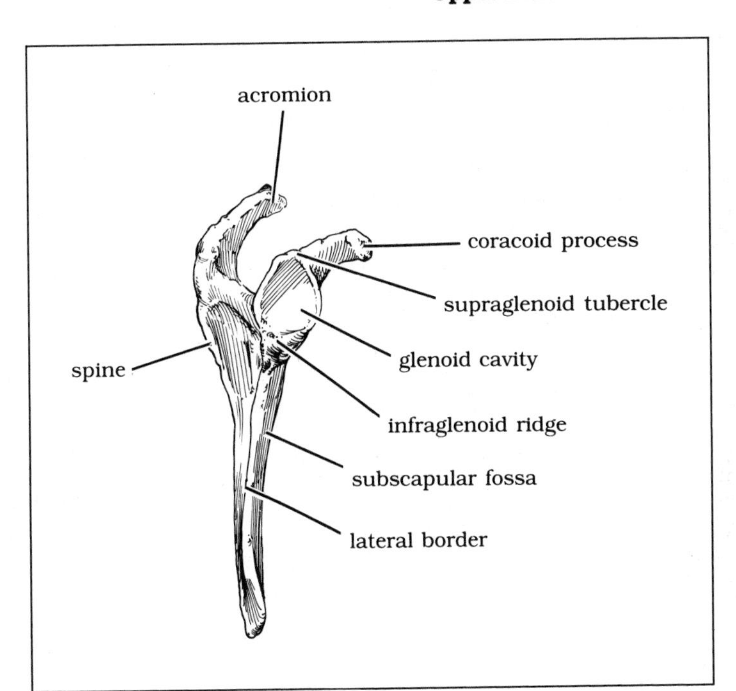

Fig. 5-7. Lateral border of the scapula.

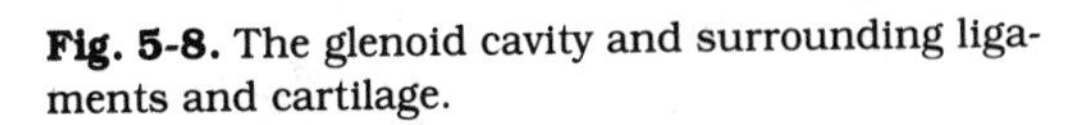
Fig. 5-8. The glenoid cavity and surrounding ligaments and cartilage.

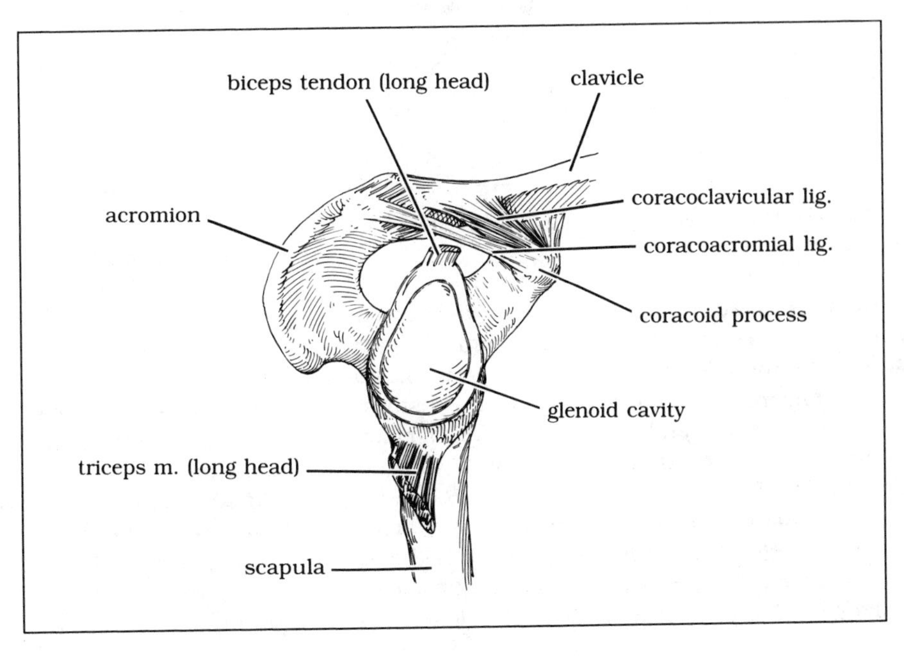

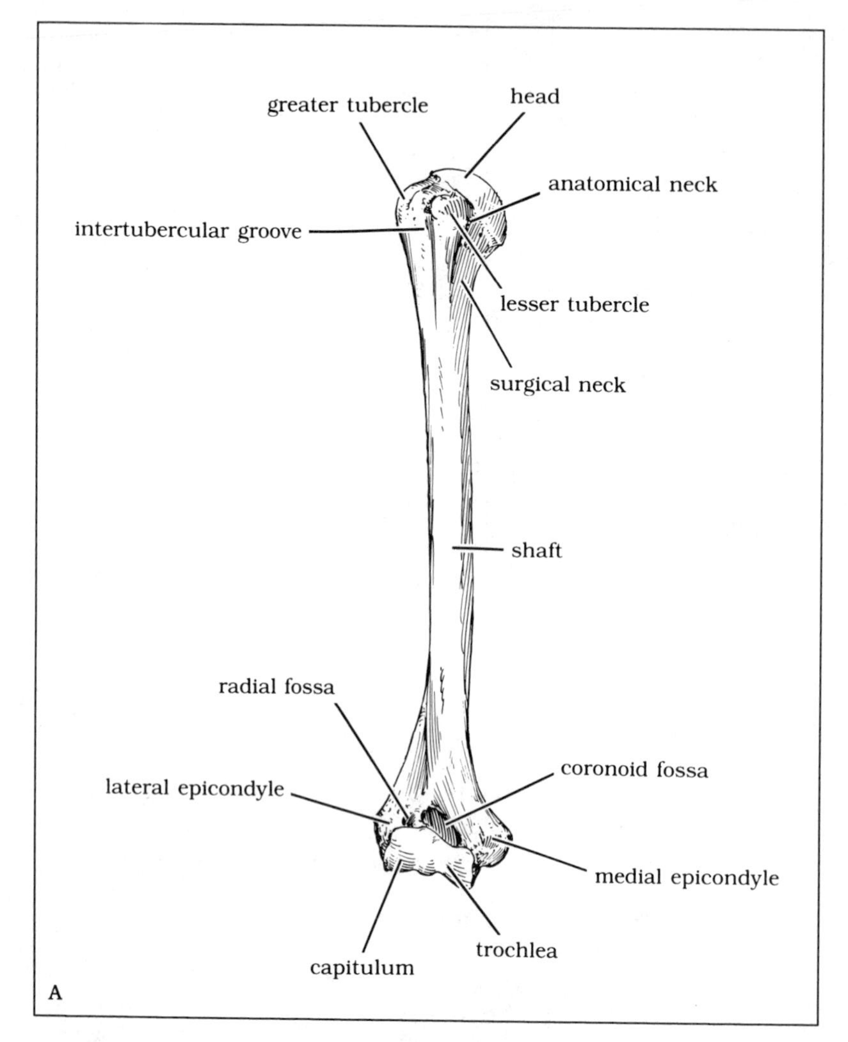

Fig. 5-9. The humerus. A. Anterior view. B. Posterior view.

muscles arise from it (see Fig. 5-6A). The coracoid process also serves as a point of attachment for the following ligaments: superior scapular ligament, coracohumeral ligament, costocoracoid ligament, and coracoclavicular ligament (both conoid and trapezoid portions) (see Figs. 5-4 and 5-8).

The **glenoid cavity** is a shallow fossa at the lateral angle of the scapula. At its top is the supraglenoid tubercle, which is the origin for the long head of the biceps muscle. The glenoid cavity articulates with the head of the humerus (see Figs. 5-3 and 5-6A).

HUMERUS

The humerus is one of the longest bones in the body. It consists of the head, anatomical neck, shaft (body), and epicondyles, and has articular surfaces at both its proximal and distal ends (Fig. 5-9). The humeral head is the smooth hemispheric region that forms the proximal articulation with the glenoid cavity. The anatomical neck is the short indented region at the base

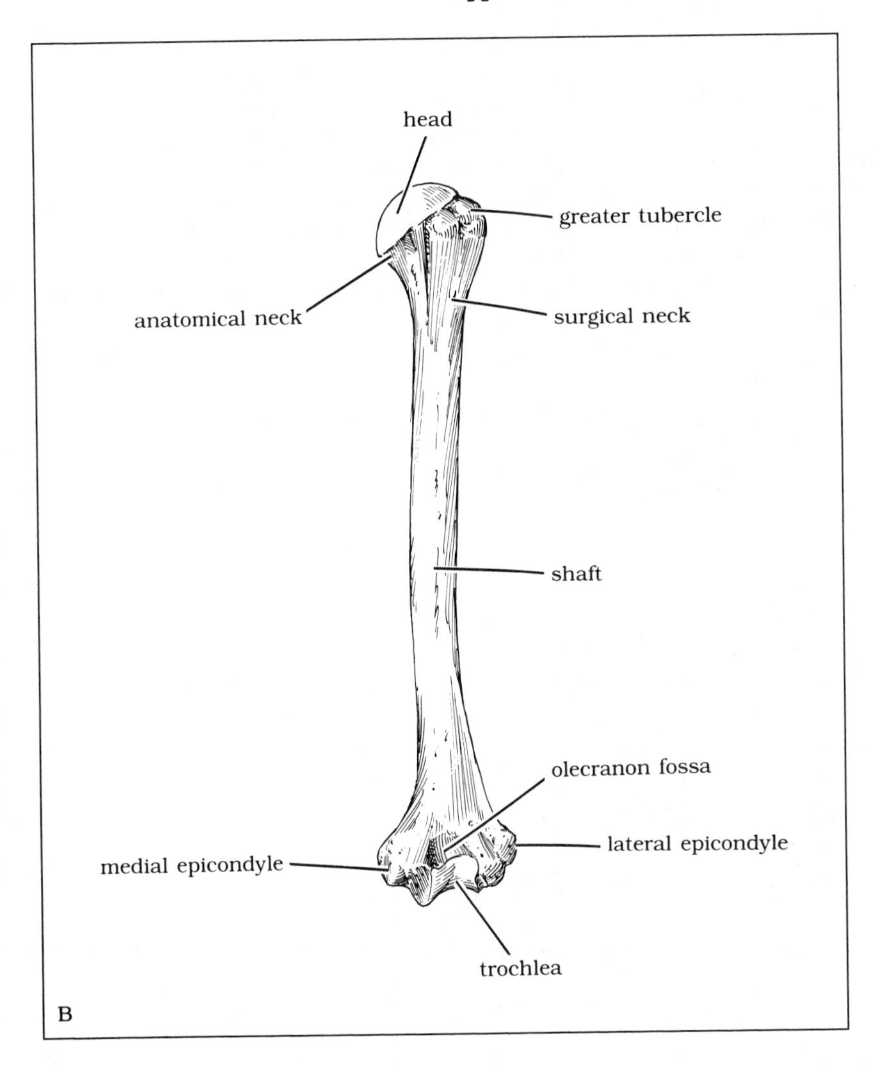

Fig. 5-9 (continued).

of the head; the articular capsule of the shoulder attaches here.

The greater tubercle of the humerus is found at the junction of the anatomical neck and the shaft on the lateral side. It is a large projection, rising almost to the top of the head, and receives the insertion of the supraspinatus, infraspinatus, and teres minor muscles. (See Figure 5-6 for origins and insertions on the humerus.) The lesser tubercle is inferior and anterior to the greater tubercle. The subscapularis muscle inserts here. Both tubercles are prolonged by downward crests. The teres major muscle inserts on the crest of the lesser tubercle, while the pectoralis major muscle inserts on that of the greater tubercle (see Fig. 5-4B). The two tubercles and their crests are separated by the intertubercular groove, in which the tendon of the long head of the biceps muscle lies. In addition the latissimus dorsi muscle inserts here (see Fig. 5-9A).

The surgical neck of the humerus is a narrow region of the shaft just below the tubercles. The deltoid tuberosity is a projection at approximately the middle of the shaft, just below the tubercles on the lateral aspect. The deltoid muscle inserts here.

On the posterior surface of the shaft, the groove for the radial nerve and the deep brachial artery begins at the base of the deltoid tuberosity and runs obliquely downward and forward.

The inferior third of the shaft is flared outward both medially and laterally, forming the medial and lateral supracondylar ridges. They end in their respective epicondyles (see Fig. 5-9).

The deltoid and coracobrachialis muscles insert on the humeral shaft. The brachialis, brachioradialis, and medial and lateral heads of the triceps muscle arise from the shaft.

The **epicondyles** are articular widenings at the distal end of the humerus. Each one is specially adapted for articulation with the ulna (medial epicondyle) or the radius (lateral epicondyle). The medial epicondyle is the larger of the two, projecting above the elbow. It contains a groove for

Fig. 5-10. The radius and ulna. A. Anterior view. B. Posterior view.

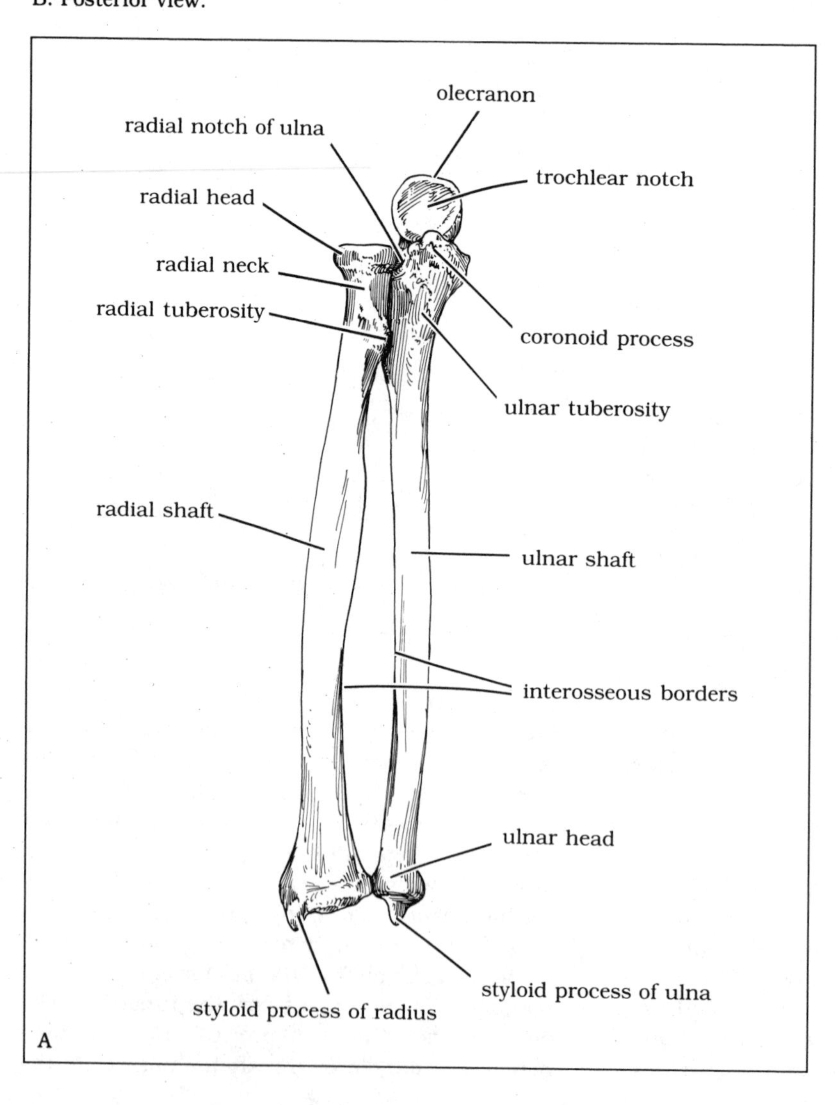

the ulnar nerve. The trochlea, a spool-shaped region of the medial condyle, articulates with the ulna. Above the trochlea are the coronoid fossa anteriorly and the olecranon fossa posteriorly. These fossae allow the bones to interdigitate as the elbow bends.

The lateral epicondyle articulates with the radius. The small, round capitulum, part of the lateral condyle, lies posteriorly to the radial fossa, where the radius comes to lie when the elbow is flexed.

RADIUS AND ULNA (FOREARM)

The forearm contains two long bones: the ulna and radius (Fig. 5-10). (Figure 5-11 details the origins and insertions on these bones.) These

Fig. 5-10 (continued).

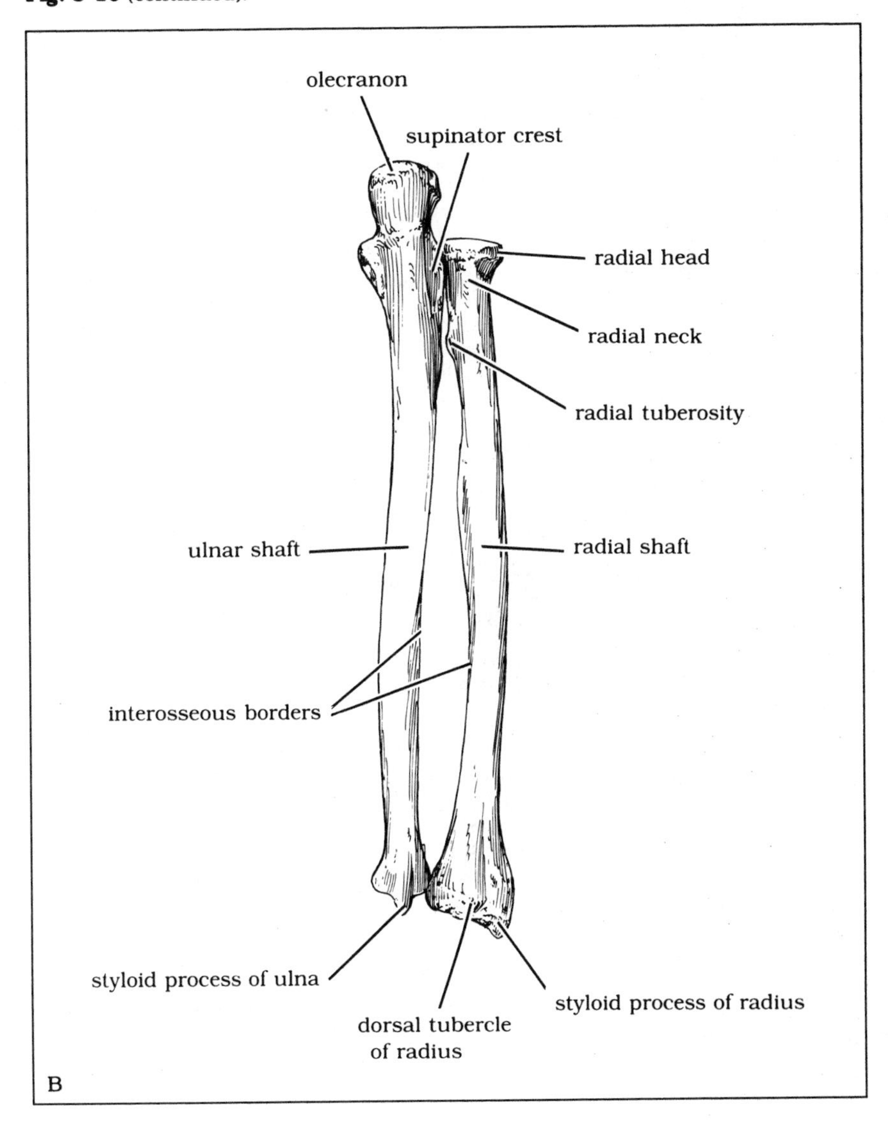

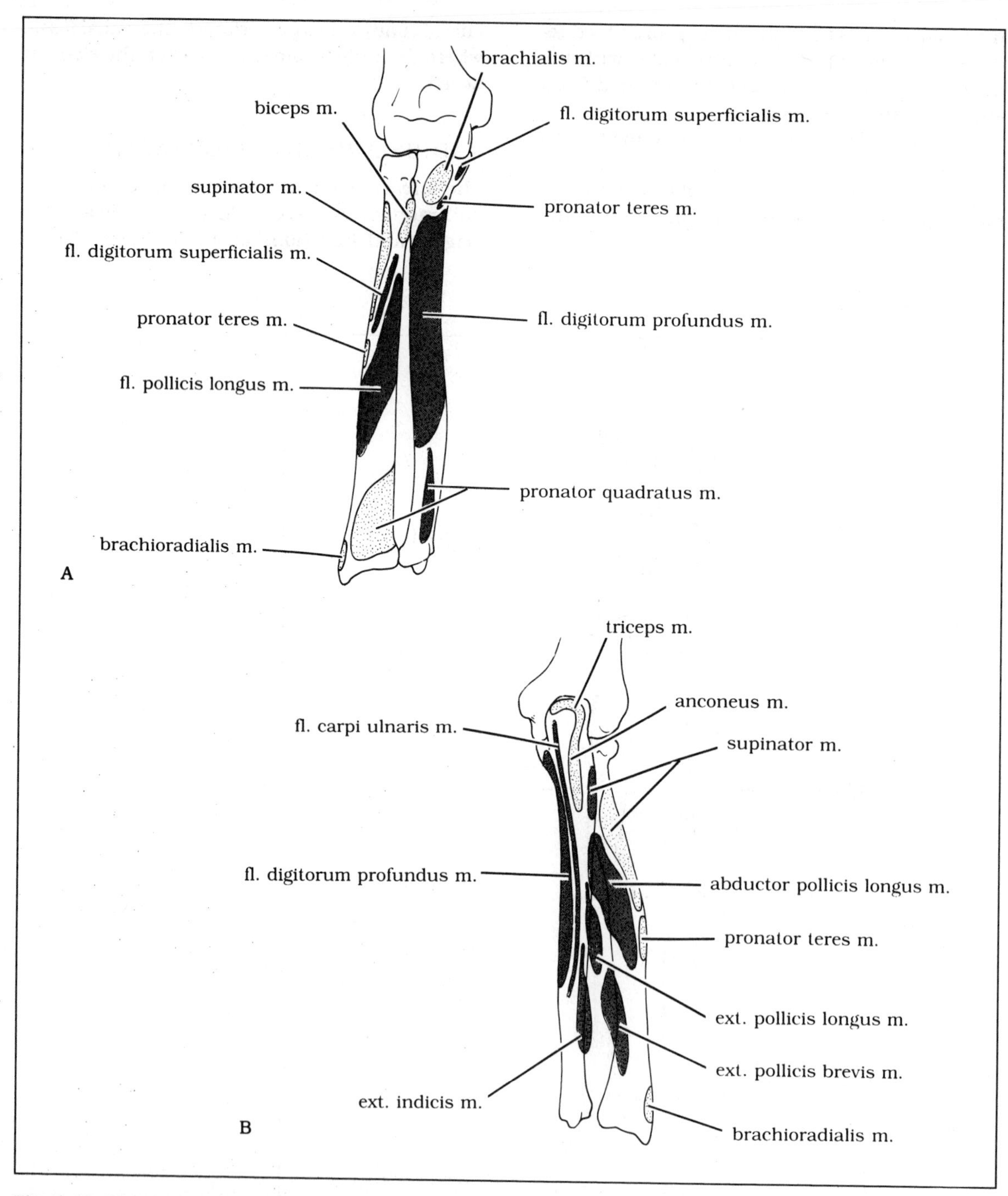

Fig. 5-11. Muscle origins (shaded areas) and insertions on the radius and ulna. A. Anterior view. B. Posterior view.

bones articulate with each other, both proximally and distally, and with the humerus. In addition, the radius articulates with the carpal bones of the wrist. The humeroulnar articulation at the elbow and the radiocarpal articulation at the wrist are generally considered the major proximal and distal articulations of the forearm. These articulations, along with the radioulnar articulations and the muscles that attach the two bones, allow for the great flexibility of motion evinced by the forearm and hand in pronation and supination.

The ulna is the medial long bone of the forearm. The thickened proximal region articulates with the humerus and with the proximal radius. Distally the ulna is reduced, articulating only with the distal radius, not with any of the carpal bones. The proximal end of the ulna has the following bony landmarks: trochlear notch, olecranon, coronoid process, and radial notch. The **trochlear notch** is a depression in the ulna into which the trochlea of the humerus fits. It is divided by two ridges: the first ridge divides the notch into medial and lateral portions, and the second ridge is at the junction of the anterior and the posterior projections from the trochlear notch.

The **olecranon** is the proximal posterior ulnar projection (Fig. 5-12). On its posterior surface it receives the triceps muscle and the capsule of the elbow joint. The anterior surface of the olecranon is the smooth posterior projection of the trochlear notch. When the elbow joint is fully extended the olecranon comes to lie in the olecranon fossa of the humerus.

The **coronoid process** is the proximal anterior ulnar projection. The upper surface of the coronoid process forms the lower anterior part of the trochlear notch. The brachialis muscle inserts on the anterior surface of the coronoid process. At the junction of the coronoid process and the shaft of the ulna is the tuberosity of the ulna, which receives the oblique cord (Fig. 5-13), a ligament that helps to hold the radius and ulna together, and a further insertion of the brachialis. On the lateral side of the coronoid process is the radial notch.

The **radial notch** of the ulna is a shallow depression within which sits the circumference of the radial head. The annular ligament of the radius attaches to the edges of the radial notch (Fig. 5-14). The supinator crest is at the posterior inferior edge of the radial notch where it borders the ulnar shaft. The supinator muscle arises from this crest and an adjacent depression, the supinator fossa.

The shaft of the ulna is thickest proximally. The anterior surface of the shaft is smooth and the posterior surface is partially subcutaneous. The interosseous (lateral) margin of the shaft is sharp and connected to the radius by the interosseous membrane (see Fig. 5-13). The posterior border of the ulnar shaft ends distally in the styloid process. This process is the distal medial end of the ulna; the distal lateral end is the head of the ulna. The head of the ulna articulates with the radius at the distal radioulnar articulation. The articular surface of the radius is the ulnar notch of the radius.

The radius is the lateral long bone of the forearm. Its head, a thickened disk with a depressed top, is wide at the top and narrow at the bottom. The top articulates with the capitulum of the humerus, and the sides of the head articulate with the radial notch of the proximal ulna. The annular ligament (see Figs. 5-13 and 5-14) wraps around the narrow portion, helping to hold the radius to the ulna without restricting movement.

The neck of the radius is the constricted region just below the head. At its distal edge on the medial side is the tuberosity of the radius. The biceps tendon inserts here. The shaft of the radius has a slight lateral convexity, which helps keep the radius and ulna separated during pronation and supination. The interosseous (medial) border, where the interosseous membrane attaches (see Fig. 5-13), is sharp.

The major articulations of the radius are at the wrist. The distal end of the radius is enlarged for its articulations with the ulna and the carpal bones. Laterally the body continues into the styloid process of the radius. Medially the ulnar notch of the radius articulates with the ulnar head. The dorsum of the distal radius has grooves for the extensor tendons.

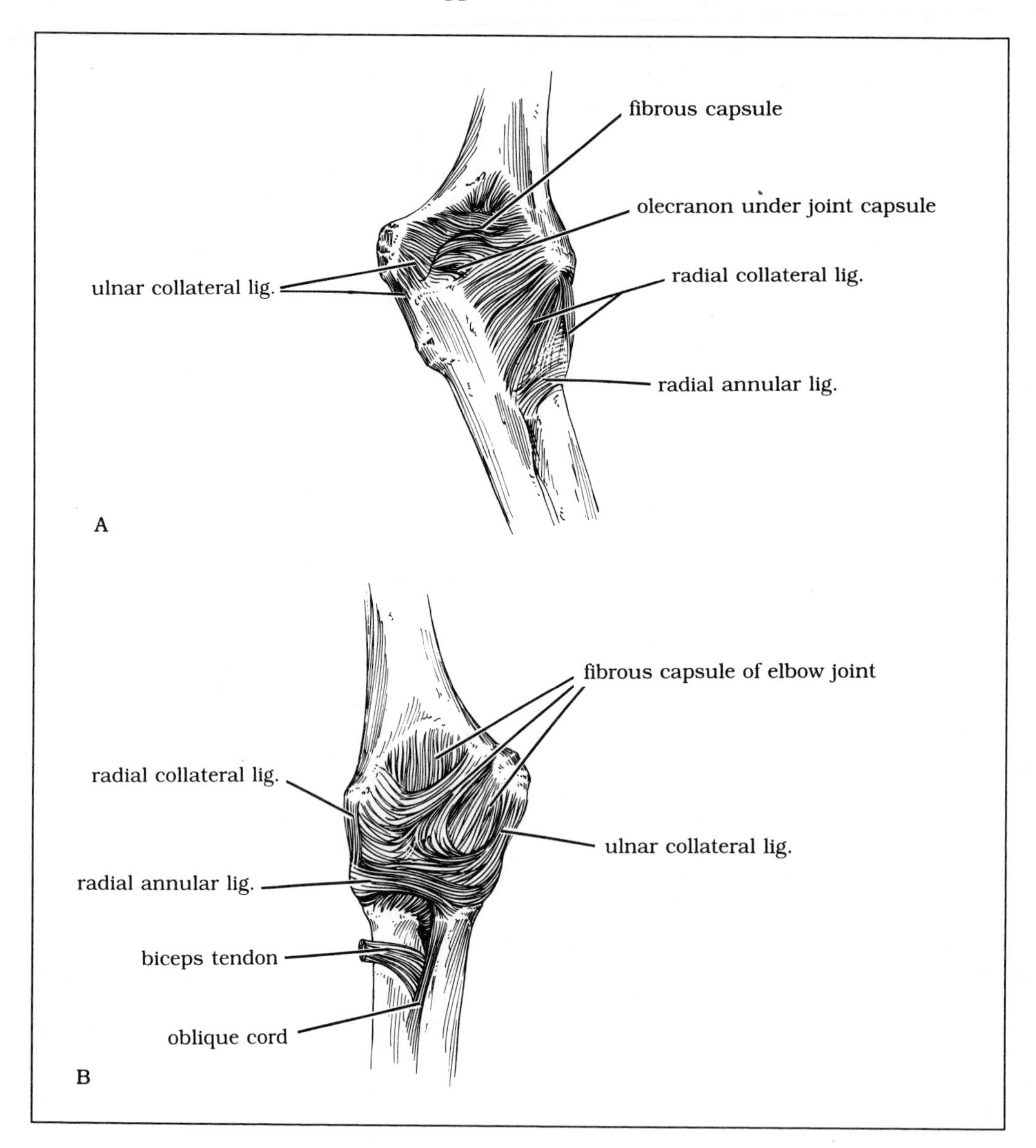

Fig. 5-12. Capsule of the elbow joint. A. Anterior view. B. Posterior view.

CARPALS

There are eight carpal (wrist) bones (Fig. 5-15). The proximal row, medially to laterally, contains the pisiform, triangular (triquetrum), lunate, and scaphoid bones. The distal row, medially to laterally, contains the hamate, capitate, trapezoid, and trapezium bones.

Each bone has six surfaces. The dorsal and palmar surfaces are nonarticular; all the other surfaces are articular, except at the ends. In all

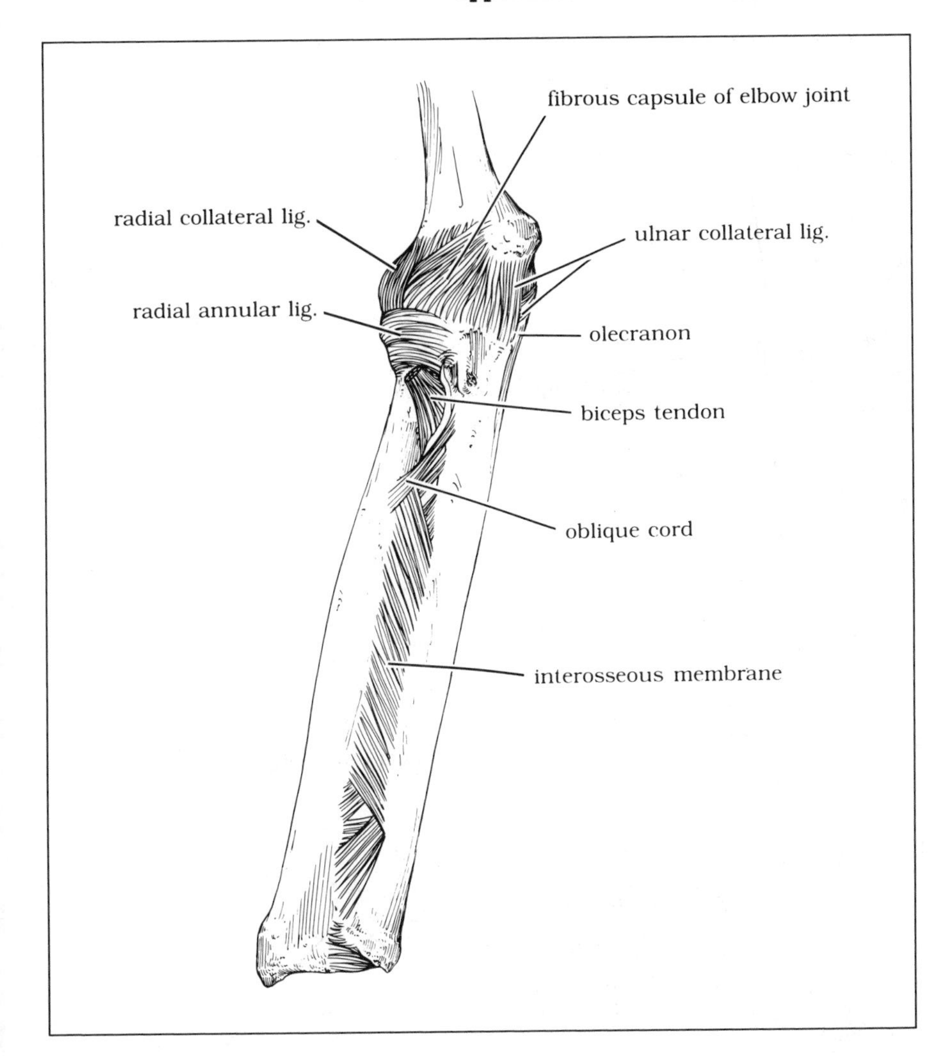

Fig. 5-13. Ligaments and membranes connecting the humerus, radius, and ulna. Anterior view.

cases the proximal articular surface is convex and the distal one is concave. The articulations of the carpal bones are as follows.

The **scaphoid** is the largest bone in the proximal row. It articulates with the radius, trapezium, trapezoid, capitate, and lunate bones.

The **lunate** is a crescent-shaped bone. It articulates with the radius, capitate, hamate, scaphoid, and triangular bones.

The pyramidal **triangular** bone articulates with the lunate, hamate, and pisiform bones.

The **pisiform** is a small pea-shaped bone. It articulates only with the triangular bone. The pisiform bone has many of the characteristics of a sesamoid bone in the tendon of the flexor carpi ulnaris muscle.

The **trapezium** is the most lateral carpal bone in the distal row. It articulates with the scaphoid,

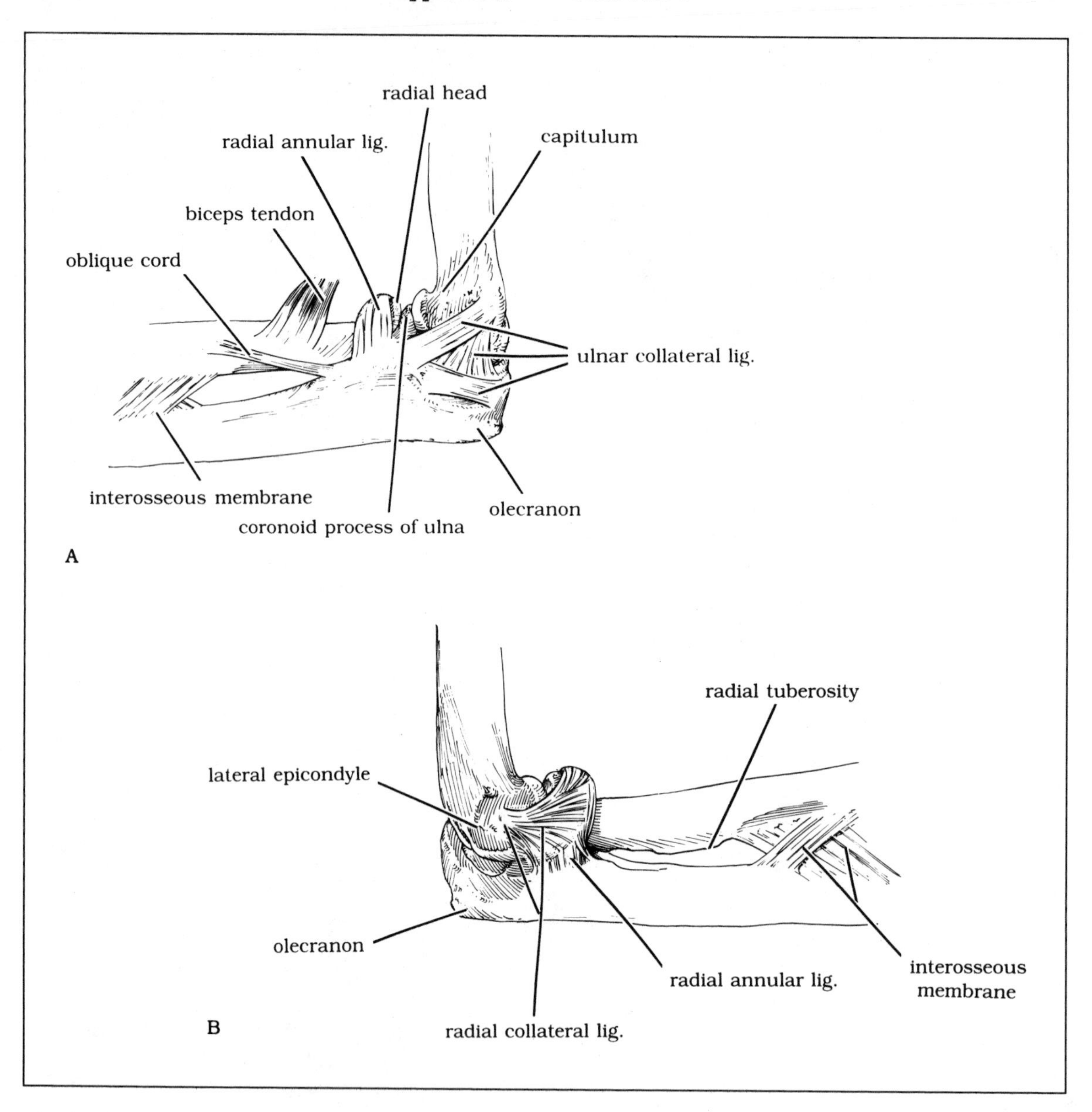

Fig. 5-14. Ligaments of the elbow joint. A. Medial view. B. Lateral view.

first metacarpal, second metacarpal, and trapezoid bones.

The **trapezoid** articulates with the scaphoid, second metacarpal, trapezium, and capitate bones.

The **capitate** is the largest carpal bone in the distal row. It articulates with the scaphoid, lunate, trapezoid, hamate, and second, third, and fourth metacarpal bones.

The **hamate** articulates with the triangular, lunate, capitate, and fourth and fifth metacarpal bones. The hamate has a hook on its palmar surface that serves as a site of attachment for several muscles (Fig. 5-16).

METACARPALS AND PHALANGES (HAND)

The hand is formed by the five metacarpal bones and the phalanges (see Figs. 5-15 and 5-16). The metacarpal bones are numbered from the radial to the ulnar side. Each metacarpal has the features of a long bone, with a head, shaft, and base. The heads of the metacarpals form the knuckles and are rounded to articulate with the base of the phalanx of the same number. The shafts are convex dorsally. Their dorsal surfaces are flattened and triangular in shape. The bases are cuboidal, with broad dorsal regions. The first metacarpal is shorter than the others and has two articular surfaces on the palmar side of the head for articulations with sesamoid bones.

There are a total of 14 phalanges. The first digit has two, while all the others have three. Each phalanx has a base, shaft, and head. The shafts taper distally. Each distal phalanx has a roughened, elevated horseshoe-shaped region on its palmar terminal surface. This region helps to support the fingertips.

Fig. 5-15. Bones of the wrist and hand. A. Dorsum. B. Palm.

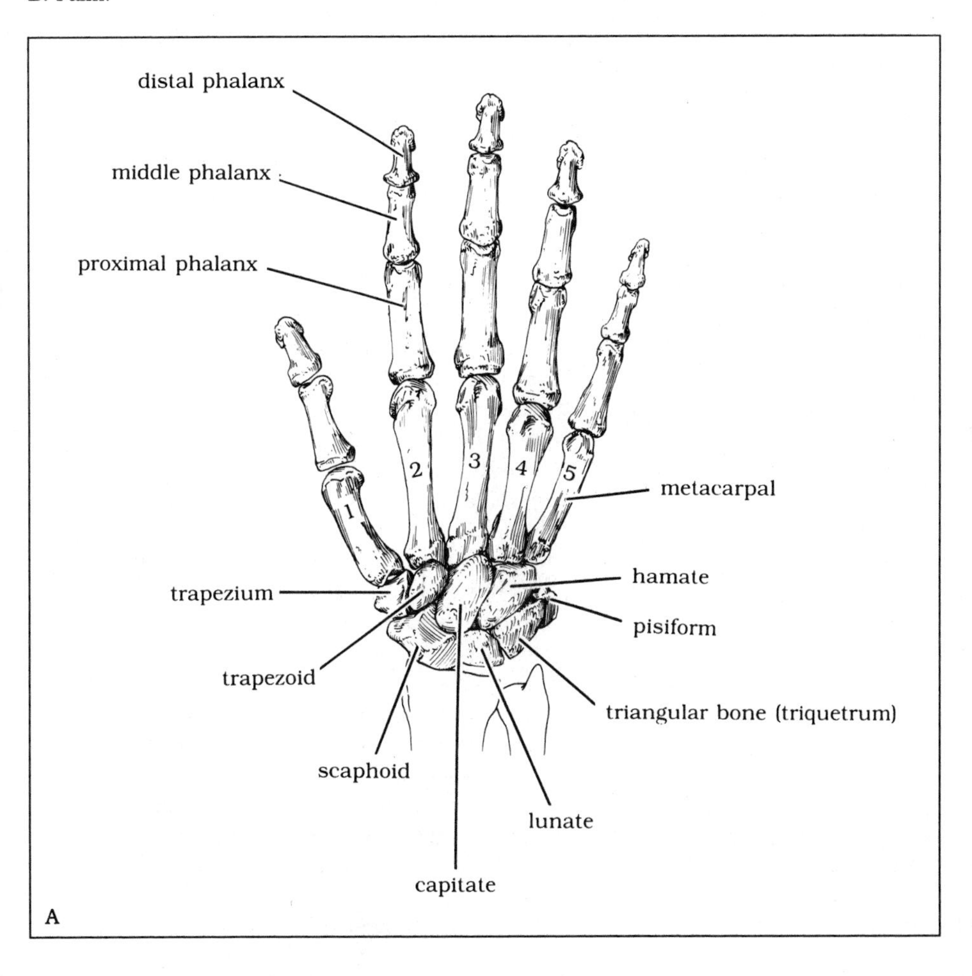

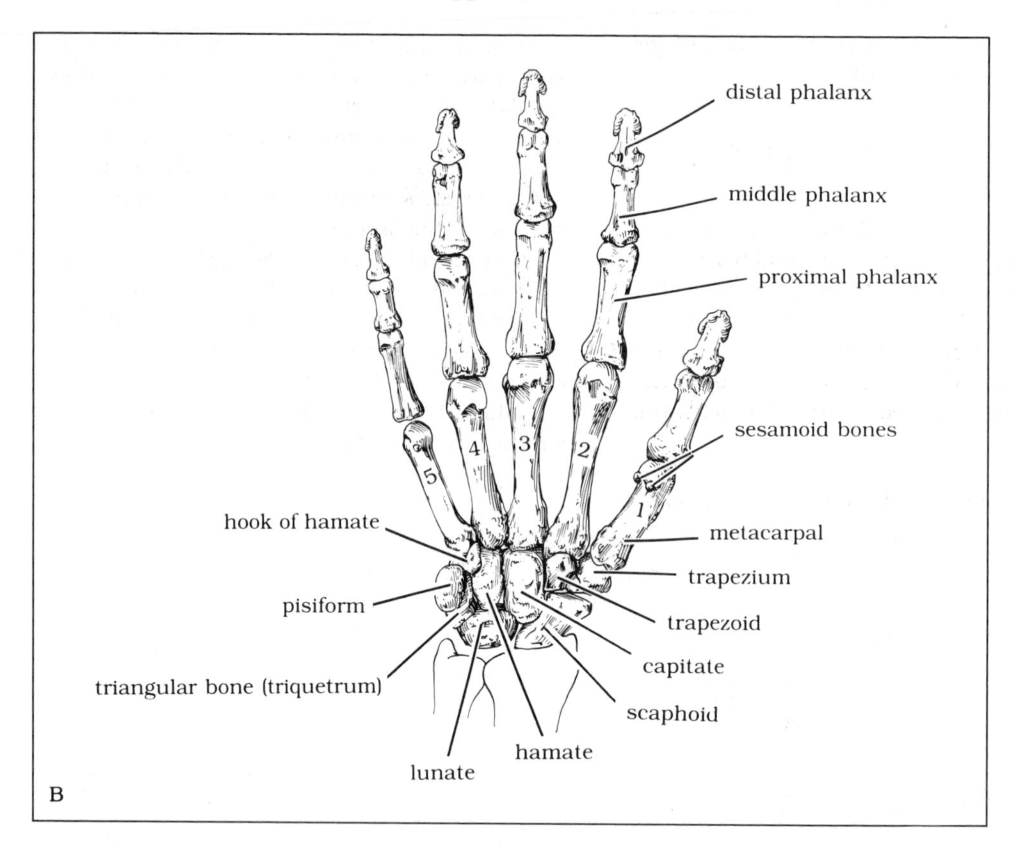

Fig. 5-15 (continued).

Back and Scapular Regions

MUSCULATURE

Back Region

The superficial musculature of the back (Fig. 5-17) connects the upper limb to the vertebral column. The muscles of this group consist of the trapezius, latissimus dorsi, levator scapulae, rhomboid major, and the rhomboid minor muscles (Table 5-1).

The **trapezius muscle** arises from the medial third of the superior nuchal line, the external occipital protuberance, the ligamentum nuchae, and the spines of vertebrae C7 to T12. It is a flat, thin muscle whose fibers converge on the shoulder to insert in upper, middle, and lower regions. The upper region inserts on the posterior border of the lateral third of the clavicle. It suspends the shoulder and holds up the point of the shoulder. The middle region of the trapezius inserts on the medial border of the acromion and the upper portion of the crest of the scapular spine. It is active in pulling and extension, drawing the shoulder back. The lower region of the trapezius inserts into a tubercle at the medial apex of the crest of the scapular spine. It is active in holding up the point of the shoulder.

The trapezius is innervated by the spinal part of the accessory (eleventh) cranial nerve and cervical nerves C3 and C4. A subtrapezial plexus formed by these nerves lies on its deep surface. Blood supply to the trapezius is from the transverse cervical and dorsal scapular arteries.

The **latissimus dorsi** is a broad, flat muscle in the lower back. It arises from the spines of vertebrae T7 to T12, the thoracolumbar fascia, the posterior third of the iliac crest, and the tenth through twelfth ribs. In addition some fibers may

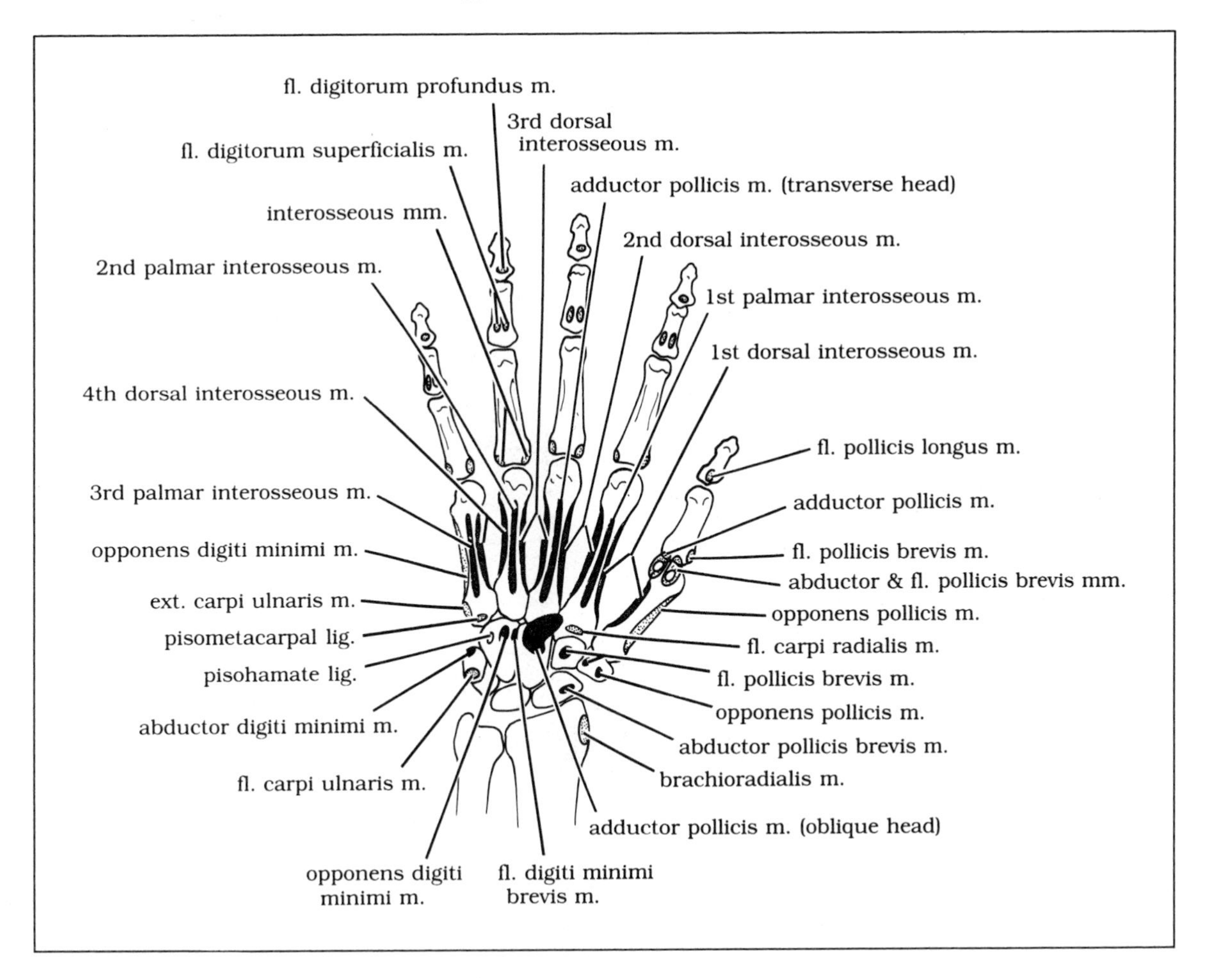

Fig. 5-16. Muscle origins (shaded areas) and insertions of the palmar hand.

arise as the muscle passes over the inferior angle of the scapula. The muscle extends laterally, narrowing and wrapping around itself to insert into the intertubercular groove of the humerus. In reaching its insertion the latissimus dorsi muscle spirals around the teres major muscle (see Fig. 5-21B), twisting so that its inferior fibers become superior. It is at this point that it also forms the posterior axillary fold. The latissimus dorsi muscle extends the humerus and pulls the arm downward and backward during medial rotation.

The muscle is innervated by the thoracodorsal (middle subscapular) nerve (C7 and C8). Blood is supplied by the thoracodorsal artery.

The **levator scapulae muscle** lies under the trapezius. It is a strap muscle that arises by separate tendons from the transverse processes of C1 to C4. It inserts onto the medial border of the scapula along the edge of the supraspinous fossa. As its name implies, the levator scapulae muscle is involved in the elevation, rotation, and support of the scapula. It receives innervation from the dorsal scapular nerve (C5) and fibers from C3 and C4. The dorsal scapular artery supplies its blood.

The **rhomboid minor muscle** arises from the lower portions of the ligamentum nuchae and

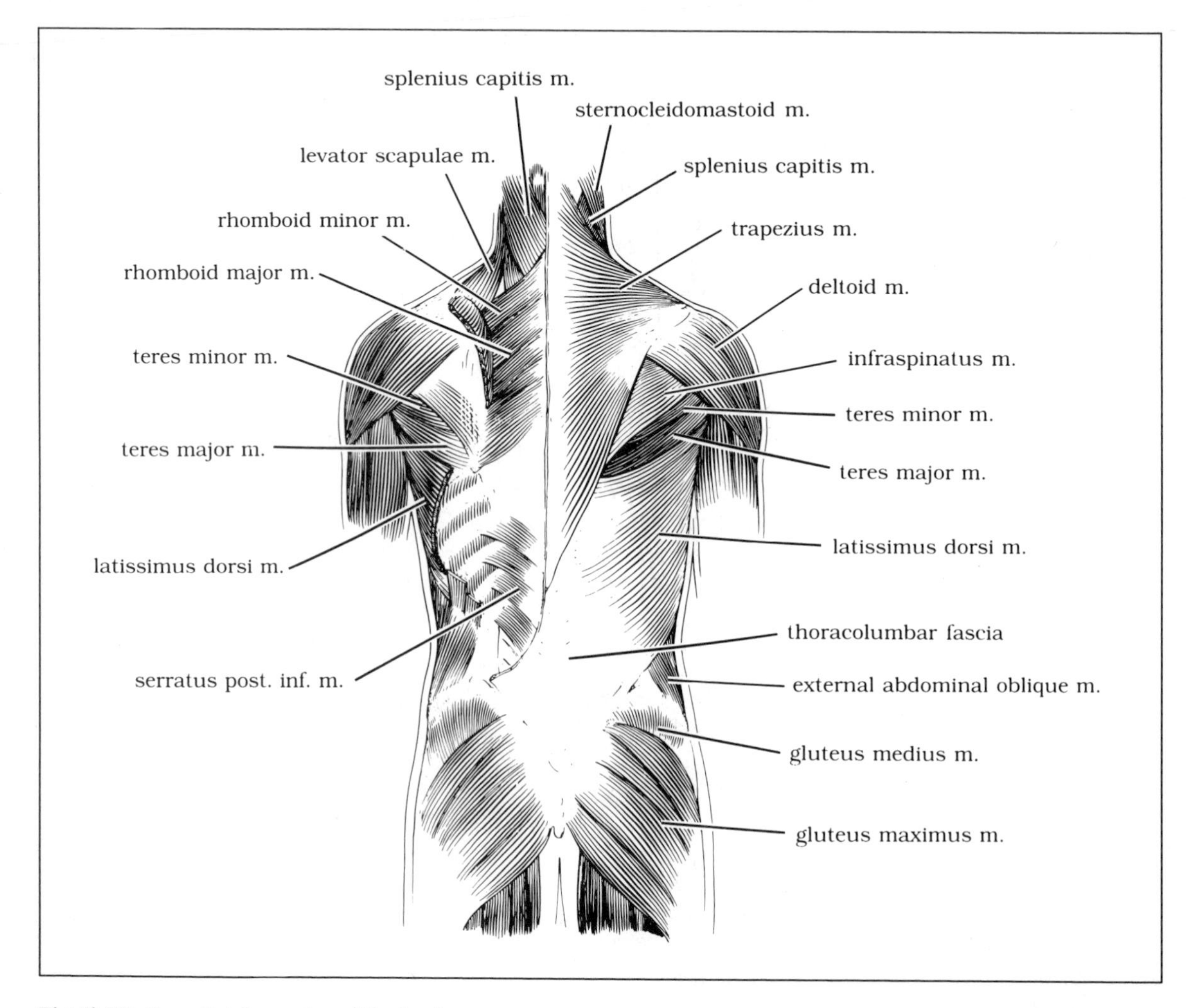

Fig. 5-17. Superficial muscles of the back.

the spines of the seventh cervical and first thoracic vertebrae. It inserts onto the medial border of the scapula at the root of the spine. The rhomboid minor muscle holds the scapula to the chest wall and draws it upward and medially. It is innervated by the dorsal scapular nerve (C5) and receives blood from the dorsal scapular artery.

The **rhomboid major muscle** arises from vertebral spines T2 to T5. It inserts along the medial border of the scapula below the root of the spine. Its actions, innervation, and blood supply are the same as those for the rhomboid minor muscle.

Scapular Region

There are six muscles in the scapular (shoulder) region: the deltoid, supraspinatus, infraspinatus, teres minor, teres major, and subscapularis muscles (see Table 5-1).

The **deltoid muscle** is the triangular, multipennate muscle that forms the curve of the shoulder (Figs. 5-18 and 5-19; see also Fig. 5-21). It arises from the lateral third of the clavicle, the lateral border of the acromion, and the lower edge of the scapular spine (see Fig. 5-5B). The fibers of the deltoid converge to insert on the deltoid tuberosity of the humerus. The action of the

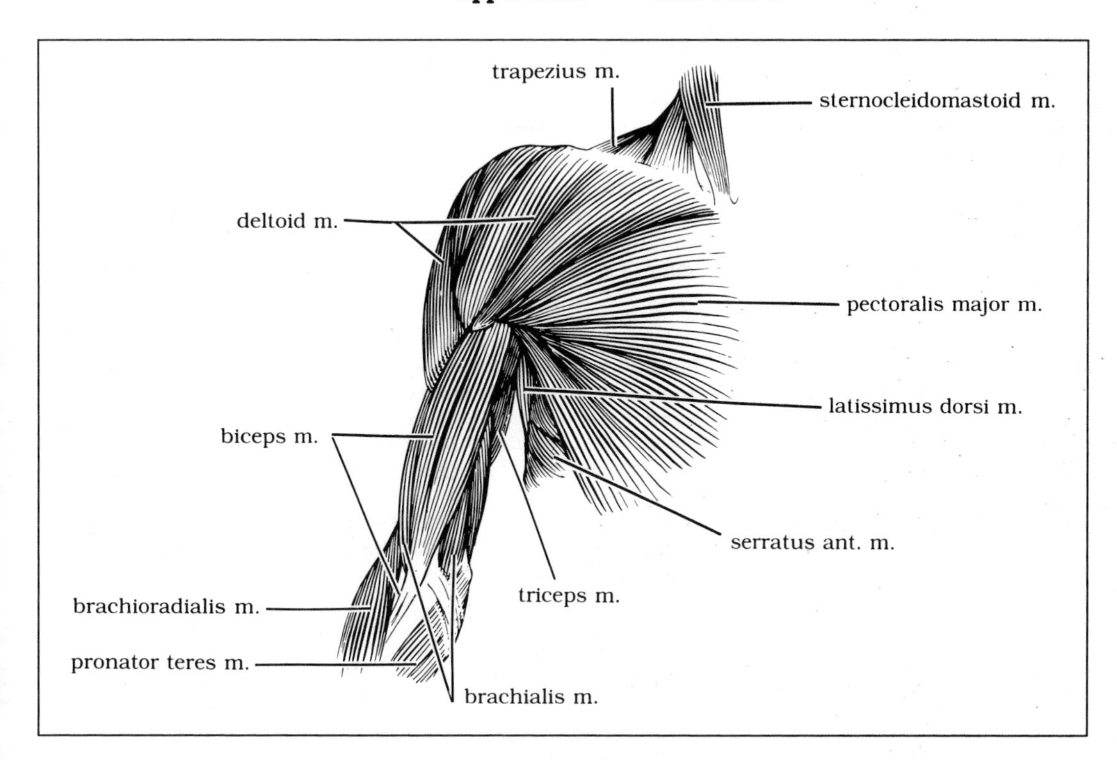

Fig. 5-18. Muscles of the shoulder and arm. Anterior view.

Fig. 5-19. Anterior musculature holding the scapula to the humerus, with the deltoid and pectoral muscles removed.

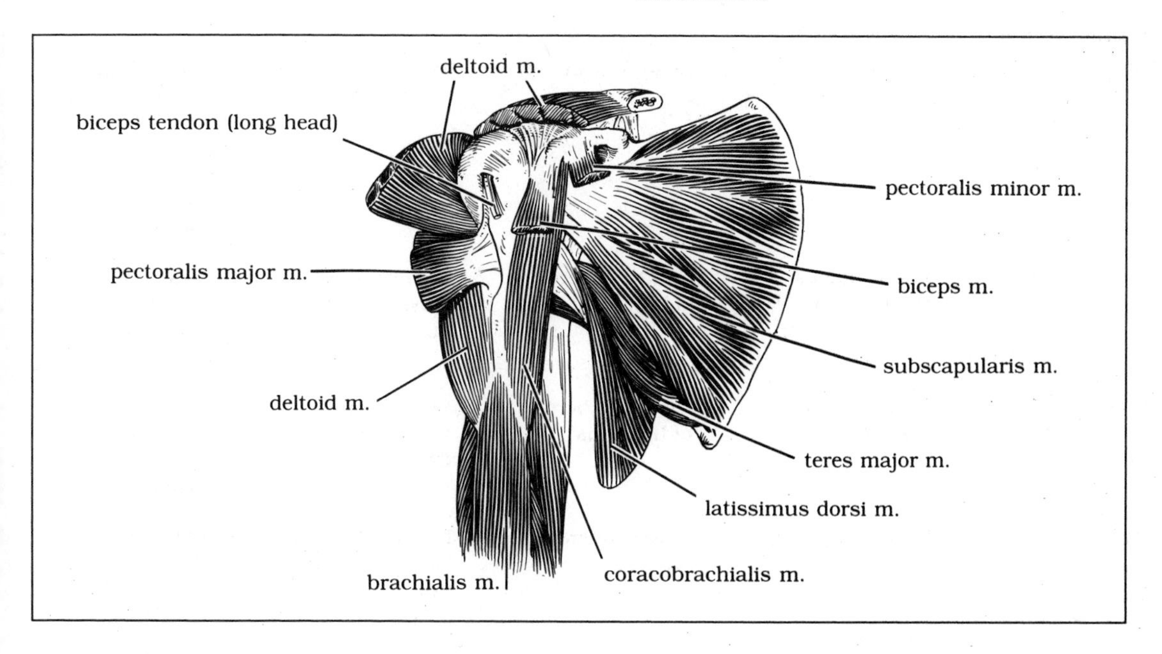

Table 5-1. Musculature of the Back and Scapular Regions

Muscle	*Origin*	*Insertion*	*Innervation*	*Action*
Back Region				
Trapezius	Superior nuchal line, external occipital protuberance, ligamentum nuchae, spines C7–T12	Lateral clavicle, acromion, crest and tubercle of spine of scapula	Accessory (eleventh) cranial nerve, C3 and C4	Suspend shoulder, hold up point of shoulder, pull shoulder back, act in extension
Latissimus dorsi	Spines T7–T12, thoracolumbar fascia, posterior iliac crest, ribs 10–12	Intertubercular groove of humerus	Thoracodorsal nerve (middle subscapular nerve; C7 and C8)	Extend humerus
Levator scapulae	Transverse processes C1–C4	Medial border of scapula at supraspinous fossa	Dorsal scapular nerve (C5)	Elevate, support, and rotate scapula
Rhomboid major	Spines T2–T5	Medial border of scapula at infraspinous fossa	Dorsal scapular nerve (C5)	Hold scapula to chest wall, draw scapula upward and medially
Rhomboid minor	Ligamentum nuchae, spines C7 and T1	Medial border of scapula at root of spine	Dorsal scapular nerve (C5)	Hold scapula to chest wall, draw scapula upward and medially
Scapular Region				
Deltoid	Lateral clavicle, acromion, spine of scapula	Deltoid tuberosity	Axillary nerve (C5 and C6)	Abduct humerus (strongest from 90 to 180 degrees); portions also involved in rotation, flexion, and extension
Supraspinatus	Supraspinous fossa	Greater tubercle of humerus	Suprascapular nerve (C5 and C6)	Abduct humerus, hold head of humerus in glenoid cavity
Infraspinatus	Infraspinous fossa	Greater tubercle of humerus	Suprascapular nerve (C5 and C6)	Laterally rotate, hold head of humerus in glenoid cavity
Teres major	Inferior angle of scapula	Lesser tubercle of humerus	Lower subscapular nerve (C5 and C6)	Adduct and medially rotate

Table 5-1 (continued).

Muscle	*Origin*	*Insertion*	*Innervation*	*Action*
Teres minor	Lateral border of scapula	Greater tubercle of humerus	Axillary nerve (C5 and C6)	Laterally rotate, hold head of humerus in glenoid cavity
Subscapularis	Subscapular fossa	Lesser tubercle of humerus	Upper and lower subscapular nerves (C5–C7)	Medially rotate, extend, hold head of humerus in glenoid cavity

deltoid is complex. The entire muscle is the primary abductor of the humerus. However, the deltoid cannot initiate abduction; it is helped, mainly by the supraspinatus muscle. The greatest strength of the deltoid is exhibited between 90 and 180 degrees of abduction. In addition the clavicular portion of the muscle is involved in flexion and medial rotation of the arm, while the scapular portion is involved in extension and lateral rotation. The deltoid muscle is innervated by the axillary nerve (C5 and C6). It receives its ma-

Fig. 5-20. Posterior musculature holding the scapula to the humerus, with the deltoid muscle removed.

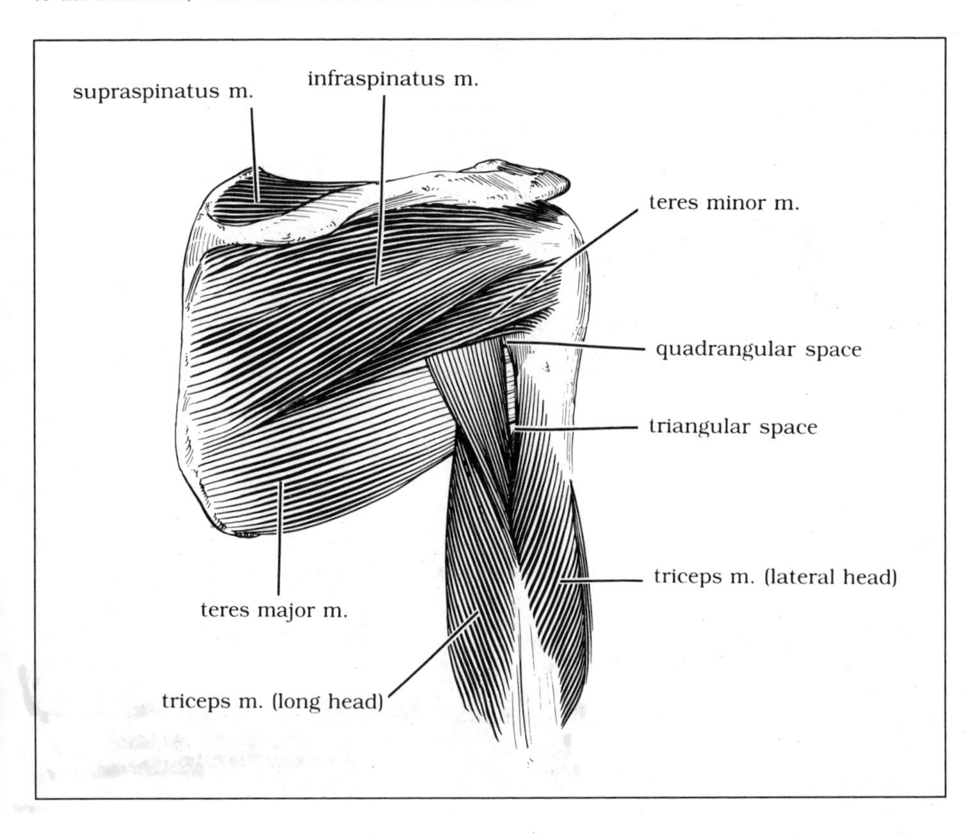

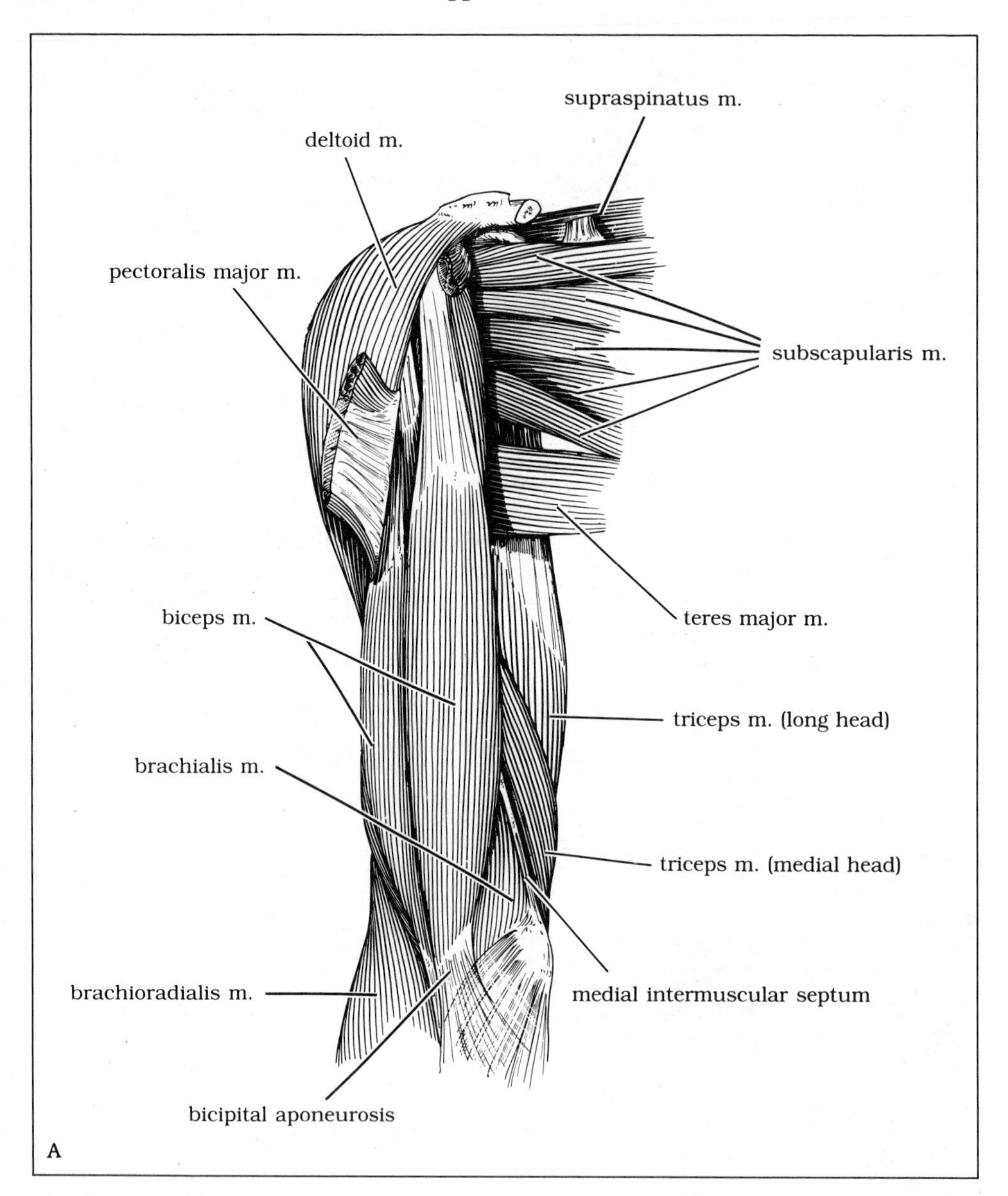

Fig. 5-21. Musculature of the arm. A. Anterior view. B. Posterior view.

jor blood supply via the posterior circumflex humeral artery.

Beneath the deltoid muscle is the **subdeltoid** (or subacromial) **bursa.** This bursa separates the deltoid muscle, the supraspinatus tendon, and the shoulder joint capsule.

The **supraspinatus muscle** arises from the supraspinous fossa and the dense fascia covering the muscle (Fig. 5-20). Its tendon blends with the capsule of the shoulder joint and inserts onto the greater tubercle of the humerus. The supraspinatus is the first muscle to act in abduction of the humerus, being the primary helper of the deltoid muscle. In addition the supraspinatus acts,

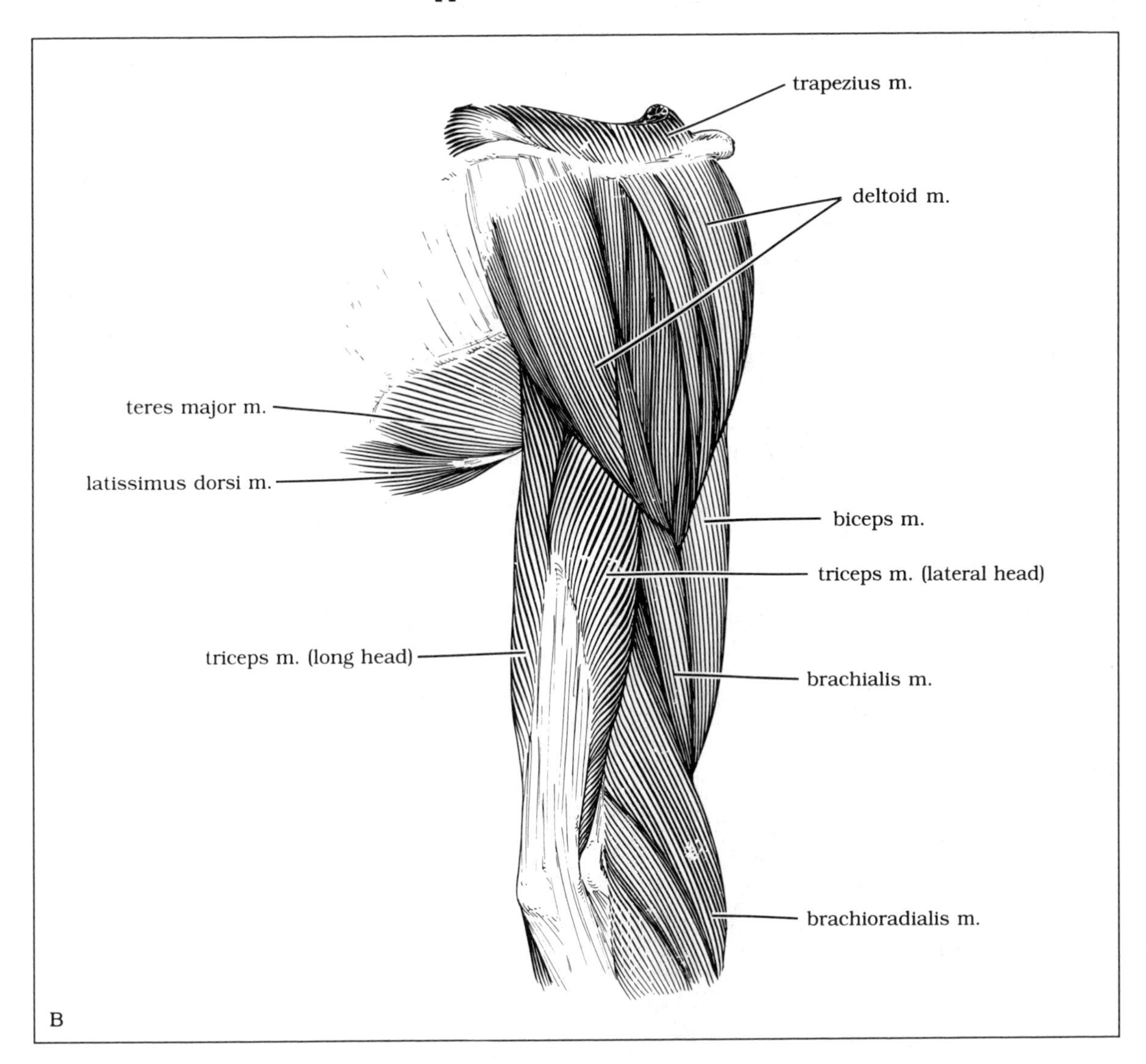

Fig. 5-21 (continued).

with the infraspinatus, subscapularis, and teres minor muscles (these four muscles form the rotator cuff), to hold the head of the humerus in the glenoid cavity. The suprascapular nerve (C5 and C6) and artery supply the supraspinatus muscle.

The **infraspinatus muscle** (see Figs. 5-20 and 5-22B) arises from the infraspinous fossa and its overlying dense fascia. The muscle inserts on the greater tubercle of the humerus, its tendon blending with the capsule of the shoulder joint. The infraspinatus muscle acts in lateral rotation and holds the head of the humerus in the glenoid cavity. The muscle is innervated by the suprascapular nerve (C5 and C6). Blood is supplied via the suprascapular and circumflex scapular arteries.

The **teres major muscle** (see Figs. 5-19, 5-20, and 5-21) arises from the dorsum of the scapula at the inferior angle and from the adjacent intermuscular septa. The muscle inserts on the lesser

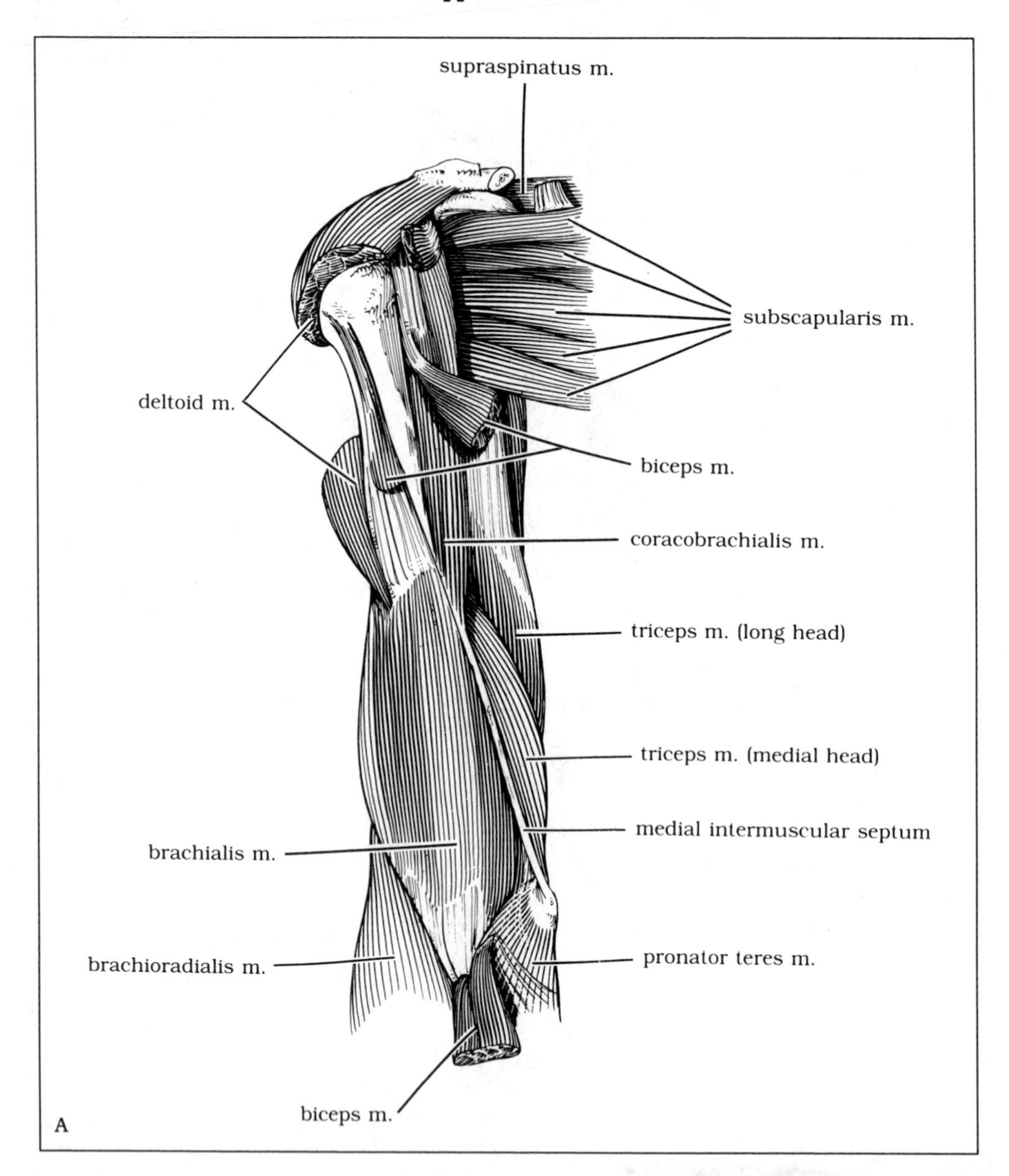

Fig. 5-22. Musculature of the arm; deep dissection. A. Anterior view. B. Posterior view.

tubercle of the humerus, its fibers twisting as they form a tendon prior to the insertion. The teres major adducts, medially rotates, and extends the arm. The lower suprascapular nerve (C5 and C6) and the circumflex scapular and posterior circumflex humeral arteries supply this muscle.

The **teres minor muscle** (Fig. 5-21) arises from the upper two-thirds of the lateral border of the scapula and from the adjacent intermuscular septa. Its tendon blends with the capsule of the shoulder joint and ends by inserting on the

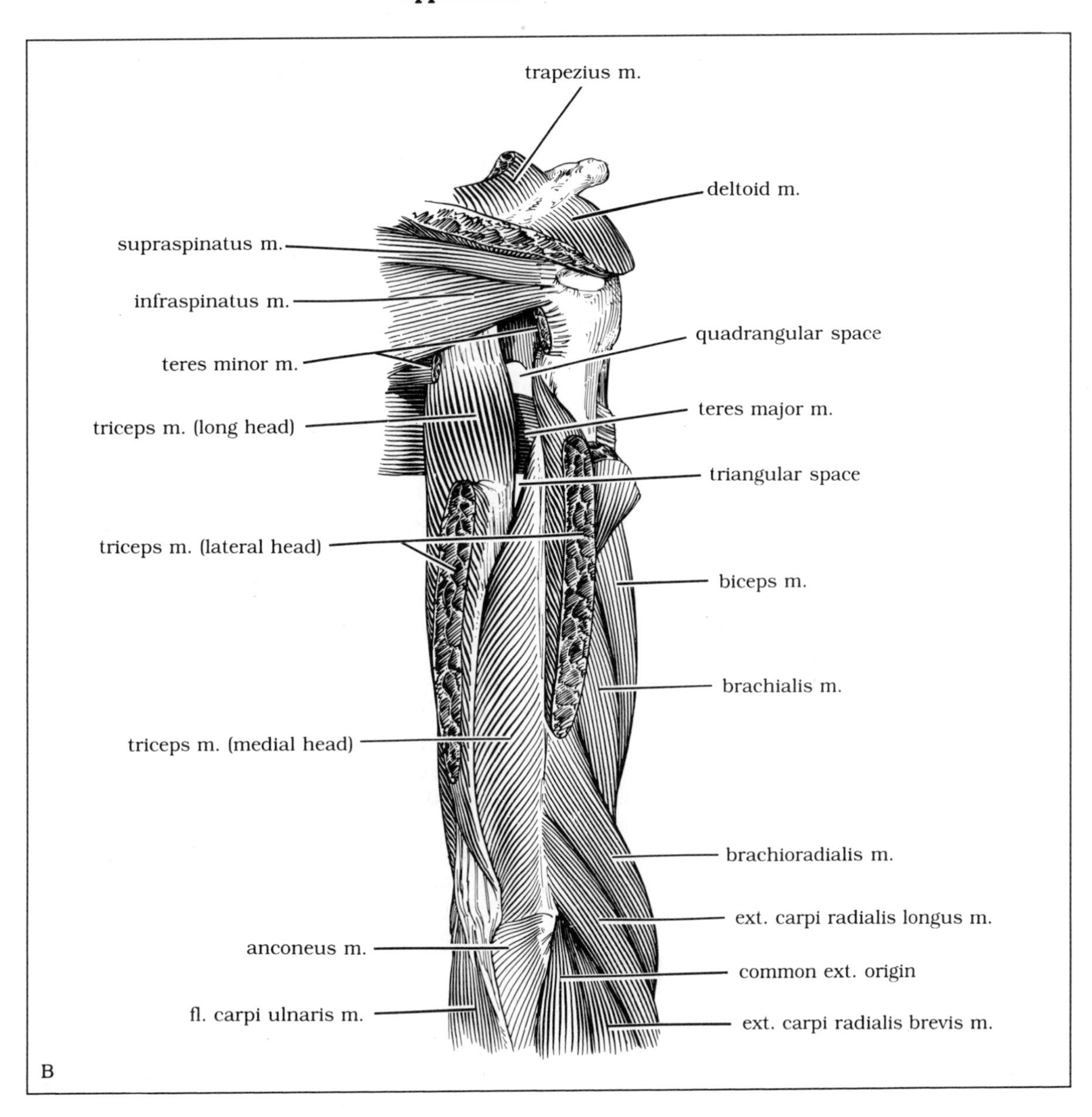

Fig. 5-22 (continued).

greater tubercle of the humerus. The teres minor muscle rotates the humerus laterally and holds the humeral head in the glenoid cavity. The axillary nerve (C5 and C6) innervates the teres minor. Blood is supplied by the circumflex scapular artery.

The **subscapularis** (Fig. 5-22A) arises from the costal (ventral) surface of the scapula and from the adjacent intermuscular septa. Its tendon blends with the capsule of the shoulder joint and ends by inserting onto the lesser tubercle of the humerus. The subscapularis has three actions. It is primarily a medial rotator of the humerus;

however, it also helps to extend the arm when it is in flexion or at rest, and to hold the humeral head in the glenoid cavity. Both the upper and lower subscapular nerves (C5 and C6) and the subscapular artery supply the subscapularis.

The **rotator cuff** is formed by the blending of the fibrous capsule of the shoulder joint with tendons of the supraspinatus, infraspinatus, subscapularis, and teres minor muscles. The cuff reinforces and strengthens the otherwise loose capsule, aiding in support of the joint at all times except in full rotation.

SPACES AND BLOOD SUPPLY

Two spaces are seen in the shoulder region (see Figs. 5-20 and 5-22B). The **quadrangular space** is formed by the borders of the humerus, teres major, triceps and teres minor muscles. The axillary nerve and posterior circumflex humeral artery pass through this space. Just medial to the quadrangular space is the **triangular space,** which is found between the triceps, teres major and teres minor muscles. The circumflex scapular vessels pass through the triangular space.

Blood is supplied to the shoulder region (see Fig. 5-25) through branches of the subclavian artery. The **suprascapular artery** is a branch of the thyrocervical trunk. It passes posteriorly, crossing over the superior transverse scapular ligament at the scapular notch to enter the supraspinous fossa. It then crosses to the notch at the neck of the scapula, passes through the notch, and terminates in the infraspinous fossa, anastomosing with the circumflex scapular artery.

The **dorsal scapular artery** is the last branch of the subclavian. It passes laterally and posteriorly through the brachial plexus (see Fig. 5-24), finally running along the medial border of the scapula. It anastomoses with the suprascapular and subscapular arteries. The dorsal scapular artery may arise in common with the superficial cervical artery. In this case the combined artery is known as the transverse cervical artery. The superficial branch of this artery follows the course of the superficial cervical artery, the deep branch the course of the dorsal scapular artery.

The subclavian artery continues past the first rib and becomes the axillary artery, which also gives off branches to the shoulder region (see pp. 174–175).

The nerves of the back and shoulder region are primarily branches of the brachial plexus (see discussion later in this chapter).

Pectoral Region and Axilla

The pectoral region is on the anterior surface of the thorax. It consists of the mammary gland and its associated structures and the pectoral muscles. Since these muscles are involved in movements of the upper limb, this region is included here rather than in Chapter 2, which dealt with the thorax.

MAMMARY GLANDS

The mammary glands are modified sweat glands that, in the female, produce and secrete milk. In the male they are rudimentary and nonfunctional. The mammary glands in the female begin as a smooth continuation of the upper chest wall (see Fig. 2-1B,D). Their lower and lateral borders are usually marked by creases. The space separating their medial borders is known as the sinus mammarum. In the nulliparous (childless) female they are conical after puberty and extend from the second or third rib to the sixth or seventh costal cartilage, and from the lateral border of the sternum to beyond the anterior axillary fold. In females who have had children they tend to be more hemispheric, with their size and extent varying according to their functional history.

The surface of the mammary glands is smooth, except for the central pigmented circular region, the areola mammae. The center of the areola is raised by a nipple. Underlying the pigmented areola is the parenchymal glandular tissue of the rudimentary milk glands known as areolar glands. The skin of the areola also contains active sebaceous glands. There is no fat in the underlying dermis.

The conical nipple contains sebaceous glands, but no fat, hairs, or sudoriferous glands. At the tip of the nipple are openings for 15 to 20 lactiferous ducts. The nipple is erected through the action of a circular layer of smooth muscle underlying the dermis. Constant contraction of the layer closes off the lactiferous duct openings; alternation of contraction with relaxation acts to pump the ducts. Beneath the areola the ducts dilate from a diameter of 1 to 2 mm to a diameter of 4 to 5 mm. This dilated area is the lactiferous sinus and represents a region for milk storage. Beyond the sinus, each duct drains one lobe of glandular (parenchymal) tissue.

The lactiferous ducts are derived from converging branches, each initially arising from alveoli. The several alveoli that drain into a particular duct branch constitute a lobule. All the lobules draining into the same lactiferous duct make up a lobe. The space between the lobes, lobules, and alveoli is filled by loose connective stromal tissue consisting of collagen fibers and fat.

The mammary glands are innervated by cutaneous fibers arising from T2 to T6. The anterior cutaneous branches of the lateral cutaneous nerves give rise to lateral mammary rami. Medial mammary rami are derived from the anterior cutaneous nerves. In addition supraclavicular fibers may innervate the upper portion of the gland.

Blood is supplied to the mammary glands from several sources. The intercostal arteries (3 to 5) and the lateral thoracic artery give rise to lateral mammary arteries. The second to the fourth perforating branches of the internal thoracic artery give rise to medial mammary arteries. In addition the pectoral branches of the thoracoacromial artery supply blood to the mammary glands. Venous drainage of the glands follows the arterial supply.

Because of the common occurrence of mammary cancer the lymphatics of this region are of particular importance. Beneath the areola lies the dense subareolar lymphatic plexus. Peripheral to the areola is the less-dense circumareolar lymphatic plexus, which is continuous with the lymphatics of the skin and the other breast.

The principal drainage of the region proceeds via the medial and lateral lymphatic trunks to the axillary lymph nodes, including the superior nodes of the pectoral group and the central and lateral groups of axillary nodes. Accessory drainage, through the underlying pectoral muscles, may proceed to the apical axillary nodes.

MUSCULATURE

Pectoral Region

There are three muscles in the pectoral region: the pectoralis major, the pectoralis minor, and the subclavius (Table 5-2). The **pectoralis major muscle** (see Figs. 5-18 and 5-23) arises from the medial half of the clavicle, the sternum, and the costal cartilages of the second to sixth ribs. The fibers converge and fold to become U-shaped, the lower fibers folding up deep to the upper ones. The lateral lower border of this fold is the anterior axillary fold. The pectoralis major inserts on the greater tubercle of the humerus.

The muscle as a whole flexes and adducts the arm. The clavicular portion elevates the shoulder and draws the arm forward and medially. The sternocostal portion drops the shoulder and draws the arm forward and medially. The pectoralis major muscle is innervated by both the lateral and medial pectoral nerves (C5 to T1). Blood is supplied via the pectoral branches of the thoracoacromial artery.

The **pectoralis minor** muscle (see Fig. 5-23) is deep to the pectoralis major. It arises from the second to the fifth ribs and inserts on the coracoid process. This muscle pulls the scapula forward, medially, and downward, helping to point the shoulder. The medial pectoral nerve (C8 and T1) innervates the pectoralis minor muscle. Blood is supplied via the pectoral branches of the thoracoacromial artery.

The **subclavius muscle** (see Fig. 5-23) arises from the first rib and passes upward to insert on the clavicle. It protects the subclavicular region and pulls the shoulder downward and forward. Innervation is accomplished through the nerve to the subclavius (C5). The clavicular branch of the thoracoacromial artery supplies blood to the subclavius muscle.

The **pectoral fascia** is the investing fascia of the pectoralis major muscle. This fascia leaves

Table 5-2. Musculature of the Pectoral and Axillary Regions

Muscle	*Origin*	*Insertion*	*Innervation*	*Action*
Pectoralis major	Clavicle, sternum, costal cartilages of ribs 2–6	Greater tubercle of humerus	Medial and lateral pectoral nerves (C5–T1)	Flex and adduct arm
Pectoralis minor	Ribs 2–5	Coracoid process	Medial pectoral nerve (C8 and T1)	Pull scapula forward, downward, and medially
Subclavius	Rib 1	Clavicle	Nerve to subclavius (C5)	Pull shoulder downward and forward
Serratus anterior	Ribs 1–8, or 9	Medial border of scapula	Long thoracic nerve (C5–C7)	Hold scapula to chest wall, rotate scapula and pull it forward

Fig. 5-23. The biceps brachii, coracobrachialis, and pectoralis minor muscles.

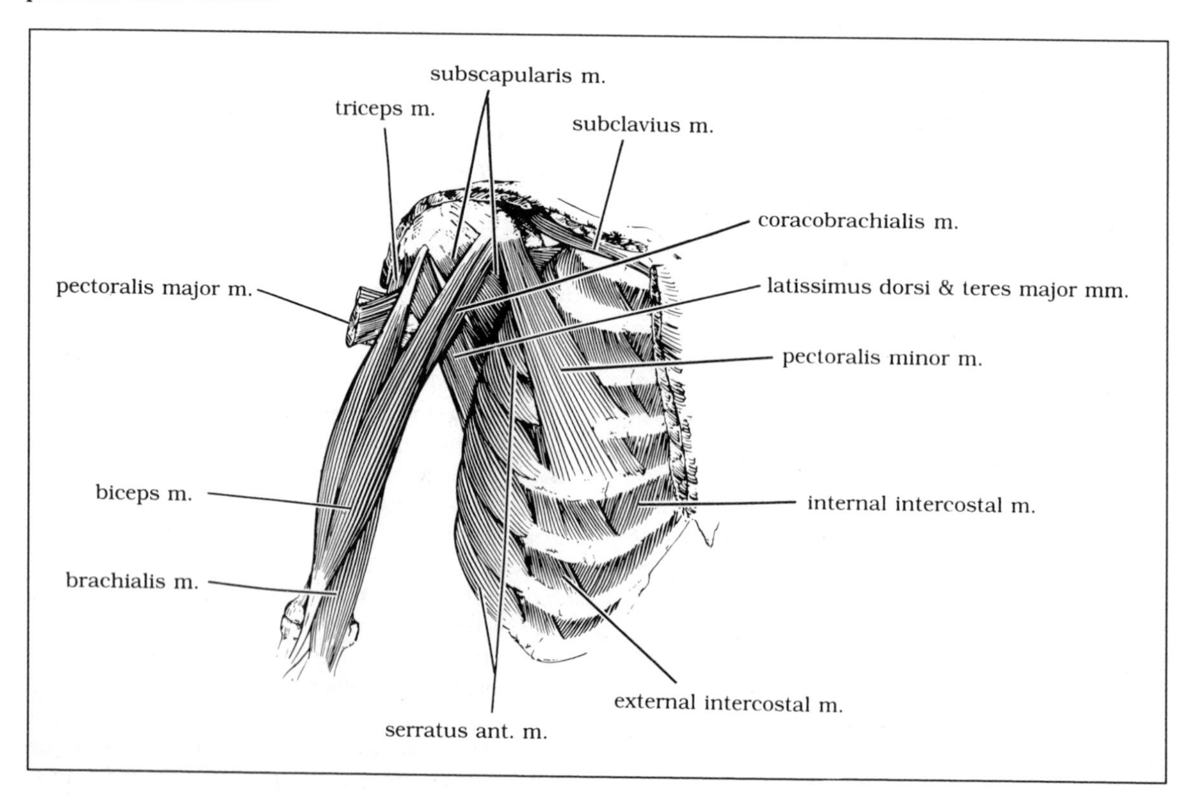

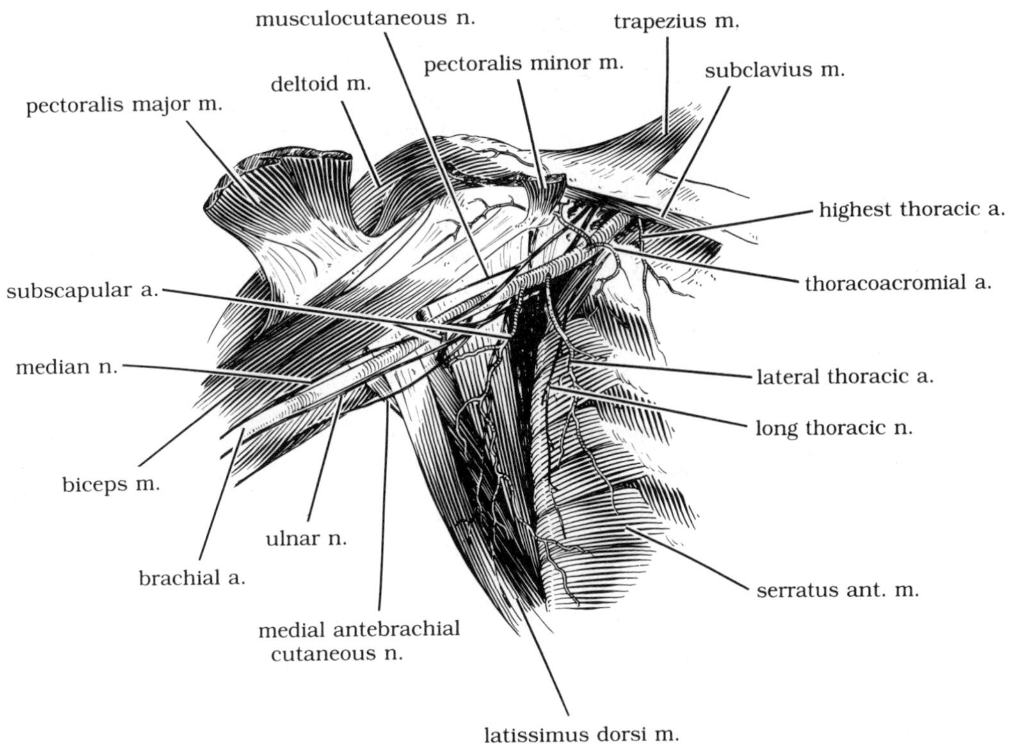

Fig. 5-24. Relationship of the axillary artery and the brachial plexus.

the muscle at its lateral border to form the axillary fascia and the floor of the axilla. Superiorly the pectoralis major muscle is separated from the deltoid muscle by the deltopectoral triangle. The cephalic vein runs through this space before penetrating the fascia to reach the axillary vein.

The **clavipectoral fascia** is the investing fascia of both the pectoralis minor and subclavius muscles. The portion of the fascia between these muscles is known as the costocoracoid membrane. This membrane is pierced by the cephalic vein, thoracoacromial artery, and lateral pectoral nerve. Where it is thickened the membrane is known as the costocoracoid ligament. After leaving the inferior lateral border of the pectoralis minor muscle, the clavipectoral fascia continues on as the suspensory ligament of the axilla. This ligament ends by blending with the axillary fascia at the floor of the axilla.

Axilla

The axilla is a pyramidal region, the base of which is formed by the axillary fascia and skin. The apex of the axilla is at the junction of the first rib, clavicle, and scapula. The anterior wall is formed by the pectoralis muscles; the posterior wall by the subscapularis, teres minor, and latissimus dorsi; the lateral wall by the humerus; and the medial wall by the serratus anterior muscle. The anterior axillary fold contains the pectoralis major muscle, the posterior fold the latissimus dorsi and teres major muscles.

The axilla (Fig. 5-24) contains the axillary artery and its branches, the axillary vein and its tributaries, a portion of the brachial plexus, and the axillary lymph nodes. The artery, vein, and nerves are contained in the axillary sheath.

The **serratus anterior muscle** (see Fig. 5-23 and Table 5-2), which forms the medial wall of the axilla, arises by fingerlike projections from the first to the eighth or ninth rib. The portion that arises from the fourth to eighth ribs interdigitates with the origin of the external abdominal oblique muscle. The serratus anterior inserts on the medial border of the scapula. It keeps the scapula close to the chest wall and draws it forward. In elevation of the shoulder the serratus anterior may also rotate the scapula. It is supplied by the lateral thoracic artery.

BLOOD SUPPLY

The **axillary artery** (Fig. 5-25; see also Fig. 5-24) begins as the subclavian artery passes the distal border of the first rib; it ends at the lower border of the teres major, where it becomes the brachial artery. It is approximately 12 cm long and 1 cm in diameter. The axillary artery, along with the axillary vein and the brachial plexus, are contained in a dense connective tissue sheath, the axillary sheath.

For didactic reasons the axillary artery is usually divided into three regions. The first region has one branch, the second two, and the third three. The first region of the axillary artery lies between the first rib and the pectoralis minor.

Fig. 5-25. The arterial supply to the axilla and arm.

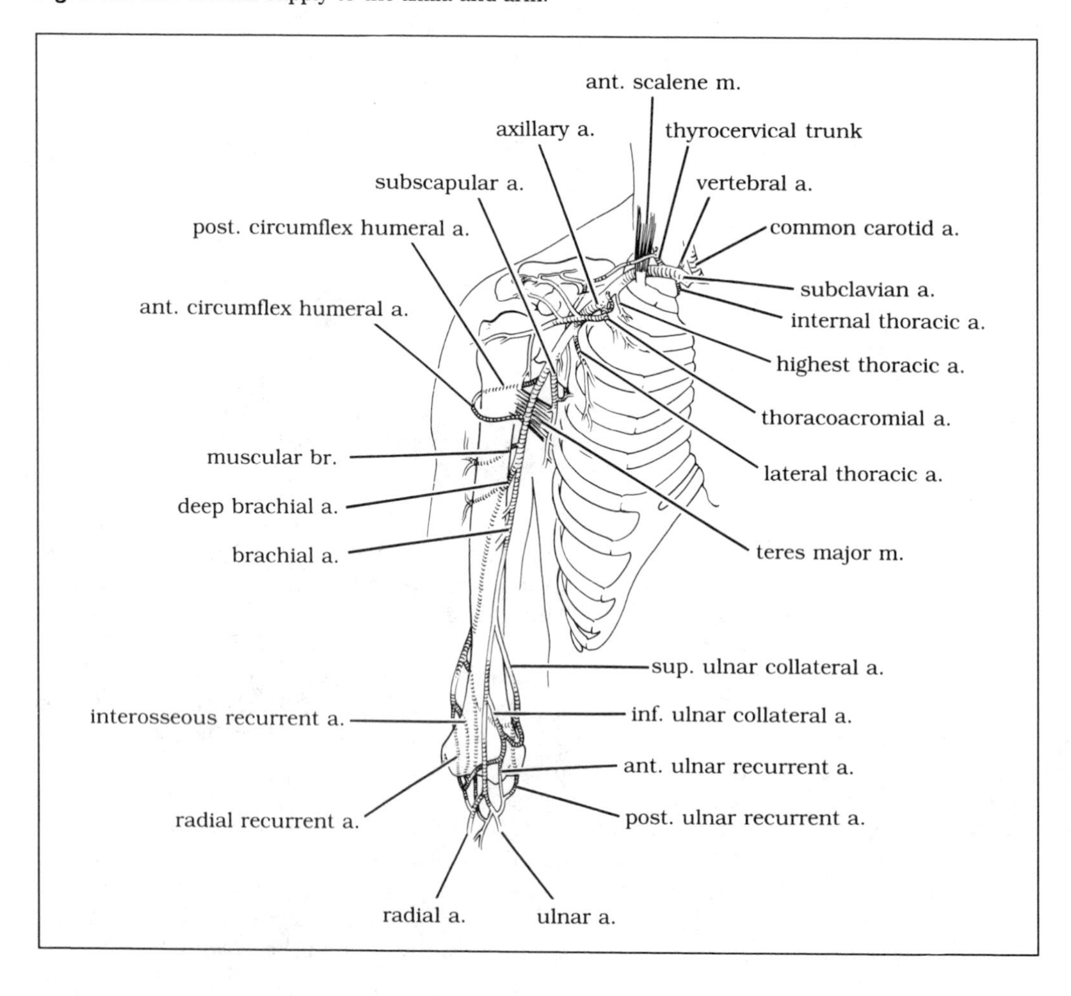

Table 5-3. Branches of the Axillary Artery

Region	*Number of Branches*	*Branch*	*Structures Supplied*
1	1	Highest thoracic	Intercostal spaces 1 and 2, serratus anterior
2	2	Thoracoacromial	Acromion, deltoid, pectoralis major and minor, mammary gland, subclavius, sternoclavicular joint
		Lateral thoracic	Pectoralis major and minor, serratus anterior, mammary gland
3	3	Subscapular	Subscapularis, teres major, teres minor, infraspinatus, supraspinatus, latissimus dorsi, serratus anterior
		Anterior circumflex humeral	Coracobrachialis, biceps (long head), pectoralis major
		Posterior circumflex humeral	Deltoid, humerus, triceps (long and lateral heads)

The **highest (superior) thoracic artery** arises here, supplying the first two intercostal spaces and the serratus anterior.

The second part of the axillary artery lies behind the pectoralis minor. The thoracoacromial and lateral thoracic arteries arise from this portion. The **thoracoacromial artery** has four branches—acromial, deltoid, pectoral, and clavicular—supplying, respectively, the acromion and deltoid muscle, the deltoid and pectoralis major muscles, the pectoralis major and minor muscles, and the subclavius muscle.

The **lateral thoracic artery** runs inferiorly, supplying the pectoral muscles, serratus anterior, and the mammary gland.

The third part of the axillary artery runs from the pectoralis minor to the teres major. The three branches from this region are the subscapular artery, anterior circumflex humeral artery, and posterior circumflex humeral artery.

The **subscapular artery** supplies the subscapularis, teres major, and serratus anterior muscles. In addition it gives rise to the circumflex scapular artery and the thoracodorsal artery. The **circumflex scapular artery** passes through the triangular space to supply the teres minor, infraspinatus, and (occasionally) supraspinatus muscles. The **thoracodorsal artery** supplies the latissimus dorsi.

The **anterior circumflex humeral artery** passes anteriorly around the surgical neck of the humerus to supply the coracobrachialis and pectoralis major and the long head of the biceps muscles.

The **posterior circumflex humeral artery** passes posteriorly around the surgical neck of the humerus and through the quadrangular space with the axillary nerve. This artery supplies the deltoid, long and lateral heads of the triceps, and humerus.

Table 5-3 provides a summary of the branches of the axillary artery. Important anastomotic connections among the branches are discussed later in this chapter.

The **axillary vein** is formed by the junction of the medial and lateral brachial veins at the level of the teres major muscle. The superficial basilic vein joins the medial brachial vein just prior to this junction. The axillary vein accompanies the axillary artery in its course, lying anterior and inferior to it.

The tributaries of the axillary vein are, in general, identical to the branches of the axillary artery. However, the thoracoacromial vein, without

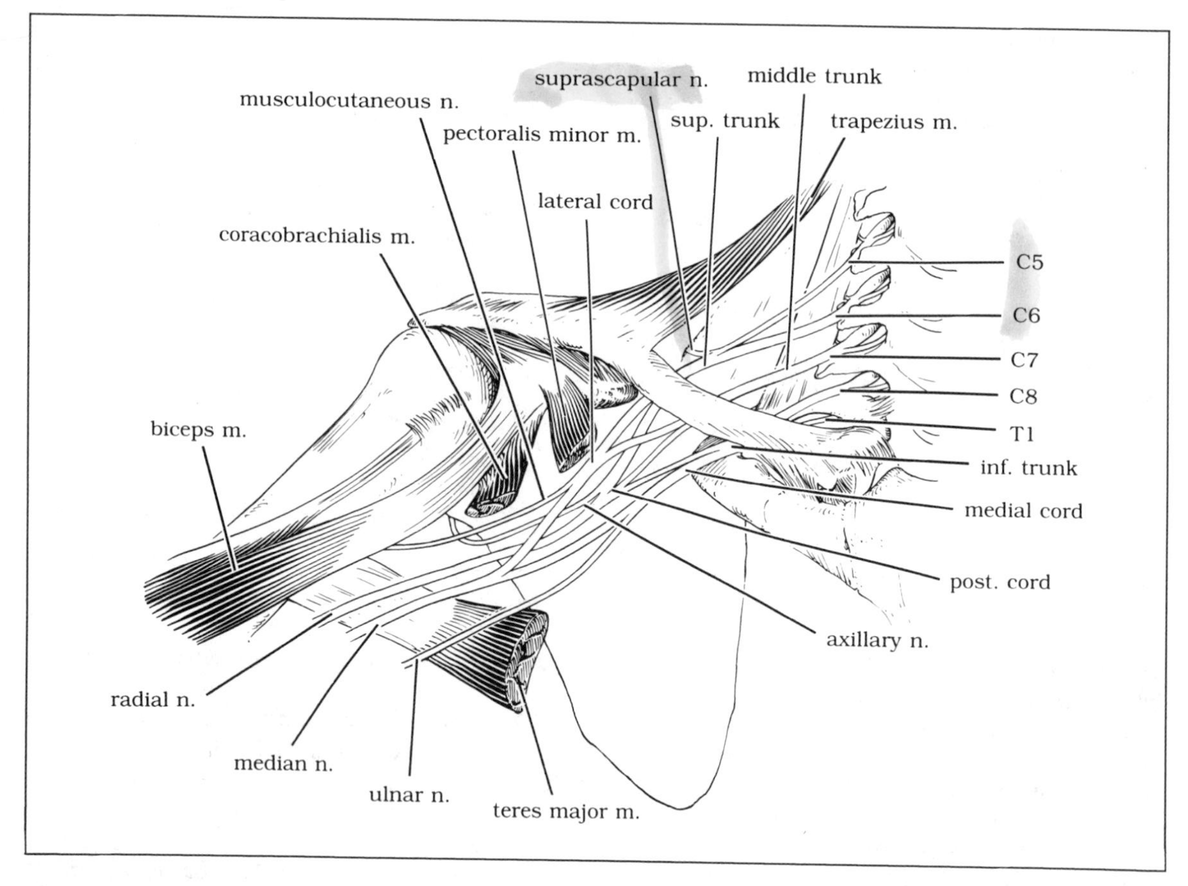

Fig. 5-26. Nerves of the brachial plexus.

its pectoral input, usually joins the cephalic vein. The pectoral tributaries empty into the subclavian vein.

The cephalic vein and the costoaxillary veins do not have arterial counterparts. The cephalic vein, a superficial vein of the upper limb, joins the axillary vein by passing through the costocoracoid membrane at the deltopectoral triangle. The costoaxillary veins drain into the lateral thoracic vein. They form connections between the intercostal veins and the lateral thoracic vein, and therefore serve as a shunt between the azygos and caval venous systems.

BRACHIAL PLEXUS

The brachial plexus (Fig. 5-26 and Table 5-4) innervates the upper limb. It arises from the ventral rami of five spinal nerves (C5–T1) and contains muscular, sensory, and autonomic fibers.

The roots pass out of the vertebral column and come to lie between the anterior and middle scalene muscles, picking up sympathetic fibers at this point. From here they pass outward and downward (except for T1, which passes upward) to reach the axillary region.

There are five levels of organization in the brachial plexus: roots, trunks, divisions, cords, and terminal nerves. Thus the terminal nerves are a complex of portions of various nerve roots. In learning this complex structure an understanding of the organization and distribution of the

Table 5-4. Brachial Plexus

Roots	*Trunks*	*Divisions*	*Cords*	*Nerves*
C5 and C6	Superior	Anterior	Lateral	Nerve to subclavius, suprascapular, lateral pectoral, musculocutaneous, median
		Posterior	Posterior	Dorsal scapular, long thoracic, upper and lower subscapular, axillary, radial
C7	Middle	Anterior	Lateral	Lateral pectoral, musculocutaneous, median
		Posterior	Posterior	Long thoracic, middle subscapular, radial
C8 and T1	Inferior	Anterior	Medial	Medial brachial cutaneous, medial antebrachial cutaneous, medial pectoral, median, ulnar
		Posterior	Posterior	Middle subscapular, radial

fibers will aid the student in determining the final composition of the various nerves.

The roots of C5 and C6 fuse to become the superior trunk. The middle trunk is formed by C7. The roots of C8 and T1 fuse to become the inferior trunk.

Each trunk divides into an anterior (flexor, preaxial) and a posterior (extensor, postaxial) division. All the posterior divisions join together to form the posterior cord. The anterior divisions of the superior and middle trunks (C5–C7) fuse to become the lateral cord. The medial cord is formed by the anterior division of the inferior trunk (C8 and T1).

Each cord is named in relation to the axillary artery at its second part, deep to the pectoralis minor muscle. Finally both the medial and lateral cords contribute to the median nerve. In addition, the lateral cord gives rise to the musculocutaneous nerve, while the medial cord gives rise to the ulnar nerve. The posterior cord divides into axillary and radial nerves.

The five terminal nerves of the brachial plexus are the median, ulnar, musculocutaneous, axillary, and radial nerves. Additional nerves are given off the brachial plexus at various stages in the distribution of the fibers. These nerves fall into two general categories: supraclavicular, or those given off above the clavicle, and infraclavicular, or those given off below the clavicle.

The four supraclavicular nerves are the dorsal scapular nerve, the long thoracic nerve, the suprascapular nerve, and the nerve to the subclavius. The dorsal scapular nerve (posterior division C4 and C5) pierces the middle scalene muscle to reach and supply the levator scapulae and rhomboid muscles. The long thoracic nerve (posterior division C5–C7) pierces the middle scalene muscle and descends to run along the serratus anterior muscle, which it supplies (see Fig. 5-24). The suprascapular nerve (posterior division C4–C6) passes through the suprascapular notch to reach the supraspinatus, and then continues through the notch of the neck of the scapula to reach the infraspinatus. The nerve to the subclavius (anterior division C4–C6) supplies the subclavius muscle.

The seven infraclavicular branches are the subscapular nerves (upper, middle, and lower), the lateral pectoral nerve, the medial pectoral nerve, the medial brachial cutaneous nerve, and the medial antebrachial cutaneous nerve (see Fig. 5-24). The upper and lower subscapular nerves (posterior division C5 and C6) supply the subscapularis; the lower nerve also supplies the teres major. The middle subscapular nerve (posterior division C7 and C8), also known as the thoracodorsal nerve, supplies the latissimus dorsi. The lateral pectoral nerve arises from the lateral cord (anterior division C5–C7) and supplies the pectoralis major. The medial pectoral nerve arises from the medial cord (anterior division C8 and T1) and supplies both the pectoralis major and minor. The medial and lateral pectoral nerves are connected by a communicating loop. The medial brachial cutaneous nerve arises from the medial cord (anterior division C8 and T1) and supplies cutaneous innervation to the skin of the medial lower third of the arm. The medial antebrachial cutaneous nerve also arises from the medial cord (anterior division C8 and T1); in the forearm it supplies cutaneous innervation to both the anterior and posterior medial forearm.

The axillary nerve is the only terminal branch of the brachial plexus to end in the axilla. Therefore it is considered below. The other terminal nerves are considered in greater detail later in this chapter. Briefly, however, the radial nerve supplies all extensor (postaxial) innervation below the shoulder; the musculocutaneous nerve supplies all flexor (preaxial) innervation in the arm; the median nerve is the primary nerve of the flexor forearm and secondary nerve of the flexor hand; the ulnar nerve is the primary nerve of the flexor hand and the secondary nerve of the flexor forearm.

The axillary nerve is a terminal nerve of the posterior cord. It arises at the lower border of the subscapularis muscle, crosses the upper border of the teres major muscle, and enters the quadrangular space, where it is accompanied by the posterior circumflex humeral artery. It supplies the shoulder joint, the deltoid muscle, and the teres minor muscle. It also gives rise to the lateral brachial cutaneous nerve, which supplies cutaneous innervation to the middle third of the lateral arm.

In considering branches of the brachial plexus, the associations between the anterior trunk divisions and the anterior flexor muscles and the posterior trunk divisions and the posterior extensor muscles are straightforward in the arm and forearm. Around the shoulder, however, the separation is not so immediately visible. It should, therefore, be remembered that those muscles that arise or insert on the scapula (except its coracoid process) are innervated by nerves from posterior divisions and those muscles that arise or insert on the clavicle or coracoid process are innervated by nerves from anterior divisions (see Fig. 5-6).

Rule of Five

The anatomy of the brachial plexus is extremely complicated, but the student may be aided in understanding it through the use of the Rule of Five.

1. The brachial plexus arises from *five* nerve roots beginning with C5.
2. There are *five* organizational categories (roots, trunks, divisions, cords, nerves).
3. There are *five* terminal nerves (axillary, radial, musculocutaneous, median, ulnar).
4. There are *five* nerves from every trunk to cord complex:
   a. medial cord to medial pectoral, medial brachial cutaneous, medial antebrachial cutaneous, ulnar, median.
   b. posterior cord to upper, middle, and lower subscapular; axillary; radial.
   c. lateral cord to lateral pectoral, musculocutaneous, median, suprascapular, nerve to subclavius.
5. The fifth rule is the exception. There are *two* nerves off the roots: the dorsal scapular and long thoracic nerves.

LYMPHATICS

There are five lymph node subgroups in the axilla defined in relation to the walls of the axilla: lateral, pectoral, subscapular, central, and apical. The lateral nodes drain the upper limb. They empty into the central and apical nodes. The pectoral nodes drain the anterolateral thoracic wall and mammary gland. They also empty into the central and apical nodes. The subscapular nodes drain the posterior thoracic wall, scapular region, and lower posterior neck. They empty into the central nodes.

The central nodes lie centrally within the axillary fat. They receive lymph from all other axillary nodal groups except the apical nodes. The central axillary nodes drain into the apical nodes. The apical axillary nodes receive all upper limb drainage, including independent drainage from the mammary glands and cephalic vein lymph channels. The subclavian lymphatic trunk begins here and continues on to empty into either the

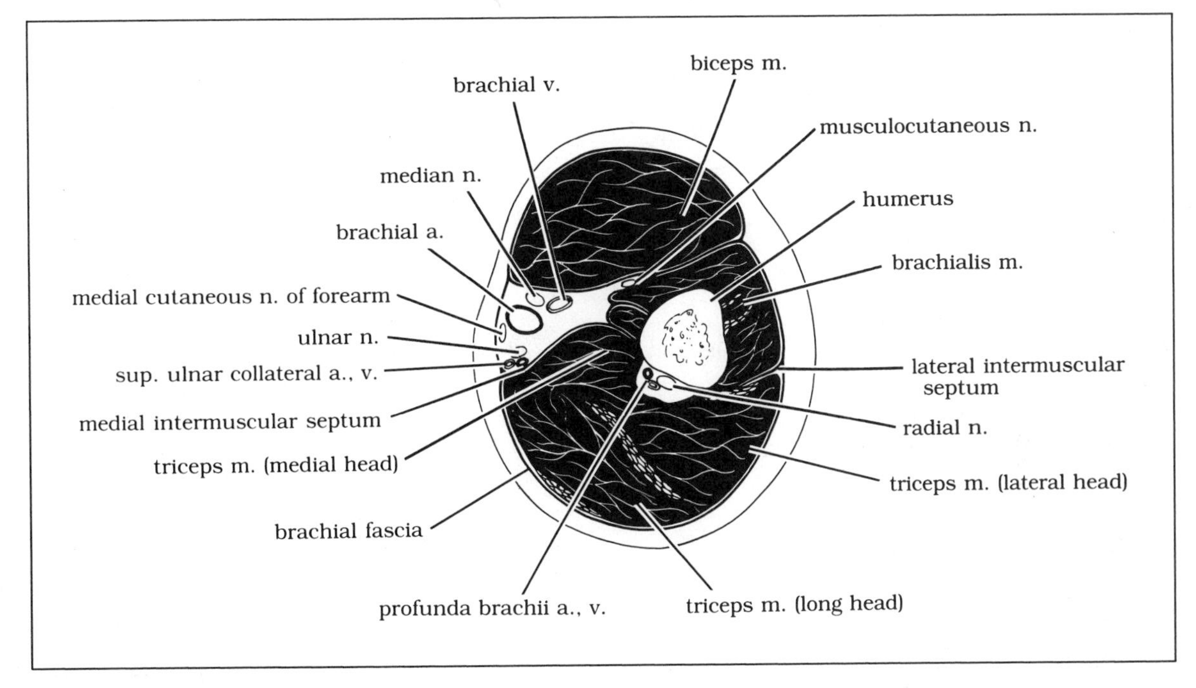

Fig. 5-27. Cross section of the arm.

thoracic duct on the left, or the junction of the internal jugular and subclavian veins on the right.

Arm

The arm is contained within a sleeve of investing fascia known as the brachial fascia. Two intermuscular septa (medial and lateral) attach to it, separating the arm into flexor (anterior, preaxial) and extensor (posterior, postaxial) compartments (Fig. 5-27).

MUSCULATURE

The flexor muscles are the coracobrachialis, biceps brachii, and brachialis muscles; the extensor muscles are the triceps brachii and anconeus (Table 5-5). All the flexor muscles are innervated by the musculocutaneous nerve, all the extensors by the radial nerve.

The **coracobrachialis muscle** (see Figs. 5-22A and 5-23) arises on the coracoid process and inserts on the humerus. It flexes the arm at the shoulder joint. The musculocutaneous nerve passes through this muscle as it innervates it.

The **biceps brachii muscle** (see Figs. 5-18, 5-21, and 5-23) arises from two heads. The origin of the long head is the supraglenoid tubercle of the scapula, and the origin of the short head is the coracoid process. The long head passes within the joint capsule, exiting it as it passes through the intertubercular groove. The two heads unite and end by inserting on the radial tuberosity of the radius. In addition the muscle gives off an aponeurosis (the bicipital aponeurosis) that runs medially to blend with the antebrachial fascia over the ulna. The biceps brachii flexes the shoulder and elbow joints and supinates the forearm.

The **brachialis muscle** (see Figs. 5-21, 5-22, and 5-30) arises from the anterior surface of the humerus. It inserts on the tuberosity of the ulna and flexes the elbow joint.

The **triceps brachii muscle** (see Figs. 5-21, 5-22, and 5-30) arises from three heads. The long head arises from the infraglenoid ridge of the scapula; the lateral head arises from the poste-

Table 5-5. Musculature of the Arm

Muscle	*Origin*	*Insertion*	*Innervation*	*Action*
Coraco-brachialis	Coracoid process	Humerus	Musculocutaneous nerve	Flex arm
Biceps brachii	Long head—supraglenoid tubercle of scapula; short head—coracoid process	Radial tuberosity of radius, bicipital aponeurosis	Musculocutaneous nerve	Flex arm, flex and supinate forearm
Brachialis	Anterior humerus	Tuberosity of ulna	Musculocutaneous nerve	Flex forearm
Triceps brachii	Long head—infraglenoid ridge of scapula; lateral head—posterior humerus; medial head—posterior humerus	Olecranon	Radial nerve	Extend and adduct arm, extend forearm
Anconeus	Lateral epicondyle of humerus	Olecranon	Radial nerve	Extend forearm

rior humerus, lateral to and above the radial groove; and the medial head arises from the posterior humerus below the radial groove. The tendon of the long head helps to form the triangular and quadrangular spaces. The triceps brachii muscle inserts on the upper surface of the olecranon. The muscle extends the elbow. In addition the long head helps to extend and adduct the arm at the shoulder.

The **anconeus** (see Fig. 5-31) is a small muscle that arises from the lateral epicondyle of the humerus. It inserts on the olecranon and extends the forearm at the elbow.

BLOOD SUPPLY AND LYMPHATICS

The axillary artery becomes the brachial artery (see Fig. 5-25) as it passes the teres major muscle. The **brachial artery** gives rise to the deep brachial, nutrient humeral, inferior and superior ulnar collateral, and some muscular arteries. It ends at the cubital fossa by dividing into radial and ulnar arteries.

The **deep brachial artery** is the largest branch of the brachial artery. It accompanies the radial nerve to the posterior compartment, traveling with it in the radial groove. The deep brachial artery supplies the muscles of the posterior compartment and gives off branches that anastomose at both the shoulder and elbow joints. The **nutrient humeral artery** supplies the humerus. The **superior ulnar collateral artery** travels with the ulnar nerve as it passes posteriorly. At anastomoses around the elbow. The **inferior ulnar collateral artery** arises just above the medial epicondyle. It anastomoses around the elbow joint and also supplies some of the proximal forearm.

The veins of the arm follow the same course as their corresponding arteries. The cubital lymph nodes drain via two or three lymphatic trunks to the lateral and central axillary nodes.

INNERVATION

The median and ulnar nerves pass through the arm but do not supply any structures. The median nerve passes through the medial arm (Fig.

5-28A); the ulnar nerve is initially anterior and medial, but midway through the arm it becomes posterior and medial (Fig. 5-28B).

The **musculocutaneous nerve** (see Figs. 5-24, 5-26, and 5-28A) supplies the coracobrachialis, biceps brachii, and brachialis muscles. It passes lateral to the axillary artery initially and then pierces the coracobrachialis muscle to run between the biceps and brachialis muscles. It supplies only cutaneous fibers to the forearm as the lateral antebrachial cutaneous nerve. It also innervates the humerus and the elbow joint.

The **radial nerve** (see Figs. 5-27 and 5-28C) supplies the triceps brachii and anconeus muscles in the arm and the elbow joint. It passes diagonally across the back of the humerus in the radial groove, the deep brachial artery accompanying it in this course. The nerve and artery thus separate the medial and lateral heads of the triceps. After leaving the groove the nerve becomes anterior and lies between the brachialis and brachioradialis muscles. It finally leaves the arm by passing in front of the lateral epicondyle.

Cubital Fossa

The cubital fossa (see Fig. 5-30) is the soft tissue region in front of the elbow. Its superior margin is the epicondyles of the humerus, and its in-

Fig. 5-28. A. Distribution of the musculocutaneous and median nerves. B. Distribution of the ulnar nerve. C. Distribution of the radial nerve.

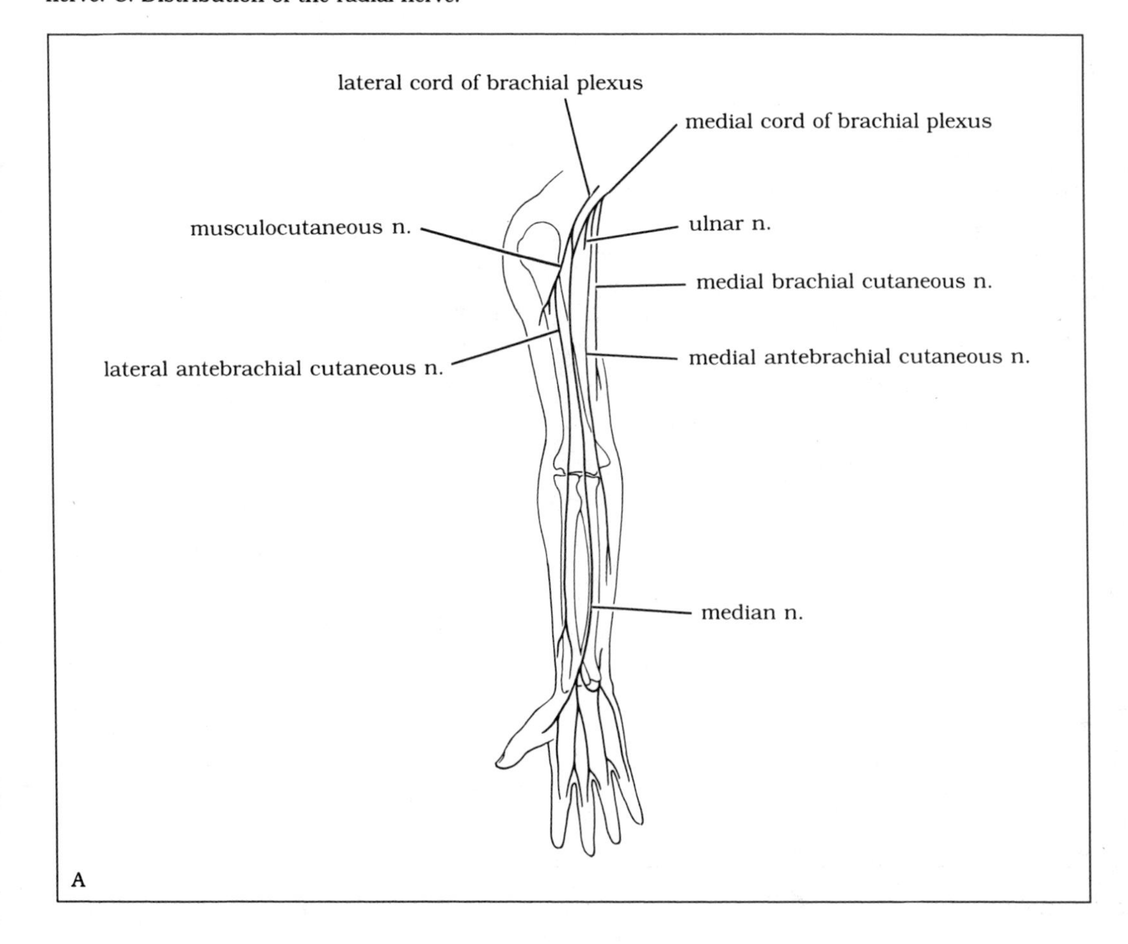

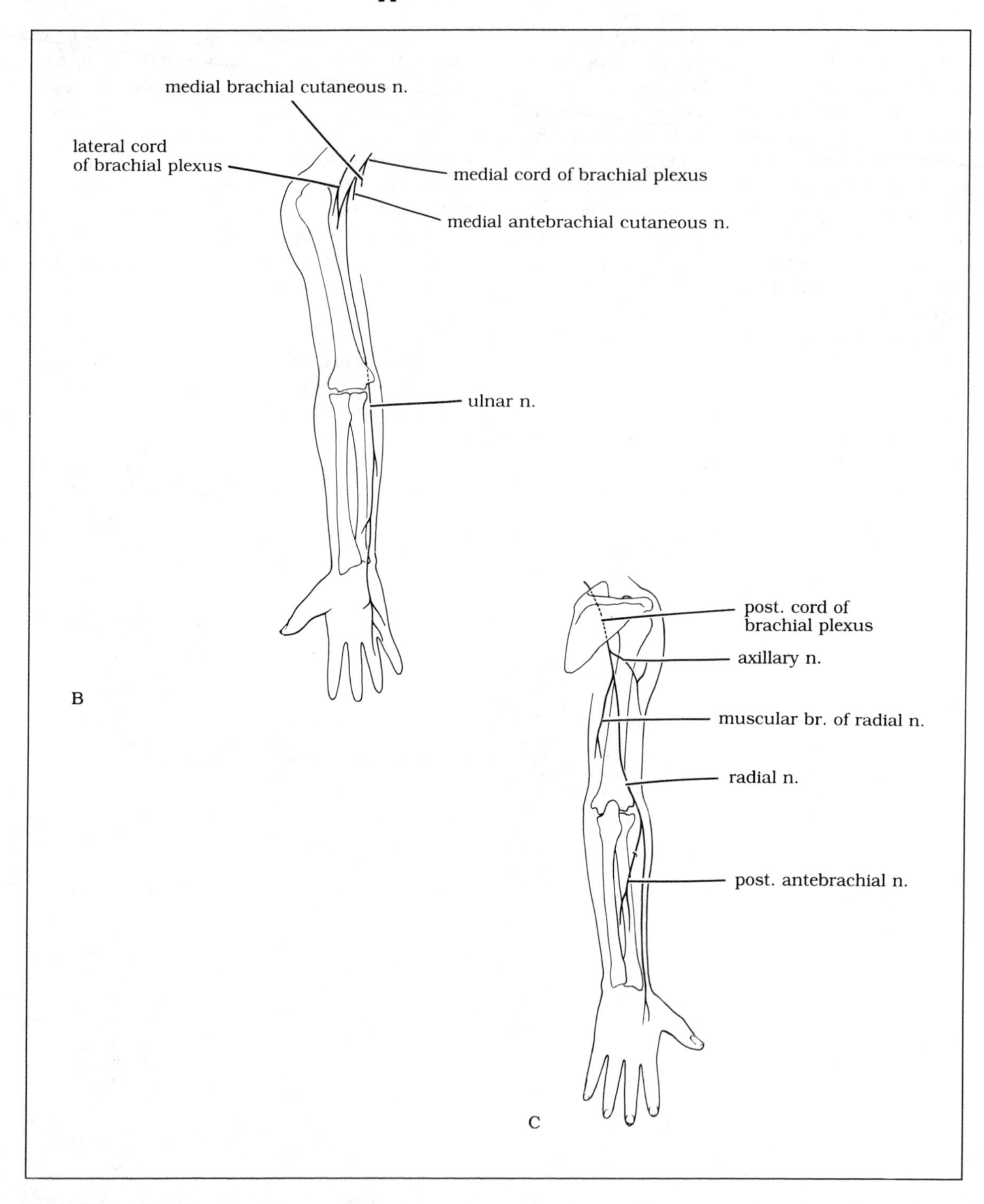

Fig. 5-28 (continued).

ferior margin is the pronator teres and brachioradialis muscles. The floor of the cubital fossa is formed by the brachialis and supinator muscles. The cubital fossa contains the biceps tendon, bicipital aponeurosis, median nerve, brachial artery (with its division to the radial and ulnar arteries), and brachial vein (with its radial and ulnar tributaries). The cubital fossa is crossed by the median cubital vein, cephalic vein, and lateral and medial antebrachial cutaneous nerves.

Forearm

The brachial fascia of the arm continues into the forearm as an investing fascial sleeve known as the antebrachial fascia. The **antebrachial fascia** attaches directly to the subcutaneous border of the ulna; it is attached to the radius by a fascial septum. This configuration, plus the interosseous membrane connecting the two bones, divides the forearm into two compartments: an anterior, preaxial flexor compartment and a posterior, postaxial extensor compartment (Fig. 5-29). At the wrist the antebrachial fascia is thickened and strengthened by the flexor and extensor retinacula.

MUSCULATURE

In considering the action of all muscles in the body, but in particular the fine actions at the forearm and the wrist, the student must understand that muscles normally act not alone but rather synergistically to balance each other out, thus allowing for the fine positioning particularly evident in hand motions (Table 5-6).

Flexor Muscles

The muscles of the flexor forearm are innervated mainly by the median nerve; the ulnar nerve supplies only one and a half muscles in this group. The flexor muscles fall into three groups: those that flex the hand at the wrist, those that flex the digits, and those that rotate the radius on the ulna. The brachioradialis acts to flex the elbow, but is innervated by the radial nerve and is therefore discussed with the extensor muscles.

The muscles of the flexor forearm may be designated superficial, intermediate, or deep. The superficial and intermediate layers all arise in common from the medial epicondyle of the humerus, the intermuscular septa, and the overlying fascia. Because of this common origin these muscles also have some slight action at the elbow joint. The muscles of the superficial group are (medially to laterally) the pronator teres, flexor carpi radialis, palmaris longus, and flexor carpi ulnaris (Fig. 5-30A). The flexor digitorum superficialis is the only muscle in the intermediate group (Fig. 5-30B). The deep group contains the flexor digitorum profundus, flexor pollicis longus, and pronator quadratus (Fig. 5-30C). (See Figure 5-11 for origins and insertions of these muscles.)

The **pronator teres muscle** arises from both the common flexor origin on the humerus and the coronoid process of the ulna. It inserts on the middle of the lateral radius. As its name implies its action is pronation. The pronator teres is innervated by the median nerve, which enters the forearm by passing between the muscle's two heads.

The **flexor carpi radialis** arises from the common flexor tendon. Its tendon passes under the flexor retinaculum, just medial to the radial artery, to insert on the base of the second metacarpal bone. The flexor carpi radialis flexes the radial side of the hand and, in conjunction with the extensor carpi radialis, abducts the hand. This muscle is innervated by the median nerve.

The **palmaris longus** arises from the common flexor tendon. It passes over the flexor retinaculum, inserting into it and into the palmar aponeurosis. The palmaris longus flexes the wrist and tenses the palmar fascia in grabbing and holding. It is innervated by the median nerve.

The **flexor carpi ulnaris** arises from the common flexor tendon, the medial border of the olecranon, and the posterior upper two-thirds of the ulna. This muscle inserts into the pisiform bone and by ligamentous extensions into the hook of the hamate and the fifth metacarpal. The flexor carpi ulnaris flexes the ulnar side of the hand and, with the extensor, adducts the hand. It is innervated by the ulnar nerve.

Table 5-6. Musculature of the Forearm

Muscle	*Origin*	*Insertion*	*Innervation*	*Action*
Pronator teres	Common flexor tendon, coronoid process of ulna	Middle of radius	Median nerve	Pronation
Flexor carpi radialis	Common flexor tendon	Base of metacarpal 2	Median nerve	Flex radial side of hand, or abduct hand
Palmaris longus	Common flexor tendon	Palmar aponeurosis	Median nerve	Flex wrist, tense palmar fascia
Flexor carpi ulnaris	Common flexor tendon, olecranon, posterior ulna	Pisiform	Ulnar nerve	Flex ulnar side of hand, or adduct hand
Flexor digitorum superficialis	Common flexor tendon, ulnar collateral ligament, coronoid process, upper anterior border of radius	Second phalanx of digits 2–5	Median nerve	Flex digits
Flexor digitorum profundus	Ulna	Base of distal phalanx of digits 2–5	Median and ulnar nerves	Flex digits
Flexor pollicis longus	Anterior radius	Base of distal phalanx of thumb	Median nerve	Flex thumb
Pronator quadratus	Distal anterior ulna	Distal anterior radius	Median nerve	Pronation
Brachioradialis	Supracondylar ridge	Styloid process of radius	Radial nerve	Flex forearm
Extensor carpi radialis longus	Supracondylar ridge	Dorsum of metacarpal 2	Radial nerve	Extend radial side of hand, or abduct hand
Extensor carpi radialis brevis	Common extensor tendon	Dorsum of metacarpal 3	Radial nerve	Extend radial side of hand, or abduct hand
Extensor digitorum	Common extensor tendon	Digits 2–5	Radial nerve	Extend digits
Extensor digiti minimi	Common extensor tendon	Digit 5	Radial nerve	Extend digit 5
Extensor carpi ulnaris	Common extensor tendon and posterior border of ulna	Base of metacarpal 5	Radial nerve	Extend ulnar side of hand, or adduct hand

Table 5-6 (continued).

Muscle	*Origin*	*Insertion*	*Innervation*	*Action*
Supinator	Lateral epicondyle of humerus, supinator crest and fossa of ulna, annular ligament of radius, radial collateral ligament of elbow	Upper lateral radius	Radial nerve	Supination
Abductor pollicis longus	Middle of radius, posterior ulna, interosseous membrane	Base of metacarpal 1	Radial nerve	Abduct thumb
Extensor pollicis longus	Middle of ulna, interosseous membrane	Base of distal phalanx of thumb	Radial nerve	Extend thumb
Extensor pollicis brevis	Distal posterior radius, interosseous membrane	Base of proximal phalanx of thumb	Radial nerve	Extend thumb
Extensor indicis	Lower ulna	Second digit, with tendon of extensor digitorum	Radial nerve	Extend digit 2

Fig. 5-29. Cross section of the forearm.

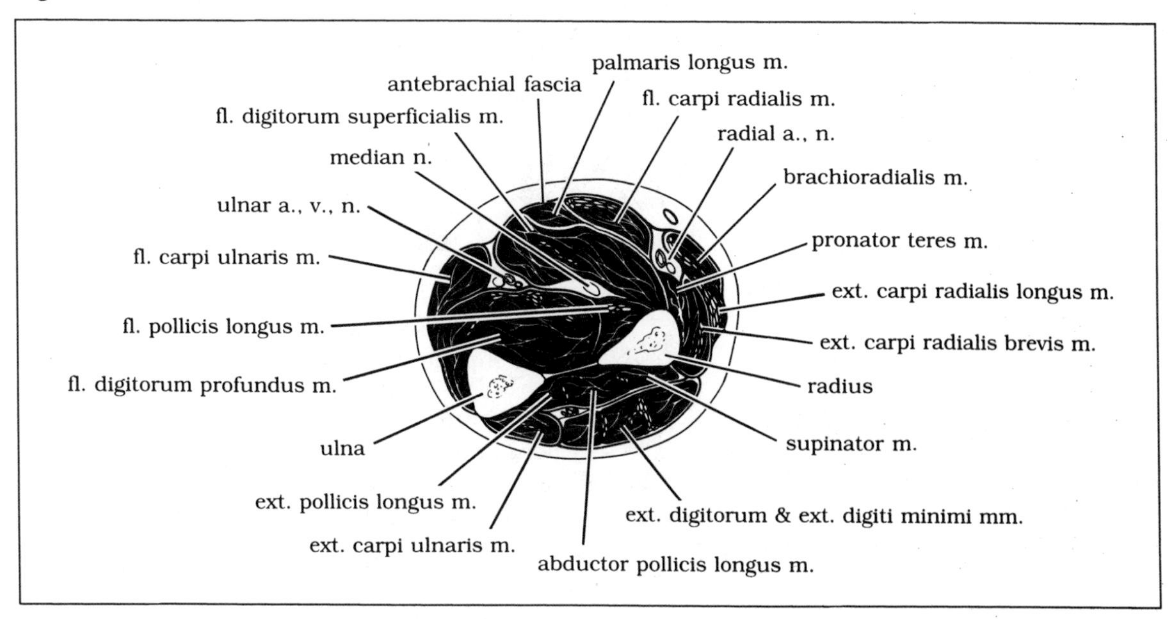

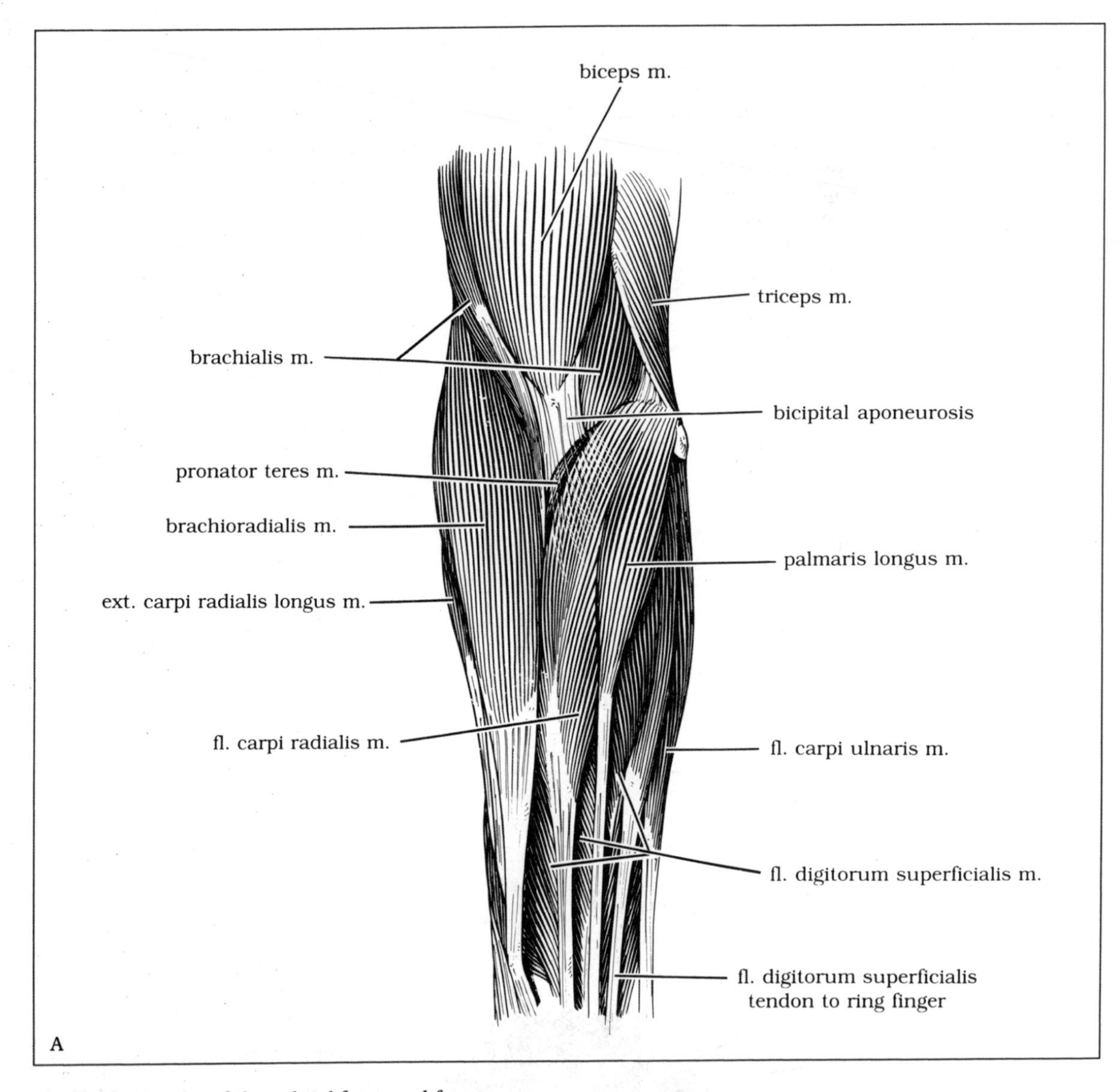

Fig. 5-30. Muscles of the cubital fossa and forearm. A. Superficial dissection. B. Deep dissection. C. Deepest dissection.

The **flexor digitorum superficialis** arises from two heads. The first originates from the common flexor tendon, the ulnar collateral ligament of the elbow, and the medial border of the coronoid process. The second originates from the upper two-thirds of the anterior border of the radius. The muscle then unites and splits into two planes. The superficial plane provides tendons for the third and fourth digits, the deep plane the tendons for the second and fifth digits. The four tendons pass under the flexor retinaculum in a common synovial sheath. In the hand they are contained in a fibrous sheath with the tendons of the deep flexor muscle. The tendons of the superficial flexor split within their digital

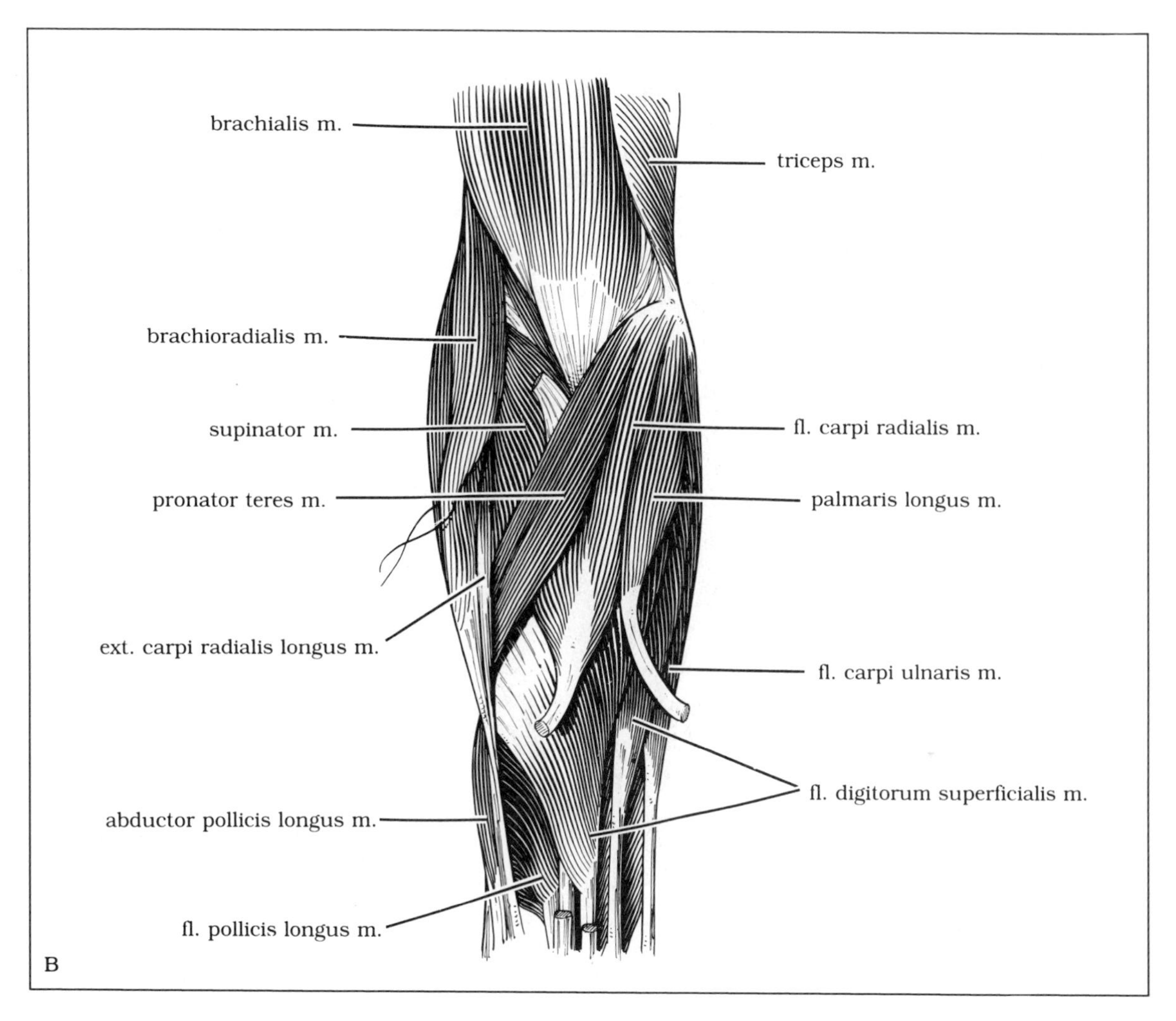

Fig. 5-30 (continued).

sheaths to allow the deep flexor tendons to continue past them. They then reunite to insert on the second phalanx of the appropriate digit (digits 2–5). The flexor digitorum superficialis is innervated by the median nerve and flexes the digits at the proximal interphalangeal joint.

The **flexor digitorum profundus** arises from the posterior ulna with the flexor carpi ulnaris, from the proximal two-thirds of the median ulna, and from the interosseous membrane. This muscle also splits into four tendons that pass under the flexor retinaculum. These tendons come to lie in fibrous digital sheaths with the accompanying superficial tendon, pass through the split in the superficial tendon, and insert on the base of the distal phalanx. The flexor digitorum profundus is innervated by both the median and ulnar nerves. It flexes the digits at the distal interphalangeal joint.

The **flexor pollicis longus** arises from the anterior radius and the interosseous membrane. Its tendon passes under the flexor retinaculum to insert into the base of the distal phalanx of the thumb. It is innervated by the median nerve and flexes the thumb at its distal joint.

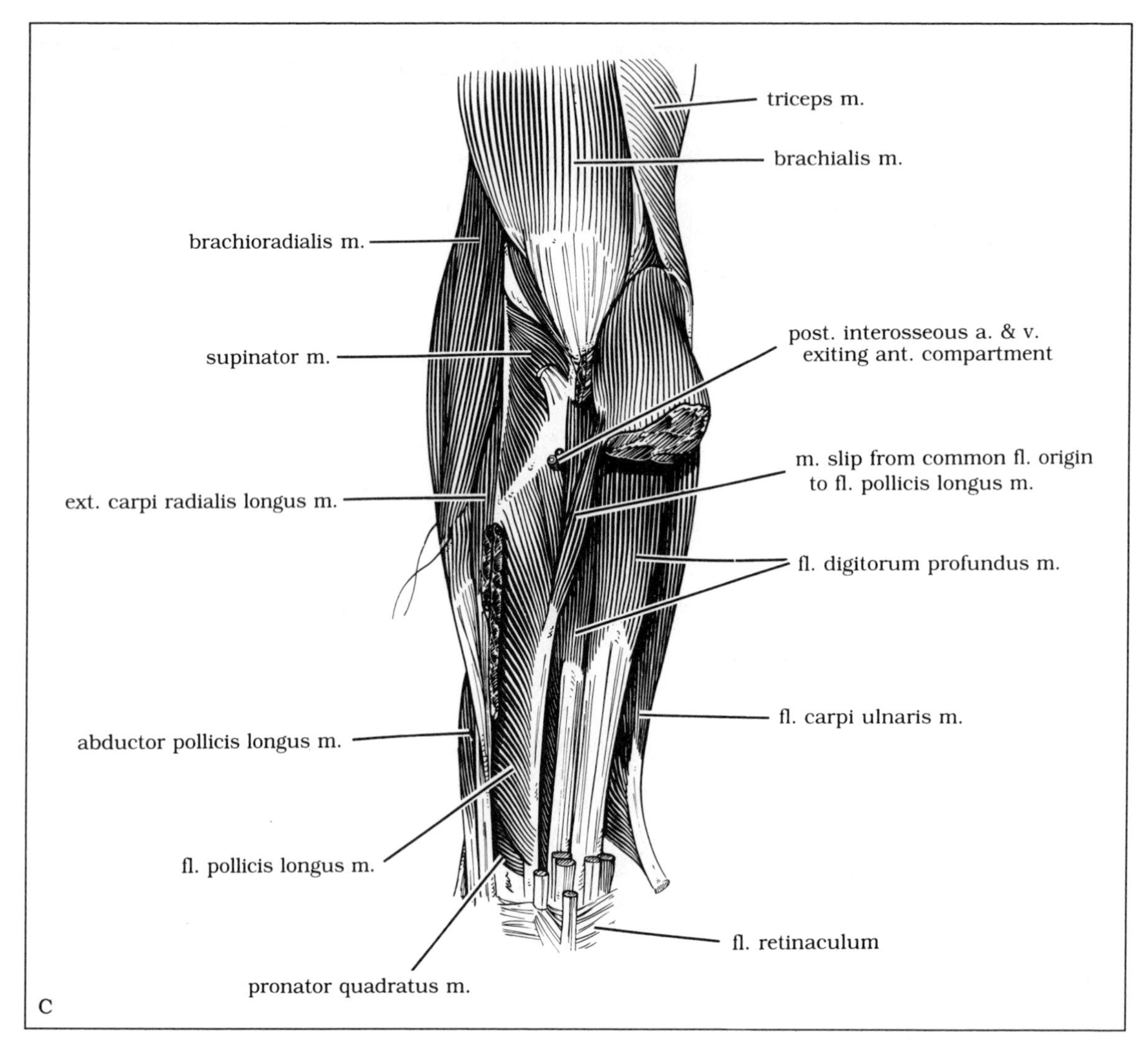

Fig. 5-30 (continued).

The **pronator quadratus** arises from the anterior medial distal fourth of the ulna and inserts on the anterior distal fourth of the radius. The pronator quadratus rotates the radius on the ulna (pronation). It is innervated by the median nerve.

Extensor Muscles

The extensor muscles of the forearm are all innervated by the radial nerve. These muscles fall into three groups: those that extend the hand at the wrist, those that extend the digits, and those that extend the thumb. In addition, the supinator, which acts in rotation of the radius on the ulna, is included here. Finally, because of its innervation by the radial nerve, the brachioradialis muscle, which flexes the elbow, is considered with the extensors.

The extensor muscles may be divided into superficial and deep groups. The superficial group contains, laterally to medially, the brachio-

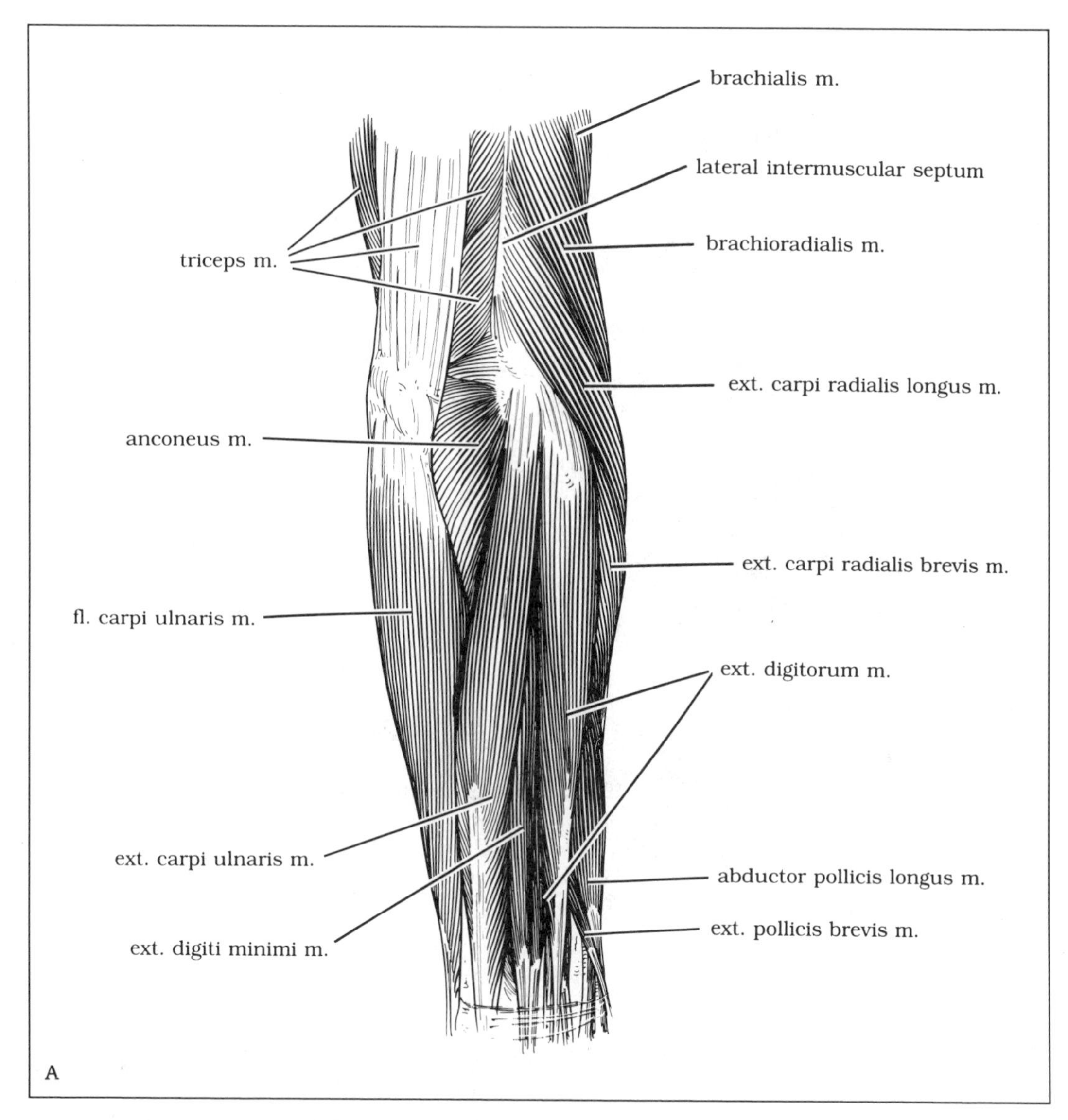

Fig. 5-31. Muscles of the posterior forearm.
A. Superficial dissection. B. Deep dissection.
C. Deepest dissection.

radialis, extensor carpi radialis longus, extensor carpi radialis brevis, extensor digitorum, extensor digiti minimi, and extensor carpi ulnaris (Fig. 5-31A). All of these muscles except the first two arise from the lateral epicondyle of the humerus via a common tendon. The brachioradialis and extensor carpi radialis longus arise from the supracondylar ridge of the humerus. The muscles of the deep extensor group are the supinator, abductor pollicis longus, extensor pollicis longus, extensor pollicis brevis, and extensor indicis (Fig. 5-31B,C).

The **brachioradialis** muscle arises from the supracondylar ridge and inserts on the lateral side of the base of the styloid process of the radius

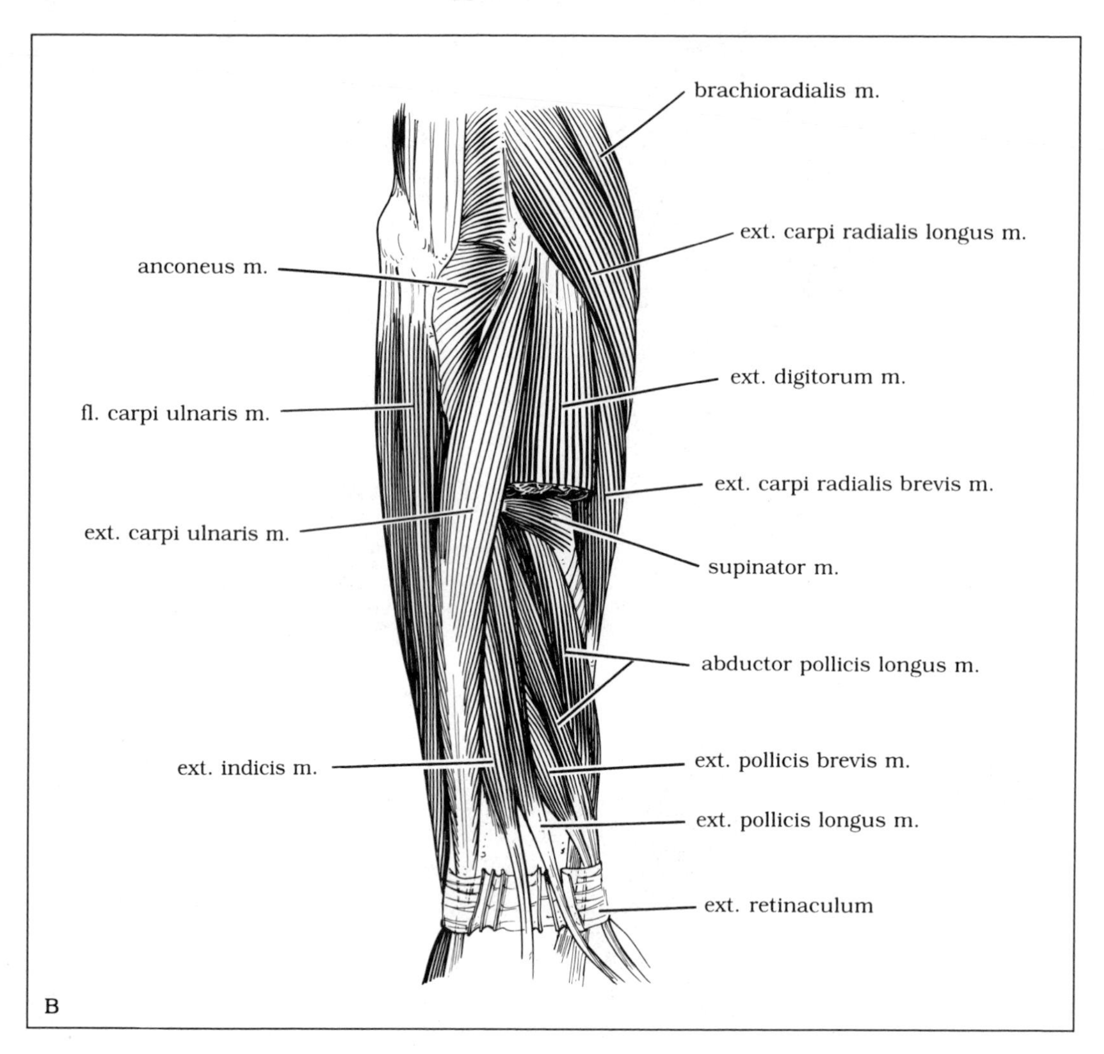

Fig. 5-31 (continued).

(see Figs. 5-30 and 5-31A,B). It flexes the forearm.

The **extensor carpi radialis longus** arises from the supracondylar ridge and inserts on the dorsum of the second metacarpal. It extends the wrist on the radial side and, in conjunction with the flexor, abducts the hand.

The **extensor carpi radialis brevis** arises from the common extensor tendon and inserts on the dorsum of the third metacarpal. It acts in conjunction with the extensor carpi radialis longus, with which it passes under the extensor retinaculum.

The **extensor digitorum** arises from the common extensor tendon and divides into four tendons that pass under the extensor retinaculum together. These tendons then flatten out over the metacarpophalangeal joints to become the extensor expansion (see Fig. 5-33). Each expansion divides into three slips, the central one going to the middle phalanx and the lateral ones to the distal phalanx. The tendons of the lumbricals and interossei join the lateral tendinous slips. The muscle extends the digits.

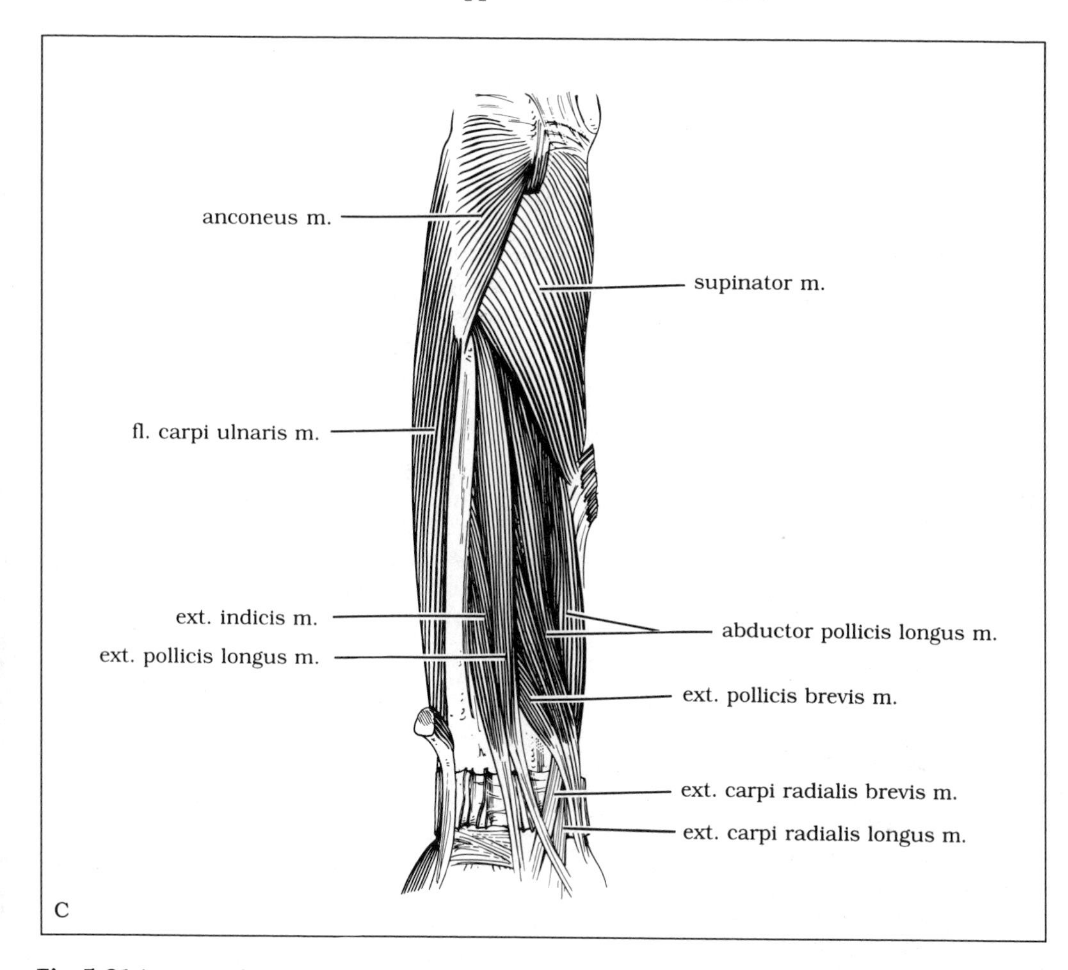

Fig. 5-31 (continued).

The **extensor digiti minimi** arises from the common extensor tendon. It inserts by joining with the extensor expansion of the fifth digit. The muscle extends the fifth digit.

The **extensor carpi ulnaris** arises from both the common extensor tendon and the middle two-fourths of the posterior border of the ulna. It ends by inserting into the base of the fifth metacarpal. The muscle, with the extensor carpi radialis, extends the hand at the wrist and, with the flexor carpi ulnaris, adducts the hand.

The **supinator** has a multiple origin from the lateral epicondyle of the humerus, the radial collateral ligament of the elbow, the annular ligament of the radius, and the supinator crest and fossa of the ulna. Its fibers go downward and laterally to wrap around the radius and end by inserting on the lateral surface of the upper third of the radius. The muscle supinates the radius on the ulna. The deep radial nerve runs between the superficial and the deep layers of this muscle.

The **abductor pollicis longus** arises from the middle third of the radius, the lateral posterior ulna, and the intervening interosseous membrane. It inserts on the radial side of the base of the first metacarpal. The muscle abducts the thumb.

The **extensor pollicis longus** arises from the middle of the ulna, below the origin of the abductor pollicis longus, and from the adjacent interosseous membrane. It inserts into the base of the distal phalanx of the thumb and extends the thumb.

The **extensor pollicis brevis** arises from the third fourth of the radius and the adjacent interosseous membrane. It inserts into the base of the proximal phalanx of the thumb and, with the extensor pollicis longus, extends the thumb.

The **extensor indicis** arises from the ulna below the extensor pollicis longus. At the level of the second metacarpal it joins with the tendon of the extensor digitorum to the second digit. This muscle extends the second digit.

The **anatomical snuff box** is a depression formed on the dorsum of the hand by abduction and extension of the thumb. It is bordered by the tendons of the extensor pollicis longus and brevis. The radial artery is found at the bottom of the snuff box.

BLOOD SUPPLY AND LYMPHATICS

The brachial artery ends at the elbow by dividing into radial and ulnar arteries (Fig. 5-32). The radial artery gives rise to the following branches: radial recurrent, palmar carpal, superficial palmar, and muscular.

The **radial recurrent artery** arises below the elbow. It ascends on the supinator and ends by anastomosing with the radial collateral artery.

The **palmar carpal artery** arises at the distal edge of the pronator quadratus. It anastomoses with the palmar carpal branch of the ulnar artery, the anterior interosseous artery, and branches from the deep palmar arch.

The **superficial palmar artery** leaves the radial artery at the wrist. It passes over the thenar eminence to anastomose with the superficial palmar arch.

The muscular branches supply all the muscles of the radial forearm.

The ulnar artery gives rise to the following branches: anterior and posterior ulnar recurrents, common interosseous, palmar carpal, dorsal carpal, and muscular.

The **anterior ulnar recurrent artery** arises just below the elbow. It ascends to anastomose with the superior and inferior ulnar collateral arteries.

The **posterior ulnar recurrent artery** arises with, or just below, the anterior ulnar recurrent artery. It ascends behind the medial epicondyle and anastomoses with the superior and inferior ulnar collateral arteries and the interosseous recurrent artery.

The **common interosseous artery** passes deep to reach the upper edge of the interosseous membrane. It then divides into anterior and posterior interosseous arteries. The **anterior interosseous artery** descends on the interosseous membrane to the upper edge of the pronator quadratus. It then passes through the membrane to anastomose with the posterior interosseous artery. The **posterior interosseous artery** descends on the posterior surface of the interosseous membrane. At its beginning it gives off an interosseous recurrent artery that anastomoses with the middle collateral branch from the deep brachial artery. The posterior interosseous artery anastomoses at the distal forearm with the anterior interosseous artery.

The **palmar carpal artery** passes from the ulnar artery to the deep palmar arch.

The **dorsal carpal artery** passes to the dorsum of the hand to help form the dorsal arterial arch.

The muscular branches supply the muscles of the ulnar forearm.

The veins of the forearm follow the same pattern as the arteries and carry similar names. Lymphatic drainage of the forearm is accomplished through deep and superficial channels to the cubital nodes.

INNERVATION

The **ulnar nerve** enters the forearm between the two heads of the flexor carpi ulnaris (see Fig. 5-28). It then continues distally on the lateral side of this muscle. In the distal half of the forearm it is accompanied by the ulnar artery. On reaching the wrist the ulnar nerve divides into deep and superficial branches. In the forearm the ulnar nerve innervates the flexor carpi ulnaris and the medial part of the flexor digitorum profundus.

The **median nerve** enters the forearm between the two heads of the pronator teres. It then passes deep to the flexor digitorum superficialis

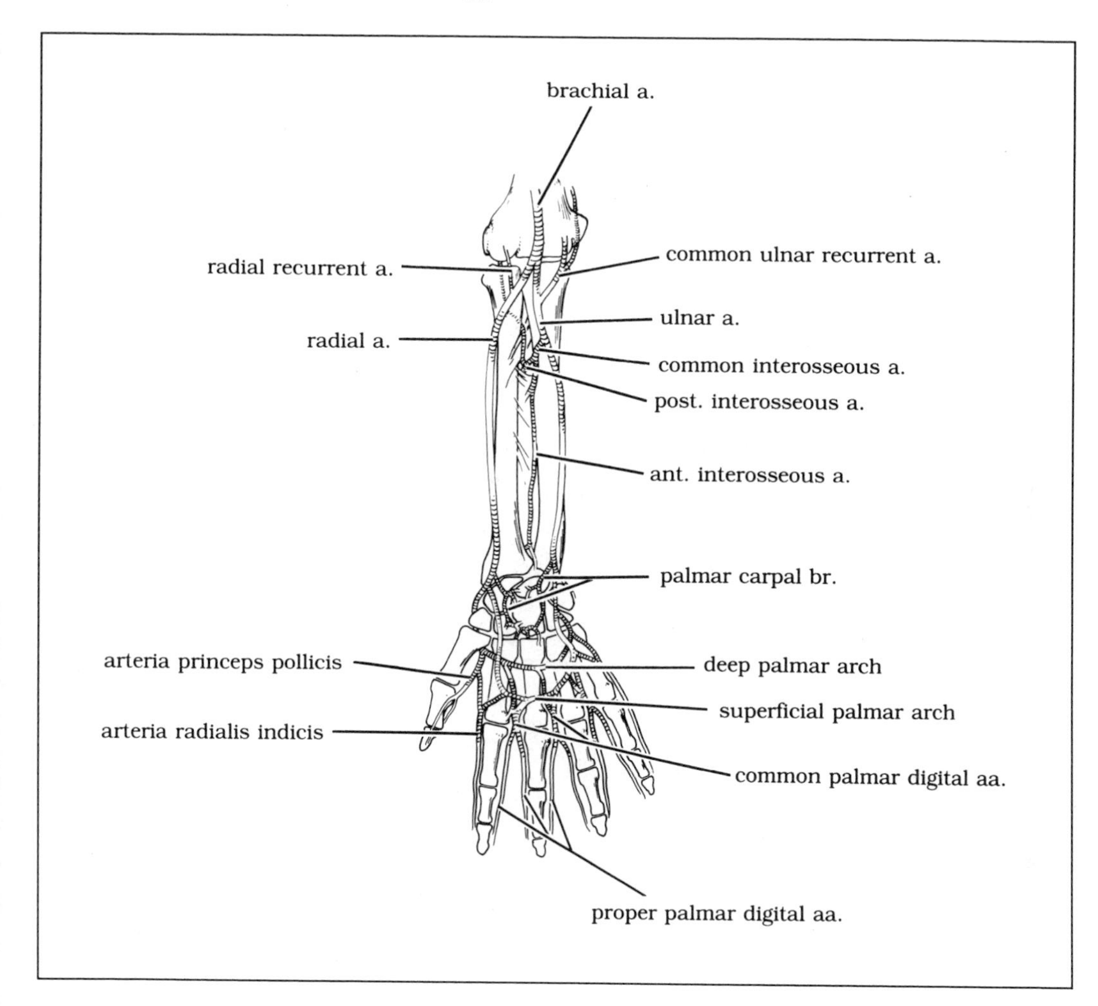

Fig. 5-32. Arterial supply to the forearm and palm.

to reach the wrist, where it passes under the flexor retinaculum on the radial side of the palmaris longus tendon. The median nerve innervates all the muscles of the flexor forearm not supplied by the ulnar nerve.

The **radial nerve** supplies all the muscles of the extensor forearm. At the level of the lateral epicondyle in the arm the radial nerve divides into superficial and deep branches. The superficial branch lies deep to the brachioradialis, in company with the radial artery. In the distal third of the forearm the nerve passes to the dorsum of the forearm and becomes subcutaneous. It ends by supplying cutaneous innervation to the dorsum of the hand and the digits. The deep branch is the muscular branch. It passes under the brachioradialis and extensor carpi radialis longus muscles to reach the supinator, which it pierces. It leaves the supinator and travels with the posterior interosseous artery as the posterior interosseous nerve. This nerve terminates at the wrist.

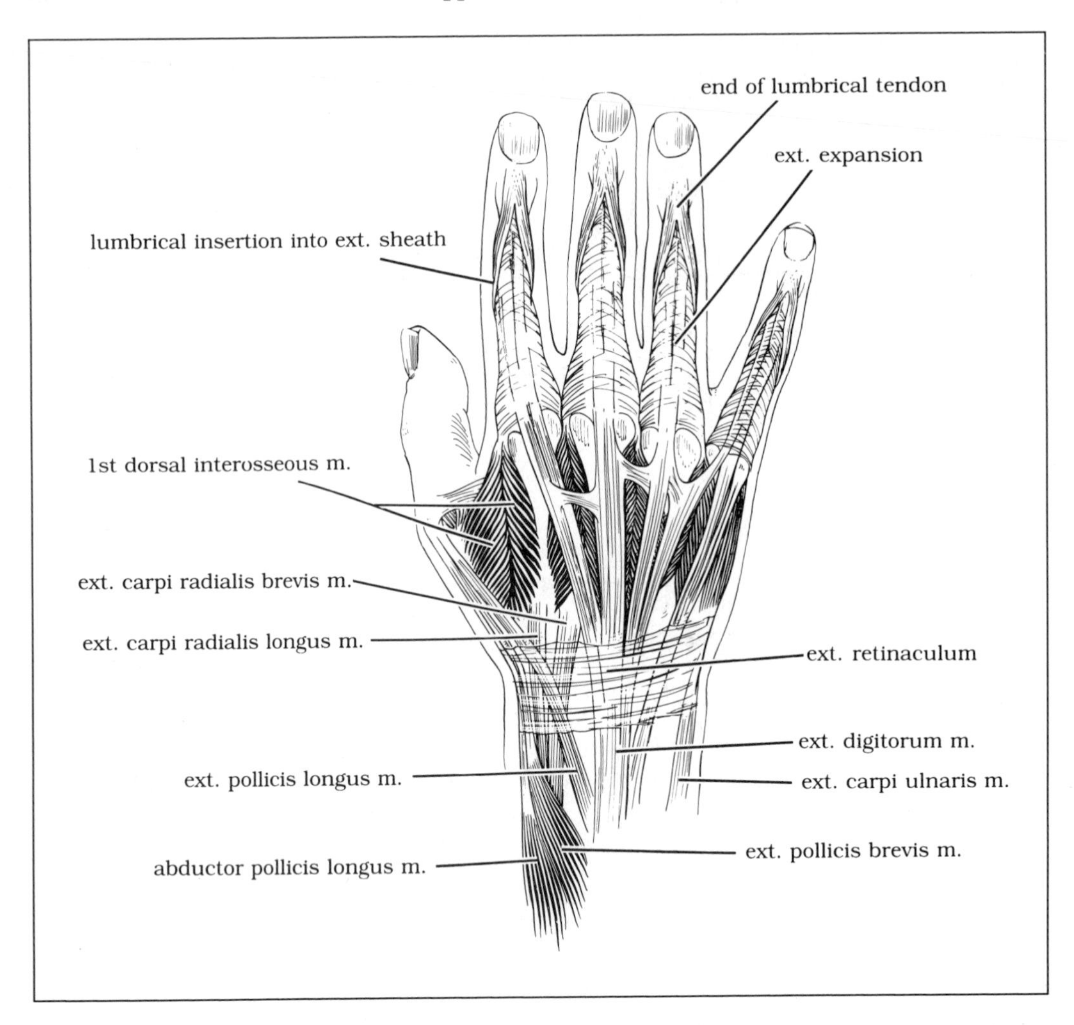

Fig. 5-33. Dorsum of the hand.

Wrist and Hand

WRIST

The wrist is the intermediate region between forearm and hand lying over the carpal bones. Tendons from the forearm muscles cross the wrist, lying in fibro-osseous canals lined by synovial sheaths. Most of these tendons lie under the retinacula at the wrist.

The **extensor retinaculum** is a thickening of the antebrachial fascia on the dorsum of the wrist. Six compartments containing nine tendons underlie it. The **flexor retinaculum** is a ligament that stretches across the concavity formed by the carpal bones. Only two synovial compartments underlie the flexor retinaculum. The contents of the compartments are as follows.

Extensor retinaculum (laterally to medially)

- Compartment 1—abductor pollicis longus, extensor pollicis brevis tendons
- Compartment 2—extensor pollicis longus tendon
- Compartment 3—extensor carpi radialis longus and brevis tendons
- Compartment 4—extensor digitorum, extensor indicis tendons

Compartment 5—extensor digiti minimi tendon
Compartment 6—extensor carpi ulnaris tendon

Flexor retinaculum (laterally to medially)
Compartment 1—flexor carpi radialis tendon
Compartment 2—flexor pollicis longus, flexor digitorum superficialis, and flexor digitorum profundus tendons, median nerve

Each compartment is lined by at least one synovial sheath. These sheaths may extend beyond the retinacular compartments into the hand, and onto the digits.

In addition to the tendons of the forearm muscles, vessels and nerves of the forearm also cross over the wrist. The median nerve passes under the flexor retinaculum on the radial side of the palmaris longus tendon. The ulnar nerve passes over the flexor retinaculum on the lateral side of the pisiform bone, the ulnar artery and vein traveling with it.

The deep radial nerve ends at the wrist, innervating the intercarpal joints. The radial artery

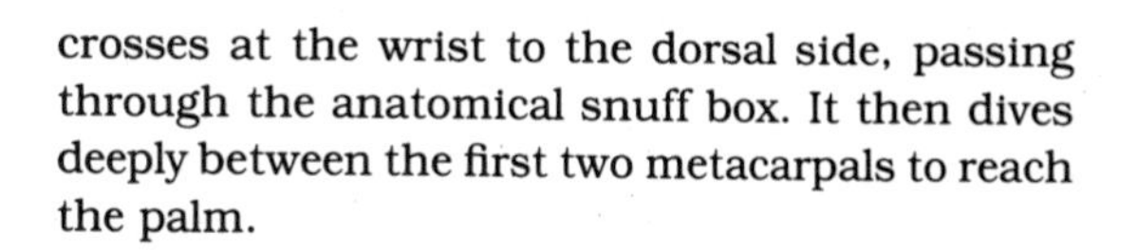

crosses at the wrist to the dorsal side, passing through the anatomical snuff box. It then dives deeply between the first two metacarpals to reach the palm.

HAND

Dorsum

The **dorsal fascia** of the hand is continuous with the extensor retinaculum and antebrachial fascia. Because there are no muscles intrinsic to the dorsum of the hand, the fascia contains only tendons and vessels (Fig. 5-33). The nerves in the dorsum are all cutaneous. The dorsal venous arch is formed in the subcutaneous fascia from veins draining dorsal structures.

The radial artery gives off two branches on the dorsum before it dives deeply to reach the palm of the hand. The branches are the dorsal carpal branch and the first dorsal metacarpal branch (Fig. 5-34). The **dorsal carpal branch** runs medially under the extensor tendons and forms the dorsal carpal arch. This arch gives rise to the second through the fourth dorsal metacarpal arteries. The **metacarpal arteries** divide into dorsal digital arteries, which proceed out along the digits.

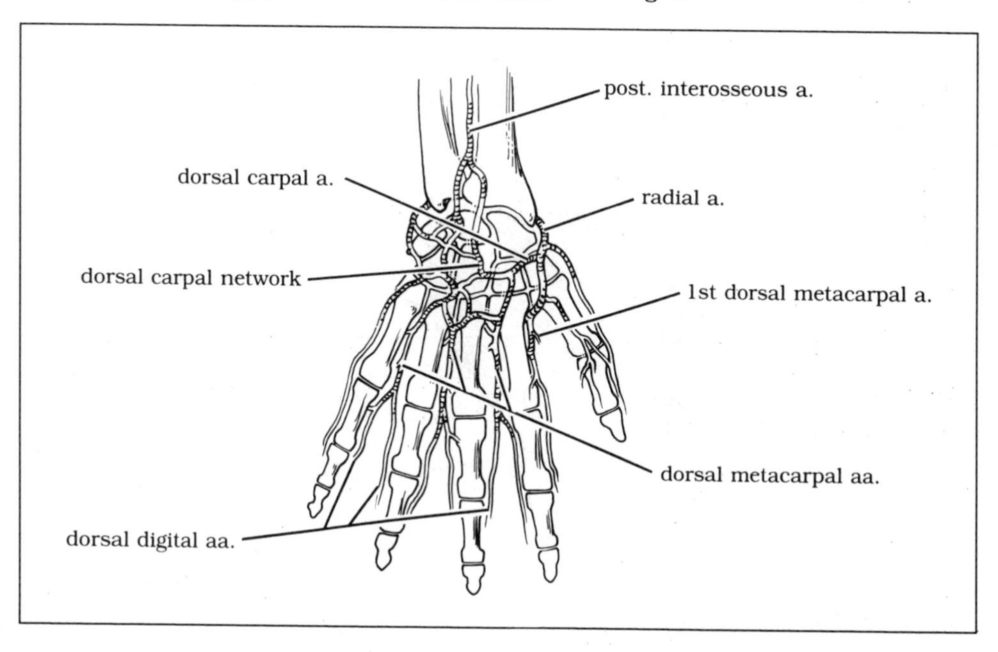

Fig. 5-34. Arterial supply to the dorsum of the hand.

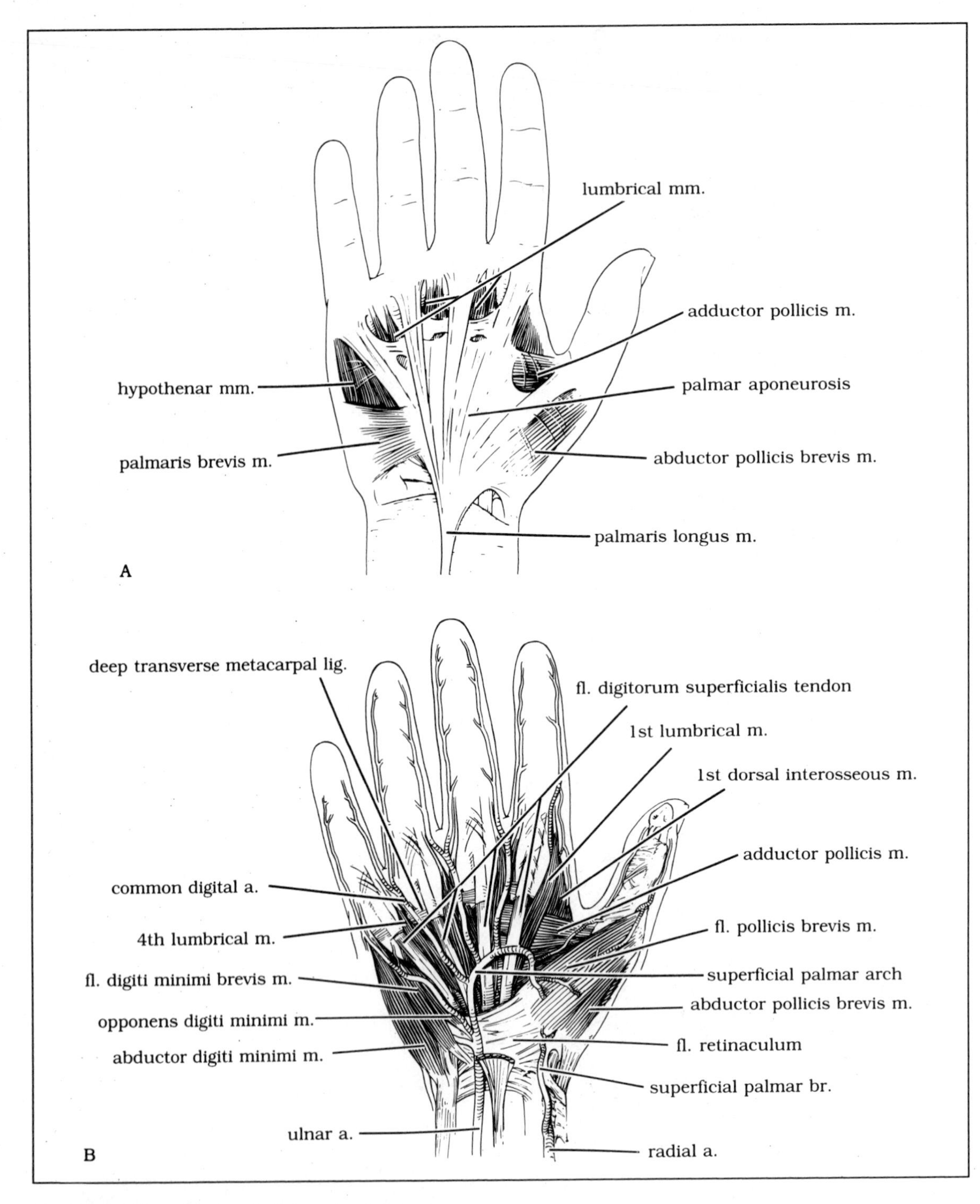

Fig. 5-35. Palm of the hand. A Superficial dissection. B. Intermediate dissection. C. Deeper dissection. D. Deepest dissection.

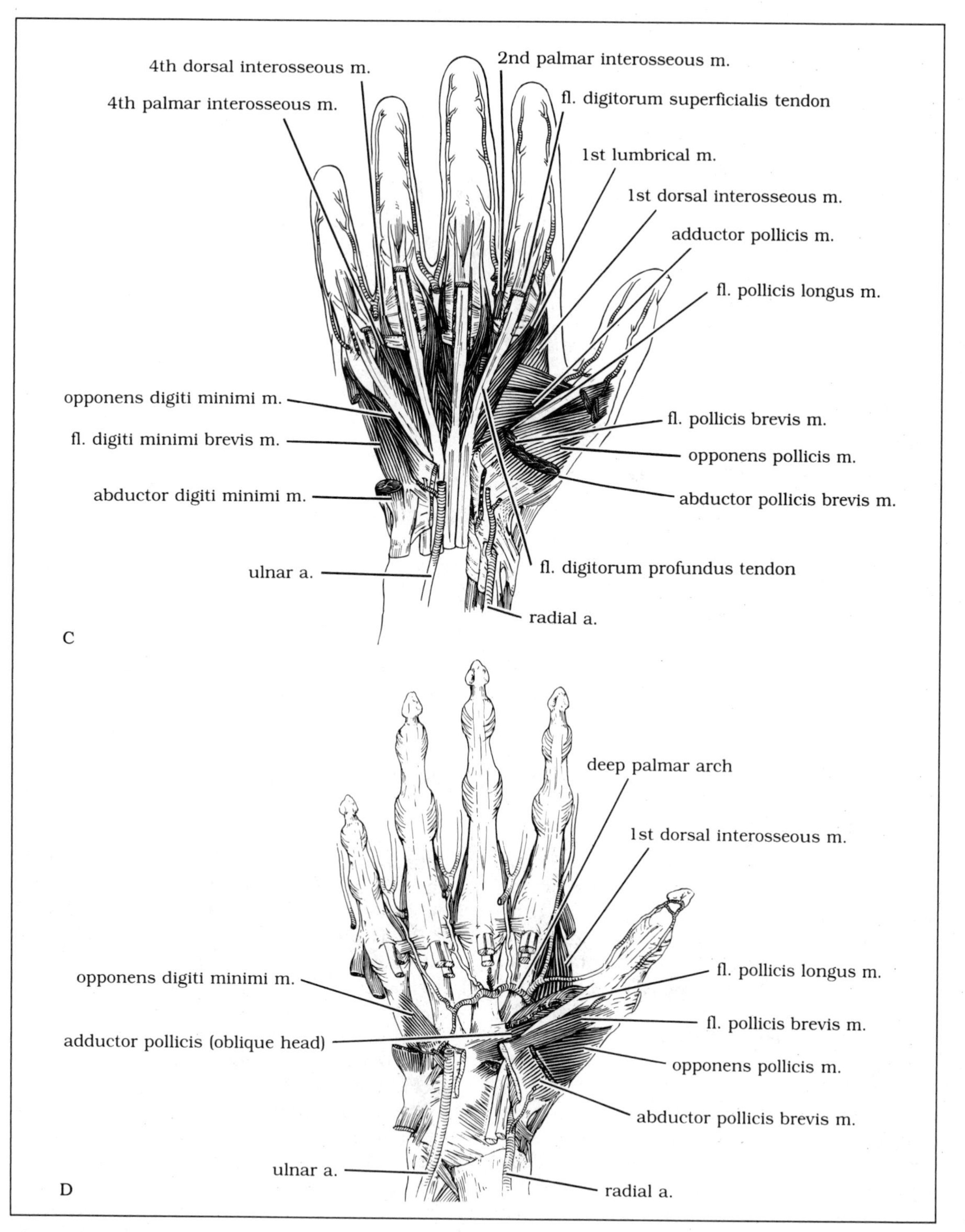

Fig. 5-35 (continued).

Table 5-7. Musculature of the Palm of the Hand

Muscle	*Origin*	*Insertion*	*Innervation*	*Action*
Abductor pollicis brevis	Scaphoid, trapezium, flexor retinaculum	Base of proximal phalanx of thumb	Median nerve	Abduct thumb
Flexor pollicis brevis	Trapezium, flexor retinaculum	Base of proximal phalanx of thumb	Median nerve	Flex thumb
Opponens pollicis	Trapezium, flexor retinaculum	Metacarpal 1	Median nerve	Oppose thumb
Abductor digiti minimi	Pisiform	Base of proximal phalanx of digit 5	Ulnar nerve	Flex digit 5
Flexor digiti minimi	Hamate, flexor retinaculum	Base of proximal phalanx of digit 5	Ulnar nerve	Flex digit 5
Opponens digiti minimi	Hamate, flexor retinaculum	Metacarpal 5	Ulnar nerve	Oppose digit 5
Lumbricals (4)	Tendon of flexor digitorum profundus	Dorsal extensor hood	Median nerve (muscles 1 and 2); ulnar nerve (muscles 3 and 4)	Flex metacarpophalangeal joint, extend interphalangeal joint
Adductor pollicis	Capitate, base of metacarpals 2 and 3	Base of the proximal phalanx of thumb	Ulnar nerve	Adduct thumb
Palmar interossei (4)	Metacarpal bone of digit	Extensor hood and proximal phalanx of digit	Ulnar nerve	Adduct digit
Dorsal interossei (4)	Both sides of metacarpal space	Extensor hood and proximal phalanx of digit	Ulnar nerve	Abduct digit

Palm

Underlying the skin of the palm is the **palmar aponeurosis,** which covers the palm and sends slips into the digits. The palmar aponeurosis is a connective tissue thickening that contains the insertions of the palmaris longus and brevis and extensions of the flexor retinaculum and subcutaneous fascia (Fig. 5-35A).

Below the aponeurosis the palm may be divided into four compartments: thenar, hypothenar, central, and interosseous-adductor. These compartments contain the muscles intrinsic to the palm of the hand (Table 5-7) and the tendons of the extrinsic muscles. All the intrinsic muscles are preaxial flexor muscles and are innervated either by the median nerve (thenar compartment and first two lumbrical muscles) or by the ulnar nerve (all the remaining muscles) (Figs. 5-35B,C,D and 5-36). (See Fig. 5-16 for origins and insertions.) The thenar and hypothenar compartments each contain abductor, opponens, and short flexor muscles. The abductors and flexors insert at the base of the proximal phalanx. Each opponens inserts along the shaft of its metacarpal.

The **thenar** (thumb) **compartment** contains three muscles, all innervated by the median nerve. The abductor pollicis brevis arises from two heads from the flexor retinaculum, scaphoid, and trapezium. The flexor pollicis brevis arises

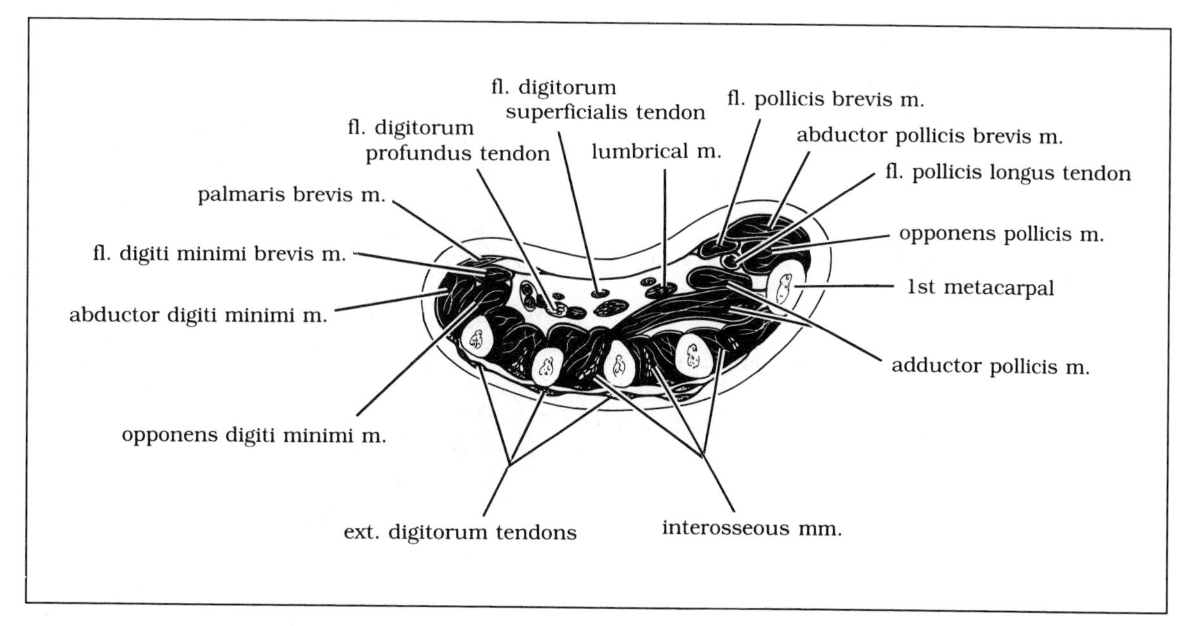

Fig. 5-36. Cross section of the hand.

from the flexor retinaculum and trapezium. The opponens pollicis arises from the flexor retinaculum and trapezium deep to the abductor pollicis brevis. The insertion of these muscles is described above.

The **hypothenar** (fifth digit) **compartment** contains three muscles, all innervated by the ulnar nerve. The abductor digiti minimi arises from the tendon of the flexor carpi ulnaris and the pisiform. The flexor digiti minimi arises from the flexor retinaculum and hamate. The opponens digiti minimi arises from the flexor retinaculum and the hamate, deep to the abductor and flexor muscles. The insertions of these muscles are described above.

The **central compartment** contains the tendons and sheaths of the flexor digitorum superficialis and profundus and the four lumbrical muscles. The lumbrical muscles (Fig. 5-37) arise from the tendons of the flexor digitorum profundus on their radial side. (They may also originate from the contiguous tendon.) They insert into the dorsal extensor hood. The lumbricals flex the metacarpophalangeal joints and extend the interphalangeal joints. The first two lumbricals (to the second and third digits) are innervated by the median nerve. The last two lumbricals (to the fourth and fifth digits) are innervated by the ulnar nerve.

The **interosseous-adductor compartment** is the deepest compartment. It contains the adductor pollicis and the eight interossei. The adductor pollicis has two heads, oblique and transverse. The oblique head arises from the capitate and base of the second and third metacarpals. The transverse head arises from the third metacarpal. The two heads unite and insert on the ulnar side of the base of the proximal phalanx. The adductor pollicis is innervated by the ulnar nerve.

The four **dorsal interosseous muscles** (there is none for the first digit) are bipennate muscles arising from adjacent metacarpal bones. They insert in the extensor hood and first phalanx of their digit (see Figs. 5-33, 5-35C, and 5-37). They abduct the digit and are innervated by the ulnar nerve.

The four **palmar interosseous muscles** (there is none for the third digit) are unipennate muscles arising from the metacarpal bone of their digit (see Figs. 5-35C and 5-37). They insert into

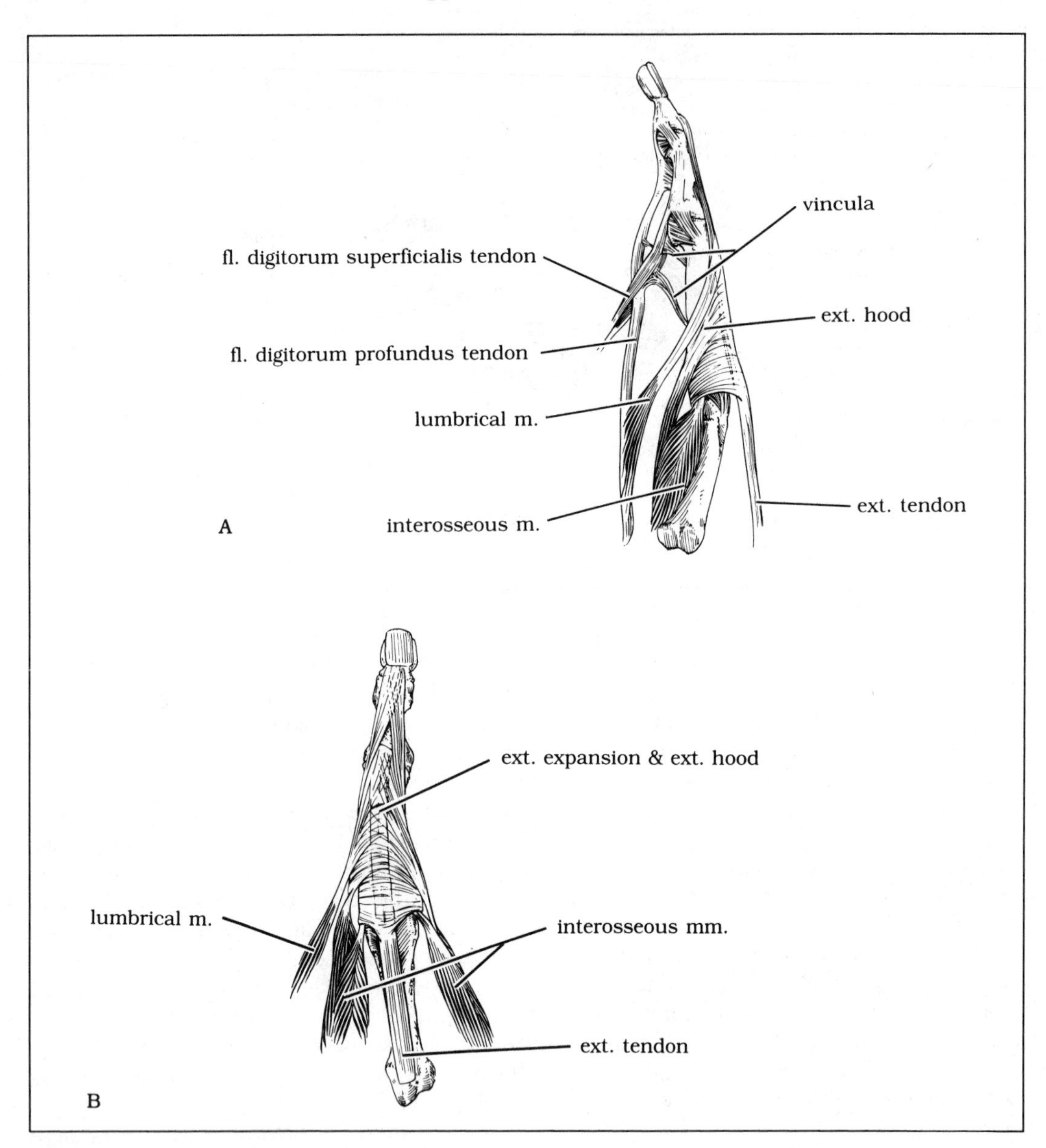

Fig. 5-37. Muscles and tendons of a digit. A. Lateral view. B. Posterior view.

the extensor hood and first phalanx of their digit. They are innervated by the ulnar nerve and adduct their digit.

Blood Supply and Innervation

The **superficial palmar arch** is formed primarily by the ulnar artery as it curves laterally across the palm of the hand. The superficial palmar arch usually anastomoses with the superficial palmar branch of the radial artery. The arch gives rise to palmar digital arteries (see Fig. 5-

32). The ulnar artery also gives rise to a palmar carpal branch, a dorsal carpal branch, and a deep palmar branch, each of which anastomoses with branches of the radial artery.

The **deep palmar arch** is formed by the radial artery as it enters the palm between the first two metacarpals. The arch usually anastomoses with the deep palmar branch of the radial artery. It gives rise to recurrents (which return to supply the wrist), perforating branches (which supply the dorsum), and palmar metacarpal branches (which anastomose with the palmar digitals of the superficial arch to supply the digits).

The radial artery has additional branches in the hand: the superficial palmar (to the superficial arch), dorsal carpal, palmar carpal, and first dorsal metacarpal branches and the arteria princeps pollicis (on the palmar surface).

There are no branches of the radial nerve in the palmar aspect of the hand. The median nerve supplies the thenar compartment and the first two lumbricals, the ulnar nerve supplying the remaining muscles of the hand.

Coordinated Movements of the Fingers

Gross movements of the fingers are controlled by flexion and extension of the muscles of the forearm. Fine movements are regulated by the intrinsic musculature of the hand.

The interosseous and lumbrical muscles act in interphalangeal extension, with the former helping to bring about metacarpophalangeal flexion. The long muscles of the thumb act in flexion (flexor pollicis longus), extension (extensor pollicis longus and brevis), and abduction (abductor pollicis longus). The short muscles aid in fine-tuning the thumb's movements, particularly flexion, opposition, abduction, and adduction. To bring the thumb into position to touch the finger pads of the second to fourth digits on the same hand, abduction, flexion, and medial rotation (produced by the opponens) are involved.

By carefully and minutely adjusting position, flexion, and extension of both thumb and digits the precision grip is achieved. Fine digital adjustments usually involve the lumbrical and interosseous muscles. In the power grip the flexor pollicis brevis is used, along with the digital and long thumb flexor, to tightly surround an object.

Movements and Joints of the Upper Limb

MOVEMENTS

The movements of the upper limb (Table 5-8) include the normal motions of flexion, extension, abduction, and adduction. In addition there are four motions found in the upper limb not generally found elsewhere: rotation and circumduction (at the shoulder joint), and supination and pronation (at the radioulnar joints). In the hand abduction and adduction are considered to occur around the axis formed by the third digit. Further explanation of all these motions may be found in the Chapter 1.

JOINTS

The **sternoclavicular joint** is the articulation between the upper limb and the trunk. This joint has the movement capabilities of a ball-and-socket joint, with the clavicle being able to be moved up, down, backward, or forward or to be rotated. The joint is formed by the bony ends of the sternum and clavicle, with an articular disk interposed between them. The articular disk forms the articular surface. The articular capsule is reinforced by ligaments in front, behind, and above. Below, the clavicle is also linked by strong ligaments to the first rib.

The **acromioclavicular joint** (see Fig. 5-4A) is a synovial joint that holds the clavicle to the scapula. The cavity of the joint is partially separated by an articular disk. The joint itself is enclosed by a loose articular capsule. The structure of the joint allows the acromion to glide forward or backward or to be turned up or down. In displacement due to trauma the joint structure tends to allow the acromion to slip under the clavicle.

The **scapulohumeral joint** (see Figs. 5-4B and 5-8) is a ball-and-socket joint and allows the greatest freedom of movement in the body. This flexibility can be attributed to the glenoid cavity surface's being approximately one-third the size

Table 5-8. Muscles Responsible for Movements of the Arm

Muscle	*Flexion*	*Extension*	*Abduction*	*Adduction*	*Medial Rotation*	*Lateral Rotation*	*Suspension of Shoulder*	*Rotation of Scapula*
Shoulder and Pectoral Regions								
Trapezius		+					+	
Latissimus dorsi		+		+	+			
Levator scapulae							+	
Rhomboid major							+	
Rhomboid minor							+	
Deltoid	+	+	+		+	+		
Supraspinatus			+					
Infraspinatus						+		
Teres major		+		+	+			
Teres minor						+		
Subscapularis		+			+			
Pectoralis major	+			+	+			
Pectoralis minor								+
Subclavius							+	
Serratus anterior								+
Arm								
Coracobrachialis	+							
Biceps	+							
Triceps		+		+				

of the humeral head. The cavity is deepened by the fibrocartilagenous glenoid labrum. As a result of the joint's freedom of motion, it has decreased stability.

The humerus is held in place by the articular cuff, which consists of the following muscles: supraspinatus, infraspinatus, teres minor, and subscapularis. The tendons of the muscles blend to reinforce the loose articular capsule. In addition the tendon of the biceps muscle, which arises within the joint, helps to hold the humeral head down in the joint. The glenohumeral and coracohumeral ligaments also strengthen the articular capsule.

The **elbow** actually consists of three joints (humeroulnar, humeroradial, and proximal radioulnar) all in one articular capsule (see Figs. 5-12, 5-13 and 5-14). The humeroulnar and humeroradial joints constitute a hinge joint, or ginglymus. The characteristics of this joint are (1) a loose capsule in front of the hinge, (2) strong collateral ligaments on the sides, (3) muscles grouped to the sides, and (4) alternating concave and convex surfaces.

The collateral ligaments of this ginglymus are the radial and ulnar collateral ligaments, which act to prevent side-to-side motion while allowing flexion and extension. The grouping of the muscles to the side, in the common flexor and extensor tendons, keeps them from interfering in the hinge actions. The trochlea of the humerus, which articulates with the trochlear notch of the radius, and the capitulum of the humerus, which articulates with the head of the radius, provide the alternating concave and convex surfaces.

The **proximal radioulnar articulation** consists of the radial notch of the ulna and the annular ligament. This ligament forms a ring in which the head of the radius rotates, allowing pronation and supination.

The **distal radioulnar articulation** (Fig. 5-38) is a pivot joint between the ulnar notch of the radius and the head of the ulna. The joint is surrounded by a weak articular capsule and separated by an articular disk. (This disk also participates in the radiocarpal joint.) It allows the hand to rotate during pronation and supination by attaching the hand to the radius and allowing the radius to rotate on the ulna. There is no ulnar carpal articulation.

The **radiocarpal joint** is between the radius and the proximal carpal bones. It is enclosed by an articular capsule. The movements allowed at this joint are flexion, extension, abduction, adduction, and circumduction.

The carpal bones are held together tightly by many ligaments (see Fig. 5-38). The intervening synovial cavity is continuous between all the bones and between the carpometacarpal joints (except for the first) and intermetacarpal joints (except for the first). The carpometacarpal joint and intermetacarpal joint for the thumb have separate capsules and cavities.

The metacarpophalangeal joints and interphalangeal joints are each contained in loose articular capsules. The capsules are strengthened by strong collateral and palmar ligaments.

Injury to the Nerves of the Upper Limb

Injury to the nerves of the upper limb can result in dramatic distortions in appearance and use. Damage to the fifth cervical nerve causes paralysis of the deltoid, biceps, brachialis, and brachioradialis, with occasional inclusion of the supraspinatus, infraspinatus, and supinator. The arm is rotated medially and the forearm is extended and pronated. The arm hangs at the side.

Damage to the inferior trunk, or precursor roots C8 and T1, paralyzes the hand's intrinsic muscles and the wrist and finger flexors.

Damage to specific nerves and the level of damage may be diagnosed by observation of the muscles affected relative to other muscles innervated by the nerve in question (see Table 5-4).

Collateral Circulation of the Upper Limb

In the region of the scapula the dorsal scapular and suprascapular arteries anastomose with the circumflex scapular and subscapular arteries. Around the acromion the thoracoacromial, anterior and posterior humeral circumflex, and acromial branches of the suprascapular artery all anastomose. The intercostobrachial anastomosis connects the intercostal arterial system with the lateral thoracic, thoracodorsal, and thoracoacromial arteries.

In the arm the ascending branch of the deep brachial artery anastomoses with the descending branch of the posterior circumflex humeral artery.

Around the elbow the superior and inferior ulnar collaterals (from the brachial artery) and the radial and middle collaterals (from the deep brachial artery) anastomose with the anterior and posterior ulnar recurrents, the interosseous recurrent (all from the ulnar artery), and the radial recurrent (from the radial artery) arteries.

At the wrist the palmar carpal branches of the

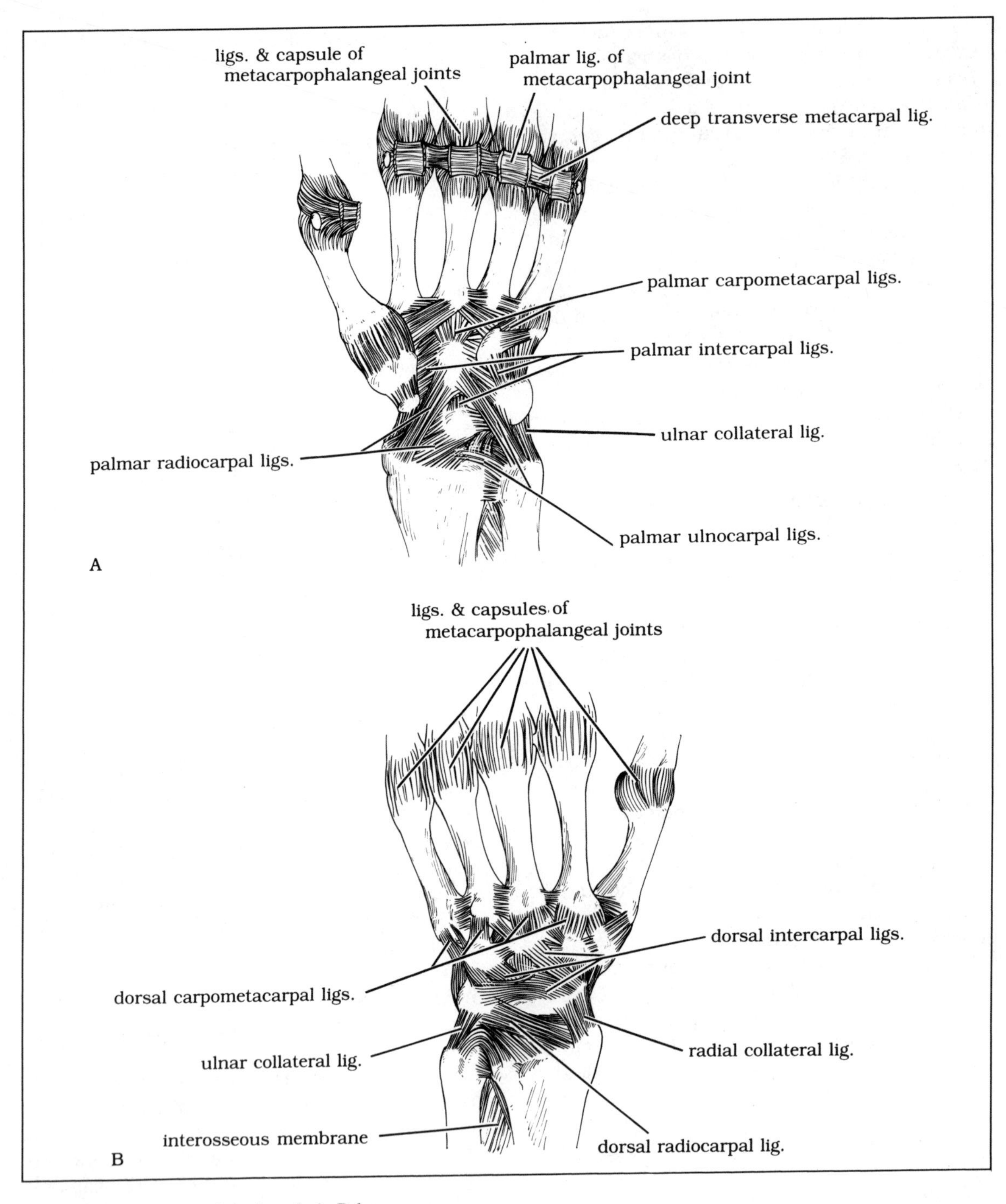

Fig. 5-38. Ligaments of the hand. A. Palm. B. Dorsum.

radial and ulnar arteries anastomose with the anterior interosseous (from the ulnar artery) and carpal recurrent (from the deep palmar artery) arteries. The dorsal carpal branches of the radial and ulnar arteries also anastomose with the anterior interosseous arteries.

In the hand the superficial and deep palmar arches have extensive anastomotic connections.

National Board Type Questions

Select the one best response for the following.

1. After ligation of the radial artery in the midforearm, most of the blood that reaches the region in the hand supplied by the cutaneous branch of the radial nerve would pass through the
   A. dorsal carpal branch of the ulnar artery
   B. anterior interosseous artery
   C. ulnar artery via the superficial palmar arch
   D. radial recurrent artery

Select the response most closely associated with each numbered item. (The headings may be used once, more than once, or not at all.)

A. Axillary nerve
B. Musculocutaneous and axillary nerves
C. Median nerve
D. Ulnar nerve
E. Radial nerve

Which of the above nerves was injured in each of the following cases?

2. A patient fractures an elbow. The position of the hand at rest shows that there is an extension of the metacarpophalangeal joints of the fourth and fifth fingers and a flexion at the interphalangeal joints of these fingers. This gives the appearance of a "claw" deformity.
3. A patient suffers a wrist laceration. Seen some weeks later a wasting of the thenar eminence (remember that muscles atrophy after denervation) and weakness of flexion of the index and middle fingers at the metacarpophalangeal joints are observed. There is also a loss of sensation over the palmar side of the thumb.
4. In pulling injuries of the arm, the force of downward traction often falls on roots C5 and C6. After such a pull a patient's arm hangs limply by the side with the forearm pronated and the palm facing backward. Examination reveals a paralysis of the deltoid, brachialis, and biceps.
5. In a car accident, the proximal end of the humerus is fractured. Following mending of the bone the patient cannot abduct the arm.

Which nerve is associated with which function or test?

A. Median nerve
B. Ulnar nerve
C. Radial nerve
D. Musculocutaneous nerve
E. Axillary nerve

6. Impairment can be tested by asking the patient to grasp a piece of paper tightly between the fingers.
7. It is involved in flexion of the terminal interphalangeal joint of the fifth digit.
8. Adduction of the thumb is lost or weakened.

Select the response most closely associated with each numbered item.

A. Deep palmar arterial arch
B. Superficial palmar arterial arch
C. Both
D. Neither

9. Located between the palmar aponeurosis and the long flexor tendons of the medial four digits.
10. Most proximally situated in the hand
11. Located between the palmar aponeurosis and the flexor retinaculum

---

For the following, select

A. if only *1, 2, and 3* are correct
B. if only *1 and 3* are correct
C. if only *2 and 4* are correct
D. if only *4* is correct
E. if *all* are correct

12. Which of the following describe(s) the anatomical relationship(s) of the axillary artery?
    1. It lies posterior to the teres major muscle.
    2. It lies superficial to the posterior cord of the brachial plexus.
    3. Its anterior and posterior circumflex humeral branches encircle the anatomical neck of the humerus.
    4. It runs between the lateral and medial cords of the brachial plexus.
13. The scapula
    1. extends as low as T10 when in normal anatomical position
    2. rotates on the thoracic wall for the arm to be fully raised above the head
    3. is downward rotated by the efforts of a cranial nerve
    4. articulates with the clavicle at the acromioclavicular joint, but its strongest union with the clavicle is the coracoclavicular ligament
14. With reference to the bones of the forearm, which of the following statement(s) is (are) true?
    1. The olecranon process of the ulna provides attachment for the triceps muscle.
    2. The coronoid process of the ulna provides attachment for the brachialis muscle.
    3. The radial tuberosity provides attachment for the tendon of the biceps brachii muscle.
    4. The styloid process of the radius provides attachment for the brachioradialis muscle.
15. The radius articulates with the
    1. Trochlea of the humerus
    2. Capitate bone
    3. Lateral epicondyle of the humerus
    4. Scaphoid bone
16. The radial artery pulse can be palpated
    1. Between the tendons of the extensor pollicis longus and the extensor pollicis brevis
    2. Along the border of the palmaris longus tendon
    3. Where the artery lies on the anterior surface of the distal end of the radius, lateral to the tendon of the flexor carpi radialis muscle
    4. On the radial side of the pisiform bone
17. Transection (cutting) of the ventral rami of C3 to C6 will result in
    1. Severe weakening to total loss of abduction of the humerus
    2. Loss of sensation on the skin of the entire arm
    3. Almost total paralysis of the diaphragm
    4. Total paralysis of intrinsic hand muscles
18. If the median nerve were severed proximal to the elbow, which of the following movements would be abolished?
    1. Flexion at the interphalangeal joint of the thumb
    2. Adduction at the carpometacarpal joint of the thumb
    3. Pronation
    4. Supination
19. A surgical procedure requires exposure of the middle third of the humerus. An anteromedial approach to the humerus during operation could result in damage to which of the following?
    1. Brachial artery
    2. Radial nerve
    3. Basilic vein
    4. Radial artery
20. Regarding the shoulder joint,
    1. The tendon of the long head of the biceps runs within the fibrous capsule of the shoulder joint.
    2. The rotator cuff provides no reinforcement inferiorly.
    3. The humerus is most likely to dislocate when it is abducted.
    4. Dislocation of the humerus may be associated with injury of the axillary nerve.
21. Choose only the correct statement(s) regarding joints and their movements.

1. Rotation occurs at the acromioclavicular joint.
2. Circumduction occurs at the articulation between the trapezium and the first metacarpal.
3. Pronation and supination involve the distal radioulnar joint.
4. Abduction and adduction occur at the metacarpophalangeal joints.

22. True statement(s) regarding the superficial veins of the upper limb include(s) the following:
    1. The cephalic vein drains blood into the axillary vein.
    2. The basilic vein runs up the ulnar side of the forearm.
    3. The cephalic vein passes between the deltoid and pectoralis major muscles.
    4. The basilic vein passes medial to the biceps brachii muscle.
23. The lateral cord of the brachial plexus contributes to the following nerve branches:
    1. Radial nerve
    2. Median nerve
    3. Ulnar nerve
    4. Musculocutaneous nerve

Annotated Answers

1. B. Although its name suggests that it supplies only the ventral surface, the anterior interosseous artery has an important distribution to the dorsum of the hand and anastomoses with the posterior interosseous artery and dorsal carpal branch of the radial artery.
2. D. The ulnar nerve, which passes dorsal to the medial epicondyle of the humerus, supplies the third and fourth lumbricals and all the interossei. These flex the metacarpophalangeal joints (balancing the long extensors) and extend the interphalangeal joints (balancing the long flexors).
3. C. The muscles of the thenar eminence are exclusively innervated by the median nerve. The weakness in flexion of the metacarpophalangeal joint of the second and third digits (but not *loss* of flexion) indicates that at least some of the muscles involved with this nerve are injured. This finding, in conjunction with the loss of the thenar muscles, would again indicate median nerve involvement, as the lumbricals of the second and third digits are innervated by the median nerve.
4. B.
5. A. The axillary nerve winds around the humerus in close association with the surgical neck (proximal shaft) and is the sole innervation to the deltoid.
6. B. Pressing the fingers together is adduction. The adductors of the fingers are the palmar interossei; *all* the interossei are innervated by the ulnar nerve.
7. B. The flexor digitorum profundus, which inserts on the terminal phalanges and therefore crosses the terminal interphalangeal joints, is partially innervated by the ulnar nerve. That portion innervated by it is associated with the tendons of the fourth and fifth digits.

9. B., 10. A., and 11. D. The superficial palmar arch, formed mainly by the ulnar artery, is the most distal of the two arches and lies between the palmar aponeurosis and the long flexor tendons. The deep arch lies on the metacarpal bones and is *crossed* by the long flexor tendons.

12. C. Remember that the cords of the brachial plexus are named in accordance with their relationship to the axillary artery. In addition it is important to understand the difference between the anatomical and surgical neck of the humerus.
13. C.
14. E.
15. D.
16. B. After passing over the anterior surface of the distal end of the radius, the radial artery winds around to the dorsum of the hand and becomes superficial between the tendons of the extensor pollicis longus and brevis.
17. B. The dermatomes of the upper limb include components from each spinal nerve

that makes up the brachial plexus—C5 to T1. Why would the diaphragm be involved? See Chapters 2 and 8.

18. B. First of all, would it alter the result if the median nerve were severed at its origin from the brachial plexus or just proximal to the elbow? (No. Understand why?) The strongest supinator is the biceps (musculocutaneous), while the adductor of the thumb is innervated by the ulnar nerve.
19. B. The radial nerve leaves to wind around the humerus in association with the deep brachial artery, while the radial artery *usually* does not arise until the level of the bend of the elbow.
20. E.
21. E. An important rotary movement occurs at the acromioclavicular joint (palpate this joint as you go through the motion of throwing a ball overhand). The thumb can be extended, abducted, flexed, and adducted (circumduction). Circumduction, along with internal rotation, is involved with the opposition movement of the thumb.
22. E.
23. C. The lateral cord, composed of anterior division fibers (which eliminates the radial—why?) from the upper (C5–C6) and middle (C7) trunks, contributes to the musculocutaneous and median nerves.

6 Lower Limb

Objectives

After reading this chapter, you should know the following:

Surface anatomy of the lower limb, including the superficial organization and subcutaneous vessels and nerves

Bones, joints, and articulations of the lower limb

Fascia, musculature, and movements of the thigh and gluteal regions

Structure, contents, and nerves and vessels of the popliteal fossa

Fascia, compartments, musculature, and movements of the leg

Musculature, movements, and retinacula of the foot

Ligaments, articulations, and movements of the hip, knee, and ankle joints

Blood supply, lymphatics, and innervation of the lower limb

Surface Anatomy

SUPERFICIAL ORGANIZATION

The lower limb (Figs. 6-1 and 6-2) consists of four anatomical regions: the thigh, buttocks (gluteal region), leg, and foot. In addition there are three major joints associated with the region: the hip, knee, and ankle joints. The proximal boundary of the lower limb is as follows: the pubic symphysis, inguinal ligament, iliac crest, dorsum of the sacrum, coccyx, sacrotuberous ligament, ischial tuberosity, and ischiopubic ramus. (See Chapter 4 for a description of the above-mentioned bones; see also Figures 4-1 and 4-2.)

The **thigh** extends from the iliac crest to the knee. At its highest point it reaches the level of the fourth lumbar vertebra (see Fig. 6-3). Surface features of the thigh include the iliac crest, inguinal ligament, gluteal fold, ischial tuberosity (visible when the hip is flexed), and greater trochanter of the femur. The gluteal fold represents the lower border of the gluteus maximus muscle.

The **femoral triangle** (Fig. 6-3; see also Fig. 6-13) is found below the middle third of the inguinal ligament. The triangle contains the femoral artery, vein, and nerve and the superficial inguinal lymph nodes. The lateral border of the triangle is formed by the sartorius muscle as it passes diagonally across the thigh. Lateral and inferior to the sartorius is the quadriceps femoris muscle mass (combined rectus femoris and vastus muscles) (see Figs. 6-1A and 6-2A). The medial border of the femoral triangle is formed by the adductor muscles, which occupy the medial side of the thigh.

The posterior thigh muscles are the hamstring muscles (the semitendinosus, semimembranosus, and biceps femoris), which form the palpable cords at the back of the knee joint. The lateral wall of the thigh is formed by the tensor fasciae latae muscle and the iliotibial tract.

The bony prominence on the anterior surface

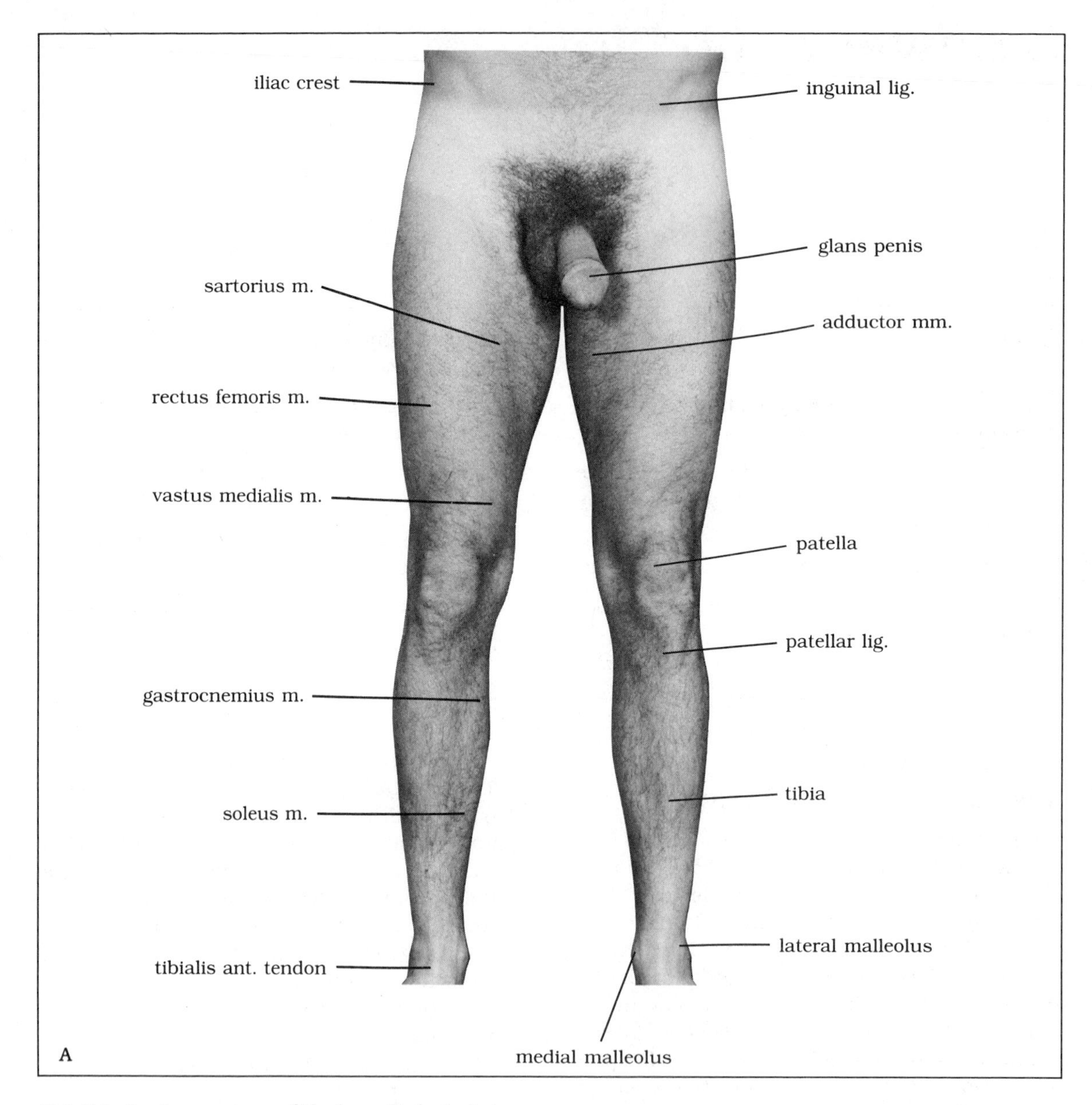

Fig. 6-1. Surface anatomy of the lower limb. A. Anterior view. B. Posterior view. C. Lateral and medial view of the feet and ankles.

of the knee joint is the **patella** (see Figs. 6-1A, 6-2A and 6-7), a sesamoid bone in the tendon of the quadriceps femoris muscle, the inferior portion of which is the patellar ligament. Medial and lateral to the patella, the condyles of the femur and tibia and the epicondyle of the femur are palpable. The posterior surface of the knee joint is the **popliteal fossa** (see Figs. 6-1B and 6-2B). It is bounded superiorly by the tendons of the hamstring muscles (the tendons of the semimembranosus and semitendinosus medially, the tendon of the biceps femoris laterally) and inferiorly by the gastrocnemius muscle.

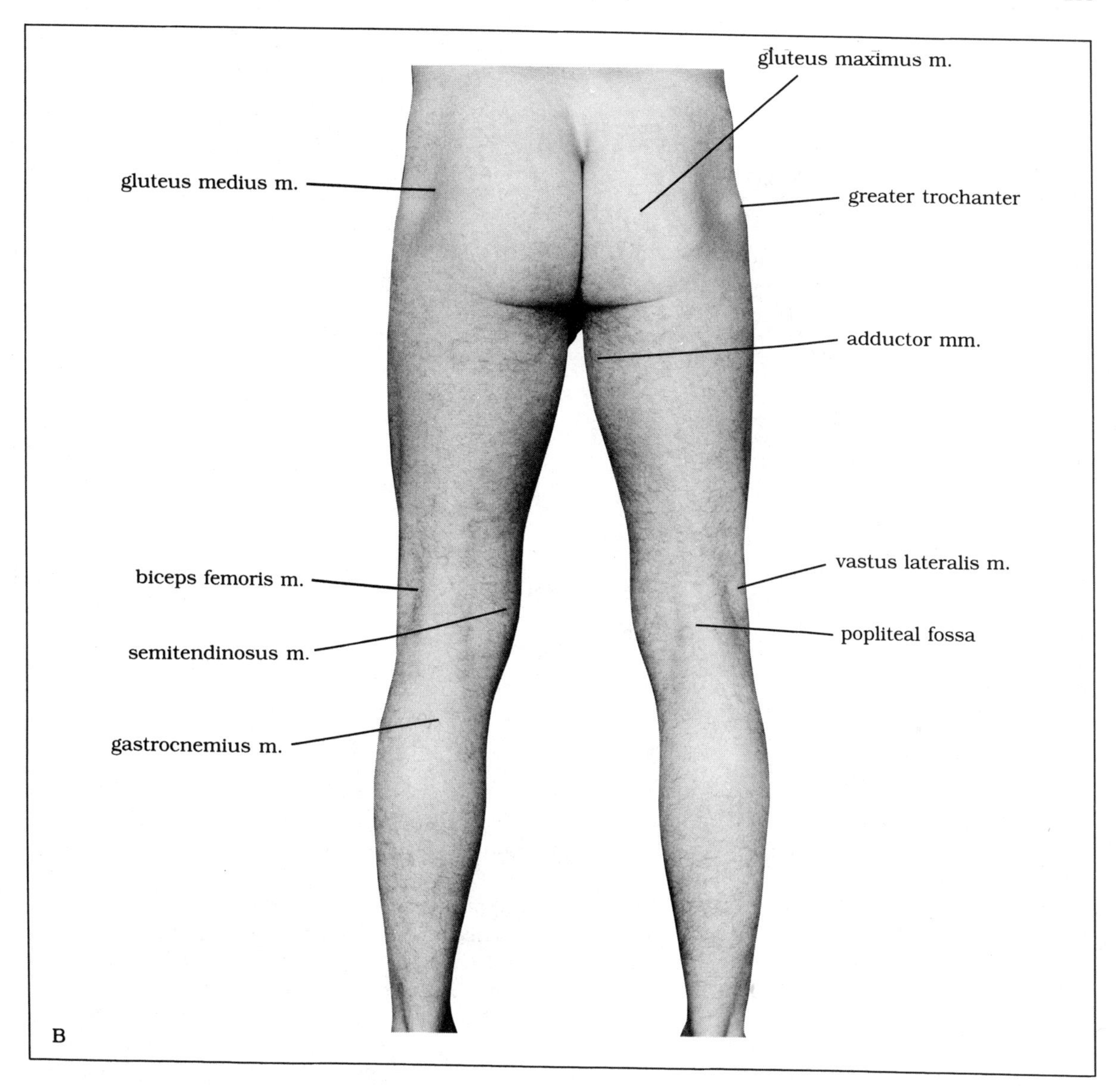

Fig. 6-1 (continued).

The **tibia** is palpable on the anterior surface of the leg. The sharp edge of the tibia, the anterior margin, forms a crest that proximally forms the tibial tuberosity, the point of insertion of the patellar ligament. Laterally the head of the **fibula** is palpable at the knee.

The **calcaneal tendon** is located at the base of the leg on the posterior surface (see Fig. 6-2B). This tendon is formed by the soleus and gastrocnemius muscles, two of the muscles of the posterior compartment. The anterior and lateral muscle groups contain the extensor and peroneal muscles.

The **medial** and **lateral malleoli** may be palpated at the ankles. The medial malleolus is a part of the tibia, the lateral malleolus is a part of the fibula. The tendons of the peroneal muscles (peroneus longus and brevis) pass posteriorly to the lateral malleolus.

By plantar flexing the foot the head of the **talus**

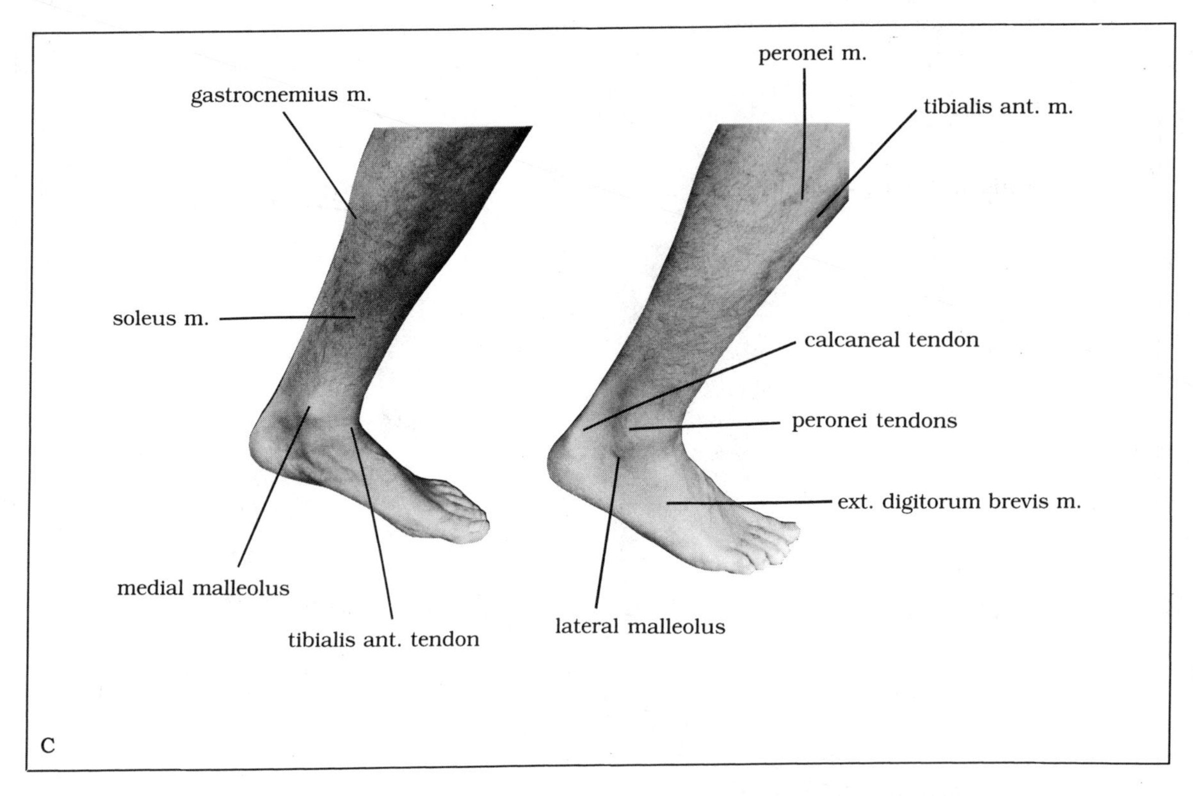

Fig. 6-1 (continued).

may be palpated medial to the lateral malleolus. The tuberosity of the navicular bone forms a prominence below and forward of the medial malleolus. At the posterior end of the foot the calcaneal tuberosity may be palpated. Along the lateral margin of the foot the tuberosity at the base of the fifth metatarsal bone may be felt.

SUBCUTANEOUS VESSELS AND NERVES

The superficial veins of the lower limb arise from a venous arch on the dorsum of the foot (Fig. 6-4). The medial end of the arch becomes the **great saphenous vein.** This vein turns upward anterior to the medial malleolus and ascends and moves posteriorly to pass behind the medial condyle of the femur. The vein then travels anteriorly to end at the femoral triangle, entering it through the saphenous opening and emptying into the femoral vein. The great saphenous vein receives tributaries from the dorsum of the foot, the ankle, the anterior leg, the calf, and the anterior and lateral thigh. In addition it receives the superficial circumflex iliac and superficial external pudendal veins (see Chaps. 3 and 4).

The **small saphenous vein** arises from the lateral side of the venous arch on the dorsum of the foot. The vein ascends behind the lateral malleolus and continues posteriorly to the midcalf level, where it ends by joining the popliteal vein in the popliteal fossa. The small saphenous vein receives tributaries from the lateral foot, the heel, and the back of the leg.

The small saphenous vein is accompanied by the sural nerve; the great saphenous vein is accompanied by the saphenous nerve in the leg and the anterior femoral cutaneous nerve in the thigh (see Fig. 6-4). The superficial nerves of the lower limb include the cluneal, posterior and lateral femoral cutaneous, femoral, saphenous, lateral sural cutaneous, sural, ilioinguinal, subcos-

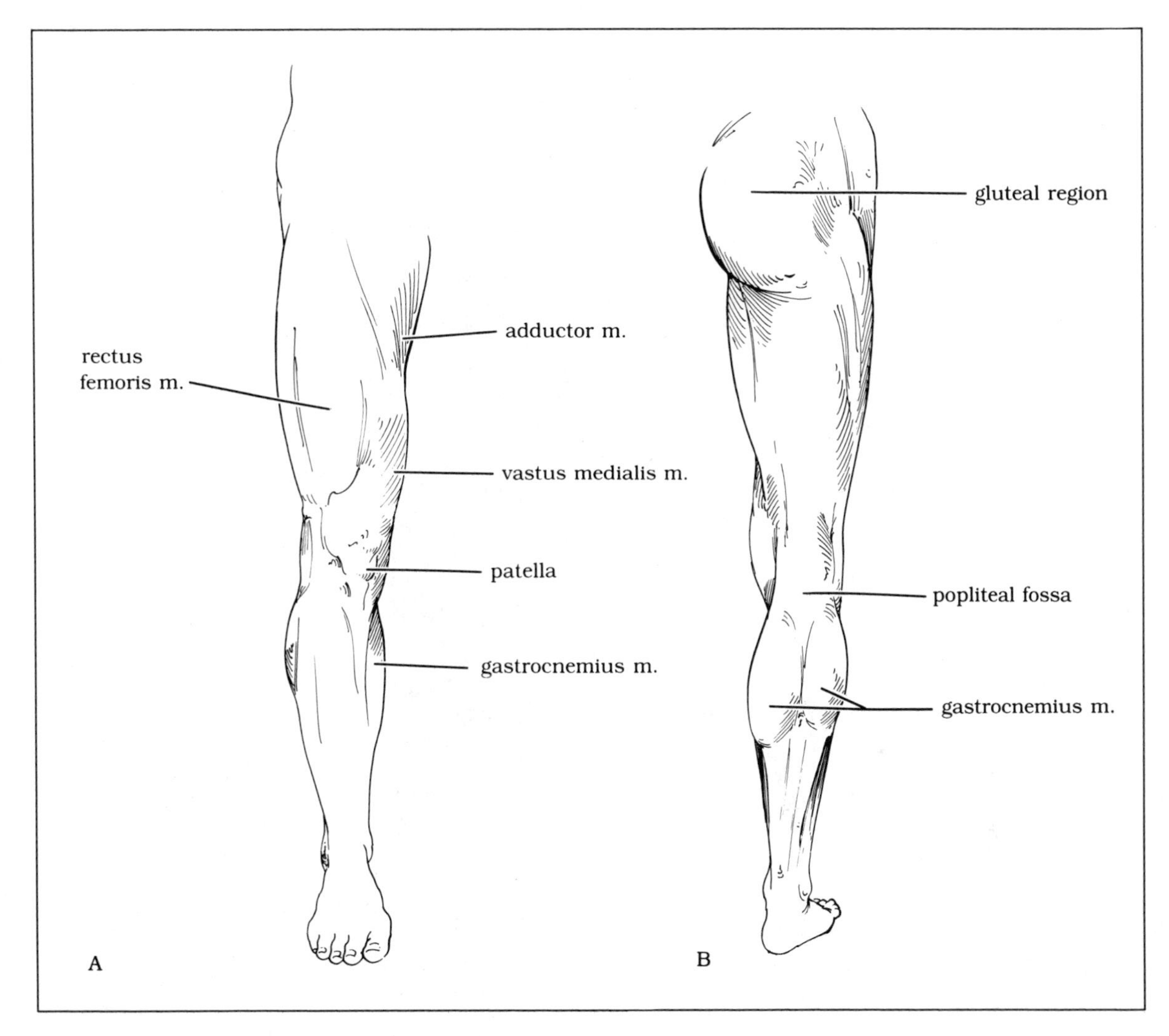

Fig. 6-2. Schematic drawing of surface anatomy of the lower limb. A. Anterior view. B. Posterior view.

tal, and iliohypogastric nerves. (Dermatome charts are shown in Figure 1-16.)

The gluteal region is supplied by the **cluneal nerves** (superior, middle, and inferior) and by perforating cutaneous nerves from sacral segments S2 and S3. The superior cluneal nerve is a branch of the dorsal ramus of the third lumbar nerve, the middle cluneal nerve is a branch of the third sacral nerve, and the inferior cluneal nerve is a branch of the posterior femoral cutaneous nerve.

The **posterior femoral cutaneous nerve,** a branch of the sacral plexus, supplies nerves to the gluteal region, the back of the thigh, and the popliteal fossa.

The **lateral femoral cutaneous nerve** is a branch of the lumbar plexus. It innervates the skin of the lateral thigh.

The **femoral nerve** supplies cutaneous branches to the distal three-fourths of the front

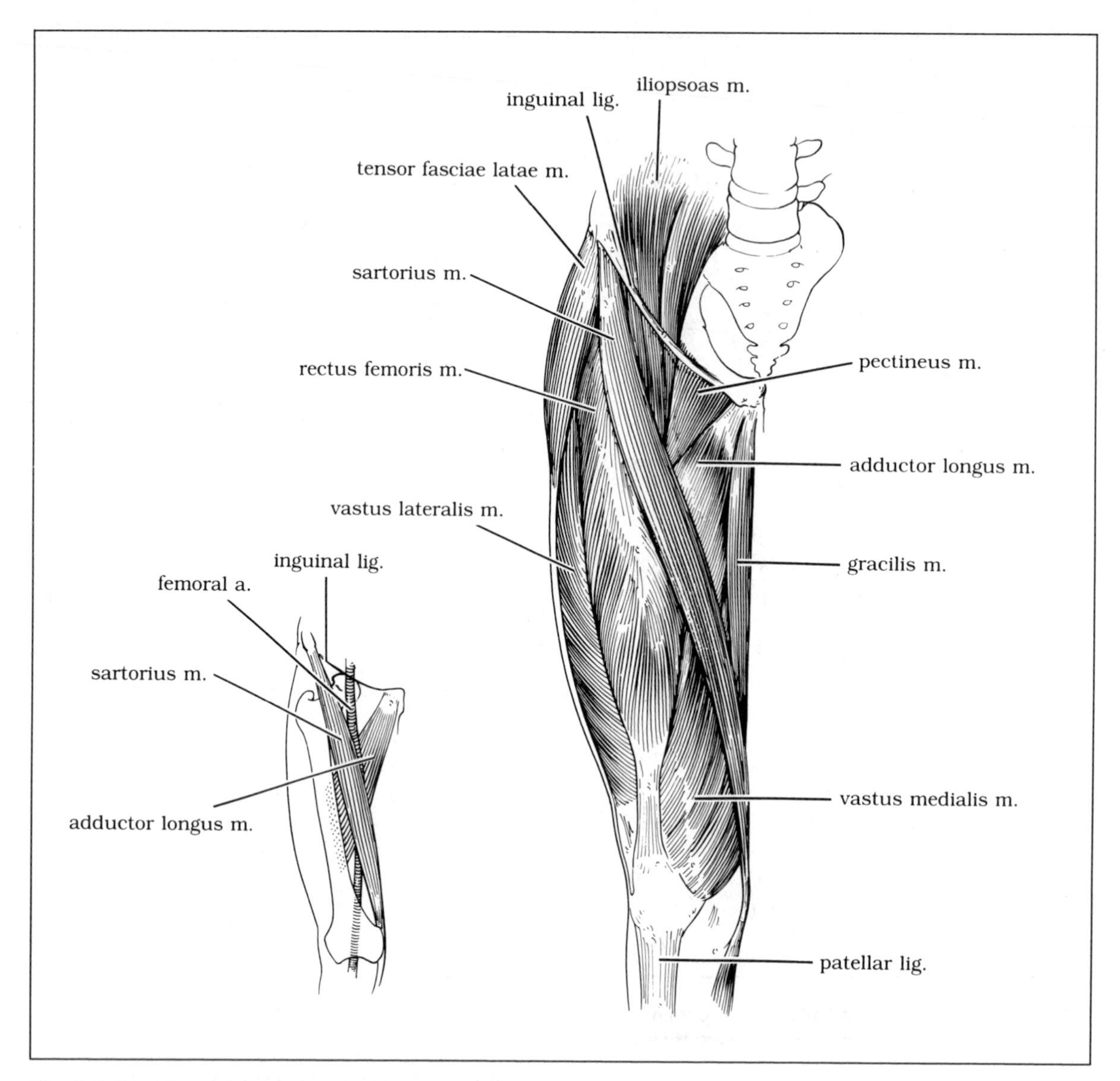

Fig. 6-3. Muscles of the anterior thigh. Inset shows boundaries of the femoral triangle and course of the femoral artery.

of the thigh, the proximal fourth being supplied by the femoral branch of the genitofemoral nerve.

The upper medial thigh from the femoral triangle to the scrotum in the male and labia majora in the female is innervated by the anterior scrotal branch of the ilioinguinal nerve. The skin of the upper lateral thigh is innervated by cutaneous branches of the subcostal and iliohypogastric nerves.

The **saphenous nerve** (see Fig. 6-4A), a branch of the femoral nerve, innervates the skin of the medial leg and the posterior half of the dorsum and medial surface of the foot. It travels with the great saphenous vein.

The **lateral sural cutaneous nerve** (see Fig. 6-

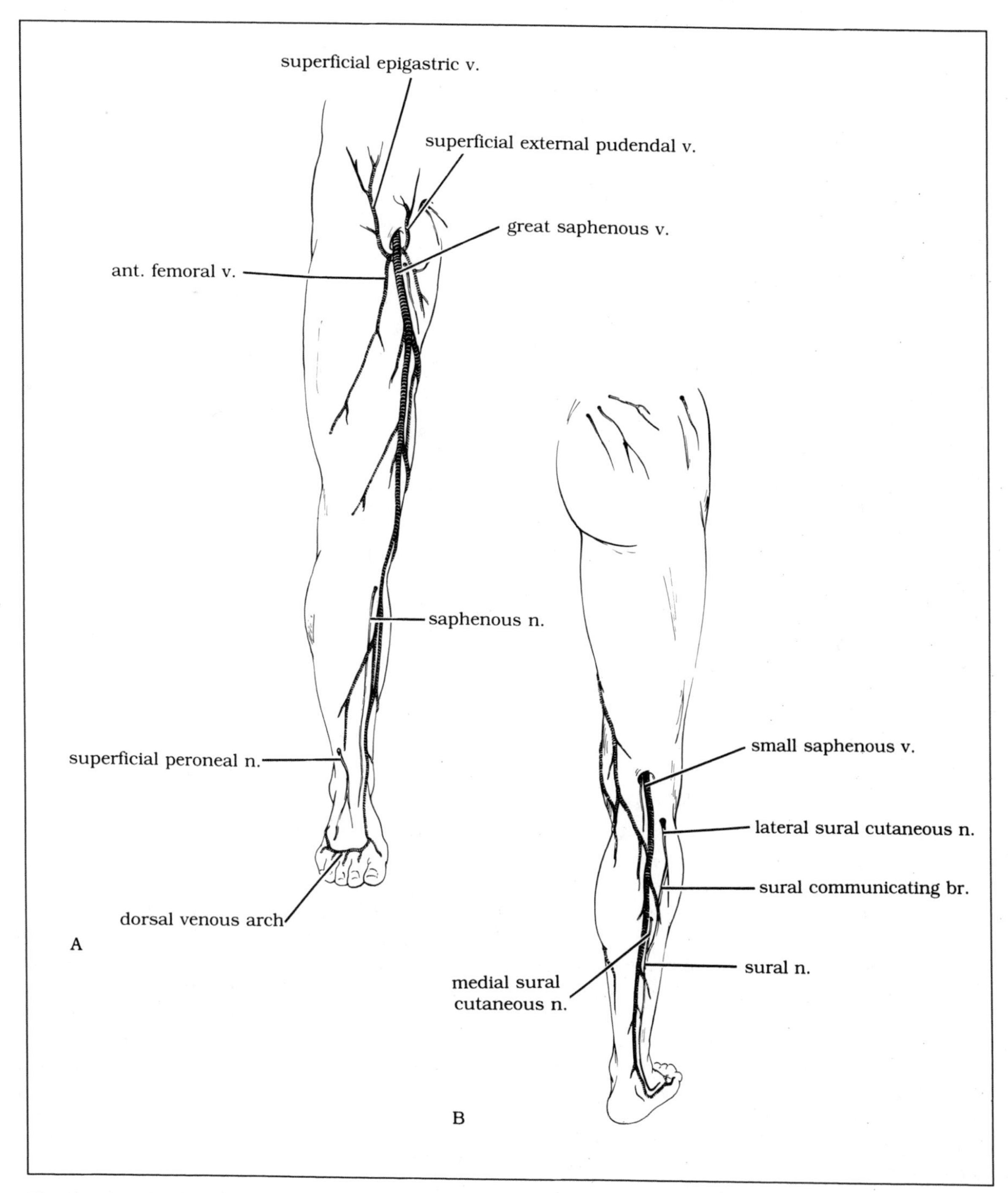

Fig. 6-4. Superficial veins of the lower limb.
A. Anterior view. B. Posterior view.

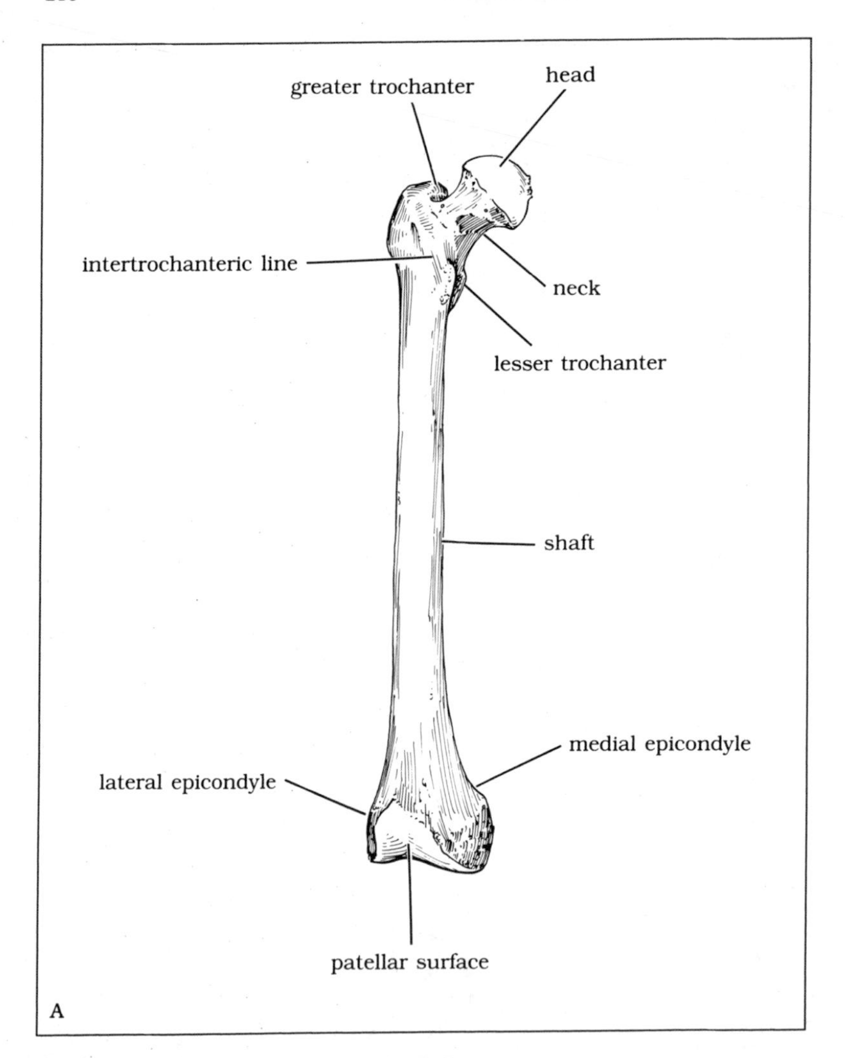

Fig. 6-5. The femur. A. Anterior view. B. Posterior view.

4B), from the common peroneal nerve, supplies the skin over the proximal two-thirds of the lateral posterior leg.

The **sural nerve** (see Fig. 6-4B), formed by the medial sural cutaneous nerve and a communicating branch from the lateral sural cutaneous nerve, innervates the distal third of the lateral posterior leg, the ankle, and the heel. The sural nerve accompanies the small saphenous vein through much of its course. Numerous other cutaneous nerves supply the ankle and foot.

The superficial lymphatics of the lower limb follow the great or small saphenous veins, primarily the former. The lymphatics following the small saphenous vein end in popliteal lymph nodes, the ones following the great saphenous vein end in the superficial inguinal nodes. Lymphatics of the lower limb are discussed at the end of the chapter.

Bones

The bones of the pelvis are discussed in Chapter 4. The bones of the lower limb include the femur,

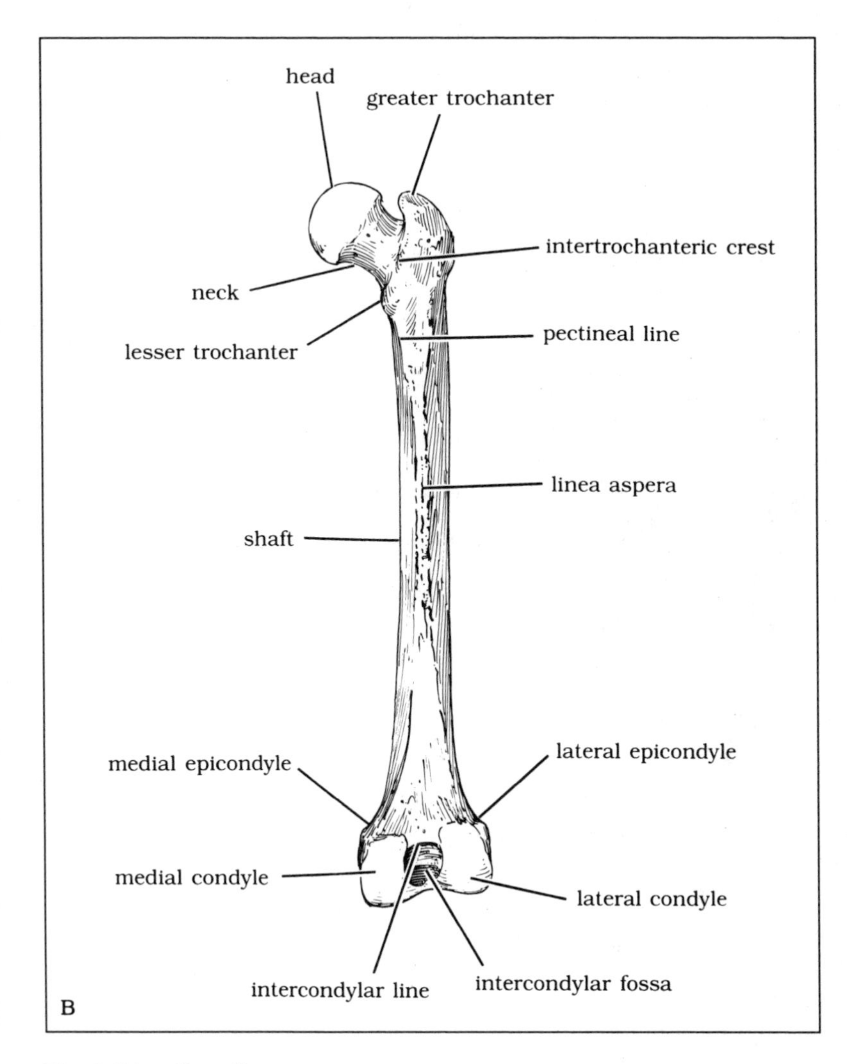

Fig. 6-5 (continued).

tibia, and fibula and the tarsal, metatarsal, and phalangeal bones.

FEMUR

The femur is the longest, strongest bone in the body. It consists of a head, neck, and shaft and a condylar (distal) region (Fig. 6-5). The **head** of the femur is globular in shape. Near the center of the globe is a pit, the fovea capitis femoris, which is where the round ligament of the head of the femur attaches. The head of the femur articulates with the bones of the pelvic girdle; its surface is therefore smooth and has a coating of cartilage.

The **neck** of the femur connects the shaft and head. In the adult male the neck protrudes from the shaft at an angle of approximately 125 degrees; in the female the angle is closer to a right angle. The neck is pyramidal in shape, with its superior surface shorter than its inferior, due to the angle the neck forms with the shaft. The superior surface of the neck ends in the greater

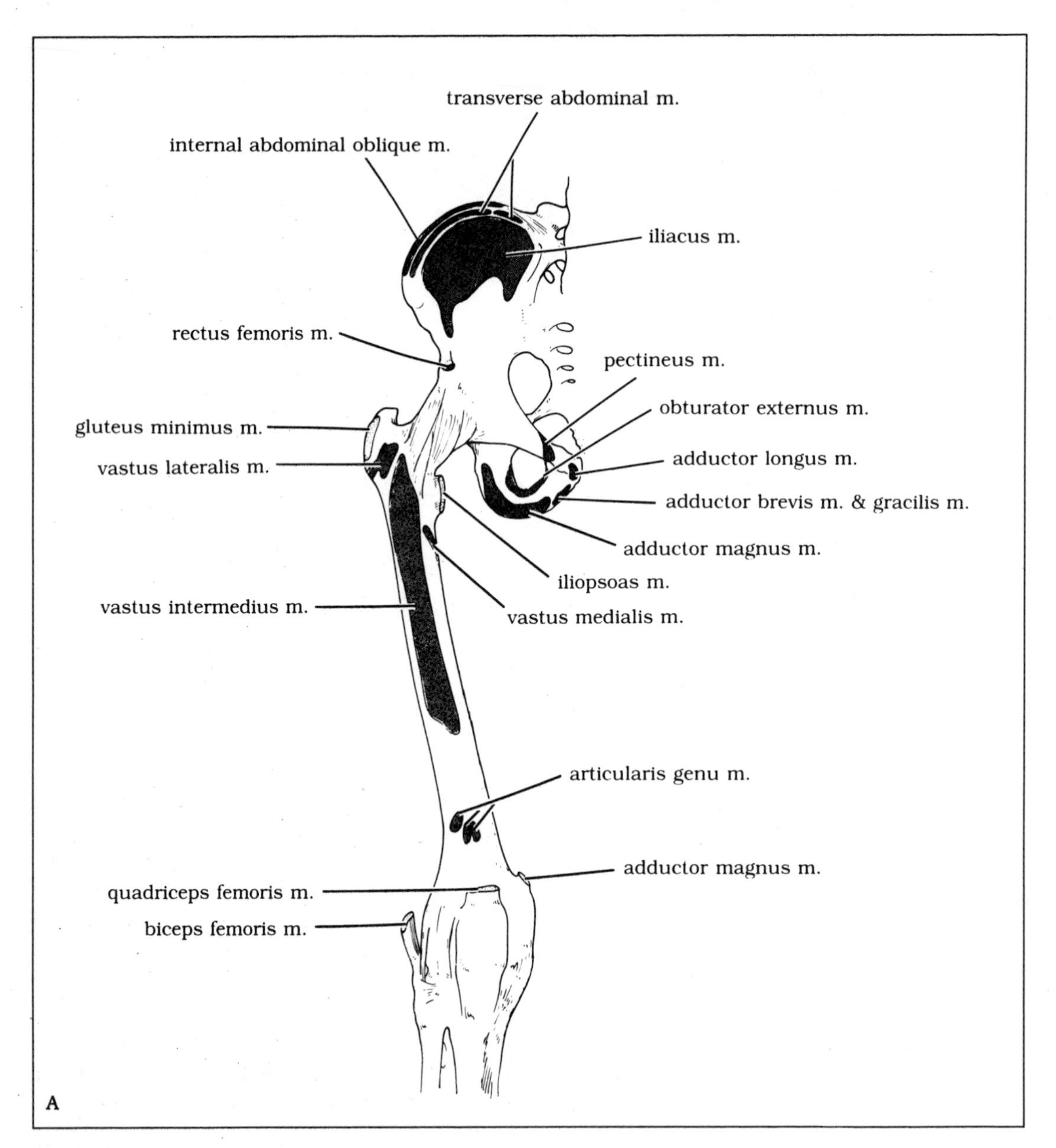

Fig. 6-6. Muscle origins (shaded areas) and insertions on the femur and portions of the pelvis. A. Anterior view. B. Posterior view.

trochanter of the femur and the inferior surface ends in the lesser trochanter.

The **greater trochanter** of the femur is a quadrilateral eminence jutting out at the junction of the neck and shaft. The lateral surface of the greater trochanter is broad and convex. The gluteus medius muscle inserts on this surface (Fig. 6-6). At the base of the medial surface there is a depression, the trochanteric fossa, in which

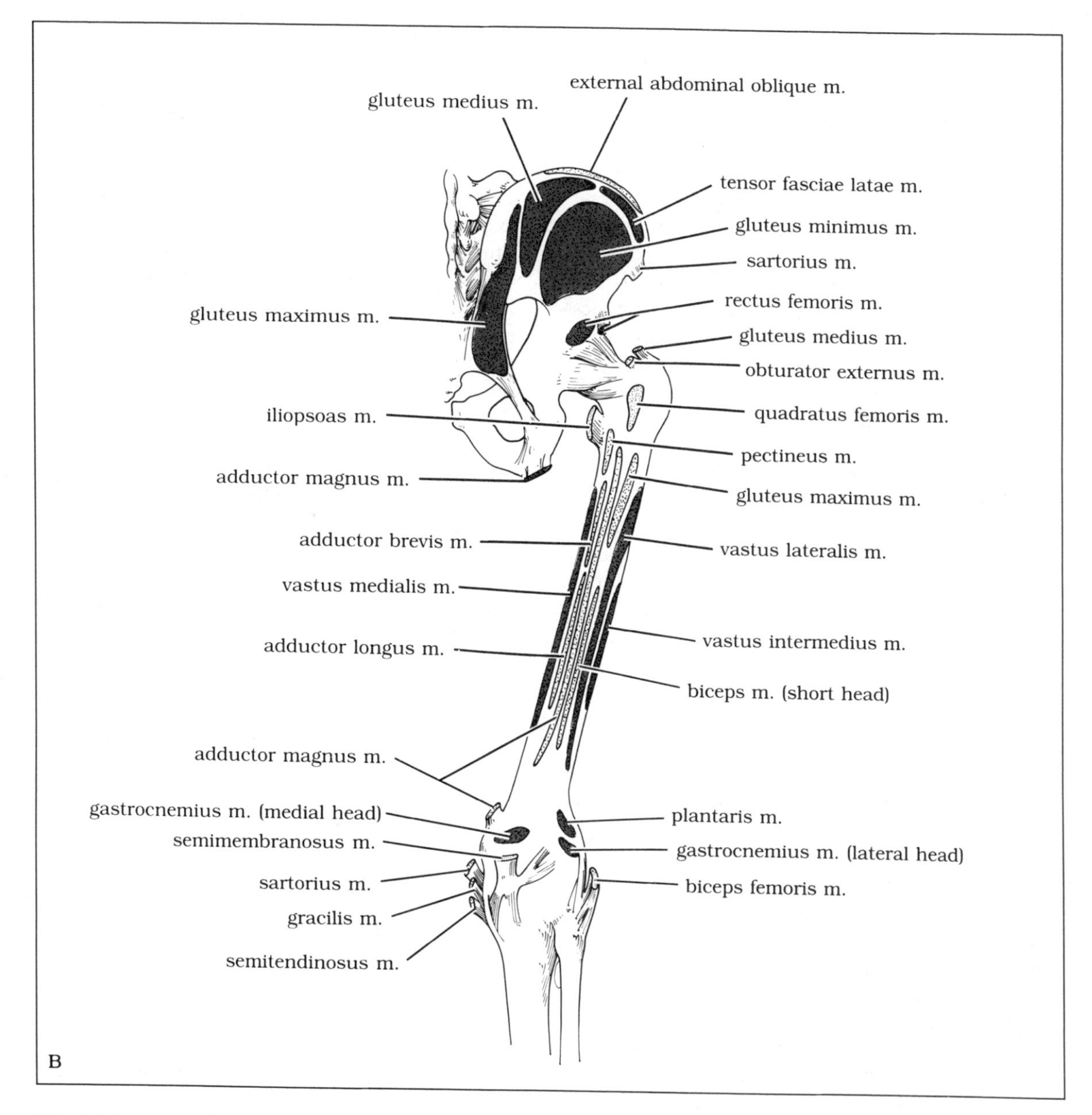

Fig. 6-6 (continued).

the obturator externus muscle inserts. Just anterior to the trochanteric fossa is the insertion of the obturator internus muscle. The piriformis muscle inserts on the superior border of the greater trochanter, and the vastus lateralis muscle arises from the lateral portion of the inferior border. The gluteus minimis muscle inserts on the lateral part of the anterior border. The quadrate tubercle, on the intertrochanteric crest, is where the quadratus femoris inserts.

The **lesser trochanter** of the femur is an eminence that projects posteromedially from the base of the neck. The iliopsoas tendon inserts on the apex of the lesser trochanter.

The **intertrochanteric crest** is a ridge that runs between the two trochanters on the posterior surface. This ridge is continuous with the posterior border of the greater trochanter and receives some fibers of insertion of the quadratus femoris muscle.

The **intertrochanteric line** runs between the two trochanters on the anterior surface. The iliofemoral ligament attaches along the proximal part of the line. The distal portion serves as the origin for part of the vastus medialis muscle. The intertrochanteric line ends in the linea aspera at the shaft of the femur.

The femoral **shaft** is almost cylindrical for much of its length. Along its posterior border there is a longitudinal ridge, the **linea aspera.** The proximal portion of the linea aspera consists of three ridges. The lateral ridge (the gluteal tuberosity) serves as the site of insertion of the gluteus maximus. The medial ridge is continuous with the intertrochanteric line via the roughened spiral line. Along its length is the origin of part of the vastus medialis muscle. The intermediate ridge, the pectineal line, serves as the origin of the pectineus muscle.

Distally the linea aspera continues as two ridges. The medial ridge (the medial supracondylar line) descends to the medial condyle of the femur, ending at the adductor tubercle, where fibers of the adductor magnus muscle insert. The lateral ridge is the lateral supracondylar line. The vastus lateralis muscle arises from this ridge. The adductor magnus inserts along the length of the linea aspera. In the proximal portion its insertion is lateral; in the distal portion it inserts medially. Between the vastus lateralis and adductor magnus muscles the gluteus maximus inserts and the short head of the biceps femoris arises. Between the vastus medialis and adductor magnus muscles the iliacus, pectineus, adductor longus, and adductor brevis muscles insert.

The proximal three-fourths of the anterior surface of the shaft of the femur serves as the origin of part of the vastus intermedius. The rest of the muscle arises from the proximal three-fourths of the lateral surface. Along the medial surface of the femur, in conjunction with the medial ridge of the linea aspera, fibers of the vastus medialis arise.

The distal portion of the femur consists of two expanded prominences: the **femoral condyles.** The condyles are separated posteriorly by the intercondylar fossa, in which lies the cruciate ligaments. At the distal end of each condyle is an articular surface for the femorotibial joint. Anteriorly the condyles are joined. This anterior junction of the condyles serves as the site of articulation with the patella.

Superior to the condyles are two prominences: the **epicondyles.** The medial epicondyle serves as the site of attachment of the tibial (medial) collateral ligament. Its proximal part is the adductor tubercle. The lateral epicondyle is the site of attachment of the fibular (lateral) collateral ligament. Posterior and proximal to this ligament are the origins of the gastrocnemius muscle (lateral head) and plantaris.

Fig. 6-7. The patella. A. Anterior view. B. Posterior view.

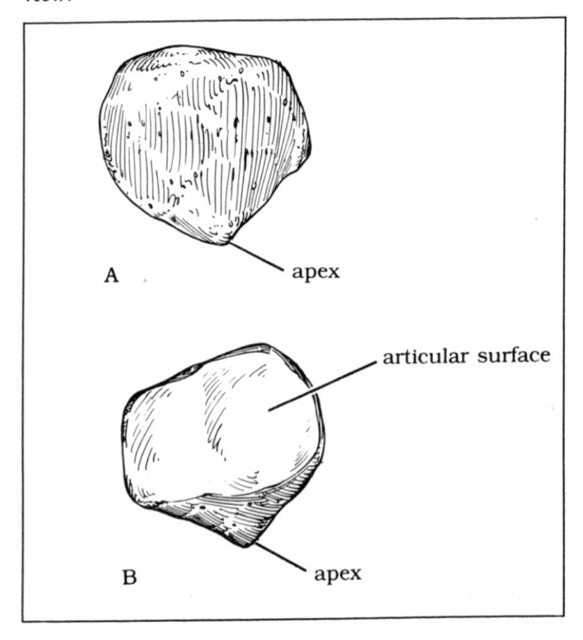

PATELLA

The patella is a sesamoid bone in the tendon of the quadriceps femoris muscle (Fig. 6-7). It serves two functions: to increase the leverage of the quadriceps femoris muscle by increasing the angle of insertion and to protect the knee joint. Its anterior surface is covered by a bursa. Inflammation of this bursa results in the condition known as housemaid's knee. The posterior surface of the patella is divided by a vertical ridge that provides a facet for articulation with each of the femoral condyles. The proximal **base** of the patella receives fibers of the quadriceps tendon, as do the medial and lateral borders. The distal **apex** of the patella serves as the origin of the patellar ligament.

TIBIA

The tibia is the second longest bone in the body (Fig. 6-8). It is expanded at both its proximal (condylar) and distal (malleolar) ends. The proximal expansion results in two **tibial condyles** that correspond with those of the femur. The superior surface of each of the condyles is organized into two smooth facets for articulation with the femur. The medial articular facet is oval and concave; the lateral articular facet is circular and convex in an anterior-to-posterior direction. Surrounding the facets are the two menisci. This configuration results in only the centers of the two facets truly articulating with the femoral condyles. Between the articular facets is an **intercondylar eminence.** Posterior and anterior to this eminence are the sites of attachment of the

Fig. 6-8. The tibia and fibula. A. Anterior view. B. Posterior view.

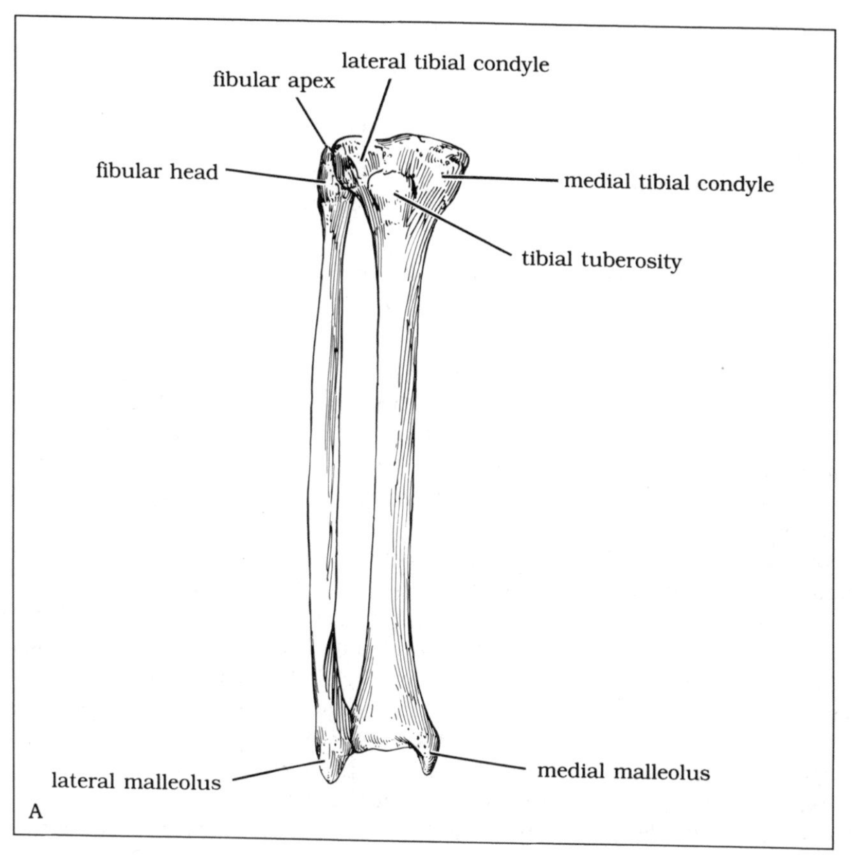

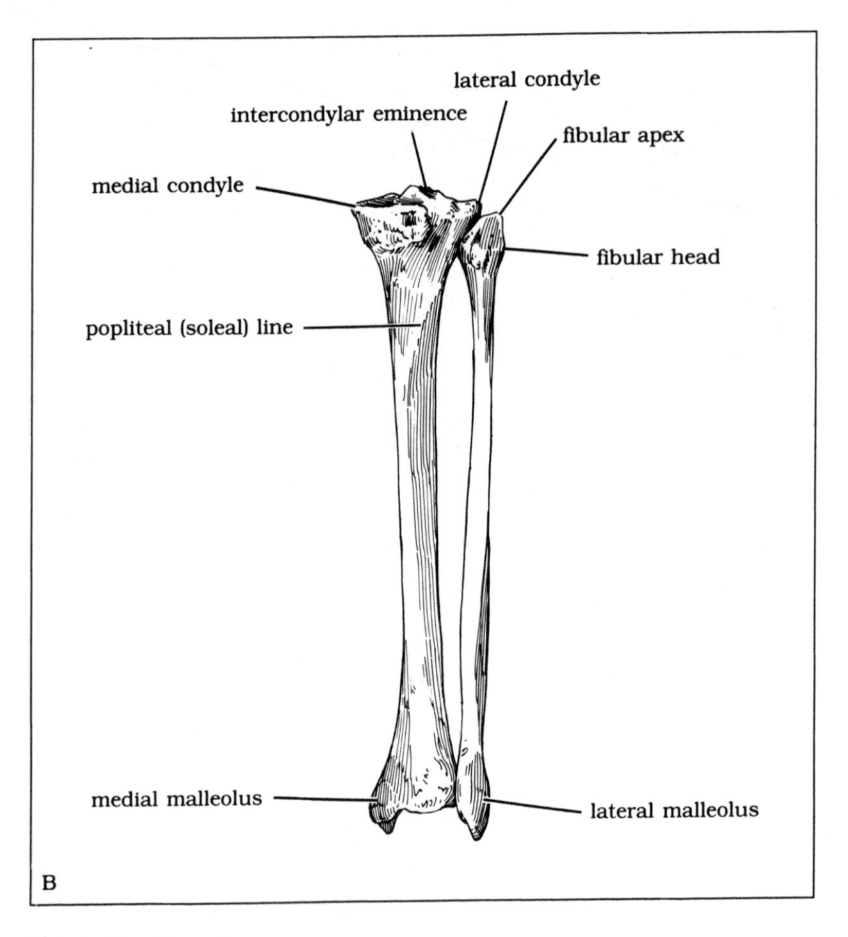

Fig. 6-8 (continued).

cruciate ligaments. On either side of this eminence there is a tubercle.

The anterior surface of the proximal tibia is a flattened, triangular region, at the apex of which is the **tibial tuberosity.** The patellar ligament attaches at the tibial tuberosity. On the posterior surface the condyles are separated by a deep groove, the **intercondylar fossa.** On the back of the medial condyle there is a groove for the attachment of the semimembranosus muscle. There is a facet for articulation with the fibula on the posterior inferior surface of the lateral condyle.

The **shaft** of the tibia has three surfaces and three borders. The medial tibial surface is subcutaneous and is palpable the length of the leg. Fibers of the tendons of the sartorius, gracilis, and semitendinosus muscles insert on it (Fig. 6-9A). The lateral tibial surface serves as the origin of the tibialis anterior along its proximal two-thirds. The posterior tibial surface has a ridge, the popliteal (soleal) line, that extends distally. The line marks the limit of the insertion of the popliteus muscle (Fig. 6-9B). It also serves as the origin of part of the soleus muscle. The flexor digitorum longus and tibialis posterior muscles also arise from the shaft of the tibia.

The anterior border of the tibia extends from the tibial tuberosity to the medial malleolus. It is subcutaneous in its course. It serves as the origin for some of the deep fascia of the leg. The lateral border of the tibia is very thin. Facing the

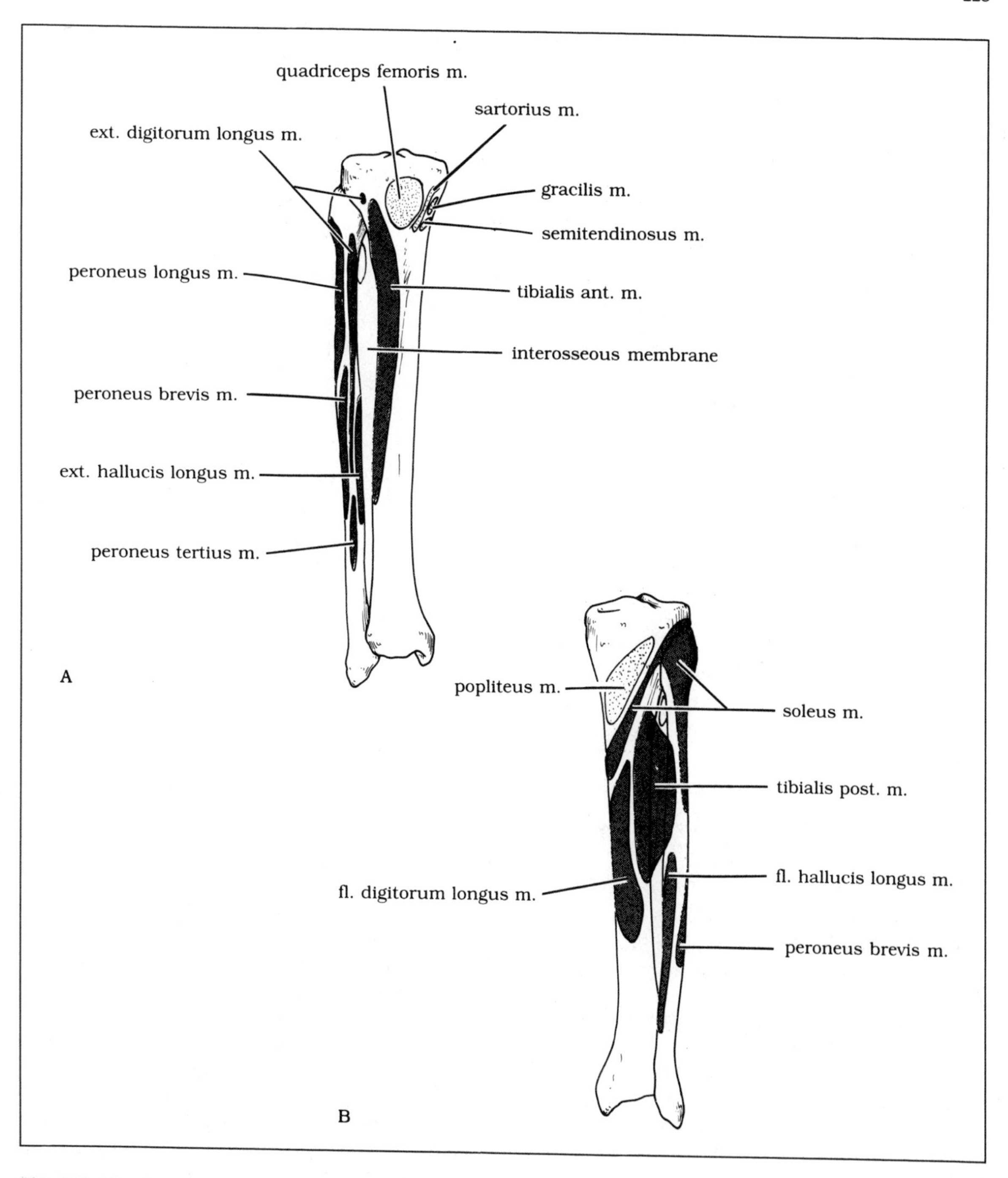

Fig. 6-9. Muscle origins (shaded areas) and insertions on the tibia and fibula. A. Anterior view. B. Posterior view.

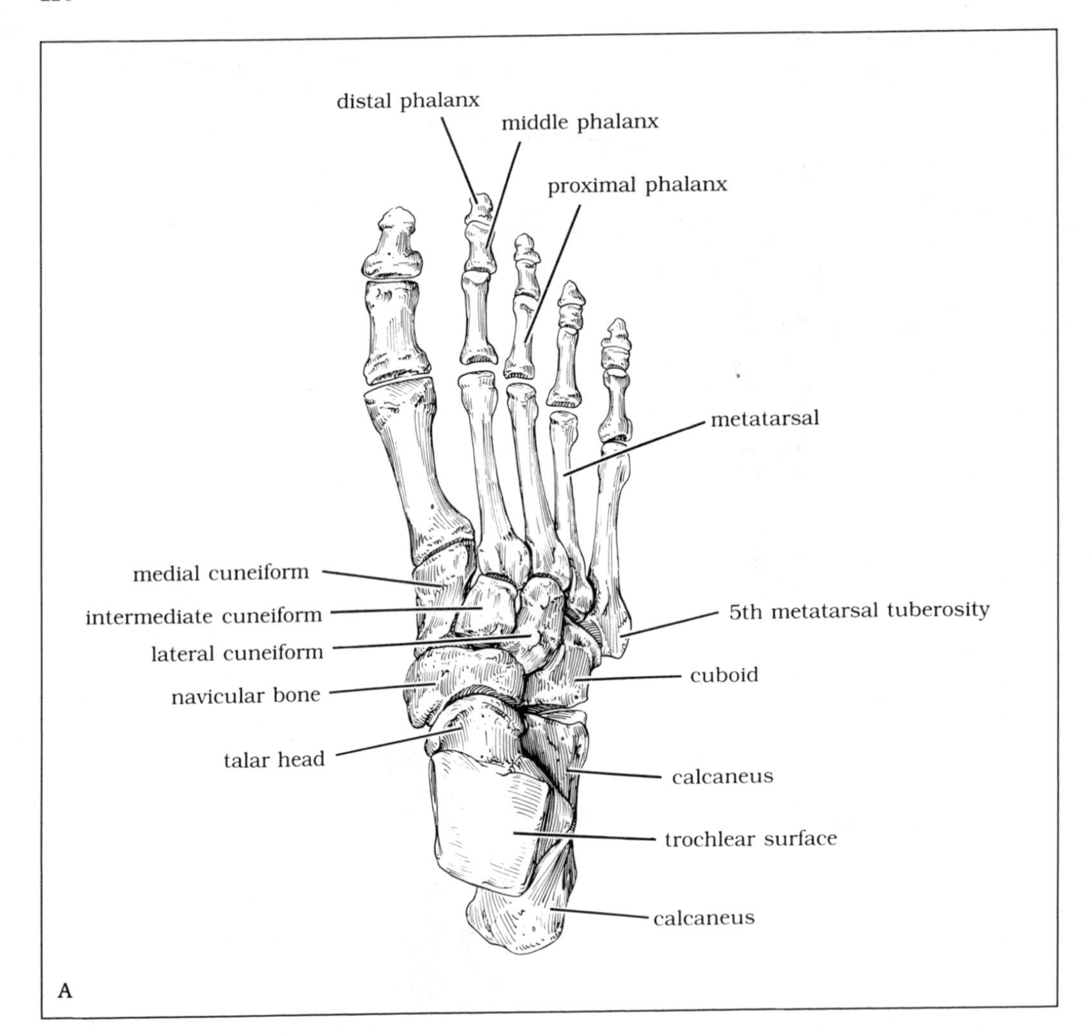

Fig. 6-10. Bones of the foot. A. Dorsum. B. Plantar.

fibula, it serves as the site of attachment of the interosseous membrane. The border extends from the proximal fibular articular facet to the distal articulation of the tibia with the fibula. The medial tibial border extends from the medial condyle of the tibia to the medial malleolus. The tibial collateral ligament attaches to the medial tibial border. In addition its middle third gives origin to some of the fibers of the soleus and flexor digitorum longus muscles.

The distal portion of the tibia is extended on its medial side as the **medial malleolus.** The inferior surface of the distal portion of the tibia is smooth, as it serves as the site of articulation with the body of the talus of the foot. The anterior surface is covered by the extensor tendons and the articular capsule.

The lateral surface of the distal portion of the tibia is marked with the fibular notch, into which an interosseous ligament between the tibia and fibula inserts. In addition the notch serves as the site of attachment of the anterior tibiofibular ligament.

The medial surface of the distal tibia is prolonged as the medial malleolus. The medial sur-

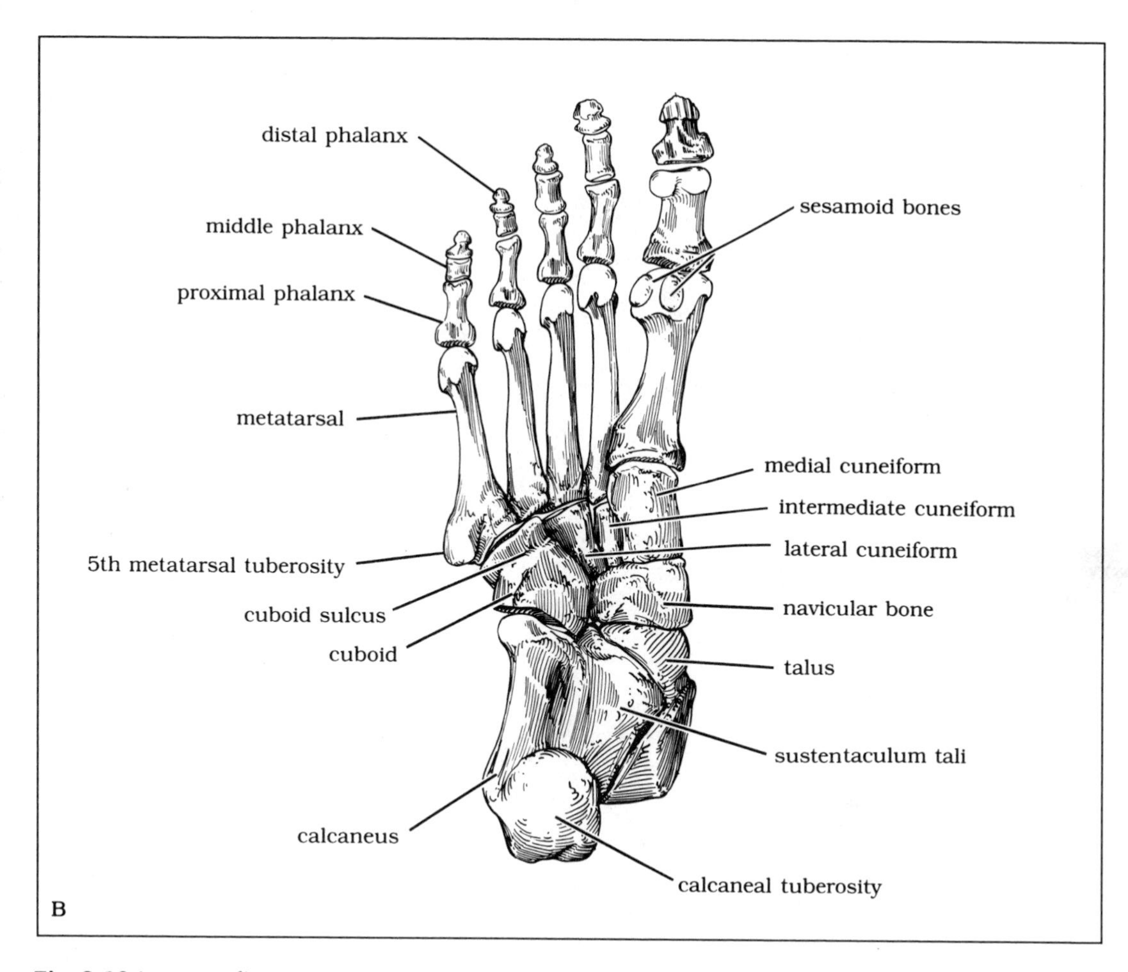

Fig. 6-10 (continued).

face of the malleolus is subcutaneous. The lateral surface of the malleolus articulates with the medial side of the talus. The posterior side has a groove for the tendons of the tibialis posterior and flexor digitorum longus muscles. The flexor hallucis longus tendon may also contact the malleolus here.

FIBULA

The fibula is the shorter, thinner bone of the leg (see Fig. 6-8). Its proximal end is below the knee joint; therefore the fibula does not participate in this joint.

The **head** of the fibula articulates with the lateral condyle of the tibia. On the lateral side of the head there is a prominence, the apex, that receives the tendon of the biceps femoris muscle and the fibular collateral ligament at the knee. Anteriorly the fibular head serves as the origin of part of the peroneus longus muscle; posteriorly some of the fibers of the soleus muscle arise (see Fig. 6-9).

The **body** of the fibula has three borders. The anterior border gives rise to the intermuscular fascia separating the anterior and lateral muscle groups. The posterior border serves as the origin of a septum separating posterior and lateral muscle groups. The interosseous (medial) border serves as the fibular attachment for the interos-

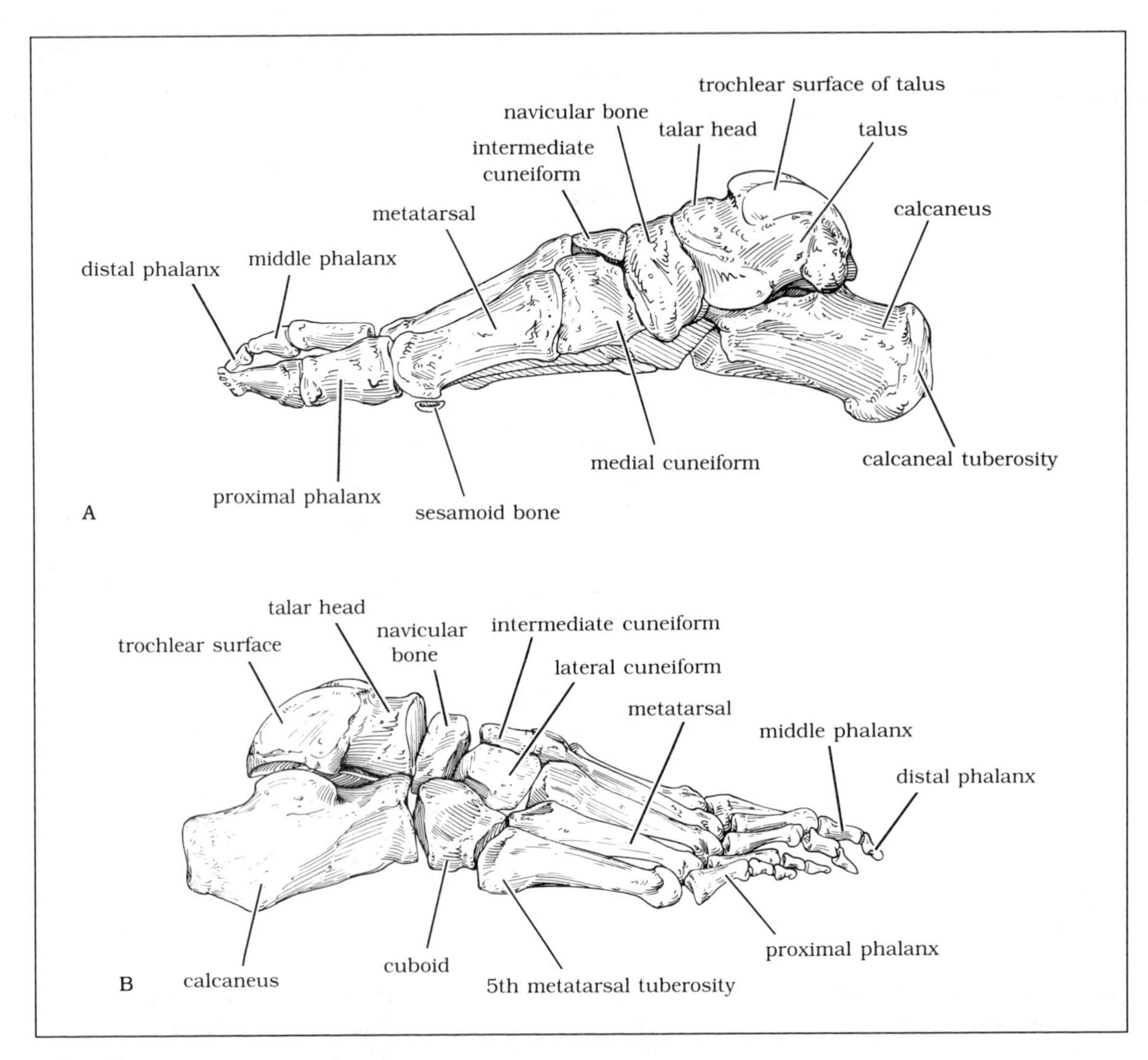

Fig. 6-11. Bones of the foot. A. Medial view. B. Lateral view.

seous membrane, which separates anterior and posterior muscle groups.

The **lateral malleolus,** which projects below the medial malleolus, is the expanded distal border of the fibula. Its articular (medial) surface is broad and oval shaped, articulating with the talus of the foot. Posteriorly the lateral malleolus has a groove for the passage of the tendons of the peroneus longus and peroneus brevis.

TARSUS, METATARSUS, AND PHALANGES (FOOT)

The foot consists of three parts: the tarsus, the metatarsus, and the phalanges (Figs. 6-10, 6-11, and 6-12). The tarsus contains the talus, calcaneus, cuboid, navicular, and three cuneiform bones. The **talus** is the proximal portion of the tarsus and acts to support the tibia. The body of the talus articulates superiorly with the malleoli and inferiorly with the calcaneus. The head

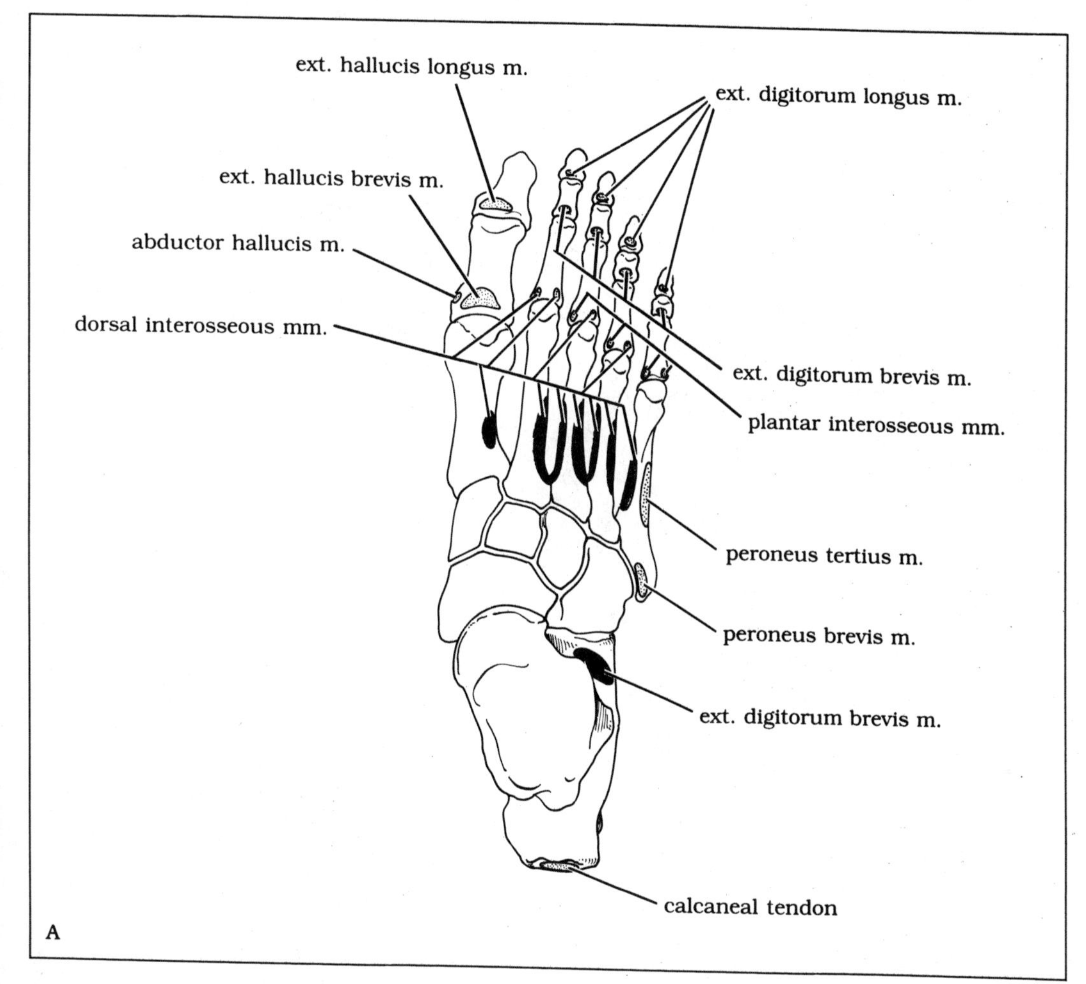

Fig. 6-12. Muscle origins (shaded areas) and insertions on the foot. A. Dorsum. B. Plantar.

of the talus articulates anteriorly with the navicular.

The **calcaneus** is the largest tarsal bone; it forms the heel of the foot. Its superior surface articulates with the talus, and its anterior surface articulates with the cuboid bone. On the medial surface of the calcaneus there is a process, the sustentaculum tali.

The **cuboid bone** is on the lateral side of the foot, proximal to the fourth and fifth metatarsal bones. On its medial side is the navicular. This bone is proximal to the three small cuneiforms.

The **cuneiforms** (medial, lateral, and intermediate) articulate with the cuboid, navicular, and metatarsal bones.

The **metatarsal bones** are five bones numbered one to five medially to laterally. Each bone has a base that articulates with the proximal tarsal bones, a slender body, and a head that articulates with its corresponding proximal phalanx. The first metatarsal is the shortest and thickest. The second metatarsal is the longest. The fifth metatarsal has a tuberosity on its lateral base.

The **phalanges** make up the digits of the toes.

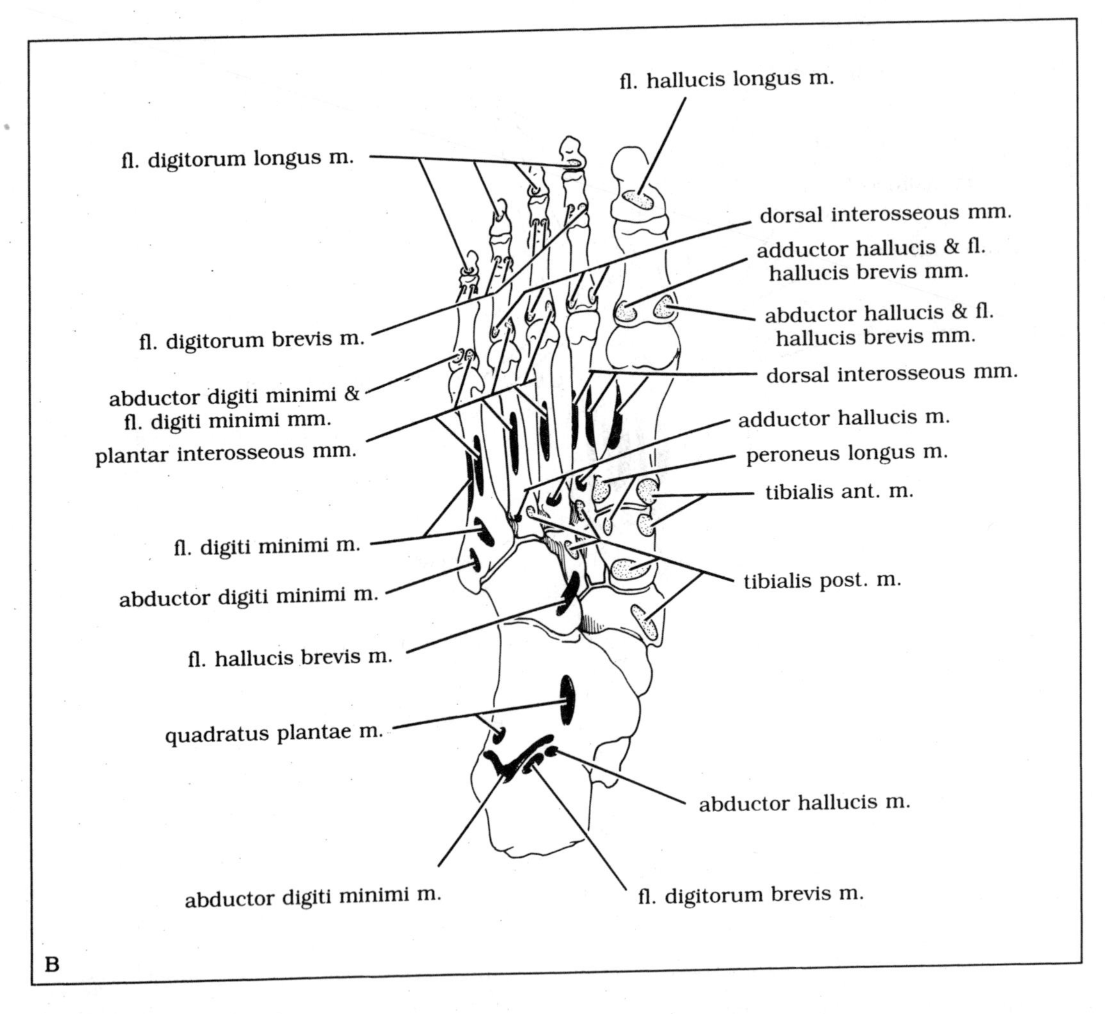

Fig. 6-12 (continued).

There are two phalanges in the first toe and three in each of the others.

In addition the foot has two constant sesamoid bones in the tendon of the flexor hallucis brevis at the metatarsophalangeal joint.

Thigh and Gluteal Region

FASCIAL COVERINGS AND PLANES

The thigh and gluteal region are covered with a varying amount of subcutaneous fat. This layer of fat is thickest over the buttocks. The superficial lymph nodes, subcutaneous blood vessels, and cutaneous nerves of the region lie between this fat layer and the membranous subcutaneous layer.

The deep fascial layer of the region is a stockinglike tough membranous layer known as the **fascia lata** (see Figs. 6-17 and 6-18A). Superiorly, attachments of the fascia lata are the iliac crest, inguinal ligament, pubic tubercle, pubic crest and symphysis, ischiopubic ramus, ischial tuberosity, sacrotuberous ligament, sacrum, and coccyx.

The fascia lata is thinnest in the gluteal region and thickest and strongest as the iliotibial tract, which arises from the iliac crest. The tensor fasciae latae muscle inserts into the tract, and the tract also receives fibers from the gluteus maximus. The iliotibial tract ends on the lateral tibial condyle, blending with the tendons of the vastus lateralis and biceps femoris muscles. Between the iliac crest and the gluteus maximus there is another region of the fascia lata, a thickening called the gluteal fascia (or gluteal aponeurosis).

There are two fascial planes within the thigh formed by the medial and lateral intermuscular septa. The septa extend from the femur to the fascia lata.

The saphenous opening is a break in the fascia lata approximately 4 cm below the inguinal ligament, lateral to the pubic tubercle. At this point the fascia lata consists of two laminae, deep and superficial. The superficial lamina forms the free sharp lateral border of the saphenous opening known as the falciform margin.

FEMORAL TRIANGLE

The femoral triangle is a subfascial region in the anterior upper third of the thigh (Fig. 6-13). The triangle contains the femoral vessels, femoral nerve, and lymph nodes. The femoral triangle is limited above by the inguinal ligament. Laterally the border is the sartorius muscle, medially the adductor long muscle. The apex of the triangle is the point at which the sartorius crosses the adductor longus. The floor of the femoral triangle consists of the pectineus and adductor longus muscles medially and the iliopsoas muscle laterally.

The femoral artery bisects the femoral triangle. It enters the triangle at the midinguinal point and leaves at the apex. Superiorly the femoral vein lies medial to the artery. As the vessels descend, the vein twists on the artery so that at the apex of the triangle the vein is posterior to the artery. The femoral nerve enters the femoral triangle between the iliacus and the psoas muscles, lateral to the artery. The nerve divides shortly after entering the femoral triangle; only the saphenous nerve and the nerve to the vastus medialis continue with the artery through the triangle.

The femoral sheath, an extension of the transversalis fascia, surrounds the femoral vessels and medial lymph nodes for 2 to 3 cm into the femoral triangle. The femoral sheath is divided into three compartments. The lateral compartment contains the artery, the middle compartment the vein, and the medial compartment the deep inguinal lymph nodes or fat or both. The femoral nerve is not contained within the sheath. A femoral hernia is one that usually involves the medial compartment; through this compartment the hernia may communicate with the saphenous opening.

FEMORAL (THIGH) MUSCULATURE

The femoral muscles are discussed in terms of four regions: anterior, medial, gluteal and lateral, and posterior regions.

Anterior Femoral Musculature

The anterior femoral muscles are the sartorius, quadriceps femoris (rectus femoris, vastus lateralis, vastus intermedius, and vastus medialis), and articularis genu. All of these muscles are innervated by the femoral nerve (L2–L4) (Table 6-1).

The **sartorius** (see Fig. 6-3) is the longest muscle in the body. It arises at the anterior superior iliac spine. The muscle crosses the front of the thigh diagonally to insert on the medial side of the tibia below the tuberosity. The tendon of insertion forms a common aponeurosis with the tendons of the gracilis and semitendinosus muscles. This common tendon is known as the pes anserinus (see Fig. 6-6B). The pes is separated from the underlying tibia by a bursa. Because the muscle crosses the thigh diagonally, contraction produces flexion, abduction, and lateral rotation of the thigh and flexion of the leg.

The **quadriceps femoris** consists of four mus-

Table 6-1. Femoral Musculature

Muscle	*Origin*	*Insertion*	*Innervation*	*Action**
Anterior				
Sartorius	Anterior superior iliac spine	Medial tibia (pes anserinus)	Femoral nerve (L2–L4)	Flex, abduct, laterally rotate thigh; flex leg
Rectus femoris	Anterior inferior iliac spine, acetabular rim	Quadriceps tendon, tibial tuberosity	Femoral nerve (L2–L4)	Flex at hip
Vastus lateralis	Greater trochanter, lateral lip of linea aspera, gluteal tuberosity, trochanteric line	Quadriceps tendon, lateral tibial condyle	Femoral nerve (L2–L4)	Extend leg at knee
Vastus intermedius	Anterior and lateral upper two-thirds of femur, lateral lip of linea aspera	Quadriceps tendon	Femoral nerve (L2–L4)	Extend leg at knee
Vastus medialis	Medial lip of linea aspera, intertrochanteric line	Quadriceps tendon, medial tibial condyle	Femoral nerve (L2–L4)	Extend leg at knee
Articularis genu	Lower fourth of anterior femur	Synovial membrane	Femoral nerve (L2–L4)	Proximal movement of articular capsule
Medial				
Gracilis	Inferior pubic ramus, pubic symphysis	Medial tibia (pes anserinus)	Anterior obturator nerve (L2–L4)	Adduct, flex leg
Pectineus	Pecten pubis	Pectineal line of femur	Femoral nerve (L2–L4), accessory obturator nerve, when present	Adduct, flex
Obturator externus	External part of pubic rami, ischial ramus, obturator membrane	Trochanteric fossa of femur	Obturator nerve (L2–L4)	Laterally rotate
Adductor longus	Superior pubic ramus	Medial lip of linea aspera	Anterior obturator nerve (L2–L4)	Adduct, flex, medially rotate
Adductor brevis	Inferior pubic ramus	Pectineal line, medial lip of linea aspera	Anterior obturator nerve (L2–L4)	Adduct, flex, laterally rotate

Table 6-1 (continued).

Muscle	*Origin*	*Insertion*	*Innervation*	*Action**
Adductor magnus	Inferior pubic ramus, ischial ramus	Medial gluteal ridge, linea aspera	Posterior obturator nerve (L2–L4)	Adduct, flex, medially rotate
	Ischial tuberosity	Medial epicondyle	Tibial nerve (L4 and L5)	Extend, medially rotate
Gluteal and lateral				
Gluteus maximus	Posterior gluteal line of ilium, sacrum and coccyx, sacrotuberous ligament, gluteal and erector spinal aponeuroses	Iliotibial tract, gluteal tuberosity	Inferior gluteal nerve (L5, S1, and S2)	Extend, laterally rotate
Gluteus medius	Ilium between posterior and anterior gluteal lines, gluteal aponeurosis	Greater trochanter	Superior gluteal nerve (L4, L5, and S1)	Abduct, medially rotate
Gluteus minimus	Ilium between anterior and inferior gluteal lines, margin of greater sciatic notch	Greater trochanter	Superior gluteal nerve (L4, L5, and S1)	Medially rotate, abduct, flex
Tensor fasciae latae	Outer iliac crest	Iliotibial tract	Superior gluteal nerve (L4, L5, and S1)	Flex, medially rotate thigh; extend, laterally rotate knee
Piriformis	Anterior sacrum	Greater trochanter	S2	Laterally rotate, abduct, extend
Obturator internus	Pubic rami, ischial ramus, circumference of obturator foramen, obturator membrane	Greater trochanter	Nerve to obturator internus (L5, S1, and S2)	Laterally rotate, help extend and abduct
Gemelli				
Superior	Spine at ischium	Greater trochanter	Nerve to obturator internus	Laterally rotate
Inferior	Ischial tuberosity	Greater trochanter	Nerve to quadratus femoris	Laterally rotate

Table 6-1 (continued).

Muscle	*Origin*	*Insertion*	*Innervation*	*Action**
Quadratus femoris	Ischial tuberosity	Vertical line from intertrochanteric crest	Nerve to quadratus femoris (L4, L5, and S1)	Laterally rotate
Posterior				
Semitendinosus	Ischial tuberosity	Medial tibia (pes anserinus)	Tibial nerve (L5, S1, and S2)	Extend thigh; flex and medially rotate leg
Semimembranosus	Ischial tuberosity	Medial tibial condyle	Tibial nerve (L5, S1, and S2)	Extend thigh; flex and medially rotate leg
Biceps femoris				
Long head	Ischial tuberosity, sacrotuberous ligament	Head of fibula	Tibial nerve (S1–S3)	Extend thigh; flex and laterally rotate leg
Short head	Lateral lip of linea aspera	Lateral tibial condyle	Peroneal nerve (L5, S1, and S2)	Flex and laterally rotate leg

*Action of thigh, unless otherwise stated.

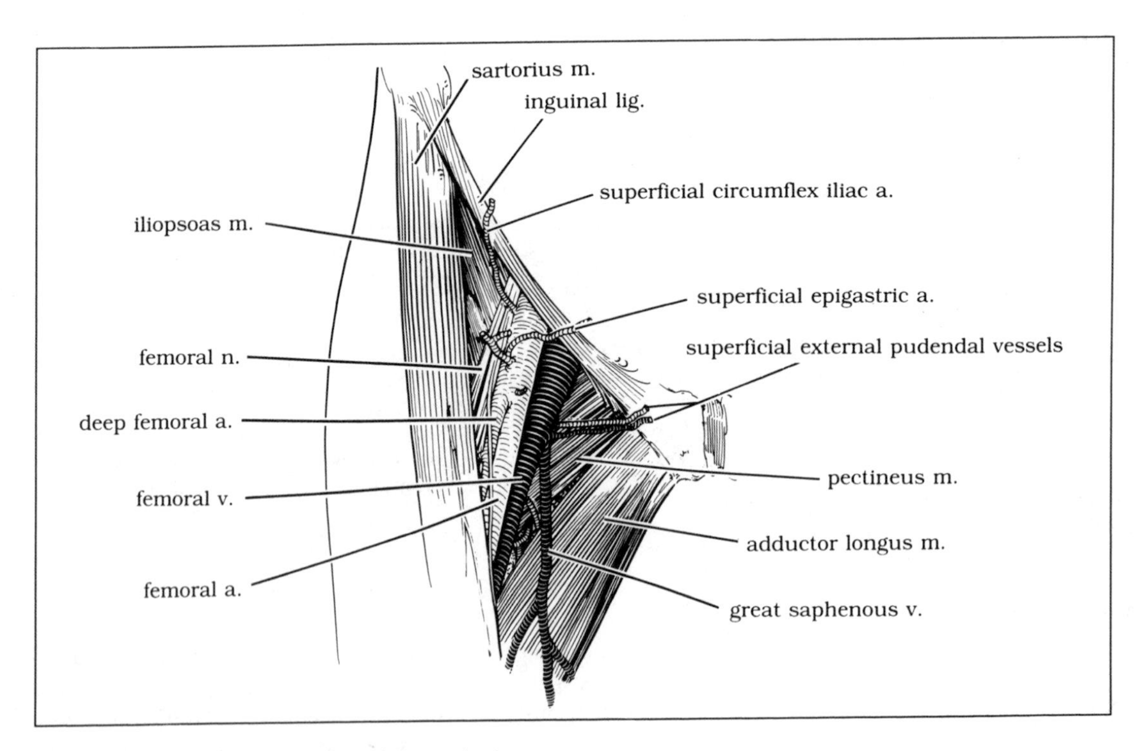

Fig. 6-13. The femoral triangle and its contents.

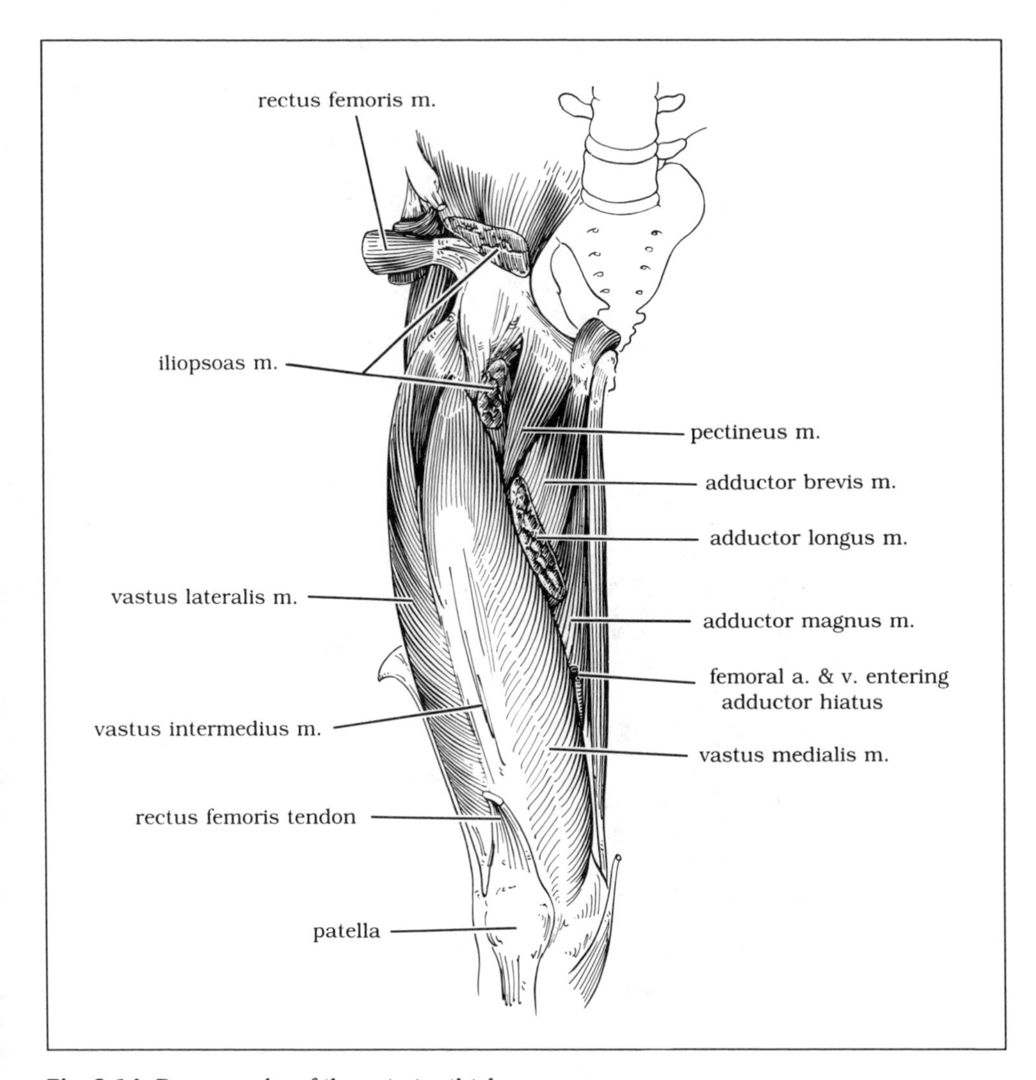

Fig. 6-14. Deep muscles of the anterior thigh.

cles: the rectus femoris (see Fig. 6-3) and the three vasti (Fig. 6-14). The **rectus femoris** has two points of origin (see Fig. 6-6). The straight head arises from the anterior inferior iliac spine. The reflected head arises from a groove above the acetabular rim. The tendon of the rectus femoris (part of the common quadriceps tendon) inserts into the tibial tuberosity.

The **vastus lateralis** (see Fig. 6-17) arises from the upper trochanteric line, the anterior and inferior borders of the greater trochanter, the gluteal tuberosity, the lateral lip of the linea aspera, and the lateral intermuscular septum (see Fig. 6-6). The muscle inserts into the superolateral patella and the lateral tibial condyle.

The **vastus intermedius** arises from the anterior and lateral upper two-thirds of the femur, the lower half of the lateral lip of the linea aspera, and the lateral intermuscular septum (see Fig. 6-6). The muscle inserts via the common tendon, its fibers blending with those of the other muscles.

The **vastus medialis** (see Fig. 6-15) arises from

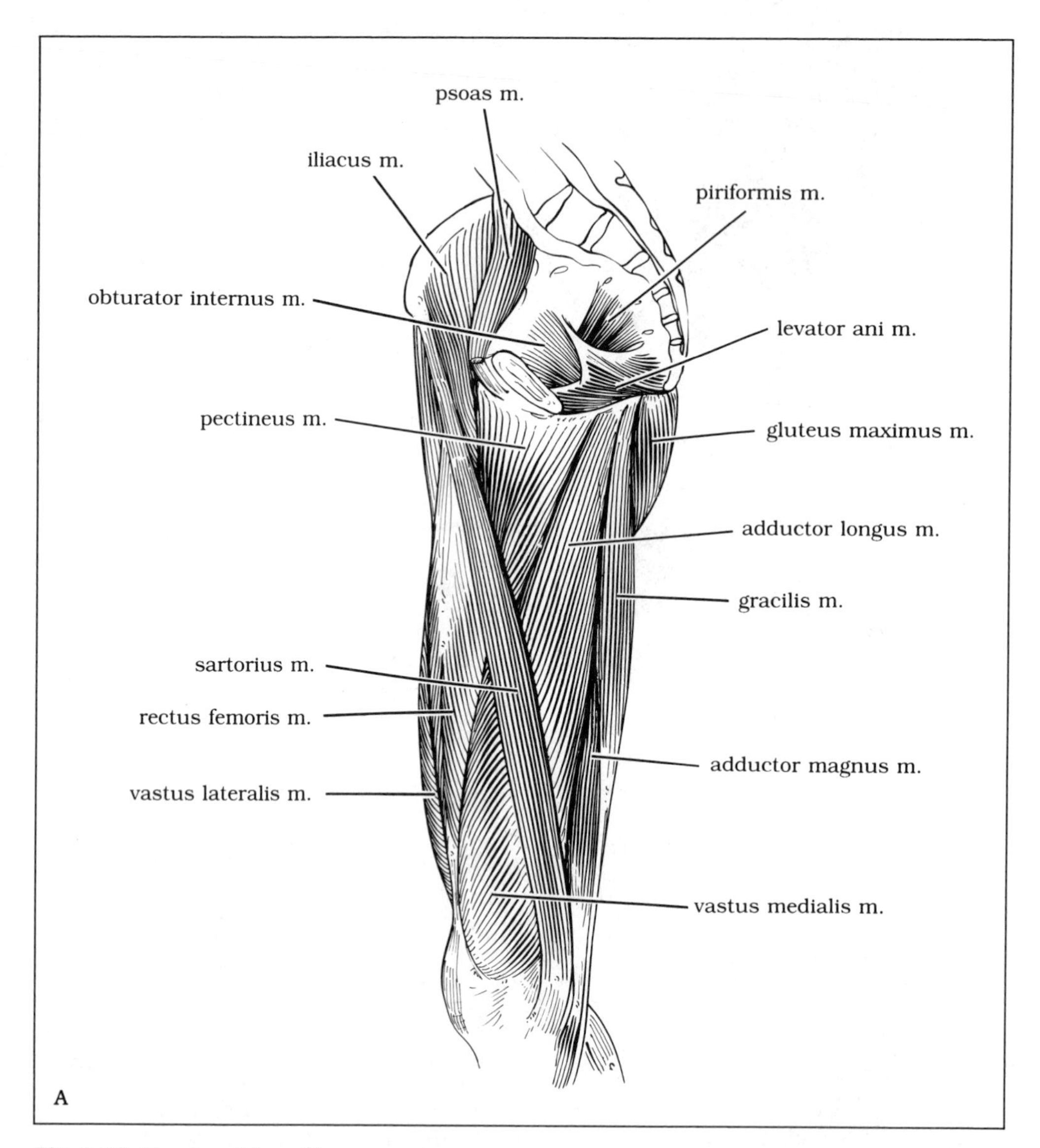

Fig. 6-15. Muscles of the adductor region.
A. Superficial view. B. Deep view.

the medial lip of the linea aspera, the distal intertrochanteric line, and the medial intermuscular septum (see Fig. 6-6). The muscle inserts into the superomedial patella and medial tibial condyle.

The patella is a sesamoid bone in the common tendon of the quadriceps femoris (see Fig. 6-14). It serves as a lever to increase the angle of insertion. The quadriceps muscle extends the leg at the knee. In addition, since the rectus femoris arises proximal to the hip joint, it flexes the thigh at the hip.

The **articularis genu** is a small muscle arising from the lowest fourth of the femur (see Fig. 6-6A). The muscle inserts into the upper synovial membrane at the knee and draws the articular capsule proximally. Innervation is provided by a

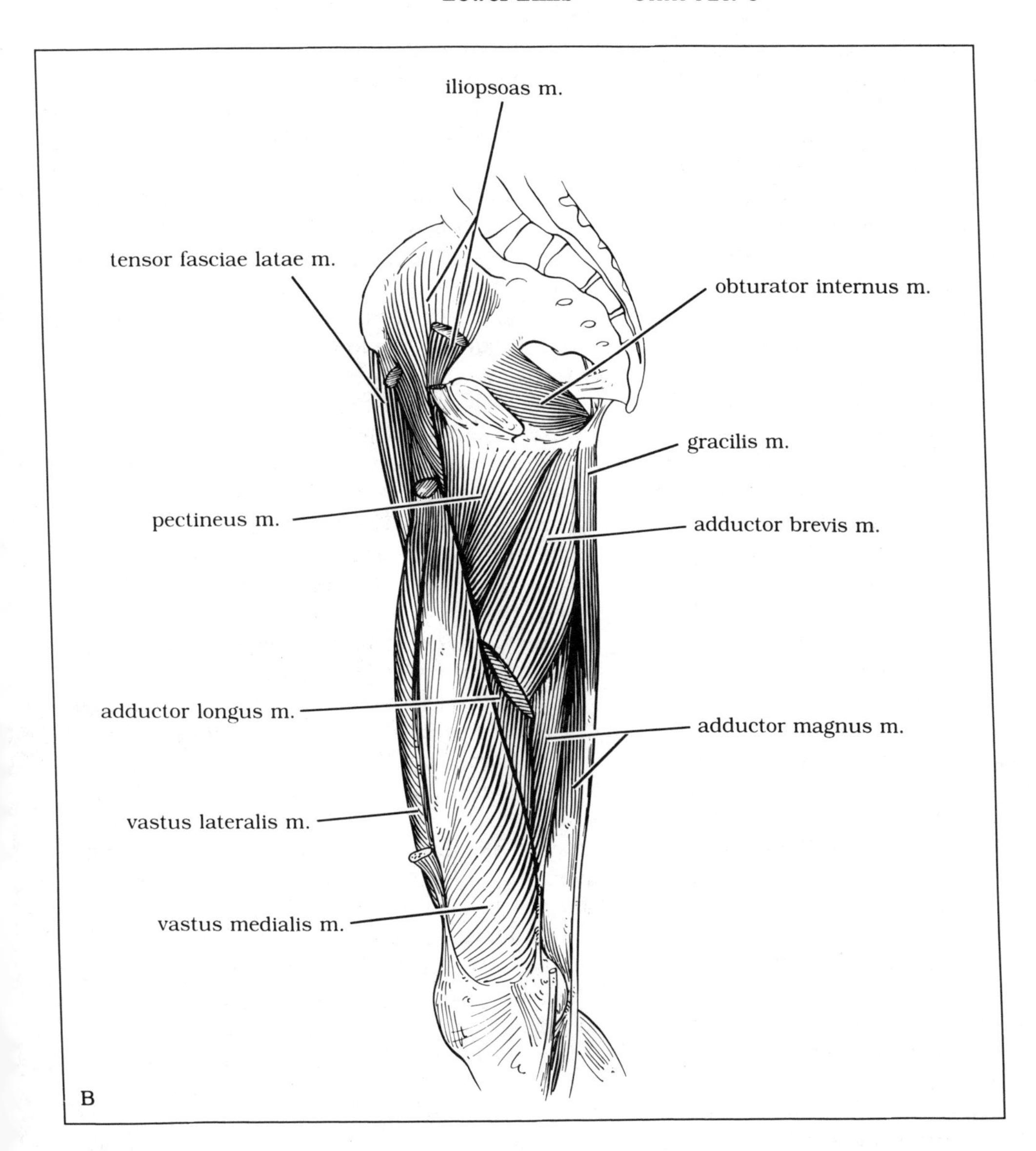

Fig. 6-15 (continued).

branch of the femoral nerve to the vastus intermedius.

Medial Femoral Musculature

The medial femoral muscles are the gracilis, pectineus, obturator externus, adductor longus, adductor brevis, and adductor magnus. All are innervated by the obturator nerve, except the pectineus, which receives femoral innervation.

The **gracilis** (Fig. 6-15; see also Fig. 6-3) is a long, thin superficial muscle, the most medial one of the thigh. It arises from the inferior pubic ramus and the pubic symphysis. The gracilis inserts onto the medial tibia as part of the pes anserinus, between the fibers of the sartorius and semitendinosus (see Fig. 6-5B). The gracilis adducts the thigh and helps to flex the leg at the knee. It is innervated by the anterior division of the obturator nerve (L2–L4).

The **pectineus** (see Figs. 6-3 and 6-14) forms the medial floor of the femoral triangle. It arises from the pecten pubis and inserts into the pectineal line of the femur (see Fig. 6-6). The pectineus is involved in adduction of the thigh, and also assists in flexion of the thigh at the hip joint. It is innervated by the femoral nerve (L2–L4) and, when present, by a branch of the accessory obturator nerve.

The **obturator externus muscle** arises from the external part of the superior and inferior pubic rami, ischial ramus, and obturator membrane. The muscle inserts into the trochanteric fossa of the femur, reaching this point by passing in back of the neck of the femur (see Fig. 6-6). Part of the synovial membrane of the hip joint acts as a bursa in separating the muscle's tendon and the neck of the femur. The obturator externus muscle laterally rotates the thigh. It is innervated by the obturator nerve (L2–L4).

The **adductor longus** (see Figs. 6-3, 6-13, and 6-15) forms the medial edge of the femoral triangle. The muscle arises from the medial portion of the superior pubic ramus; it inserts on the middle third of the medial lip of the linea aspera (see Fig. 6-6). The adductur longus adducts the thigh and in addition helps to flex and perhaps medially rotate the thigh. It is innervated by a branch of the anterior division of the obturator nerve (L2—L4).

The **adductor brevis** (see Figs. 6-6, 6-14, and 6-15B) arises from the inferior pubic ramus. It inserts into the lower two-thirds of the pectineal line of the upper half of the medial lip of the linea aspera. The adductor brevis adducts the thigh and helps to flex and laterally rotate it, too. The muscle is innervated by the anterior division of the obturator nerve (L2–L4).

The **adductor magnus** (see Figs. 6-6, 6-14, and 6-15) may be considered two muscles: an upper, adductor minimus, portion whose fibers run horizontally and a lower portion whose fibers run vertically. The muscle arises from the lower portion of the inferior pubic ramus, the ischial ramus, and the ischial tuberosity. The fibers of the upper, adductor minimus, portion insert into the medial gluteal ridge and the upper part of the linea aspera. The fibers of the lower portion continue, inserting along the linea aspera and finally ending in the adductor tubercle of the medial epicondyle of the femur. These last fibers, which arise specifically from the ischial tuberosity and insert into the medial epicondyle by a round tendon, are known as the ischiocondylar portion of the adductor magnus.

Between the insertion of the fibers into the linea aspera and the insertion into the adductor tubercle there is a gap, the **adductor hiatus** (see Fig. 6-14), that allows for the passage of the femoral vessels. The insertion of the adductor magnus into the linea aspera is also pierced by the perforating branches of the deep femoral artery.

The adductor magnus adducts the thigh. The adductor minimus portion also assists in flexion and may assist in medial rotation of the thigh. The ischiocondylar portion of the muscle is actually a hamstring muscle and extends and medially rotates the thigh. The posterior division of the obturator nerve (L2–L4) innervates the adductor magnus, except for the ischiocondylar portion, which is innervated by the tibial division of the sciatic nerve (L4, L5).

The **adductor canal** (Fig. 6-16; see also Fig. 6-14) is a passage for the femoral vessels through the middle third of the thigh. The canal begins at the apex of the femoral triangle and ends at the adductor hiatus, where the vessels enter the posterior compartment of the thigh. The anterolateral boundary of the canal is the vastus medialis. The anteromedial boundary is the sartorius. The posteromedial boundary is formed by the adductor longus and adductor magnus. The femoral vessels are accompanied through the canal by the saphenous nerve and the nerve to the vastus medialis, branches of the femoral nerve. Within the adductor canal the femoral artery is anterior to the femoral vein. Deep within the canal the deep femoral vessels may be found.

Gluteal and Lateral Femoral Musculature

The muscles of the gluteal and lateral femoral region are the gluteus maximus, gluteus medius, gluteus minimus, tensor fasciae latae, piriformis, and quadratus femoris and the obturator internus, superior gemellus, and inferior gemellus muscles.

The **gluteus maximus** (Figs. 6-17 and 6-18A) is a broad, heavy, coarse, quadrilateral-shaped muscle, the most prominent muscle in the buttocks. The muscle arises (see Fig. 6-6B) from the posterior gluteal line of the ilium and the bone both above and behind it, from the posterior surface of the sacrum and coccyx, from the aponeurosis of the erector spinae muscle group, from the sacrotuberous ligament, and from the gluteal fascia, which covers the gluteus medius. The superficial fibers of the gluteus maximus insert into the iliotibial tract, the deep fibers into the gluteal tuberosity. The gluteus maximus extends the thigh and laterally rotates it. In addition, through its insertion into the iliotibial tract the muscle braces the knee when the knee is extended. The gluteus maximus is innervated by the inferior gluteal nerve (L5, S1, and S2). The muscle is associated with three bursae: one over the greater trochanter, one separating it from the origin of the vastus lateralis, and an inconstant one separating it from the ischial tuberosity.

The **gluteus medius** (Fig. 6-18B,C) lies under the gluteus maximus and the gluteal fascia. The

Fig. 6-16. Cross section of the thigh, above the origin of the biceps femoris.

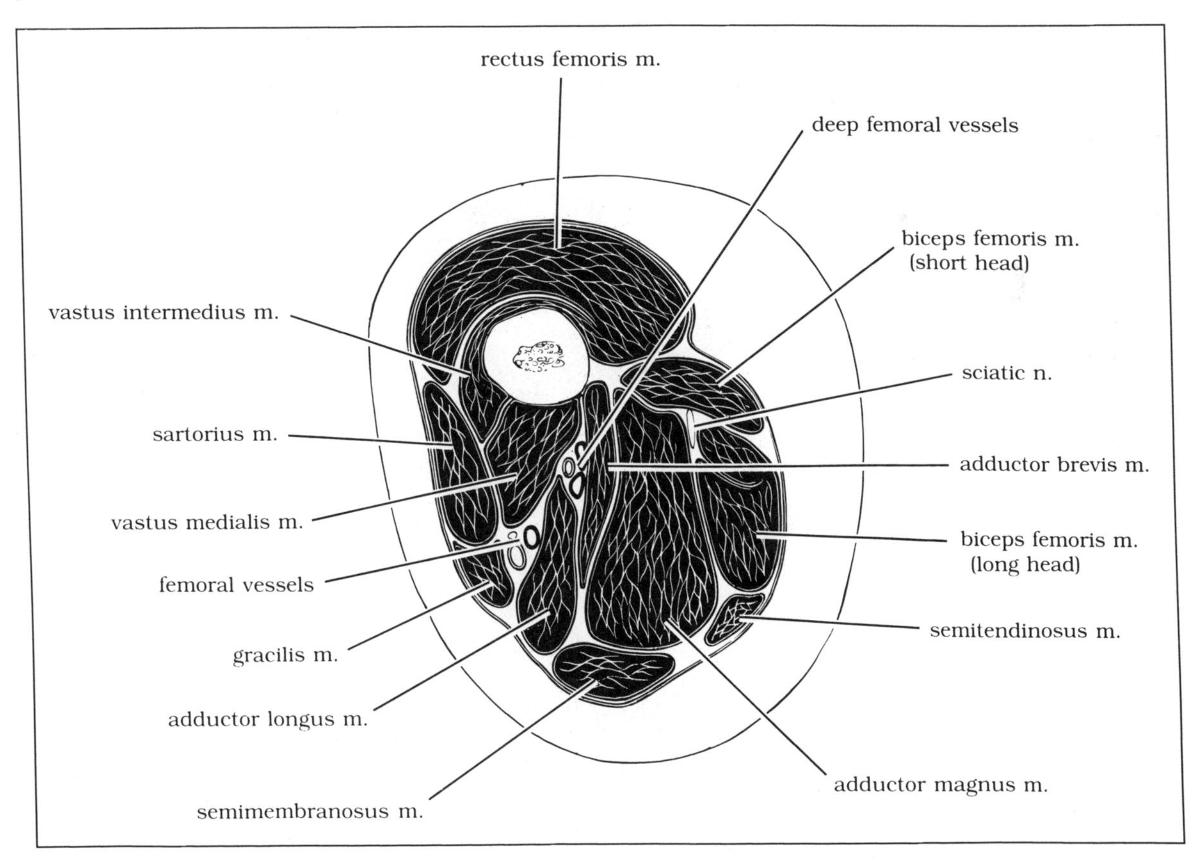

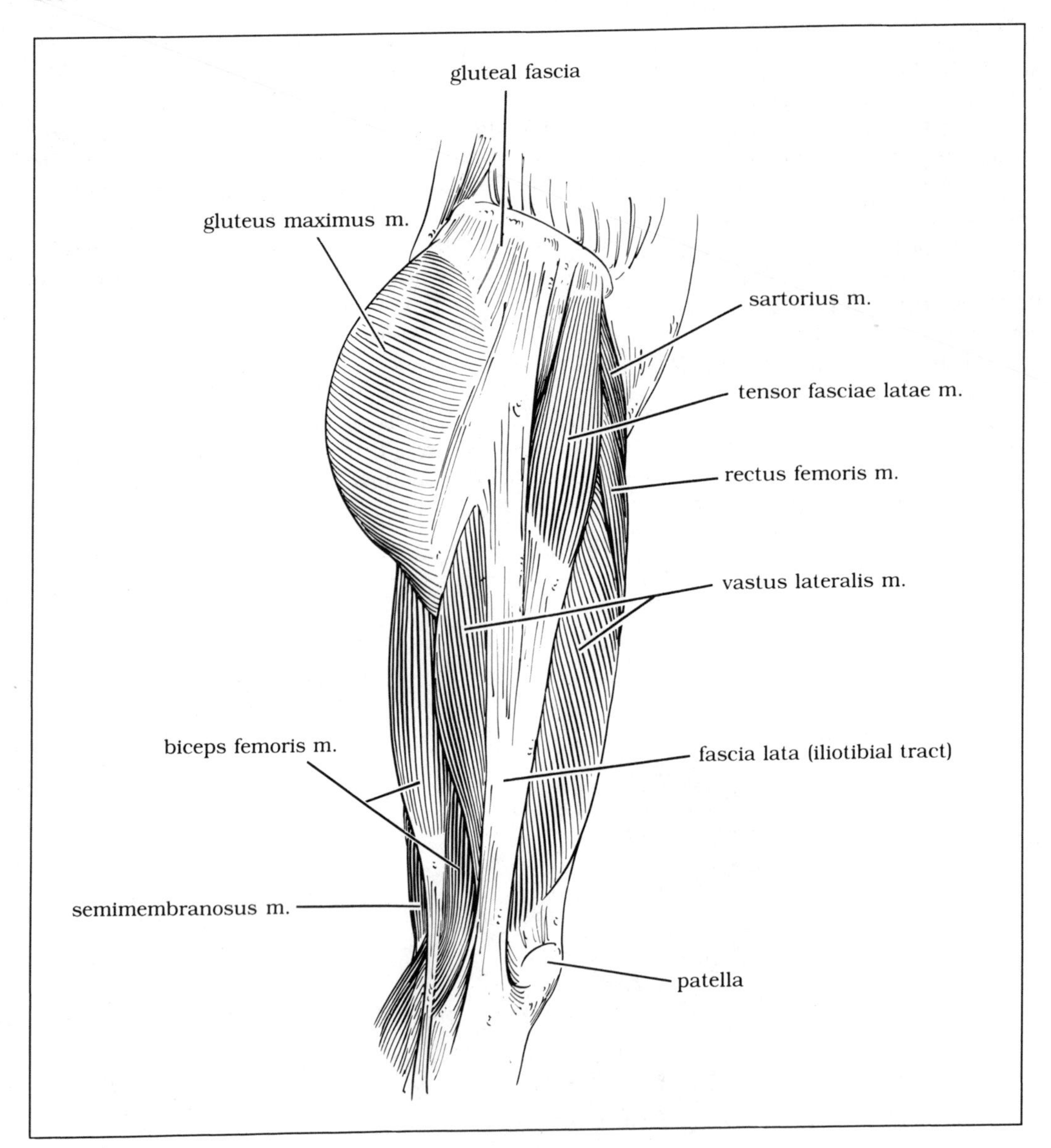

Fig. 6-17. The lateral thigh.

gluteus medius arises from the external surface of the ilium between the posterior and anterior gluteal lines and from the gluteal fascia; it inserts into the lateral surface of the greater trochanter, from which it is separated by a bursa before the insertion (see Fig. 6-6B). The gluteus medius medially rotates the thigh and abducts it. The anterior part of the muscle flexes and medially rotates the thigh, the posterior part extends and laterally rotates it. The gluteus medius is innervated by the superior gluteal nerve (L4, L5, and S1).

The **gluteus minimus** (see Fig. 6-18C) is the deepest of the gluteal muscles. It arises from the outer surface of the ilium between the anterior and inferior gluteal lines and from the margin of the greater sciatic notch; the muscle inserts into the anterior border of the greater trochanter, from which it too is separated by a bursa prior to insertion (see Fig. 6-6). The gluteus minimus medially rotates, abducts, and flexes the thigh. The muscle is supplied by the superior gluteal nerve (L4, L5, and S1). The gluteus minimus and gluteus medius are separated by a fascial plane that contains the superior gluteal vessels and nerves (see Fig. 6-18C).

The **tensor fasciae latae** (see Figs. 6-3 and 6-17) is the only other muscle of the group to arise from the external surface of the ilium—specifically, the anterior part of the iliac crest, the anterior superior iliac spine, and the notch below it (see Fig. 6-6B). The muscle inserts into the iliotibial tract approximately one-third of the way down the thigh. The tensor fasciae latae flexes the thigh and also helps to medially rotate it. It also acts to extend and laterally rotate the leg. The muscle is innervated by the superior gluteal nerve (L4, L5, and S1).

The **piriformis** arises within the pelvis, by three digitations, from the anterior sacrum (see Figs. 6-15A and 6-18B,C). The muscle passes out of the pelvis, through the greater sciatic foramen, to insert into the superior border of the greater trochanter. The piriformis laterally ro-

Fig. 6-18. Muscles of the gluteal region.
A. Superficial dissection. B. Intermediate dissection.
C. Deep dissection.

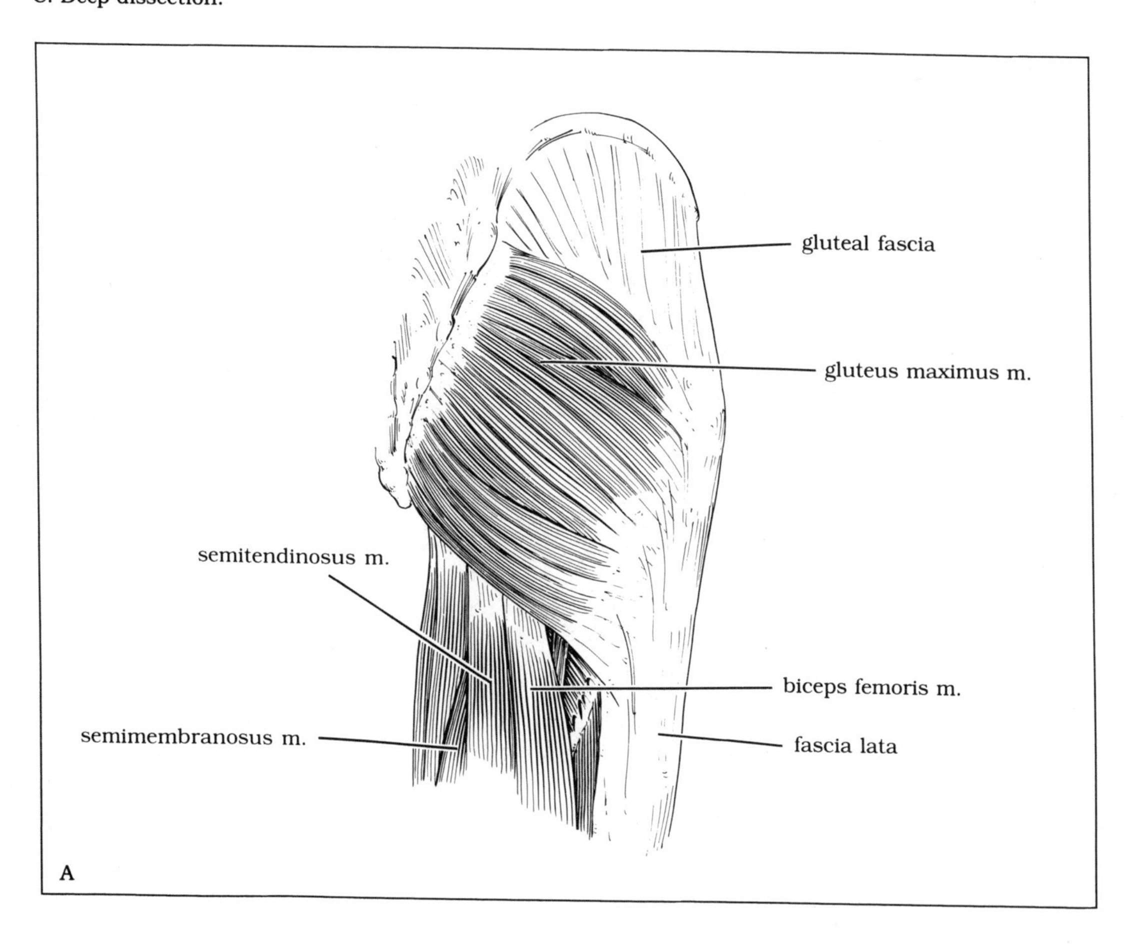

tates, extends, and abducts the thigh. It is innervated by branches of the second sacral nerve.

The **obturator internus** (see Figs. 6-15 and 6-18B,C) also arises within the pelvis from the superior and inferior pubic rami, the ischial ramus, the circumference of the obturator foramen, and the surface of the obturator membrane. The muscle exits the pelvis through the lesser sciatic foramen, passes across the capsule of the hip joint, and inserts into the anterior part of the medial surface of the greater trochanter. There is a bursa between the tendon of the muscle and the capsule of the hip joint. The obturator internus muscle laterally rotates the thigh. When the hip joint is flexed the muscle also extends and abducts the thigh. Innervation is accomplished by the nerve to the obturator internus (L5, S1, and S2).

The **obturator membrane** is a fibrous sheet that closes off the obturator foramen. The sheet is pierced by the obturator vessels and nerves, which exit the pelvis by passing through the obturator canal.

The **superior gemellus** (see Fig. 6-18B,C) arises from the outer surface of the spine of the ischium and inserts with the obturator internus. The **inferior gemellus** (see Fig. 6-18B,C) arises

Fig. 6-18 (continued).

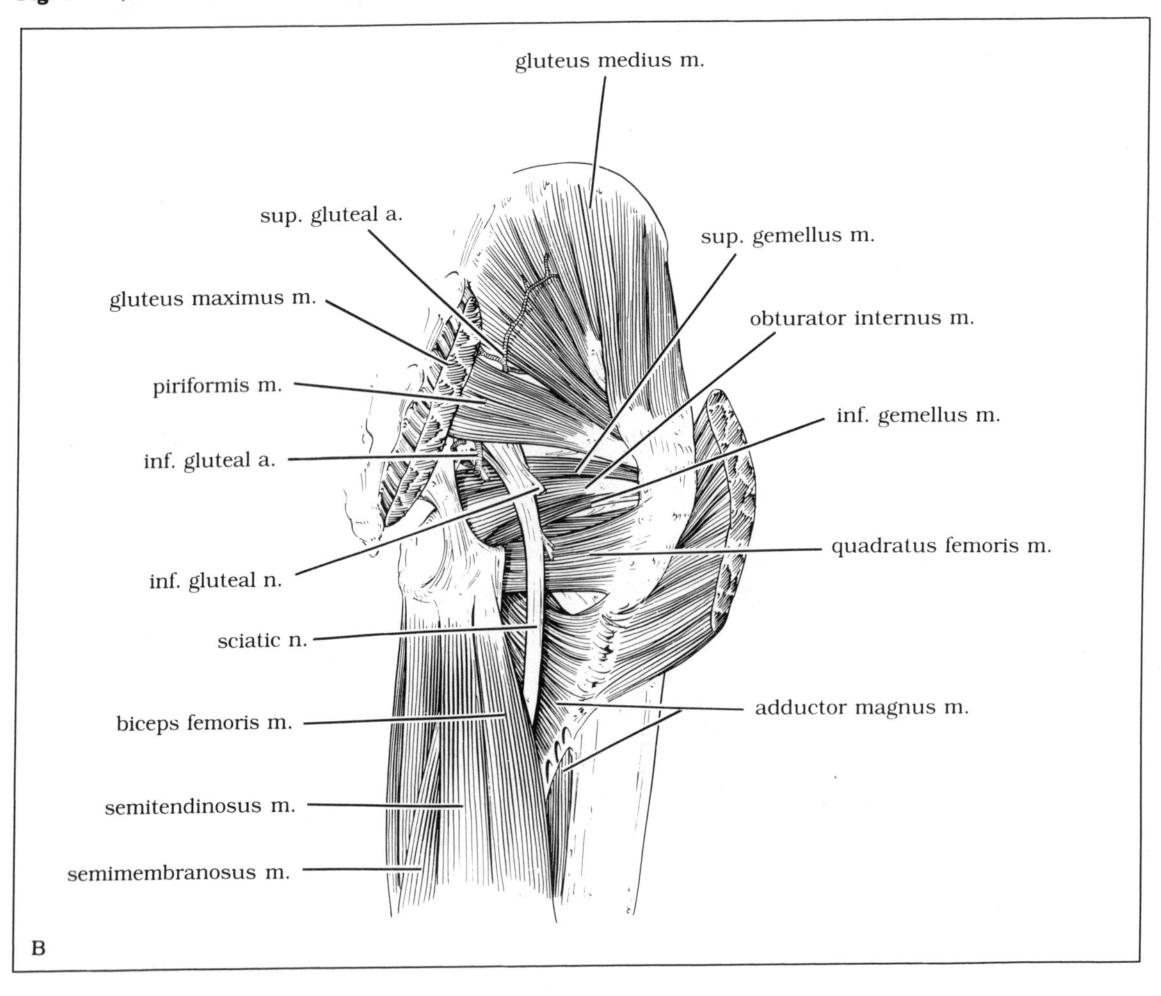

from the ischial tuberosity and also inserts with the obturator internus. The two muscles laterally rotate the thigh. The superior gemellus is innervated by a branch of the nerve to the obturator internus. The inferior gemellus is innervated by a branch of the nerve to the quadratus femoris.

The **quadratus femoris** (see Fig. 6-18B,C) is a quadrilateral muscle between the inferior gemellus and the adductor magnus. The quadratus femoris arises from the external edge of the ischial tuberosity and inserts along a line that descends vertically from the intertrochanteric crest (see Fig. 6-6). Along with the gemelli and obturator internus, the quadratus femoris acts as a lateral rotator of the thigh. The muscle is innervated by its own nerve, the nerve to the quadratus femoris (L4, L5, and S1).

Posterior Femoral Musculature

The posterior femoral muscles (the hamstrings) are the semitendinosus, semimembranosus, and biceps femoris muscles (Fig. 6-19). The ischiocondylar portion of the adductor magnus may also be considered a posterior femoral mus-

Fig. 6-18 (continued).

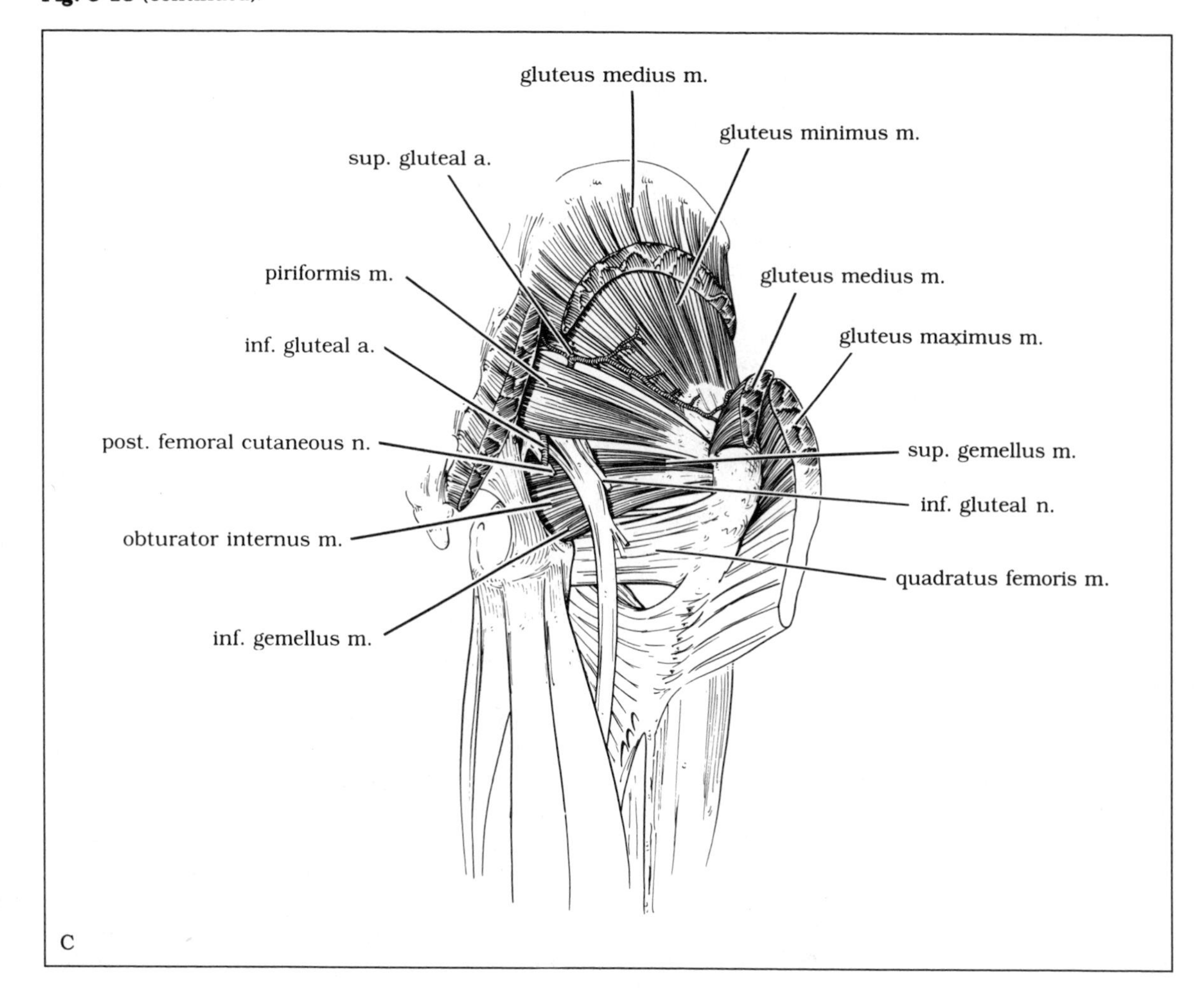

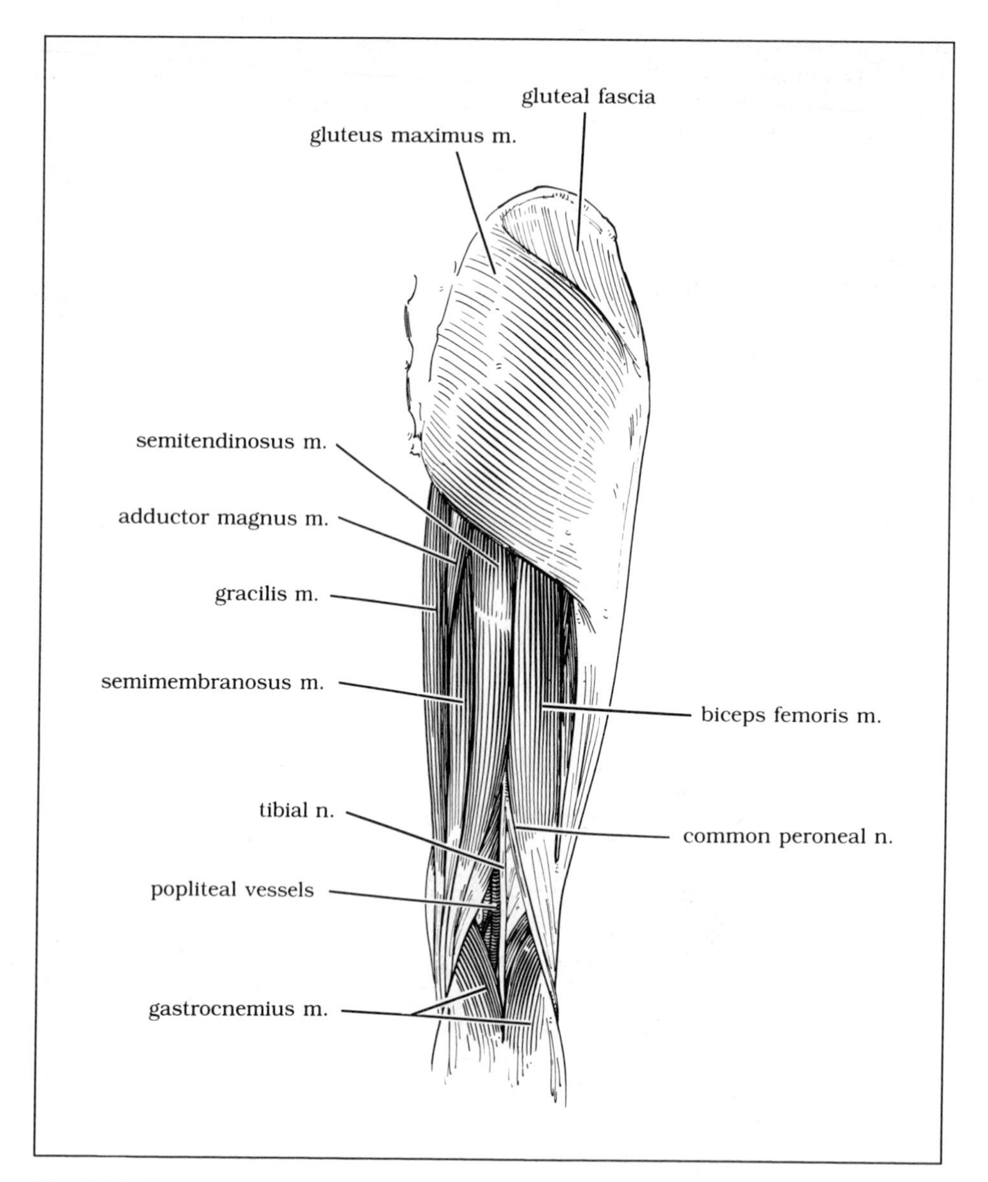

Fig. 6-19. Muscles of the posterior thigh.

cle. All the muscles of this group, except the short head of the biceps femoris, are innervated by the tibial division of the sciatic nerve.

The **semitendinosus** (see Fig. 6-18A,B) arises along with the long head of the biceps femoris from the lower medial portions of the ischial tuberosity. The muscle inserts into the upper medial tibia as part of the pes anserinus (see Fig. 6-6B). In its descent the muscle forms the medial margin of the popliteal fossa. The semitendinosus extends the thigh and flexes and medially rotates the leg. Innervation is supplied by the tibial division of the sciatic nerve (L5, S1, and S2).

The **semimembranosus** (see Fig. 6-19), the other medial hamstring muscle, arises from the upper and outer portions of the ischial tuberosity. The muscle inserts into the posteromedial portion of the medial tibial condyle (see Fig. 6-6B). The semimembranosus, with the semitendinosus, extends the thigh and flexes and medially rotates the leg. The muscle is innervated by the tibial division of the sciatic nerve (L5, S1, and S2).

The **biceps femoris** is the lateral hamstring muscle (see Figs. 6-17 and 6-19). The long head of the biceps femoris arises from the lower and inner part of the ischial tuberosity and from the inferior part of the sacrotuberous ligament (see Figs. 6-6 and 6-18A,B). The short head of the biceps femoris arises from the lateral lip of the linea aspera. The muscle inserts into the lateral side of the head of the fibula and the lateral tibial condyle. It flexes the leg and rotates it laterally. In addition the long head helps in extension of the thigh. The long head of the biceps is innervated by the tibial division of the sciatic nerve (S1–S3). The short head of the biceps is innervated by the peroneal division of the sciatic nerve (L5, S1, and S2).

Actions of the Femoral Muscles

The anterior femoral muscles are primarily concerned with extension of the leg and have only secondary, and weak, action as flexors of the thigh. They also act to stabilize the knee joint and guide and control the motion of the patella. The medial femoral muscles are involved in adduction of the thigh and lateral rotation. They are involved in the maintenance of posture through synergistic actions.

The lateral femoral and gluteal muscles are extensors and abductors of the hip and lateral rotators of the thigh; the gluteal muscles, particularly the maximus, are also involved in abducting the thigh. All of the gluteal muscles are involved in walking, particularly the gluteus medius and minimus. These muscles act in concert when one foot is off the ground to keep the hip of the unsupported side from sagging. Impairment of their action (either congenitally, through nerve damage, or by fracture) results in a sinking of the unsupported side of the pelvis when weight is applied to the affected side: the so-called Trendelenburg sign. Thus, damage to these muscles results in considerable gait abnormality. The remaining smaller lateral muscles are involved in lateral rotation of the thigh and in postural support.

The posterior femoral muscles flex the leg at the knee and extend the thigh at the hip. They also have some rotational capabilities (biceps femoris, laterally; semimembranosus and semitendinosus, medially). These muscles are also active in maintaining balance when the body weight shifts forward.

BLOOD SUPPLY

The gluteal region is served by the superior and inferior gluteal arteries (see Fig. 6-18B,C). The **superior gluteal artery** is a continuation of the posterior division of the internal iliac artery. The artery exits the pelvis through the greater sciatic foramen above the piriformis. There it divides into superficial branches that supply the gluteus maximus and deep branches. The deep branches of the superior gluteal artery run between the gluteus medius and minimus to supply these muscles, the tensor fasciae latae, and the hip joint.

The **inferior gluteal artery** is a terminal branch of the anterior division of the internal iliac artery. The vessel exits the pelvis below the piriformis and then descends deep to the gluteus maximus, giving off branches to this muscle, the hamstring muscles, the sciatic nerve, and the anastomoses around the hip joint. The internal pudendal artery also runs through the gluteal region (see Chap. 4).

The **obturator artery** is the first branch of the anterior division of the internal iliac artery. The artery enters the thigh through the obturator canal, which is a passageway through the obturator foramen. On entering the thigh the artery almost immediately divides into anterior and posterior divisions. The anterior division supplies blood to the obturator externus and internus muscles, and the adductors and gracilis. It also forms an anastomosis with the medial femoral circumflex artery. The posterior division of the obturator artery supplies the adductor magnus and the hamstrings. Branches are also sent to the hip joint and to anastomoses with the inferior gluteal artery.

Femoral Artery and Its Branches

The femoral artery is the major artery of the lower limb (Fig. 6-20). It is the direct continuation of the external iliac artery, which becomes the femoral artery after it passes under the inguinal ligament (see Fig. 6-13). The femoral artery lies in the femoral triangle between the femoral vein and the femoral nerve. From here it runs through the adductor canal (see Figs. 6-14 and 6-16) to reach the adductor hiatus. After passing through the hiatus the femoral artery becomes the popliteal artery (see Fig. 6-19). Its branches are the superficial epigastric, superficial circumflex iliac, superficial external pudendal (see Chap. 2), deep external pudendal (see Chap. 4), deep femoral, and descending genicular arteries.

The **deep femoral artery** is the main branch of the femoral artery. It arises from the lateral side of the femoral artery approximately 5 cm below the inguinal ligament. Within the femoral triangle the deep femoral artery gives off medial and lateral femoral circumflex arteries and muscular branches. It then comes to lie deep to the femoral artery and vein on the medial side of the femur (see Fig. 6-16). The deep femoral artery descends, crossing the tendon of the adductor brevis, to end deep to the adductor longus as the fourth perforating artery. It gives off the first three perforating arteries within the adductor canal (see Fig. 6-20).

The **medial femoral circumflex artery** leaves the femoral triangle by passing between the pectineus and the iliopsoas muscles. The artery then passes backward to the neck of the femur. It supplies the adductor muscles and the acetabulum. It also divides into ascending and descending branches, the ascending division going to the trochanteric fossa and the descending division going to the hamstrings.

The **lateral femoral circumflex artery** leaves the femoral triangle by passing laterally, deep to the sartorius and rectus femoris muscles. It then divides into three branches: ascending, transverse, and descending branches. The ascending branch goes to the gluteal region, where it anastomoses with the superior gluteal artery. The transverse branch winds around the femur to anastomose with the medial femoral circumflex artery, the inferior gluteal artery, and the first perforating artery. The descending branch travels down on the vastus lateralis, which it supplies, to anastomose with the descending genicular and superior lateral genicular arteries.

The **perforating arteries** pass posteriorly through the tendon of the adductor magnus to supply the hamstring muscles. The first perforating artery supplies the adductor brevis, adductor magnus, biceps femoris, and gluteus maximus. It anastomoses with the inferior gluteal, femoral circumflex, and second perforating arteries. The second perforating artery supplies the hamstring muscles and the femur. It anastomoses with the first and third perforating arteries. The third perforating artery supplies the hamstring muscles.

The **descending genicular artery** arises from the femoral artery just before the adductor hiatus. It divides into saphenous and articular branches. The saphenous branch runs with the saphenous nerve. The articular branch supplies the vastus medialis and anastomoses with the medial superior genicular, anterior tibial recurrent, lateral femoral circumflex, and lateral superior genicular arteries.

The veins of the thigh follow the course and carry the name of their respective arteries.

INNERVATION

The obturator nerve and the femoral nerve are the two nerves of the anterior and lateral thigh (Table 6-2). The **obturator nerve** (L2, L3, and L4) enters the thigh through the obturator canal in the obturator foramen. After entering the thigh it divides into anterior and posterior divisions. The anterior division, running above the adductor brevis, gives off an articular branch to the hip joint and muscular branches to the adductor longus, adductor brevis, and gracilis muscles. The posterior division runs beneath the adductor brevis and gives an articular branch to the knee joint and muscular branches to the adductor magnus and adductor brevis and the obturator externus muscles.

The **accessory obturator nerve** (L3 and L4) is an inconstant nerve. When present, it exits the pelvis by passing over the pubic ramus and then travels deep to the pectineus, which it supplies.

Fig. 6-20. Distribution of the femoral artery and its branches in the thigh.

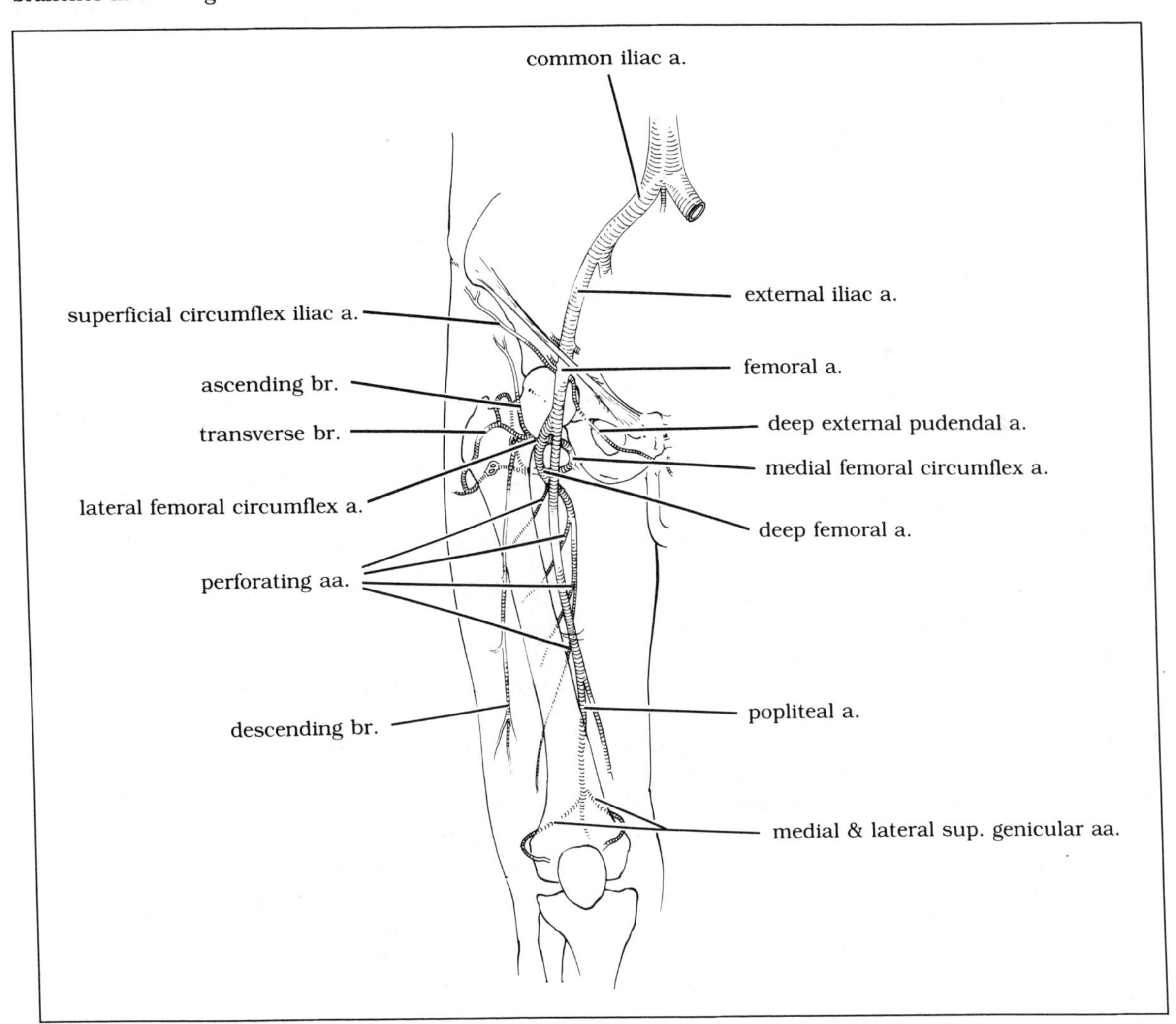

In addition it supplies an articular branch to the hip joint.

The **femoral nerve** (L2, L3, and L4) is the principal nerve to the anterior thigh (Fig. 6-21A). The nerve passes under the inguinal ligament to enter the femoral triangle. Within the femoral triangle it branches, supplying articular branches to the hip and knee joints and muscular branches to the sartorius, rectus femoris, vasti, and pectineus muscles.

The gluteal region is supplied by several small nerves (Fig. 6-21B). The **superior gluteal nerve**

Table 6-2. Nerves of the Thigh

Nerve	*Origins*	*Divisions*	*Innervations*
Obturator	L2, L3, and L4	Anterior	Adductor longus, adductor brevis, gracilis, hip joint
		Posterior	Adductor magnus, adductor brevis, obturator externus, knee joint
Accessory obturator	L3 and L4		Pectineus, hip joint
Femoral	L2, L3, and L4		Sartorius, rectus femoris, vasti, pectineus (when there is no accessory obturator), knee joint, hip joint
Superior gluteal	L4, L5, and S1		Gluteus medius, gluteus minimus, tensor fasciae latae
Inferior gluteal	L5, S1, and S2		Gluteus maximus
Piriformis	S1 and S2		Piriformis
Obturator internus	L5, S1, and S2		Obturator internus, superior gemellus
Quadratus femoris	L4, L5, and S1		Quadratus femoris, inferior gemellus, hip joint
Sciatic	L4–S3	Tibial	Biceps femoris (long head), semitendinosus, semimembranosus, adductor magnus, hip joint, knee joint
		Common peroneal	Biceps femoris (short head), hip joint, knee joint

(L4, L5, and S1) exits the pelvis above the piriformis in company with the superior gluteal artery. The nerve supplies the gluteus medius, gluteus minimus, and tensor fasciae latae. The **inferior gluteal nerve** (L5, S1, and S2) exits the pelvis below the piriformis accompanied by the inferior gluteal artery and supplies the gluteus maximus. The **nerve to the piriformis muscle** (S1 and S2) enters this muscle within the pelvis. The **nerve to the obturator internus** (L5, S1, and S2) exits the pelvis with the internal pudendal vessels. In the gluteal region the nerve innervates the superior gemellus; in the perineum it innervates the obturator internus. The **nerve to the quadratus femoris** (L4, L5, and S1) exits the pelvis through the sciatic foramen. This nerve supplies the quadratus femoris and the inferior gemellus muscle and sends an articular branch to the hip joint.

The **sciatic nerve** (L4–S3) is the principal nerve to the posterior thigh and to the leg and foot (see Figs. 6-18B,C and 6-21B). The sciatic is actually two nerves in a common sheath: The **tibial nerve** is the larger one and is located on the medial side; the **common peroneal nerve** is on the lateral side. The sciatic nerve exits the pelvis through the greater sciatic foramen and descends down the middle of the thigh. It usually separates into two nerves in the lower third of the thigh, but may separate at any level. Before separating, the sciatic nerve sends articular branches to the hip joint and to the hamstring and the adductor magnus (ischiocondylar part) muscles. The branches to all these muscles, except the short head of the biceps femoris, come from the tibial division. The short head of the biceps is supplied by the common peroneal division.

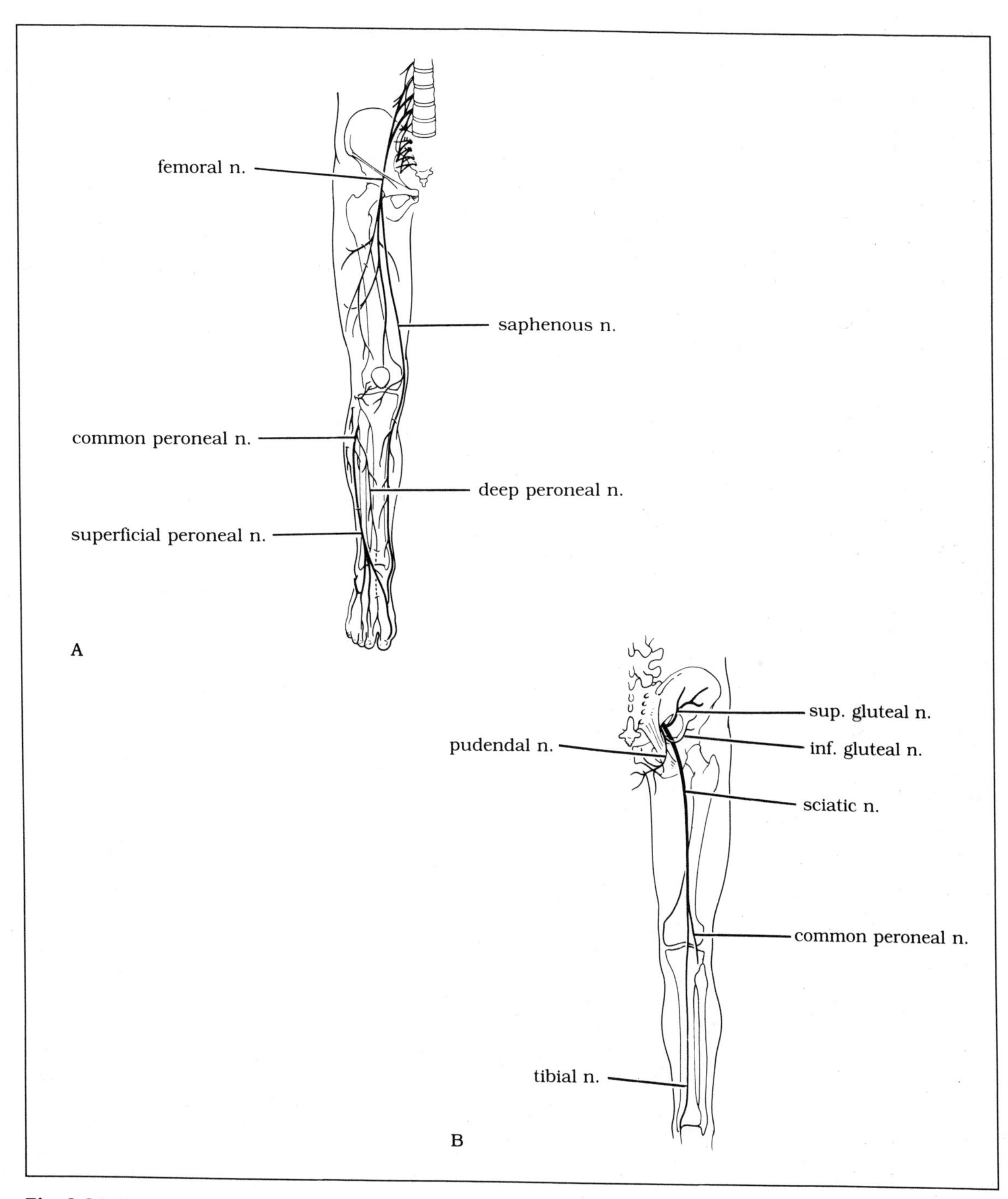

Fig. 6-21. Nerves of the lower limb. A. Anterior view. B. Posterior view.

Popliteal Fossa

The popliteal fossa (see Fig. 6-19) lies behind the knee at the back of the leg and thigh. The fossa is diamond shaped; the divergent hamstring muscles form the upper borders, the biceps femoris forms the lateral margin, and the semitendinosus and semimembranosus form the medial margin. The inferior borders of the popliteal fossa are formed by the two heads of the gastrocnemius muscle.

The popliteal fossa is crossed by the tibial and common peroneal terminal branches of the sciatic nerve and by the popliteal vessels. The remaining space in the fossa is filled with fat, connective tissue, and the lymph nodes that accompany the popliteal vessels. The popliteal fossa is covered by dense connective tissue, the popliteal fascia.

The tibial nerve descends vertically through the midline of the fossa (see Figs. 6-19 and 6-21B). Within the fossa it gives off genicular branches to the knee and sural branches to the gastrocnemius and plantaris muscles. The common peroneal nerve (see Figs. 6-19 and 6-21B) descends along the lateral margin of the fossa in company with the tendon of the biceps femoris and enters the leg by curving around the neck of the fibula. The common peroneal nerve gives off genicular branches to the knee and the lateral sural cutaneous nerve within the popliteal fossa.

The **popliteal vessels** are the direct continuations of the femoral vessels after the latter pass through the adductor hiatus (see Figs. 6-19 and 6-20). In the popliteal fossa the artery lies anterior to the vein, against the capsule of the knee joint. The branches of the popliteal artery are the medial and lateral superior genicular, medial and lateral inferior genicular, middle genicular, and sural arteries. The popliteal artery terminates as the anterior and posterior tibial arteries in the leg.

The **superior genicular arteries** branch from the popliteal artery at the top of the popliteal fossa and travel around to the front of the femur to form a ring by anastomotic connections with each other. In addition the superior genicular arteries form anastomoses with the descending branch of the lateral femoral circumflex artery, the inferior genicular arteries, and the descending genicular artery. The **middle genicular artery** arises in the middle of the popliteal fossa and travels forward to penetrate the knee joint and supply blood to the synovial membrane and the cruciate ligaments.

The **inferior genicular arteries** (see Fig. 6-27), like the superior genicular arteries, form a ring by anastomotically connecting with each other. In addition the arteries form anastomoses with the superior genicular arteries, the descending genicular artery, and the anterior tibial recurrent artery.

There are two **sural arterial branches** of the popliteal artery, one going to each head of the gastrocnemius.

The **popliteal vein** and its branches follow the course of their like-named artery. In addition the popliteal vein receives the small saphenous vein within the popliteal fossa.

Leg

The subcutaneous structures of the leg are held in place by a deep fascial layer, the crural fascia. The **crural fascia** is continuous with the fascia lata of the thigh and serves the same purpose, that being to hold the structures of the leg tightly in place. The crural fascia blends into the periosteum of the tibia in the region where that bone is subcutaneous.

In addition the leg is divided into three compartments by three fascial septa: the anterior intermuscular septum, the posterior intermuscular septum, and the interosseous membrane (Fig. 6-22). The two intermuscular septa extend from the fibula to the crural fascia and enclose the lateral muscular compartment. The third fascial septum, the interosseous membrane, extends from the tibia to the fibula. The interosseous membrane separates the anterior muscular and posterior muscular compartments. The posterior muscular compartment is further divided into deep and superficial regions by the transverse intermuscular septum. This septum ex-

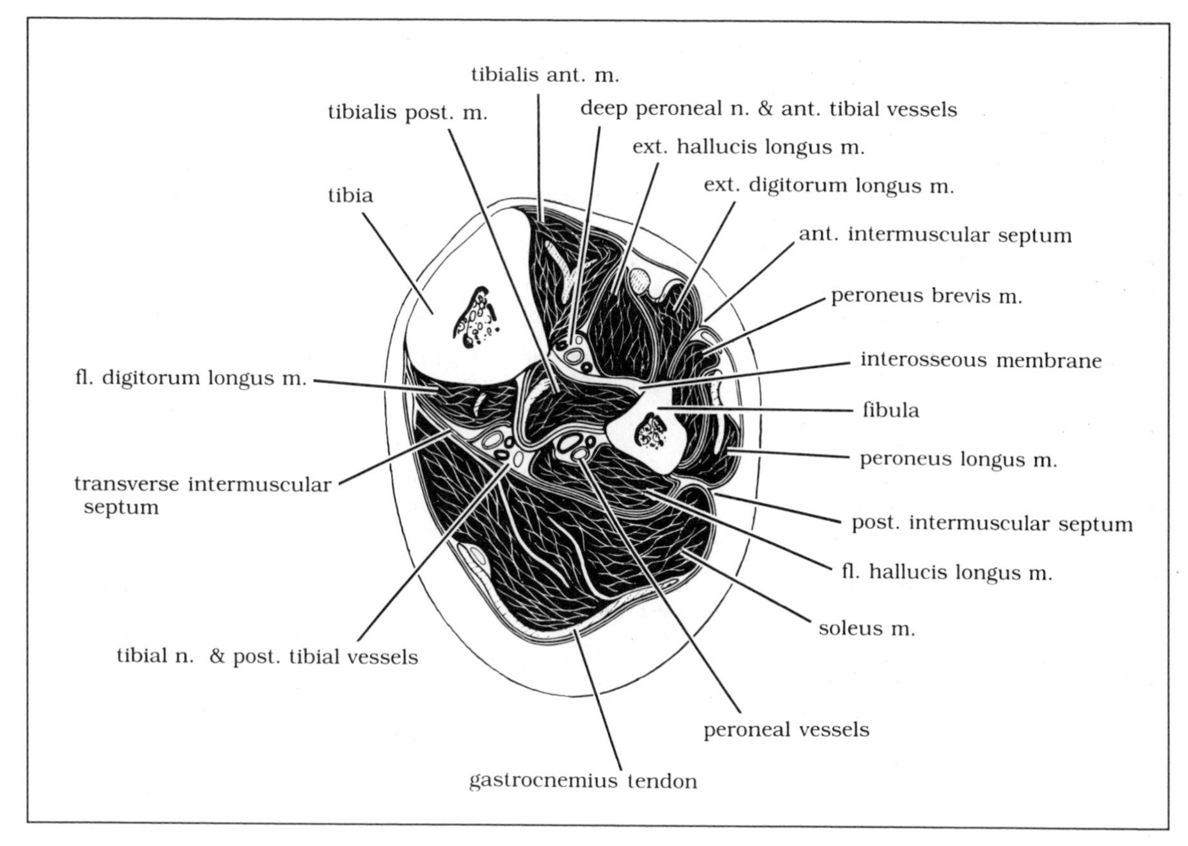

Fig. 6-22. Cross section of the leg.

tends from the posterior intermuscular septum to the tibia (see Fig. 6-22).

The crural fascia at the ankle joint is reinforced by five retinacula: the superior extensor retinaculum, the Y-shaped inferior extensor retinaculum, the flexor retinaculum, the superior peroneal retinaculum, and the inferior peroneal retinaculum (see Fig. 6-26).

MUSCULATURE

The muscles of the anterior, lateral, and posterior compartments of the leg are summarized in Table 6-3.

Anterior Muscular Compartment

The muscles of the anterior muscular compartment are the tibialis anterior, extensor hallucis longus, extensor digitorum longus, and peroneus tertius (Fig. 6-23; see also Fig. 6-24). All the muscles in the anterior compartment are innervated by the deep peroneal nerve.

The **tibialis anterior** is the most medial muscle in this compartment. It arises from the lateral tibial condyle, the proximal half of the lateral tibia, and the interosseous membrane. The tendon of the tibialis anterior passes under the extensor retinacula, in the most medial compartments, to insert on the medial and plantar surfaces of the medial cuneiform and the base of the first metatarsal bone. The tibialis anterior dorsiflexes and supinates (adducts and inverts) the foot.

The **extensor hallucis longus** arises from the middle half of the anterior fibula and the interosseous membrane and inserts on the distal phalanx of the great toe (hallux). The extensor

Table 6-3. Musculature of the Leg

Name	*Origin*	*Insertion*	*Innervation*	*Action*
Anterior Compartment				
Tibialis anterior	Lateral tibial condyle, lateral tibia, surrounding fascia	Medial cuneiform, metatarsal 1	Deep peroneal nerve (L4, L5, and S1)	Dorsiflex and supinate (adduct and invert) foot
Extensor hallucis longus	Anterior fibula, interosseous membrane	Distal phalanx of toe 1	Deep peroneal nerve (L4, L5, and S1)	Extend toe 1, dorsiflex and supinate foot
Extensor digitorum longus	Lateral tibial condyle, anterior fibula, surrounding fascia	Phalanges 2 and 3 of toes 2–5	Deep peroneal nerve (L4, L5, and S1)	Extend toes, dorsiflex and pronate (abduct and evert) foot
Peroneus tertius	Distal fibula, surrounding fascia	Base of metatarsal 5	Deep peroneal nerve (L4, L5, and S1)	Dorsiflex and pronate foot
Lateral Compartment				
Peroneus longus	Head and lateral fibula, surrounding fascia	Base of metatarsal 1, medial cuneiform	Superficial peroneal nerve (L4, L5, and S1)	Plantar flex and pronate foot
Peroneus brevis	Lateral fibula, surrounding fascia	Base of metatarsal 5	Superficial peroneal nerve (L4, L5, and S1)	Plantar flex and pronate foot
Posterior Compartment				
Gastrocnemius	Medial and lateral femoral condyles, posterior femur and capsule of knee joint	Calcaneal tendon to calcaneus	Tibial nerve (S1 and S2)	Plantar flex foot, flex leg, supinate foot
Plantaris	Lateral linea aspera	Posterior calcaneus	Tibial nerve (L4, L5, and S1)	Plantar flex foot, flex leg
Soleus	Posterior surface of head of fibula, proximal third of tibia, popliteal line	Calcaneal tendon to calcaneus	Tibial nerve (S1 and S2)	Plantar flex foot
Popliteus	Lateral femoral condyle	Tibia proximal to popliteal line	Tibial nerve (L4, L5, and S1)	Flex and medially rotate leg
Flexor hallucis longus	Distal two-thirds of posterior fibula	Base of terminal phalanx of great toe	Tibial nerve (L5, S1, and S2)	Flex great toe, plantar flex and supinate foot
Tibialis posterior	Posterior interosseous membrane and adjacent bones	Tuberosity of navicular bone	Tibial nerve (L5 and S1)	Plantar flex and supinate foot
Flexor digitorum longus	Middle third of posterior tibia	Base of distal phalanx of toes 2–5	Tibial nerve (L5 and S1)	Flex phalanges 2–5, plantar flex and supinate foot

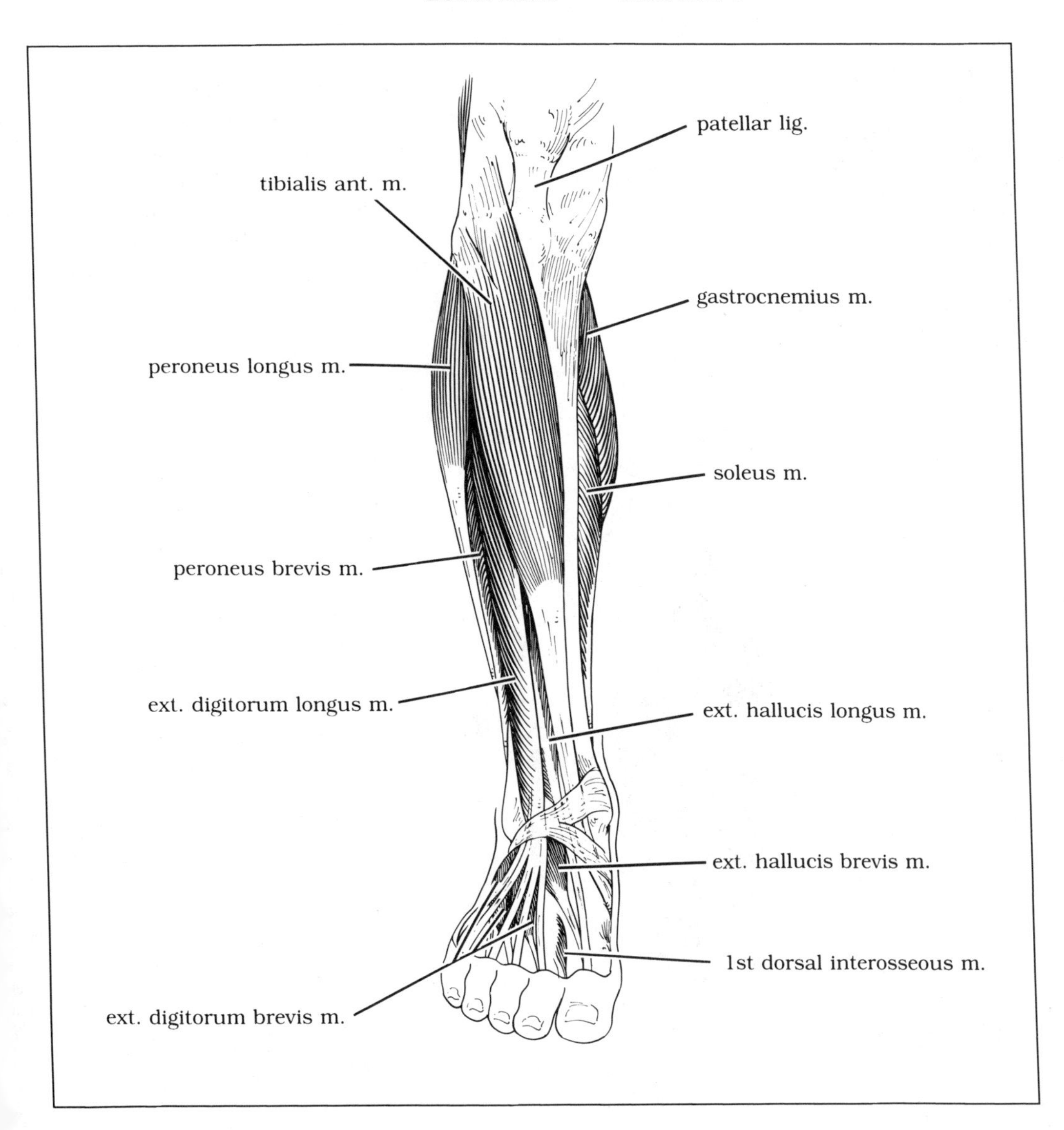

Fig. 6-23. Muscles of the anterior compartment of the leg.

hallucis longus extends the great toe and dorsiflexes and supinates the foot.

The **extensor digitorum longus** arises from the lateral tibial condyle, the first three-fourths of the anterior fibula, the interosseous membrane, and the surrounding fascia. Its tendon passes under the superior extensor retinaculum and then divides into four tendons at, or just distal to, the inferior extensor retinaculum. The tendinous slips insert onto the second and third phalanges at the second through fifth toes. The extensor digitorum longus extends the toes and dorsiflexes and pronates (abducts and everts) the foot.

The **peroneus tertius** is the most lateral muscle in the anterior compartment. The muscle arises from the distal third of the fibula and its surrounding fascia. Its tendon passes under the extensor retinacula with the tendon of the extensor digitorum longus to insert on the dorsal surface of the base of the metatarsal bone of the fifth toe. The peroneus tertius muscle dorsiflexes and pronates the foot.

Lateral Muscular Compartment

The lateral muscular compartment consists of two muscles, the peroneus longus and peroneus brevis muscles, that are innervated by the superficial peroneal nerve (Fig. 6-24).

The **peroneus longus** arises from the head and the proximal two-thirds of the lateral fibula and its surrounding fascia (see Fig. 6-9A). Its tendon passes behind the lateral malleolus, beneath

Fig. 6-24. Muscles of the lateral compartment of the leg.

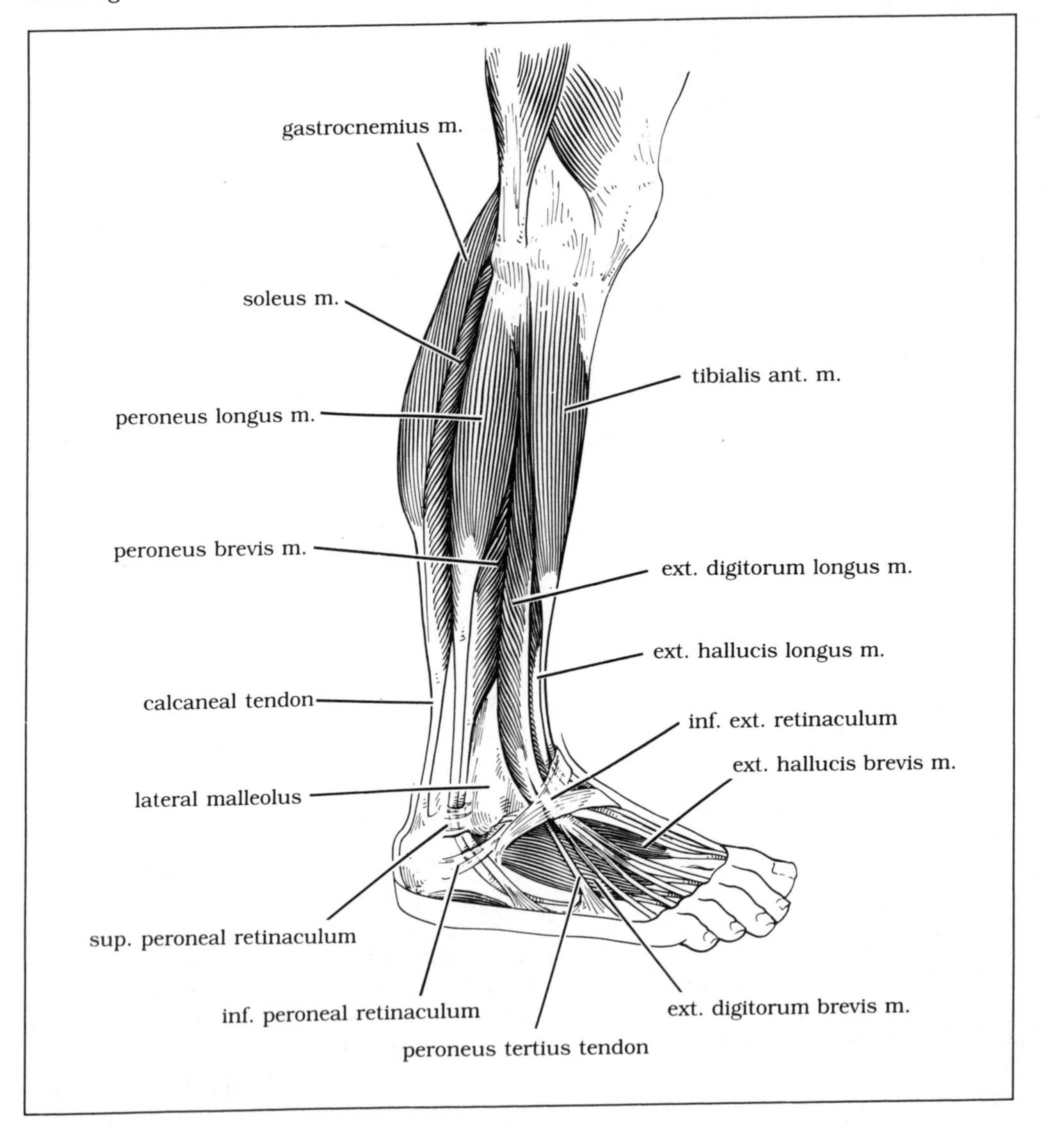

the peroneal retinacula, and across the sole of the foot to the medial side, where it inserts on the medial cuneiform bone and the base of the first metatarsal bone. The peroneus longus pronates and plantar flexes the foot.

The **peroneus brevis** arises from the distal two-thirds of the lateral fibula and from the surrounding fascia. Its tendon passes behind the lateral malleolus and deep to the peroneal retinacula to insert on the tuberosity at the base of the fifth metatarsal bone. The peroneus brevis plantar flexes and pronates the foot.

Posterior Muscular Compartment

The posterior muscular compartment of the leg is divided into superficial and deep regions by the transverse intermuscular septum (Fig. 6-25). The superficial muscles are the gastrocnemius and plantaris and the soleus muscles. The deep muscles are the popliteus, flexor hallucis longus, flexor digitorum longus, and tibialis posterior muscles. Figure 6-9 shows origins and insertions of these muscles.

The **gastrocnemius** is the most superficial and longest muscle in the calf. It arises by two heads—one from each femoral condyle—from the posterior femur and from the posterior capsule of the knee joint. The muscle inserts, along with the soleus muscle, via the calcaneal tendon into the posterior calcaneus. The gastrocnemius plantar flexes and supinates the foot and flexes the leg. It is innervated by the tibial nerve (S1 and S2).

The **plantaris** is a small muscle between the gastrocnemius and the soleus. It arises from the distal part of the lateral linea aspera. The tendon of the plantaris crosses the calf diagonally to insert into the posterior calcaneus, medial to the calcaneal tendon. The muscle plantar flexes the foot and flexes the leg. It is innervated by the tibial nerve (L4, L5, and S1).

The **soleus muscle** is a broad flat muscle. It arises from the posterior head of the fibula, the proximal posterior third of the fibula, the popliteal line, and the medial third of the tibia. It inserts, along with the gastrocnemius, via the calcaneal tendon. The soleus muscle plantar flexes the foot. It is innervated by the tibial nerve (S1 and S2).

The **calcaneal tendon,** the tendon of insertion of the gastrocnemius and the soleus muscle, is the thickest and strongest tendon in the body and is about 15 cm long. The tendon begins in the middle of the leg and inserts into the middle of the posterior calcaneus.

The **popliteus** forms the distal floor of the popliteal fossa. The muscle arises on the anterior groove of the lateral femoral condyle and passes backward to insert along the medial two-thirds of the posterior tibia proximal to the popliteal line. It flexes the leg and rotates it medially. It is innervated by the tibial nerve (L4, L5, and S1).

The **flexor hallucis longus** arises from the inferior two-thirds of the posterior fibula. The tendon of this muscle travels in a groove across the distal posterior tibia, behind the talus, and under the sustentaculum tali. From here it passes deep to the tendon of the flexor digitorum longus, winding from lateral to medial as it does so. The tendon of the flexor hallucis longus then passes between the two heads of the flexor hallucis brevis to insert on the base of the terminal phalanx of the great toe. The muscle flexes the great toe and plantar flexes and supinates the foot. It is innervated by the tibial nerve (L5, S1, and S2).

The **tibialis posterior** is the deepest muscle of the calf. It arises from the proximal posterior interosseous membrane, the posterior lateral tibia, and the proximal two-thirds of the posterior medial fibula. Its origin is separated by the anterior tibial vessels. The tendon of the tibialis posterior lies in a groove behind the medial malleolus with the tendon of the flexor digitorum longus. It passes under the flexor retinaculum and inserts on the foot into the tuberosity of the navicular bone. The tibialis posterior supinates and plantar flexes the foot. It is innervated by the tibial nerve (L5 and S1).

The **flexor digitorum longus** arises from the middle third of the posterior body of the tibia. Its tendon passes with that of the tibialis posterior behind the medial malleolus. On the plantar sur-

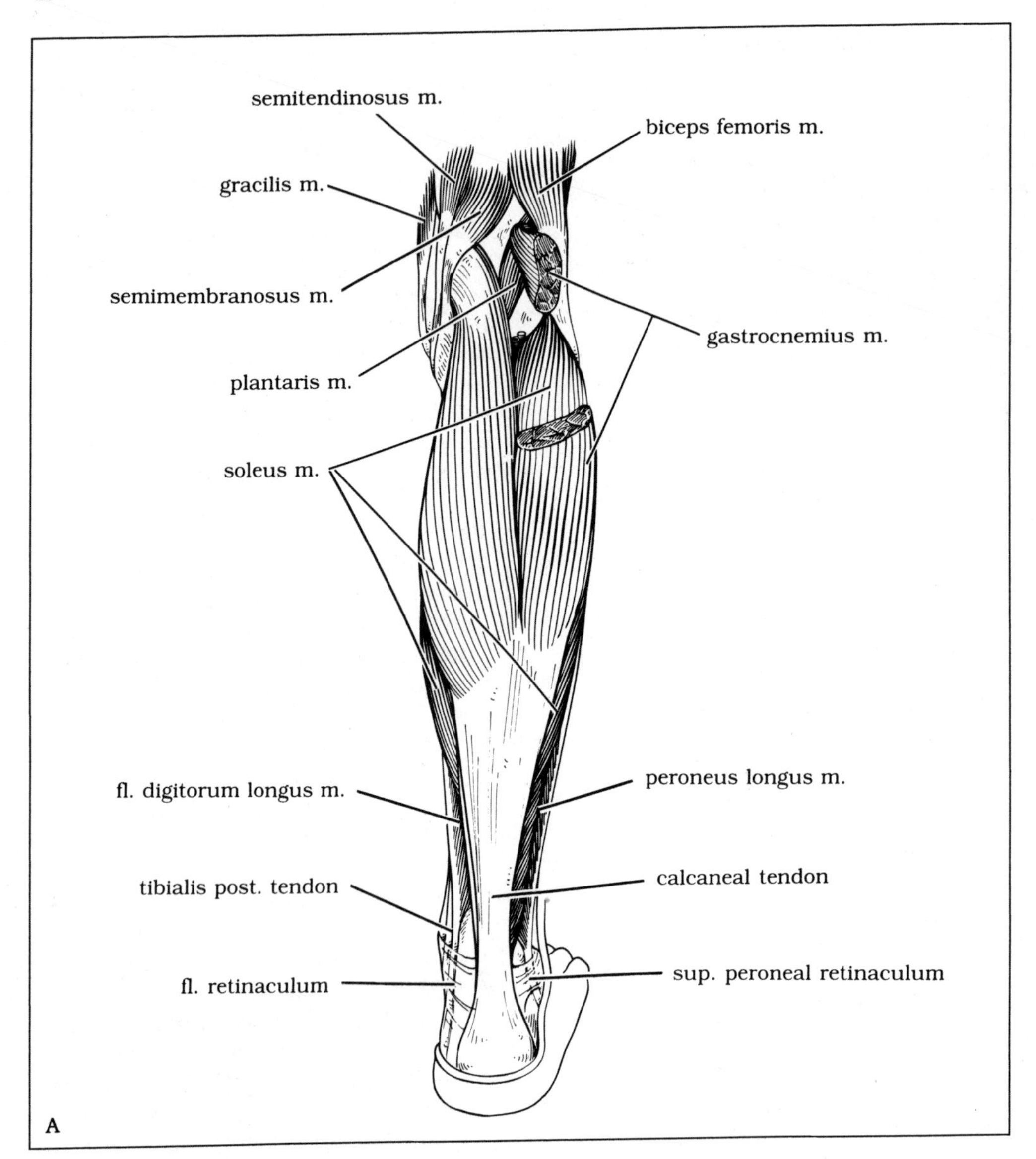

Fig. 6-25. Muscles of the posterior compartment of the leg. A. Superficial dissection. B. Deep dissection.

face of the foot it receives a tendinous slip from the flexor hallucis longus and the muscular fibers of the quadratus plantae. The tendon then divides into four tendons that insert onto the base of the distal phalanx of the second to fifth toes. In getting to their points of insertion the tendons pass through a slit in the tendons of the flexor digitorum brevis over the proximal phalanx. The flexor digitorum longus flexes the toes and plantar flexes and supinates the foot. It is innervated by the tibial nerve (L5 and S1).

Actions of the Leg Muscles

The anterior leg muscles are all active in dorsiflexion of the foot. They are therefore active in

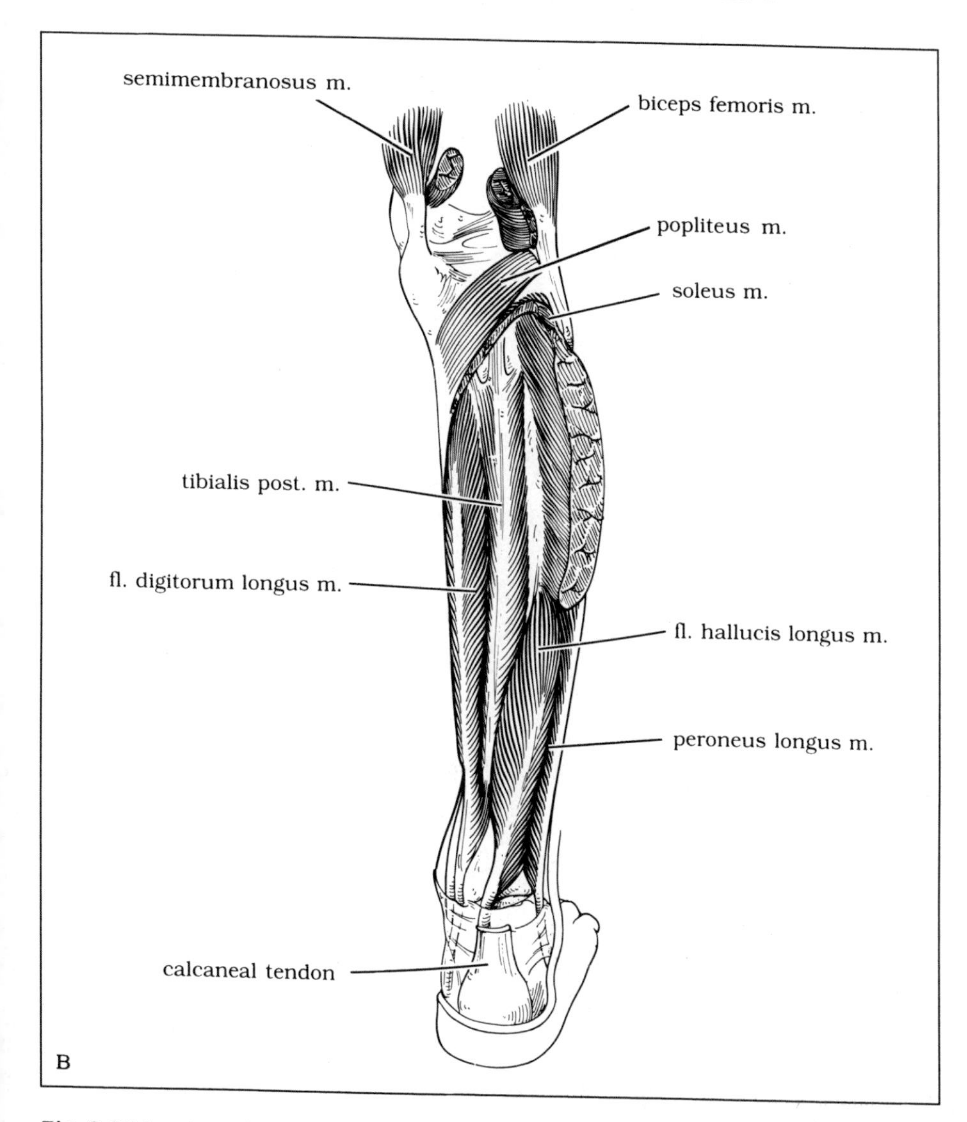

Fig. 6-25 (continued).

walking and running, particularly through take-off and toeing. The lateral leg muscles act in walking by helping to maintain the arch of the foot during takeoff and tip-toeing.

The superficial posterior leg muscles are active in maintaining upright stance. In addition, by plantar flexing the foot, they provide the major propelling force for walking and running. The deep muscles act in part to maintain the foot when it is carrying a load. They aid in the distribution of that load over the metatarsals, which are the fulcrum upon which movement occurs. Thus they are active in both takeoff and the maintenance of balance. They may also assist in plantar flexion, particularly when excessive power is required.

RETINACULA

There are five retinacula around the ankle joint (Fig. 6-26). The **superior extensor retinaculum**

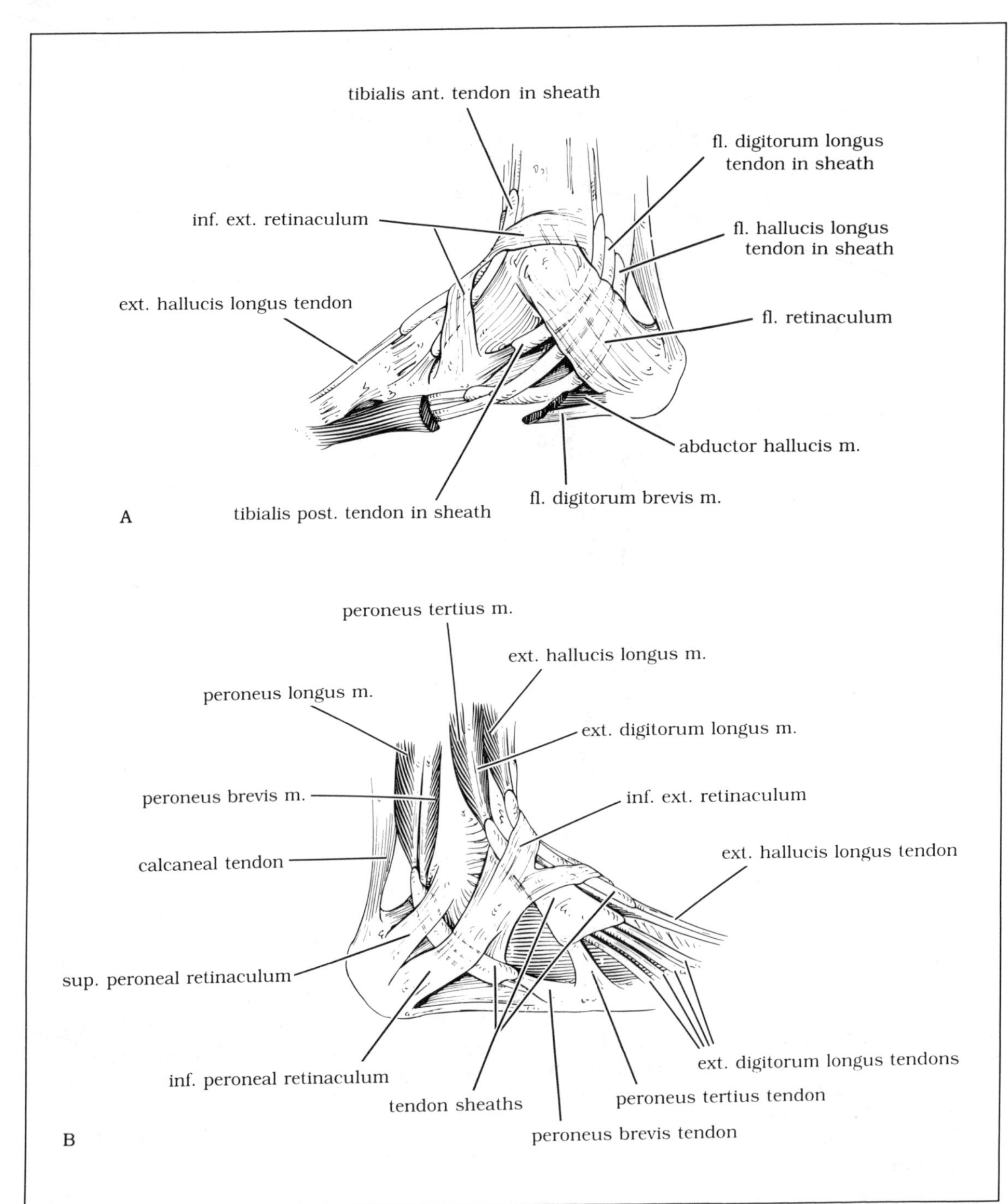

Fig. 6-26. The ankle. A. Medial view. B. Lateral view.

extends from the distal fibula to the tibia. It holds down the tendons of the extensor digitorum longus, extensor hallucis longus, peroneus tertius, and tibialis anterior muscles. In addition the anterior tibial vessels and the deep peroneal nerve pass deep to it.

The **inferior extensor retinaculum** is Y shaped. The stem of the Y is attached laterally to the calcaneus, and the limbs diverge, one proximally and medially to the medial malleolus, the other distally and medially to the plantar aponeurosis. This retinaculum also holds down the tendons of the extensor digitorum longus, extensor hallucis longus, peroneus tertius, and tibialis anterior and the anterior tibial vessels and deep peroneal nerve.

The **flexor retinaculum** extends from the medial malleolus to the calcaneus. It holds down the tendons of the tibialis posterior, flexor digitorum longus, and flexor hallucis longus and the posterior tibial vessels and tibial nerve.

The **superior peroneal retinaculum** extends from the lateral malleolus to the lateral calcaneus. It holds down the tendons of the peroneus longus and brevis. The **inferior peroneal retinaculum** also holds down these tendons. It is a continuation of the inferior extensor retinaculum to the lateral calcaneus.

BLOOD SUPPLY

The popliteal artery divides just distal to the popliteal fossa into anterior and posterior tibial arteries (Fig. 6-27). The **anterior tibial artery** proceeds anteriorly between the heads of the tibialis posterior and over the interosseous membrane to enter the anterior compartment. The artery then descends on the interosseous membrane to the ankle, where it becomes the **dorsalis pedis artery.** The anterior tibial artery runs with the deep peroneal nerve. The branches of the anterior tibial artery are the posterior tibial recurrent, fibular, anterior tibial recurrent, anterior medial malleolar, anterior lateral malleolar, and muscular branches.

The **posterior tibial recurrent artery** branches off the anterior tibial artery before it goes over the interosseous membrane. It ascends to the knee, where it anastomoses with the inferior genicular artery.

The **fibular artery** proceeds around the neck of the fibula.

The **anterior tibial recurrent artery** branches off the anterior tibial artery just after it passes through the interosseous membrane. It ascends to the knee, where it anastomoses with the descending genicular and both inferior genicular arteries.

The **anterior medial malleolar artery** branches off the anterior tibial artery 5 cm before the ankle. It passes medial to the medial malleolus, where it anastomoses with the posterior tibial and medial plantar branches and with the medial calcaneal artery.

The **anterior lateral malleolar artery** passes lateral to the ankle joint to anastomose with branches of the peroneal and lateral tarsal arteries.

Muscular branches of the anterior tibial artery supply blood to the muscles of the anterior and lateral compartments.

The **posterior tibial artery** descends behind the tibia to the ankle, where it divides to become the medial and lateral plantar arteries. The artery travels in company with the tibial nerve. The branches of the posterior tibial artery are the peroneal, nutrient, posterior medial malleolar, communicating, medial calcaneal, and unnamed muscular arteries.

The **peroneal artery** descends on the fibular side of the posterior leg. It provides muscular branches to the soleus, tibialis posterior, flexor hallucis longus, and peronei muscles. Additional branches include the following: the nutrient artery to the fibula; the perforating artery, which pierces the interosseous membrane and anastomoses with the anterior lateral malleolar and lateral tarsal arteries; the communicating artery, which joins the communicating branch of the posterior tibial artery; the posterior lateral malleolar artery, which anastomoses around the lateral malleolus; and the lateral calcaneal arteries, which

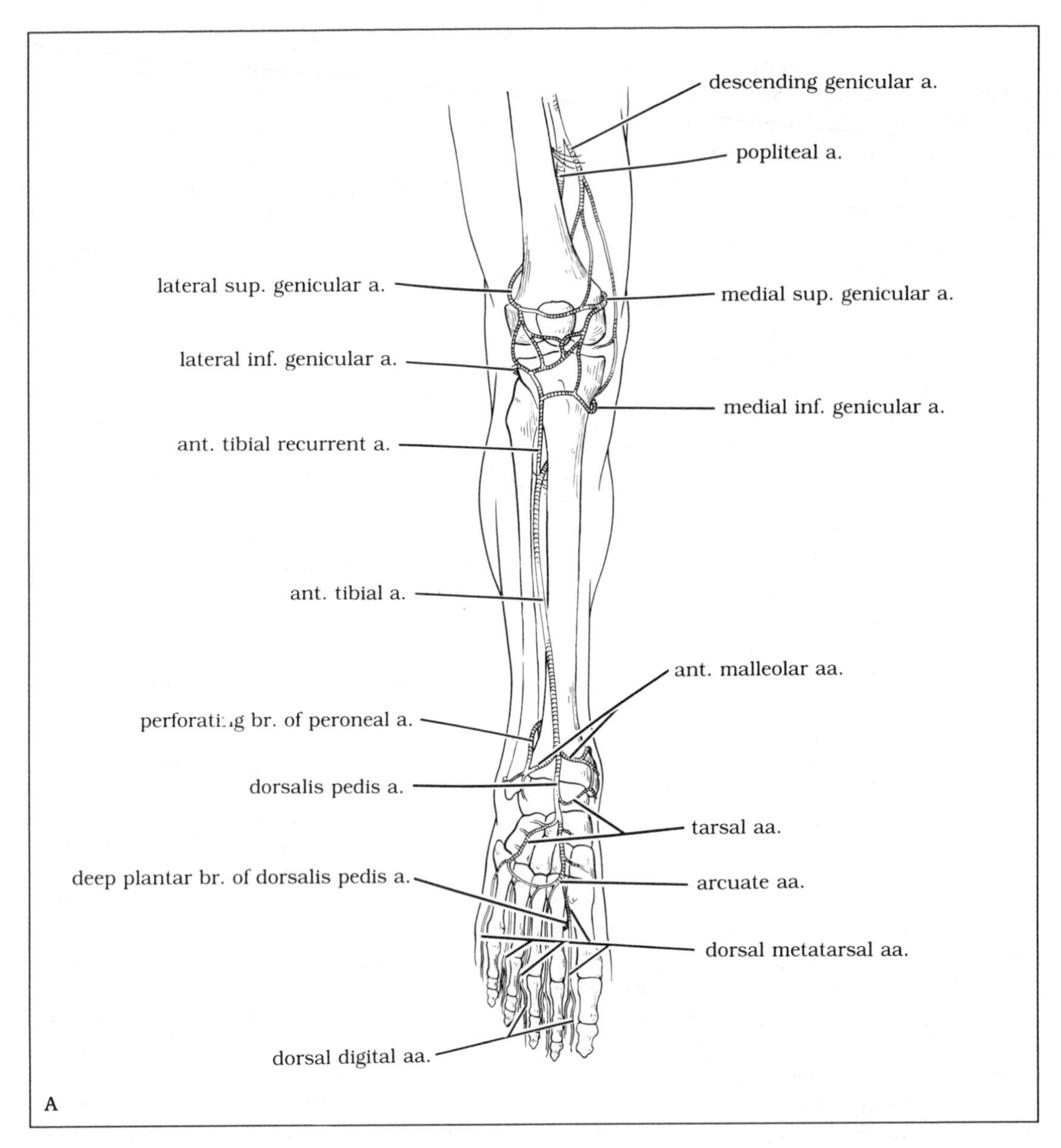

Fig. 6-27. Arteries of the leg and foot. A. Anterior view. B. Posterior and plantar views.

supply the lateral side of the heel and anastomose with the lateral malleolar and medial calcaneal arteries.

The **nutrient arterial branch of the posterior tibial artery** is the nutrient artery to the tibia.

The **posterior medial malleolar artery** supplies the medial ankle and joins the anastomoses around the medial malleolus.

The **communicating branch of the posterior tibial artery** connects with the communicating branch of the peroneal artery.

The **medial calcaneal branches of the posterior tibial artery** supply the skin and subcutaneous

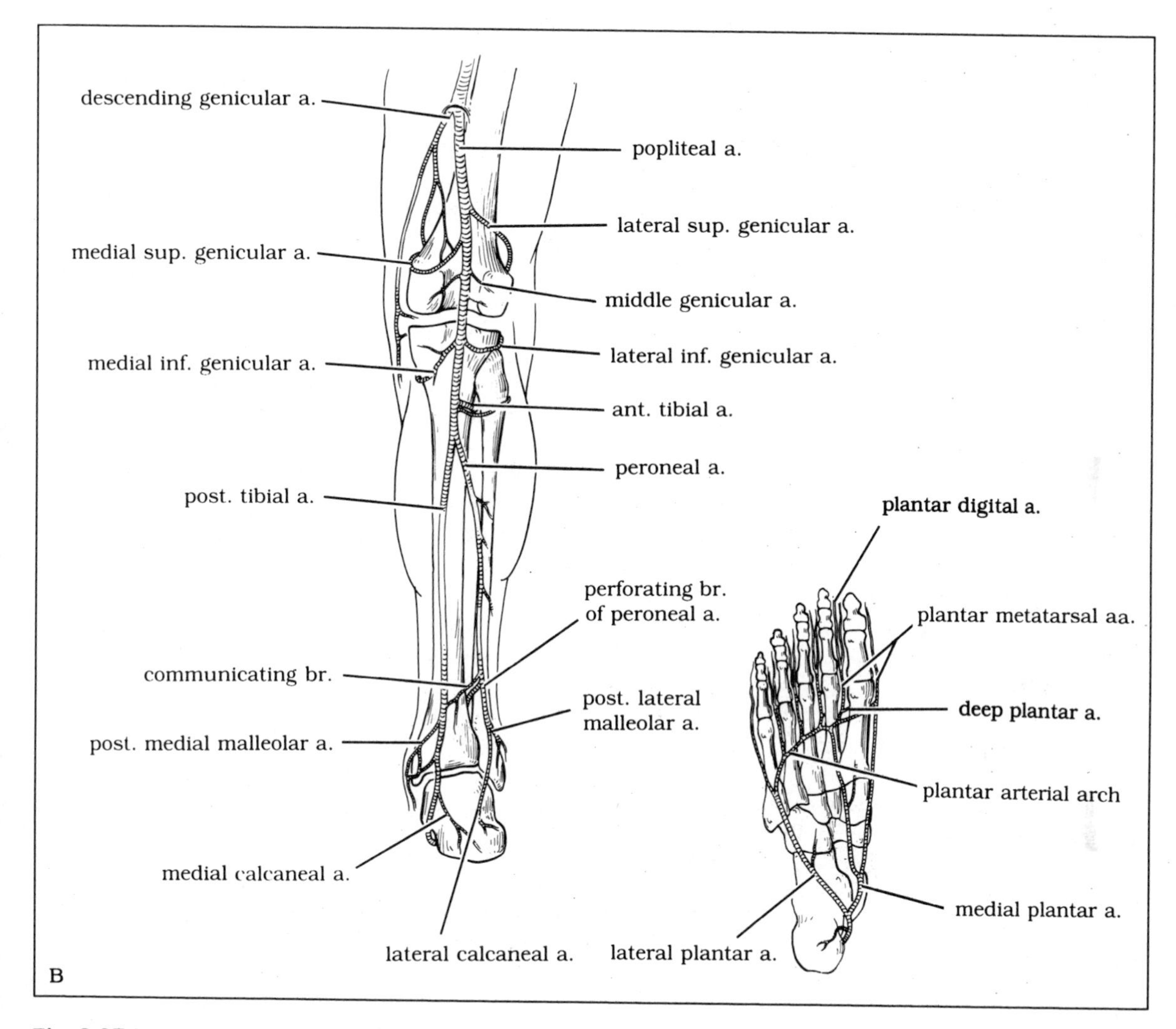

Fig. 6-27 (continued).

tissue on the medial side of the heel and the muscles on the medial side of the foot. They anastomose with the medial malleolar and lateral calcaneal arteries.

The veins of the region follow the arterial pattern.

INNERVATION

The sciatic nerve splits in the popliteal fossa to form the tibial and common peroneal nerves (see Fig. 6-21). The tibial nerve is formed from ventral branches of ventral rami of spinal nerves L4 to S3. The common peroneal nerve is formed from dorsal branches of ventral rami of spinal nerves L4 to S2.

The **tibial nerve** continues down the midline of the posterior leg, under the soleus muscle, in company with the posterior tibial artery. As it passes under the flexor retinaculum it divides into medial and lateral plantar nerves. The tibial nerve supplies articular branches to the knee and ankle joints, muscular branches to the muscles of the posterior leg, cutaneous innervation to the back of the leg via the medial sural cutaneous nerve (which connects to the sural nerve), and innervation to the skin of the heel and medial sole of the foot via the medial calcaneal nerve.

The **common peroneal nerve** crosses the popliteal fossa to the medial side of the biceps femoris and thence to the head of the fibula. It then winds around the neck of the fibula, passing deep to the peroneus longus, where it divides into deep peroneal and superficial peroneal nerves. Before splitting it sends articular branches to the knee joint and the peroneal communicating branch to join the medial sural cutaneous nerve in forming the sural nerve.

The **deep peroneal nerve** passes deep, to lie on the anterior surface of the interosseous membrane in company with the anterior tibial artery. The nerve descends, passing under the extensor retinacula, to the ankle, where it divides into medial and lateral terminal branches that innervate the foot. The deep peroneal nerve supplies muscular branches to the muscles of the anterior leg and articular branches to the ankle joint. The **superficial peroneal nerve** descends between the peronei and extensor digitorum longus, becoming superficial in the lower third of the leg. It supplies muscular branches to the peroneus longus and brevis, and cutaneous branches to the lower leg. It ends as medial and intermediate dorsal cutaneous nerves in the dorsum of the foot.

Foot

The deep fascia of the dorsum of the foot is continuous above with the extensor retinacula, and to the sides with the deep fascia of the sole. The latter is thickened over the center of the sole to form the **plantar aponeurosis.** This structure arises from the tuberosity of the calcaneus. Near the toes it diverges into digital slips, inserting into the skin at the junction of the toe and the sole and into the sheaths of the flexor tendons.

DORSUM

The dorsum of the foot contains one intrinsic muscle, the extensor digitorum brevis (Fig. 6-28). The **extensor digitorum brevis** arises from the distal, lateral, and superior portions of the calcaneus; it then passes diagonally across the dorsum of the foot, dividing into four tendons. The first (sometimes known as the tendon of the **extensor hallucis brevis**) inserts into the base of the first phalanx of the great toe. The second through fourth slips insert into the lateral side of the corresponding tendons of the extensor digitorum longus muscle (see Fig. 6-12A). The extensor digitorum brevis extends the proximal phalanges of the first four toes. It is innervated by the deep peroneal nerve (L5 and S1).

The **deep peroneal nerve** (see Fig. 6-21A) divides at the end of the inferior extensor retinaculum into medial and lateral branches. The medial branch gives rise to two dorsal digital branches to the adjacent sides of the first two toes. The lateral branch runs deep to the extensor digitorum brevis muscle. It innervates this muscle and the tarsal joints. Both nerves innervate the metatarsophalangeal joints.

The artery of the dorsum of the foot is the **dorsalis pedis artery** (see Fig. 6-27A), which crosses the dorsum of the foot to the first metatarsal space, where it divides into the first dorsal metatarsal and deep plantar arteries. In addition it gives rise to the medial and lateral tarsal arteries and the arcuate artery.

The **first dorsal metatarsal artery** supplies the adjacent sides of the first and second toes.

The **deep plantar artery** descends to the sole between the heads of the first dorsal interosseous muscle. There it unites wth the lateral plantar artery to form the plantar arterial arch.

The **lateral tarsal artery** runs laterally and distally, deep to the extensor digitorum brevis, supplying both this muscle and the underlying joints. The artery anastomoses with the arcuate artery, the anterior lateral malleolar artery, the lateral plantar artery, and the perforating branch of the peroneal artery.

The **medial tarsal artery** supplies the medial foot, anastomosing with the medial malleolar arteries.

The **arcuate artery** forms an arch across the dorsum of the foot at the base of the metatarsal bones. It gives rise to the dorsal digital arteries and anastomoses with the lateral tarsal and lateral plantar arteries.

SOLE

The plantar foot is divided into four regions: the medial (great toe) region, the lateral (small toe) region, the central region, and the interosseous adductor region. The sole of the foot is covered by a layer of strong, dense connective tissue, the plantar aponeurosis (Fig. 6-29). (See Figure 6-12B for muscle origins and insertions on the sole of the foot.)

The medial region contains the abductor hallucis and flexor hallucis brevis. The **abductor hallucis** muscle forms the medial side of the sole. It arises from the medial process of the calcaneal tuberosity and inserts into the medial side of the base of the proximal phalanx of the great toe. It

Fig. 6-28. The dorsum of the foot. A. Superficial dissection. B. Intermediate dissection. C. Deep dissection.

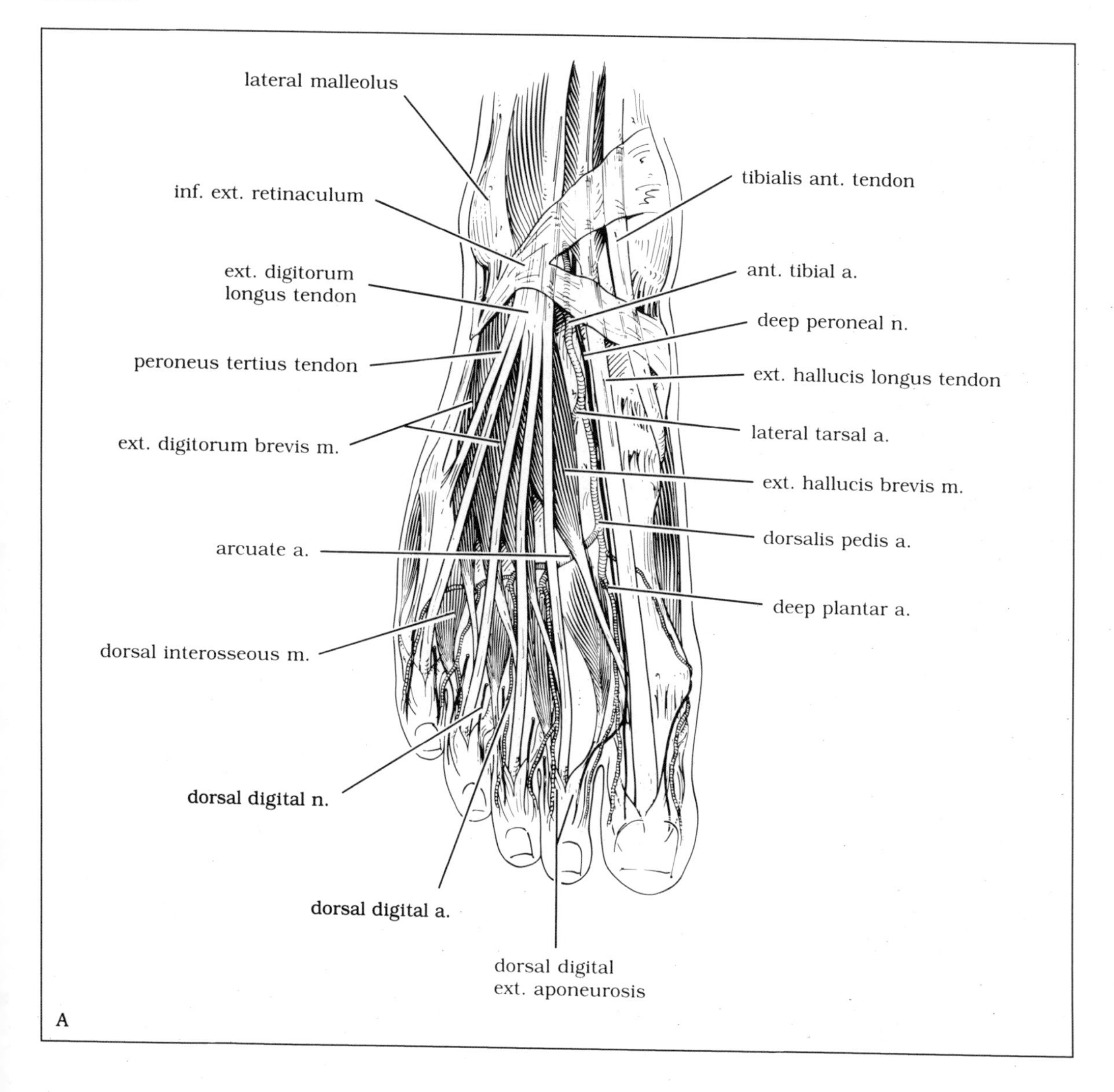

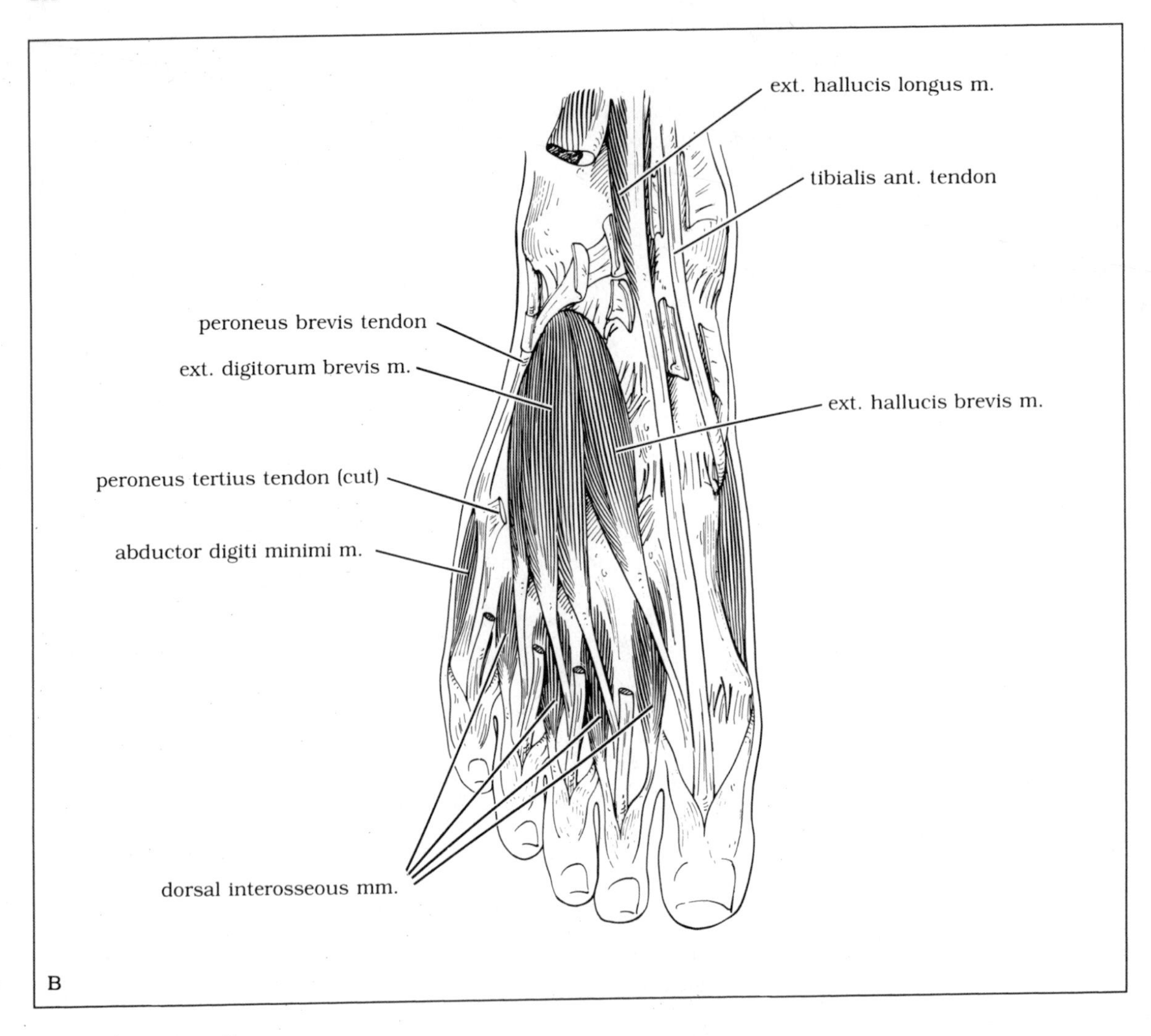

Fig. 6-28 (continued).

abducts the first toe and is innervated by the medial plantar nerve (L4 and L5).

The **flexor hallucis brevis** is a two-bellied muscle with the tendon of the flexor hallucis longus muscle lying in between. It arises from the cuboid and lateral cuneiform bones and the tendon of the tibialis posterior. It then divides in two, inserting into the medial and lateral sides of the base of the proximal phalanx of the great toe. Each of its tendons has a sesamoid bone in it. The medial tendon blends with the tendon of the abductor hallucis; the lateral with the tendon of the adductor hallucis. The muscle flexes the great toe at its proximal phalanx. It is innervated by the medial plantar nerve (L4, L5, and S1).

The lateral region contains the **abductor digiti minimi** muscle and the flexor digiti minimi brevis muscle. The abductor forms the lateral border of the sole. It arises from the calcaneal tuberosity and inserts into the lateral side of the base of the proximal phalanx of the small toe. It abducts the small toe and is innervated by the lateral plantar nerve (S1 and S2). The **flexor**

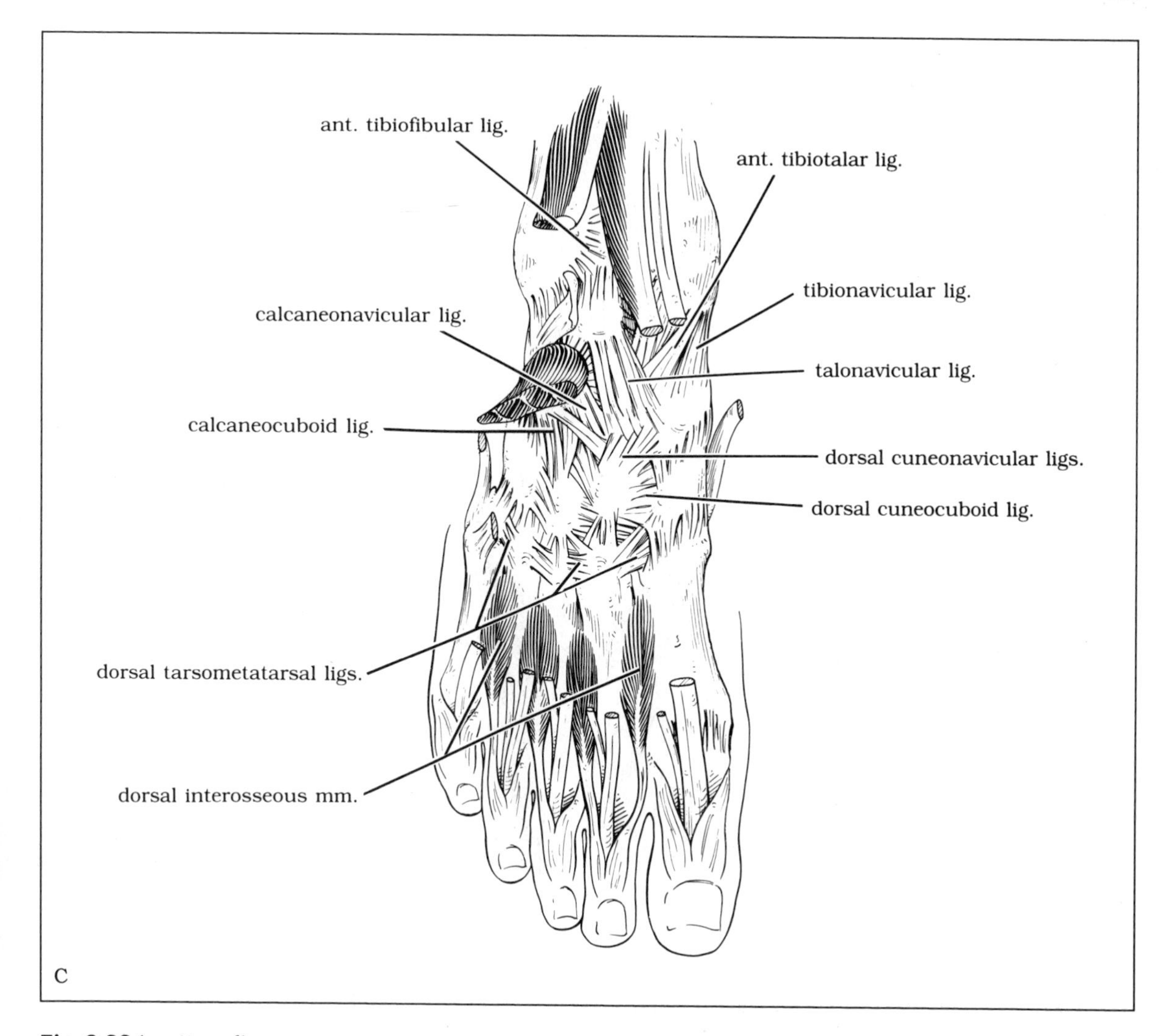

Fig. 6-28 (continued).

digiti minimi brevis arises from the base of the fifth metatarsal bone. It inserts into the base of the first phalanx of the small toe on the lateral side. It flexes the small toe at the proximal phalanx. The lateral plantar nerve (S1 and S2) innervates it.

The central region of the sole of the foot contains the flexor digitorum brevis muscle, the quadratus plantae muscle, the tendon of the flexor digitorum longus muscle, and the lumbricals.

The **flexor digitorum brevis** lies just below the plantar aponeurosis. The muscle arises from the medial process of the calcaneal tuberosity. As it passes distally over the sole of the foot it divides into four tendons. Each tendon in turn splits to allow for passage of the flexor digitorum longus tendon and then reunites. These tendons then divide again to insert into the two sides of the second phalanx of the corresponding toe. The muscle flexes the second phalanx of the four small toes. The medial plantar nerve (L4 and L5) innervates it.

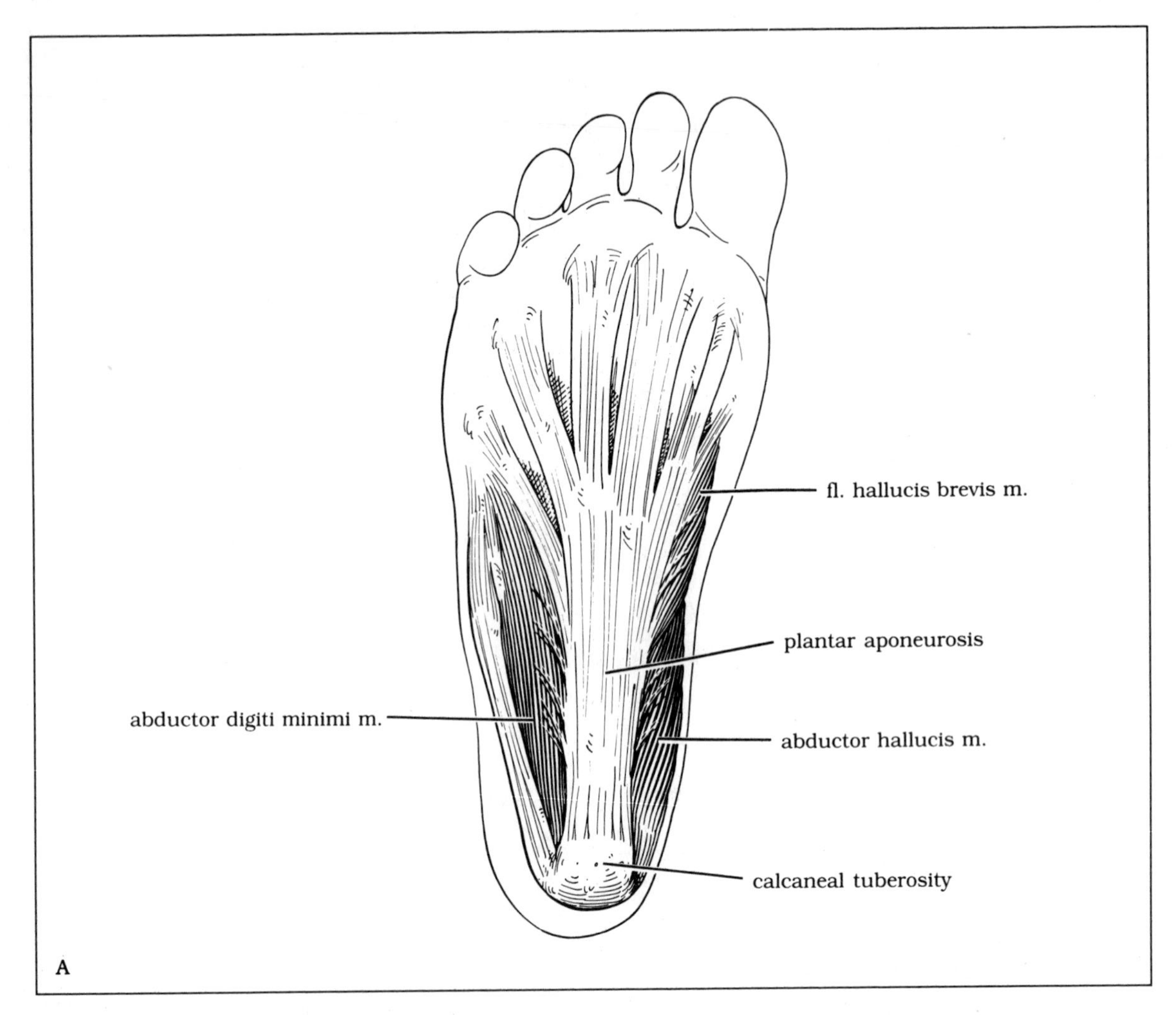

Fig. 6-29. The plantar foot. A. Superficial dissection. B. Intermediate dissection. C. Deep dissection. D. Deepest dissection.

The **quadratus plantae** arises by two heads from the medial and lateral calcaneus. The heads are separated by the long plantar ligament. The muscle inserts into the lateral margin and dorsal and plantar surfaces of the tendon of the flexor digitorum longus. Through this tendon it flexes the terminal phalanges of the second to fifth toes. It is innervated by the lateral plantar nerve (S1 and S2).

Each of the four **lumbrical muscles** arises from two tendons of the flexor digitorum longus. The lumbrical muscles pass on the medial side of the toes to insert into the tendinous sheaths of the extensor digitorum longus (see Fig. 6-28B). The lumbrical muscles flex the proximal phalanx and extend the distal phalanges of the four small toes. The first lumbrical muscle is innervated by the medial plantar nerve (L4 and L5), the others by the lateral plantar nerve (S1 and S2).

The interosseous adductor region contains the adductor hallucis muscle and the dorsal and

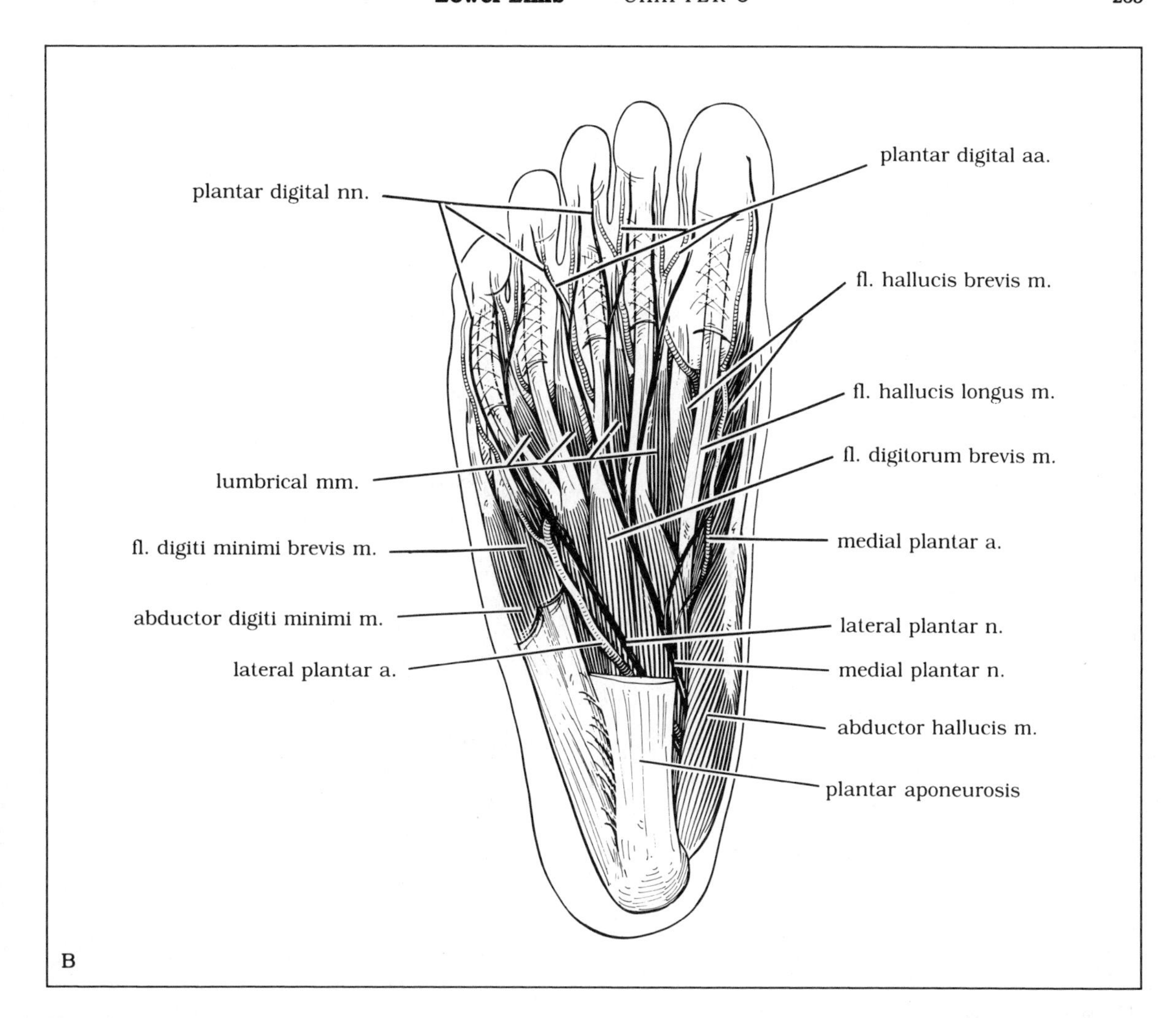

Fig. 6-29 (continued).

plantar interossei muscles. The **adductor hallucis** is a two-headed muscle. The oblique head arises from the bases of the second through fourth metatarsals. The transverse head arises from the plantar metatarsophalangeal ligaments of the third through fifth toes. The two heads insert into the lateral side of the base of the proximal phalanx of the first toe. They adduct the great toe and are innervated by the lateral plantar nerve (S1 and S2).

The **interossei muscles** act around an axis through the midline of the second digit. The four dorsal interossei abduct the toes, and the three plantar interossei adduct the toes. Both sets of muscles also flex the proximal phalanx and extend the distal phalanges of their corresponding toes. They are innervated by the lateral plantar nerve (S1 and S2).

The four dorsal interossei are bipennate muscles, arising by two heads from the adjacent sides of the metatarsal bones. They insert into the base of the proximal phalanx and into the extensor tendon expansions. The first inserts into the medial side of the second toe, the second through fourth into the lateral side of the second through

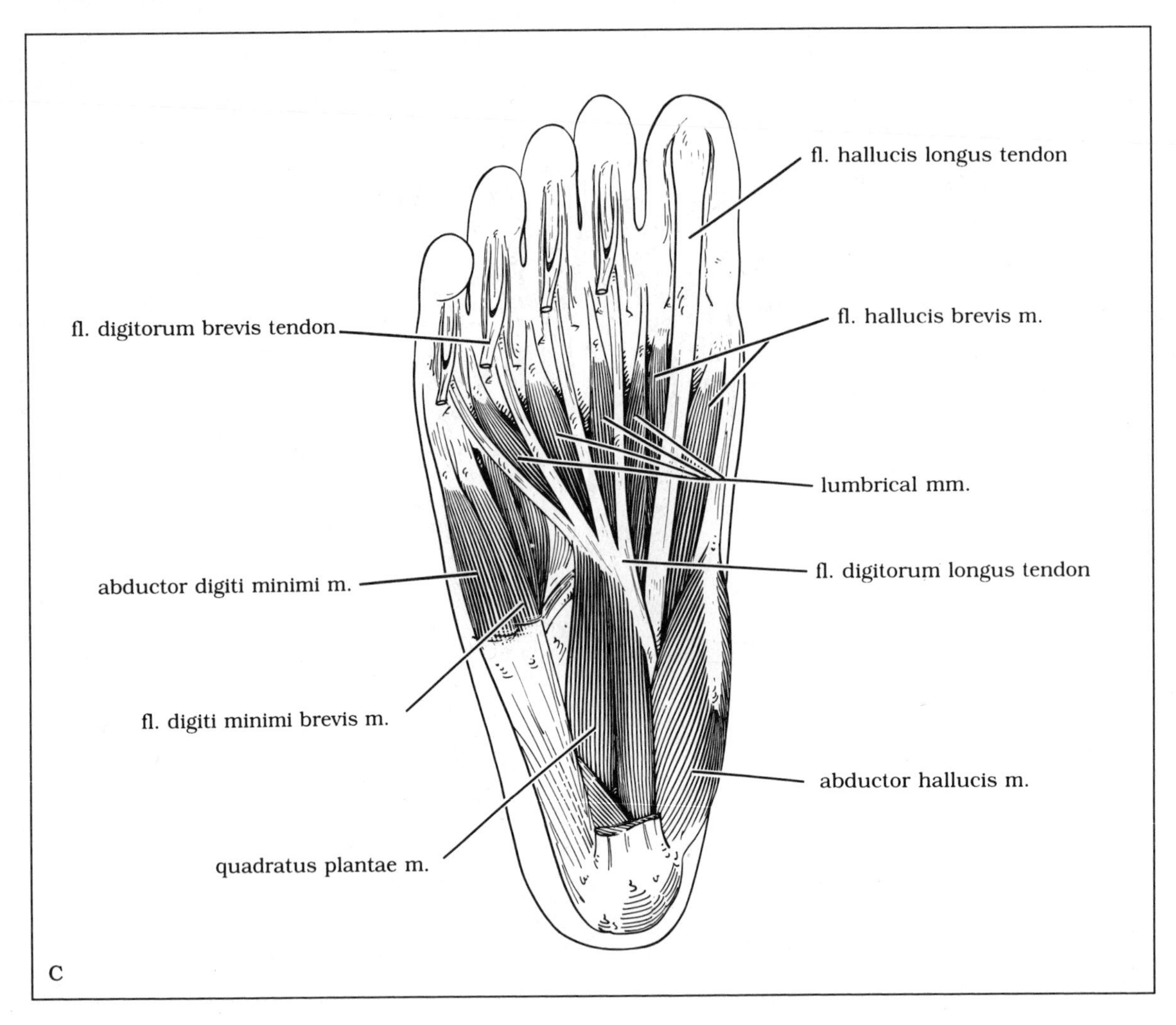

Fig. 6-29 (continued).

fourth toes. The plantar interossei arise from the base and medial side of the third through fifth metatarsal bones. They insert into the medial side of the proximal phalanx and the extensor tendon expansions of the same toe.

ACTIONS OF THE MUSCLES OF THE FOOT

The dorsal muscle of the foot, the extensor digitorum brevis, aids in extending the phalanges of the toes. The plantar muscles are active in maintaining the arch of the foot as the heel comes off the ground during walking or running. In addition, as the foot acts as a lever during takeoff, the muscles serve to increase the structural rigidity of the foot, most likely resulting in more powerful movement.

BLOOD SUPPLY AND INNERVATION

The two arteries of the sole of the foot, the medial and lateral plantar arteries, are branches of the posterior tibial artery (see Figs. 6-27B and 6-29B,D). The **medial plantar artery** lies first

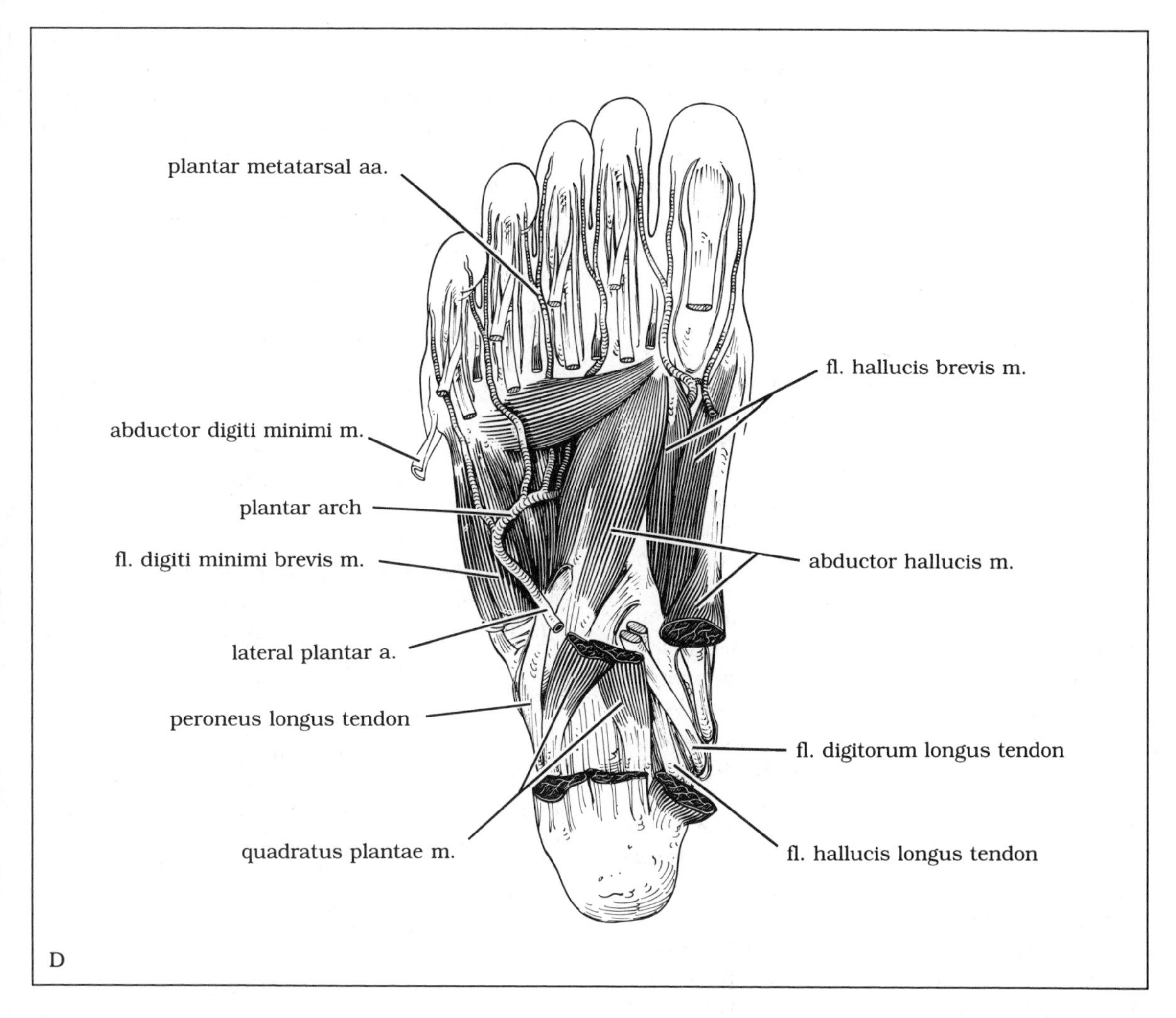

Fig. 6-29 (continued).

under the abductor hallucis and then between it and the flexor digitorum brevis. It sends digital branches to the first three interdigital clefts. The medial plantar artery anastomoses with the digital branches of the first plantar metatarsal artery and the three plantar metatarsal arteries of the plantar arch.

The **lateral plantar artery** crosses the sole of the foot diagonally to the fifth metatarsal. There it sinks deeply and passes medially as the plantar arch, deep to the adductor hallucis muscle. At the first interosseous space it anastomoses with the deep plantar ramus of the dorsalis pedis artery. The lateral plantar artery gives off the four **plantar metatarsal arteries,** which in turn divide into the digital arteries.

The **medial plantar nerve,** a branch of the tibial nerve, travels with the medial plantar artery. It innervates the abductor hallucis and flexor digitorum brevis muscles and is articular to tarsal and metatarsal joints. The nerve also supplies the flexor digitorum brevis muscle and the first lumbrical muscle.

The **lateral plantar nerve** also arises from the tibial nerve. It courses with the lateral plantar vessels to innervate the abductor digiti minimi, flexor digiti minimi brevis, and quadratus plantae and the interossei muscles. In addition it supplies all the lumbricals except the first.

The veins of the sole of the foot accompany their respective artery and are named accordingly.

Joints of the Lower Limb

HIP JOINT

At the hip joint, the spheroidal head of the femur sits in the pocket of the coxal bone known as the acetabulum (see Fig. 4-4). It is a synovial joint. The head of the femur is covered by articular cartilage that is thickest above and thinnest below. The cartilage does not cover the foveal pit, where the ligamentum capitis femoris inserts into the head of the femur.

The **acetabulum** has a horseshoe-shaped articular surface that is also thickest above. Below, the horseshoe is closed by the **transverse acetabular ligament.** Within the fossa, but outside the synovial membrane, is a fat pad. The fossa is deepened by the **acetabular labrum,** a fibrocartilaginous ring, on its bony edge. The labrum's thin free edge surrounds the head of the femur and helps to hold it in the acetabulum. Vessels and nerves enter the fossa by passing under the transverse acetabular ligament.

The **capsule** of the hip joint is a strong structure with fibers that run in two directions. Most fibers run from the pelvis to the femur, winding around the joint as they do so, so that in full extension they provide a "screw-in" effect for greater stability. Other fibers encircle the joint (the zona orbicularis), particularly in the posterior regions.

In addition, three accessory ligaments help to stabilize the hip joint (Fig. 6-30). The strong **iliofemoral ligament** stabilizes the joint anteriorly. It is shaped like an inverted Y, extending from the anterior inferior spine of the ilium to the intertrochanteric line of the femur. It is taut in full extension. The **pubofemoral ligament** adds stability medially and inferiorly, extending from the acetabular rim of the pubis and the obturator crest to the neck of the femur. This ligament tightens in extension and limits abduction. The **ischiofemoral ligament** stabilizes the joint posteriorly, extending from the ischial part of the acetabulum in a spiral to reach the posterior neck of the femur. The ischiofemoral ligament aids in limiting adduction. The capsule is weakest between the iliofemoral and pubofemoral ligaments. It is strengthened here by the presence of the iliopsoas tendon.

The **ligamentum capitis femoris** extends from the acetabular notch to the fovea of the femur. It serves as a route by which vessels and nerves reach the femoral head and aids in limiting adduction.

The arteries supplying the hip joint are the medial and lateral femoral circumflex, and the inferior and superior gluteal arteries and the posterior division of the obturator artery (see Fig. 6-20). Venous drainage follows the arterial supply. Innervation is supplied by the femoral nerve, the superior gluteal nerve, the anterior division of the obturator nerve, and (when present) the accessory obturator nerve.

Movement at the hip is limited as follows: Flexion is limited when the thigh touches the abdomen; extension is limited by the iliofemoral ligament; abduction is limited by the pubofemoral ligament; and adduction is limited by the ischiofemoral ligament, the ligamentum capitis femoris, and the opposite limb.

KNEE JOINT

The knee is primarily a weight-bearing joint that allows movement in only one direction. In actuality, the knee is three joints—the femoropatellar, medial tibiofemoral, and lateral tibiofemoral joints—with an interconnected joint cavity.

The femur and tibia expand at the joint, enhancing its stability. The wheellike femoral condyles articulate with the concave tibial condyles (Fig. 6-31). The **menisci** are two horseshoe-shaped fibrocartilages on top of the articular surfaces of the tibial condyles (Fig. 6-32). They aid in deepening the articular surfaces, protecting the articular cartilage, and increasing the efficiency of lubrication of the joint.

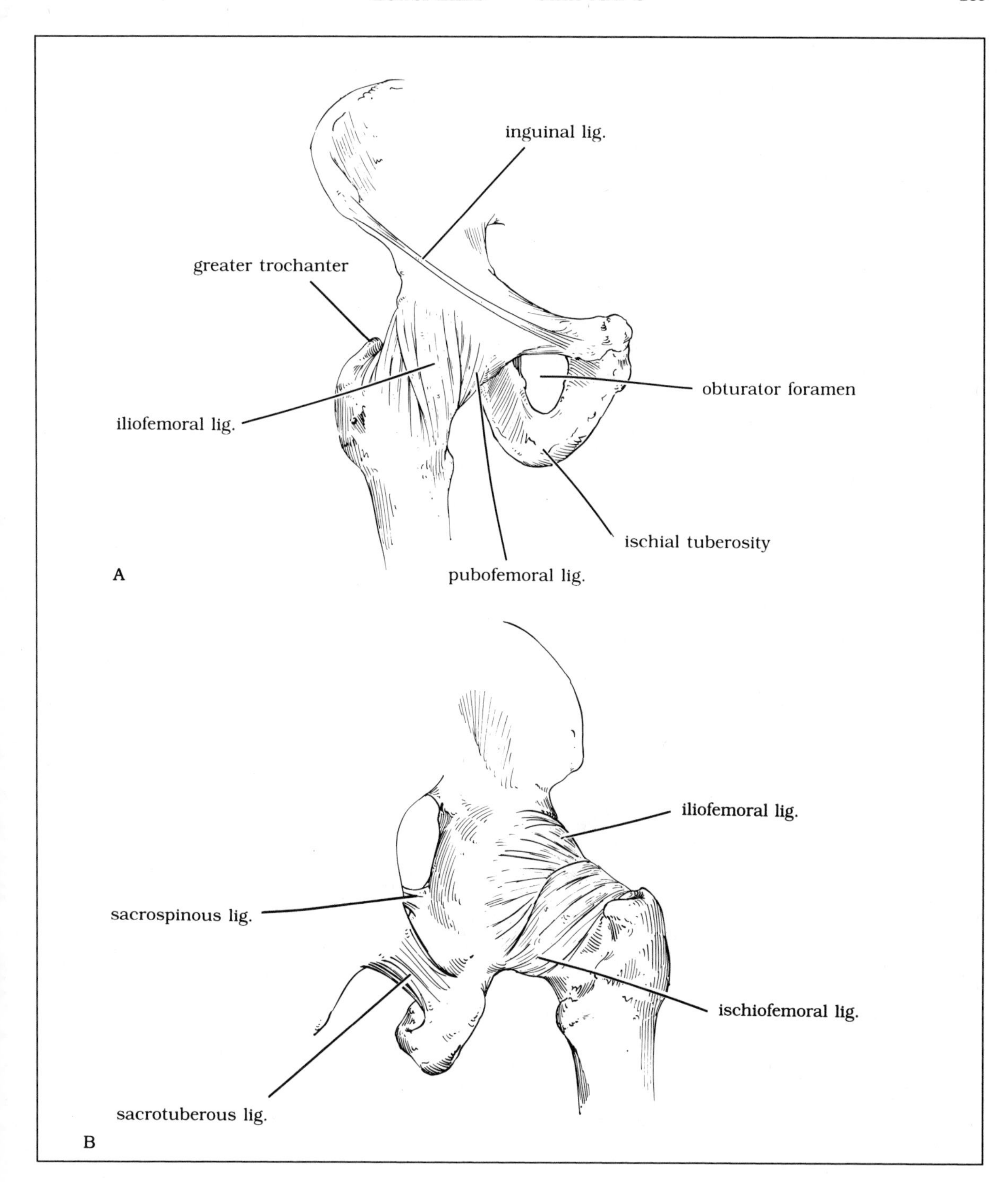

Fig. 6-30. The hip joint. A. Anterior view.
B. Posterior view.

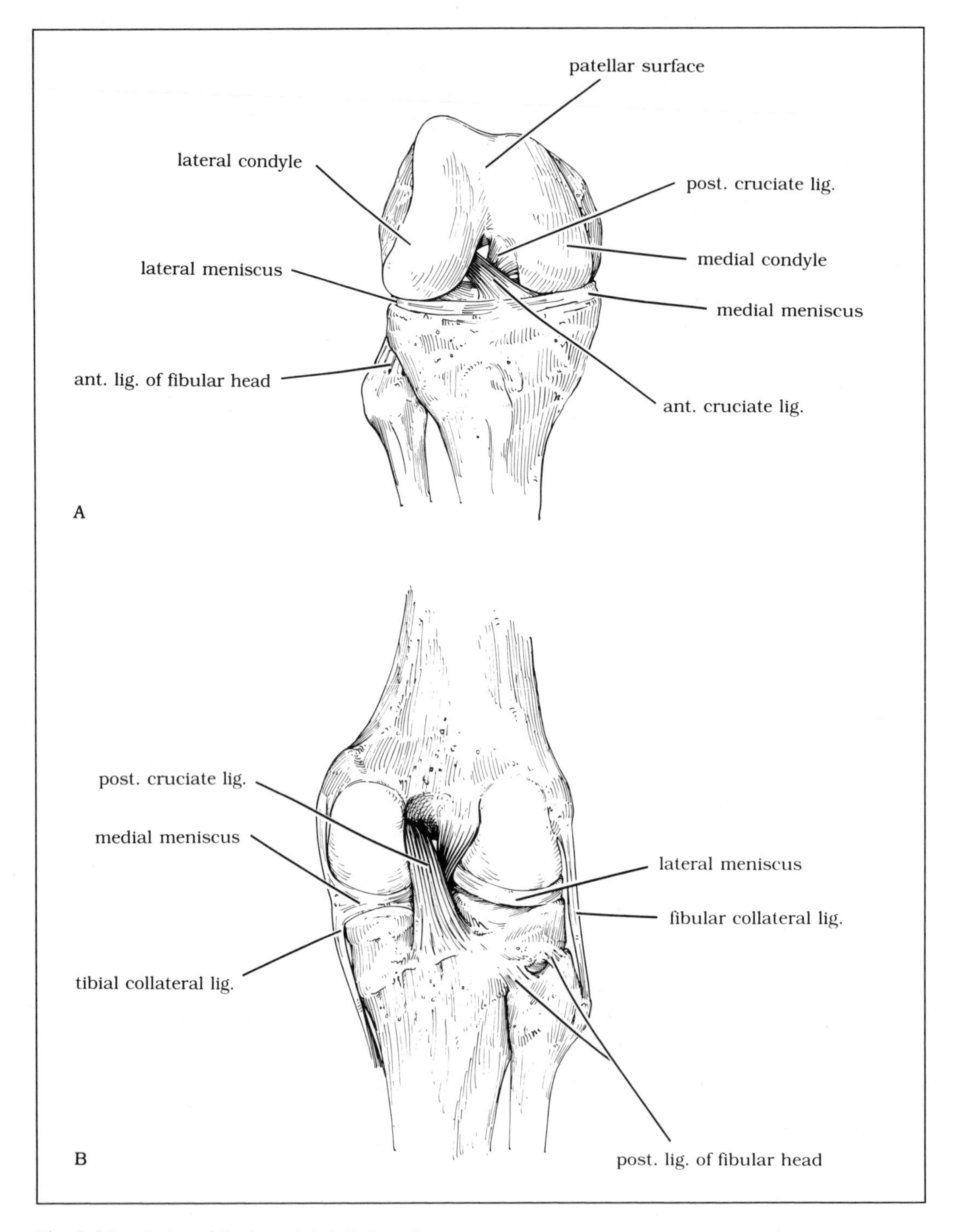

Fig. 6-31. Interior of the knee joint. A. Anterior view.
B. Posterior view.

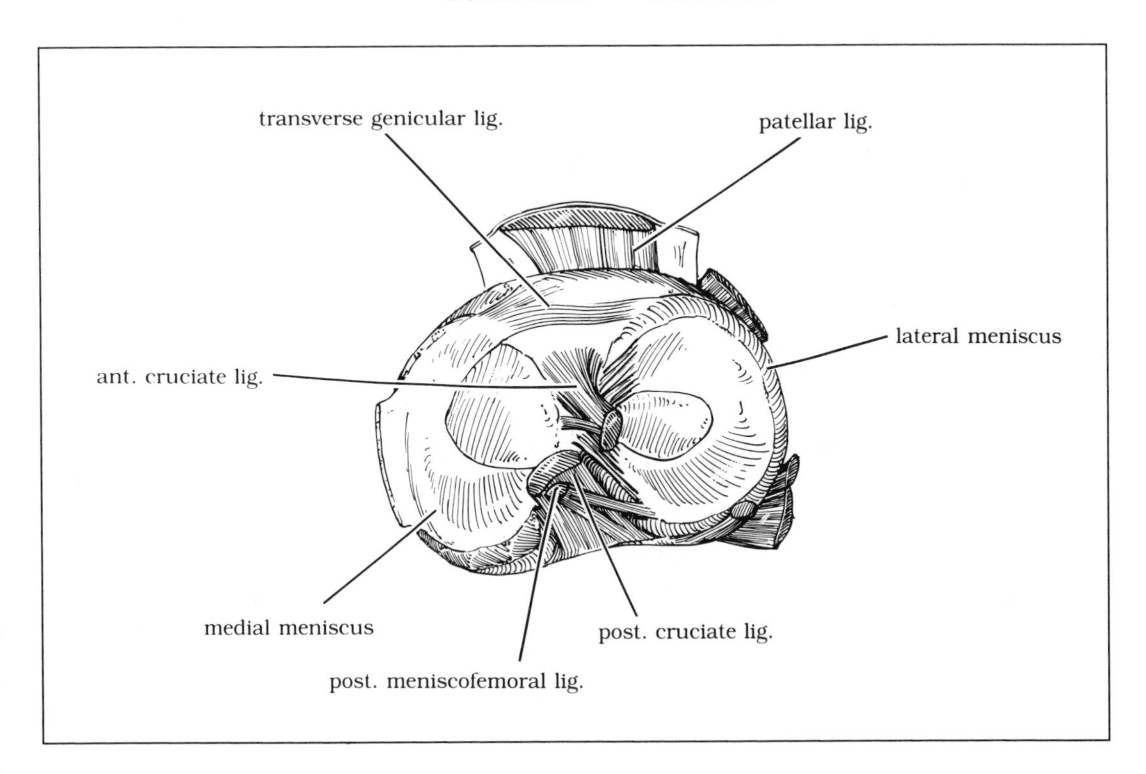

Fig. 6-32. The knee joint from above, with the femur removed.

All articular surfaces are covered by articular cartilage. The articular capsule is attached to, and considerably strengthened by, the ligaments and aponeuroses surrounding it: the iliotibial tract, fascia lata, popliteal ligaments, and quadriceps tendon (patellar ligament) (Fig. 6-33). The joint is held together by ligaments. Two collateral ligaments, both situated behind the vertical plane of the joint, prevent hyperextension. The **tibial collateral ligament** (Fig. 6-34A) extends from the femur to the tibia on the medial side. It is a flat, broad ligament and is attached to the medial meniscus. The **fibular collateral ligament** (Fig. 6-34B), on the lateral side, attaches to the femur above and to the fibula below. The ligament is a strong, round band and the tendon of the popliteus muscle passes deep to it.

The two **cruciate ligaments** (see Figs. 6-31 and 6-32) are within the capsule of the knee joint but are outside the synovial cavity, as they are surrounded by synovial membrane. The ligaments are named after their location on the tibia. The two cross each other, the anterior one going from the front of the tibia to the back of the femur, and the posterior one going from the back of the tibia to the front of the femur. The ligaments are tightest in full extension. The anterior ligament prevents forward movement of the tibia, and the posterior ligament prevents backward movement.

The articular cavity is lined with **synovial membrane.** This membrane is continuous around the condyles, behind the patella, and between the femur and the quadriceps tendon. The two cruciate ligaments are surrounded by this membrane and they and the infrapatellar fold of membrane serve to separate the two tibiofemoral joints. The infrapatellar fat pad lies within the infrapatellar fold.

There are three bursae associated with the patella: the suprapatellar, prepatellar, and infrapatellar bursae. In addition there are six bur-

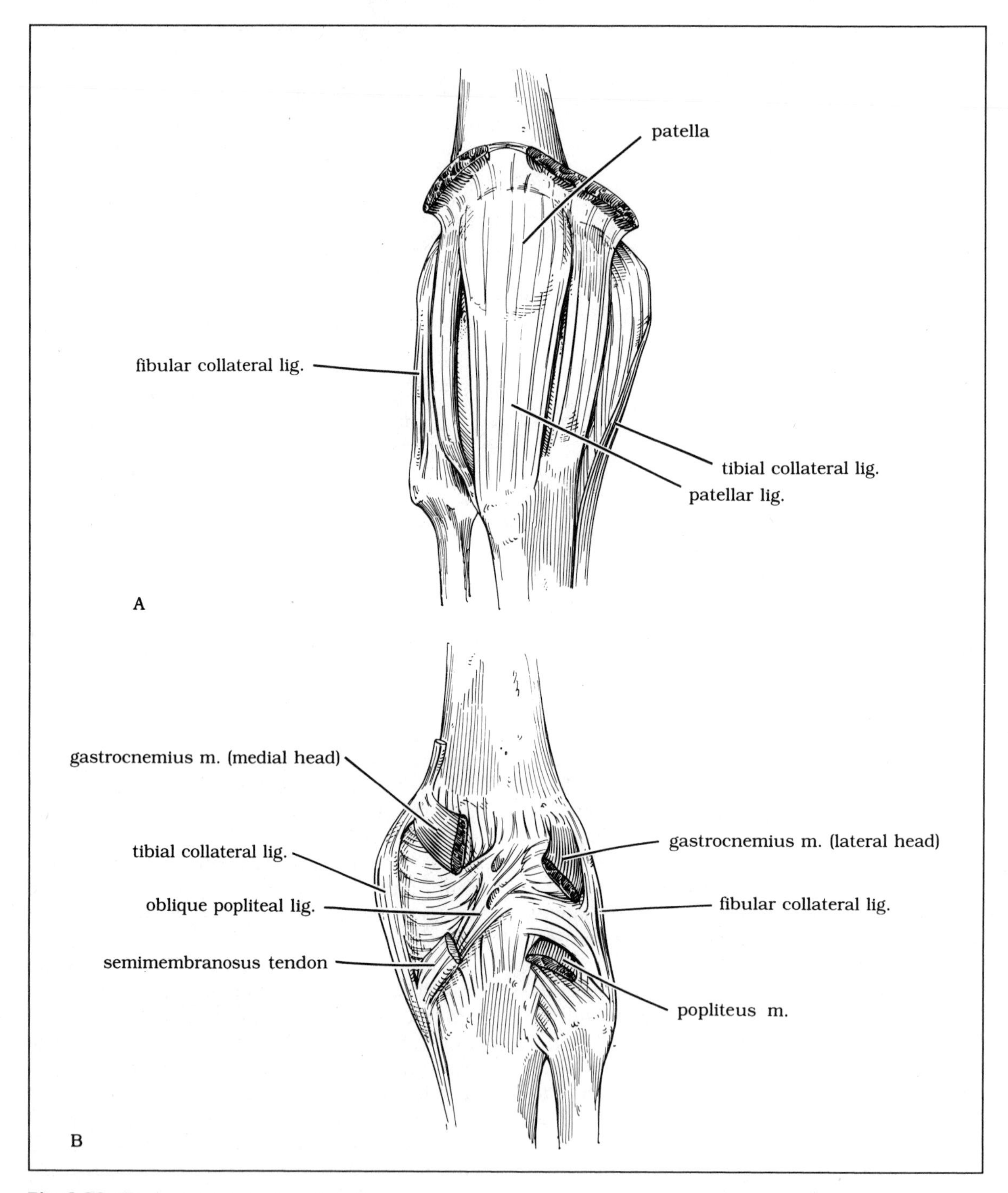

Fig. 6-33. The knee joint. A. Anterior view.
B. Posterior view.

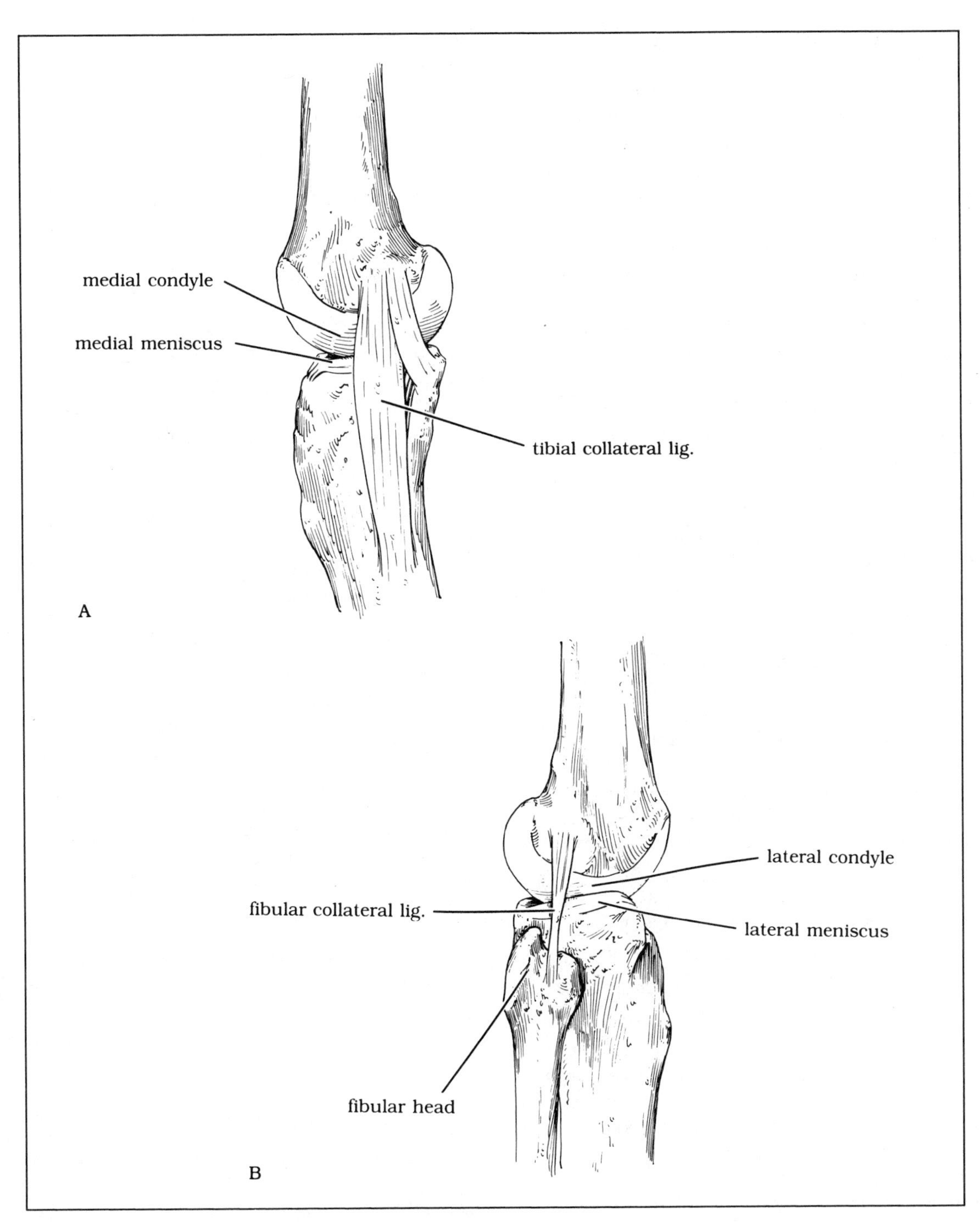

Fig. 6-34. The knee joint. A. Medial view.
B. Lateral view.

sae associated with surrounding muscles. The semimembranosus muscle has a bursa underlying it. The other five bursae underlie the pes anserinus or the tendons of the following muscles or muscle groups: biceps femoris, lateral head of gastrocnemius, medial head of gastrocnemius, and popliteus.

The arteries composing the **genicular anastomosis** are the descending branch of the lateral femoral circumflex artery, the descending genicular branch of the femoral artery, the medial and lateral superior genicular arteries, the medial and lateral inferior genicular arteries, the middle genicular artery, the posterior and anterior tibial recurrent arteries, and the circumflex fibular artery. Lymph from the knee joint drains to the popliteal and inguinal nodes.

The innervation of the knee is supplied by the femoral nerve, the posterior division of the obturator nerve, the tibial nerve, and the common peroneal nerve (see Fig. 6-21).

TIBIOFIBULAR ARTICULATIONS

There are two tibiofibular articulations. The proximal joint (see Fig. 6-31) is strengthened by the anterior and posterior ligaments of the head of the fibula. There is little movement at this joint. At the distal articulation, the tibiofibular syndesmosis, the bones are held together by the anterior and posterior tibiofibular ligaments and by the interosseous ligament, which is a stronger continuation of the interosseous membrane. There is slight movement at this joint. The tibia and the fibula are also held together by the interosseous membrane.

ANKLE JOINT

The ankle joint is a hinge joint. The cap is formed by the medial and lateral malleoli, which fit over the trochlea of the talus. The **capsule** of the joint is weak in front and behind, but strengthened on the sides by collateral ligaments. The **medial collateral ligament,** or deltoid ligament, is a strong triangular ligament. The three **lateral collateral ligaments** are the anterior talofibular, posterior talofibular, and middle ligaments. The middle (calcaneofibular) ligament attaches below to the calcaneus, the others to the talus.

Blood is supplied to the ankle joint by the malleolar arteries. Innervation is provided by the tibial and deep peroneal nerves.

JOINTS OF THE FOOT

The **intertarsal joints** of the foot are the subtalar (talus-calcaneus), talocalcaneonavicular, transverse tarsal (talonavicular and calcaneocuboid), cuneonavicular, intercuneiform, and cuneocuboid joints (see Fig. 6-28C). The last three have a common cavity and capsule, which also includes the tarsometatarsal and intermetatarsal joints for the second and third metatarsals. All the intertarsal joints have interosseous, dorsal, and plantar ligaments.

The **tarsometatarsal joints** connect the following bones of the foot: the first metatarsal with the medial cuneiform, the second metatarsal with all the cuneiforms, the third metatarsal with the lateral cuneiform, the fourth metatarsal with the lateral cuneiform and cuboid, and the fifth metatarsal with the cuboid.

The **intermetatarsal joints** all contain articular cavities, except the joint between the first and the second metatarsals, which are held together only by interosseous fibers.

The **metatarsophalangeal** and **interphalangeal joints** all are surrounded by joint capsules that are strengthened by plantar and collateral ligaments.

The joints and ligaments of the foot all provide resiliency and flexibility. Its arched structure, strengthened by the strong plantar and interosseous ligaments, gives the foot its shock-absorbing quality, allowing it both to distribute body weight and to maintain its spring at the same time.

Lymphatics of the Lower Limb

The terminal nodes of the lower limb are the deep and superficial inguinal nodes. The **superficial inguinal nodes** drain the gluteal region, the abdominal wall below the umbilicus, the external genitalia, the anal and perianal regions, and the superficial structures of the lower limb. The superficial inguinal nodes drain to the external iliac nodes.

The **deep inguinal nodes** drain the glans penis or clitoris, the lymph vessels accompanying the femoral vessels, and the deep structures of the lower limb. The deep inguinal nodes also drain to the external iliac nodes.

The **popliteal nodes,** lying in the popliteal fossa, drain the knee joint, the lymph vessels accompanying the tibial vessels, and portions of the lateral superficial calf. These nodes drain to the deep inguinal nodes.

Lymphatic vessels in the lower limb accompany the arterial and venous vessels, including the great and small saphenous veins.

National Board Type Questions

Select the one best response for each of the following

1. Regarding the knee joint
   A. The tibial collateral ligament, if stretched, could damage the medial meniscus because of its attachment to it.
   B. Allows flexion and abduction.
   C. The patella is surrounded by bursae.
   D. Is unaffected by action of the popliteus muscle.
   E. Contains cruciate ligaments named for their attachment to the femur.
2. Regarding cutaneous sensation all of the following are true **except:**
   A. The gluteal region is supplied by the cluneal nerves.
   B. The gluteal region is supplied by the posterior femoral cutaneous nerves.
   C. The gluteal region is supplied by dorsal rami.
   D. The gluteal region is supplied by sacral nerve 3.
   E. The gluteal region is supplied by the superior gluteal nerves.
3. Following a skiing injury a tightly fitting cast was applied to the right leg (knee to ankle) of a young woman. When the cast was removed the patient exhibited sensory loss on the lateral surface of the right leg and the dorsum of the right foot. She also had markedly reduced ability to dorsiflex and evert her right foot and extend the toes on this foot. These signs are consistent with a damaged
   A. Tibial nerve
   B. Femoral nerve
   C. Common peroneal nerve
   D. Superficial peroneal nerve
   E. Deep peroneal nerve
4. The chief ligament preventing forward sliding of the femur on the tibia (i.e., backward sliding of the tibia on the femur) is the
   A. Tibial collateral ligament
   B. Fibular collateral ligament
   C. Oblique popliteal ligament
   D. Anterior cruciate ligament
   E. Posterior cruciate ligament
5. Which of the following quadriceps muscles can act as a flexor of the hip joint?
   A. Vastus lateralis
   B. Vastus medialis
   C. Vastus intermedius
   D. Rectus femoris
   E. None of the above

Select the response most closely associated with each numbered item. (The headings may be used once, more than once, or not at all.)

The *chief* action of each of the numbered muscles is

A. Abduction of thigh
B. Adduction of thigh
C. Flexion of thigh
D. Extension of thigh
E. Lateral rotation of thigh

6. Gluteus minimus
7. Gluteus medius
8. Gracilis
9. Iliopsoas
10. Gluteus maximus

Select the response most closely associated with each numbered item.

A. Hamstring muscle group
B. Quadriceps femoris muscle group

C. Both
D. Neither

11. Flexion of the knee joint
12. Innervated by the obturator nerve
13. All of its component muscles are attached to the pelvis
14. Blood supply from branches of the femoral artery
15. Most of its component muscles attach to the femur

For the following, select

A. if only *1, 2, and 3* are correct
B. if only *1 and 3* are correct
C. if only *2 and 4* are correct
D. if only *4* is correct
E. if *all* are correct

16. The head and neck of the femur are supplied by which of the following arteries?
    1. Obturator
    2. Lateral femoral circumflex
    3. Medial femoral circumflex
    4. Internal pudendal
17. Which of the following bone(s) articulate with the calcaneus?
    1. Talus
    2. Fibula
    3. Cuboid
    4. Tibia
18. The fibrous capsule of the hip joint
    1. Is thickened anteriorly by the iliofemoral ligament
    2. Where weakest is strengthened by the tendon of the iliopsoas
    3. Limits extension, abduction, and adduction
    4. Attaches to the base of the femoral head
19. The muscles that form a sling around the transverse arch of the foot by inserting into the first cuneiform and first metatarsal bones are
    1. Extensor hallucis longus
    2. Tibialis anterior
    3. Peroneus brevis
    4. Peroneus longus
20. If the tibial nerve were severed at the popliteal fossa, which of the following would be affected?
    1. Flexor hallucis longus
    2. Abductor hallucis
    3. Soleus
    4. Tibialis anterior
21. Structure(s) coursing behind the medial malleolus and covered by the flexor retinaculum is (are) the
    1. Tendon of flexor hallucis longus
    2. Posterior tibial artery
    3. Tendon of tibialis posterior
    4. Saphenous vein
22. Which of the following are in the femoral sheath?
    1. Femoral artery
    2. Saphenous vein
    3. Femoral vein
    4. Femoral nerve

Annotated Answers

1. A.
2. E. The gluteal region receives sensory innervation from several nerves, but the large motor nerves to the muscles of the region do not carry a sensory component.
3. C. The injury must be to the common peroneal nerve because areas involved include the sensory and motor field of the superficial peroneal nerve and the motor component of the deep peroneal.
4. E. To test the integrity of the cruciate ligaments one may take the patient's flexed knee and while holding the thigh pull the leg forward and backward. If the tibia slides backward (excessively, as compared to the other limb) one could suspect damage to the posterior cruciate ligament.
5. D. The rectus femoris is a component of the quadriceps group that spans both the hip and knee joints. Thus it can act on both joints.
6. A. and 7. A. These muscles act as powerful abductors of the hip joint. However, one should remember that this abduction can be considered as an antagonistic movement to

the adduction that tends to occur when one walks and at a point in the gait cycle one limb is in contact with the ground and one limb is free. (Stand on one leg and try to keep your balance without shifting (abducting) your weight over the stance limb.)

8. B.
9. C.
10. D.
11. A. The hamstrings take part in flexion of the knee because they attach on the tibia or fibula and pass dorsal to the axis of the knee joint.
12. D.
13. A.
14. C. Remember that the femoral artery supplies both the anterior and posterior thigh via the profunda femoris.
15. B.
16. A. The first three arteries supply the joint and are involved in the so-called cruciate anastomosis. The pudendal supplies structures in the perineum.
17. B.
18. A. The fibrous capsule of the hip joint, which reaches to the neck of the femur, serves an important function in support by limiting movements, and is associated with one of the strongest ligaments in the body, the iliofemoral.
19. C. The sling is formed by an inverter, tibialis anterior, and an everter, peroneus longus.
20. A.
21. A. Remember that the saphenous vein is always found anterior to the medial malleolus. An incision to find this vein, for purposes of catheterization for example, should never be made posterior to the malleolus.
22. B.

7 Head

Objectives

After reading this chapter, you should know the following:

Structure of the skull
- Features of the cranial and facial bones
- Location and contents of all foramina, fissures, canals, and fossae

Tissue layers and musculature of the scalp

Muscles of facial expression

Location and function of the parotid gland

Structure, components, and functions of the nose, including the nasal cavities, paranasal sinuses, and nasopharynx

Structure of the orbital region
- Extraocular muscles and conjugate eye movements
- Tissues and musculature of the eyelid
- Components and functions of the lacrimal apparatus
- Layers of the eye
- Components and functions of the refractive apparatus
- Sensory and autonomic innervation of the eye and the lacrimal gland

Structure, components, and functions of the ear, including conduction of vibration and special sensory receptive organs

Structure, components, and functions of the mouth

Musculature and taste buds of the tongue

Structure, components, and functions of the oral pharynx

Classification and functions of the teeth

Structure and function of the temporomandibular joint

Contents, musculature, and functions of the temporal and infratemporal fossae

Structures and musculature involved in mastication

Arterial supply to the head, including branches of the external and internal carotid arteries and the vertebral artery

Venous drainage of the head, including the cerebral sinuses

Blood supply, lymphatics, and innervations to the regions of the head

Innervations of the 12 cranial nerves
- Differences between motor and sensory innervation
- Differences between visceral and somatic innervation
- Differences between spinal and cranial nerves
- Categories of the cranial nerves
- General functions and pathways of the cranial nerves
- Effects of cranial nerve injuries

Skull

The skull consists of 22 bones: 8 cranial bones, which enclose the brain, and 14 facial bones, which are found at the anterior and lateral aspects of the skull:

Cranial Bones	*Facial Bones*
Frontal	Maxillae (2)
Parietal (2)	Mandible
Occipital	Zygomas (2)
Temporal (2)	Nasal (2)
Sphenoid	Palatine (2)
Ethmoid	Lacrimal (2)
	Vomer
	Inferior nasal conchae (2)

The bones of the skull are joined together primarily at immobile joints, the sutures; only the mandible, which articulates with the remainder of the skull at the temporomandibular joint, and the joints between the auditory ossicles are moveable.

The skull contains approximately 85 openings (foramina, fissures, and canals) that permit connections between the central nervous system within the cranial cavity and portions of the central nervous system external to the cranial cavity, or permit nerves and vessels to enter the bones and interior of the skull. The skull is covered both inside and out with a layer of connective tissue, the periosteum. The dura mater forms the periosteum of the inner table of cranial bone.

EXTERIOR ASPECT OF THE SKULL

Superior Aspect

When viewed from above the skull is oval, wider posteriorly than anteriorly (Fig. 7-1). The superior aspect of the skull is also called the skull cap, or calvaria. Four bones—frontal, right and left parietal, and occipital—and three sutures—sagittal, lambdoid, and coronal—can be seen from this view. The junction point between the sagittal and coronal sutures is the **bregma,** and that between the lambdoid and sagittal sutures is the **lambda.** These junction points indicate the positions of the anterior and posterior fontanelles, respectively, in the infant skull; the anterior fontanelle is usually present into the second year, and the posterior fontanelle is obliterated by the second month after birth.

The **vertex,** or the highest point on the skull, is along the sagittal suture near its midpoint, several centimeters behind the bregma. Occasionally, a few centimeters in front of the lambda and adjacent to the sagittal suture, the **parietal foramen** may be found; the parietal emissary vein runs through this foramen.

Posterior Aspect

The back of the skull consists of portions of the occipital and parietal bones, the mastoid portion of the temporal bone, and the sagittal and lambdoid sutures (Fig. 7-2). Inferiorly the lambdoid suture meets the parietomastoid and occipitomastoid sutures at a point called the **asterion.** Commonly, the **mastoid foramen,** which transmits an emissary vein, is identified near the lambdoid suture.

The dominant structure in the occipital bone is the **external occipital protuberance,** which is found in the midline between the lambda and the foramen magnum and is readily palpable in vivo. The **superior nuchal line** delineates the upper end of the neck, and in some skulls another bony landmark, the **highest nuchal line,** can be seen just above the superior nuchal line.

Lateral Aspect

The lateral aspect of the skull contains five cranial bones—the frontal, parietal, occipital, and temporal bones and the greater wing of the sphenoid bone—and the zygoma of the face (Fig. 7-3). The lateral aspect also includes the temporomandibular joint, the temporal and infratemporal fossae, the pterygopalatine fossa, the pterygomaxillary fissure, and the inferior orbital fissure (Table 7-1).

The **pterion** is the point in the temporal fossa where the frontal and parietal bones, the greater wing of the sphenoid bone, and the squamous

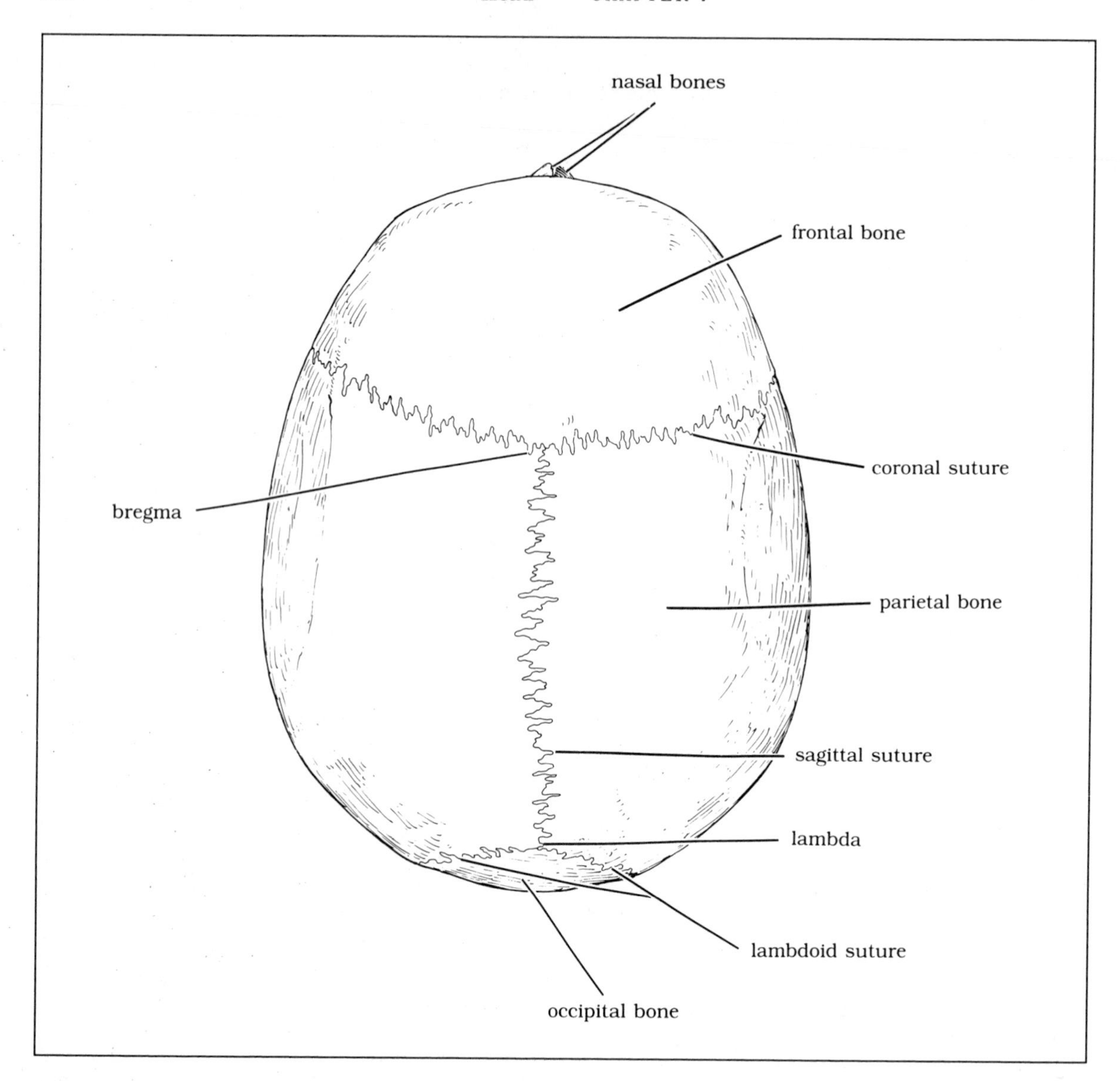

Fig. 7-1. Superior aspect of the skull.

portion of the temporal bone approach each other; under this point in the cranial cavity is the middle meningeal artery and the lateral fissure of the brain. The **squamous suture** arches back from the pterion and lies between the temporal squama and the most inferior portion of the parietal bone. This suture is continuous posteriorly with the parietomastoid suture. The coronal and lambdoid sutures are also seen in the lateral aspect.

The **temporomandibular joint** is the articulation between the mandible and the temporal bone. The mandibular fossa and articular tubercle of the temporal bone form the articular surface on the skull; the condyle is the mandibular articular surface. (This joint is discussed in detail later in this chapter.)

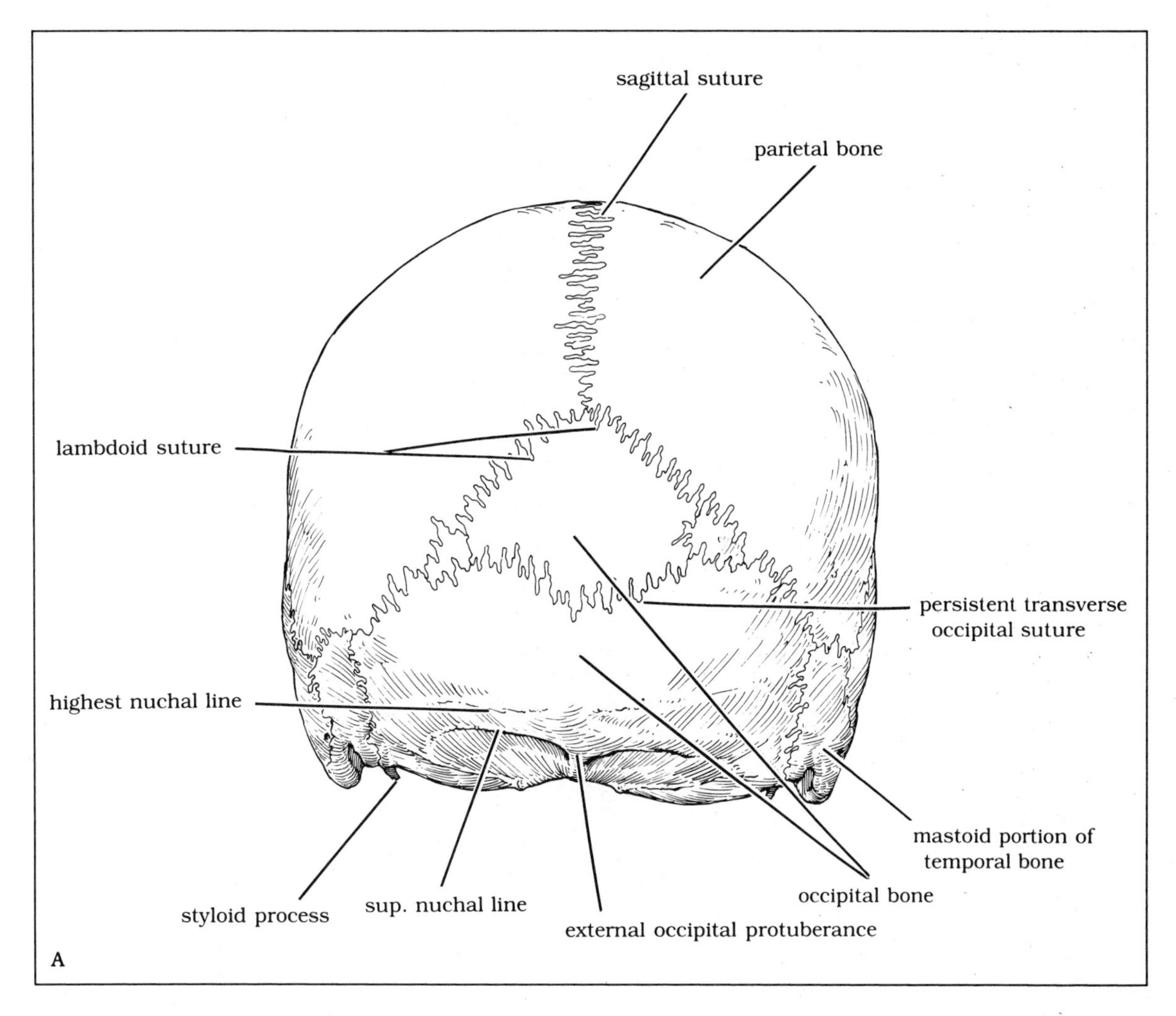

Fig. 7-2. Posterior aspect of the skull. A. Diagram. B. X-ray film. (1 = parietal bone; 2 = occipital bone; 3 = foramen magnum; 4 = nasal septum; 5 = mandible; 6 = greater wing of the sphenoid bone; 7 = mastoid air cells; 8 = coronal suture; 9= sagittal suture.)

The temporal and infratemporal fossae are separated from one another by the infratemporal crest on the greater wing of the sphenoid bone and by a ridge that continues posteriorly across the squamous portion of the temporal bone to the anterior root of the zygomatic process. (These fossae are covered in detail later in this chapter.)

The **temporal fossa** contains the temporalis muscle and its vessels and nerves, as well as the zygomaticotemporal nerve. The fossa is bounded by the temporal line superiorly, the zygomatic arch inferiorly, and the frontal and zygomatic bones anteriorly. The **temporal line** begins at the zygomatic process of the frontal bone and arches superiorly and posteriorly across the frontal and parietal bones, ending by joining the supramastoid crest of the temporal bone. Portions of the frontal and parietal bones, the greater wing of the sphenoid bone, and the squamous part of the temporal bones form the floor of the fossa. The

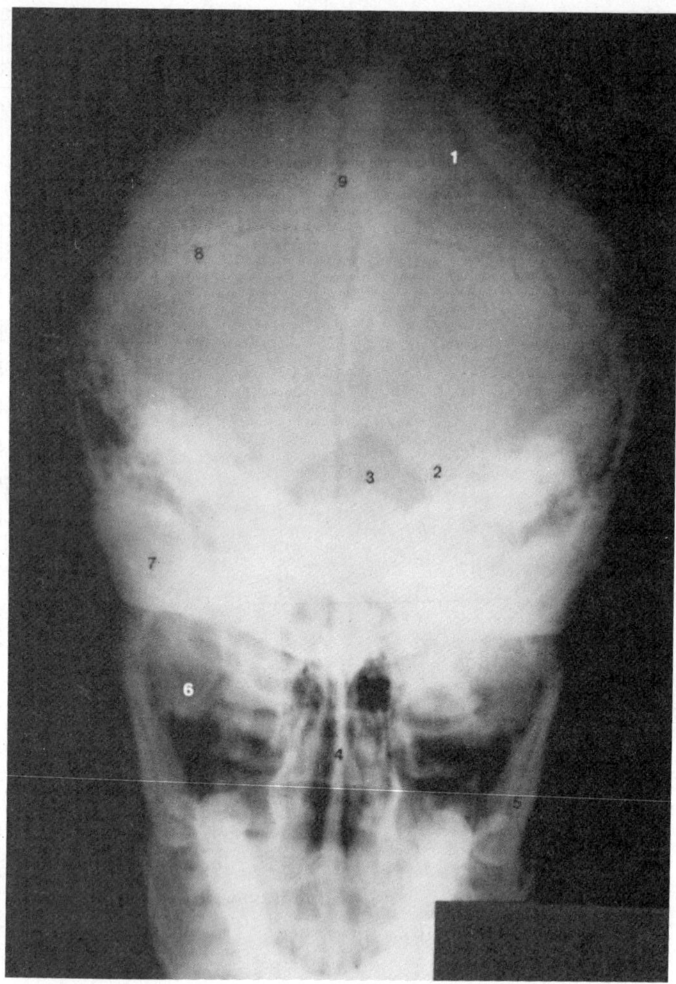

B

Fig. 7-2 (continued).

fossa communicates with the orbital cavity through the inferior orbital fissure.

The **infratemporal fossa** is found in the space between the zygomatic arch and the skull, posterior to the maxillae. It contains the lower portion of the temporalis muscle, the lateral pterygoid muscle, and a portion of the medial pterygoid muscle; the maxillary artery and its branches; the pterygoid venous plexus; and branches of the maxillary and mandibular nerves and the chorda tympani. The roof of the fossa is formed by the infratemporal surface of the greater wing of the sphenoid bone: its medial border is the lateral pterygoid plate of the sphenoid bone; its lateral margin is the ramus and coronoid process of the mandible. The foramen ovale, foramen spinosum, and alveolar canals open into the infratemporal fossa. The horizontal inferior orbital fissure, through which the fossa communicates with the orbit, and the vertical pterygomaxillary fissure meet at right angles at the superomedial angle.

The **pterygomaxillary fissure** is located be-

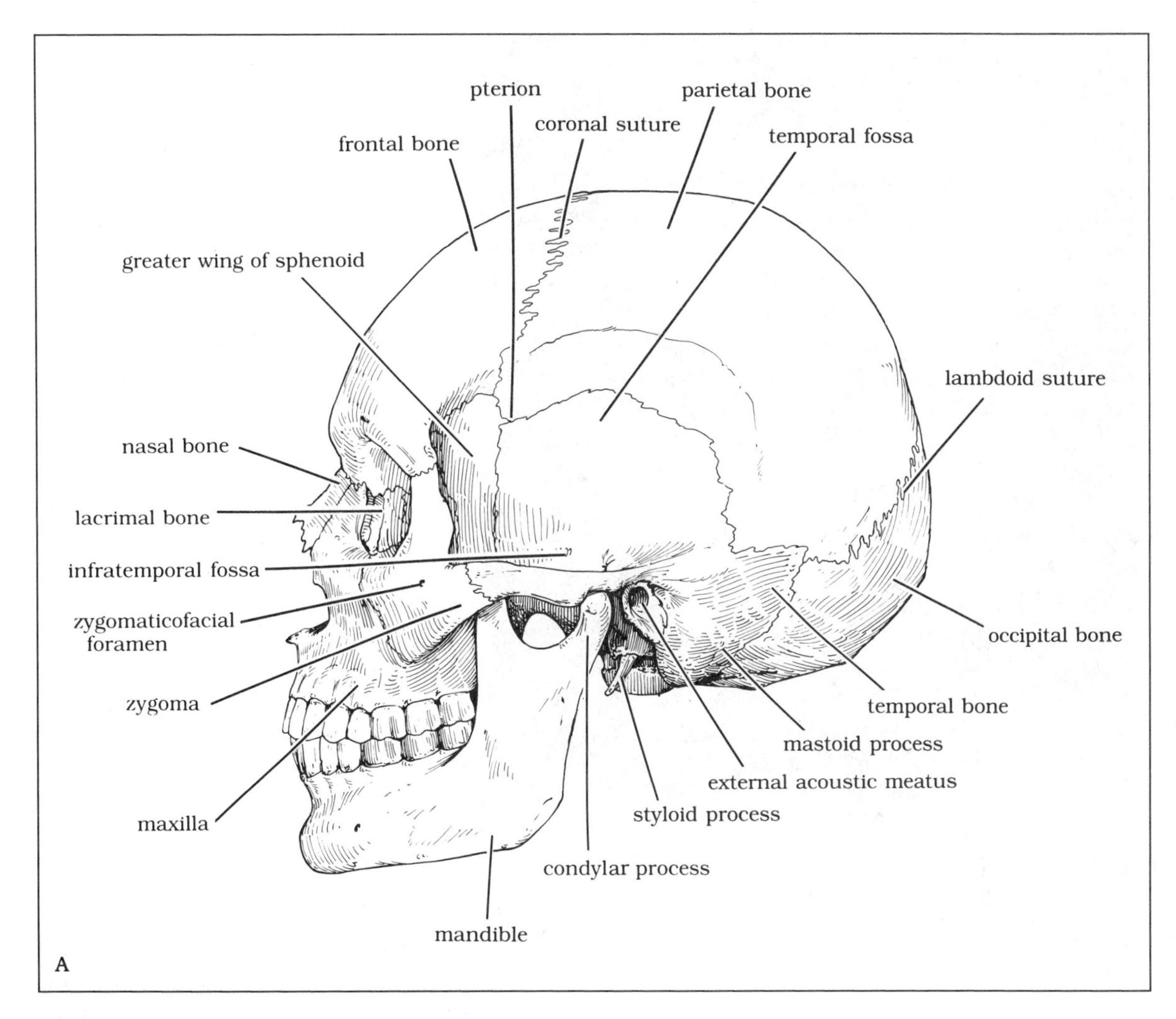

Fig. 7-3. Lateral aspect of the skull. A. Diagram. B. X-ray film. (1 = superior sagittal sinus; 2 = middle fossa; 3 = posterior fossa; 4 = atlas; 5 = axis; 6 = sella turcica; 7 = anterior fossa.)

tween the maxilla and the lateral pterygoid plate. It serves as the communication between the infratemporal and pterygopalatine fossae, and it carries the maxillary artery.

The **pterygopalatine fossa** is a small space at the junction of the inferior orbital and pterygomaxillary fissures below the apex of the orbit. It lies between the pterygoid plates of the sphenoid bone and the palatine bone and communicates with the orbit through the inferior orbital fissure and with the nasal cavity through the sphenopalatine foramen. The pterygopalatine fossa contains branches of the maxillary nerve, the pterygopalatine ganglion, and the terminal portion of the maxillary artery. Five foramina open into this fossa: posteriorly, the foramen rotundum, pterygoid canal, and pharyngeal canal; medially, the sphenopalatine foramen; and laterally, the pterygomaxillary fissure.

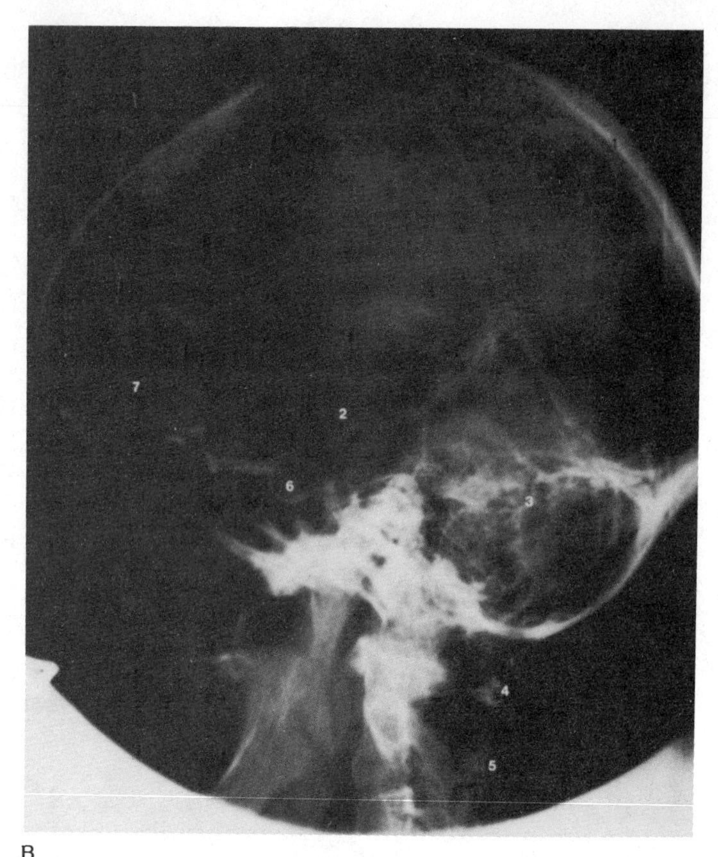

B
Fig. 7-3 (continued).

Communications of the Temporal and Infratemporal Fossae

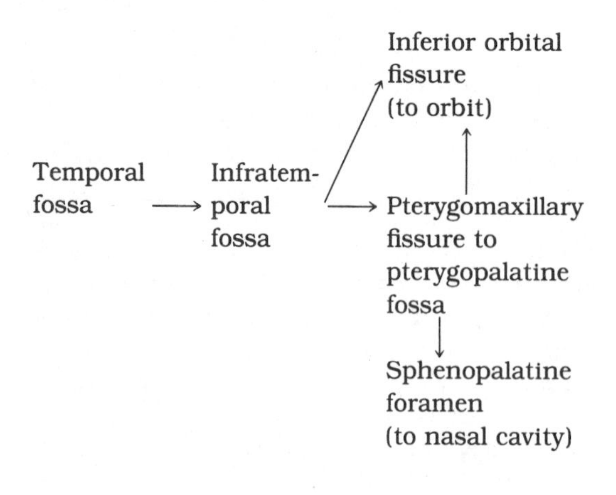

Anterior Aspect

The prominent structures on the anterior aspect of the skull are the forehead, bony orbits, cheeks, bony nose, and upper and lower jaws with their associated teeth (Fig. 7-4). Table 7-2 summarizes the openings seen in this aspect of the skull.

The **forehead** is formed by the frontal bone, which by six years of age has fused into a single bone. Inferiorly it articulates with the nasal bones at a junction called the **nasion.** The zone between the superciliary ridge (eyebrows) and above the nasion is the **glabella.**

The **orbits** are two bony cavities that contain the eyes and their muscles, nerves, blood vessels, glands, and adipose tissue. The cavity that forms the orbit is conical, narrowing posteriorly, and consists of a roof, a floor, medial and lateral

Table 7-1. Major Foramina, Canals, Fissures, and Fossae in the Lateral Aspect of the Skull

Opening	*Bone*	*Contents*
External acoustic meatus	Temporal	
Mastoid foramen	Temporal and occipital	Mastoid emissary vein
Zygomaticotemporal foramen	Zygomatic	Zygomaticotemporal nerve
Pterygomaxillary fissure	Sphenoid and maxillary	Maxillary artery, vein, and nerve branches
Pterygopalatine fossa	Sphenoid and palatine	Pterygopalatine ganglion
Inferior orbital fissure	Sphenoid, maxillary, and palatine	Maxillary nerve, branches from pterygopalatine ganglion, maxillary artery and vein
Posterior superior alveolar foramen	Maxillary	Posterior superior alveolar nerves and vessels
Temporal fossa	Temporal	Temporalis muscle
Infratemporal fossa	Temporal and sphenoid	Muscles of mastication, maxillary mandibular nerve, artery, and pterygoid venous plexus

Table 7-2. Foramina, Canals, Fissures, and Fossae in the Anterior Aspect of the Skull

Opening	*Bone*	*Contents*
Supraorbital foramen	Frontal	Supraorbital nerves (V_1) and vessels
Infraorbital foramen	Maxillary	Infraorbital nerves (V_2) and vessels
Zygomaticofacial foramen	Zygomatic	Zygomaticofacial nerve (V_2)
Optic canal	Sphenoid	Optic nerve, ophthalmic artery
Superior orbital fissure	Sphenoid and frontal	Oculomotor, trochlear, trigeminal (V_1) and abducens nerves; orbital branches of middle meningeal and ophthalmic veins
Anterior ethmoid foramen	Ethmoid	Anterior ethmoid nerve, artery, and vein
Posterior ethmoid foramen	Ethmoid	Posterior ethmoid nerve, artery, and vein
Zygomatico-orbital foramen	Zygomatic	Zygomaticotemporal and zygomaticofacial nerves (V_2)
Mental foramen	Mandibular	Mental nerves and vessels

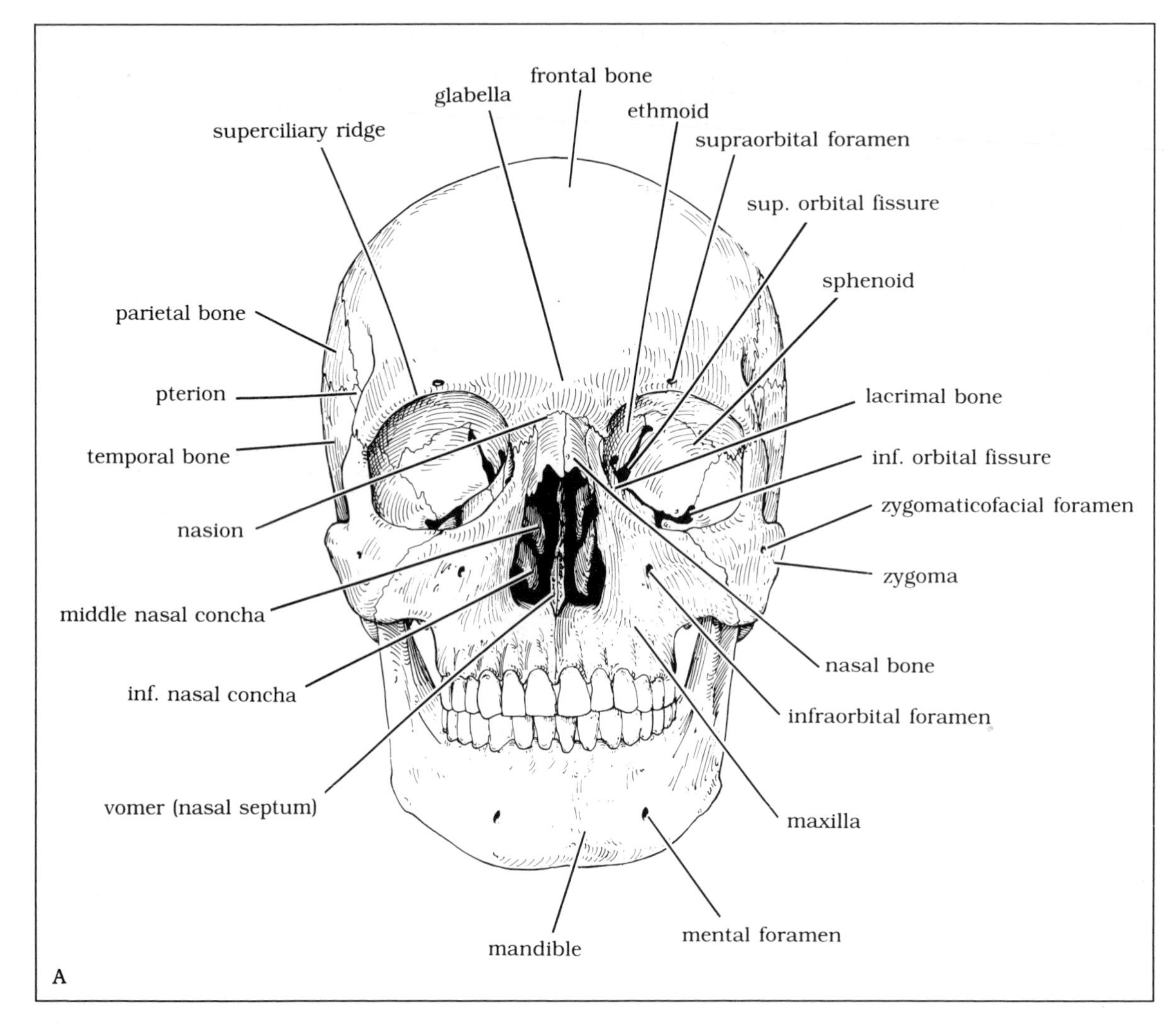

Fig. 7-4. Anterior aspect of the skull. A. Diagram. B. X-ray film. (1 = sagittal suture; 2 = frontal sinus; 3 = orbit; 4 = nasal septum; 5 = nasal conchae; 6 = maxilla; 7 = mandible; 8 = teeth; 9 = nasal bone; 10 = ethmoid air cells; 11 = maxillary sinus; 12 = supraorbital ridge; 13 = frontal bone.)

walls, a base, and an apex (Table 7-3) and contains nine openings that carry nerves and blood vessels in and out of the orbit (Table 7-4).

The roof of the orbit is formed anteriorly by the orbital plate of the frontal bone and posteriorly by the lesser wing of the sphenoid bone. The lacrimal fossa and the trochlear fossa are found medially.

The floor of the orbit is formed by the orbital surface of the maxilla and the orbital processes of the zygomatic and palatine bones. Medially, the lacrimal fossa is found. The **infraorbital groove,** which leads into the infraorbital canal, is seen in the middle of the floor—the infraorbital nerve and vessels are found in the canal.

The medial wall of the orbit is formed by the frontal process of the maxilla, the lacrimal bone, the orbital lamina of the ethmoid bone, and a

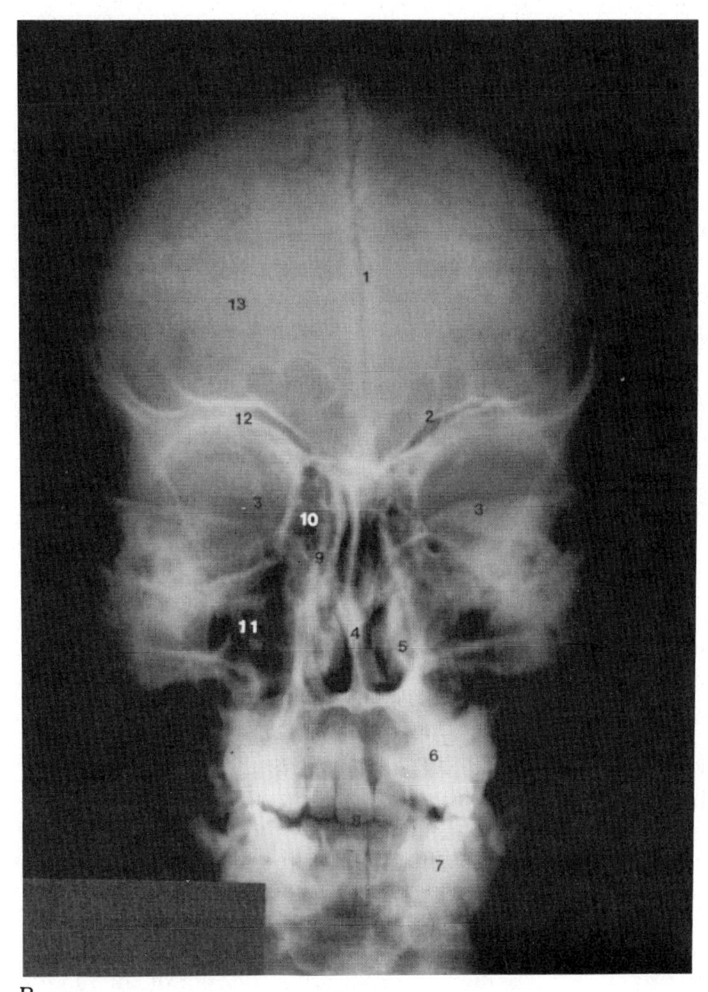

B

Fig. 7-4 (continued).

portion of the body of the ethmoid bone. Anteriorly the **lacrimal fossa** contains the lacrimal sac. Seven sutures are found in the medial wall (see Table 7-3). In the frontoethmoid suture are found the anterior ethmoidal foramen, which carries the anterior ethmoid branch of the nasociliary nerve and anterior ethmoidal blood vessels, and the posterior ethmoidal foramen, which carries the posterior ethmoidal nerve and blood vessels. The dura of the middle cranial cavity enters the orbit through the optic canal and divides into a periosteal portion, the **periorbita** (an outer layer), and an inner layer that covers the optic (second cranial) nerve and is continuous with the fascia in the eye.

The lateral wall of the orbit is formed by the orbital process of the zygoma and the orbital surface of the greater wing of the sphenoid bone. The **superior orbital fissure** is located between the roof and the lateral wall near the apex of the orbit. Through this fissure, the oculomotor (third cranial), trochlear (fourth cranial), ophthalmic division (V_1) of the trigeminal (fifth cranial), and abducens (sixth cranial) nerves and nerve fibers from the carotid plexus and orbital branches of the middle meningeal artery enter the orbit. The superior ophthalmic vein and a

Table 7-3. Bony Sutures in the Orbit

Region of Orbit	*Sutural Bones*
Roof	Frontal, lesser wing of sphenoid
Floor	Maxillae, zygomatic Maxillae, orbital process of palatine
Medial wall	Lacrimal, maxillae Lacrimal, ethmoid Sphenoid, ethmoid Frontal, maxillae Frontal, lacrimal Frontal, ethmoid Sphenoid, frontal
Lateral wall	Sphenoid, zygomatic

recurrent branch from the lacrimal artery leave the orbit through this fissure.

The base of the orbit is formed by the bones at the orbital margin, the zygoma and the maxilla. In the apex is found the optic canal, which carries the optic nerve and ophthalmic artery.

The prominence of the **cheek** is formed by the zygomatic bone, which is found at the inferior and lateral surface of the orbit and sits on the maxilla. The lateral portion of the zygomatic bone forms the prominent cheek bone of the face, its orbital portion forms the lateral surface of the orbit, and its temporal surface extends into the temporal fossa. The frontal process of the zygomatic bone articulates with the zygomatic process of the frontal bone, while the temporal process of the zygomatic bone articulates with the zygomatic process of the temporal bone. Laterally the zygomatic bone contains the zygomaticofacial foramen for passage of the zygomaticofacial nerve.

The bony part of the **external nose** is formed by the nasal bones and the maxillae and is bounded by the piriform apertures (anterior nasal openings in the skull). The apertures are formed superiorly by the nasal bones and laterally and inferiorly by the maxillae. The nasal cavity, which can be seen internal to the piriform apertures, is divided into right and left fossae by the nasal septum. The cartilaginous flexible portion of the external nose is anchored, in the piriform aperture, to the nasal septum. In the midline the lower margin of the piriform aperture contains the anterior nasal spine, a bony spur formed by the maxillae. The anterior portion of the nasal septum is formed by cartilage, while the posterior portions are formed by parts of the ethmoid and vomer bones. The lateral wall of the nasal cavity contains three or four plates of bones, the **conchae,** with spaces, the **meatuses,** beneath them.

The **jaws** consist of the maxillary (upper) and mandibular (lower) portions from each side. Each bone is united at the midline. The lower teeth are found in the alveolar part of the mandible, the upper teeth in the alveolar part of the maxilla.

Inferior (Basilar) Aspect

When the mandible is removed the inferior (basilar) aspect of the skull is seen to consist of the palatine processes of the maxilla, the palatine bones, the vomer, the sphenoid bone (inferior surfaces of the greater wings, the spinous process, and part of the body), the temporal bones (squamous, tympanic, and petrous portions), and the occipital bone (Fig. 7-5). The bony (hard) palate is considered below. All of the other bones in the inferior aspect of the skull are discussed separately later in this chapter.

The **bony palate** forms the roof of the mouth and the floor of the nasal cavity. Behind and above the bony palate are the posterior openings into the nasal fossae: the choanae. The bony palate is formed anteriorly by the palatine processes of the maxillae and posteriorly by the horizontal portions of the palatine bones. The bones are joined together by a cruciate suture. The palate's anterior and lateral boundaries are the alveolar processes of the maxillae containing the 16 teeth in the upper jaw. Depressions are seen in the palate for the palatine glands.

At the posterior and lateral margins of the palatine bones are the **greater palatine foramina** for the transmission of the greater palatine nerves and blood vessels (Table 7-5). Posterior to the

Table 7-4. Foramina, Canals, and Fissures of the Orbit

Regional Orbit	*Opening*	*Communication*	*Contents*
Roof (superior)	Optic canal	Anterior cranial fossa	Cranial nerve II, ophthalmic artery
Lateral wall	Superior orbital fissure	Middle cranial fossa	Cranial nerves III, IV, V_1, VI; nerve fibers from cavernous plexus; orbital branches of middle meningeal artery; lacrimal artery; superior ophthalmic vein
Floor (inferior)	Inferior orbital fissure	Infratemporal fossa, pterygopalatine fossa	Cranial nerve V_2, zygomatic nerves, infraorbital artery
	Zygomatico-orbital foramen	Zygomaticotemporal, zygomaticofacial	Zygomaticofacial nerve and vessels
Medial wall	Anterior ethmoid foramen	Anterior cranial fossa	Anterior ethmoid nerves and vessels
	Posterior ethmoid foramen	Anterior cranial fossa	Posterior ethmoid nerve and vessels
	Canal for nasolacrimal duct	Nasal cavity	Nasolacrimal duct

Table 7-5. Major Foramina, Canals, Fissures, and Fossae in the Inferior Aspect of the Skull

Opening	*Bone*	*Contents*
Incisive foramen	Maxillary	Nasopalatine nerve, greater palatine vessels
Greater and lesser palatine foramina and canals	Palatine and maxillary	Greater and lesser palatine nerves and vessels
Foramen ovale	Sphenoid	Cranial nerve V_3, accessory meningeal artery
Foramen spinosum	Sphenoid	Middle meningeal artery and meningeal ramus
Stylomastoid foramen	Temporal	Cranial nerve VII and stylomastoid artery
Mastoid foramen	Temporal and occipital	Emissary vein
Foramen lacerum	Sphenoid and temporal	Emissary vein, medial end of internal carotid artery
Carotid canal	Temporal	Internal carotid artery, carotid sympathetic plexus
Jugular foramen	Occipital and temporal	Internal jugular vein; cranial nerves IX, X, and XI
Hypoglossal canal (anterior condylar canal)	Occipital	Cranial nerve XII
Posterior condylar canal	Occipital	Emissary vein
Foramen magnum	Occipital	Medulla, spinal cord, vertebral arteries, cranial nerve XI, spinal root
Sphenopalatine foramen	Sphenoid and palatine	Sphenopalatine artery and vein Posterior superior nasal nerve
Pterygoid canal	Sphenoid	Vidian nerve (nerve of the pterygoid canal), pterygoid artery

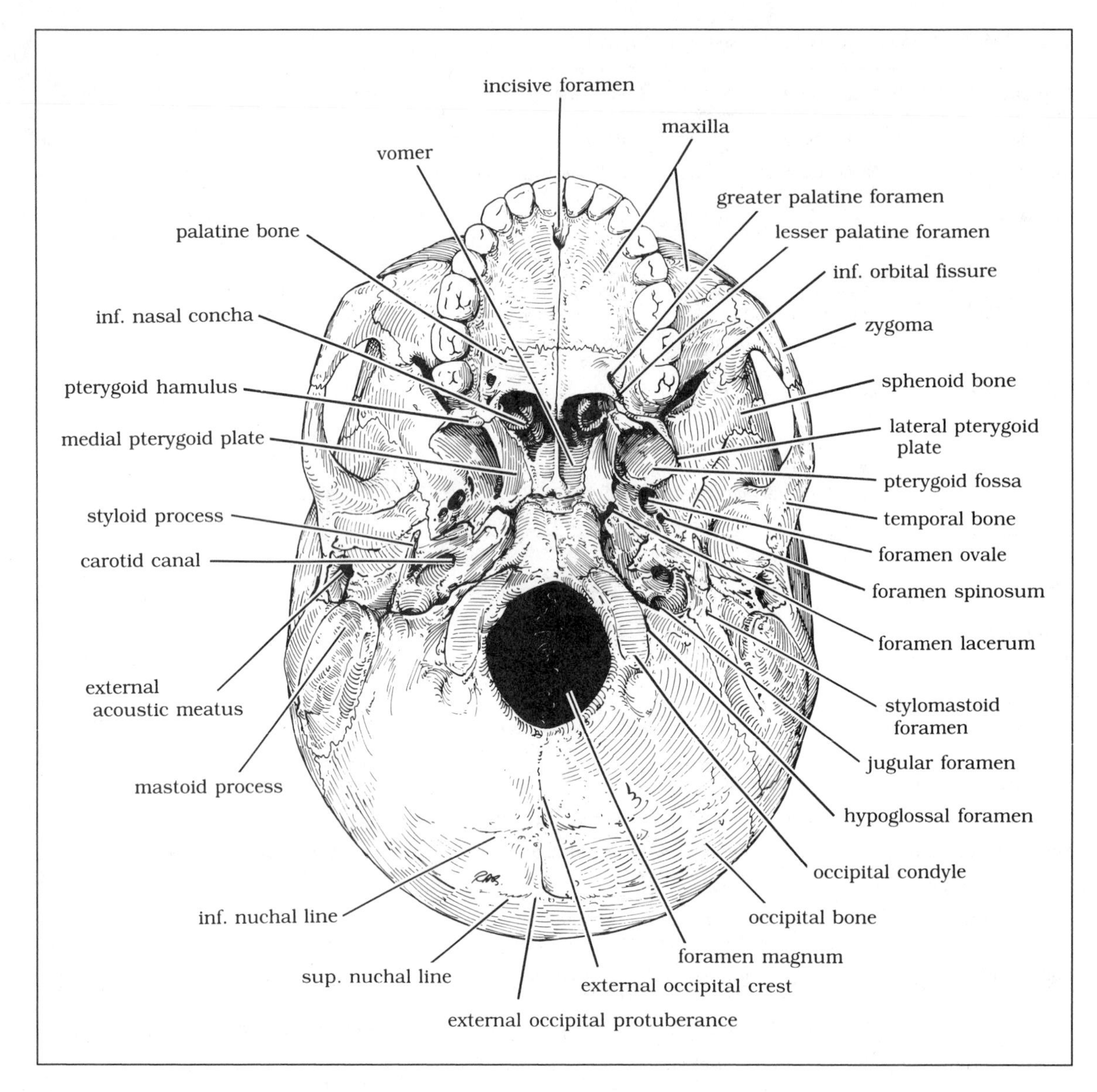

Fig. 7-5. External surface of the base of the skull.

greater foramina and in the pyramidal portion of the palatine bone are usually several smaller foramina, the **lesser palatine foramina,** that contain the lesser palatine nerves and blood vessels. Anteriorly, in the maxillary bone behind the incisor teeth, there is a depression, the **incisive fossa,** through which the nasopalatine nerves pass from the nasal cavity.

The posterior margin of the bony palate gives origin to the **soft palate**. In the midline the posterior nasal spine projects posteriorly from the palatine bone and provides the insertion for the musculature of the uvula.

INTERIOR ASPECT OF THE SKULL

The interior of the skull is best revealed by a section that removes the skull cap, or **calvaria.** The calvaria forms the roof of the cranial cavity; the floor is formed by the superior or inner surface of the base of the skull.

Calvaria

The internal aspect of the calvaria is marked in the midline by the sagittal suture between the parietal bones, anteriorly by the coronal suture between the frontal and parietal bones, and posteriorly by the lambdoid suture between the parietal and occipital bones. In some skulls indentations by the gyri of the cerebrum can also be seen.

In the midline is a longitudinal groove, the **sagittal groove,** that begins at the frontal crest and extends posteriorly; the superior sagittal sinus is found in this groove and the falx cerebri (the process of the dura mater separating the cerebral hemispheres) is attached at its margins. Adjacent to the sagittal groove, granular pits that contain the arachnoid granulations are found. Parietal foramina that carry the parietal emissary veins are also present. The inner aspect of the calvaria is also marked by the impressions created by branches of the meningeal vessels that are especially prominent in the parietal bone. In a cross section through the bone, spaces that in vivo contain the diploic veins are seen.

Floor of the Cranial Cavity

The floor of the cranial cavity is divided into three fossae: the anterior, middle, and posterior cranial fossae (Fig. 7-6 and Table 7-6). These fossae are organized stepwise, with the highest level being the anterior fossa and the lowest level being the posterior fossa.

ANTERIOR CRANIAL FOSSA

The floor of the anterior cranial fossa is formed by three bones: the orbital plate of the frontal bone, the cribriform plate of the ethmoid bone, and the lesser wing and anterior part of the body of the sphenoid bone.

The anterior fossa contains the frontal lobes of the cerebrum and the olfactory (first cranial) nerve and associated olfactory bulb and tract. The impressions of the base of the frontal lobe are evident on the orbital surface of the frontal bone. The **crista galli** is a midline process in the ethmoid bone. The frontal crest and the crista galli give attachment to the falx cerebri. The **foramen cecum** is a pit between the frontal crest and the crista galli; a vein connecting the nasal cavity to the superior sagittal sinus is usually found in this foramen. On the sides of the crista galli, in the cribriform plate of the ethmoid bone, the olfactory groove is pierced by 10 to 20 foramina transmitting filaments of the olfactory nerve from the nasal mucosa to the olfactory bulbs, which are found in the olfactory groove. Branches of the nasociliary nerve (anterior and posterior ethmoid nerves) are found rostrally on the cribriform plate. In the lateral walls of the olfactory groove are the internal openings for the anterior and posterior ethmoid foramina.

The orbital portion of the frontal bone articulates posteriorly with the lesser wing of the sphenoid bone. The lesser wing presents a sharp posterior border, the sphenoidal ridge, that forms one of the two steps in the cranial fossa. The **sphenoidal ridge** is a border above and overhanging the middle fossa. The medial border of the sphenoidal ridge is the anterior clinoid process, which gives attachment to the tentorium cerebelli, the process of dura mater that covers the cerebellum.

MIDDLE CRANIAL FOSSA

The middle cranial fossa is bordered anteriorly by the posterior surfaces of the lesser wings of the sphenoid bone, the anterior clinoid processes, and the ridges forming the margin of the chiasmatic groove; posteriorly by the superior margin of the petrous portion of the temporal bone and the dorsum sellae; and laterally by the squamous portion of the temporal bone, the sphenoidal angle of the parietal bone, and the greater wings of the sphenoid bone.

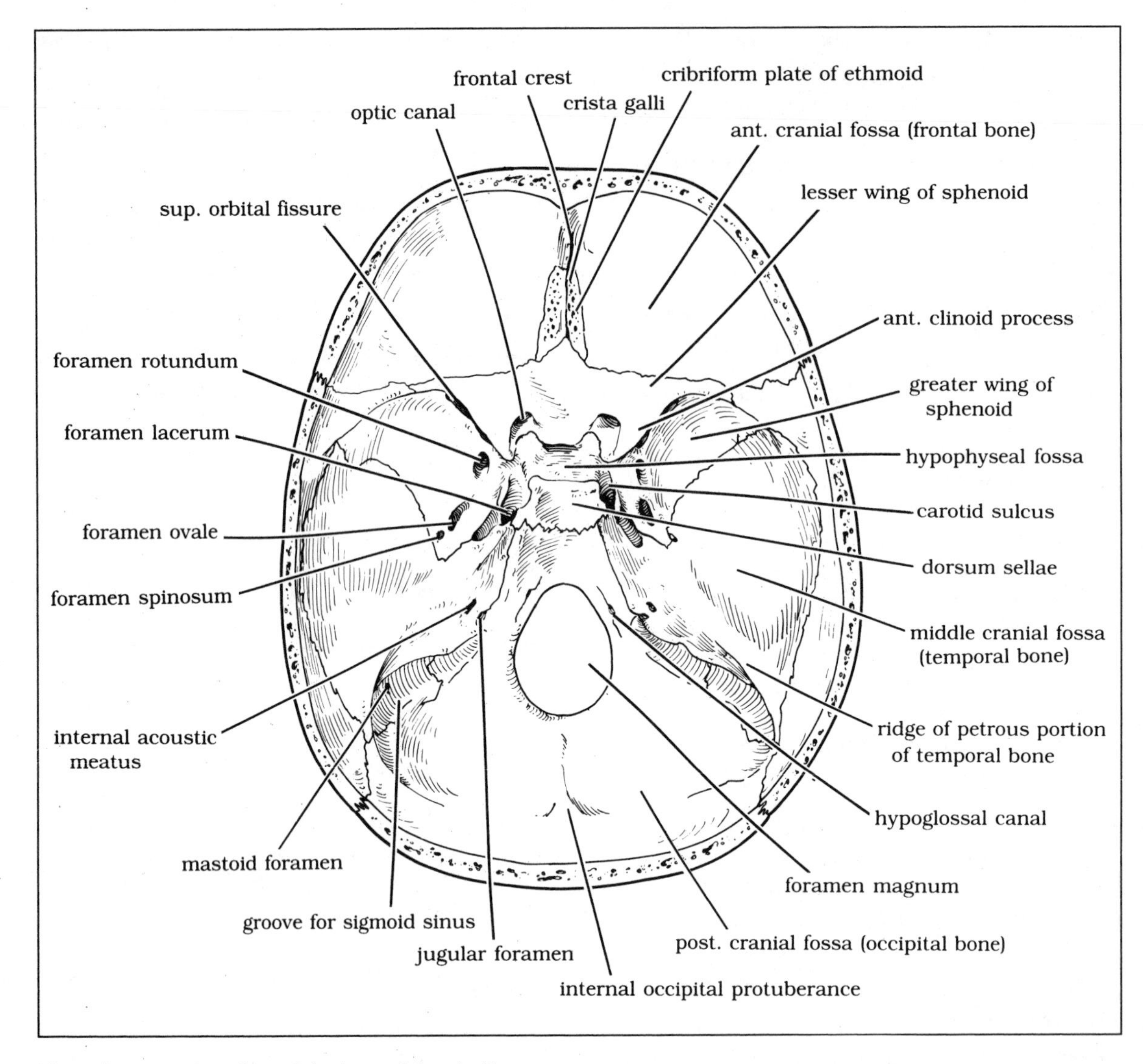

Fig. 7-6. Internal surface of the base of the skull.

The floor of the middle fossa consists of a smaller median and a larger lateral region. The superior portion of the body of the sphenoid bone forms the median region, which contains the pituitary gland in the tuberculum sellae and the chiasmatic groove. The limbus sphenoidalis limits the median region of the fossa anteriorly at the anterior margin of the chiasmatic groove. The chiasmatic groove continues laterally into the optic canal, which contains the optic nerve and ophthalmic artery. The optic canal is bounded by the body of the sphenoid bone and the roots of its lesser wing. The anterior clinoid processes are posterior to the optic canal. Just behind the chiasmatic groove is the sella turcica, which consists of a tubercle, the tuberculum sellae; the hypophyseal fossa, which contains the hypophysis; and posteriorly, a quadrilateral plate, the dorsum sellae, that contains the posterior clinoids laterally. The posterior clinoids receive the tentorium cerebelli. On the sides of the sella turcica are the carotid grooves, which begin at the foramen lacerum, run superiorly and anteriorly to lie medial to the anterior clinoids, and continue into the superior orbital fissure. The internal carotid and the cavernous sinus are found in the carotid groove. A groove for the abducens nerve is seen below the posterior clinoid process. Commonly a middle clinoid process arises from the lateral surface of the tuberculum sellae.

The lateral region of the middle cranial fossa is formed by the greater wing of the sphenoid bone and the squamous and petrous portions of the temporal bone. Its anterior limit is the sharp posterior border of the lesser wing of the sphenoid bone. Its posterior limit is the upper border of the petrous portion of the temporal bone. The anterior portion of the temporal lobe of the cerebral hemisphere is found in the anterior and lateral portions of the middle cranial fossa. Grooves for the anterior and posterior branches of the middle meningeal vessels are also seen.

The superior orbital fissure is located between the greater and lesser wings of the sphenoid; it transmits the oculomotor, trochlear, and abducens nerves and the ophthalmic branch of the trigeminal nerve. It thus connects the middle

Table 7-6. Major Foramina, Canals, Fissures, and Fossae in the Interior Aspect of the Skull

Opening	*Bone*	*Contents*
Anterior Cranial Fossa	Frontal, ethmoid, and sphenoid	Frontal lobes
Anterior and posterior ethmoid foramina	Ethmoid	Anterior and posterior ethmoid nerves and vessels
Foramina in cribriform palate	Ethmoid	Cranial nerve I rootlets (olfactory filaments)
Middle Cranial Fossa	Sphenoid, temporal, and parietal	Temporal lobe, pituitary gland
Optic canal	Sphenoid	Cranial nerve II, ophthalmic artery
Carotid canal	Sphenoid	Internal carotid artery and plexus
Foramen rotundum	Sphenoid	Cranial nerve V_2
Foramen ovale	Sphenoid	Cranial nerve V_3, accessory meningeal artery
Foramen spinosum	Sphenoid	Middle meningeal artery
Foramen lacerum	Sphenoid and temporal	Emissary vein, end of internal carotid artery
Posterior Cranial Fossa	Sphenoid, occipital, temporal, and parietal	Cerebellum, pons, medulla
Internal acoustic meatus	Temporal	Cranial nerves VII and VIII
Jugular foramen	Occipital and temporal	Internal jugular vein; cranial nerves IX, X and XI
Hypoglossal canal (anterior condylar canal)	Occipital	Cranial nerve XII
Posterior condylar canal	Occipital	Emissary vein
Foramen magnum	Occipital	Medulla, spinal cord, and vertebral artery

fossa with the orbit. The foramen rotundum (despite its name, usually oval), found below the medial end of the superior orbital fissure, transmits the maxillary nerve (the second division of the trigeminal) from the trigeminal ganglion to the pterygopalatine fossa. The foramen ovale is found posterior and lateral to the foramen rotundum; the mandibular branch (V_3) of the trigeminal nerve passes into the infratemporal fossa through this foramen. A sphenoidal emissary foramen, which provides a connection between the cavernous sinus and pterygoid venous plexus, is usually present medial to the foramen ovale.

The foramen spinosum, which carries the middle meningeal vessels, is located lateral and posterior to the foramen ovale. A groove for the vessels continues on to the floor of the fossa in the squamous and petrous portions of the temporal bone. The hiatus of the canal for the greater petrosal nerve is usually seen in the petrous portion of the temporal bone; the greater petrosal nerve and the petrosal branch of the middle meningeal artery are found here. Behind the hiatus on the anterior surface of the petrous portion of the temporal bone, a prominence, the arcuate eminence, is produced by the underlying anterior semicircular canal. The foramen lacerum (a ragged foramen, really a canal) lies medial to the foramen ovale. The internal carotid artery with its accompanying sympathetic plexus runs in the superior part of the foramen lacerum. The carotid canal opens medially into the foramen lacerum and the carotid artery crosses the foramen lacerum to enter the carotid groove on the sphenoid bone. A bony spur, the lingula, from the sphenoid bone is commonly seen in the lateral wall of the foramen lacerum. The deep petrosal and greater petrosal nerves are also found in the foramen lacerum; these nerves unite to form the nerve of the pterygoid canal.

POSTERIOR CRANIAL FOSSA

The posterior cranial fossa is the largest and deepest of the three cranial fossae and is entered by stepping down from the petrous ridge of the temporal bone. It is formed by the following bones: the posterior surface of the dorsum sellae and the clivus of the sphenoid bone, the occipital bone, the petrous and mastoid portions of the temporal bone, and the mastoid portion of the parietal bone.

The posterior fossa contains the cerebellum, pons, and medulla, and is separated from the overlying occipital lobes of the cerebrum by a dural fold, the **tentorium cerebelli.** The tentorium attaches anteriorly to the upper border of the petrous portion of the temporal bone and posterior clinoid processes and posteriorly and laterally to the transverse groove in the occipital bone.

The foramen magnum forms the lowest part of the posterior cranial fossa (see pp. 295–296). The jugular tubercle is seen in the anterior margin of the foramen and the hypoglossal canal is found below it carrying the hypoglossal (twelfth cranial) nerve. The jugular foramen is found lateral to the jugular tubercle, between the lateral part of the occipital bone and the petrous portion of the temporal bone. The internal acoustic meatus lies just superior to the jugular foramen. The facial (seventh cranial) and vestibulocochlear (eighth cranial) nerves and the internal auditory artery are transmitted through the internal acoustic meatus.

In front of the foramen magnum is seen a grooved region where the basilar portion of the occipital bone meets the body of the sphenoid bone and continues into the dorsum sellae above. This region is called the clivus, meaning "slope." The medulla and pons are found just posterior to the clivus. The superior margin of the petrous portion of the temporal bone receives the tentorium cerebelli and is grooved for the superior petrosal sinus.

Posterior to the foramen magnum a midline ridge, the internal occipital crest, leads to the internal occipital protuberance. The cerebellar fossae, containing the cerebellum, are separated by the internal occipital crest. The tentorium cerebelli extends laterally from the crest between the underlying cerebellar hemispheres and overlying occipital lobes. The internal occipital protuberance gives attachment to the falx cerebri, tentorium cerebelli, and falx cerebelli (a process of dura mater between the cerebellar hemispheres).

The posterior cranial fossa is bounded laterally by the deep grooves for the transverse sinus. These grooves run laterally from the internal occipital protuberance to the mastoid region of the parietal bone, to the mastoid region of the temporal bone, and then to the jugular process of the occipital bone, and end in the jugular foramen. The transverse sinus turns downward and becomes the sigmoid sinus over the mastoid region of the temporal bone. This sinus continues into the jugular foramen, where it forms the internal jugular vein.

Near the sigmoid groove, in the mastoid region of the temporal bone, vascular openings can commonly be found. The surfaces of the occipital and temporal bones are usually grooved by posterior meningeal vessels.

CRANIAL BONES

Each cranial bone consists of a thin inner table and a thicker outer table of compact bone joined by a middle spongy layer, the diploë.

Frontal Bone

The frontal bone has three portions: the squamous portion, which forms the convexity of the forehead and contains the **frontal eminence,** the **superciliary ridges** superior to the supraorbital margins, and the **frontal sinus;** the orbital portion, which forms the superior wall of the orbit and the floor of the anterior cranial fossa; and the nasal portion, which articulates with the nasal bones and the frontal process of the maxillae (see Figs. 7-3 and 7-4). The frontal bone also has two openings: the **supraorbital foramen,** or supraorbital notch, allows for the passage of the supraorbital nerve and artery; the **frontal** (supratrochlear) **notch** allows for the passage of the supratrochlear artery and the medial branch of the supraorbital nerve.

Parietal Bones

The parietal bones are paired and lie superiorly between the frontal and occipital bones, forming much of the roof and lateral walls of the cranium (see Figs. 7-1, 7-2, and 7-3). Each bone is a quadrilateral plate with the following margins: the sagittal margin, which forms the **sagittal suture** with the opposite sagittal margin; the frontal margin, which forms the **coronal suture** with the frontal bone; the occipital margin, which forms the **lambdoid suture** with the occipital bone; and the squamous margin, which articulates with the greater wing of the sphenoid bone and the squamous and petrous portions of the temporal bone.

On the external surface of the parietal bone are the **superior** and **inferior temporal lines,** which are the lines of insertion for the temporal fascia and origin for the temporalis muscle, respectively. The **parietal eminence,** the most convex part of the parietal bone, lies superior to the temporal lines. The internal surface is marked by a depression formed in vivo by the cerebral convolutions (gyri of the cerebrum).

Occipital Bone

The occipital bone forms the posterior portion of the base of the skull and is marked by several bony prominences and the foramen magnum (see Fig. 7-2). The occipital bone consists of four portions that fuse by the sixth year: the basilar, paired lateral, and squamous portions.

The basilar portion of the occipital bone forms the anterior boundary of the **foramen magnum,** a large ovoid opening with its largest diameter oriented anteroposteriorly (see Fig. 7-5). The posterior cranial fossa communicates with the vertebral canal and the brain connects with the spinal cord through this foramen. The following structures pass through the foramen magnum: the medulla and spinal cord, the spinal roots of the spinal accessory (eleventh cranial) nerve, the meningeal branches of the first to third cervical nerves, the vertebral arteries with the sympathetic plexuses, the anterior and posterior spinal arteries, the meninges, the alar ligaments of the dens, the cruciform ligaments of the atlas, the tectorial membranes of the posterior longitudinal ligaments, and the atlanto-occipital membrane.

On the inferior surface of the basilar portion of the occipital bone, in the midline about a centimeter anterior to the foramen magnum, is the **pharyngeal tubercle,** where the superior constrictor and pharyngeal raphe attach to the occipital bone. This tubercle marks the junction of the pharynx in front with the bones and muscles of the neck in back and inferiorly. The longus capitis and rectus capitis muscles insert in front of the tubercle.

The **foramen lacerum,** which carries the internal carotid artery, is found on the lateral margin of the basilar portion of the occipital bone between the greater wing and body of the sphenoid bone and the petrous portion of the temporal bone (see following discussion of the temporal bone).

The lateral portions of the occipital bone contain the **occipital condyles,** paired prominences on the sides of the foramen magnum that articulate with the lateral masses of the atlas so that the weight of the head is transmitted to the vertebral column. The articular surface is convex. The **condylar fossa** lies behind it and receives the superior facet of the atlas when the head is moved posteriorly with extension of the neck. A **condylar canal** may be found in the fossa, carrying emissary veins connecting the transverse sinus with the deep veins of the neck.

The **hypoglossal canal** is found at the anterior medial margin of the base of each condyle. The hypoglossal nerve and a meningeal branch from the ascending pharyngeal artery are found in this canal. The anterior and posterior **atlanto-occipital capsule,** from the lateral mass of the atlas, attaches to the margin of the foramen magnum and the alar tubercle on the medial surface of the condyle.

The **jugular process** extends laterally from the condyle to the temporal bone and has a concave anterior border, the **jugular notch,** that marks the posterior part of the jugular foramen (see discussion of temporal bone). The rectus capitis lateralis muscle inserts on the jugular process, and the transverse process of the atlas lies below the jugular process.

The squamous portion lies in the inferior and posterior aspect of the skull (see Fig. 7-2). The **superior nuchal line** extends from the external occipital protuberance and demarcates the inferior from the posterior aspect; this line gives attachment to the galea aponeurotica and the occipitalis, trapezius, sternocleidomastoid, and splenius capitis muscles. The **external occipital crest** extends anteriorly from the external protuberance to the foramen magnum; the nuchal ligament (ligamentum nuchae), from the external occipital protuberance to the superior processes of C7, attaches to this crest and helps to support the head. The **inferior nuchal lines** extend laterally from the external occipital crest. The convex bony part of the squamous portion between the superior and inferior nuchal lines is called the **planum occipitale,** and the semispinalis capitis and obliquus capitis superior muscles insert here medially and laterally, respectively. The part of the squamous portion below and anterior to the inferior nuchal line is the **planum nuchae,** and the rectus capitis posterior major and minor muscles insert here, laterally to medially.

Temporal Bones

Each temporal bone is situated at the side and base of the cranium and contains the vestibular and cochlear organs (see Fig. 7-3). The bone consists of five portions—squamous, tympanic, styloid, mastoid, and petrous—and articulates with five bones—the occipital, parietal, and sphenoid bones, the mandible, and the zygoma. The first four parts form prominent portions on the lateral aspect of the skull; all are visible on the inferior aspect of the skull.

The squamous portion of the temporal bone is thin and scalelike and forms the anterior and superior parts of the temporal bone. The parietal bone articulates inferiorly with the squamous portion. The **zygomatic process** arises from the squamous portion by two roots: The posterior root runs into the upper border of the zygoma above the external acoustic meatus and is continuous with the supramastoid crest and posterior glenoid tubercle; the anterior root is broad and continuous with the inferior border and ends as the **articular tubercle of the zygoma,** which is where the lateral temporomandibular ligaments attach. The zygomatic process extends forward from the squamous portion and articulates with the zygomatic bone, completing the **zygomatic arch.** The lower border of the cerebral hemispheres is found in the cranial cavity at a level with the zygomatic arch. The superficial layer of the temporal fascia is attached to the superior border of the arch, and the masseter originates from the medial and inferior border of the arch.

The head of the mandible is found in the **mandibular (glenoid) fossa** internal to the articular tubercle of the zygoma with the mouth closed, and resting on the articular tubercle with the mouth open. An articular disk is found between the condyle of the mandible and the mandibular fossa (see discussion of the temporomandibular joint). The margin of the mandibular fossa and the articular tubercle receives the attachment of the capsule for the temporomandibular joint.

Behind the mandibular fossa is the **external acoustic** (auditory) **meatus.** The most lateral portion of the meatus is cartilaginous and is not found within the skull (see discussion on the ear later in this chapter). The medial end of the acoustic meatus is separated from the tympanic cavity by the tympanic membrane in vivo, but in a skeletal preparation the tympanic membrane has been removed and the middle ear can be seen. The **supramental triangle,** or mastoid fossa, lies above and posterior to the external acoustic meatus and posterior root of the zygoma; the mastoid antrum of the temporal bone lies internal to the triangle.

The **tympanic portion** of the temporal bone is a curved plate that fuses with the mastoid and petrous portions and forms a sheath for the styloid process. Superiorly it forms the floor and anterior wall of the external acoustic meatus. Anteriorly it is separated from the mandible by the parotid gland. The **squamotympanic fissure** separates the tympanic portion from the squamous portion of the temporal bone and forms the anterior wall of the bony auditory canal. The squamotympanic fissure continues anteriorly as the petrosquamous fissure and posteriorly as the petrotympanic fissure. The chorda tympani exits the skull through the petrotympanic fissure.

The **styloid portion** of the temporal bone is a slender pointed projection of variable length directed downward and forward from the inferior surface of the temporal bone, under the tympanic plate. Two ligaments (stylohyoid and stylomandibular) and three muscles (styloglossus, stylohyoid, and stylopharyngeus) attach to this process. The hyoid bone is suspended from the skull by the stylohyoid ligaments, muscles, and other structures of each side. The parotid gland covers the process laterally. The styloid process develops from the second pharyngeal arch.

The **stylomastoid foramen** is located just posterior and lateral to the styloid process. The facial nerve exits the temporal bone through this foramen between the styloid and mastoid processes.

The **mastoid portion** of the temporal bone forms the posterior part of the temporal bone and consists of a roughened surface and an anterior process, the mastoid process. The mastoid portion in the adult contains several air spaces, the **mastoid air cells,** that communicate with the middle ear through the **mastoid antrum.** The occipital bellies of the occipitofrontalis and auricularis posterior muscles originate from its roughened surface.

The **mastoid processes** are in line with the foramen magnum and receive the most anterior portions of the insertion from the sternocleidomastoid, splenius capitis, and longissimus capitis muscles. Medial to the mastoid process is the **mastoid notch,** where the posterior belly of the digastric muscle arises. Just inferior and medial to the mastoid notch is the groove that carries the occipital artery. The anterior surface of the mastoid process is separated from the tympanic plate by the **tympanomastoid fissure,** which carries the auricular branch of the vagus (tenth cranial) nerve.

The **petrous portion** of the temporal bone contains the inner ear and portions of the middle ear. It is pyramid shaped, its apex extending medially and forward and articulating with the occipital and sphenoid bones. The **jugular foramen** is a large opening between the petrous portion of the temporal bone and the occipital bone; it is bounded anteriorly by bone and the carotid canal, laterally by the styloid process, medially by the hypoglossal canal, and posteriorly by the transverse process of the atlas. The jugular foramen can be divided into three portions: an anterior compartment containing the inferior petrosal sinus; an intermediate portion containing rootlets for the glossopharyngeal (ninth cranial), vagus, and spinal accessory nerves; and a posterior portion containing the sigmoid sinus and meningeal branches of the occipital and ascending pharyngeal arteries.

The **jugular fossa** is found medial to the styloid process. Several small openings are seen in the jugular fossa: laterally, the mastoid canal carries the auricular branch of the vagus nerve, and anteriorly, the tympanic canal carries the tympanic nerve from the glossopharyngeal nerve.

The **carotid canal** in the petrous portion of the temporal bone is found close to the internal ear and contains the internal carotid artery with its sympathetic plexus. This artery enters the skull anterior to the exit of the internal jugular vein. The quadrilateral area of the petrous portion of the temporal bone lies between the carotid canal and the foramen lacerum. The levator veli palatini muscle originates from the quadrilateral area. The cartilaginous portion of the auditory tube is located in a groove between the quadrilateral area and the greater wing of the sphenoid bone.

The **foramen lacerum** is an opening between the petrous portion of the temporal bone, the body and the greater wing of the sphenoid bone, and the basilar portion of the occipital bone. The internal carotid artery runs in the superior part of the foramen lacerum after emerging from the carotid canal. The inferior part of the foramen is covered with cartilage, and running through it are the small nerve of the pterygoid canal and a small meningeal branch of the pharyngeal artery with a small emissary vein connecting the cavernous sinus to the pterygoid plexus. The **pterygoid canal** (vidian canal) passes from the anterior margin of the foramen lacerum to the pterygopalatine fossa and carries an artery and nerve. The artery in the pterygoid canal arises from the pterygopalatine portion of the maxillary artery; the nerve is formed by the union of the greater and deep petrosal nerves at the foramen lacerum and also receives a small branch from the otic ganglion. The nerve exits the canal, crosses the pterygopalatine fossa, and enters the pterygopalatine ganglion.

Sphenoid Bone

The sphenoid bone is found at the base of the skull and articulates with four bones at the sphenozygomatic, sphenofrontal, sphenoparietal, and sphenosquamosal (sphenotemporal) sutures. It joins in the formation of the floor of the anterior, middle, and posterior cranial fossae; the temporal and infratemporal fossae; the nasal cavity; and the orbit. It is irregular in shape and consists of a central part, the body; two lateral expansions, the greater and lesser wings; and the inferiorly placed pterygoid processes. The body is cuboidal and hollow, containing the sphenoid sinus. The superior surface of the body has a spine at its anterior end, the sphenoid crest, and also contains the **sella turcica,** a depression surrounding the **hypophyseal fossa,** which in vivo contains the hypophysis cerebri. The posterior boundary of the sella is formed by the **dorsum sellae,** a quadrilateral plate that is continuous with the basilar part of the occipital bone and supports the pons and the basilar artery. The superior angles of the dorsum sellae contain the **posterior clinoid processes.**

The **lesser wings of the sphenoid bone** are two thin triangular plates of bone that extend horizontally and laterally from the anterosuperior part of the body. The lesser wings arise by thin anterior and thick posterior roots; between the roots is the optic canal.

The **greater wings of the sphenoid bone** arise from the inferior lateral surface of the body and lie between the squamous and petrous portions of the temporal bones. They form the roofs of the infratemporal fossae and are lateral to the lateral pterygoid processes. Anteriorly the greater wing is limited by the inferior orbital fissure. The inner (cerebral) surface of the greater wing is smooth and concave and enters into the formation of the middle cranial fossa. The **foramen rotundum** (for the maxillary nerve), the large **foramen ovale** (for the mandibular nerve and emissary vein from the cavernous sinus), and the **foramen spinosum** (for the middle meningeal artery and meningeal branch of the mandibular nerve) are seen medially and laterally in the greater wing. These foramen are found from medial to lateral with the rotundum most medial, the ovale next, and the laterally placed spinosum. Commonly, a small aperture, the **sphenoidal emissary foramen,** is found anteromedial to the foramen ovale; the sphenoidal emissary vein runs in this foramen, connecting the cavernous sinus and pterygoid venous plexus.

The **spine of the sphenoid bone** is found on the greater wing behind the foramen spinosum; the tensor veli palatini muscle and the sphenomandibular and pterygospinous ligaments attach to this bony spine. The auriculotemporal nerve is found lateral to the spine, and the chorda tympani is seen medially.

The cartilaginous portion of the auditory tube is found in a groove medial to the foramina ovale and spinosum. The mandibular nerve and the middle meningeal arteries lie lateral to the tube at the base of the external surface of the skull.

The **pterygoid processes** descend from the junction of the greater wings and body of the sphenoid bone. Each pterygoid process, one on each side, consists of two plates or laminae, the medial and lateral pterygoid plates, that contain the pterygoid fossae between them.

The **lateral pterygoid plate** gives origin on its medial and lateral surfaces to portions of the lateral and medial pterygoid muscles, respectively, and forms the medial boundary of the infratemporal fossa. The **medial pterygoid plate** contains at its posterior border the pterygoid hamulus, which anchors the pterygomandibular ligament. The **scaphoid fossa** is found on the lateral side at the base of the medial pterygoid plate. The tensor veli palatini muscle originates in this fossa and its tendon hooks around the pterygoid hamulus.

Ethmoid Bone

The ethmoid bone is cuboid in shape and found at the anterior part of the base of the cranium, where it contributes to the floor of the anterior cranial fossa and walls of the orbits and nasal cavity. It consists of a cribriform plate, a perpendicular plate, and the ethmoid labyrinth. The **cribriform plate** is found in the median plane and contains on its superior surface a thick process, the **crista galli,** which gives origin to the falx cerebri. On the sides of the crista galli, the cribriform plate is grooved to contain the olfactory bulbs and is pierced by many (approximately 20) foramina, which contain the olfactory nerve filaments, which end in the olfactory bulbs. The **perpendicular plate** is found at right angles to the cribriform plate and forms the superior and posterior third of the nasal septum. The **ethmoid labyrinth** is suspended from the lateral part of the cribriform plate and contains the superior and middle nasal conchae (which form a portion of the lateral wall of the nasal cavity) and the ethmoid air cells.

FACIAL BONES

The facial bones are the maxillae, mandible, zygomas, nasal bones, palatine bones, lacrimal bones, vomer, and inferior nasal conchae.

Maxillae

The maxillae, which constitute the upper jaw, are formed by the fusion of two bones that are united at the midline at the intermaxillary suture. Each maxilla consists of a body and four maxillary processes that extend from its body. The body is pyramidal in shape and contains the large maxillary sinus. It presents a nasal surface or base that forms some of the inferior lateral wall of the nasal cavity, an orbital surface that forms the bulk of the floor of the orbit, an infratemporal surface that forms the anterior wall of the infratemporal fossa, and an anterior surface that is covered by some of the facial muscles. On the anterior surface of the maxilla, below the orbital surface, is the **infraorbital foramen,** which carries the infraorbital nerve and artery.

The four maxillary processes are the frontal, zygomatic, palatine, and alveolar processes. The **frontal process** extends superiorly and articulates with the nasal part of the frontal bone. Its lateral surface serves as an attachment site for the orbicularis oculi muscle, the medial palpebral ligament, and the levator labii superioris alaeque nasi muscle. The **zygomatic process** extends superolaterally from the body and articulates with the zygomatic bone. The **palatine process** projects transversely from the body and articulates with the palatine process of the opposite side to form the anterior three-fourths of the bony palate. A large pit, the **incisive fossa,** containing the incisive foramen, is found in the midline where the palatine processes of the maxillae articulate. The **alveolar processes** extend from the inferior surface of the body and contain eight sockets on each side for the upper teeth.

Mandible

The mandible, the lower jaw, is the largest and strongest bone in the face (Fig. 7-7). It articulates with the skull in the mandibular fossa of the squamous part of the temporal bone. The mandible consists of a single body and paired rami.

The body is curved and horseshoe shaped. Its external surface has a faint median ridge, the **symphysis menti** (symphysis mandibulae), indicating where the fusion of the two halves of the body occurred (fusion usually occurs in the second year). Inferior to the symphysis the body expands into the prominent **mental protuberance,** which is limited laterally by the mental tubercles. The incisive fossae are found on either side of the symphysis, the mentalis muscle and a portion of the orbicularis oris muscle originating here. The **mental foramen,** where the mental branch of the trigeminal nerve and the mental vessels exit, lies lateral to the protuberance, usually inferior to the second premolar tooth. The **oblique line** is a ridge extending laterally and superiorly from the mental tubercle to the anterior border of the ramus. The depressor labii inferioris and depressor anguli oris muscles attach to this line; the buccinator muscle originates superior to the upper end of the oblique line from the first to the third molar. The upper surface of the body, the alveolar part, contains the alveolar arch with its hollowed out 16 dental sockets and in vivo is covered on the anterior and posterior surfaces by the oral mucous membrane. The lower edge of the mandible is termed the base; the platysma muscle attaches to here. At the point where the base of the mandible joins the ramus, the groove for the facial artery is found.

On the internal surface of the symphysis the **mental spines** are seen. The genioglossus muscle originates from the superior spines, and the geniohyoid muscle from the inferior spines. Adjacent to the midline on both sides is a roughened depression, the **digastric fossa,** that serves as the site of attachment of the anterior belly of the digastric muscle. An oblique ridge called the **mylohyoid line** is seen extending posteriorly from above the digastric fossa toward the third premolar. The mylohyoid muscle originates from this line. Below the mylohyoid line is a fossa for the submandibular gland; the fossa for the sublingual gland lies just lateral to the mental spines.

The **ramus** of the mandible is quadrilateral in shape. Its lateral surface is flat and serves as the place of insertion of the masseter muscle. Inferiorly there are several oblique **masseteric tuberosities,** ridges that receive insertions from some of the masseter tendons. The medial surface of the mandible, near the middle, contains the **mandibular foramen,** which leads into the **mandibular canal.** The canal traverses the bone and ends up at the mental foramen on the anterior surface; it contains the blood vessels and nerves for the lower teeth, which enter the teeth through small openings. The **pterygoid tuberosity** is found inferior to the mandibular foramen; the medial pterygoid muscle inserts here. Above and in front of this tuberosity is the **mylohyoid groove,** which ends at the mandibular foramen and contains the mylohyoid vessels and nerves.

The inferior border of the ramus forms the angle of the jaw and continues on to the posterior border. The posterior border is thick and

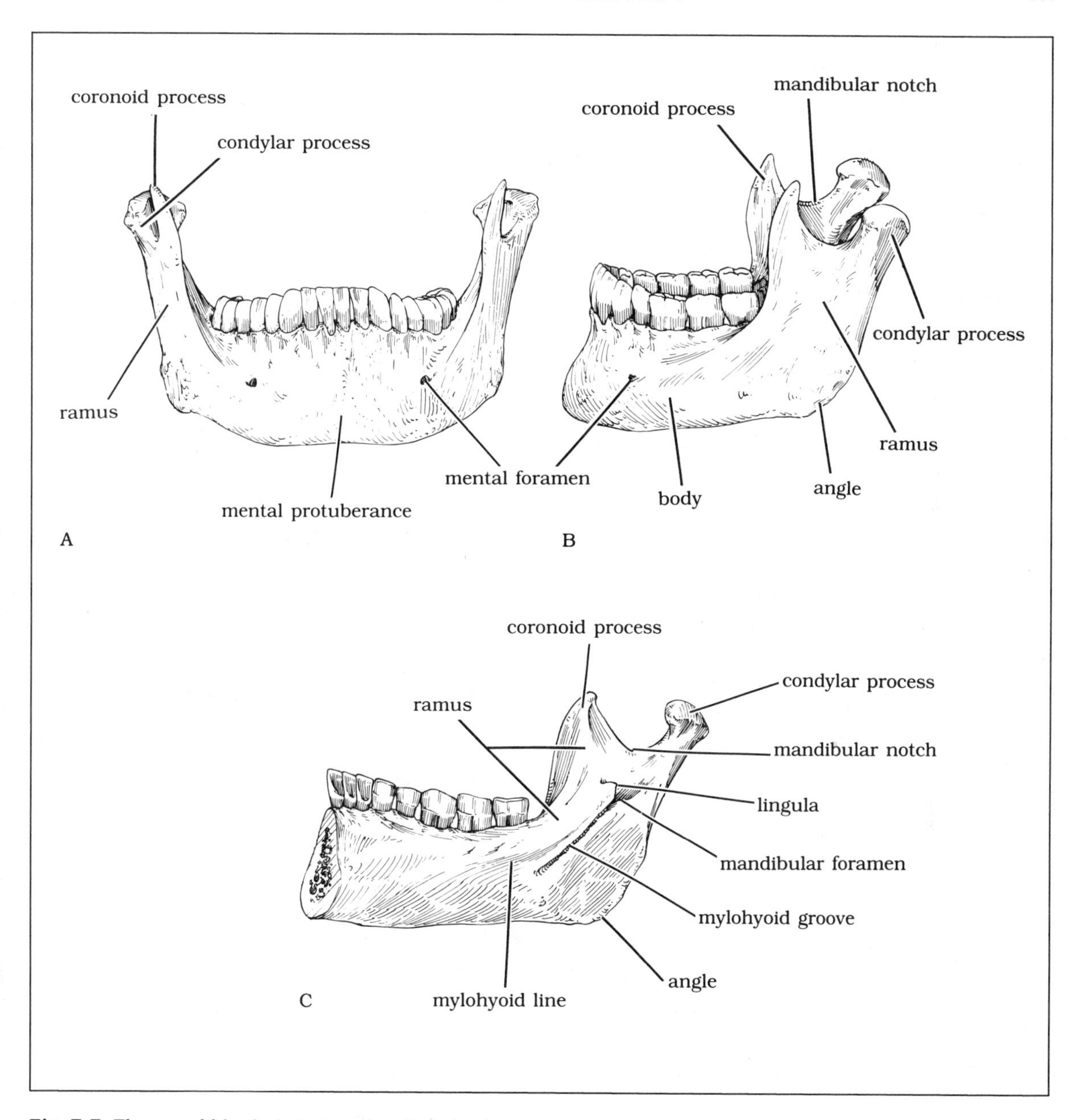

Fig. 7-7. The mandible. A. Anterior view. B. Lateral view. C. Medial view.

rounded and covered in vivo by the parotid gland. The anterior border is continuous with the body. The superior border contains two processes—the anterior coronoid process and the posterior condylar process—separated by the deep mandibular notch.

The **coronoid process** is flattened; the temporalis muscles insert in part on its lateral surface, but primarily on its medial surface. The inner aspect of the coronoid process continues inferiorly as a ridge to the inner side of the last molar; the temporalis muscles insert along this ridge.

The **condylar process** consists of a head and a narrow neck. The head is covered by fibrocartilage in vivo and articulates with the articular disk between the mandible and the temporal bone (within the temporomandibular joint). The neck is flattened and has a depression anteriorly, the **pterygoid depression,** for insertion of the lateral pterygoid muscles. At the lateral end of the condyle is a tubercle for attachment of the lateral temporomandibular ligament.

The **mandibular notch** separates the coronoid and condylar processes and allows the passage of the masseteric nerves and blood vessels.

Zygomas

The paired zygomatic bones form the prominence of the cheek and articulate between the zygomatic processes of the temporal bone at the zygomaticotemporal suture, the frontal bone at the zygomaticofrontal suture, and the maxillae. The zygoma is quadrilateral, with frontal and temporal processes. The zygomatic muscles originate on its external surface. The lateral surface of the zygoma is convex and contains the **zygomaticofacial foramen** for the zygomaticofacial branch of the trigeminal nerve.

Nasal Bones

The nasal bones consist of two small and oblong bones in the superior part of the face. Each articulates with its opposite nasal bone at the internasal suture to form the dorsum of the nose. The superior border articulates with the nasal part of the frontal bone, while the inferior border receives the nasal cartilage. The lateral border articulates with the frontal process of the maxilla.

Palatine Bones

The L-shaped palatine bones form the posterior portion of the bony palate (the horizontal plate, which also serves as the boundary of the choanae, the posterior nasal apertures), the lateral wall of the nasal cavity, and the posterior inferior wall of the orbit (the perpendicular plate).

Lacrimal Bones

The lacrimal bones are the smallest facial bones and are thin and delicate. Each forms a portion of the anterior medial wall of the orbit. The lacrimal bone is quadrilateral in shape, and its anterior surface lodges the lacrimal groove, which contains the lacrimal sac.

Vomer

The vomer is found in the midline. It fuses anteriorly with the ethmoid bone and inferiorly with the maxillae and the palatine bones. Its anterior portion forms the posterior inferior portion of the nasal septum, separating the two openings in the nasal cavity; the **nasopalatine groove** for the nasopalatine nerve and blood vessels is seen here. (The ethmoid bone forms the anterior superior portion of the nasal septum.) Two lateral expansions, the **alae of the vomer**, articulate with the medial pterygoid processes of the sphenoid bone laterally and with the medial pterygoid plates just below the body of the sphenoid bone superiorly.

Inferior Nasal Conchae

The paired inferior nasal conchae are slender and covered with mucous membrane. They attach superiorly to the lateral wall of the nasal cavity, separating the inferior meatus from the middle meatus of the nose.

Scalp and Face

SCALP

The scalp begins at the supraorbital ridges, continues posteriorly to the external occipital protuberance and laterally to the temporal lines, and is continuous with the fascia of the temporalis muscles. Five layers form the scalp:

1. **S**kin The skin on the scalp is similar to skin elsewhere on the body, but may have more hair.
2. **C**onnective tissue The dense connective tissue of the scalp contains nerves, arteries, and veins. When the scalp is cut, this layer tends to hold the wound closed, but the vessels bleed heavily.
3. **A**poneurosis The galea aponeurotica is a tendon between the frontalis, auricularis superior, and occipitalis muscles that cover the vault of the skull. This tendon is attached posteriorly to the external occipital protuberance, superiorly to the nuchal line, laterally over the temporalis fascia to the zygomatic arch, and anteriorly to the fascia of the frontalis muscle.
4. **L**oose connective tissue The subaponeurotic tissue separates the galea aponeurotica from the periosteum of the skull, the pericranium.
5. **P**ericranium The pericranium is the external periosteum of the skull.

The skin layer and the dense connective tissue layer over the temple are the same as those over the scalp. The aponeurosis is thinner, with a tough fascia, the temporalis fascia, covering the temporalis muscle and joining it to the pericranium. The pericranium, temporalis muscle, and underlying bone are supplied by the same vessels.

Musculature

The **occipitofrontalis muscle** is the main muscle in the scalp (see Fig. 7-8). It arises by two frontal (frontalis) and two occipital (occipitalis) bellies. The frontalis arises from the galea aponeurotica and inserts in adjacent muscles and skin in the nose and eyebrow. The occipitalis arises from the lateral portion of the highest nuchal line on the occipital bone and from the mastoid portion of the temporal bone and inserts in the galea aponeurotica. The occipitofrontalis muscle moves the scalp forward and backward and elevates the eyebrows. Branches of the facial nerve innervate this muscle. The occipitofrontalis muscle, a small temporoparietal muscle, and the galea aponeurotica constitute the epicranius muscle.

Arterial Supply

The arterial supply to the scalp comes from the internal carotid (supratrochlear and supraorbital branches) and the external carotid (superficial temporal, posterior auricular, and occipital branches) arteries. The **supratrochlear** and **supraorbital arteries** are branches of the ophthalmic artery and distribute together to the forehead, anastomosing with each other and the same arteries from the opposite side.

The **superficial temporal artery** is a terminal branch of the external carotid and supplies the parotid region, external ear, and face (see Fig. 7-33). It crosses the zygomatic process of the temporal bone and divides into parietal and temporal branches, which freely anastomose with the other arteries. The auriculotemporal nerves run with this artery. The **posterior auricular artery** (see Fig. 7-33) branches off the external carotid in the upper part of the neck and runs under the parotid gland with the posterior auricular nerve to the scalp posterior to the ear. The **occipital artery** (see Fig. 7-33) supplies the posterior part of the scalp after running deep to the sternocleidomastoid muscle; as it runs through the trapezius muscle it divides into many branches on the scalp. Posteriorly, it is accompanied by the greater occipital nerve. The occipital artery also gives off a branch that anastomoses with branches of the subclavian artery and two branches that supply the meninges via the mastoid and parietal foramina.

Venous Drainage

The veins of the scalp are named according to their accompanying arteries and drain the area that their artery supplies, with the exception of the ophthalmic vein. (The ophthalmic vein has no valves and serves to connect the cavernous sinus and facial veins in either direction.) The supratrochlear and supraorbital veins join to form the facial vein (see Fig. 7-40). The superficial temporal vein drains the side of the scalp and joins the maxillary vein to form the retromandibular vein. The posterior auricular vein drains the region behind the ear and joins the inferior portions of the retromandibular vein to form the beginning of the external jugular vein. The occipital vein drains much of the posterior parietal and occipital region and joins veins in the suboccipital triangle. Emissary veins exit through the mastoid and parietal foramina, connecting to the dural sinuses.

Innervation

The nerves of the scalp arise inferiorly and are found in the dense connective tissue layer. The trigeminal and cervical nerves supply the scalp from the forehead to the ear. **Trigeminal branches** to the scalp include the following.

The ophthalmic branch gives rise to the supratrochlear nerve (to the front of the forehead) and the supraorbital nerve (to the vertex of the scalp).

The maxillary branch gives rise to the zygomaticotemporal branch to the temple.

The mandibular branch gives rise to the auriculotemporal nerve to the external ear, external auditory meatus, tympanic membrane, and temple.

Cervical nerves supplying the scalp include two branches of the cervical plexus, which is formed by the primary ventral rami of nerves C1 to C4, and the greater occipital nerve and the third occipital nerve, which are derived from the dorsal rami of nerves C2 and C3, respectively.

The lesser occipital nerve (ventral ramus of C2) supplies the superior part of the external ear and the skin above and behind this region.

The great auricular nerve (ventral rami of C2 and C3) supplies the posterior surface of the external ear and the skin posterior to it.

The greater occipital nerve (dorsal ramus of C2) supplies the posterior scalp up to the vertex, to meet the supraorbital and auriculotemporal nerves.

The third occipital nerve (dorsal ramus of C3) supplies the skin over the external occipital protuberance.

FACE

The face is the area inferior to the supraorbital ridges continuing down to the lower edge of the jaw and posteriorly to the ear. The anterior, or central, part of the face consists of the nose, eyes, and mouth, and most movement in the face is associated with these areas and the lower jaws and the ears. The area in the face bounded by the mouth, eyes, and ear contains the masseter muscle and the parotid gland. Below this muscle and gland is the buccal fat pad. The parotid duct from the parotid gland to the mouth crosses the masseter and enters the oral cavity near the upper second molar. Five branches of the facial nerve originate deep to the gland and cross through its substance and over the masseter muscle to supply the muscles of facial expression. The coiled facial artery from the external carotid artery supplies blood to the anterior part of the face. It arises in the neck and ascends around the edge of the mandible. It then proceeds laterally to the corner of the mouth and nose. The superficial temporal artery supplies the posterior part of the face as described above.

Muscles of Facial Expression

All the muscles of facial expression (Fig. 7-8) are superficially located and act on the skin. They are considerably variable in size, shape, and strength and tend to fuse with adjacent muscles. All of the muscles of facial expression are controlled by the facial nerve. These muscles can be grouped as follows: (1) muscles around the mouth (orbicularis oris, risorius, depressor anguli oris, depressor labii inferioris, mentalis, transversus menti, zygomaticus major, zygomaticus minor, levator labii superioris, levator anguli oris, and buccinator); (2) muscles around the nose (dilator naris anterior and posterior, depressor septi, procerus, nasalis pars transversa and pars alaris, and levator labii superioris alaeque nasi); (3) muscles around the orbit (orbicularis oculi and corrugator supercilii); (4) muscles around the ear (anterior, posterior, and superior auricularis); (5) muscles of the scalp (occipitofrontalis) (discussed earlier in this chapter); and (6) muscles of the neck (platysma) (discussed in Chap. 8).

MUSCLES AROUND THE MOUTH

The **orbicularis oris** encircles the mouth and has no bony attachments. Its fibers blend with those of most of the other muscles around the mouth, most actions around the mouth being a result of combined functions of the oris and its associated muscles. The superficially located **risorius** (smile muscle) extends from the corner of the mouth and pulls the corners of the mouth laterally. The **depressor anguli oris** originates from the exterior surface of the mandible on the oblique ridge and inserts on the corners of the mouth to pull the mouth down. The **depressor labii inferioris** originates above the depressor anguli oris on the mandible and inserts into the lower lip and pulls the lip down. The **mentalis** arises from the mandible medial to the mental foramen in the mental fossa and inserts into the skin on the chin; it raises the skin of the chin as the lower lip is protruded. The **transversus menti** is not found in all cases; it is small and is usually continuous with the depressor anguli oris muscle, acting as a sling for the chin.

The **zygomaticus major** arises from the zygomatic bone and inserts into the orbicularis oris at the angle of the mouth. The **zygomaticus minor** parallels the origin and insertion of the zygomaticus major, lying medial to it. These muscles act together to lift up the corners of the mouth in the act of smiling. The **levator labii superioris** arises from the maxilla above the infraorbital foramen and inserts into the upper lip. The **levator anguli oris** arises from the maxilla below the infraorbital foramen and inserts on the corner of the mouth deep to the zygomaticus and levator labii superioris muscles. These muscles assist in pulling up the lip in the act of smiling or grimacing.

The **buccinator** originates from the lateral surface of the maxillae and mandible and from the pterygomandibular raphe. It inserts into the muscles around the mouth, dividing into a superior portion for the upper lip and an inferior portion for the lower lip. This muscle pulls the cheeks against the teeth to prevent food from accumulating in the space between the cheeks and teeth during mastication.

MUSCLES AROUND THE NOSE

The **procerus** is an inferior extension of the frontalis muscle and its action is to wrinkle the skin of the nose. The **nasalis** consists of transverse and alar portions. The **nasalis pars transversa** arises from the maxilla above the fossae for the incisors and inserts on the dorsum of the nose; it compresses the sides of the nose. The **nasalis pars alaris** originates with the pars transversa and inserts on the lateral side of the nostril; it assists the pars transversa in decreasing the nasal aperture.

The **depressor septi** originates on the incisor fossa of the maxilla and inserts on the ala and septum of the nose; it pulls the nose inferiorly. The **levator labii superioris alaeque nasi** arises from the frontal process of the maxilla and inserts on the ala of the nose. The **dilator naris,** located on the lateral margin of the nostril, assists the levator labii in dilating the nostril.

MUSCLES AROUND THE ORBIT

The **orbicularis oculi** consists of three divisions: the palpebral, orbital, and lacrimal portions. The palpebral portion forms a series of curves that arise from the medial palpebral ligaments in the medial corner of the eye and insert at the lateral palpebral raphe in the lateral corner of the eye. The orbital portion is thicker and completely surrounds the palpebral portion. The lacrimal portion arises from the orbital surface of the lacrimal bone and inserts on the tarsal plate. The palpebral portion closes the eye in blinking and the orbital portion closes the eye tightly in the act of winking or trying to get dirt out of the eye.

The **corrugator supercilii** are found deep to the eyebrows. They originate from the medial end of the superciliary arch and insert into the skin of the eyebrow laterally. They act to form a wrinkle on the forehead.

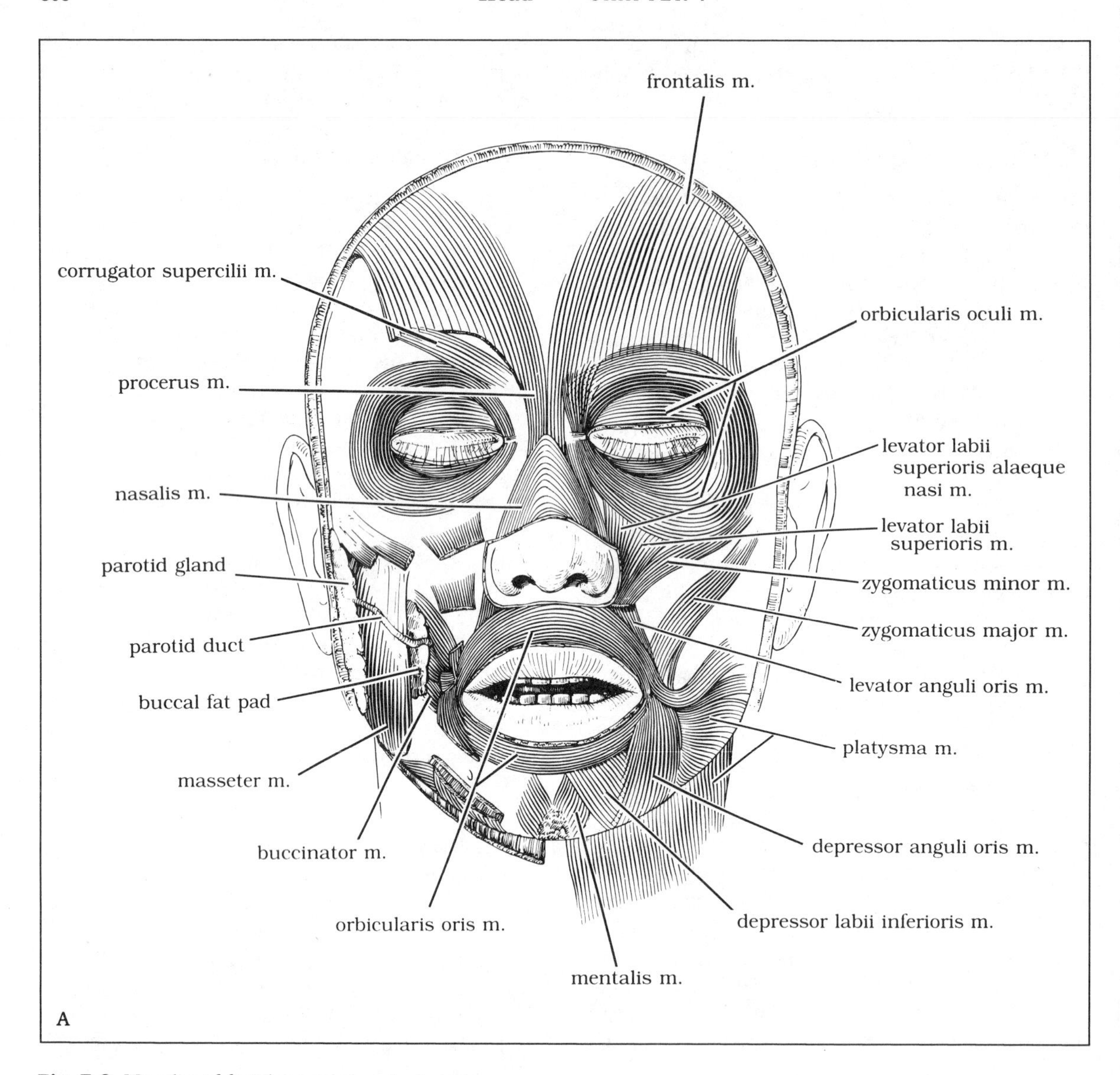

Fig. 7-8. Muscles of facial expression. A. Anterior view. B. Lateral view.

MUSCLES AROUND THE EAR

There are three muscles associated with the ear: the auricularis anterior, auricularis superior, and auricularis posterior muscles. They extend the ear out from the scalp anteriorly, posteriorly, and superiorly with the aid of the aponeurosis and connective tissue. These muscles are important in mammals more able than humans to move their ears (e.g., elephants, dogs, cats). In humans these muscles are more for wiggling one's ears for entertainment and may also assist in adjusting the position of the ears to varying levels of sound.

Arterial Supply

The main arteries to the face are the facial artery and the transverse facial artery (see Fig. 7-33).

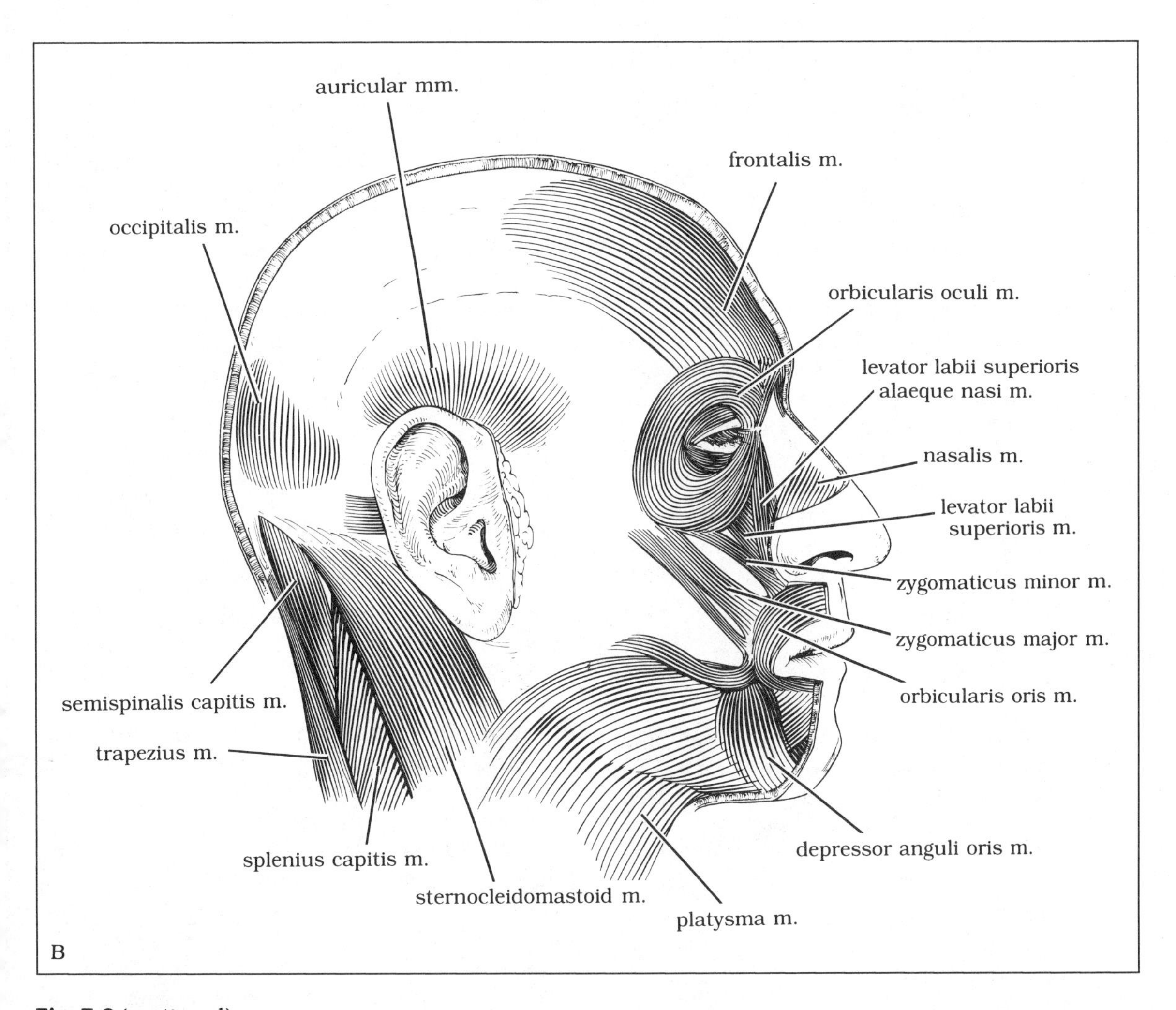

Fig. 7-8 (continued).

The **facial** artery originates from the external carotid and passes obliquely deep to the posterior belly of the digastric and stylohyoid. It becomes superficial and winds around the inferior border of the mandible of the anterior margin of the masseter and enters the face, being covered by only the platysma. It passes over the cheek lateral to the angle of the mouth and terminates as the **angular artery** along the margin of the nose. The angular artery anastomoses with the dorsal nasal branch of the ophthalmic artery. The facial artery is coiled and as a result is not torn when the mandible is opened and closed.

The facial artery has three branches. The **inferior labial branch** extends on to the lower lip to supply the muscles and mucous membranes there. The **superior labial branch** courses medially to the upper lip between the orbicularis oris muscle and the mucous membranes, and terminates in the septum of the nose. The **lateral nasal branch** runs on the lateral surface of the nose and supplies the ala and dorsum of the nose. Muscular branches are named for each of the muscles of facial expression around the nose and mouth supplied by the facial artery. (The facial artery is discussed in detail under Arterial Supply to the Head.)

The **transverse facial artery**, a branch of the superficial temporal artery, lies under the cheek and crosses the masseter, lying between the zygoma and the parotid gland.

Venous drainage

The **facial vein** starts at the medial angle of the eye, where the supratrochlear and supraorbital veins unite (see detailed discussion under Venous Drainage of the Head; see also Fig. 7-40). It runs superficial to the muscles at the angle of the mouth and posterior to the artery. The facial vein usually joins with the anterior division of the retromandibular vein to form the common facial vein, which crosses the submandibular gland under the platysma and empties into the internal jugular vein. This vessel is external to the mandible, so it is not affected by the movements associated with the lower jaw.

Innervation

The trigeminal sensory nerves to the scalp and temples are described earlier in the chapter. The trigeminal nerve divides into the following seven branches to supply the face (see Fig. 7-43):

Ophthalmic division (V_1)
- Infratrochlear nerve
- Lacrimal nerve
- External nasal nerve

Maxillary division (V_2)
- Zygomaticofacial nerve
- Infraorbital nerve

Mandibular division (V_3)
- Buccal nerve
- Mental nerve

The **infratrochlear nerve** enters the orbital region above the medial angle of the eye, supplying the skin on the medial angle of the eye and adjacent bridge of the nose. The **lacrimal nerve** enters the lateral part of the eyelid and supplies the upper lid. The **external nasal nerve** innervates the anterior part of the nose. The **zygomaticofacial nerve** exits through the zygomaticofacial foramen to supply skin on the zygomatic bone. The **infraorbital nerve** exits through the infraorbital foramen and divides into the nasal branch, which supplies the skin of the nose; the superior labial branch, which supplies the upper lip; and the inferior palpebral branch, which supplies the lower lid. The **buccal nerve** is found on the buccinator and supplies the corner of the mouth and cheek and the mucous membranes in the mouth. The **mental nerve** exits through the mental foramen in the mandible and supplies the chin.

LYMPHATICS OF THE SCALP AND FACE

The lymphatics of the scalp and face form a ring (pericervical collar) at the junction of the head and neck, beginning at the chin with the submental nodes and then continuing laterally to the submandibular nodes and superior buccal nodes. At the angle of the mandible are the superficial cervical nodes. There are also associated nodes in front of the ear and parotid gland (preauricular or parotid nodes), behind the ear (postauricular nodes), and on the posterior margin of the head (occipital nodes).

The lymphatic drainage of the scalp and face parallels the arterial supply. The scalp and the back of the head drain through lymphatics associated with the occipital artery to the occipital nodes. The lymphatics that drain the mouth, lips, chin, and nose parallel the course of the facial artery and empty into the superficial cervical, submandibular, submental, and buccal nodes. The anterior surface of the ear and the parotid regions are drained by lymphatics associated with the superficial temporal artery and the preauricular nodes. Finally, the lymphatics draining the posterior surface of the ear follow the posterior auricular artery and empty into the postauricular nodes.

PAROTID GLAND

There are three major salivary glands in the head and neck: the parotid, sublingual, and submandibular glands. The largest is the parotid gland (Fig. 7-9), which is found in the lateral face overlying the mandible. Its shape on the surface is roughly triangular. Its superior border lies along the lower edge of the zygomatic arch; its posterior border extends inferiorly along the posterior border of the ramus of the mandible, adjacent to the external acoustic meatus, until the sternocleidomastoid muscle. Its anterior border runs along a diagonal between the most inferior point and the anterior edge of the gland at the zygomatic arch. The deep portion of the parotid continues behind the mandible.

The parotid gland is contained in the parotid fascia, a continuation of the superficial layer of cervical fascia. The fascia both contains and invests the gland. A thickened region forms the stylomandibular ligament.

The **parotid duct** extends anteriorly from the anterior border, crosses the masseter and buccal fat pad, penetrates the buccinator, and terminates in the interior of the mouth opposite the second upper molar (Fig. 7-10).

Branches of the facial nerve to the muscles of facial expression exit the stylomastoid foramen deep to the gland. They then traverse the gland, breaking up as they do so into several interconnected branches before leaving the gland to reach the appropriate muscles.

Nose

The nose, which consists of the external nose and the nasal cavity, provides an open airway to admit air into the lungs; filters, warms, and moistens the air entering the lungs; and contains the receptive organ for smell.

EXTERNAL NOSE

The external nose has a free tip and a bridge that is attached to the nasal, frontal, and maxillary bones of the face. Inferiorly, there are two openings, the nostrils, or nares. The **nostrils** are separated medially by the nasal septum and are bounded laterally by the ala of the nose. Branches of the ophthalmic artery and facial ar-

Fig. 7-9. The parotid gland and duct.

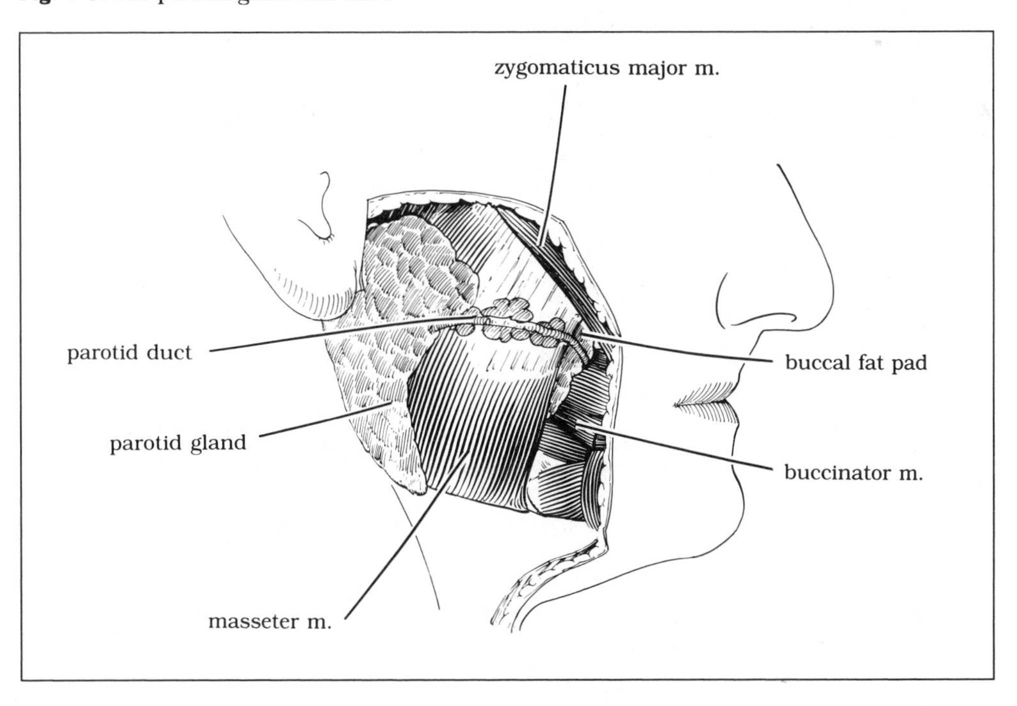

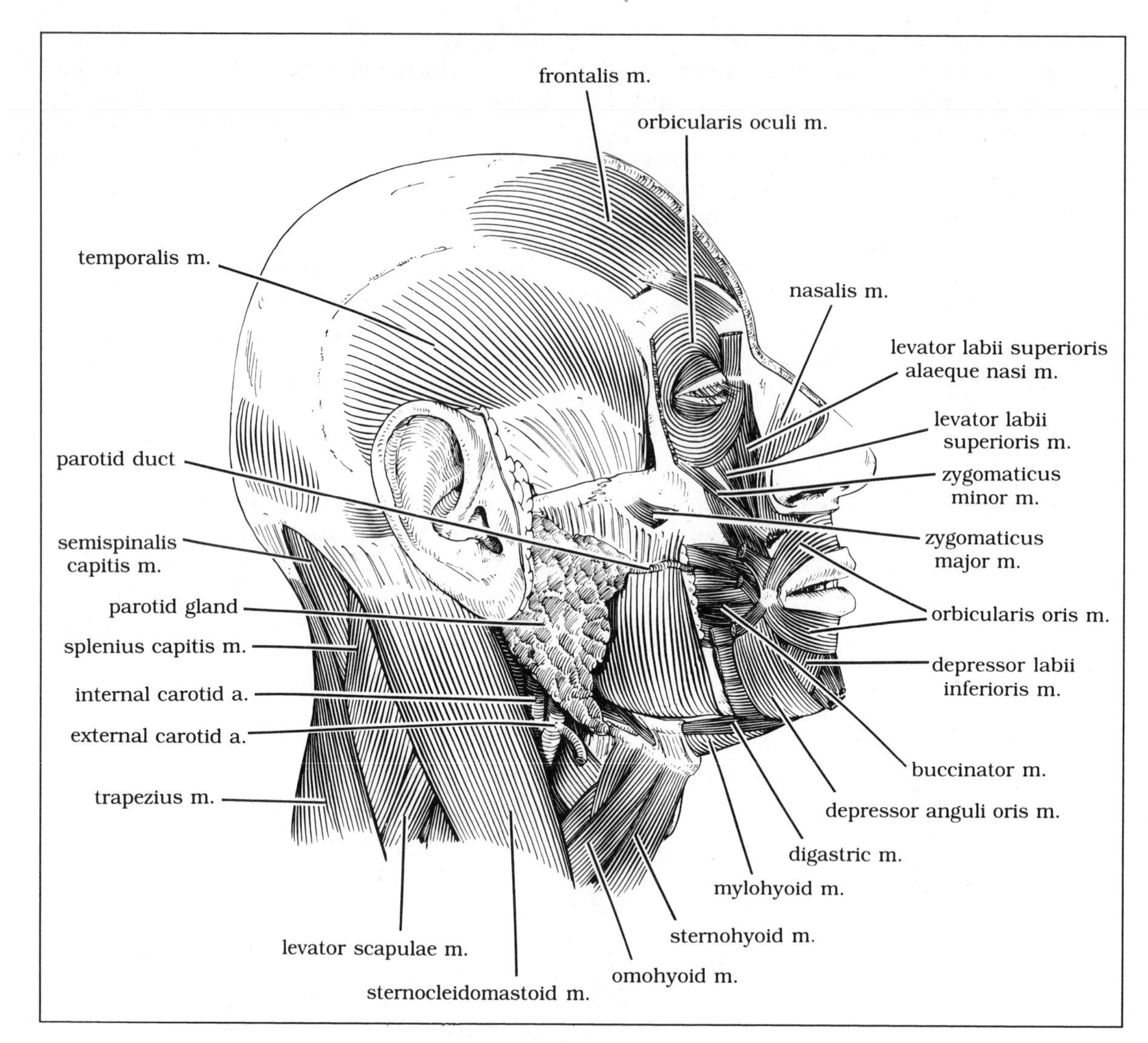

Fig. 7-10. The lateral face, with the muscles of facial expression removed to show the parotid gland and masseter muscle.

tery supply this region. Terminal branches of the ophthalmic and maxillary nerves supply the nose.

INTERNAL NOSE

The internal nose consists of the nasal cavity and the paranasal sinuses, which are embedded in the bones of the face.

Nasal Cavity

The nasal cavity, divided into two nasal fossae, begins at the nares and continues posteriorly to the posterior nasal apertures, the choanae. It is subdivided into the vestibule, respiratory region, and olfactory region. The nasal cavities are continuous with the nasopharynx via the choanae. Medially the choanae are separated by the vomer; laterally they are bounded by the body of the sphenoid bone (medial pterygoid plate) and the horizontal portion of the palatine bone.

The roof of the nasal cavity is formed, from front to back, by the nasal cartilages and the vomer, the nasal and frontal bones, the cribriform plate of the ethmoid bone, the body of the sphenoid bone, and the palatine bone. The floor is horizontal and wider than the roof and is formed, from front to back, by the palatine process of the maxillae and the horizontal plate of the palatine bone.

Its medial wall is formed by the **nasal septum.** Anteriorly the nasal septum is cartilaginous; posteriorly it consists of bony elements from the perpendicular plate of the ethmoid bone and the vomer. The septum usually deviates toward one side.

The **lateral wall** of the nasal cavity (Fig. 7-11) contains the medial projections of the nasal conchae with their underlying meatuses. Parts of seven bones form the lateral wall—the nasal, maxillary, lacrimal, and ethmoid (superior and middle conchae) bones; the inferior conchae; the perpendicular plate of the palatine bone; and the medial pterygoid plate of the sphenoid bone.

Usually superior and middle conchae are present, and occasionally the highest (supreme) concha is also present (see Fig. 7-41). The space above the **superior nasal concha** is the **sphenoethmoid recess,** which opens into the sphenoid sinus. The highest concha, when present, is found in this recess. The **superior meatus** is under the overhanging superior concha; the posterior ethmoid cells open into the superior meatus. The **middle meatus** lies inferior to the **middle nasal concha.** The middle meatus contains the openings into the maxillary frontal and ethmoid sinuses and the anterior and middle ethmoid air cells.

The **inferior nasal concha** is a separate bone at the lower border of the lateral wall of the nasal cavity. Its inferior border is free and its upper border articulates, from front to back, with the maxilla and the lacrimal, ethmoid, and palatine bones. The **inferior meatus** is covered by the inferior nasal concha and lies above the bony palate and receives the opening of the nasolacrimal duct from the orbit.

The **vestibule** is the dilatation just inside the nasal openings. It is lined with skin, hairs, and

Fig. 7-11. The nasal wall and palate.

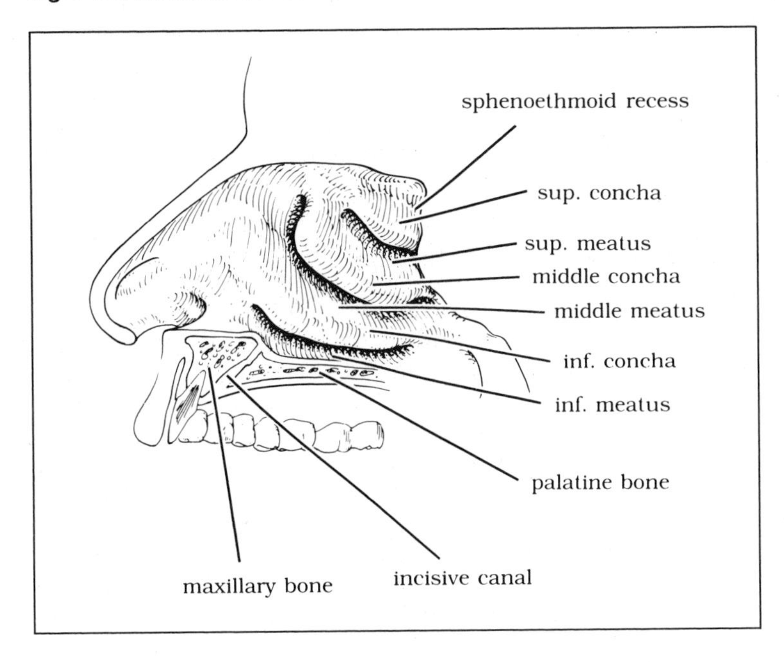

sebaceous and sweat glands. The vestibule connects with the respiratory region posteriorly. It is limited superiorly by a ridge, the lumen nasi, where the skin becomes continuous anteriorly with the membrane in the vestibule and the mucous membrane of the paranasal sinuses, nasopharynx, and nasolacrimal duct (which connects with the conjunctiva). In the mucosa of the nasal conchae is an extensive venous plexus that warms and moistens the incoming air. The inferior conchae in the respiratory region are especially important in both warming and cooling the air. When the middle conchae are removed or elevated the features of the lateral wall of the middle meatus are better seen. A rounded elevation, the **bulla ethmoidalis,** and below it a cleft, the **hiatus semilunaris,** are its principal features. The prominence of the bulla ethmoidalis is caused by the bulging of the middle ethmoid sinus, which opens above it. The hiatus semilunaris leads upward into the ethmoid inferior dilatation, where the anterior ethmoid sinuses open into the middle meatus. In addition, the frontal sinus opens into the most anterior portion of the hiatus, and the maxillary sinus into the center of the hiatus.

The **olfactory region** is found superiorly and posteriorly, bounded by the superior nasal conchae and the upper part of the nasal septum (see Fig. 7-41). The bipolar receptive cells of the olfactory system are found here in the mucous membrane. The membrane is pigmented and contains the pseudostratified columnar epithelium that contains the olfactory cells. The olfactory nerve originates from the epithelial cells. The nerves consist of about 20 bundles that pass superiorly through the foramina on the cribriform plate of the ethmoid bone to form the olfactory nerve, which then synapses in the overlying olfactory bulb.

BLOOD SUPPLY AND INNERVATION

The sphenopalatine branch of the maxillary artery and the anterior ethmoid branch of the ophthalmic artery provide blood to the nasal cavity. The veins form a plexus under the mucosa and run with the anterior ethmoid artery.

The sensory innervation of the anterior region in the nasal cavity comes from the anterior ethmoid branch of the ophthalmic nerve, while the posterior region is innervated by nasal, nasopalatine, and palatine branches of the maxillary nerve.

Sympathetic innervation comes from nerves in the upper thoracic cord that synapse in the superior cervical ganglion. The postganglionic sympathetics first join the plexus on the internal carotid and continue on to the deep petrosal nerve and the nerve of the pterygoid canal, finally reaching the pterygopalatine ganglion. These fibers are vasoconstrictive.

The preganglionic parasympathetics originate in the intermediate branch of the facial nerve, join the greater petrosal nerve, and synapse in the pterygopalatine ganglion. The postganglionic fibers stimulate vasodilatation and secretion of glands in the nasal mucosa. All these fibers distribute with the branches of the maxillary nerve.

Paranasal Sinuses

The paranasal sinuses are found within the maxillae and the frontal, sphenoid, and ethmoid bones, and they are named for these bones. The walls of the sinuses are compact bone lined with mucous membrane that is similar to the mucosa in the nasal cavity. All sinuses develop from the nasal cavity; consequently they connect with the nasal cavity and drain into it. Nasal infections readily spread from the nasal cavity into the paranasal sinuses. All sinuses can be readily identified radiographically. Branches from the maxillary nerves innervate the sphenoid, ethmoid, and maxillary sinuses, while the ophthalmic nerve supplies the frontal sinus.

The **maxillary sinus** is the largest paranasal sinus and appears pyramidal in shape. Its medial wall is the lateral wall of the nasal cavity, its roof is the floor of the orbit, and its floor is the alveolar process of the maxillae. The maxillary sinus drains by several openings into the middle meatus. The **ethmoidal sinuses** consist of many small openings in the ethmoid labyrinth between the orbit and the nasal cavity: Anterior and posterior cells drain into middle and superior meatuses, respectively. The **frontal sinuses,** narrow cavities in the frontal bone, open into the middle meatus. The sphenoid sinus is located in the midline in the body of the sphenoid bone. It varies in size and opens into the sphenoethmoid recess of the nasal cavity.

NASOPHARYNX

The nasal part of the pharynx is the most superior portion of the pharynx, lying above the soft palate and posterior to the choanae. Its walls are formed laterally by bone, its ceiling by the base of the skull, and its floor by the soft palate. Anteriorly, the nasopharynx is continuous with the nasal cavity through the choanae.

The lateral wall of the nasopharynx has an opening for the **auditory tube** (eustachian tube) just behind the inferior nasal concha. The nasopharyngeal isthmus of the auditory tube is found posterior to the torus tubarius. The pharyngeal recess is formed by the cartilaginous portion of the auditory tube. A fold of mucous membrane, the **salpingopharyngeal fold,** extends inferiorly. In the fold is the salpingopharyngeal muscle, which originates from the auditory tube and inserts on the thyroid cartilage. The muscle elevates the pharynx during swallowing and is innervated by the vagus nerve.

The **pharyngeal tonsils** (lymphoid tissue) are seen in the posterior wall of the nasopharynx. The mucous membrane containing the pharyngeal tonsils extends from the base of the occipital bone to the body of the sphenoid bone.

Orbital Region

The orbits, paired structures lying in the front of the skull, contain the eyes and the blood vessels, muscles, nerves, and fat necessary to protect and maintain their functions. The dura mater enters the orbit through the optic canal and forms the periosteum of the orbit, attaching to both eyelids and separating the orbit from the exterior. The dura also fuses with the sclera. The eyeball is surrounded by fascia, and each of its muscles has its own covering of fascia. The fascia for the inferior muscles—the inferior rectus and inferior oblique muscles—are thought to form a suspensory ligament for the eyeball. There is also an extensive fat pad in the orbit, the periorbital pad, that helps cushion the eye.

EXTRAOCULAR MUSCLES

In the physiological condition the main axis of the eye is slightly adducted (slightly divergent), and the eyeballs and extraocular muscles move in tandem. Six pairs of skeletal muscles are needed to move the eyeball in all planes necessary for adequate vision: four rectus and two oblique muscles. (Table 7-7 lists the movements of the individual extraocular muscles; Table 7-8 lists the conjugate movements.) In all eye movements, the muscles work together through the circuitry in the cranial nerve nuclei of the oculomotor and trochlear nerves (in the midbrain) and the abducens nerve (in the pons) (Fig. 7-12). Only when there is a lesion in this system will the eyes move independently of each other.

Rectus Muscles

The four rectus muscles (superior, inferior, lateral, and medial) form a cone that encloses most of the structures in the orbit (Fig. 7-13). The four rectus muscles originate from a tendinous band that arises from a tubercle lateral to the supraor-

Table 7-7. Functions of Individual Extraocular Muscles*

Muscle	*Cranial Nerve*	*Action on Eyeball*
Superior oblique	IV	Depress and abduct
Inferior oblique	III	Elevate and abduct
Lateral rectus	VI	Abduct
Superior rectus	III	Elevate and adduct
Medial rectus	III	Adduct
Inferior rectus	III	Depress and adduct

*The extraocular muscles can be tested by asking the patient to look in the direction in which the single action of each muscle predominates.

Table 7-8. Conjugate Movements of Extraocular Muscles

Movement of Eyeball	*Muscles*
Abducted	Medial rectus on one side; lateral rectus on other side
Adducted	Medial recti on both sides
Elevated	Superior recti and inferior obliques
Depressed	Inferior recti and superior obliques
Elevated and abducted	Inferior oblique on one side; superior and lateral recti on other side
Depressed and abducted	Superior oblique on one side; inferior and lateral recti on other side
Rotated	Combined actions of all muscles in proper sequence

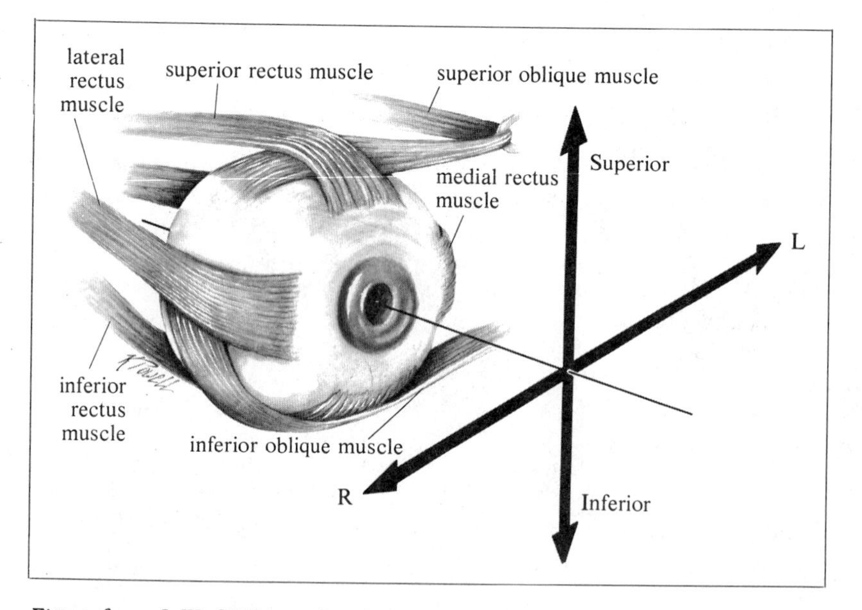

Figure from J. W. Gittinger, Jr., *Ophthalmology: A Clinical Introduction*. Boston: Little, Brown, 1984. With permission.

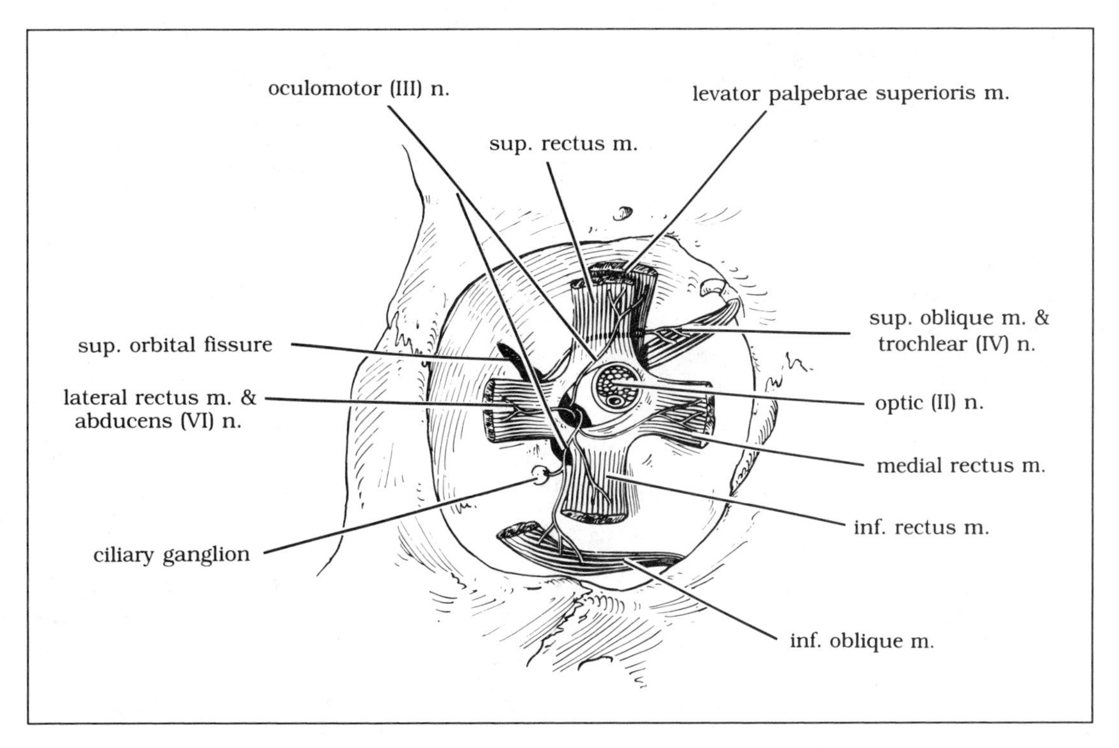

Fig. 7-12. Nerves and muscular origins within the orbit.

Fig. 7-13. Muscles of the orbit, superior view.

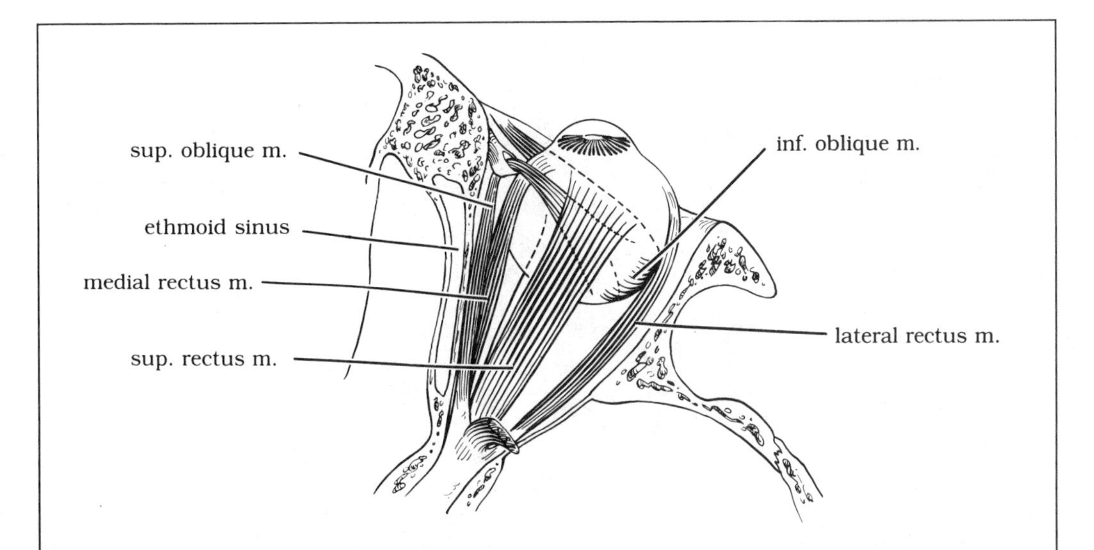

bital fissure. This tendinous band forms a ring that surrounds the optic nerve and attaches to the bone above and below the optic foramen. The muscles insert on the sclera on the anterior half of the eyeball. In the physiological condition, contraction of the rectus muscles results in rotation.

The two oblique muscles insert on the posterior half of the eyeball. The **superior oblique muscle** originates from the medial portion of the roof of the orbit. The tendon of the superior oblique muscle passes anteriorly through a fibrous ring or pulley (trochlea) in the anterior superior and medial edge of the orbit. The tendon, after going through the pulley, bends posteriorly and laterally to insert on the superior surface of the eyeball.

The **inferior oblique muscle** arises from the orbital surface of the maxillary bone and passes under the eyeball to insert on the inferior surface of the eyeball, posteriorly and laterally.

EYELIDS

The eyelids (palpebrae), two moveable folds at the most anterior end of the orbit, protect the eye and orbital contents from foreign particles and excessive light. The upper lid is larger and more mobile than the lower lid. The eyelids meet at the medial and lateral angles of the eye, the medial and lateral canthi, respectively. The palpebral fissure is the opening between the upper and lower eyelids. The free margin of each eyelid has several rows of hairs, the eyelashes or cilia. The ciliary glands (modified sweat glands) are mixed in with the eyelashes. The eyelid consists of several layers (Fig. 7-14). From outside to inside the layers are (1) the skin and subcutaneous tissue, (2) the orbicularis oculi and levator palpebrae superioris muscles, (3) the tarsal muscles and plate, and (4) the mucous membranes (palpebral part of the conjunctiva).

Musculature

The **orbicularis oculi,** an oval sphincter muscle, arises from the nasal part of the frontal bone and the medial palpebral ligament and divides into superior and inferior portions: the palpebral and orbital portions, respectively (see Fig. 7-8B). The palpebral portion is thinner than the orbital portion. The temporal and zygomatic branches of

Fig. 7-14. Muscles of the eyelid and lens.

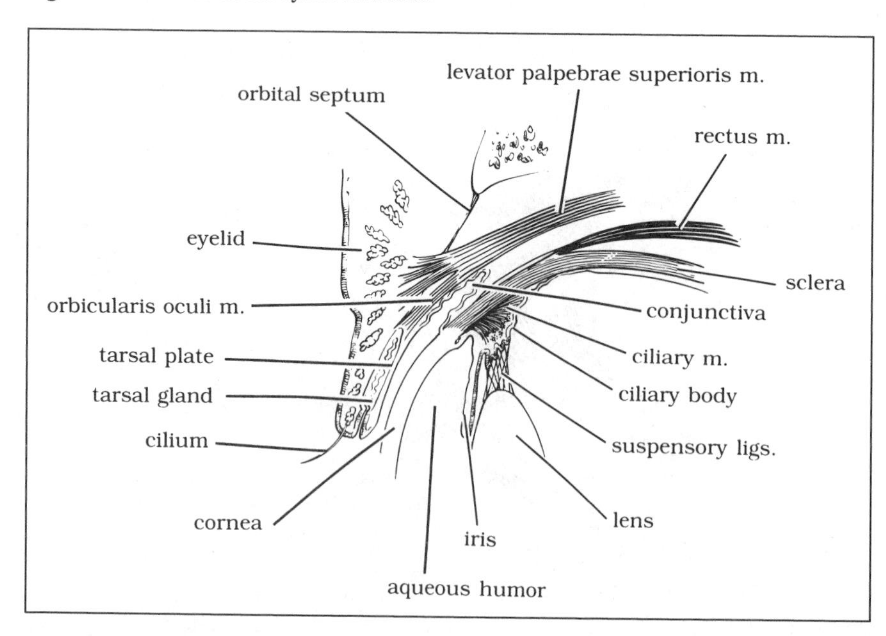

the facial nerve innervate the orbicularis oculi, which functions to close the eyelids.

The **levator palpebrae superioris** originates in the orbit from the lesser wing of the sphenoid bone above the optic canal and inserts into the fibrous aponeurosis of the upper lid (see Figs. 7-12 and 7-14). This muscle is controlled by the oculomotor nerve, and it raises the upper lid. It is the antagonist muscle to the orbicularis oculi.

The **tarsal plate** (see Fig. 7-14) is a dense connective tissue layer that contains sebaceous glands, tarsal glands, and lateral and medial palpebral ligaments. The tarsal muscles are innervated by sympathetic fibers and assist in widening the palpebral fissure. A lesion of the oculomotor nerve produces a drooping eyelid (ptosis) and a dilated pupil, while a lesion of the sympathetic chain produces ptosis and a constricted pupil. The sensory innervation to the eyelids is from the supraorbital and infraorbital branches of the trigeminal nerve.

Conjunctiva

The conjunctiva is the mucous membrane that lines the inner surface of the upper and lower eyelids and the anterior portion of the eyeball (see Fig. 7-14). The conjunctival sac is a capillary-filled area between the eyelids and the eyeball. The reflections of the conjunctiva from the upper and lower lid onto the eyeball are called the superior and inferior fornices.

The conjunctiva is divided into palpebral and bulbar portions. The **palpebral conjunctiva** lines the inner surface of the upper and lower eyelids and contains the openings of the lacrimal canaliculi. This zone is vascularized and appears reddish. The superior fornix receives the opening of the lacrimal glands.

The **bulbar conjunctiva,** which lines the anterior portion of the eyeball, is transparent and permits the sclera to show as the so-called white of the eye. At the medial angle of the eye is a fold of conjunctiva, the plica semilunaris (nictitating membrane), that helps to move foreign particles abrading the cornea out of this extremely sensitive zone. The infratrochlear, lacrimal, and ciliary nerves of the trigeminal nerve supply the conjunctiva. The anterior ciliary and medial palpebral arteries provide blood to this region.

LACRIMAL APPARATUS

The lacrimal apparatus is the mechanism for secreting tears and draining them into the nasal cavity. It consists of the lacrimal gland and its ducts, the conjunctiva, the lacrimal puncta and canaliculi, and the lacrimal sac and nasolacrimal duct (Fig. 7-15).

The **lacrimal gland** is found in the lacrimal fossa in the frontal bone at the anterolateral angle in the roof of the orbit. Tears are formed by the lacrimal gland and accessory glands. The gland is surrounded by fascia and separated from the eyeball by the levator palpebrae superioris and the lateral rectus muscles. The levator palpebrae superioris divides the gland into orbital and palpebral portions. A series of ducts leaves the gland and ends in the superior fornix of the conjunctiva. The fluid is removed by collection in the **lacrimal canaliculi,** which begin at lacrimal puncta, on papillae in the eyelids. The canaliculi connect to the lacrimal sac. The **lacrimal sac** lies in a fossa formed by the lacrimal and maxillary bones in the medial edge of the orbit. The lacrimal sac is continuous with the nasolacrimal duct, which empties directly into the inferior meatus of the nasal cavity. The **nasolacrimal duct** lies in a bony canal formed by the lacrimal bone, maxilla, and inferior nasal concha.

EYE

The eyeball is a spheroid structure occupying the anterior part of the orbit (Fig. 7-16). It is divided into anterior and posterior portions and has three layers: the sclera, choroid, and retina.

Outer Layer (Sclera and Cornea)

The sclera and cornea comprise the tough connective tissue layer of the eye. The sclera is the white of the eye and is loosely interconnected with the choroid layer. Anteriorly, the sclera is

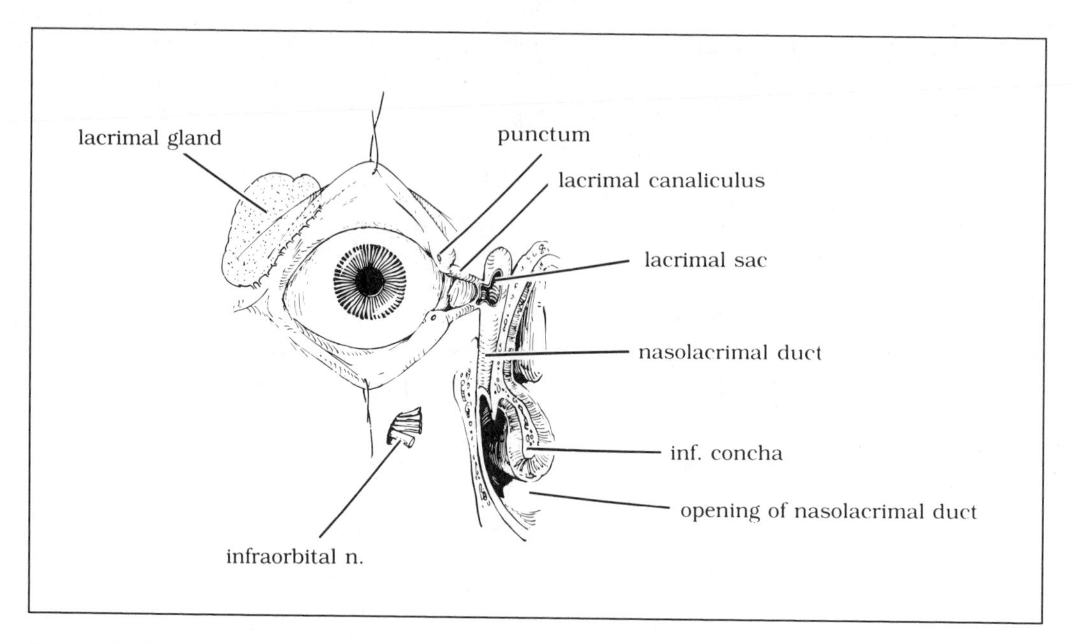

Fig. 7-15. The eye and the lacrimal apparatus.

covered by the transparent bulbar conjunctiva. The extraocular muscles insert into the sclera. The optic nerve pierces the sclera slightly medial and inferior to the posterior pole of the eyeball; the long and short ciliary nerves and vessels enter the sclera near the optic nerve. The cornea is the transparent layer over the pupillary opening; its external layer is continuous with the conjunctiva. The cornea and sclera join at the corneoscleral junction.

Middle Layer (Choroid)

The **choroid** is the vascular tunic of the eye. There are two layers of blood vessels: a superficial venous layer and a deeper arterial layer. The choroid is loosely connected to the sclera and firmly attached to the retina.

The **ciliary body** is the anterior continuation of the choroid and contains the ciliary muscles and processes (see Fig. 7-14). The ciliary muscles consist of smooth muscle that can be longitudinally or obliquely oriented or oriented in both directions; their contractions help to accommodate the eye for far and near vision. Zonular fibers (ciliary zonule), or suspensory ligaments, from the ciliary processes suspend the lens.

The **iris** is a circular pigmented diaphragm that lies in front of the lens. Its periphery is attached to the ciliary body and its center is free; the space bordered by the free edge is the pupil. The **sphincter pupillae** is an annularly oriented smooth muscle found at the free edge of the iris. The **dilator pupillae** is more anterior and is radially oriented in the iris. Parasympathetic fibers control the sphincter pupillae and produce constriction (miosis), while sympathetic fibers innervate the dilator and produce dilatation. The iris divides the space between the lens and cornea into anterior and posterior chambers, both of which are filled with aqueous humor. The **anterior chamber** lies in front of the iris and behind the cornea. The **posterior chamber** lies behind the iris and in front of the lens.

Inner Layer (Retina)

The inner layer of the eye consists of an outer pigmented layer and an inner neuronal layer. Its most anterior portion at the ora serrata continues on to the surface of the ciliary body and posterior surface of the iris. The neuronal portion becomes continuous with the optic nerve at the optic disk. The retinal veins and arteries enter and leave the retina at the optic disk. The central artery of the retina divides into four major divisions that supply the four retinal quadrants.

The neuronal layer consists of three layers: the outermost light-receptive rods and cones, the bipolar cell layer, and the ganglion cell layer, which gives origin to the cranial II nerve.

Refractive Apparatus

Anteriorly to posteriorly, the refractive apparatus of the eye consists of the cornea, the aqueous humor, the lens, and the vitreous body. The **aqueous humor** fills the anterior and posterior chambers and then is absorbed at the iridocorneal angle by the ciliary veins.

The **lens** is biconvex and is surrounded by a capsule that forms an elastic covering. It is suspended from the ciliary processes by the suspensory ligaments; changing the tension on these ligaments changes the focal length. Normal tension flattens the lens so that there is minimal refraction of light rays (far accommodation). Near accommodation (increased refraction and convergence of light rays) is accomplished by contracting the ciliary muscles, which draw the ciliary body anteriorly, releasing the tension on the suspensory ligaments and allowing the lens to assume a more spherical shape.

The **vitreous body** is a transparent gelatinous mass that fills the posterior four-fifths of the eyeball. It is adherent to the ora serrata. It contains fluid similar to the aqueous humor but it also contains collagenous fibers with a mucopolysaccharide hyaluronic acid. It is both refractive and structural, helping to maintain the oval shape of the eye.

Fig. 7-16. Cross section of the eyeball.

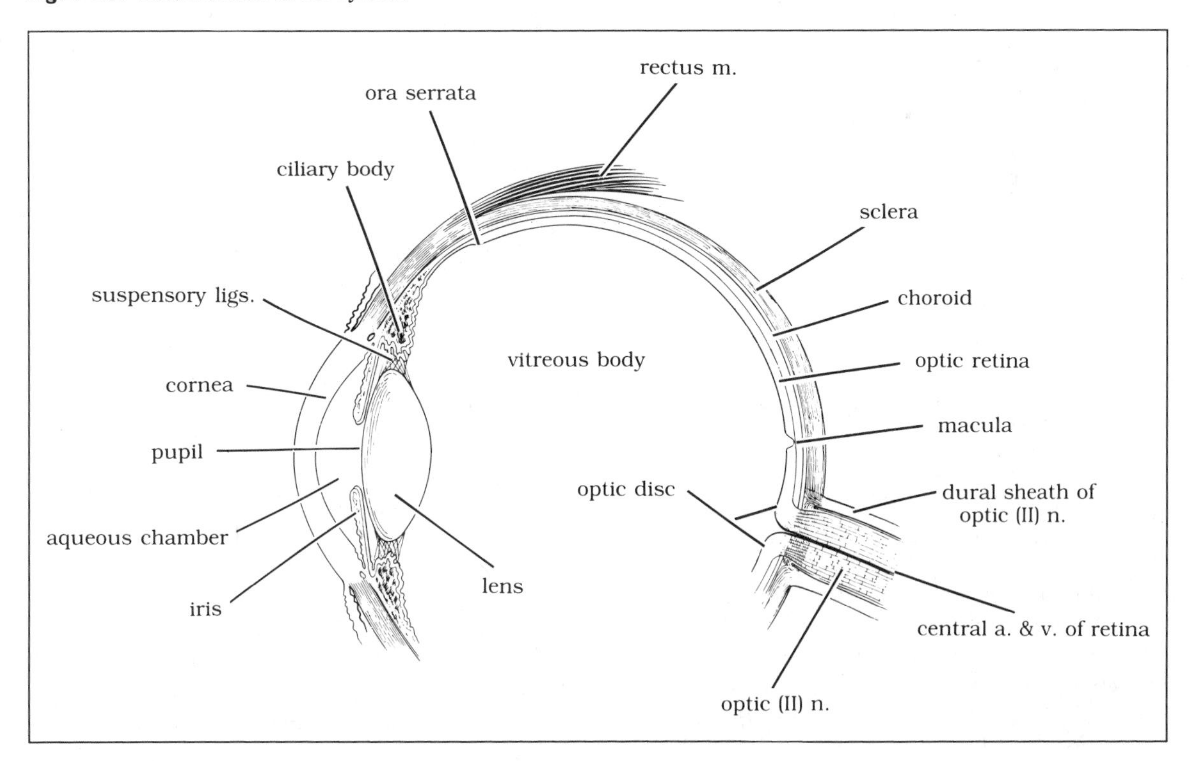

Ciliary Ganglion

The ciliary ganglion is very small and provides parasympathetic postganglionic neurons to the eye (see Fig. 7-12). It lies in the back of the orbit lateral to the optic nerve and the ophthalmic artery. The branch of the oculomotor nerve to the inferior oblique muscle provides the input to this ganglion. The **short ciliary nerves** contain the postganglionic parasympathetic fibers to the ciliary muscles and the sphincter pupillae muscle.

Sympathetic fibers course through the ciliary ganglion but do not synapse therein; they originate from the superior cervical ganglion. The short ciliary nerves also contain postganglionic sympathetic fibers to the dilator pupillae muscles.

BLOOD SUPPLY

Ophthalmic Artery

The ophthalmic artery provides the major blood supply to the orbit (Fig. 7-17). It is the first branch off the internal carotid, arising medial to the anterior clinoid processes. It enters the orbit through the optic canal, inferior to the optic nerve. In the orbit it first lies between the optic nerve and the lateral rectus muscle. It then runs obliquely toward the medial wall of the orbit, where it turns forward between the superior oblique and medial rectus muscles. At the front of the orbit it divides into a supratrochlear and a dorsal nasal artery.

The branches of the ophthalmic artery to the eye are the following:

The **central artery of the retina,** the first branch of the ophthalmic artery, enters the optic nerve and courses to the retina, where it divides into branches for the temporal and nasal quadrants of the retina. Its terminal branches are end arteries.

The two **long posterior ciliary arteries** arise within the orbit, pierce the sclera, and supply the ciliary body and iris.

The **short posterior ciliary arteries** (there are several of them) pierce the sclera and supply the choroid.

The **anterior ciliary arteries** pierce the sclera and supply the iris. They arise from muscular branches.

The branches of the ophthalmic artery to the eyelid, scalp, and nose are the following:

The **medial palpebral arteries** supply arcades in the eyelids and form posterior conjunctival arteries.

The **lacrimal artery** to the lacrimal glands, conjunctiva, and eyelids has two branches. The **recurrent meningeal branch** passes through the superior orbital fissure and anastomoses with the middle meningeal artery. The **lateral palpebral arteries** supply the arcades in the upper and lower eyelids.

The **supraorbital artery** exits through the supraorbital notch to supply the upper eyelids and scalp.

The **supratrochlear artery** passes through the supratrochlear notch (or foramen) and supplies the forehead and scalp.

The **dorsal nasal artery** leaves the orbit above the medial palpebral ligament and supplies the root of the nose and the lacrimal sac. It anastomoses with branches of the facial artery.

The branches of the ophthalmic artery to the cranial fossa are the following:

The **anterior ethmoid artery** joins the nerve of the same name and passes superiorly through the anterior ethmoid canal into the anterior fossa, nasal cavity, and nose.

The **posterior ethmoid arteries** may also be present and irrigate nearly the same zone as does the anterior ethmoid artery.

The muscular branches of the ophthalmic artery arise from the main trunk, the long ciliary arteries, and many other branches of the ophthalmic artery to supply the muscles (rectus, oblique, and levator) of each eye.

Ophthalmic Veins

The superior and inferior ophthalmic veins drain the orbit. They communicate with the facial vein, pterygoid plexus, and cavernous sinus. The

superior ophthalmic vein with the ophthalmic artery and exits through the superior orbital fissure, ending in the cavernous sinus. The **inferior ophthalmic vein** may be in the form of a plexus on the orbital floor and also drains into the cavernous sinus. The **central vein of the retina** exits through the optic nerve and empties into the cavernous sinus.

INNERVATION

Ophthalmic Division of the Trigeminal Nerve

The ophthalmic nerve is the principal general sensory nerve to the contents of the orbit including the eye, lacrimal gland, and eyelids. Near the superior orbital fissure, the ophthalmic nerve divides into three branches—lacrimal, frontal, nasociliary—that then enter the orbit through the superior orbital fissure.

The **lacrimal nerve** enters the orbit and runs laterally above the muscles of the eye near the lateral rectus muscle. At the front of the orbit it branches to the lacrimal gland, conjunctiva, and upper eyelid.

The **frontal nerve** also enters above the eye muscles and continues to pass anteriorly, superior to the levator palpebrae superioris muscle. The frontal nerve divides into the supraorbital and supratrochlear nerves. The supraorbital nerve is the direct continuation of the frontal nerve and continues through the orbit, exits through the supraorbital notch, and supplies the upper eyelid, forehead, scalp, and frontal sinuses. The supratrochlear nerve exits the orbit at the medial end of the supraorbital margin and supplies the upper eyelid and forehead.

The **nasociliary nerve** enters the orbit below the lacrimal and frontal nerves, lying between the superior and inferior branches of the oculomotor nerve. It continues forward under the superior rectus muscle, crossing the optic nerve and ophthalmic artery to reach the medial surface of the orbit. In the orbit the following branches are

Fig. 7-17. Blood supply to the eye and orbit.

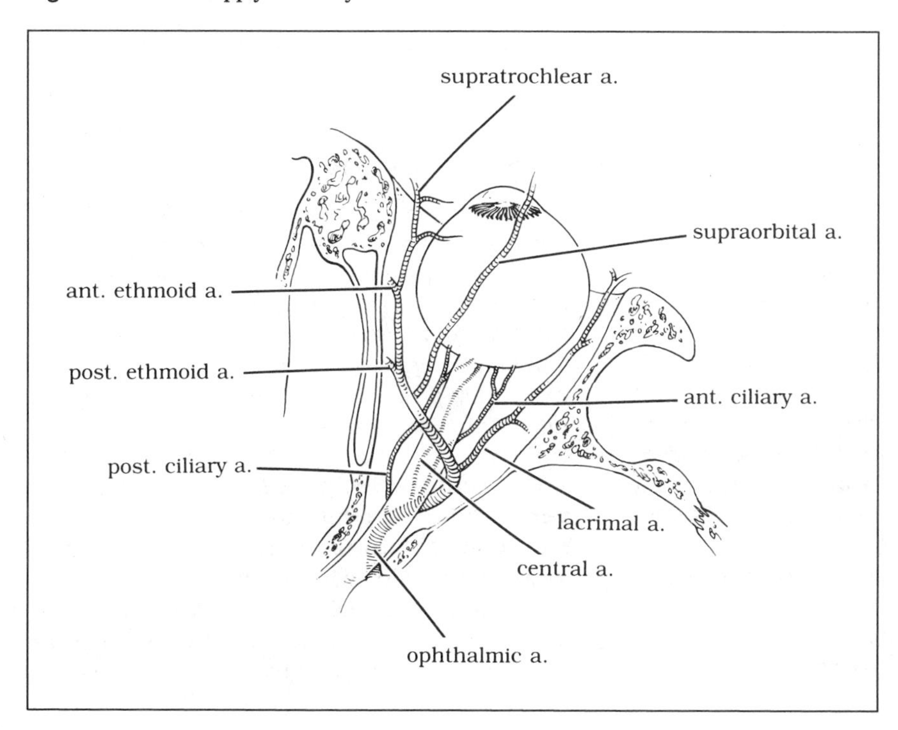

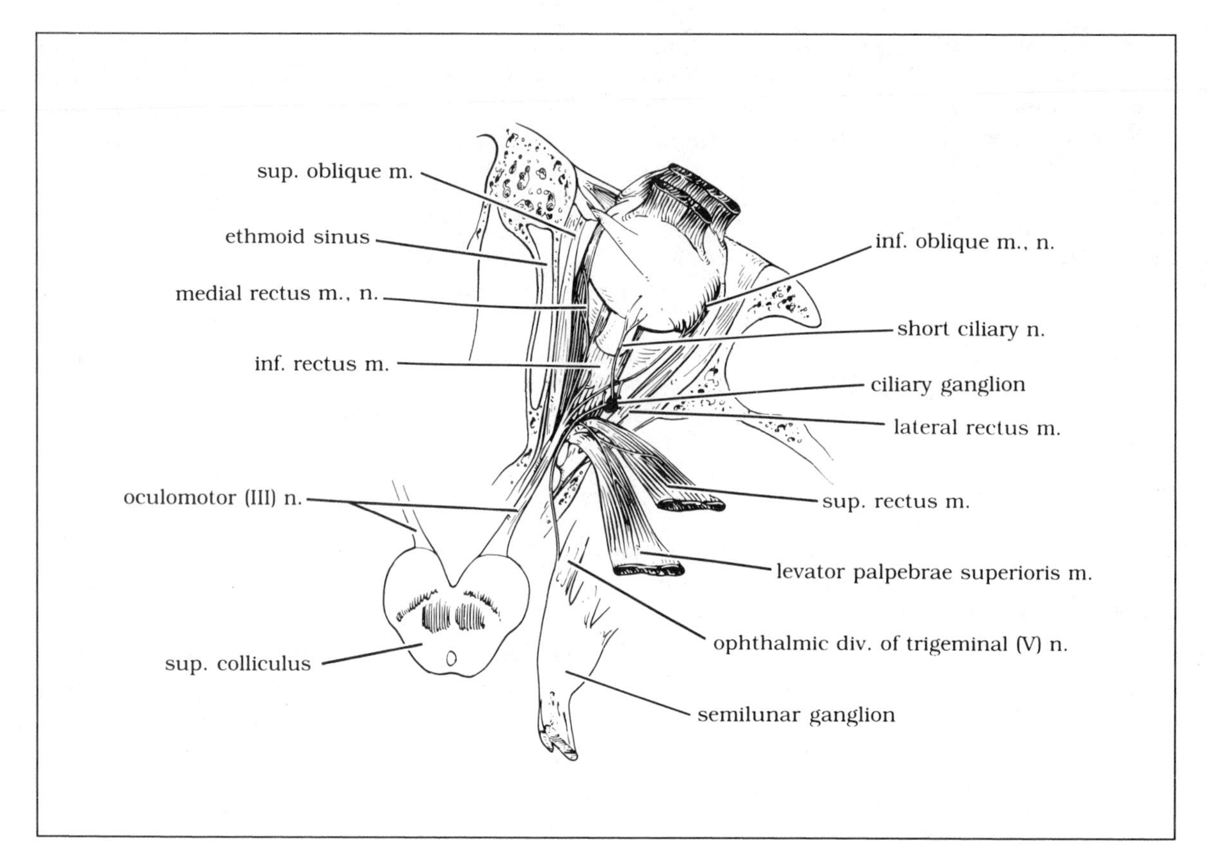

Fig. 7-18. Superior view of the orbit showing distribution of the oculomotor nerve.

found: the communicating branch to the ciliary ganglion, which transmits sensory information from the short ciliary nerves; one or two long ciliary nerves (sympathetic to dilator pupillae and afferent to cornea and uvea); and the infratrochlear nerve to the eyelids, lacrimal sac, and skin of the nose. The **anterior ethmoid nerve,** the continuation of the nasociliary nerve, passes superiorly through the anterior ethmoid foramen into the anterior cranial fossa and then into the nasal cavity. One branch, the external nasal branch, may even reach the skin over the nose. Thus, the nasociliary nerve is found in the middle cranial fossa, orbit, anterior cranial fossa, nasal cavity, and even the skin over the nose.

Other Cranial Nerves

The **optic nerve** (special somatic afferent fibers; see discussion of cranial nerves later in this chapter) is the nerve for vision. It originates in the ganglionic layers of the retina, converges on the optic disk, and pierces the retina, choroid, and lamina cribrosa sclerae to exit the eye. The nerve then proceeds posteriorly and medially, lying within the cone formed by the rectus muscles.

The **oculomotor, trochlear,** and **abducens nerves** enter the orbit through the superior orbital fissure. The oculomotor nerve divides into a superior division, which supplies the superior rectus and levator palpebrae superioris muscles, and an inferior division, which supplies the inferior rectus muscle and branches to the medial rectus muscle and the ciliary ganglion (Fig. 7-18). The trochlear nerve supplies the superior oblique muscle; the abducens nerve supplies the lateral rectus muscle (Fig. 7-19). (See Cranial Nerves later in this chapter for full coverage of the oculomotor, trochlear, and abducens nerves.)

Autonomic Innervation

THE EYE

The preganglionic sympathetic fibers originate in the upper thoracic levels of the spinal cord. They enter the sympathetic chain through the white rami communicantes and ascend in the chain, finally synapsing in the superior cervical ganglion. The postganglionic nerves leave the ganglion, join the plexus on the internal carotid, pass into the skull, and then follow the ophthalmic artery and enter the eyeball with the **short ciliary nerves.** The short ciliary nerves innervate the blood vessels and the dilator pupillae and tarsal muscles of the upper lid. The **long ciliary nerve** is one of several branches of the nasociliary nerve and provides afferents to the iris and cornea, and probably sympathetics from the superior cervical ganglion to the dilator muscles as well.

THE GLANDS

The preganglionic parasympathetic fibers that supply autonomic innervation to the lacrimal glands originate from the facial nerve, which branches into the greater petrosal nerve and the nerve of the pterygoid canal. These branches synapse in the pterygopalatine ganglion, and the postganglionic parasympathetics enter the orbit to supply the gland. The preganglionic sympathetic fibers originate from the upper thoracic levels of the spinal cord and synapse in the superior cervical ganglion. The postganglionic sympathetics join the carotid plexus on the carotid artery, continue on to the ophthalmic artery, then into the orbit, and finally on to the lacrimal artery and nerve.

Fig. 7-19. Superior view of the orbit showing distribution of the trochlear and abducens nerves.

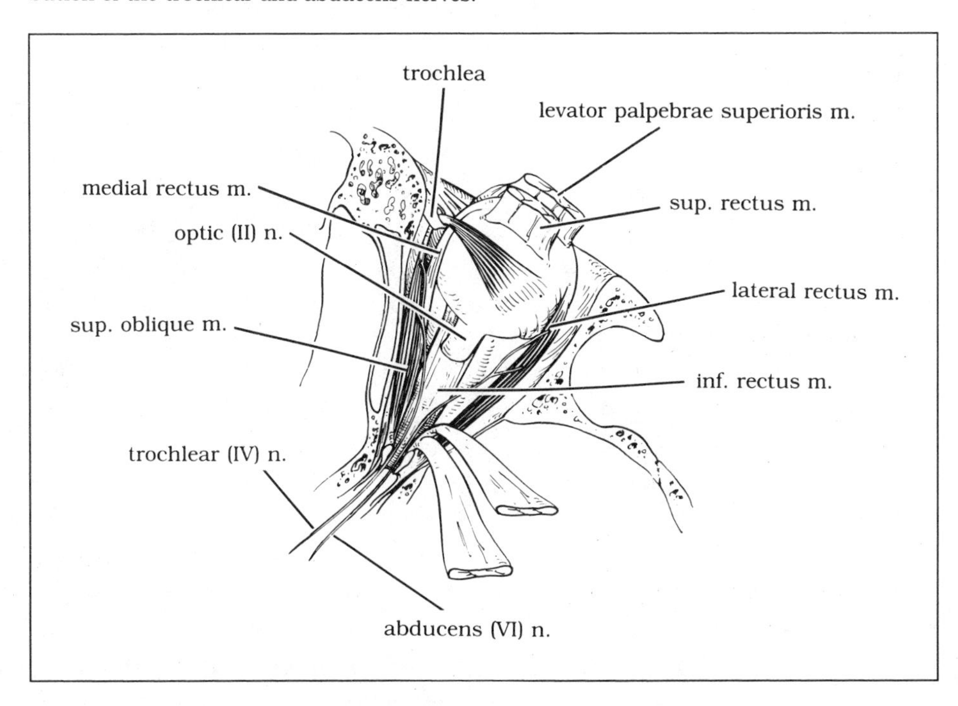

Ear

Each ear consists of an external, middle, and internal portion.

EXTERNAL EAR

The external ear consists of the auricle and the external acoustic (auditory) meatus.

The **auricle** consists of elastic cartilage covered with skin (Fig. 7-20). The margin of the auricle is the helix that ends inferiorly in the lobule. The lobule has no cartilage and consists of connective tissue and fat. The hollow in the center of the ear is the concha. The tragus is the elevation anterior to the external acoustic meatus. The auricle receives sensory innervation from the auriculotemporal branch of the trigeminal nerve, the glossopharyngeal nerve, and the great auricular and lesser occipital nerves of the cervical plexus (ventral rami of C2 and C3). Vagal fibers reach the auricle through the auricular branch of the vagus nerve. The superficial temporal and posterior auricular artery supply the auricle.

The **external acoustic (auditory) meatus** is a canal that begins at the deepest part of the conchae and continues medially to the tympanic membrane (Fig. 7-21). The meatus conducts changes in external air pressure into the middle ear cavity via the tympanic membrane and ear ossicles. The lateral part of the canal is cartilaginous, and the longer medial part is within the temporal bone. The canal is lined with skin that continues on to the tympanic membrane. Many hairs are present in the skin, and modified sweat glands (ceruminous glands) that form the ear wax are also present.

The **tympanic membrane** separates the external acoustic meatus from the middle ear. It is formed of connective tissue lined on its external surface with skin and on its inner surface, in the middle ear, with mucous membrane. The tympanic membrane lies obliquely in the meatus; its lateral surface is concave and in the center of the concavity is the umbo. The tympanic membrane is thickened around the periphery where it attaches to the tympanic portion of the temporal bone. The membrane can be divided into anterosuperior, posterosuperior, anteroinferior and posteroinferior quadrants. The superior quadrants are separated by the manubrium of the most external ear bone, the malleus. The

Fig. 7-20. The auricle of the ear.

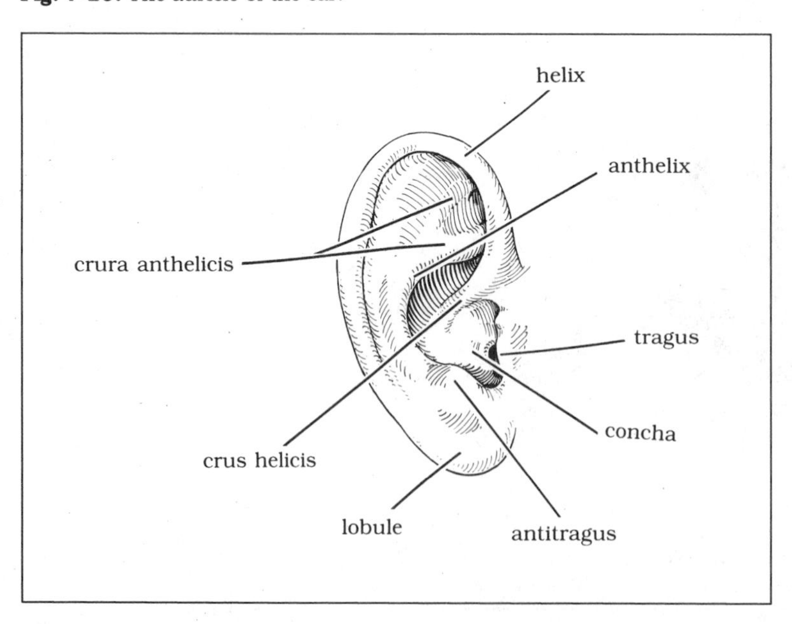

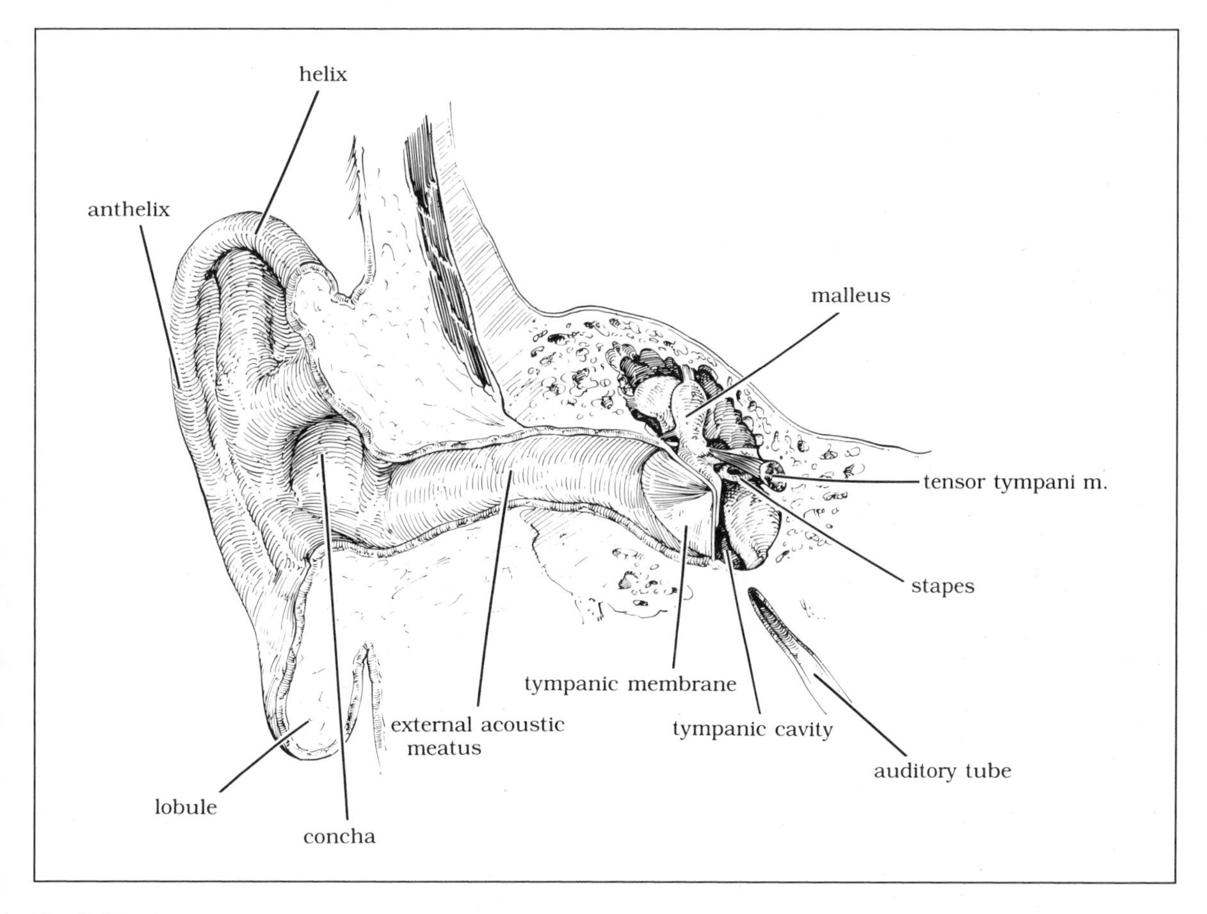

Fig. 7-21. The external ear.

most concave portion of the tympanic membrane is where the manubrium of the malleus attaches to the center of the tympanic membrane. The chorda tympani crosses the tympanic membrane on its medial surface, near to where the handle and lateral process of the malleus attach (see Fig. 7-22). Incisions through the membrane are usually made in the posteroinferior quadrant to avoid the ossicles and chorda tympani.

Three cranial nerves supply the surface of the tympanic membrane: the auriculotemporal branch of the trigeminal nerve and the vagus nerve supply the outer surface, and the tympanic branch of the glossopharyngeal nerve supplies the inner surface.

MIDDLE EAR

The middle ear is a space in the petrous portion of the temporal bone lined with mucous membranes. It contains the ossicles and is continuous anteriorly with the nasopharynx through the auditory tube and posteriorly with the mastoid cells. The middle ear bones—the malleus, incus, and stapes—conduct vibrations picked up by the tympanic membrane into the **vestibular (oval) window,** or fenestra vestibuli, of the inner ear (Fig. 7-22). The middle ear bones are attached to each other by ligaments. The tensor tympani muscle attaches to the manubrium of the malleus and the stapedius muscle inserts on the neck of the stapes (Fig. 7-23). These muscles contract and help to dampen the vibrations picked

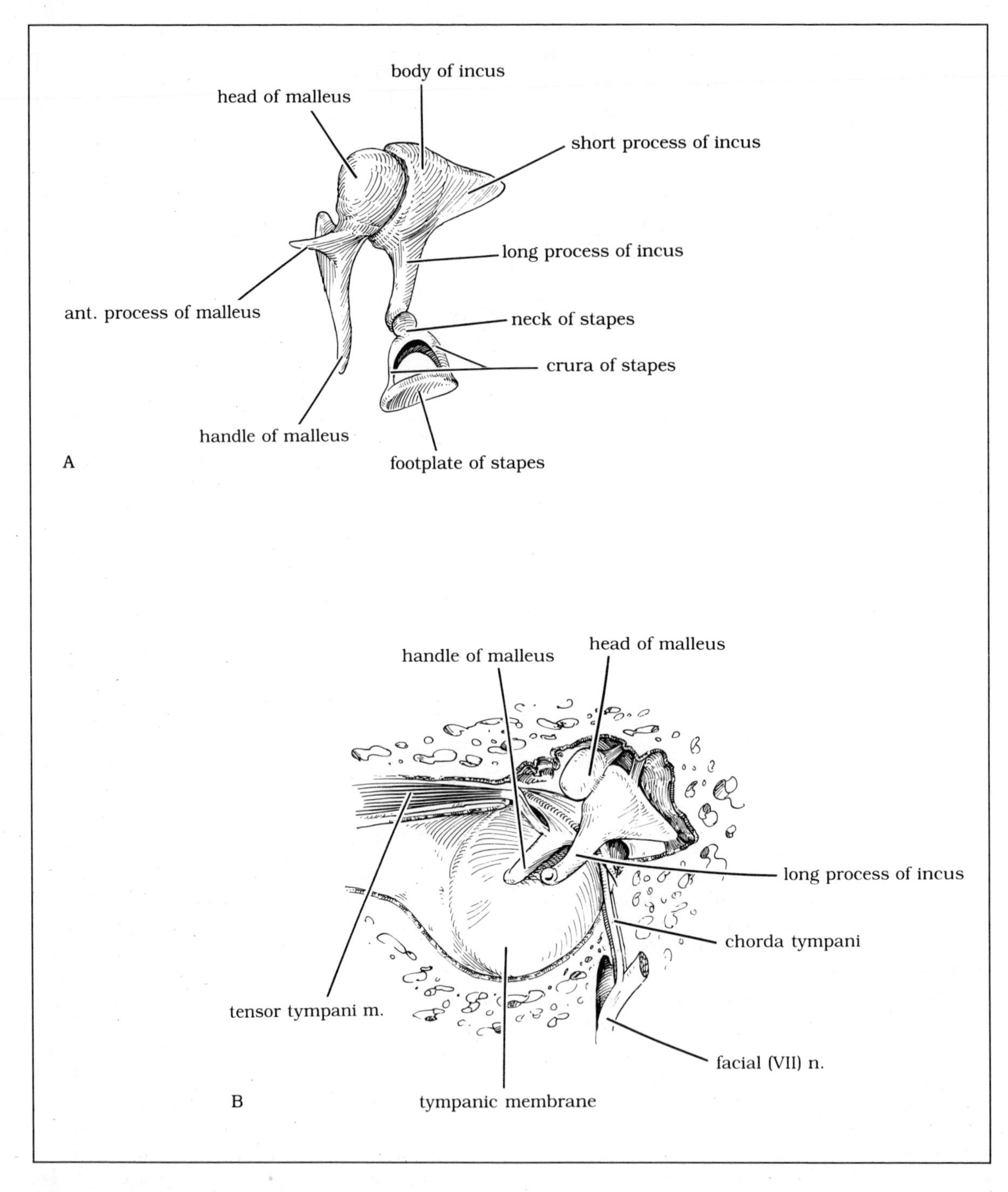

Fig. 7-22. A. Bones of the middle ear. B. The tympanic membrane and the middle ear.

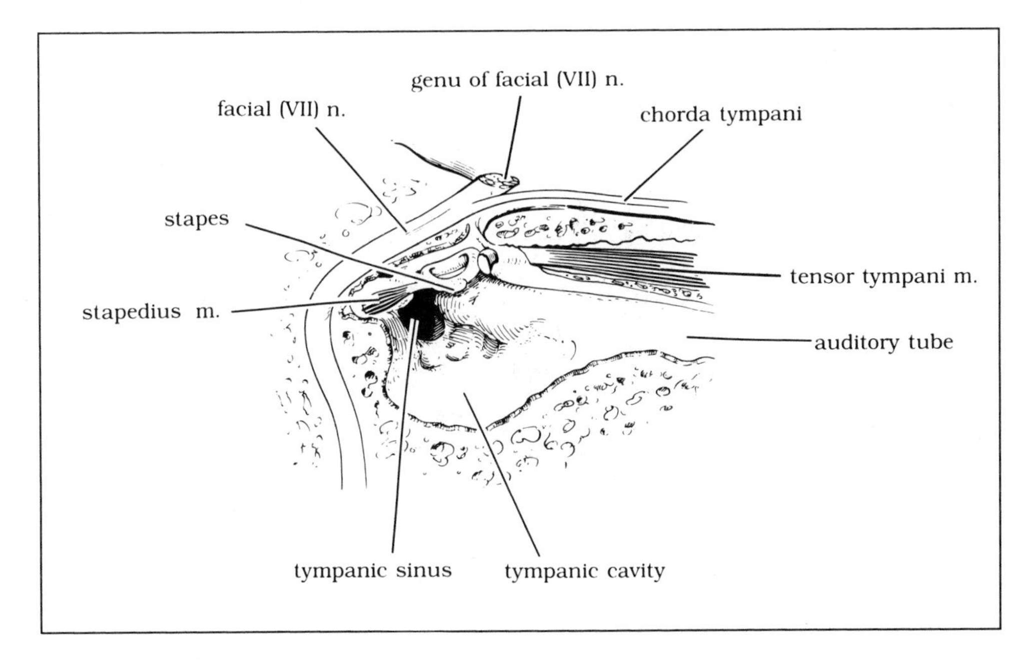

Fig. 7-23. The middle ear and the opening to the auditory tube.

up at the tympanic membrane before they are delivered to the vestibular window.

The **tensor tympani** develops in the first brachial arch, and consequently is innervated by the trigeminal nerve, while the stapedius muscle develops in the second brachial arch and is innervated by the facial nerve.

The **chorda tympani** from the facial nerve enters the middle ear posteriorly and leaves it anteriorly. On the tympanic membrane the chorda is lateral to the incus and medial to the manubrium of the malleus. In the infratemporal fossa the chorda tympani joins the lingual branch of the trigeminal nerve and carries taste fibers from the anterior two-thirds of the tongue and parasympathetic innervation to the submandibular and sublingual glands.

The middle ear has a floor, a roof, and anterior, posterior medial, and lateral walls. The roof (tegmental wall) is a thin bone, the tegmen tympani of the petrous portion of the temporal bone, that separates the middle ear from the middle cranial fossa. The head of the malleus is attached to the roof by suspensory ligaments. The floor is a thin bone that separates the middle ear from the jugular fossa (jugular wall) posteriorly and the internal carotid anteriorly. Superiorly on the posterior wall (mastoid wall) is the entrance to the mastoid cells, the aditus ad antrum; the pyramidal eminence, housing the stapedius; and the foramen through which the chorda tympani enters the middle ear. Superiorly to inferiorly, the anterior wall (carotid wall) contains the semicanal for the tensor tympani, the entrance to the auditory tube, and the foramen for the exit of the chorda tympani from the middle ear. The tympanic membrane forms the lateral wall (membranous wall).

The medial wall of the middle ear contains many structures that are related to the inner ear: most anteriorly is the promontory produced by the basal coil of the cochlea. A plexus of nerves, the tympanic plexus, covers the promontory. Behind the promontory, the vestibular window can be seen covered by the stapes. The stapedius muscle is attached to the neck of the stapes. Inferior to the vestibular window is the cochlear (round) window, or fenestra cochleae. Behind and above the cochlear window is a ridge cover-

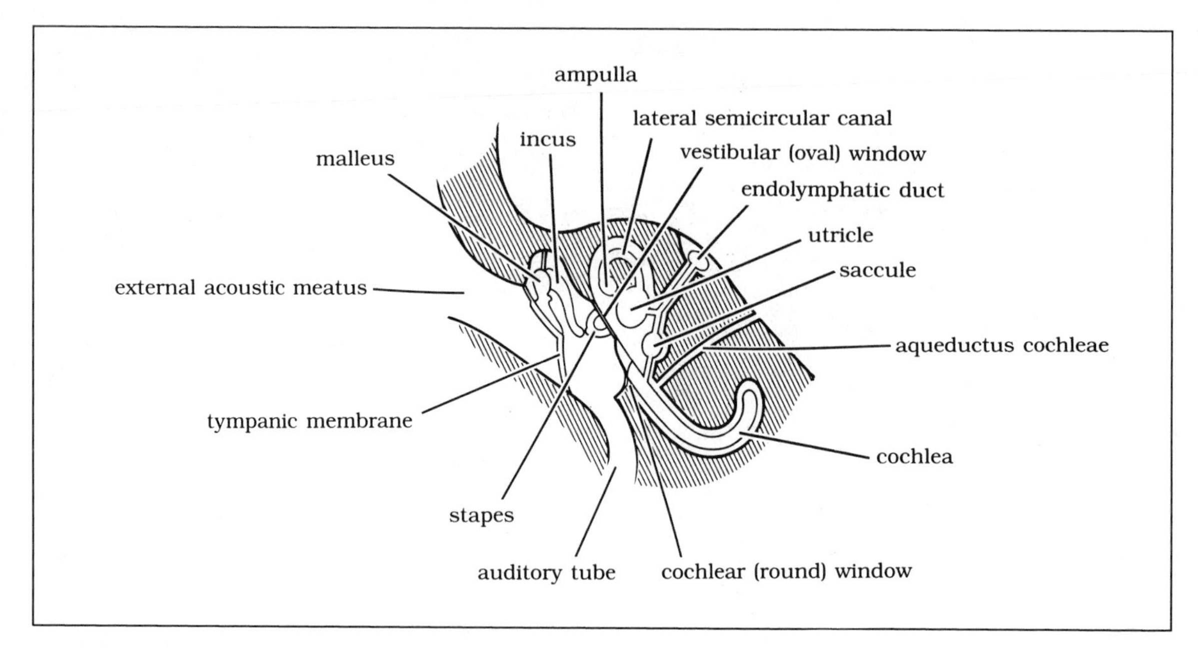

Fig. 7-24. Schematic view of the middle and inner ears.

ing the facial nerve. Behind this ridge a prominence formed by the lateral semicircular canal in the epitympanic recess is found. Behind this recess is the entrance to the mastoid air cells.

The **auditory tube** connects the middle ear with the nasopharynx. In the middle ear the tube is narrower and bony, while in the nasopharynx it is cartilaginous. The tube opens into the nasopharynx behind the nasal cavity and is lined with mucous membrane. The auditory tube serves to balance the pressure on each side of the tympanic membrane.

INNER EAR

The inner ear is found in the petrous portion of the temporal bone on the bony floor of the middle fossa, between the middle ear and the internal acoustic meatus (Fig. 7-24). It is a series of fluid-filled cavities, consisting of a membranous labyrinth within a similarly shaped bony (osseous) labyrinth. The bony labyrinth consists of the vestibule, the cochlea, and the three semicircular canals; the membranous labyrinth consists of the utricle and saccule (contained within the vestibule), the cochlear duct, and the semicircular ducts. The bony labyrinth is filled with perilymph, a fluid similar to cerebrospinal fluid but containing more protein; the membranous labyrinth is filled with endolymph.

The vestibular portion of the inner ear monitors the position of the head in space and is affected by movements in all planes. The cochlear portion responds to sound waves, whether sent up in the air through the tympanic membrane or directly through the bone. The vestibulocochlear nerve connects these sensory receptors to the brain.

Bony Labyrinth

The **vestibule** is the central portion of the bony labyrinth and contains the utricle and saccule of the membranous labyrinth. It lies anterior to the semicircular canals, immediately medial to the tympanic cavity and posterior to the cochlea. In its lateral wall are the vestibular window (oval window, or fenestra vestibuli), which is closed by the base of the stapes, and the cochlear window (round window, or fenestra cochleae), which is closed by the secondary tympanic membrane.

The three canals are the anterior, posterior, and lateral **semicircular canals** (Figs. 7-25 and 7-26). The anterior and posterior canals face laterally and posteriorly and are oriented vertically, at right angles to each other. The lateral canal is horizontally placed. Medially, each canal enlarges into an ampulla that opens into the vestibule.

The **cochlea,** named for its similarity in shape to the shell of a snail, is helical, containing two and a half turns. The base of the cochlea lies at the lateral end of the internal acoustic meatus facing the medial wall of the middle ear; its apex faces anterolaterally. The **modiolus,** the bony core of the cochlea, carries the cochlear nerve and the primary sensory cells (spiral ganglion). A bony, winding shelf (similar to a screw thread), the **osseous spiral lamina,** projects from the modiolus; the basilar and vestibular membranes extend from this lamina to the wall of the cochlea, forming between them the **cochlear duct.** The osseous spiral lamina and the cochlear duct serve to divide the cochlea into the scala vestibuli anteriorly and the scala tympani posteriorly. The footplate of the stapes fits into the vestibular window at the base of the scala vestibuli; the base of the scala tympani ends at the cochlear window. At the apex of the cochlea, the modiolus and the cochlear duct end, and the two scalae communicate at the helicotrema. The perilymphatic duct extends superiorly and approaches the dura in the floor of the middle fossa.

Fig. 7-25. The inner ear, with the semicircular canals and the cochlea.

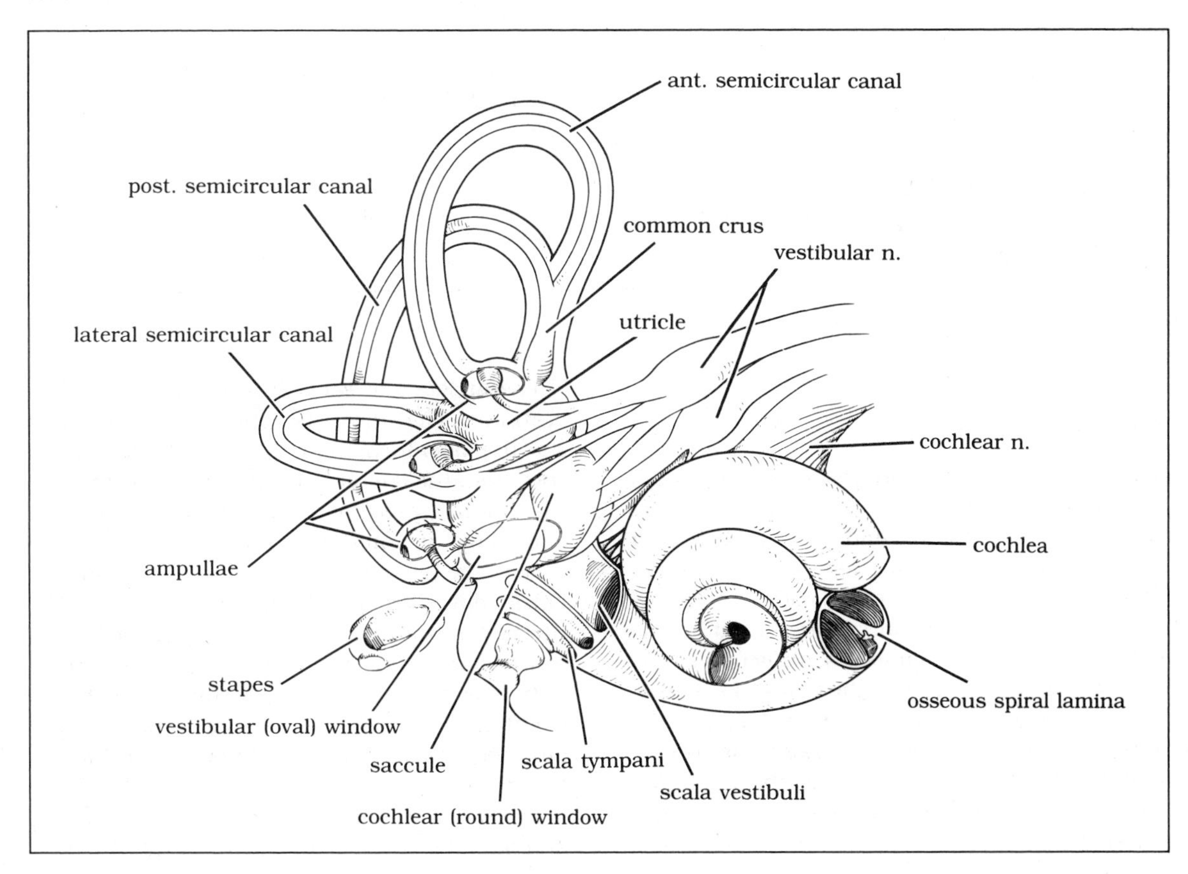

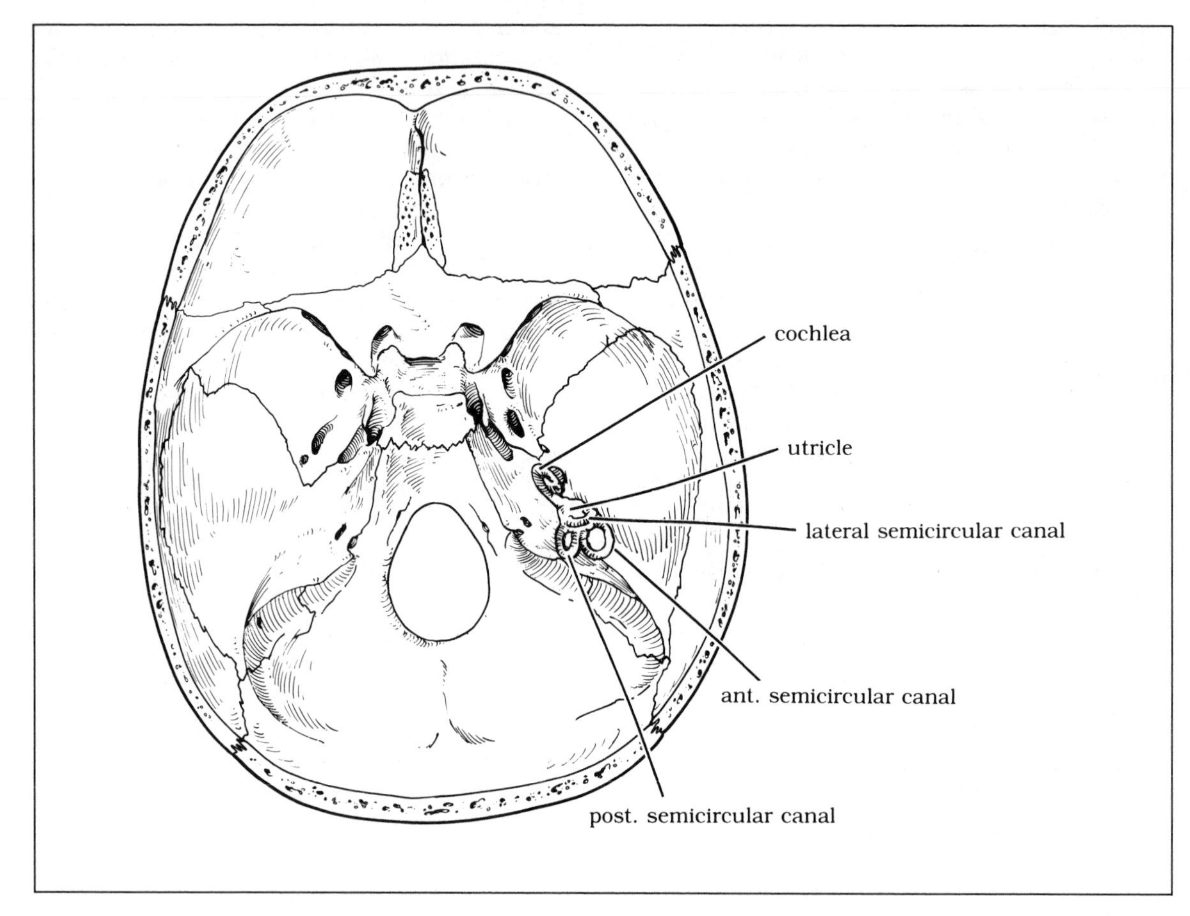

Fig. 7-26. The inner ear within the skull.

Membranous Labyrinth

The three **semicircular ducts** are arranged and named just as the bony semicircular canals are (i.e., anterior, posterior, and lateral). Each duct enlarges laterally into an ampulla that contains an ampullary crest. Neuroepithelial hair cells covered with a gelatinous cupula are found on the ampullary crest. When pulses of air generated by sound enter the inner ear, the hair cells are displaced by the movement of the gelatinous cupula, and an action potential is generated that is carried by the vestibular portion of the vestibulocochlear nerve into the pons and medulla of the brain stem.

The **utricle** and **saccule** are contained in the vestibule. The utricle (elliptical recess) receives the five openings from the semicircular ducts (two from the lateral duct, one from the anterior duct, one from the posterior duct, and one shared by the anterior and posterior ducts). The saccule (spherical recess) connects to the cochlear duct through the ductus reuniens. The utricle and saccule each have a thickened zone, called a **macula,** that contains neuroepithelial hair cells covered with a gelatinous membrane containing crystals of calcium carbonate, the otoliths. The maculae are responsive to changes in gravity.

The endolymphatic duct arises from the utricle and saccular ducts and continues into the vestibular canal, ending as the endolymphatic sac under the dura on the posterior margin of the middle cranial fossa.

The **cochlear duct** turns two and a half times, is over 30 mm long, and is centrally placed in the bony cochlea. Its posterior wall is formed by the basilar membrane, which attaches to the cochlear wall through the spiral ligament. The anterior wall of the cochlear duct is the thin vestibular membrane. (Both membranes extend from an osseous spiral lamina to the wall of the cochlea.) The stria vascularis ductus cochlearis is found in the spiral ligament and secretes the endolymph. The basilar membrane separates the cochlear duct from the scala tympani.

The spiral organ (of Corti), which is the receptive organ for hearing, lies on the basilar membrane. The receptor cells are neuroepithelial hair cells that are attached to a gelatinous membrane, the tectorial membrane. Movement of the tectorial membrane generates the transduction of sound from a mechanical signal to an electrical signal that is transmitted to the central nervous system. The apex of the cochlea responds to lower frequencies, and the basilar portion to higher frequencies.

Blood Supply

The internal ear is supplied by the labyrinthine artery from the basilar or anterior inferior cerebellar artery. The venous drainage connects to the petrosal sinuses.

Innervation

The vestibulocochlear nerve is a purely special somatic afferent fiber (see discussion under Cranial Nerves). It exits the internal acoustic meatus and enters the central nervous system in the medullopontine junction.

The **vestibular nerve** consists of five portions that originate in relationship to the hair cells in the maculae of the utricle and saccule and from the ampullary crests of the semicircular ducts. The cell bodies for this nerve are located in the vestibular ganglion in the internal acoustic meatus. The **cochlear nerve** originates in relationship to the hair cells of the spiral ganglion. Its cell bodies are found in the spiral ganglion in the spiral canal of the modiolus. The vestibular and cochlear nerves join and run in the internal acoustic meatus. The facial nerve accompanies the vestibulocochlear nerve until the ganglion of the facial nerve, the geniculate ganglion, is reached. At this point the facial nerve enters the canal for the facial nerve to emerge at the stylomastoid foramen. The vestibulocochlear nerve continues medially to the internal acoustic meatus and then exits it.

Mouth, Tongue, Teeth, and Oral Pharynx

MOUTH

The cavity of the mouth has a small external portion, the vestibule, and a larger internal portion, the oral cavity. The vestibule is the space between the internal surfaces of the lips and cheeks, and the teeth and gums. The mucous membrane that covers the lips and cheeks reflects onto the gums and forms the roof and floor of the vestibule. Labial glands open into the vestibule. The parotid duct also opens into the vestibule opposite the second molar in the upper jaw.

The oral cavity is broader than the vestibule. The boundaries of the cavity in front and on the side are the alveolar arches, teeth, and gums; the roof is formed by the palate; the floor comprises the tongue and its supporting tissues. The oral cavity communicates anteriorly with the vestibule through the teeth; posteriorly it communicates with the oropharynx.

Lips and Cheeks

The lips are mobile musculoconnective folds that form the opening into the mouth. The median part of the upper lip is called the philtrum. The inner surface of each lip is connected to the adjacent gum by a median fold of mucous membrane, the frenulum of the lip, and is lined by mucous membranes. The lips are covered with skin containing hair. They include labial glands, and the underlying orbicularis oris muscle. The cheeks are similarly organized, being covered by skin containing buccal glands and some hair and with the buccal fat pad underlying the masseter muscle. The parotid duct passes around the anterior border of the masseter muscle, through the buccal fat pad, and then through the buccinator muscle. It opens into the vestibule opposite the upper second molar. The junction between the lips and the cheeks is marked externally by the nasolabial groove, which extends from the nose to the angle of the mouth.

Palate

The palate forms the roof of the mouth and the floor of the nasal cavity and posteriorly it creates a partial division between the oral and the nasal pharynx. The palate has two parts: the anterior two-thirds is the hard, or bony, palate and the posterior third is the soft palate.

The **bony,** or **hard palate** is formed by the palatine process of the maxillae anteriorly and the horizontal plates of the palatine bones posteriorly (see Fig. 7-5). It is covered superiorly by the mucous membrane of the nasal cavity. Inferiorly it is covered by the mucoperiosteum of the bony palate, which contains blood vessels and the mucous palatine glands. A median raphe ends anteriorly in the incisive papilla. Transverse palatine folds or rugae are also seen.

The **soft palate** is a moveable fibromuscular fold suspended from the posterior end of the bony palate. It has two functions: It forms an incomplete partition between the nasopharynx and the oropharynx, and it closes off the pharyngeal isthmus during swallowing and speech. The soft palate is covered by stratified squamous epithelium and contains glands. In the midline a projection, the uvula, is seen. The soft palate is continuous laterally with the palatoglossal and palatopharyngeal arches.

There are five muscles in the soft palate: the palatoglossus, palatopharyngeus, uvulae, levator veli palatini, and tensor veli palatini muscles. These palatine muscles are described in Table 7-9. The greater palatine artery branch from the maxillary artery irrigates this zone.

TONGUE

The tongue is a muscular organ in the floor of the mouth covered by a mucous membrane. The tongue functions in taste, mastication, swallowing, and phonation. It is attached by muscles to the mandible, hyoid bone, styloid process, pharynx, and palate.

Surfaces

The tongue has four surfaces: tip and margin, dorsum, inferior surface, and root. The tip of the tongue rests against the incisor teeth and its margin lies adjacent to the teeth and gums.

The **dorsum** (Fig. 7-27) contains a U-shaped groove, the sulcus terminalis, that marks the boundary between the oral (anterior two-thirds) and pharyngeal (posterior third) parts of the dorsum. The oral part is covered by a mucous membrane with four types of papillae. The filiform papillae, the most numerous papillae, are narrow and conical and the most common papillae in the oral part of the dorsum. The fungiform papillae, which have round red heads and narrow bases, contain taste buds and are usually seen at the apex and margin of the tongue. The circumvallate (vallate) papillae are the largest papillae and are arranged in a U-shaped row just in front of the sulcus terminalis. They also contain taste buds (see discussion of the facial and glossopharyngeal nerves later in this chapter). The foliate papillae are inconstant grooves or ridges near the posterior end of the margin of the tongue.

The pharyngeal part of the dorsum faces posteriorly. Its mucous membrane is devoid of papillae and has many mucous glands and lymphatic follicles called the lingual tonsil. The mucous membrane is reflected onto the epiglottis and labial wall of the pharynx (glossoepiglottic and pharyngoepiglottic folds, respectively).

The **inferior surface** of the tongue is found in the oral cavity and is connected to the floor of the mouth by the frenulum of the tongue. Anterior

Table 7-9. Palatine Muscles

Muscle	*Origin*	*Insertion*	*Innervation*	*Action*
Palatoglossus (in palatoglossal fold)	Palatine aponeurosis in soft palate	Lateral margin of tongue	Cranial nerve XI through pharyngeal plexus	Close off pharynx
Palatopharyngeus (in palatopharyngeal fold)	Palatine aponeurosis, posterior margin of bony palate	Posterior border of thyroid (palatothyroideus) cartilage and esophagus (palatopharyngeus)	Cranial nerve XI through pharyngeal plexus	Separate oropharynx from nasopharynx
Uvulae	Posterior nasal spine of palatine bone	Mucous membrane of uvula	Cranial nerve XI through pharyngeal plexus	Elevate uvula
Levator veli palatini	Undersurface of petrous portion of temporal bone, front of carotid canal, cartilage of auditory tube	Palatine aponeurosis, contralateral levator velum	Cranial nerve XI through pharyngeal plexus	Elevate palate and draw it backward during ingestion of liquids and during speaking; elevate pharynx
Tensor veli palatini	Scaphoid fossa at root of medial pterygoid spine of sphenoid bone	Via tendon into palatine aponeurosis and membranous part of auditory tube	Cranial nerve V_3	Tighten palate during speaking and swallowing

lingual glands are seen in the musculature of the inferior surface.

The **root** of the tongue rests on the floor of the mouth and is attached by muscles to the mandible and the hyoid bone (Fig. 7-28). The nerves, blood vessels, and extrinsic muscles of the tongue enter the tongue through the root.

Musculature

The intrinsic muscles of the tongue originate and insert within the tongue and act to change the shape of the tongue. The extrinsic musculature, arising outside the tongue, moves the tongue about. The four intrinsic muscles—the superior longitudinal, inferior longitudinal, transverse, and vertical muscles—act in concert with the extrinsic muscles to extend the tongue and withdraw or curl it during eating of solids and liquids and during speaking (see Fig. 7-28). The hypoglossal (XII) nerve provides motor innervation to these muscles. The five extrinsic muscles (Fig. 7-29) are the genioglossus, hyoglossus, styloglossus, palatoglossus, and chondroglossus (variable) muscles (Table 7-10).

Blood Supply and Lymphatics

The lingual branch of the external carotid artery irrigates the tongue through the dorsal lingual branches and the arteria profunda linguae (deep lingual artery). Lingual veins receive the dorsal lingual vein and the deep lingual vein. The lingual vein runs on the lateral side of the frenulum (where it can readily be seen) and joins the sublingual vein to form the vena comitans nervi hypoglossi, which drains into the internal jugular or facial veins. The submental, submandibular, and deep cervical lymph nodes drain this region and freely anastomose in the midline.

Innervation

The motor innervation of the tongue is supplied by the hypoglossal (XII) nerve. The sensory supply for general sensation of the anterior two-thirds of the tongue arises from the lingual branch of the mandibular (V_3) nerve, and for taste from the chorda tympani of the facial (VII) nerve. Sensory innervation for general sensation of the posterior third of the tongue and the circumvallate papillae (taste buds) is supplied by the lingual branch of the glossopharyngeal (IX) nerve. Taste buds in the pharynx are supplied by the vagus (X) nerve. (See Cranial Nerves later in this chapter.)

TEETH

The function of the teeth is to prepare solid food for digestion by cutting, grinding, and mascerating it. The teeth also help to maintain the structures they are housed in, the alveolar bone. The teeth are organized into lower right and lower left, upper right and upper left quadrants.

Structure

Each tooth consists of a crown, a neck, and one to three roots. A vascularized, nerve-rich pulp extends from the crown through the root and is covered by three calcified layers: dentin, enamel, and cementum. The anatomical crown is the part of the tooth covered with enamel, while the clinical crown is the part of the tooth projecting into the oral cavity. The neck is adjacent to the crown and is surrounded by the gum. The root of the tooth is covered by cementum and lies in its body socket, or alveolus, the alveolar process of the bones forming the upper and lower jaw. The root canal opens at the tip of the root into apical foramina that permit nerves, blood vessels, and lymphatic vessels to enter the pulp. The blood supply and nerve supply to the lower and upper

Fig. 7-27. The dorsum of the tongue.

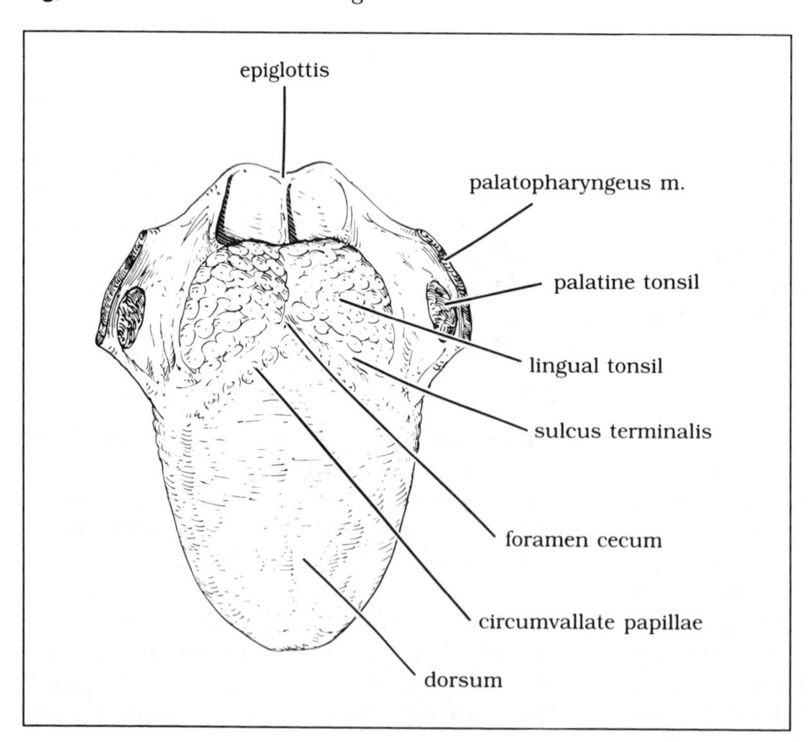

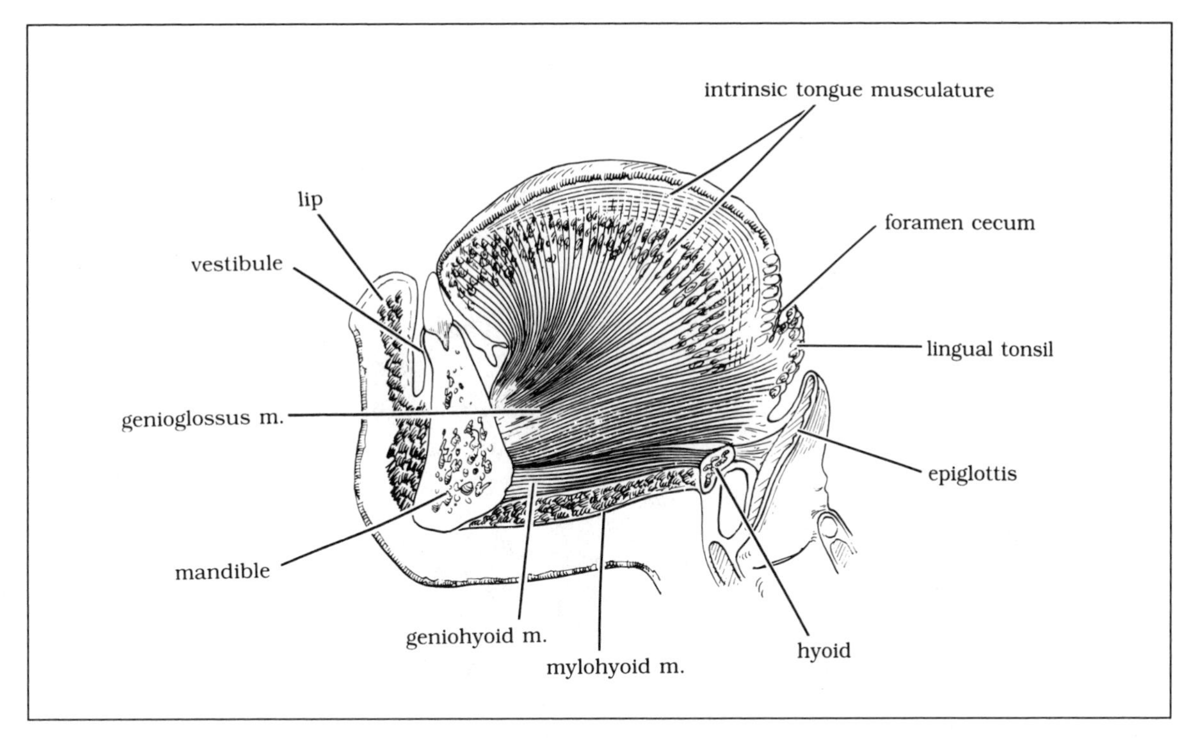

Fig. 7-28. Sagittal view of the tongue, epiglottis, and lower jaw.

jaws including the teeth are described on pages 343 to 345.

Classification

There are four types of teeth: incisors, canines, premolars, and molars (Fig. 7-30).

The eight **incisors** cut the food; the lingual surfaces of their crowns are triangular. On each side, the most medial incisor is called the central incisor and the adjacent incisor is the lateral incisor.

The **canines** are long teeth with a prominent tubercle, or cusp, on their crown. They assist the incisors in cutting the food and are also important in facial expression. There are four permanent canines.

The **premolars** are broad and have two tubercles or cusps on the crown. They are the bicuspid teeth and help in crushing food. There are eight permanent premolars.

The **molars** are the largest teeth, the first molars being the largest of all. They have three to five cusps to help in chopping solid food; the cusps can be worn down. Each upper molar has three roots and each lower molar has two roots. The third molar, the so-called wisdom tooth, may or may not be present and is variable in form. When all wisdom teeth are present, there are twelve permanent molars.

Primary Teeth

At birth there are usually no teeth present. The primary, or milk, teeth appear in the oral cavity between six months and two and a half years of age; the first teeth to erupt are the lower medial incisors at about six months of age. The other lower teeth may erupt before the upper teeth. There are twenty deciduous teeth, five in each quadrant: two incisors, one canine, and two molars. They are smaller than the permanent teeth. By 12 years of age roots of the deciduous teeth

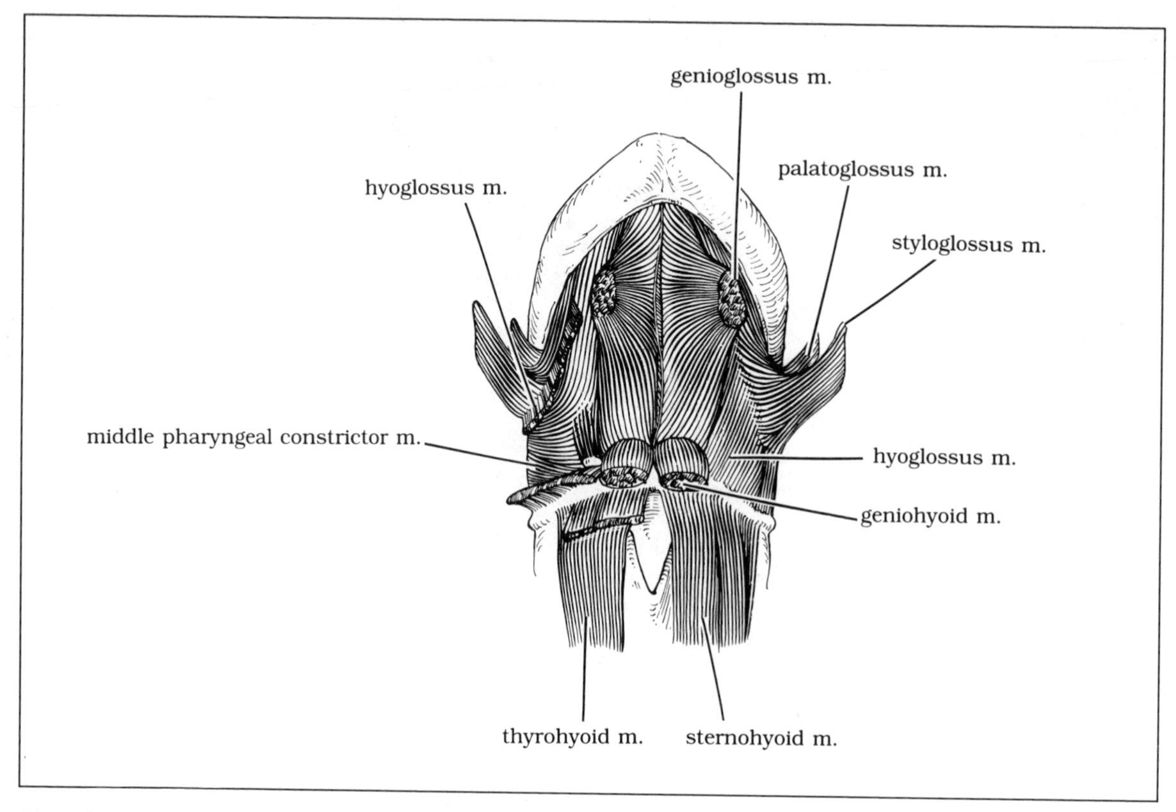

Fig. 7-29. Muscles of the tongue from below.

have been shed because the roots have been loosened by the eruption of the permanent teeth.

Permanent Teeth

There are usually 32 permanent teeth: eight incisors, four canines, eight premolars, and twelve molars. These teeth, organized in the four quadrants, are numbered from one to eight mesially to distally in similarly numbered sockets in either the lower jaw or the upper jaw, left side or right side: 1 is the mesial incisor, 2 is the lateral incisor, 3 is the canine, 4 is the mesial premolar, 5 is the lateral premolar, 6 is the first molar, 7 is the second molar, and 8 is the third molar or wisdom tooth. The first permanent tooth to erupt is the first molar (sixth socket). The teeth in each quadrant usually erupt in the following sequence: 6, 1, 2, 4, 3, 5, 7, and 8.

Occlusion

The teeth are organized into an upper and a lower arch. The upper and lower jaws' coming in contact with each other is termed occlusion. A normal occlusion depends on normal tooth alignment and development. Abnormal occlusion is termed malocclusion.

ORAL PHARYNX

The oral pharynx extends from the level of the soft palate to the hyoid bone. Above it is the nasopharynx, below is the laryngeal part of the pharynx. The anterior wall of the oral pharynx (essentially the glossopalatine arches) is the entrance to the oral cavity. Posteriorly the oral pharynx is contained by the muscular wall formed by the pharyngeal constrictors. Laterally the palatine tonsils lie between the palatoglossal and palatopharyngeal folds.

Table 7-10. Extrinsic Musculature of the Tongue

Muscles	*Origin*	*Insertion*	*Innervation*	*Action*
Genioglossus (bulk of posterior part of tongue)	Superior genial tubercle of mandible	Inferior aspect of tongue	XII	Depress and (posterior part) pull tongue forward and protrude it, prevent tongue from obstructing respiration
Hyoglossus (flat quadrilateral) concealed by mylohyoid	Greater horn and body of hyoid	Side and inferior aspect of tongue	XII	Retract tongue
Chondroglossus (variable)	Hyoid	Dorsum of tongue	XII	Retract tongue
Styloglossus	Front of styloid process, stylomandibular ligament	Side and inferior aspect of tongue	XII	Retract tongue
Palatoglossus (palatoglossal fold)	Palatine aponeurosis, posterior margin of bony palate	Side of tongue	XI through pharyngeal plexus	Elevate tongue

Fig. 7-30. Upper and lower quadrants of teeth. A. Medial view. B. Lateral view.

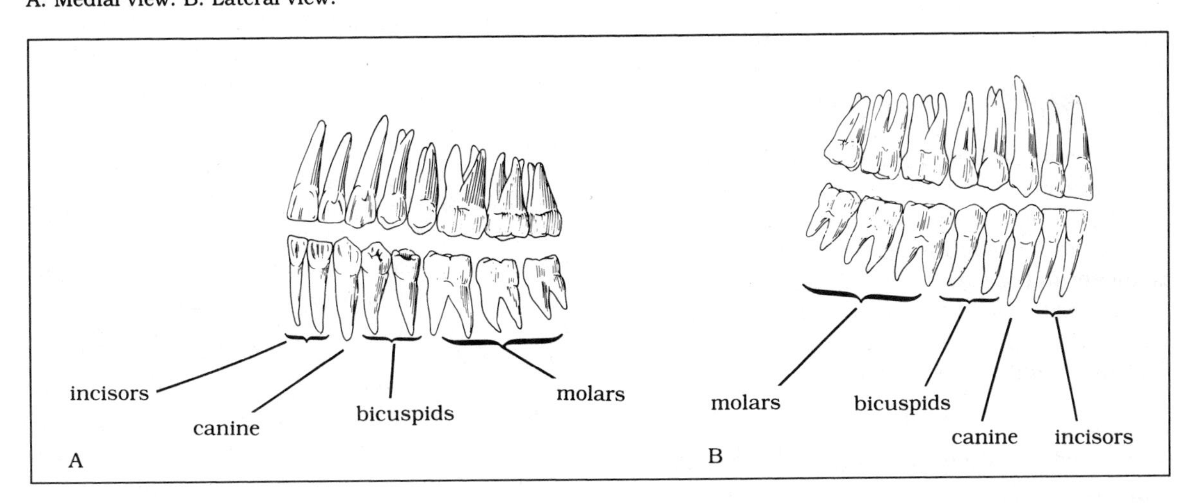

The muscles of the pharynx are the three constrictor muscles, the stylopharyngeus, the palatopharyngeus, and the salpingopharyngeus (see Chap. 8). Overlying the muscles are two layers, a strong submucosal layer known as the pharyngobasilar fascia (overlying the upper margin of the superior pharyngeal constrictor), and a layer of mucous membrane.

The oral cavity and oral pharynx are isolated from the associated respiratory structures during the act of swallowing by elevation and spreading of the soft palate and by movement of the epiglottic and associated cartilages. Food is then pushed down into the esophagus through serial contractions of the constrictor muscles.

Temporomandibular Joint

The temporomandibular joint is the articulation between the mandible and the temporal bone. The articular surfaces on the skull and the mandible are covered by fibrocartilage and separated by a dense connective tissue **articular disk** that allows for different movements at each surface. The mandibular disk interaction is one of a hinge joint; the temporal disk interaction is one of a gliding joint.

The capsule of the temporomandibular joint is thin and loose. It is strengthened laterally by the **lateral (temporomandibular) ligament** between the zygomatic process of the temporal bone and the neck of the mandible (Fig. 7-31). Two additional ligaments are associated with this joint: the sphenomandibular and stylomandibular ligaments. The **sphenomandibular ligament** is a strong ligament extending from the spine of the sphenoid bone to the lingula of the mandible. The **stylomandibular ligament** extends from the styloid process to the ramus of the mandible near the angle; it is a thickening of the parotid fascia.

The temporomandibular joint is innervated by branches of the mandibular (V_3) nerve. Its blood supply is provided by branches of the superficial temporal, middle meningeal, anterior tympanic, and ascending pharyngeal arteries.

Temporal and Infratemporal Fossae

The structures contained within the temporal and infratemporal fossae are primarily concerned with mastication. The fossae contain the temporalis, medial pterygoid, and lateral pterygoid muscles; the maxillary artery and its branches; the mandibular nerve and its branches; and the chorda tympani. The masseter muscle is also considered in the discussion of the temporal and infratemporal fossae because of its functional interactions with the other muscles of mastication.

MUSCULATURE

The **temporalis muscle** occupies the temporal fossa (Fig. 7-32). The muscle arises from the periosteum of the fossa as high as the superior temporal line. Its fibers descend deep to the zygomatic arch and converge to insert on the anterior border and medial surface of the coronoid process of the mandible. The muscle is innervated by the deep temporal branches of the mandibular (V_3) nerve.

The **masseter muscle** lies on the superficial surface of the mandible. It arises from the zygomatic process of the maxilla and the lower border of the zygomatic arch. The muscle inserts into the lateral surface of the coronoid process, ramus, and angle of the mandible. The buccal fat pad separates this muscle from the buccinator. The parotid gland and duct partly cover it. Innervation is accomplished via branches of the mandibular division of the trigeminal nerve, which reach the muscle by passing through the mandibular notch. A branch of the maxillary artery accompanies the nerve.

The **medial pterygoid muscle** parallels the masseter on the medial surface of the mandible within the infratemporal fossa. It arises from the medial surface of the lateral pterygoid plate and from the pyramidal process of the palatine bone. A small piece also arises from the tuberosity of the maxilla. The muscle inserts on the medial surface of the ramus and angle of the mandible as high as the mandibular foramen. The muscle is innervated by a branch of the mandibular nerve.

The **lateral pterygoid muscle** runs horizontally in the infratemporal fossa. It has two heads of origin. The larger head arises from the lateral surface of the lateral pterygoid plate, the smaller from the greater wing of the sphenoid bone in the

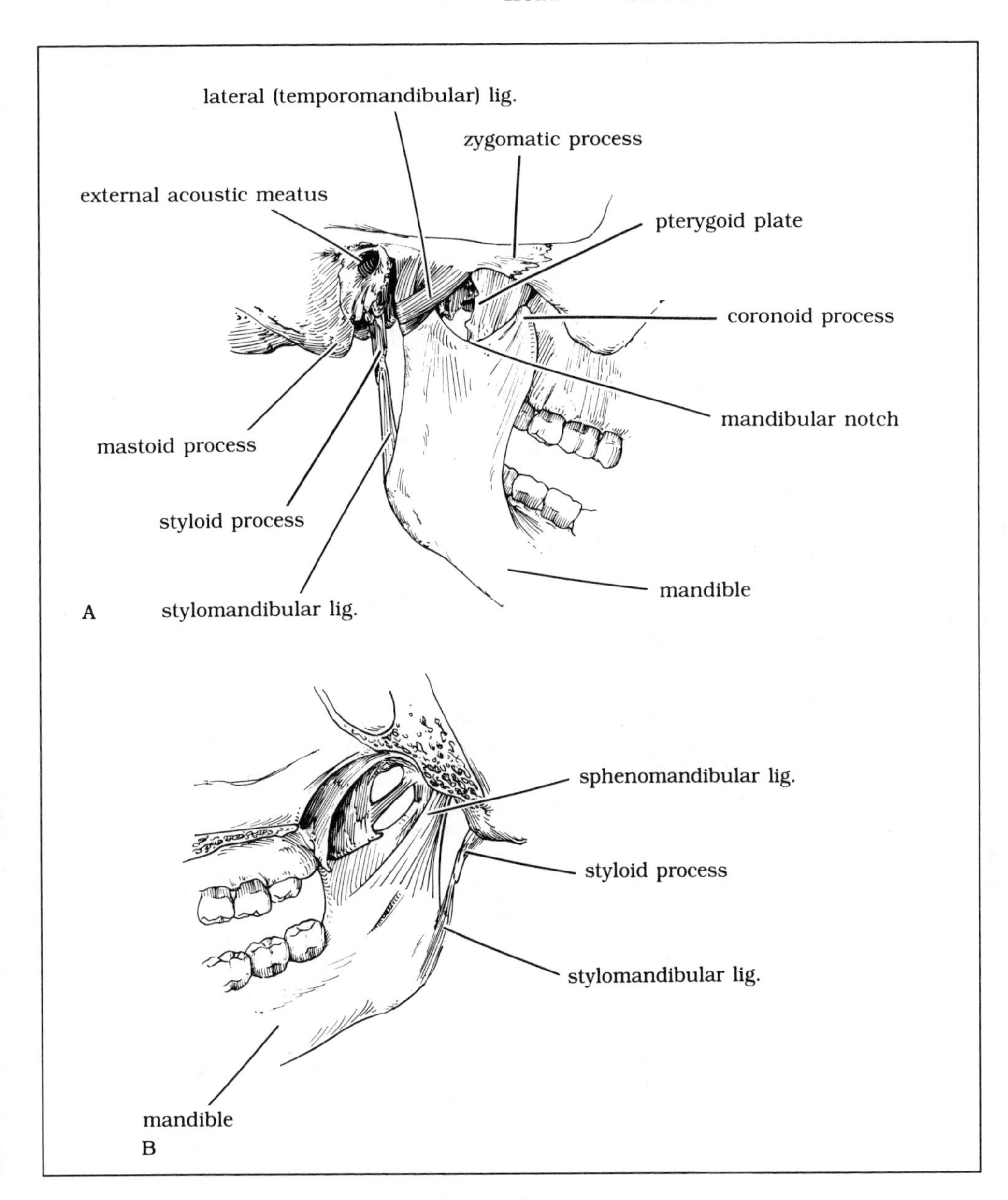

Fig. 7-31. Mandible and temporomandibular joint. A. External surface. B. Internal surface.

infratemporal fossa. The larger head inserts onto the neck of the mandible, the smaller onto the articular disk of the temporomandibular joint and the neck of the condyle. The mandibular nerve innervates this muscle.

In mastication the lower jaw moves up and down, forward and backward, and laterally. The elevators of the mandible are the masseter, temporalis, and medial pterygoid muscles; the depressors of the mandible are the lateral pterygoid muscle—in concert with the suprahyoid, mylohyoid, digastric, geniohyoid, and infrahyoid

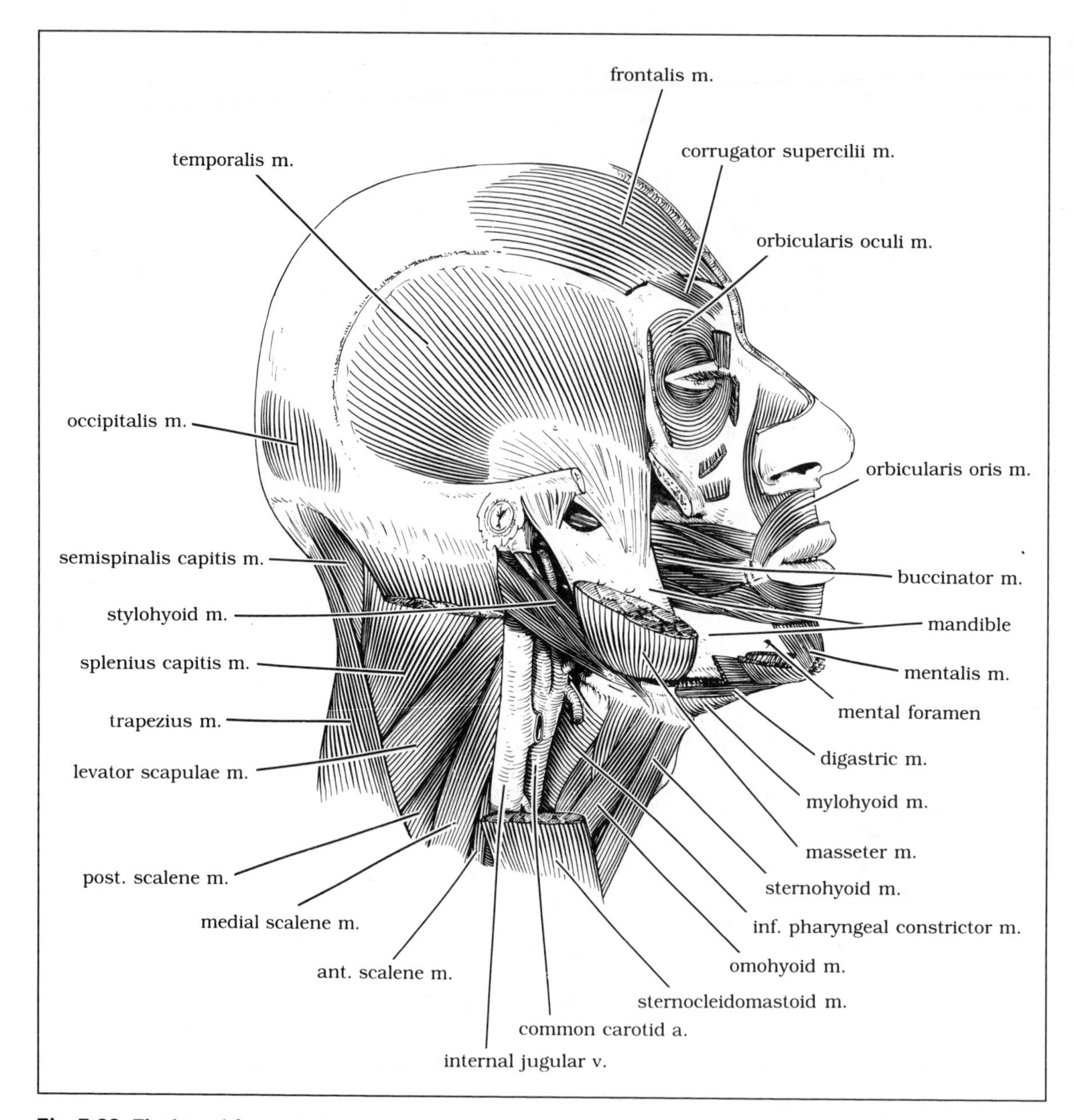

Fig. 7-32. The lateral face with the parotid gland and the masseter and sternocleidomastoid muscles removed. Deeper view.

muscles—and the sternohyoid and omohyoid muscles (the depressors of the hyoid bone) along with gravity. The protruders of the mandible are the medial and lateral pterygoid muscles assisted by the masseter muscle. The retractors of the mandible are the temporalis and digastric muscles. The pterygoid muscles produce side-to-side movement.

BLOOD SUPPLY AND INNERVATION

The **maxillary artery** is one of the two terminal branches of the external carotid artery, the other being the superficial temporal artery. The maxillary artery passes anteriorly, medial to the mandible and through the infratemporal fossa, ending at the pterygopalatine fossa by dividing into terminal branches. This artery is discussed later in this chapter, under Arterial Supply to the Head.

The **maxillary vein** lies behind the neck of the mandible. It joins with the superficial temporal vein to form the retromandibular vein. It receives tributaries from the pterygoid plexus, which lies on and around the pterygoid muscles, from the veins accompanying the maxillary artery branches, and from communications from the cavernous sinus, facial vein, and inferior ophthalmic veins.

The nerves of the temporal and infratemporal fossae are the mandibular nerve and its branches and the chorda tympani, which are described later in this chapter with the trigeminal and facial nerves, respectively.

Arterial Supply to the Head

Arterial blood is supplied to the head by the external and internal carotid arteries (branches of the common carotid artery) and the basilar artery (formed by the fusion of the two vertebral branches of the subclavian artery).

EXTERNAL CAROTID ARTERY

The external carotid artery begins opposite the upper border of the thyroid cartilage (Fig. 7-33). It passes superiorly and anteriorly, entering the space behind the neck of the mandible, where it divides into the superficial temporal and maxillary arteries. This vessel quickly diminishes in size in the neck as it gives off numerous branches. The branches of the external carotid in their order of origin are as follows:

1. Superior thyroid
2. Ascending pharyngeal
3. Lingual
4. Facial
5. Occipital
6. Posterior auricular
7. Superficial temporal
8. Maxillary

The superior thyroid, ascending pharyngeal, and lingual arteries are discussed with the neck in Chapter 8.

Facial Artery

The facial artery arises in the carotid triangle under the ramus of the mandible, passes under the digastric and stylohyoid muscles, and winds around the inferior border of the mandible at the anterior edge of the masseter to enter the face. The facial artery in the neck has eight branches:

The **ascending palatine artery** supplies the soft palate, palatine gland, superior pharyngeal constrictor muscle, palatine tonsil, and auditory tube.

The **tonsillar branch** supplies the palatine tonsil and root of the tongue.

The **glandular branches** supply the submandibular gland and adjacent muscles.

The **submental branch** supplies the mylohyoid and digastric muscles and the lip. It anastomoses with the inferior labial and mental arteries.

The **inferior labial branch** supplies the labial glands, mucous membranes, and muscles of the lower lip. It anastomoses with the opposite inferior labial branch and the mental branch of the inferior alveolar and submental arteries.

The **superior labial branch** supplies the upper lip and gives off branches to the nasal septum and ala of the nose.

The **lateral nasal** branch arises alongside the nose. It supplies the nose and anastomoses with branches of the ophthalmic and infraorbital branches of the maxillary artery.

The **angular branch** is the terminal branch of the facial artery. It reaches to the medial angle of the orbit and anastomoses with the infraorbital and dorsal nasal branches of the ophthal-

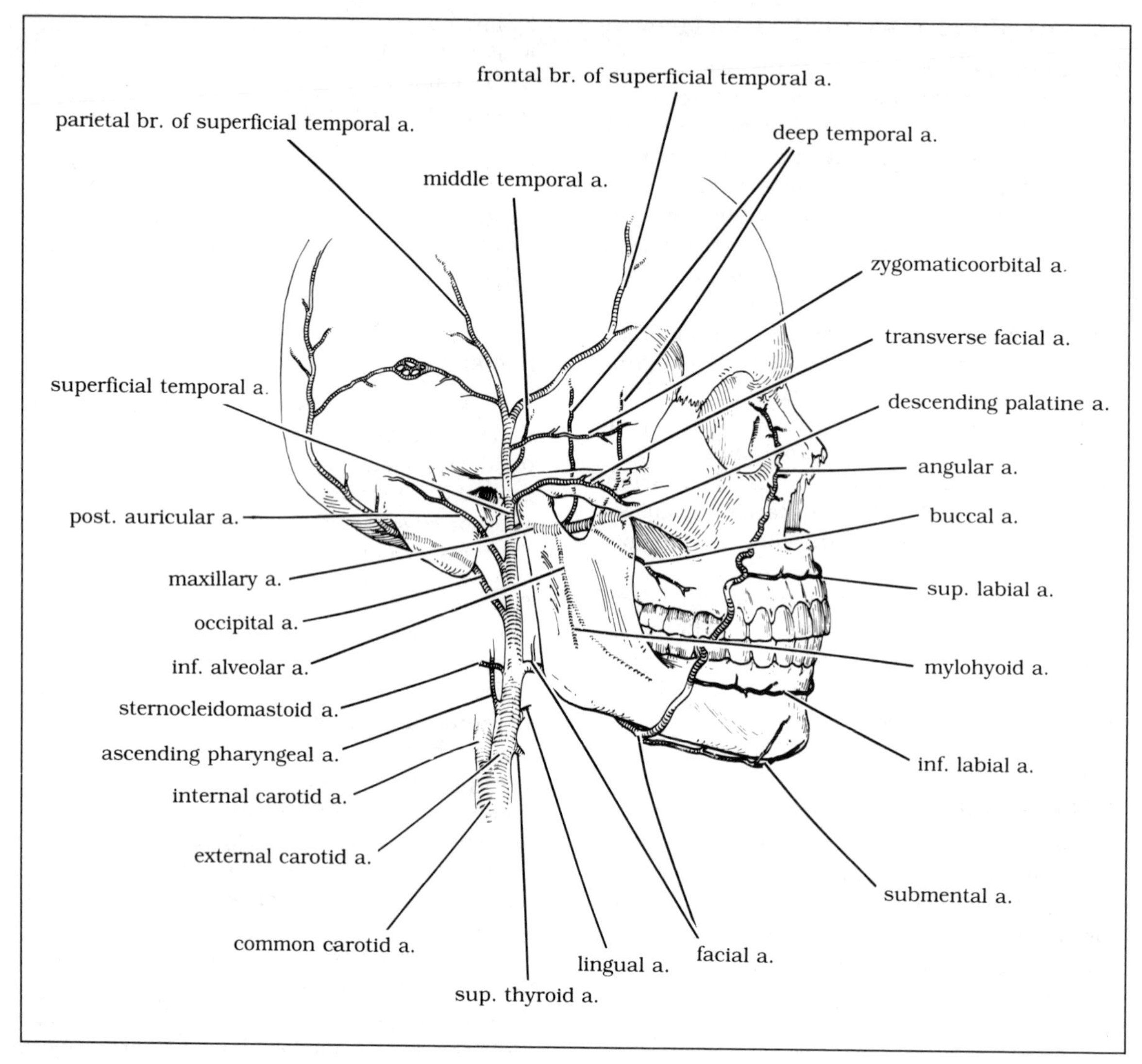

Fig. 7-33. Branches of the external carotid artery in the head.

mic artery, supplying the lacrimal sac and the orbicularis oculi muscle.

Occipital Artery

The occipital artery arises from the posterior part of the external carotid opposite the origin of the facial artery. There are six branches of the occipital:

The **mastoid branch** supplies the muscles attached to the mastoid process and the mastoid itself.

The **sternocleidomastoid branch** supplies the sternocleidomastoid muscle with the spinal accessory nerve.

The **auricular branch** supplies the back of ear and dura mater, diploë and mastoid air cells through the mastoid foramen.

The **meningeal branch** supplies the dura mater

in the posterior fossa through the jugular foramen and condylar canal.

The **descending branch** divides into a superficial and deep portion. The superficial portion supplies the splenius and trapezius muscles. The deep portion anastomoses with the vertebral artery and the deep cervical artery. These anastomotic channels provide collateral circulation after ligation of the common carotid or subclavian.

The **occipital** or terminal branches supply the back of the head, lying between the skin and the occipital muscle.

Posterior Auricular Artery

The posterior auricular artery arises near the styloid process, supplies the parotid gland, and has four branches:

The **stylomastoid artery** runs through the stylomastoid foramen into the tympanic cavity and semicircular canals.

The **auricular branch** supplies the back of the ear and ear muscles, anastomosing with the parietal and anterior auricular branches of the superficial temporal.

The **occipital branch** supplies the scalp above and behind the ear, passing over the sternocleidomastoid to supply the occipital muscles and overlying scalp.

The **posterior tympanic artery** supplies the middle and inner ear.

Superficial Temporal Artery

This artery is the smaller of the two terminal branches of the external carotid. It begins in the parotid gland and has seven branches:

The **parotid branches** supply the parotid gland.

The **transverse facial artery** supplies the parotid gland and duct, and the masseter. Overlying the masseter it anastomoses with the masseteric, buccal, and infraorbital arteries.

The **middle temporal artery** supplies the temporalis muscle, anastomosing with the deep temporal branch.

The **zygomatico-orbital artery** runs from the middle temporal artery along the superior border of the zygomatic arch to supply the orbicularis oculi.

The **anterior auricular branch** supplies the anterior portion of the ear, anastomosing with the posterior auricular artery.

The **frontal branch** supplies the forehead, anastomosing with the supraorbital artery.

The **parietal branch** supplies the scalp and muscle on the side of the head (temporal fascia).

Maxillary Artery

The maxillary artery is the larger of the two terminal branches of the external carotid and is found in the infratemporal fossa (Fig. 7-34). It arises within the parotid compartment, where it is embedded in the parotid gland, and passes deep to the neck of the mandible. It runs anteriorly between the neck of the mandible and the sphenomandibular ligament and then deep (or superficial) to the lateral pterygoid muscle and into the pterygopalatine fossa. The maxillary artery is divided into three portions: mandibular, pterygoid, and pterygopalatine.

The mandibular, or first portion, has five branches:

The **deep auricular artery** supplies the outer surface of the tympanic cavity and the temporomandibular joint. It arises in the parotid gland, and pierces the wall of the external acoustic meatus to reach the tympanic cavity.

The **anterior tympanic artery** reaches the tympanic cavity by passing through the petrotympanic fissure. It anastomoses with the artery of the pterygoid canal and the caroticotympanic arteries of the internal carotid artery.

The **inferior alveolar artery** joins the inferior alveolar nerve to enter the mandibular foramen and run in the mandibular canal to the first molar. Here it divides into mental, incisor, and dental branches. The mental branch exits through the mental foramen to supply the chin and lower lips. The incisor branches, or terminations, supply the incisor teeth. Dental branches enter the apertures in the roots of the teeth and supply the tooth pulp. The inferior

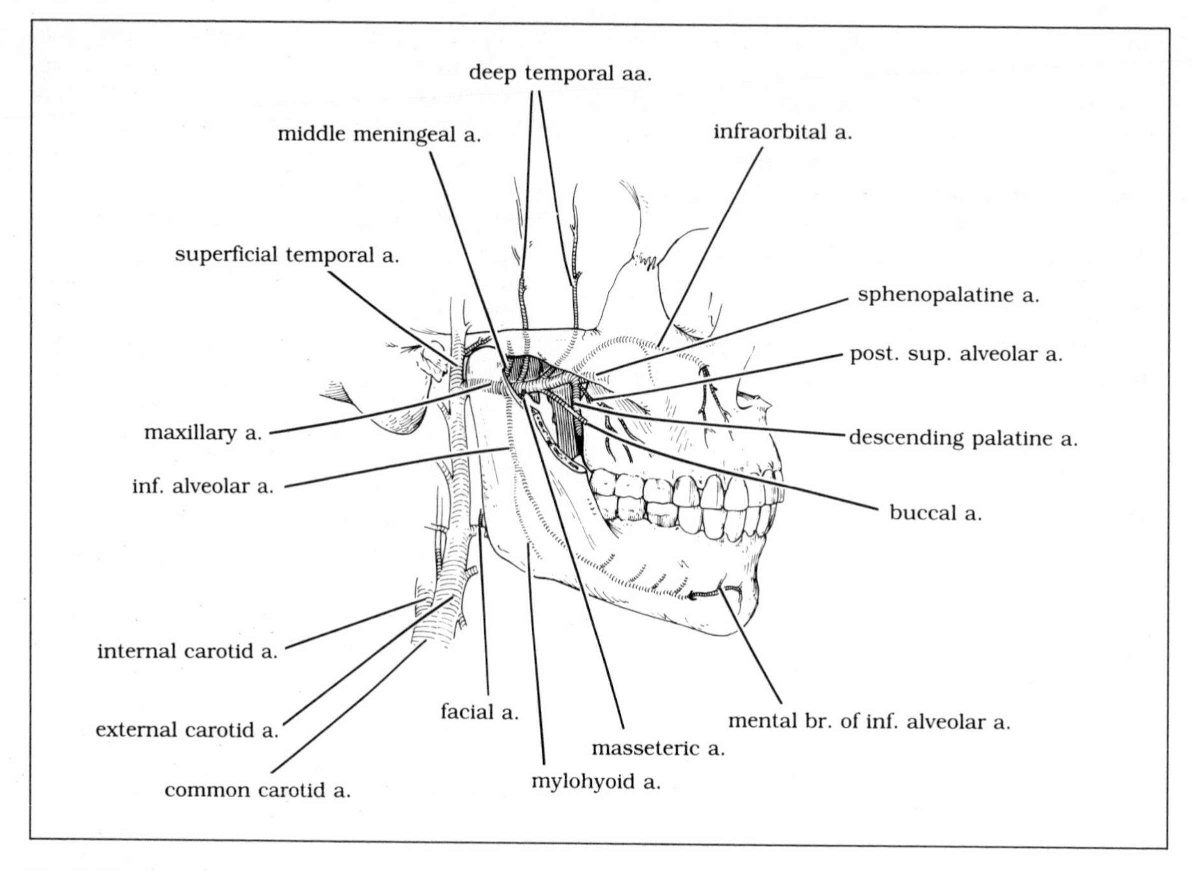

Fig. 7-34. The maxillary artery and its branches.

alveolar artery also has a lingual branch to the mucosa of the mouth and a mylohyoid branch to the mylohyoid muscle; both these branches arise prior to the artery's entering the mandibular foramen.

The **middle meningeal artery** is the largest most constant artery to the dura and is consequently a clinically significant vessel. It enters the cranial cavity through the foramen spinosum and runs on the greater wing of the sphenoid bone. In the middle cranial fossa it divides into anterior and posterior branches. These vessels supply the dura and bones in the middle and anterior fossae.

The **accessory meningeal artery** enters the cranium via the foramen ovale as a supplement to the middle meningeal artery. It supplies the trigeminal ganglion and surrounding dura mater.

The pterygoid, or second portion, originates inferior to the lateral pterygoid and often runs deep to the two heads of this muscle. Each branch of this portion travels with a branch of the mandibular division of the trigeminal nerve. This portion has four branches:

The **deep temporal arteries** (anterior and posterior) supply the temporalis muscle and pericranium.

The **pterygoid branches** are numerous short branches supplying the pterygoid muscles.

The **masseteric arteries** supply the masseter muscle and anastomose with the masseteric branches of facial arteries.

The **buccal artery** supplies the external surface

of buccinator muscle, the cheek, and the mouth; it anastomoses with branches of facial and infraorbital arteries.

The pterygopalatine, or third portion, is found in the pterygopalatine fossa and ends as the sphenopalatine artery. It has six major branches, each of which accompanies a branch of the maxillary nerve.

The **posterior superior alveolar artery** comes off as the vessel enters into the pterygopalatine fossa. There are numerous branches of this vessel entering the alveolar canal to supply the molar and premolar teeth, the lining of the maxillary sinus, the gingivae, and the buccinator muscle.

The **infraorbital artery** enters the orbit through the infraorbital fissure and runs along the infraorbital groove and canal with the infraorbital nerve. It passes through the infraorbital foramen to emerge on the face. Orbital branches arise in the infraorbital canal and supply the lacrimal gland and the inferior oblique and inferior rectus muscles. Anterior superior alveolar arteries descend in the anterior alveolar canal, supplying incisor and canine teeth. The facial branches emerge from the infraorbital foramen to supply the medial angle of the orbit and the lacrimal sac. Some run to the nose and anastomose with ophthalmic branches. Other branches descend between the levator labii and anastomose with facial and buccal arteries.

The descending palatine artery has two branches. The **greater palatine artery** arises in the pterygopalatine fossa and descends with the greater palatine nerve in the greater palatine canal. It exits from the greater palatine foramen and supplies the hard palate, gingivae, palatine glands, and mucous membranes of the mouth. The **lesser palatine arteries** arise in the pterygopalatine canal and descend through the lesser palatine canal to supply the soft palate and palatine tonsil. They are accompanied by the lesser palatine nerves.

The **artery of the pterygoid canal** enters with the nerve of the same name and supplies the upper pharynx, auditory tube, sphenoidal sinus and tympanic cavity. (Branches of both the maxillary artery and internal carotid artery are capable of filling this vessel.) The **pharyngeal artery** passes posteriorly through the pharyngeal canal to supply the roof of the pharynx, the sphenoid sinus, and the auditory tube.

The **sphenopalatine artery** is the terminal branch of the maxillary. It enters the nasal cavity via the sphenopalatine foramen and divides into a posterior lateral nasal branch supplying the meatuses and conchae and posterior septal branches to the nasal septum. The **sphenopalatine artery** accompanies the nasopalatine nerve to the nasal cavity via the sphenopalatine foramen.

INTERNAL CAROTID ARTERY

The internal carotid branch of the common carotid artery has four portions: cervical, petrous, cavernous, and cerebral. The cervical portion has no branches.

Petrous Portion

The petrous portion of the internal carotid artery has two branches: the **caroticotympanic arteries,** which enter the tympanic cavity through small foramina in the carotid canal, and the **artery of the pterygoid canal,** which is small and is found in the pterygoid canal anastomosing with the maxillary artery (see previous discussion of maxillary artery).

Cavernous Portion

The cavernous portion of the internal carotid artery has three branches: (1) the branches in the cavernous and inferior petrosal sinuses; (2) the anterior meningeal artery, which supplies the dura in the anterior cranial fossa and anastomoses with the posterior ethmoid artery; and (3) the hypophysial group of arteries, which supplies the hypophysis through numerous small branches.

Cerebral Portion

The cerebral portion of the internal carotid artery has five branches: ophthalmic, middle cerebral, anterior cerebral, anterior choroidal, and posterior communicating arteries (Fig. 7-35).

OPHTHALMIC ARTERY

The ophthalmic artery arises from the internal carotid as it leaves the cavernous sinus (see Fig. 7-17). It enters the orbital cavity through the optic canal and runs in the medial wall of the orbit, dividing into two terminal branches, the supratrochlear and dorsal nasal. The branches of the ophthalmic artery are divided into an orbital group and an ocular group.

Orbital Branches

The **lacrimal artery,** one of the ophthalmic artery's largest branches (see Fig. 7-17), arises near the optic canal and joins the lacrimal nerve before entering the lacrimal gland. It gives rise to lateral palpebral branches, terminal branches that supply the eyelid. **Zygomatic vessels** leave the lacrimal artery to enter the zygomatico-orbital foramen and then the temporal fossa. They anastomose with the deep temporal artery. A **recurrent branch** runs backward through the superior orbital fissure to the dura, anastomosing with the middle meningeal artery.

The **supraorbital artery** accompanies the supraorbital nerve supplying the skin and muscles of the forehead, anastomosing with the supratrochlear and frontal branches of the superficial temporal artery. In the orbit it supplies the rectus superior and levator palpebrae superioris muscles.

The **posterior ethmoid artery** is smaller than the anterior ethmoid artery. It enters the posterior ethmoid foramen, supplying the ethmoid air cells, and then enters the cranial cavity, giving off branches to the dura and nasal cavity. The **anterior ethmoid artery** runs with the nasociliary nerve and enters the anterior ethmoid canal, supplying the anterior and middle ethmoid air cells and the frontal sinus. It then enters the cranial cavity, supplying the dura. Nasal branches enter the nasal cavity through the crista galli.

The **medial palpebral artery** arises from the ophthalmic artery near the trochlea for the superior oblique muscle and encircles the eyelids at their free margin to form superior and inferior palpebral anastomoses.

The **supratrochlear artery,** one of the terminal branches of the ophthalmic artery, leaves the orbit at its medial angle and supplies the skin and muscle up to the forehead, anastomosing with the supraorbital artery. The **dorsal nasal artery** is the other terminal branch of the ophthalmic artery. It exits the orbit after supplying the lacrimal sac and divides into branches that supply the root and dorsum of the nose.

Ocular Branches

The **central artery of the retina** is the first branch of the ophthalmic artery and one of the smallest. It runs in the dural sheath of the optic nerve and pierces the nerve to run in the center of the optic nerve and enter the substance of the retina.

The **ciliary arteries** are divided into three groups: the long posterior ciliary, short posterior ciliary, and anterior ciliary arteries (see Fig. 7-17). There are six to twelve **short posterior ciliary arteries** that arise from the ophthalmic artery and surround the optic nerve. Running to the posterior part of the eyeball, they pierce the sclera near the entrance of the nerve and supply the choroid and ciliary processes. The paired **long posterior ciliary arteries** pierce the posterior part of the sclera and run between the sclera and choroid to the ciliary muscle. Here they divide into branches that run around the circumference of the iris, forming major and minor arterial circles.

Muscular branches arise from the trunk of the ophthalmic artery, forming superior and inferior branches. The superior vessel supplies the levator palpebrae superioris, superior rectus, and superior oblique muscles. The inferior vessel supplies the inferior, lateral, and medial rectus muscles and the inferior oblique. The muscular branch also gives rise to the anterior ciliary arteries, which pierce the sclera and supply the iris and ciliary body.

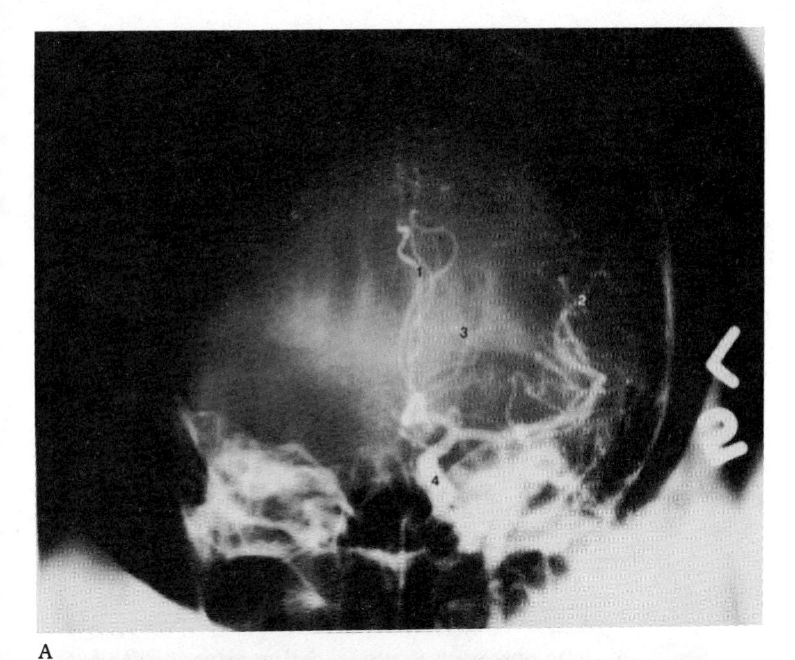

A

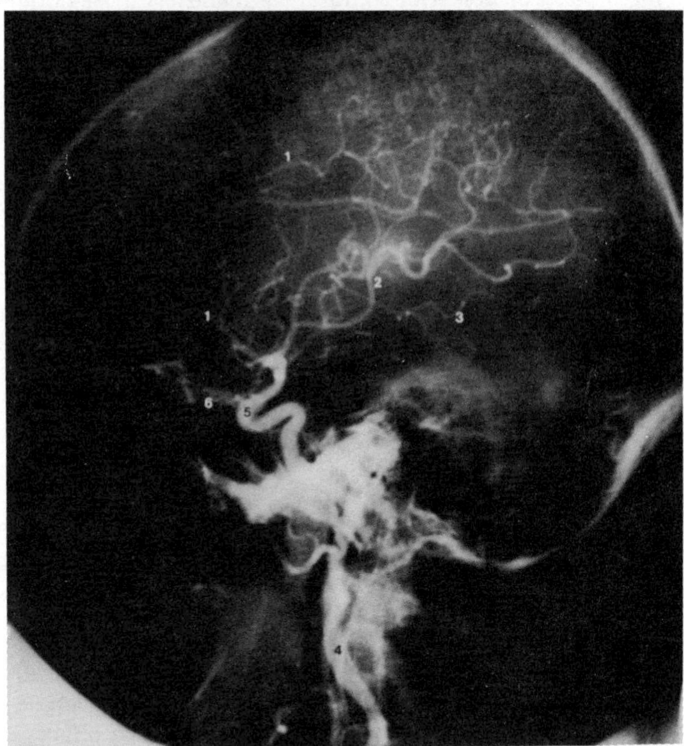

B

Fig. 7-35. X-ray films of the arterial supply to the brain. A. Anterior view. B. Lateral view. (1 = anterior cerebral artery; 2 = middle cerebral artery; 3 = posterior cerebral artery; 4 = internal carotid artery; 5 = carotid siphon; 6 = ophthalmic artery.)

MIDDLE CEREBRAL ARTERY

The middle cerebral artery is the largest branch of the internal carotid and is considered its direct continuation (see Fig. 7-35). The artery is first seen to run laterally in the lateral cerebral fissure, where it passes backward over the insular cortex. The lateral thalamostriatal branch arises from the middle cerebral artery in the most medial part of the fissure and penetrates the brain, supplying most of the thalamus, most of the caudate, the lateral portion of the globus pallidus, and the upper portion of the internal capsule. The following cortical branches of the lateral thalamostriatal artery are named for the cortical gyri they supply.

The **inferior lateral frontal branch** supplies the inferior frontal gyrus, Broca's area, and the lateral orbital gyrus.

The **ascending frontal branch** supplies the anterior pole of the frontal lobe.

The **ascending parietal branch** supplies the postcentral gyrus and much of the superior parietal lobule.

The **central branch** supplies the lower portions of the pre- and postcentral gyri.

The **parietotemporal branch** supplies the inferior parietal lobule, the supramarginal and angular gyri, and the posterior portion of the superior and middle temporal gyri.

The **temporal branches** supply the superior and middle temporal gyri.

The **medial thalamostriatal branch** penetrates the brain and supplies the lower part of the basal nuclei and the anterior limb of the internal capsule.

ANTERIOR CEREBRAL ARTERY

The anterior cerebral artery arises from the internal carotid at the medial end of the lateral cerebral fissure and passes forward to enter the longitudinal fissure, where it approaches the other anterior cerebral artery and is connected to it by the anterior communicating artery. The two anterior cerebrals run together in the longitudinal fissure, curving around the genu of the corpus callosum and running on the corpus callosum to where they anastomose with the posterior cerebral artery in the vicinity of the parieto-occipital fissure.

ANTERIOR CHOROIDAL ARTERY

The anterior choroidal artery arises from the internal carotid near the origin of the posterior communicating artery. It runs along the optic tract and cerebral peduncle to the lateral geniculate, where a major branch enters the choroid plexus in the inferior horn of the lateral ventricle and supplies the lateral geniculate, globus pallidus, and the posterior third of the internal capsule.

POSTERIOR COMMUNICATING ARTERY

The posterior communicating artery arises from the internal carotid and runs posteriorly to join the posterior cerebral artery. It is commonly very small. Branches from this vessel supply the genu of the internal capsule and anterior portions of the posterior limb, and the hypothalamus and infundibulum.

VERTEBRAL ARTERY

The vertebral artery is the first and largest branch of the subclavian artery (Fig. 7-36). It arises in the neck and runs superiorly through foramina in the transverse processes of the upper six cervical vertebrae. It penetrates the foramen in the atlas, runs posteriorly on a groove in the posterior arch of the atlas to enter the foramen magnum under the atlanto-occipital membrane, and joins the vertebral artery of the other side to form the basilar artery. The branches of the vertebral artery are covered in Chapter 8.

The **spinal branches** enter the vertebral canal through the intervertebral foramina and divide into branches for the spinal cord and its nerves and ganglia and its membranes. They anastomose with other spinal arteries and unite with branches from cranial and segmental arteries to form anastomoses on the dorsal surface of the vertebral bodies. The **meningeal branch** arises in

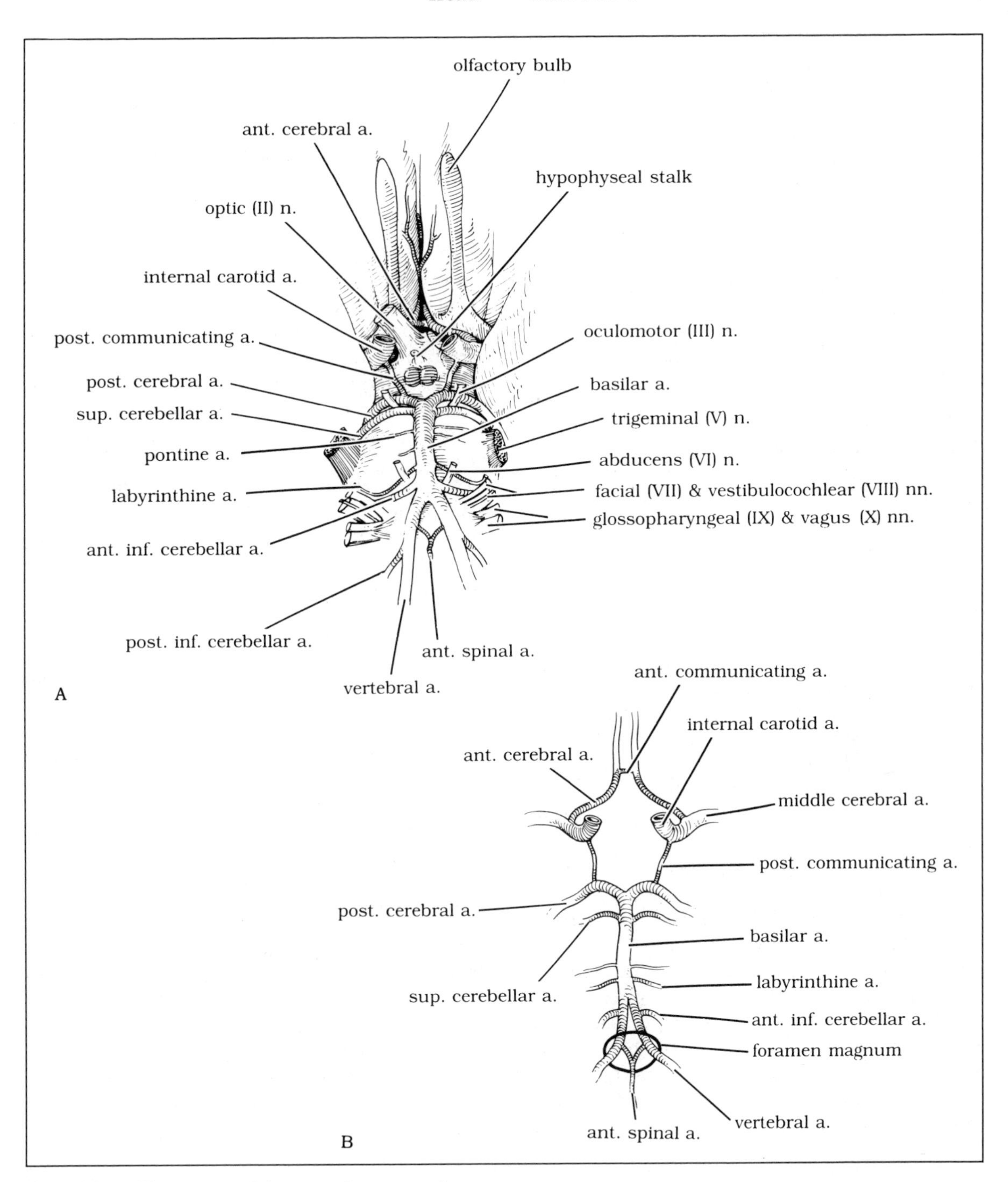

Fig. 7-36. A. The origins of the cranial nerves and their relationship to the circle of Willis. B. The arteries forming the circle of Willis.

the foramen magnum and supplies the falx cerebri and cerebellar fossa.

The **posterior spinal branch** arises from the vertebral artery at the lateral margin of the medulla, posterior to the dorsal roots of the spinal nerves; at each segment of the cord it receives a spinal branch. The posterior spinal vessel, found in this position down to the cauda equina, supplies the posterior third of the spinal cord and the dorsal roots.

The **anterior spinal artery** arises near the formation of the basilar artery and runs in the anterior median fissure. This artery anastomoses inferiorly with segmental arteries at each spinal level (cervical vertebral and ascending cervical; intercostal arteries at the thoracic level; lumbar, iliolumbar, and lateral sacral arteries at the lumbar and sacral levels). This vessel supplies the anterior two-thirds of the spinal cord and the ventral roots.

BASILAR ARTERY AND THE CEREBRAL ARTERIAL CIRCLE

After giving off the anterior spinal and posterior inferior cerebellar arteries, the two vertebral arteries unite to form the basilar artery. This artery runs in the midline of the ventral surface of the pons and gives off three paired branches, the anterior inferior cerebellar, superior cerebellar, and labyrinthine arteries. The basilar artery then ends by bifurcating into two posterior cerebral arteries.

The **posterior cerebral arteries** are connected to the internal carotid arteries by posterior communicating arteries. The terminal two **anterior cerebral arteries** are connected to each other by an anterior communicating artery. The result of these interconnections is an arterial circle (circle of Willis) surrounding the optic chiasm and mamillary bodies on the base of the brain. This vascular connection is important for equalizing pressure variations in any of the major vessels supplying the brain.

Venous Drainage of the Head

The circulation in cerebral veins and sinuses is summarized in Figure 7-37.

CEREBRAL VEINS

In the cerebral hemispheres veins are seen in the sulci and are usually external to arteries. There are superior, middle, and inferior cerebral veins. The **superior cerebral veins** drain the superior surface of the cerebrum and drain into the superior sagittal sinus. The **middle cerebral veins** drain the lateral and some of the inferior surface of the cerebral hemispheres and drain into the cavernous sinuses. The **inferior cerebral veins** drain the inferior surface of the temporal lobe and the occipital lobe and enter directly into the transverse sinus.

The internal cerebral vein is formed near the median aperture of the fourth ventricle (the foramen of Monro) by the fusion of vessels from the basal ganglia, septum, and choroid plexus. The internal cerebral veins pass posteriorly in the roof of the third ventricle and unite with the basal vein (Rosenthal's vein) to form the single great cerebral vein (Galen's vein), which joins the straight sinus.

BASAL VEINS

The paired basal veins originate near the anterior perforated substance by the union of vessels from the corpus callosum, thalamus, insula, and anterior temporal lobe. The basal veins pass around the cerebral peduncle and behind the splenium of the corpus callosum to join the internal cerebral vein and help to form the great cerebral vein.

CEREBELLAR VEINS

The superior cerebellar veins drain the superior surface of the cerebellum and join either the straight sinus or the great cerebral vein. Some of these vessels join the inferior petrosal sinus and transverse sinus, while others enter the occipital sinus. Veins from the medulla and pons terminate in the superior and inferior petrosal sinuses.

SINUSES

There are six major cranial dural sinuses (Fig. 7-38):

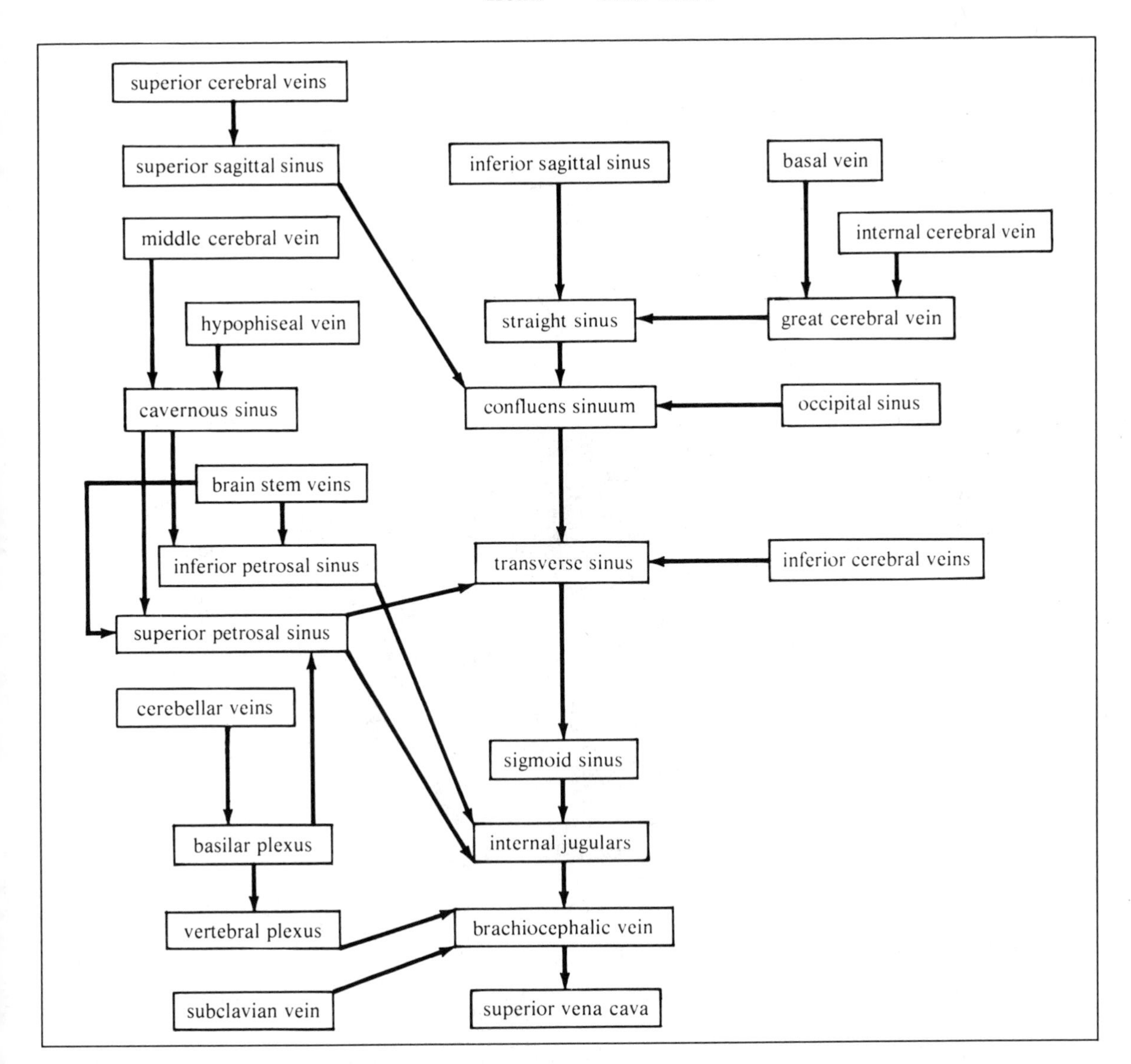

Fig. 7-37. Diagram of circulation in cerebral veins and sinuses.

Paired	*Unpaired*
Transverse sinuses	Superior sagittal sinus
Sigmoid sinuses	Inferior sagittal sinus
Cavernous sinuses	Straight sinus

The **superior sagittal sinus** is found in the attached margin of the falx cerebri. It receives the superior cerebral veins and communicates through the parietal emissary vein with the superficial temporal vein. It drains into the **confluens sinuum** (confluence of sinuses or torcular Herophili). The **inferior sagittal sinus** is found in the free margin of the falx cerebri. It also receives the superior cerebral veins and joins with the great cerebral vein to form the **straight sinus,** which continues to the confluens sinuum. At the **confluens sinuum,** the superior sagittal and straight sinuses become continuous with the right and left transverse sinuses. The **transverse sinuses** are found in the inner surface of the oc-

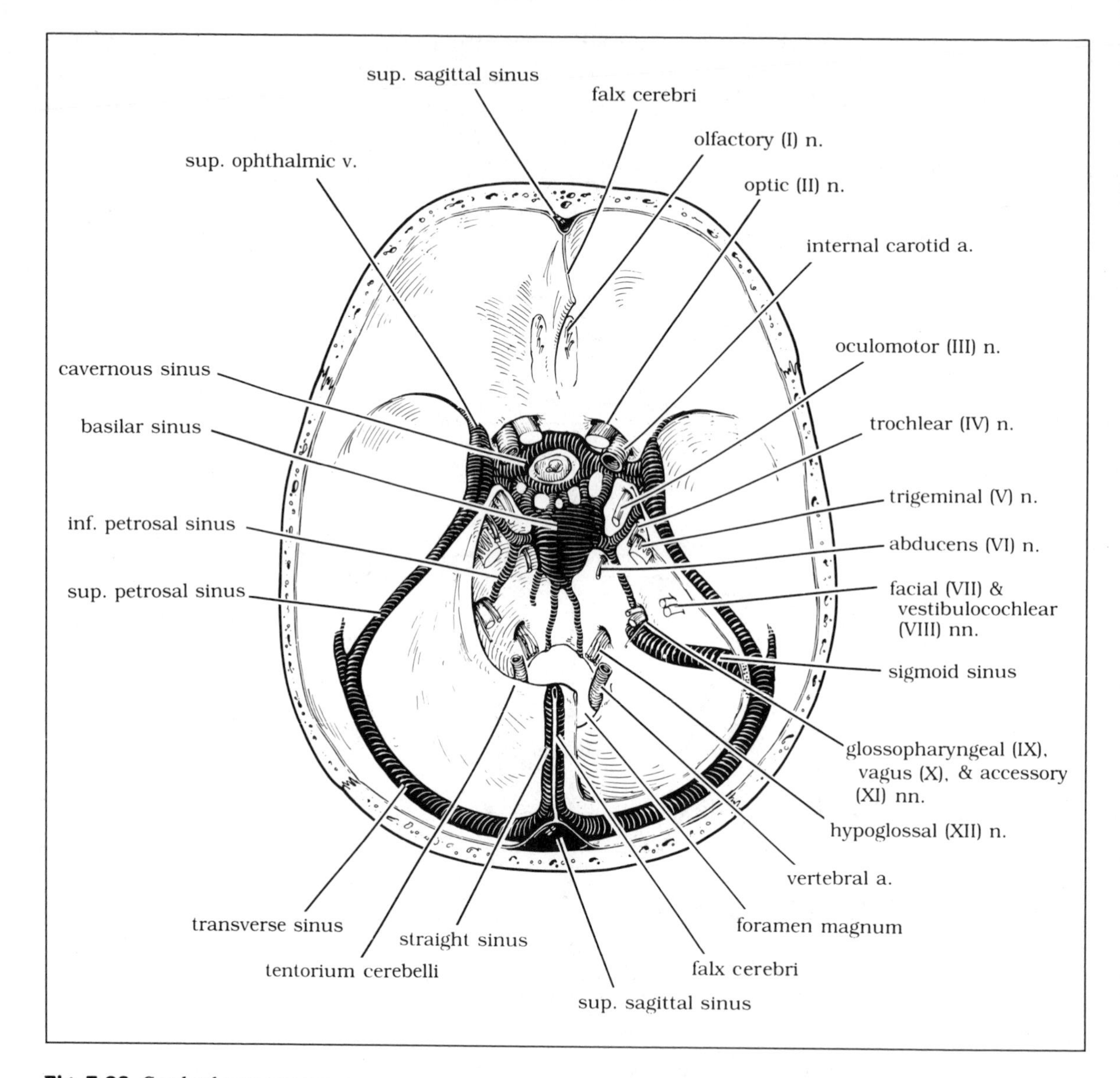

Fig. 7-38. Cerebral venous sinuses.

cipital bone and receive the superior petrosal sinuses and the inferior cerebral veins.

The **sigmoid sinuses** are continuations of the transverse sinuses and lie in the inner surface of the occipital and temporal bones. The sigmoid sinus enters the jugular fossa, where it becomes continuous with the internal jugular vein.

The **cavernous sinuses** are found in the dura around the sella turcica. Each continues from the medial portion of the superior orbital fissure to the apex of the petrous portion of the temporal bone. Rootlets of the oculomotor, trochlear, and abducens nerves and the ophthalmic and maxillary branches of the trigeminal nerve are found herein (Fig. 7-39). The internal carotid artery is also found in the cavernous sinus. The middle cerebral veins and hypophyseal plexus drain into the sinus, and the facial vein connects to it via the superior ophthalmic veins. The **superior** and **inferior petrosal sinuses** drain the cavernous sinus and connect it to the internal jugular vein. The superior petrosal sinus also connects to the transverse sinus.

An **occipital sinus** is also seen in the inner surface of the occipital bone draining the cerebellum and brain stem and emptying into the confluens sinuum.

The extensive vertebral venous system connects the intracranial sinuses with the cervical, thoracic, abdominal, and pelvic veins (see Chap. 9). The vertebral veins arise from the vessels in the basilar plexus at the base of the medulla and pons. They are initially found around the vertebral artery, and at lower levels they form a single vein that enters the brachiocephalic vein.

The **occipital veins** receive the occipital emissary vein from the straight sinus.

FACIAL VEINS

Venous drainage of the face closely parallels the arterial distribution (Fig. 7-40; see also Fig. 7-33). The facial vein is formed by the junction of the angular and palpebral veins at the lower margin of the orbit. It continues inferiorly, receiving nasal, palpebral, and labial tributaries. Just below the mandible the vein receives a communicating branch from the retromandibular vein. It terminates in the internal jugular vein. Through tributaries the facial vein is connected to the ophthalmic, infraorbital, and mental veins and the pterygoid plexus.

The **infraorbital vein,** which runs with the infraorbital artery, receives blood from superficial regions. The **mental vein** is the first tributary of the **inferior alveolar vein,** which drains the lower jaw. Both the infraorbital and the mental veins empty into the pterygoid plexus, which surrounds the pterygoid muscles in the infratemporal fossa. This venous plexus communicates with the cavernous sinus through an emissary vein; its main drainage, however, is via the maxillary vein. The **ophthalmic vein** lies within the orbit. It has no valves and so serves to connect, in either direction, the cavernous sinus and facial veins.

The **retromandibular vein** is formed by the junction of the superficial temporal vein and the maxillary vein. It descends posterior to the

Fig. 7-39. Contents of the cavernous sinus.

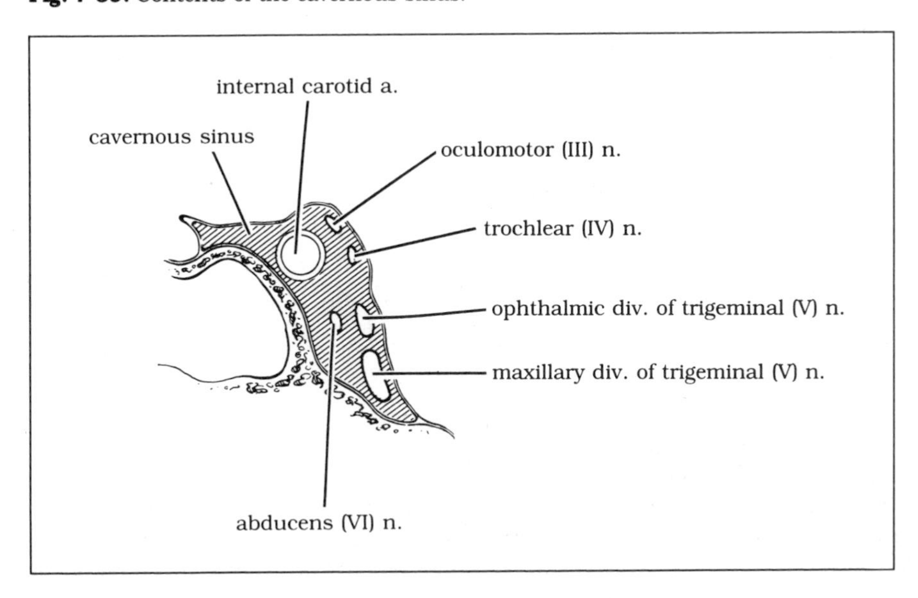

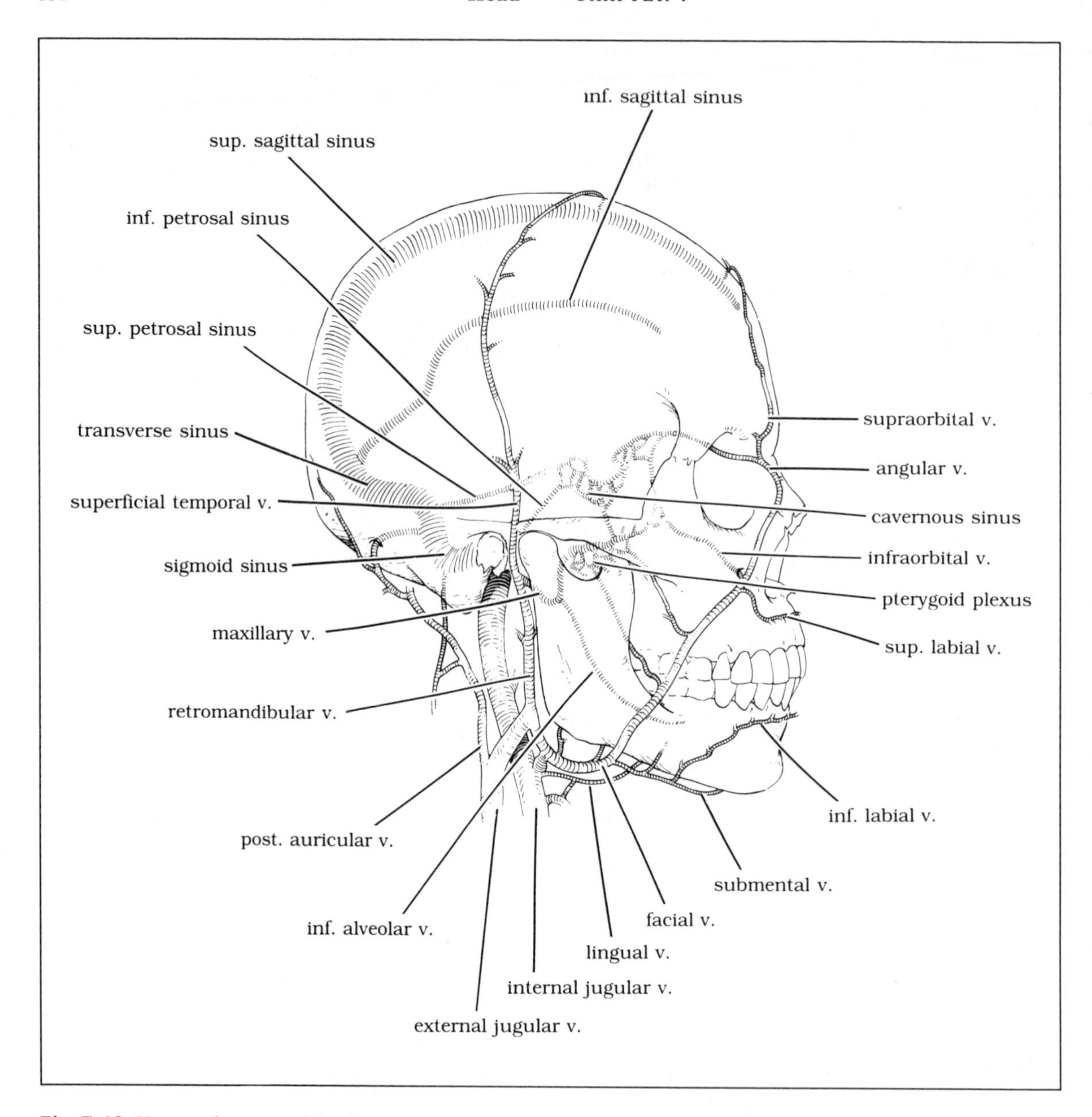

Fig. 7-40. Venous drainage of the face.

ramus of the mandible, giving off a communicating (anterior division) branch that joins the facial vein. The remaining (posterior division) vein joins with the posterior auricular vein to become the external jugular vein.

Cranial Nerves

COMPONENTS

In the spinal cord and brain stem gray matter, neurons with similar functions (e.g., motor, sensory, autonomic) are arranged in columns of cells. The classification of cranial or spinal nerves into somatic or visceral with general or special functions is called a nerve component (of Herrick).

Four functional types of "general" nerve fibers are found in the spinal cord and are involved in the innervation of skin, muscle, blood vessels, glands, and viscera. In the nerve component scheme of Herrick, the four functional types of general nerve fibers are as follows:

Motor (efferent)
1. Somatic (skeletal muscle)—General somatic efferent (GSE)
2. Visceral (smooth muscle)—General visceral efferent (GVE)

Sensory (afferent)
1. Somatic (cutaneous and proprioceptive)—General somatic afferent (GSA)
2. Visceral—General visceral afferent (GVA)

In addition to these four types of nerve fibers, the cranial nerves have three other types of fibers, which innervate the special sensory organs and muscles of branchiomeric origin (i.e., musculature thought to be derived from the pharyngeal arches and associated with the embryonic gills). The **special somatic afferent fibers** (SSA) innervate the visual, auditory, and equilibratory systems; the **special visceral afferent fibers** (SVA) provide olfactory, taste, and visceral reflex innervation; and the **special visceral efferent fibers** (SVE) innervate skeletal muscles of branchiomeric origin (muscles of mastication). These special fibers are present only in the cranial nerves. The categories of cranial nerves are as follows.

1. Nerves that provide motor innervation to skeletal muscles of presumed somite origin. The **general somatic efferent** (GSE) column consists of the motor nuclei of cranial nerves III (midbrain), IV (midbrain), VI (pons), and XII (medulla). The column is adjacent to the midline underlying the ventricular gray matter in the brain stem.

2. Nerves that provide motor innervation from the cranial portion of the parasympathetic nervous system to smooth muscle, glands, and blood vessels: the preganglionic parasympathetics. The **general visceral efferent** (GVE) column contains the Edinger-Westphal nucleus of cranial nerve III (midbrain), the superior salivatory nuclei of cranial nerve VII (pons), the inferior salivatory nucleus of cranial nerve IX (medulla), and the dorsal motor nuclei of cranial nerve X (medulla). These nuclei are found lateral to the GSE column. The axons leave the central nervous system and end in parasympathetic ganglia associated with the target organ.

3. Nerves that provide motor innervation to skeletal muscles of branchiomeric origin. The **special visceral efferent** (SVE) column consists of the motor nuclei of cranial nerves V, VII, IX, X and XI (pons and medulla). These nuclei are not found near the floor of the ventricle, but have migrated and are located in the ventrolateral margin of the tegmentum of the pons and medulla.

4. Nerves that transmit visceral and gustatory sensation. The **general** and **special visceral afferent** (GVA and SVA) column consists of the cells located peripherally in the sensory ganglia of cranial nerves VII, IX, and X. The axons of these cells enter the central nervous system, run in the fasciculus solitarius, and terminate in the solitary nucleus. The general sensation fibers from the viscera (GVA), as well as the taste fibers (SVA), synapse in this column.

5. Nerves that transmit cutaneous and proprioceptive sensations from the skin and the muscles in the head and neck. The **general somatic afferent** (GSA) column is located throughout the brain stem and upper cervical spinal levels and consists of three secondary mesencephalic nuclei in the brain stem: the mesencephalic nucleus of cranial nerve V for proprioception (midbrain), the chief sensory nucleus of nerve V for touch (midpontine), and the descending nucleus of nerve V for pain and temperature (lower pons, upper cervical level). General sensations from structures innervated by cranial nerves VII, IX, and X are conveyed by these nerves into the central nervous system, where they synapse in the descending nucleus of cranial nerve V.

6. Nerves that transmit special sensory infor-

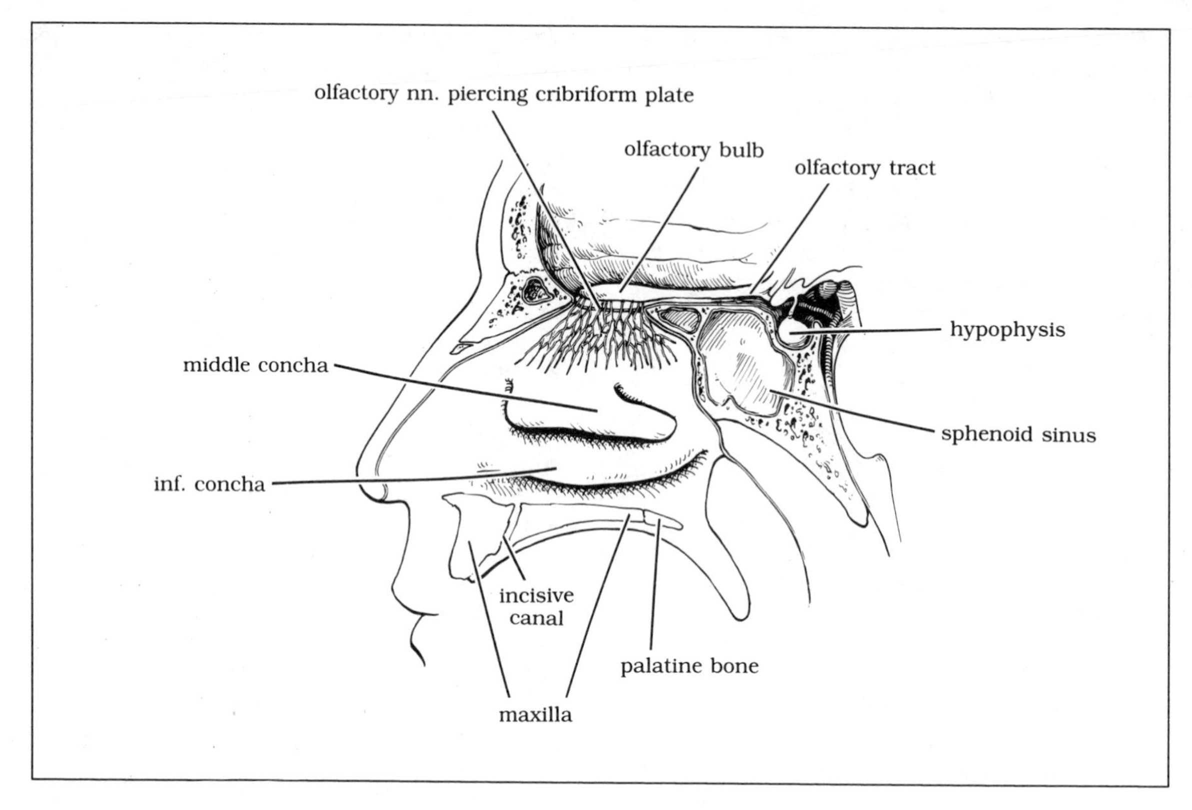

Fig. 7-41. The nasal wall and palate, showing the distribution of the olfactory nerve.

mation from the olfactory mucosa, retina, and inner ear. The **special somatic afferent** (SSA) column includes the four nuclei of the vestibular nerve and the two nuclei of the cochlear nerve, which together make up cranial nerve VIII, and cranial nerves I and II, which carry olfactory and visual sensations, respectively. The vestibular and cochlear nuclei are located in the dorsal lateral tegmentum underlying the fourth ventricle in the lower pontine and medullary levels (vestibular area and acoustic tubercle). Cranial nerve I is found attached to the inferior surface of the frontal lobe, while cranial nerve II ends in the lateral geniculate nucleus of the diencephalon.

Table 7-11 describes each cranial nerve component and its function.

GENERAL FUNCTIONS

Olfactory Nerve

The olfactory nerve (cranial nerve I; SSA) originates from receptor cells in the nasal mucosa found in the olfactory region of the nasal cavity (Fig. 7-41). These unmyelinated nerve fibers combine into bundles, pierce the cribriform plate, and end in the glomerular layer of the olfactory bulb. This is the only cranial nerve associated with the telencephalon.

Optic Nerve

The optic nerve (cranial nerve II; SSA) is the nerve for vision (Fig. 7-42). It is really a tract of fibers in the central nervous system, since it is invested with glia and not Schwann cells. The optic nerve originates in the ganglionic layers of

Table 7-11. Cranial Nerves

Cranial Nerve (number)	*Fiber Types*	*Location of Cell Bodies*	*Function*	*Path of Exit from Skull*
Olfactory (I)	SSA	Neuroepithelial cells in nasal cavity	Olfaction	Foramina in cribriform plate of ethmoid
Optic (II)	SSA	Ganglion cells in retina	Vision	Optic canal
Oculomotor (III)	GSE	Tegmentum of upper midbrain (oculomotor nucleus)	Innervation of eye movements involving all eye muscles except lateral rectus and superior oblique	Superior orbital fissure
	GVE	Tegmentum of upper midbrain (Edinger-Westphal nucleus; preganglionic nerves to ciliary ganglion)	Pupillary constriction, accommodation of lens for near vision	Superior orbital fissure
Trochlear (IV)	GSE	Tegmentum of lower midbrain	Innervation of eye movements involving superior oblique muscle	Superior orbital fissure
Trigeminal (V)	GSA	Primary trigeminal ganglion; midbrain (secondary mesencephalic nucleus); pons (sensory nucleus); pons, medulla, upper cervical levels (descending nuclei)	Transmission of cutaneous and proprioceptive sensations from skin and muscles in face, orbit, nose, mouth, forehead, teeth, and meninges and from anterior two-thirds of tongue	Superior orbital fissure (ophthalmic nerve) Foramen rotundum (maxillary nerve) Foramen ovale (mandibular nerve)
	SVE	Pons (motor nucleus)	Innervation of muscles of mastication: masseter, medial and lateral pterygoids, temporalis, mylohyoid, anterior belly of digastric, tensor veli palatini, and tensor tympani muscles	Foramen ovale (mandibular nerve)
Abducens (VI)	GSE	Tegmentum of pons	Innervation of eye movements involving lateral rectus muscle	Superior orbital fissure
Facial (VII)	SVE	Lateral margin of pons (motor nucleus)	Innervation of muscles of facial expression and platysma, extrinsic and intrinsic ear muscles, stapedius stylophyoid, and posterior digastric muscles	Internal acoustic meatus to stylomastoid foramen
	GSA	Geniculate ganglion	Anterior surface of ear and mastoid process	Petrotympanic fissure

Table 7-11 (continued).

Cranial Nerve (number)	*Fiber Types*	*Location of Cell Bodies*	*Function*	*Path of Exit from Skull*
	GVE	Superior salivatory nucleus—preganglionic to ganglia associated with these glands	Regulation of secretions from glands in nose, palate, and pharynx and lacrimal, submandibular, and sublingual glands	Petrotympanic fissure
	SVA	Primary geniculate ganglion	Transmission of gustatory sensations from taste buds in anterior two-thirds of tongue	Petrotympanic fissure
Vestibulocochlear (VIII)	SSA	Primary spiral ganglion, temporal bone; medulla (secondary cochlear nuclei)	Audition	Internal acoustic meatus
	SSA	Primary vestibular ganglion, temporal bone; medulla and pons (secondary vestibular nuclei)	Equilibrium, coordination, orientation in space	Internal acoustic meatus
Glossopharyngeal (IX)	GVE	Inferior salivatory nucleus—preganglionic to otic ganglion	Regulation of secretions from parotid gland	Jugular foramen
	SVE	Nucleus ambiguus	Swallowing, innervation of stylopharyngeus muscle	Jugular foramen
	GSA	Primary inferior ganglion	Innervation of pinna of ear	Jugular foramen
	SVA	Primary inferior ganglion, secondary nucleus solitarius	Transmission of gustatory sensations from taste buds in posterior third of tongue	Jugular foramen
	GVA	Primary inferior ganglion; secondary nucleus solitarius	Transmission of visceral sensations from intraceptive palate, posterior third of tongue, carotid body	Jugular foramen
Vagus (X)	GVE	Dorsal motor nucleus—preganglionic parasympathetic innervation	Innervation of smooth muscle in heart, blood vessels, trachea, bronchi, esophagus, stomach, intestine to left colic flexure, pancreas, liver	Jugular foramen
	SVE	Nucleus ambiguus	Innervation of larynx and pharynx for phonation and deglutition	Jugular foramen
	GSA	Primary inferior ganglion	Innervation of skin on external ear	Jugular foramen

Table 7-11 (continued).

Cranial Nerve (number)	*Fiber Types*	*Location of Cell Bodies*	*Function*	*Path of Exit from Skull*
	GVA	Primary inferior ganglion, secondary nucleus solitarius	Transmission of visceral sensation from pharynx, larynx, aortic body, thorax, abdomen	Jugular foramen
	SVA	Primary inferior ganglion, secondary nucleus solitarius	Transmission of gustatory sensation from taste buds in epiglottis and pharynx	Jugular foramen
Spinal accessory (XI)	SVE	C1–C4, nucleus ambiguus	Motor innervation of trapezius and sternocleidomastoid muscles in neck	Jugular foramen
Hypoglossal (XII)	GSE	Medulla	Motor innervation of extrinsic and intrinsic muscles of tongue, except palatoglossus	Hypoglossal canal

GSE = general somatic efferent; GVE = general visceral efferent; GSA = general somatic afferent; GVA = general visceral afferent; SSA = special somatic afferent; SVA = special visceral afferent; SVE = special visceral efferent.

the retina, converges on the optic disk, and pierces the retina, choroid, and lamina cribrosa sclerae to exit the eye. The nerve then proceeds posteriorly and medially, lying within the cone formed by the rectus muscles. It is pierced by the retinal veins and arteries and is crossed by the ophthalmic artery and nasociliary nerve. The nerve exits the orbit through the optic canal and enters the middle cranial fossa. The fibers from the nasal quadrant of the retina cross in the optic chiasm. Those from the temporal quadrant remain uncrossed and join with the nasal fibers of the opposite side. The fibers, as the optic tract, then synapse in the diencephalon (optic thalamus, or lateral geniculate nucleus) or midbrain (superior colliculus and tectal nuclei).

Oculomotor Nerve

The nucleus of the oculomotor nerve (cranial nerve III; GSE and GVE) is a purely motor nucleus and is found in the tegmentum of the midbrain at the superior collicular levels. It may be differentiated into somatic (GSE) and visceral (GVE) portions. The somatic portion consists of a paired lateral nuclear complex and an unpaired central nucleus and supplies the rectus (except lateral rectus), levator, and inferior oblique muscles. The visceral Edinger-Westphal nucleus provides the preganglionic parasympathetic fibers (GVE) to the ciliary ganglion in the orbit. The oculomotor nerve enters the orbit via the superior orbital fissure, where it divides into superior and inferior divisions that then innervate the eye muscles (see Fig. 7-12).

The **superior division** innervates the levator palpebrae superioris and the superior rectus muscles, while the inferior division supplies the medial and inferior rectus and inferior oblique muscles. The **inferior division** also sends parasympathetic fibers from the Edinger-Westphal nucleus to innervate the constrictor muscle of the pupil (via episcleral ganglia) and the ciliary muscle (via ciliary ganglia of the orbit), permitting, respectively, constriction of the pupil and an increase in the thickness of the lens for accommodation (focusing on objects within six inches of the eye).

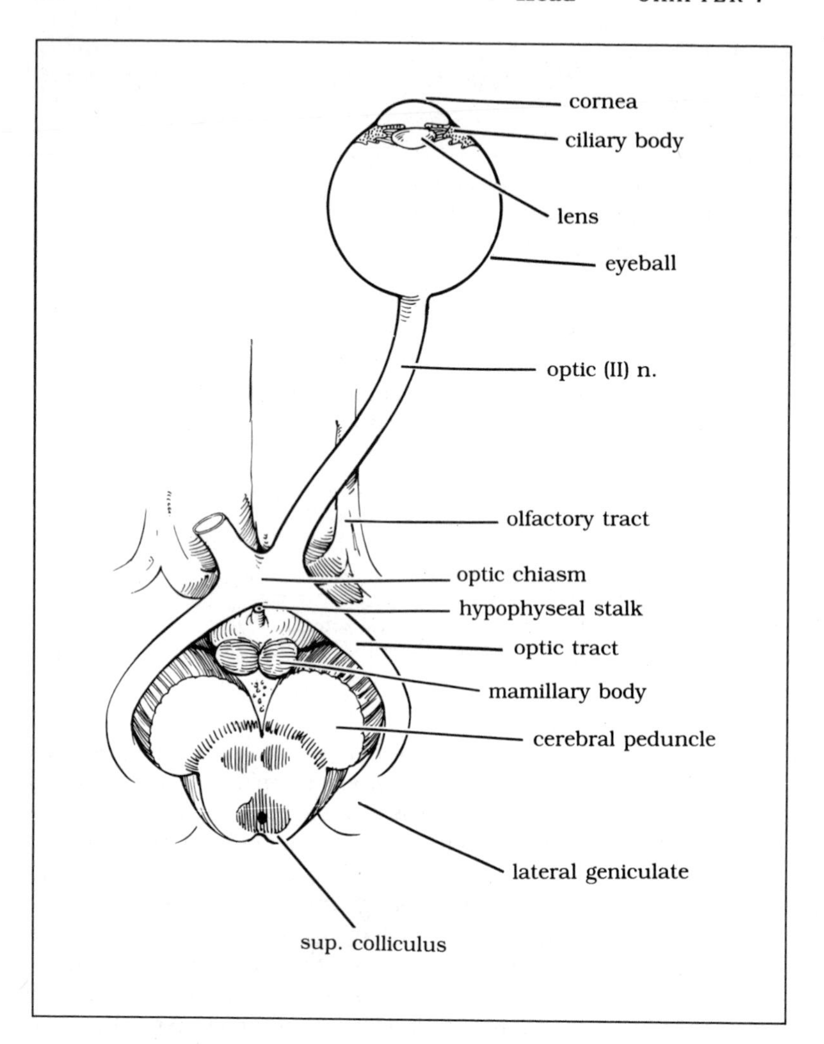

Fig. 7-42. The optic nerve, its origin, and its course.

Complete paralysis of nerve III produces ptosis of the lid, paralysis of medial and upward gaze, weakness in downward gaze, and dilatation of the pupil. The eyeball deviates laterally and slightly downward; the dilated pupil does not react to light; the lens does not accommodate. Paralysis of the intrinsic muscles (ciliary, sphincter, and dilator pupillae) is called internal ophthalmoplegia, while paralysis of the extraocular muscles (rectus and oblique) is called external ophthalmoplegia.

Trochlear Nerve

The nucleus of the trochlear nerve (cranial nerve IV; GSE) is purely motor and is located inferiorly in the midbrain. It extends throughout the inferior collicular levels and is nearly continuous with the lateral nuclear complex of nerve III.

Nerve IV is unique for two reasons: (1) It is the only cranial nerve that has rootlets on the posterior surface of the brain stem, and (2) these rootlets decussate just before they exit the brain stem. In the cavernous sinus the nerve is lateral to the rootlets of nerve III and enters the orbit through the superior orbital fissure, where it terminates in the superior oblique muscle (see Figs. 7-12 and 7-19).

In a nuclear lesion of nerve IV within the central nervous system the contralateral superior oblique muscle is paralyzed, while a lesion in the nerve roots outside the central nervous system involves the ipsilateral muscle.

Trigeminal nerve

The trigeminal nerve (cranial nerve V; GSA and SVE) is the largest cranial nerve and provides sensory fibers (GSA) to the face—including the posterior surface of the ear, the scalp up to the vertex, and the under surface of the lower jaw (Fig. 7-43A)—and motor fibers (SVE) to the muscles of mastication and the tensor tympani and veli palatini muscles (Fig. 7-43C). The sensory root is larger than the motor root. Associated with this nerve are one motor and three sensory nuclei in the brain stem and a large peripheral sensory ganglion, the semilunar or gasserian ganglion. The motor and chief sensory nuclei are found in the upper pons, deep to where the nerve enters the central nervous system.

NUCLEI IN THE BRAIN STEM

The chief sensory nucleus of nerve V receives cutaneous sensory information from the face and head and the nasal and oral cavities. The descending or **spinal nucleus** of nerve V forms the bulk of the dorsal horn in C1 to C4. Pain and temperature fibers synapse here. The **mesencephalic nucleus,** located lateral to the fourth ventricle and the cerebral aqueduct in the upper pontine and midbrain levels, is proprioceptive for the facial muscles and the muscles of mastication. This nucleus is unique in that it is the only primary sensory nucleus (dorsal root ganglion equivalent) in the central nervous system. The mesencephalic nucleus and the motor nucleus provide a two-neuron reflex arc for the jaw jerk.

PERIPHERAL BRANCHES

There are three main peripheral branches of nerve V: the ophthalmic, maxillary, and mandibular nerves. The primary cell bodies for the sensory fibers, with one exception (mesencephalic nuclei), are located in the semilunar ganglion in the floor of the middle cranial fossa.

The **ophthalmic branch** (V_1), the smallest division of nerve V, innervates the skin of the forehead and scalp to the vertex, the upper eyelid, the skin of the anterior and lateral surfaces of the nose, the eyeballs, the cornea, the ciliary body, the conjunctiva, the iris, the mucosa in the frontal and nasal sinuses, and the dura forming the cerebellar tentorium. The nerve originates from the medial part of the ganglion, passes into the lateral wall of the cavernous sinus, and divides into its terminal branches in the superior orbital fissure.

The **maxillary division** (V_2) of nerve V supplies the skin on the temples, the posterior half of the nose, the lower eyelid, the upper cheek and the upper lip, and the dura in the middle cranial fossa (middle meningeal nerve). The maxillary division also supplies the gums; molar and premolar canine teeth (superior dental plexus) of the upper jaw; and the mucous membranes of the mouth, nose, and maxillary sinus. The nerve arises from the middle of the semilunar ganglion in the cavernous sinus. It leaves the middle cranial fossa via the foramen rotundum and enters the pterygopalatine fossa, where it comes in contact with the pterygopalatine ganglion (Fig. 7-43B) before dividing into its terminal branches.

The **mandibular division** (V_3) of nerve V is the largest of the three divisions and is formed by the union of the sensory and motor roots at the inferior border of the ganglion. The motor root originates in the motor nucleus of nerve V in the pons, unites with the mesencephalic root, and exits the middle cranial fossa through the foramen ovale, where it unites with the sensory root. The motor, or masticator, nerve innervates the muscles of mastication (medial and lateral pterygoids, temporalis, and masseter), the anterior belly of the digastric muscle, and the mylohyoid muscle.

The **sensory division** of the mandibular branch of nerve V supplies the skin of the cheek,

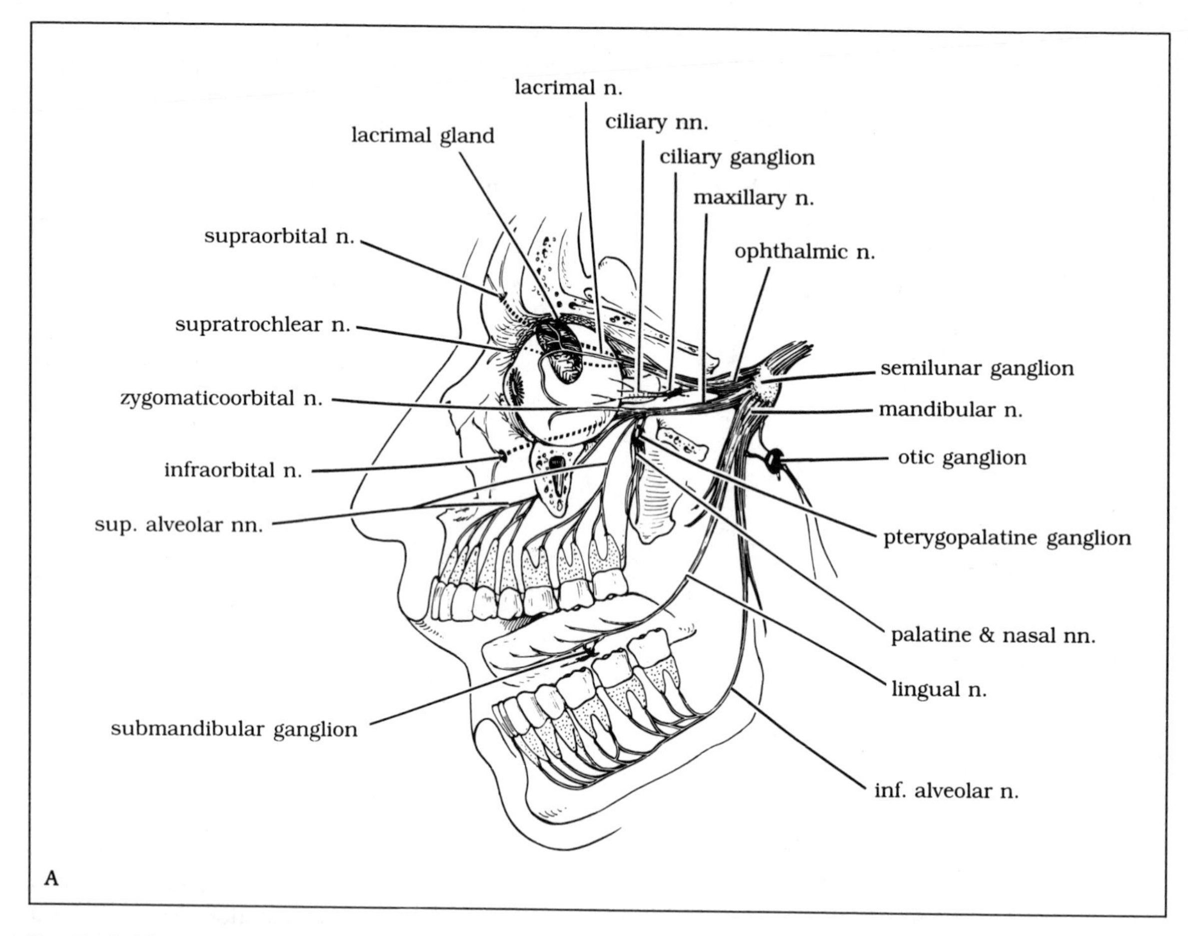

Fig. 7-43. The trigeminal nerve. A. Sensory branches. B. Branches from the pterygopalatine ganglion. C. Motor branches.

chin, lower jaw, and temporomandibular joint; the dura in the middle and anterior cranial fossa; the lower teeth and gums (inferior dental plexus); the oral mucosa; and part of the ear. Sympathetic and parasympathetic fibers to the otic and submandibular ganglia distribute with branches of the mandibular nerve.

The trigeminal nerve also innervates the tensor veli palatini and the tensor tympani muscles. The tensor veli palatini muscle tenses and draws the palate to the sides, which prevents food from entering the nasal pharynx. The tensor tympani muscle pulls on the malleus bone, which tenses the tympanic membrane and diminishes the amplitude of the vibration caused by a loud noise.

INJURY

Injury to the trigeminal nerve produces paralysis of the masticatory muscles and causes the jaw to deviate toward the side of the lesion. Injury also leads to loss of sensation of light touch, pain, and temperature in the face and to an absence of the corneal and sneeze reflexes. The jaw jerk is also absent. **Trigeminal neuralgia** (tic douloureux) is a disorder of the sensory division of nerve V characterized by recurrent paroxysms of stabbing pains along the distribution of the involved branches.

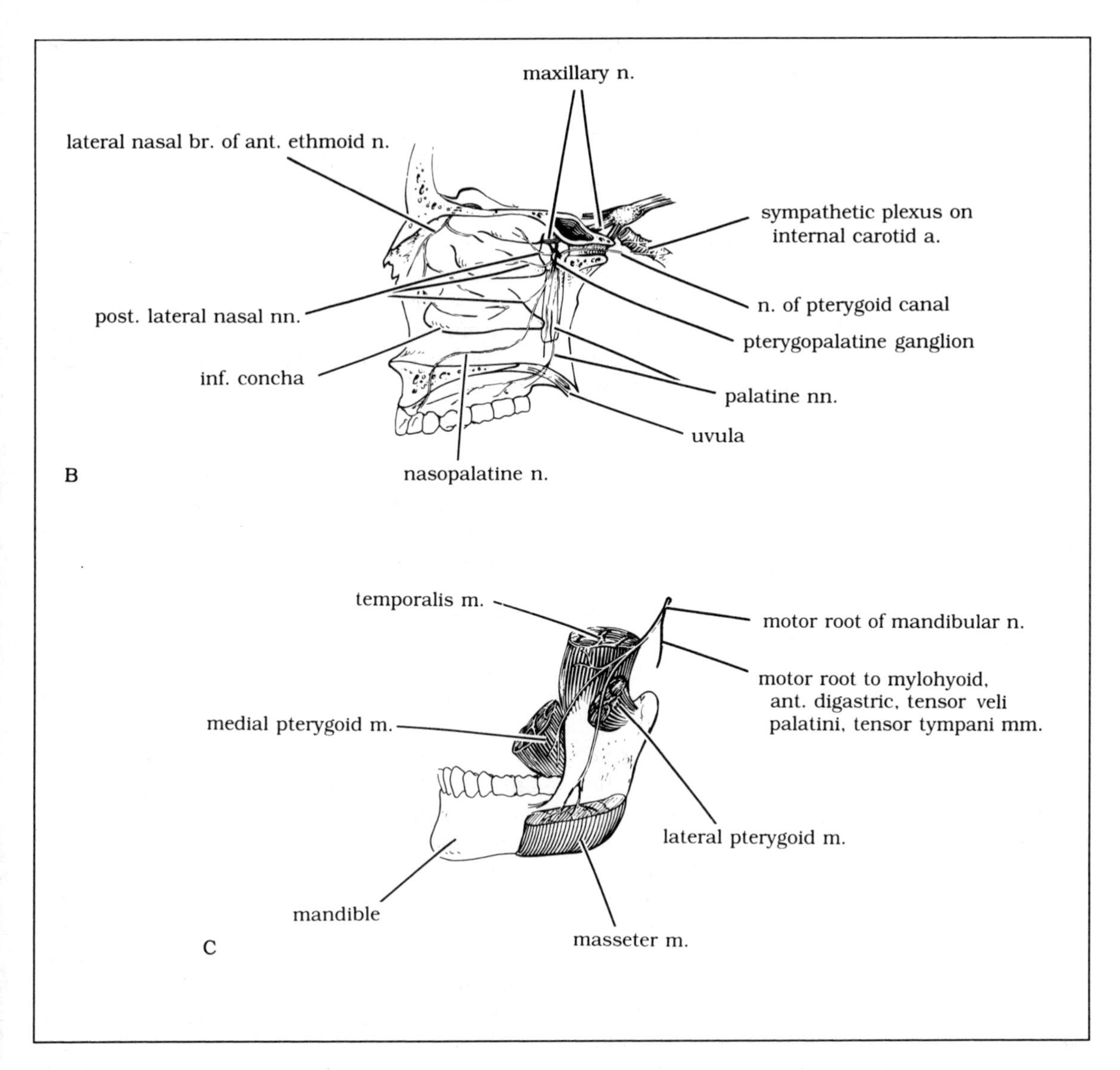

Fig. 7-43 (continued).

Abducens Nerve

The nucleus of the abducens nerve (cranial nerve VI; GSE) is a purely motor nucleus and is located in the inferior pontine levels posterior to the motor nucleus of nerve VII and surrounded by the looping fibers of nerve VII. The rootlets enter the cavernous sinus medial to nerve III and lateral to the internal carotid artery. The nerve enters the orbit via the superior orbital fissure (see Fig. 7-12) and innervates the lateral rectus muscle (see Fig. 7-19), which abducts or deviates the eye laterally. Injury to the nucleus or rootlets of nerve VI causes the eyeball to turn medially and makes it impossible to move the eye laterally on the side of the lesion.

Facial Nerve

The facial nerve (cranial nerve VII; SVE, GSA, GVE, and SVA) is primarily a motor nerve with a small sensory component; consequently the motor root is larger than the sensory root. There is one large motor nucleus and one sensory nucleus for nerve VII in the pons and one sensory ganglion peripherally.

The motor fibers supply the muscles associated with the second branchial arch (hyoid arch). The fibers originate from the motor nucleus of nerve VII (for SVE) in the lateral part of the reticular formation in the caudal pontine levels. They enter the posterior cranial fossa and pass into the internal acoustic meatus, where the motor and sensory divisions of nerves VII and VIII are separated from one another by only the arachnoid sheaths on the nerves. In the meatus, the motor and sensory roots of nerve VII combine and enter the facial canal of the temporal bone. This combined nerve continues in the facial canal until it reaches the hiatus of the facial canal, where it bends around the anterior border of the vestibule of the inner ear, forming the external genu of nerve VII. The **geniculate ganglion** of nerve VII is located at the external genu and provides the primary cell bodies of the intermediate, or sensory, root of nerve VII (Fig. 7-44A).

In the facial canal the greater petrosal nerve leaves the geniculate ganglion; passes toward the pterygoid canal, where it is joined by the deep petrosal nerve; and then becomes the nerve of the pterygoid canal. Preganglionic parasympathetic fibers in this nerve synapse in the pterygopalatine ganglion, and the postganglionic branches supply the lacrimal, nasal, palatal, and pharyngeal glands.

The sensory root fibers run through the facial canal into the internal acoustic meatus, and then enter the pons, run through the lateral part of the reticular formation, and enter the tractus solitarius. The sensory root of nerve VII carries gustatory impulses from the taste buds in the anterior two-thirds of the tongue and cutaneous sensations from the anterior surface of the external ear and mastoid process.

The motor and sensory roots continue in the facial canal between the inner ear and the middle ear and then turn and descend to the stylomastoid foramen, where they emerge from the temporal bone (Fig. 7-44B). Just prior to leaving the stylomastoid foramen, a branch, the **chorda tympani,** leaves nerve VII and enters its own small canal, then enters the tympanic cavity and runs on the medial surface of the tympanic membrane and onto the medial side of the manubrium (see Figs. 7-22B and 7-23). The nerve leaves the tympanic cavity and then emerges from the skull near the medial surface of the spine of the sphenoid bone and joins the lingual nerve at the medial surface of the lateral pterygoid muscle. The chorda tympani contains parasympathetic preganglionic fibers (GVE) and provides innervation to the submandibular gland and sublingual glands. It also carries taste sensation from the anterior two-thirds of the tongue.

The motor root innervates the stapedius muscle, the posterior belly of the digastric muscle, the stylohyoid muscle, and the muscles of facial expression: occipitofrontalis, orbicularis oculi, anterior auricular, corrugator supercilii, zygomatic, orbicularis oris, depressor labii inferioris, levator labii superioris, levator anguli oris, nasalis, mentalis, platysma, buccinator, and risorius (see Fig. 7-44B).

Injury to the facial nerve near its origin or in the facial canal produces paralysis of the facial muscles, a loss of taste in the anterior two-thirds of the tongue, and improper secretion in the lacrimal and salivary glands. The muscles will in time atrophy. Injury to nerve VII at the stylomastoid foramen produces ipsilateral paralysis of the facial muscles without affecting taste. Injury to the chorda tympani results in absence of taste in the anterior two-thirds of the tongue.

One of the more common involvements of the facial nerve is **Bell's palsy,** in which the entire nerve may be affected. This condition commonly results from swelling of the nerve somewhere in the facial canal often attributable to inflammation of the nerve. The Bell's palsy patient has sagging ipsilateral muscles in the lower half of the face and in the fold around the lips and nose and widening of the palpebral fissure. There is an absence of voluntary control of facial and platysmal musculature. When the patient smiles the lower portion of the face is pulled to the unaffected side. Saliva and food tend to collect on the affected side. When the injury is distal to the gan-

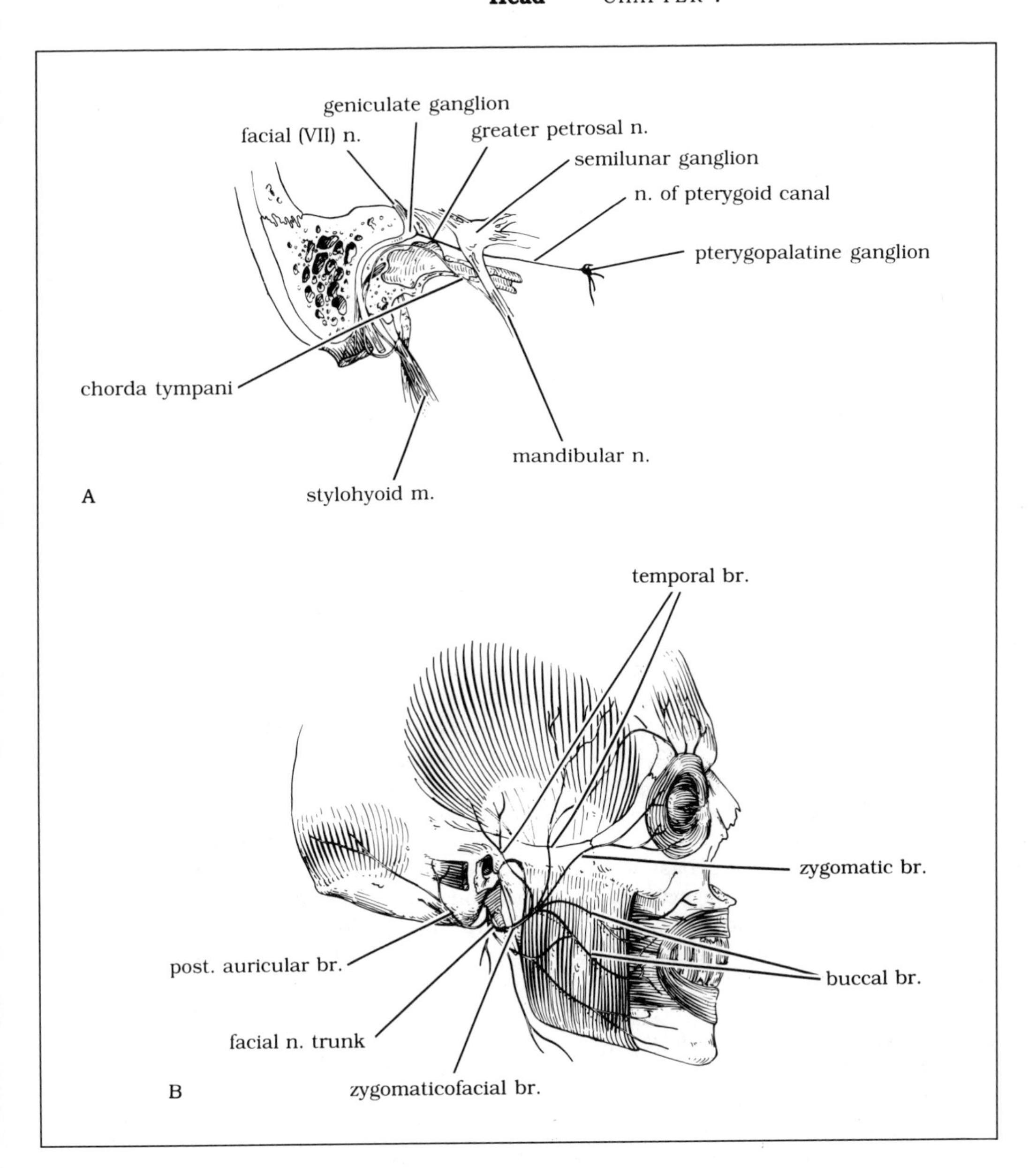

Fig. 7-44. A. The facial nerve in the facial canal. B. Distribution of the facial nerve to the muscles of facial expression.

glion there is excessive accumulation of tears, because the eyelids do not move as the lacrimal gland continues to secrete.

Vestibulocochlear Nerve

The vestibulocochlear nerve (cranial nerve VIII; SSA) is solely sensory and has two divisions: the vestibular and cochlear nerves (Fig. 7-45). The vestibular and cochlear nerves are attached to the medulla at its border with the pons lateral

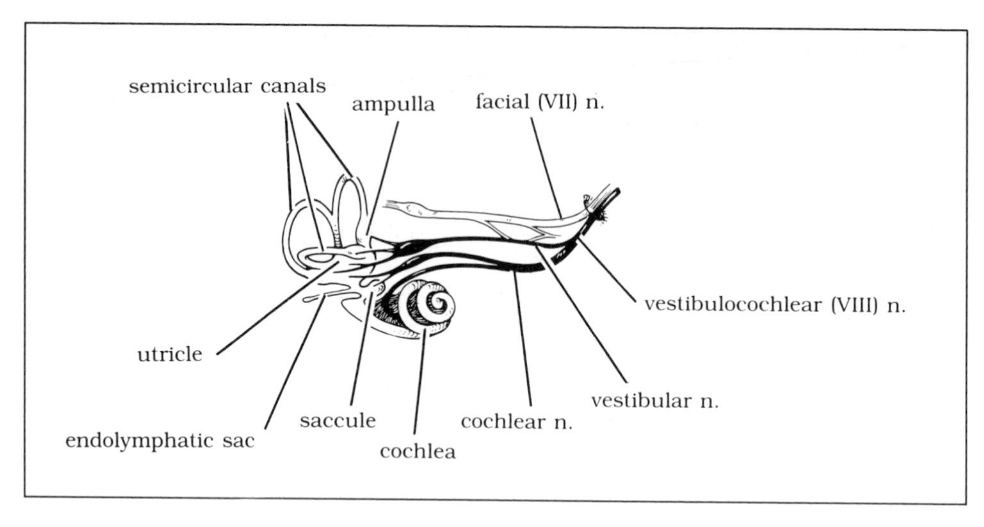

Fig. 7-45. Distribution of the vestibulocochlear nerve.

to the root of nerve VII. The cochlear nerve is smaller than the vestibular nerve and is lateral to it.

The **vestibular nerve** is concerned with equilibrium. Its primary bodies are bipolar and are located in the vestibular ganglia in the internal acoustic meatus. The fibers terminate in the four vestibular nuclei located in the ventricular floor of the medulla. The vestibular nerve has five branches that originate from the hair cells in maculae of the saccule and utricle and from the cristae of the ampullae of the three semicircular canals (see discussion of the ear earlier in this chapter; see also Fig. 7-25).

The **cochlear division** of nerve VIII is concerned with hearing and originates from the spiral ganglion in the petrous portion of the temporal bone. The peripheral nerve process originates from the hair cells in the organ of Corti, and the central nerve process terminates on the ventral or dorsal cochlear nuclei or other nuclei in the auditory pathway.

Glossopharyngeal Nerve

The glossopharyngeal nerve (cranial nerve IX; GVE, GSA, SVE, GVA, and SVA) is a mixed nerve and has sensory and motor nuclei in the medulla (Fig. 7-46). The nerve exits the skull via the jugular foramen, where it lies anterolateral to the vagus nerve. The two sensory ganglia of the glossopharyngeal nerve, the superior and inferior ganglia, are located in the jugular foramen.

The **sensory cell bodies** are located in the superior and inferior ganglia and consist of GSA fibers from the pinna of the ear; SVA fibers carrying gustatory sensations from the taste buds in the posterior third of the tongue and cutaneous sensations; and GVA fibers from the pharynx, auditory tube, middle ear, palatine tonsils, and carotid sinus. The general and special visceral axons enter the medulla and run in the tractus solitarius, synapsing in the nucleus solitarius.

The **motor fibers** of nerve IX (SVE) originate from the most superior part of the nucleus ambiguus in the medulla and control the stylopharyngeus muscle. The preganglionic parasympathetic (GVE) fibers originate from the inferior salivatory nucleus in the medulla and run to the otic ganglion via nerve IX and via the tympanic plexus and lesser petrosal nerve below the foramen ovale. The postganglionic fibers pass via the auriculotemporal nerve to the parotid gland.

Lesions restricted to the nuclei or nerve roots of nerve IX are rare. Taste is lost on the posterior third of the tongue and the gag reflex is absent ipsilaterally. There is no response when the posterior pharyngeal wall and soft palate are stimulated. The function of the parotid gland may also be impaired and can be examined by placing a highly seasoned food on the tongue and seeing if there is a copious flow from the duct.

Vagus Nerve

The vagus nerve (cranial nerve X; GVE, GSA, SVE, GVA, and SVA) has two peripheral sensory ganglia and one sensory and two motor nuclei in the medulla and also has the most extensive distribution of any cranial nerve (Fig. 7-47). Its rootlets are located in the medulla, in line with the fibers of nerve IX but inferior to them. They exit the cranial cavity via the jugular foramen. The superior ganglion of nerve X lies within the jugular foramen; the inferior ganglion is found just as the nerve leaves it. In the neck the vagus nerve is found in the carotid sheath close to the common carotid artery and internal jugular vein.

MOTOR DIVISION

The dorsal motor nucleus (GVE) of the vagus nerve forms the vagal trigone on the floor of the

Fig. 7-46. Distribution of the glossopharyngeal nerve.

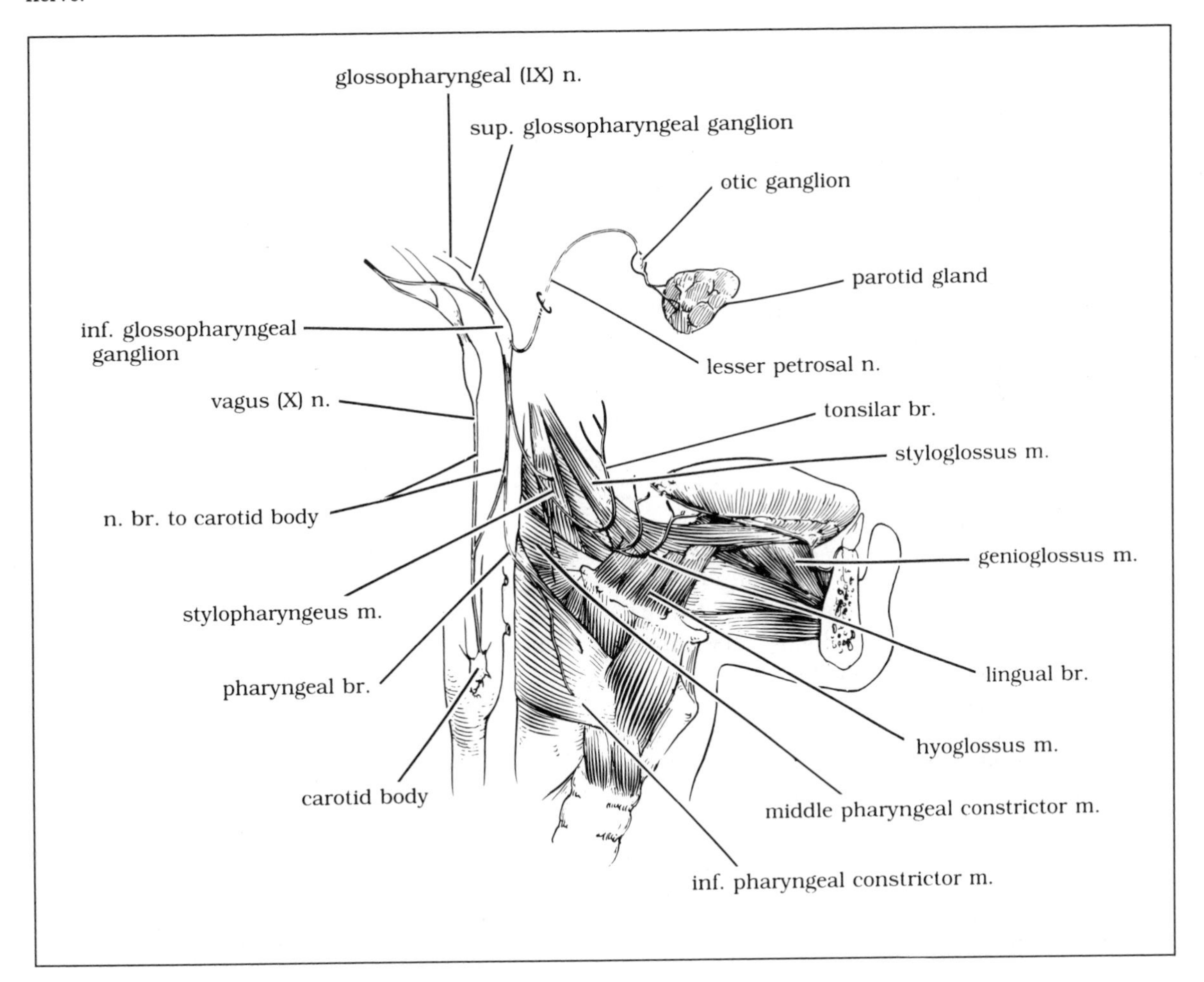

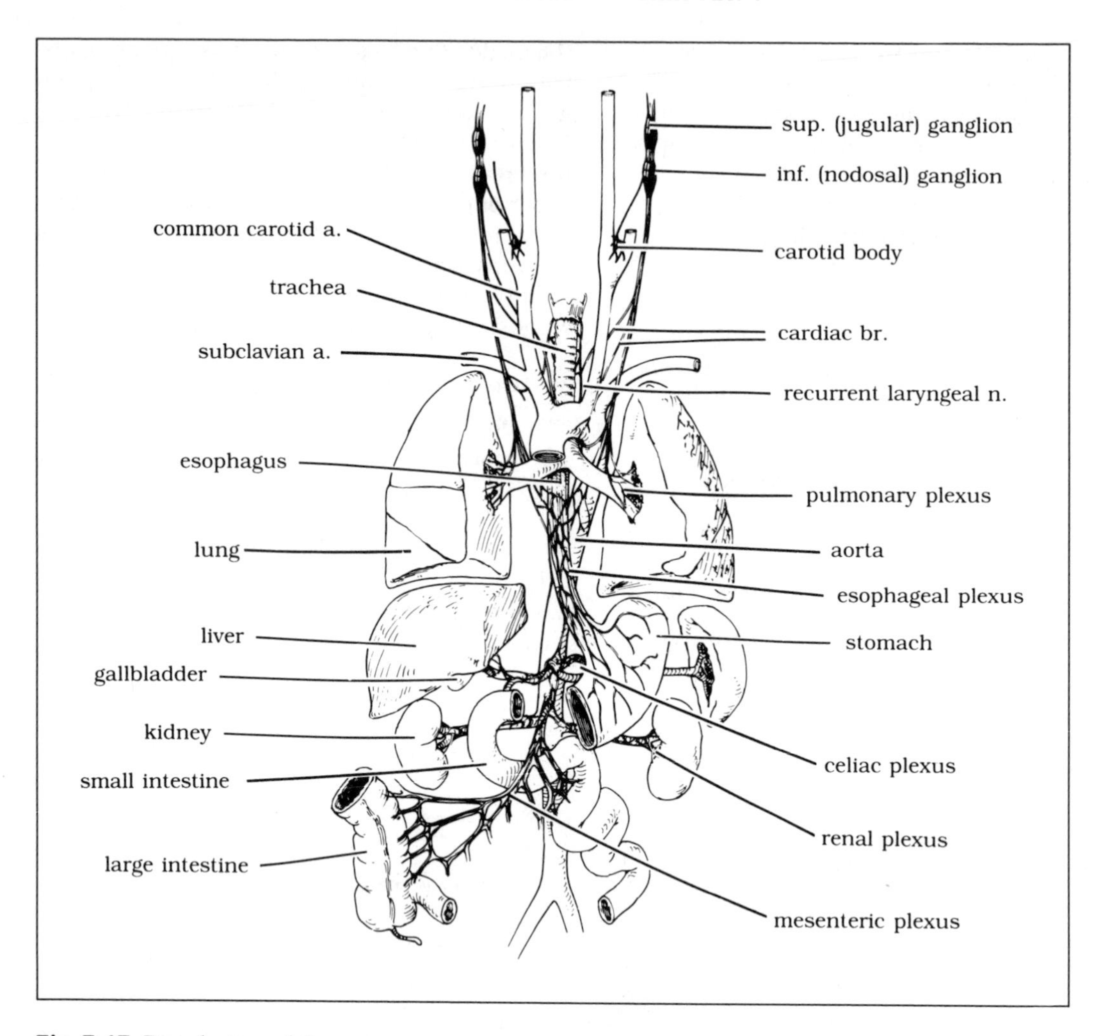

Fig. 7-47. Distribution of the vagus nerve.

fourth ventricle in the medullary levels. It provides preganglionic parasympathetic innervation to the ganglia in the walls of the pharynx, trachea, bronchi, esophagus, stomach, small intestine, ascending portions of the large intestine, and heart. The other motor division of the vagus nerve originates from the posterior two-thirds of the nucleus ambiguus (SVE) in the medulla and connects to the striate muscles of the soft palate and pharynx and the intrinsic muscles of the larynx and is thus important in swallowing and speech. The following muscles are innervated by the ambiguus portion of nerve X: the superior, middle, and inferior constrictors of the pharynx and the palatoglossus, palatopharyngeus, salpingopharyngeus, levator veli palatini, posterior cricoarytenoid, arytenoid, cricothyroid, lateral cricoarytenoid, and thyroarytenoid muscles.

SENSORY DIVISION

The primary cell bodies for the sensory fibers of nerve X are located in the inferior ganglion and carry general sensation (GSA) from the skin on the external ear and gustatory information (SVA) from taste buds in the epiglottis and pharynx, cutaneous sensation from the base of the tongue and epiglottis, and general visceral sensation from all the structures receiving motor innervation from the vagus: the pharynx, larynx, heart, trachea, bronchi, esophagus, small and large intestines, and dura of the sigmoid sinus. The fibers enter the medulla and synapse in the nucleus solitarius.

INJURY

Bilateral involvement of the vagal nuclear complex in the medulla usually has a fatal result because of the complex's distinct relationship to the pneumotaxic, apneustic, and medullary respiratory centers. Unilateral lesions of the vagus nerve may or may not produce autonomic dysfunction; the heart rate and respiratory and gastronomic tracts appear to function normally, but there may be minimal difficulty in swallowing. Injury to pharyngeal branches produces difficulty in swallowing. Lesions of the superior laryngeal nerve produce anesthesia of the upper part of the larynx and paralysis of the cricothyroid muscle. The voice is weak and the laryngeal muscles tire easily with these lesions. Interruptions of the recurrent laryngeal nerves produce paralysis of the vocal cords and hoarseness and dysphonia. Bilateral involvement of both recurrent laryngeal nerves produces aphonia and inspiratory stridor.

Although the vagus nerve is important in many visceral reflexes—vomiting, swallowing, coughing, sneezing, sucking, hiccuping, yawning, and the carotid sinus reflex—any abnormalities in these responses usually involve structures in addition to the vagus nerve.

Accessory Nerve

The accessory nerve (cranial nerve XI; SVE) is a motor nerve that originates from the medulla and spinal cord; it is consequently divided into cranial (medullary) and spinal portions (Fig. 7-48).

Fig. 7-48. Schematic drawing of the spinal root of the accessory nerve.

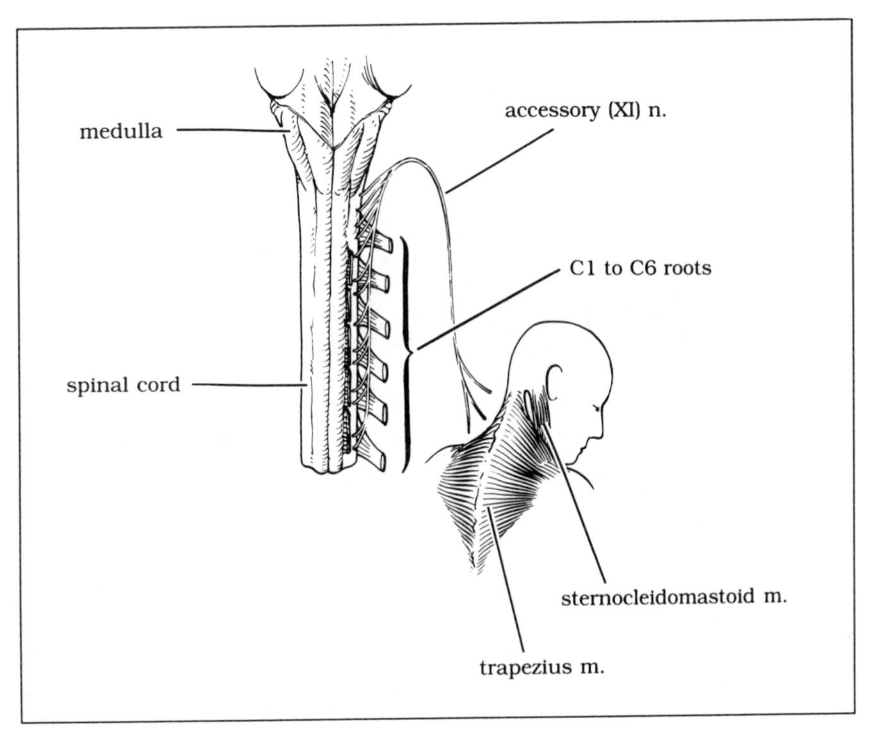

The **cranial root** arises from the most posterior part of the nucleus ambiguus, exits in a postolivary position, enters the jugular foramen, and then combines with the spinal root. Functionally it is best included with the vagus. The **spinal portion** arises from the nucleus of nerve XI in the ventrolateral part of the ventral horn in the upper four cervical segments. The fibers pass superiorly through the foramen magnum, combine with the cranial roots, and then separate from them to distribute to the sternocleidomastoid and trapezius muscles. General sensations from these muscles are carried by branches of the trigeminal nerve and upper cervical nerves.

Injury to the nucleus ambiguus in the medulla causes paralysis and atrophy of the trapezius and sternocleidomastoid muscles and weakness in movements of the head to the opposite side and in shrugging.

Hypoglossal Nerve

The hypoglossal nerve (cranial nerve XII; GSE) is solely a motor nerve and originates from the hypoglossal nucleus, which forms the hypoglossal trigone in the medullary floor of the fourth ventricle (Fig. 7-49). The fibers are found on the medulla in a preolivary position. The fibers exit through the hypoglossal canal and supply the intrinsic muscles of the tongue and all the extrinsic muscles of the tongue (styloglossus, hypoglossus, and genioglossus) except the palatoglossus. General sensations from the tongue musculature are carried by branches of the trigeminal nerve and upper cervical nerves.

Unilateral injury to the hypoglossal nerve or its nucleus produces atrophy and paralysis in the ipsilateral tongue so that the tongue protrudes toward the side of the lesion. Bilateral paralysis of the nucleus or nerve causes difficulty in eating and dysarthria.

Fig. 7-49. Distribution of the hypoglossal nerve and the ansa cervicalis.

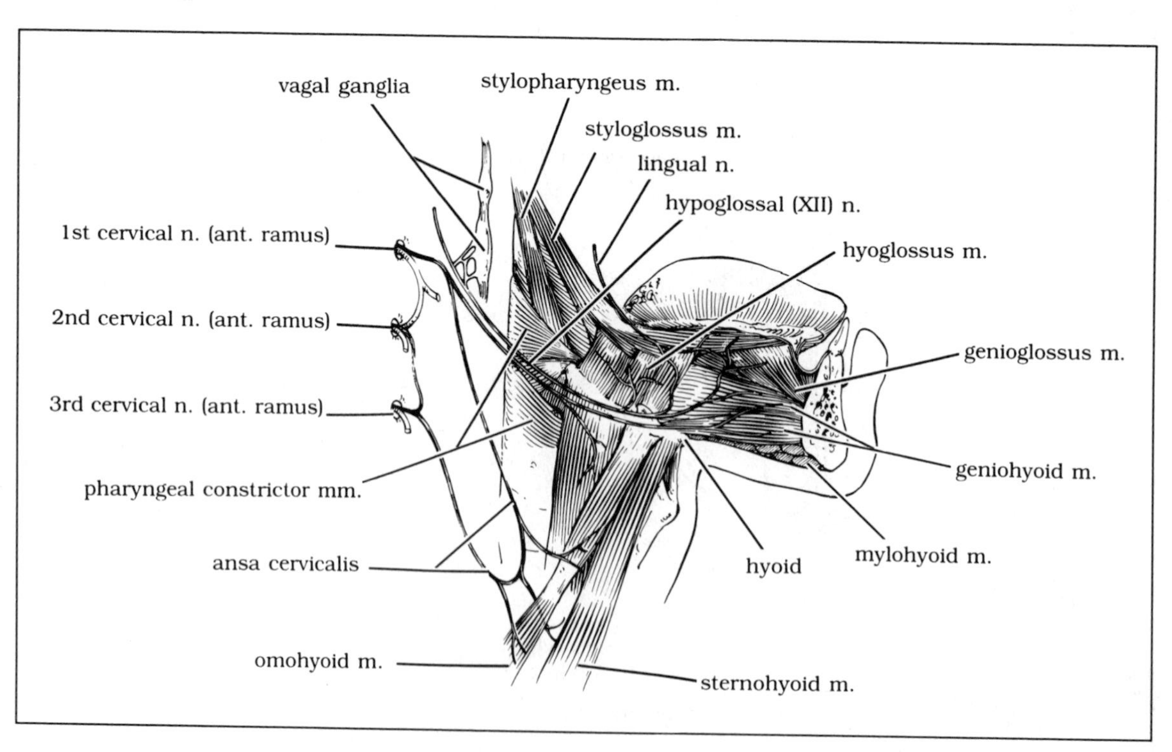

National Board Type Questions

Select the one best response for each of the following.

1. With respect to the oculomotor (III), trochlear (IV), and abducens (VI) nerves,
   A. Only one supplies a single voluntary muscle.
   B. All three supply only a single muscle each.
   C. All three nerves contain parasympathetic fibers.
   D. One of these nerves supplies four of the six extrinsic eye muscles.
   E. All three supply intrinsic eye muscles.
2. Which nerve may be injured during an incision of an abscessed submandibular gland, resulting in deformity of the mouth?
   A. Hypoglossal (XII) nerve
   B. Lingual nerve
   C. Marginal mandibular nerve
   D. Inferior alveolar nerve
   E. Buccal nerve
3. Which of the following openings is in the sphenoid bone?
   A. Foramen magnum
   B. Hypoglossal canal
   C. Cribriform plate
   D. Optic foramen
   E. Stylomastoid foramen
4. Deviation of the protruded (stuck-out) tongue to the right could be due to damage to the
   A. Geniculate ganglion
   B. Right lingual nerve
   C. Left lingual nerve
   D. Right hypoglossal (XII) nerve
   E. Left hypoglossal (XII) nerve
5. The ostium of the nasolacrimal duct
   A. Is adjacent to the sphenopalatine foramen
   B. Opens on the lateral nasal wall under cover of the middle concha
   C. Opens on the lateral nasal wall under cover of the superior concha
   D. Opens into the bulla ethmoidalis
   E. Opens on the lateral nasal wall under cover of the inferior concha
6. Taste is mediated by all but the
   A. Facial (VII) nerve
   B. Vagus (X) nerve
   C. Glossopharyngeal (IX) nerve
   D. Lesser petrosal nerve
   E. Chorda tympani
7. Cell bodies responsible for the sensation of taste in the anterior two-thirds of tongue are found in the
   A. Semilunar ganglion
   B. Vagal ganglion
   C. Pterygopalatine ganglion
   D. Geniculate ganglion
   E. Submandibular ganglion

Select the response most closely associated with each numbered item. (The headings may be used once, more than once, or not at all.)

A. Anterior cranial fossa
B. Middle cranial fossa
C. Posterior cranial fossa
D. Both anterior and middle cranial fossae
E. Both middle and posterior cranial fossae

8. Basilar venous plexus
9. Foramen lacerum
10. Trigeminal (V) nerve
11. Abducens (VI) nerve
12. Greater petrosal nerve

A. Ciliary ganglion
B. Geniculate ganglion
C. Otic ganglion
D. Semilunar ganglion
E. Submandibular ganglion

The above ganglia are located within which of the following areas:

13. Inside the petrous portion of the temporal bone
14. Orbital cavity
15. Middle cranial fossa
16. Immediately inferior to the foramen ovale

A. Sensory innervation by V_1
B. Sensory innervation by V_2
C. Sensory innervation by V_3
D. Sensory innervation by IX
E. Sensory innervation by X

17. Conchae of nose
18. Middle ear cavity
19. Piriform recess
20. Inside of cheeks
21. Antrum (anterior region) of nose

Select the response most closely associated with each numbered item.

A. Sphenoid bone
B. Maxilla
C. Both
D. Neither

22. Articulates with the nasal bone
23. Articulates with the occipital bone
24. Articulates with the mandible
25. Articulates with the zygoma

For the following, select

A. if only *1, 2, and 3* are correct
B. if only *1 and 3* are correct
C. if only *2 and 4* are correct
D. if only *4* is correct
E. If *all* are correct

26. Which of the following are correct relationships of the optic chiasm?
    1. Anterior to the superior orbital fissures
    2. Medial to the carotid arteries
    3. Inferior to the sphenoid sinus
    4. Anterior to the infundibular stalk
27. Which of the following arteries would be concerned in a cerebrovascular accident involving the entire motor cortex?
    1. Basilar artery
    2. Middle cerebral artery
    3. Posterior cerebral artery
    4. Anterior cerebral artery
28. Blood flows from the internal carotid artery directly into the
    1. Lingual artery
    2. Middle cerebral artery
    3. Posterior cerebral artery
    4. Ophthalmic artery
29. The middle meningeal artery
    1. Is a branch of the maxillary artery
    2. Enters the cranium through the foramen rotundum
    3. Is the largest of the meningeal arteries
    4. Lies between the arachnoid and the pia mater
30. Which of the following are part of the temporal bone or are contained within it?
    1. Mastoid process
    2. Styloid process
    3. Internal acoustic (auditory) meatus
    4. External acoustic (auditory) meatus
31. Which of the following nerves leave(s) the brain case via the superior orbital fissure?
    1. Oculomotor (III) nerve
    2. Trochlear (IV) nerve
    3. Abducens (VI) nerve
    4. Ophthalmic branch of the trigeminal (V) nerve
32. Which of the following items drain(s) into the middle meatus?
    1. Frontal sinus
    2. Nasolacrimal duct
    3. Anterior ethmoid sinus
    4. Posterior ethmoid sinus
33. Bone(s) to which the falx cerebri is attached include the
    1. Frontal bone
    2. Occipital bone
    3. Parietal bone
    4. Sphenoid bone
34. Which of the following bone(s) contribute(s) surfaces to the orbital cavity?
    1. Frontal bone
    2. Zygoma
    3. Maxilla
    4. Ethmoid bone
35. An acoustic neurofibroma is a tumor of the vestibulocochlear (VIII) nerve. This tumor generally occurs within the internal auditory canal and often compresses the facial (VII) nerve as well as the vestibulocochlear (VIII) nerve. Which of the following symptoms

could occur as a result of compression of the facial (VII) nerve in this location?
1. Paralysis of the facial muscles (loss of muscle function
2. Drying of the eye (loss of lacrimal gland function)
3. Drying of the nasal cavity (loss of nasal gland function)
4. Complete absence of salivation (loss of saliva production)

36. When a teenager "pops" a pimple, occasionally a bacillary embolus is carried into the facial vein and may eventually lodge in the cavernous sinus, leading to a serious infection of the brain. Which of the following veins anastomose with the facial vein and provide a route for the embolism to reach the cranial cavity?
1. Parietal emissary veins
2. Pterygoid plexus of veins
3. Diploic veins
4. Ophthalmic veins

37. Select only the muscle(s) that receive(s) its (their) motor innervation from the mandibular division (V_3) of the trigeminal (V) nerve.
1. Tensor tympani
2. Temporalis
3. Masseter
4. Platysma

38. Choose only the correct statement(s) regarding the trigeminal (V) nerve.
1. Its sensory ganglion is located within the middle cranial fossa
2. Its ophthalmic division passes through the wall of the cavernous sinus
3. It supplies motor fibers to the lateral pterygoid muscle
4. Its mandibular division passes through the pterygopalatine fossa

39. The parasympathetic innervation of the parotid gland
1. Originates in cranial nerve IX
2. Travels via the tympanic nerve
3. Travels via the lesser petrosal nerve
4. Is relayed via the auriculotemporal nerve

40. Select only the correct statement(s):
1. Cranial nerve VII supplies motor fibers to the muscles of facial expression.
2. The chorda tympani contains preganglionic parasympathetic fibers to the submandibular ganglion.
3. The facial (VII) nerve innervates the posterior belly of the digastric muscle.
4. The chorda tympani contains taste fibers that innervate the anterior two-thirds of the tongue.

41. The vagus (X) nerve has which of the following functional components?
1. Special somatic afferent
2. Special visceral afferent
3. General somatic efferent
4. General visceral efferent

42. Which of the following drain directly into the internal jugular vein?
1. External jugular vein
2. Middle thyroid vein
3. Inferior thyroid vein
4. Sigmoid sinus

Annotated Answers

1. D. Intrinsic eye muscles are innervated by the autonomic system, the parasympathetic component being delivered only by the oculomotor nerve. The abducens and trochlear nerves are so named because of a unique characteristic of the single muscle they each innervate.
2. C. Remember that one of the motor branches of the facial nerve, the marginal mandibular, runs superficial to the digastric triangle, home of the submandibular gland.
3. D.
4. D. The hypoglossal nerve supplies all the tongue muscles. The genioglossus muscle protrudes the tongue. The direction of pull (forward and lateral to medial) means that if the right nerve was injured, the left genioglossus would pull the tongue forward and to the right.
5. E.
6. D. and 7. D. Cell bodies in the geniculate ganglion of the facial nerve convey taste sen-

sation from the anterior two-thirds of the tongue via the chorda tympani branch of the facial nerve. The glossopharyngeal and the vagus nerves also have taste fibers.

8. C.
9. B.
10. E. and 11. E. The trigeminal nerve is attached to the pons in the posterior cranial fossa and passes to the middle cranial fossa, where it divides into its three main branches. The abducens also arises in the posterior cranial fossa. It then pierces the dura to enter the cavernous sinus. Although the nerve cannot be seen at this point, by definition the cavernous sinus is in the middle cranial fossa, so that this nerve is also in the middle cranial fossa.
12. B.
13. B. and 14. A. Remember that the peripheral ganglia of the parasympathetic system in the head are associated with (attached to) other nerves and not within the cranial cavity. Thus the ciliary ganglia are in the orbit (associated with the oculomotor and nasociliary nerves) and the otic ganglia are intimately associated with the mandibular branch of the trigeminal as that nerve emerges through the foramen ovale. (Now, with what is the submandibular ganglion associated?)
15. D.
16. C.
17. B., 18. D., 19. E., 20. C., and 21. A. The sensory innervation of the nose is divided between the second branch (maxillary division) of the trigeminal nerve, which supplies the conchae on the lateral wall, and the first branch (ophthalmic division), which sends fibers to the anterior region of the nose. The ninth and tenth cranial nerves also have sensory components, with the ninth supplying the middle ear cavity via a tympanic branch (pain from a middle ear infection?) and the tenth supplying the pharyngeal mucosa, which includes the area of the piriform recess (see Chap. 8).
22. B.
23. D.
24. D.
25. C.
26. C.
27. C. The cortical branches of the middle cerebral artery supply the motor cortex with the exception of the "leg" area, which is supplied by the cortical branches of the anterior cerebral artery.
28. C. The middle cerebral artery can be considered the terminal branch of the internal carotid. The ophthalmic artery arises from the internal carotid as it emerges from the cavernous sinus.
29. B.
30. E.
31. E.
32. B. Remember that the nasolacrimal duct opens into the inferior meatus.
33. A.
34. E.
35. A. At this level the seventh cranial nerve has motor fibers to the facial muscles and the preganglionic parasympathetic fibers to the pterygopalatine ganglion, which then sends postganglionic fibers to the lacrimal gland and glands of the nasal cavity. Although the submandibular gland and lingual gland would also be affected, complete loss of saliva production is avoided because the parotid is still innervated (by what nerve?).
36. C. A direct route to the cavernous sinus from the facial vein can follow the valveless ophthalmic veins, which begin just deep to the eyelids, and the pterygoid plexus of veins, which links to the cavernous sinus through the foramen ovale and emissary foramina.
37. A.
38. A. The semilunar ganglion of the trigeminal contains the sensory cell bodies (like a dorsal root ganglion). Remember also that a cavernous sinus thrombosis (see Question 36) can involve the ophthalmic nerve.
39. E. The ninth cranial nerve contains preganglionic parasympathetics that eventually reach the parotid gland by first traveling in the tympanic nerve (What other types of fi-

bers are in this nerve?), then the lesser petrosal, and are eventually distributed to the gland via the auriculotemporal nerve (part of what cranial nerve?).

40. E.
41. C. Remember that since the vagus nerve has two ganglia, it must have some sensory function, and that it does *not* innervate skeletal muscle (it is important to understand the difference between skeletal muscle and muscle of branchiomeric origin).
42. C.

8 Neck

Objectives

After reading this chapter, you should know the following:

Surface anatomy and superficial organization of the neck

Fascia and compartments of the neck

Bony structures of the neck

Musculature of the neck, including musculature involved with the back and support of the head

Location, composition, and function of the neck's endocrine (thyroid and parathyroid) and salivary glands

Composition, structure, and function of the cervical viscera (pharynx, larynx, esophagus, trachea) and their involvement in speech

Organization, components, and functions of the cervical plexus and cervical sympathetic chain and ganglia

Blood supply, lymphatics, and innervations of the neck

Surface Anatomy

Anteriorly the neck is the region of the body extending from the chin to the sternum (Fig. 8-1). The line of the neck continues, superiorly, along the line of the lower jaw and across the mastoid process and ends at the greater occipital protuberance. Inferiorly the line of the neck extends along the clavicle and across the soft structures covered in part by the trapezius and ends at the seventh cervical vertebra.

The prominent structures of the neck are the hyoid bone, thyroid cartilage, trachea, and sternocleidomastoid muscles (Fig. 8-2). The hyoid bone lies at the angle between the chin and the anterior neck. The greater cornua of the hyoid are palpable at the sides of the neck.

The thyroid cartilage (laryngeal prominence) lies just below the hyoid bone in the midline. In men it forms the prominent Adam's apple. Below the thyroid cartilage, the cricoid cartilage and then the trachea may be palpated. In thin subjects the trachea may also be visible.

Lying laterally in the neck are the two sternocleidomastoid muscles. They cross the lateral neck at a diagonal to converge at the jugular notch of the sternum in the midline. The external jugular vein is often visible crossing the sternocleidomastoid to the base of the neck.

Triangles

For purposes of discussion, the neck, both superficially and deep, is often divided into triangles, and these triangles are often subdivided (Fig. 8-3). The largest triangles are the anterior and posterior cervical triangles formed by the sternocleidomastoid crossing the neck. The limits of the two **anterior cervical triangles** are the midline, sternocleidomastoid, and man-

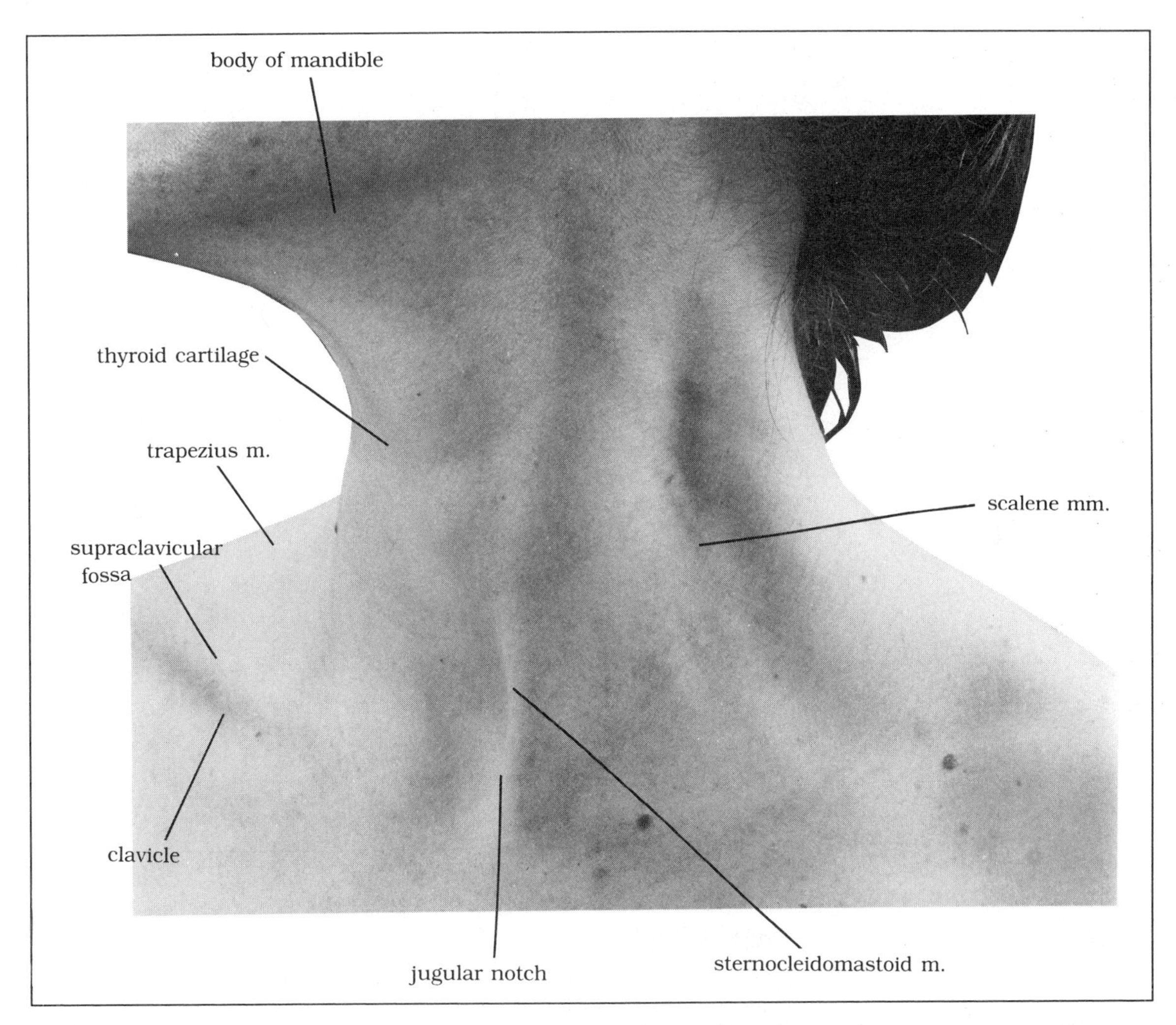

Fig. 8-1. Surface anatomy of the neck.

dible. The two **posterior cervical triangles** are limited by the sternocleidomastoid, clavicle, and trapezius.

Each anterior triangle is divided again: The region above the hyoid contains the submental and submandibular triangles. The **submental triangle** is limited by the hyoid and the anterior belly of the digastric. The **submandibular triangle** is bounded by the mandible and the two bellies of the digastric. The submandibular gland is found within this triangle.

Below the hyoid each anterior triangle contains the muscular triangle and the carotid triangle. The **muscular triangle,** containing the infrahyoid muscles and thyroid glands, is limited by the midline and the omohyoid and sternocleidomastoid muscles. The **carotid triangle** is bounded by the posterior belly of the digastric and the sternocleidomastoid and omohyoid muscles. This triangle provides entrance to those structures associated with the carotids.

Each posterior cervical triangle contains within it an **omoclavicular triangle.** This triangle is limited by the inferior belly of the omohyoid, the clavicle, and the sternocleidomastoid.

Outside the neck (but discussed within this

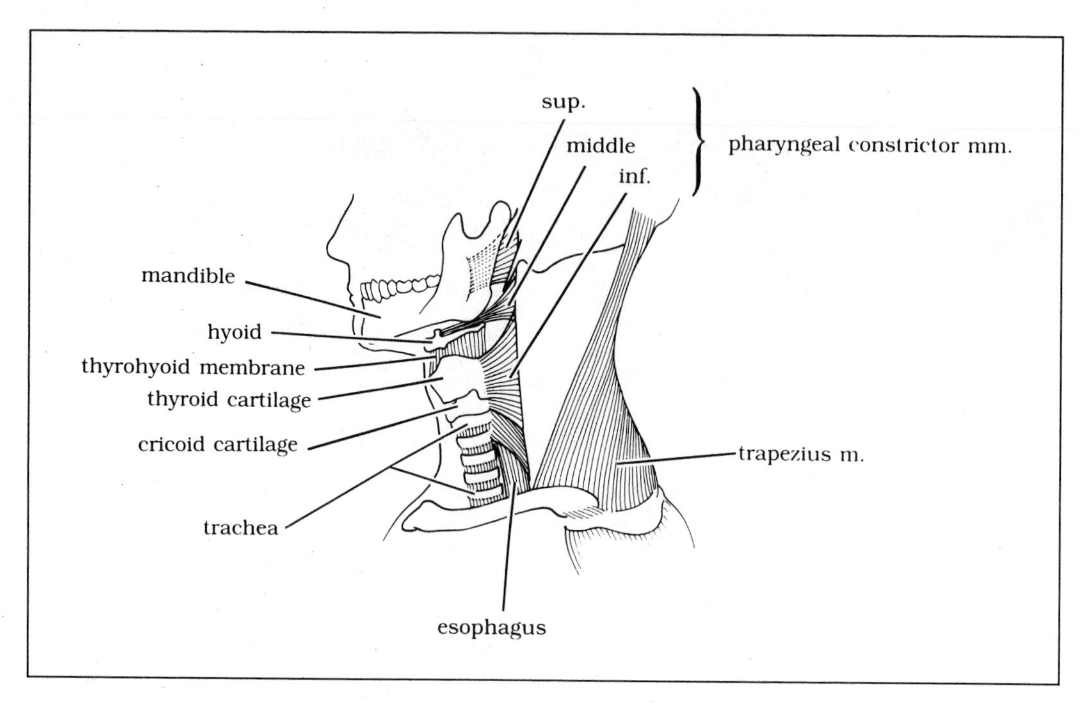

Fig. 8-2. Schematic structures of the neck.

Fig. 8-3. Triangles of the neck.

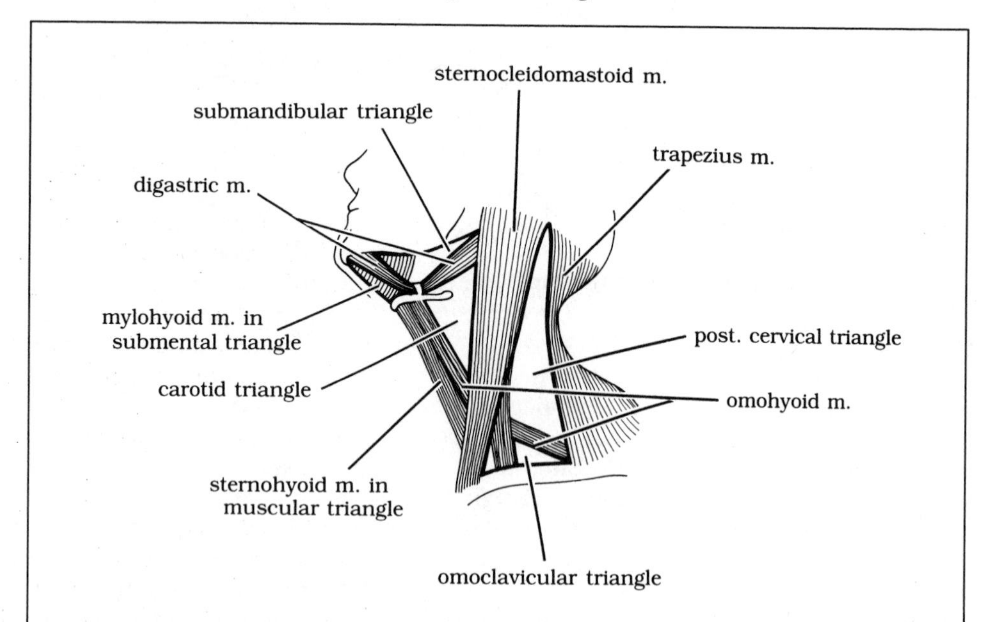

chapter), is the **suboccipital triangle.** This triangle, primarily underlying the trapezius muscle, contains the deep posterior region of the neck.

Superficial Organization

SUPERFICIAL STRUCTURES

The skin of the neck is thin and mobile. It receives the insertion of subcutaneous muscle, the platysma, as does the skin of the face. The **platysma** is a thin broad muscle overlying the subcutaneous veins, nerves, and fascia (see Table 8-1). The muscle arises over the pectorals and deltoids and ascends to insert into the skin along the lower edge of the mandible, and into the mandible itself. Some of its fibers may travel farther upward to blend around the lower mouth with the muscles of facial expression. The platysma is innervated by a cervical branch of the facial (seventh cranial) nerve.

The veins of the superficial neck form an interconnected plexus, extending from superficial to deep levels. The primary veins are the external and anterior jugulars, both of which lie under the platysma. The **external jugular vein** is formed by the joining of the posterior division of the retromandibular vein and the posterior auricular vein just below the angle of the mandible (see Fig. 8-6). The vein descends directly downward, crossing the sternocleidomastoid and piercing the cervical fascia approximately 2 cm above the clavicle. It empties into the subclavian vein.

The external jugular vein receives several tributaries. The posterior external jugular vein (which drains the skin and fascia of the posterior neck) reaches the external jugular vein at the inferior margin of the sternocleidomastoid. The suprascapular and transverse cervical veins (which drain the deep structures of the shoulder), the anterior jugular vein, and branches of the facial vein are all tributaries of the external jugular vein.

The **anterior jugular vein** may be either a single vessel running along the midline or a set of two vessels running on either side of the midline. The vein descends to a level approximately 3 cm above the sternum before piercing the fascia to run laterally behind the sternocleidomastoid over to the external jugular vein. The anterior jugular vein usually receives a branch from the facial vein. If two anterior jugular veins are present, they are usually connected to each other through a branch just above the jugular notch, at which level the veins turn laterally to reach the external jugular.

The **dorsal rami** of the cervical nerves have a superficial presence beginning with the second cervical nerve. The second cervical nerve's cutaneous branch is the **greater occipital nerve.** This nerve travels with the occipital artery on the back of the scalp. The third through sixth cervical nerves supply the skin along the back of the neck superiorly to inferiorly in numerical order. The branch of the third cervical nerve is known as the **occipitalis tertius nerve.**

The **cervical plexus** is formed by the ventral rami of the second through fourth cervical nerves. The branches (see Fig. 8-5) emerge from behind the middle of the posterior border of the sternocleidomastoid to reach the skin of the neck and face. In order of appearance they are the lesser occipital (C2 and C3), great auricular (C2 and C3), transverse cervical (C2 and C3), and supraclavicular (C3 and C4) nerves.

The **lesser occipital nerve** travels upward along the sternocleidomastoid to supply the region above and behind the ear and the upper portion of the auricle. The **great auricular nerve** crosses the sternocleidomastoid to run anterior and superior to the ear. It supplies the skin over the mastoid process, the lower auricle, and the region in front of the ear over the parotid gland. The **transverse cervical nerve** proceeds anteriorly, supplying the skin of the anterior cervical triangle.

The three **supraclavicular nerves** descend along the sternocleidomastoid to pierce the platysma near the clavicle. The medial branch supplies the base of the neck and the adjacent region just over the sternum. The intermediate branch covers the middle clavicle and descends, over the pectorals, as low as the third rib. The lateral branch extends to the shoulder as far as the first two-thirds of the deltoid.

FASCIA, FASCIAL PLANES, AND COMPARTMENTS

Because the neck is cylindrical, each fascial sheath essentially compartmentalizes the neck. The first sheath is the **superficial layer of cervical fascia.** This layer, containing all but the most superficial structures, is a single layer all around the neck except where it splits to contain the trapezius and sternocleidomastoid muscles. It is continuous with the parotid fascia at the face and the pectoral and deltoid fasciae of the chest. This layer of fascia blends at the margins of the neck with the periosteum of the clavicle, acromion, occiput, and scapular spine. The area between the layers containing the sternocleidomastoid muscle extends upward above the manubrium to contain the link between the anterior jugular veins. This area is known as the suprasternal space.

Deep to the superficial layer is a second fascial cylinder known as the **prevertebral fascia.** This layer contains the vertebrae and deep cervical muscles (scalenes, levator scapulae, and splenius). At the sides of the neck it forms the floor of the posterior cervical triangle.

Between these two cylinders lie several other fascial layers. The most anterior is the **infrahyoid fascia,** which consists of two layers of fascia in concentric semicircles, both extending from the hyoid to the sternum. The superficial layer contains both the sternohyoid and omohyoid; a portion of it forms the sling for the midtendon of the omohyoid. The deep layer contains the sternothyroid and thyrohyoid.

The next fascial layers are the ones containing the cervical viscera. The **pretracheal layer** encloses the larynx, trachea, thyroid gland, and parathyroids. The **buccopharyngeal layer** contains the pharynx and esophagus.

The last fascial layer to be considered is the **carotid sheath.** This fibrous sheath contains, separates, and invests the carotid arteries (common and internal), internal jugular vein, vagus (tenth cranial) nerve, and the superior root of the ansa cervicalis. The sheath attaches above to the jugular foramen and carotid canal, and below to the sheaths of the great vessels and the scalene fascia. The cervical sympathetic trunk lies behind the carotid sheath.

The fascial layers divide the neck into compartments. The superficial cervical layer contains the entire neck. The prevertebral layer contains the vertebrae and muscles. The visceral layer contains the viscera. The carotid layer contains the neurovascular structures. The interfascial spaces are potential spaces. The largest one is the retropharyngeal space, lying between the buccopharyngeal and prevertebral layers and the carotid sheath.

Bony Structures

The vertebrae are discussed in Chapter 9. The only other bone in the neck is the hyoid. The **hyoid** is the only bone in the body not directly connected with any other; instead it hangs within muscular slings. It consists of a U-shaped body (see Fig. 8-4A) with two horns, the greater and lesser cornua, on each side. The central body serves as a site of attachment for the geniohyoid, mylohyoid, sternohyoid, and omohyoid muscles. The greater horns, which extend posteriorly, serve as a site of attachment for the hyoglossus, thyrohyoid, stylohyoid, and middle constrictor muscles. In addition the thyrohyoid membrane and the sling for the digastric attach here. The lesser horns arise from nodes at the junction of the body and greater horn and extend superiorly. The middle constrictor muscle and stylohyoid ligament attach to them.

The hyoid serves to anchor most muscles of the neck. It functions in movements of the tongue and lower jaw by supplying a stable structure against which the muscles may pull.

Anterior Cervical Triangle

MUSCULATURE

The four superficial muscles in each anterior cervical triangle constitute the infrahyoids (Fig. 8-4 and Table 8-1): the sternohyoid, sternothyroid, thyrohyoid, and omohyoid. The **sternohyoid** is the most central and superficial muscle. It arises from the medial posterior clavicle and posterior

Table 8-1. Musculature of the Anterior Cervical Triangle

Muscle	*Origin*	*Insertion*	*Innervation*	*Action*
Platysma	Skin over pectorals and deltoids	Skin at lower edge of mandible	Cervical branch of facial nerve	Muscle of facial expression
Sternohyoid	Clavicle, manubrium	Body of hyoid	Ansa cervicalis	Lower hyoid
Sternothyroid	Rib 1, manubrium	Thyroid cartilage	Ansa cervicalis	Lower thyroid cartilage and larynx
Thyrohyoid	Thyroid cartilage	Greater horn of hyoid	Ansa cervicalis	Lower hyoid or raise larynx
Omohyoid	Superior border of scapula	Body of hyoid	Ansa cervicalis	Lower hyoid
Longus colli	Vertebral column	Atlas	Nerves C2–C6	Bend neck forward, flex and rotate neck
Longus capitis	Vertebral column	Occipital bone	Nerves C1–C3	Flex head
Rectus capitis anterior	Atlas	Occipital bone	Nerves C1 and C2	Flex head
Rectus capitis lateralis	Atlas	Occipital bone	Nerves C1 and C2	Bend head to same side

manubrium and inserts into the lower border of the body of the hyoid. The sternohyoid, in concert with the thyrohyoid and omohyoid, pull down the hyoid. It is innervated through the superior root of the ansa cervicalis.

The **sternothyroid** lies deep to the sternohyoid. It extends from the posterior manubrium and cartilage of the first rib to the thyroid cartilage. Its action is to pull down the thyroid cartilage and the larynx. Innervation is accomplished through the ansa cervicalis.

The **thyrohyoid** lies deep to the sternohyoid. It is a continuation of the sternothyroid, arising from the thyroid cartilage and inserting into the greater horn of the hyoid. It is innervated through the ansa cervicalis. The thyrohyoid acts either to raise the larynx when the hyoid is fixed or to lower the hyoid.

The **omohyoid** is a superficial two-bellied muscle lying laterally in the neck. Its inferior belly (which lies in the posterior cervical triangle) arises from the superior border of the scapula next to the transverse ligament. In its middle the muscle is pulled close to the clavicle through a fascial sling. The superior belly ascends to insert on the lower border of the body of the hyoid. It is innervated through the ansa cervicalis. The muscle acts in concert with the thyrohyoid and sternohyoid to draw down the hyoid.

The deep muscles of the triangle lie on the anterior surface of the vertebral column (see Fig. 8-6). They are the longus colli, longus capitis, rectus capitis anterior, and rectus capitis lateralis.

The slips of the **longus colli** arise from the vertebral column as low as the third thoracic vertebra. They insert as high as the atlas. The muscle bends the neck forward (in concert with its mate on the other side), flexes the neck laterally, and

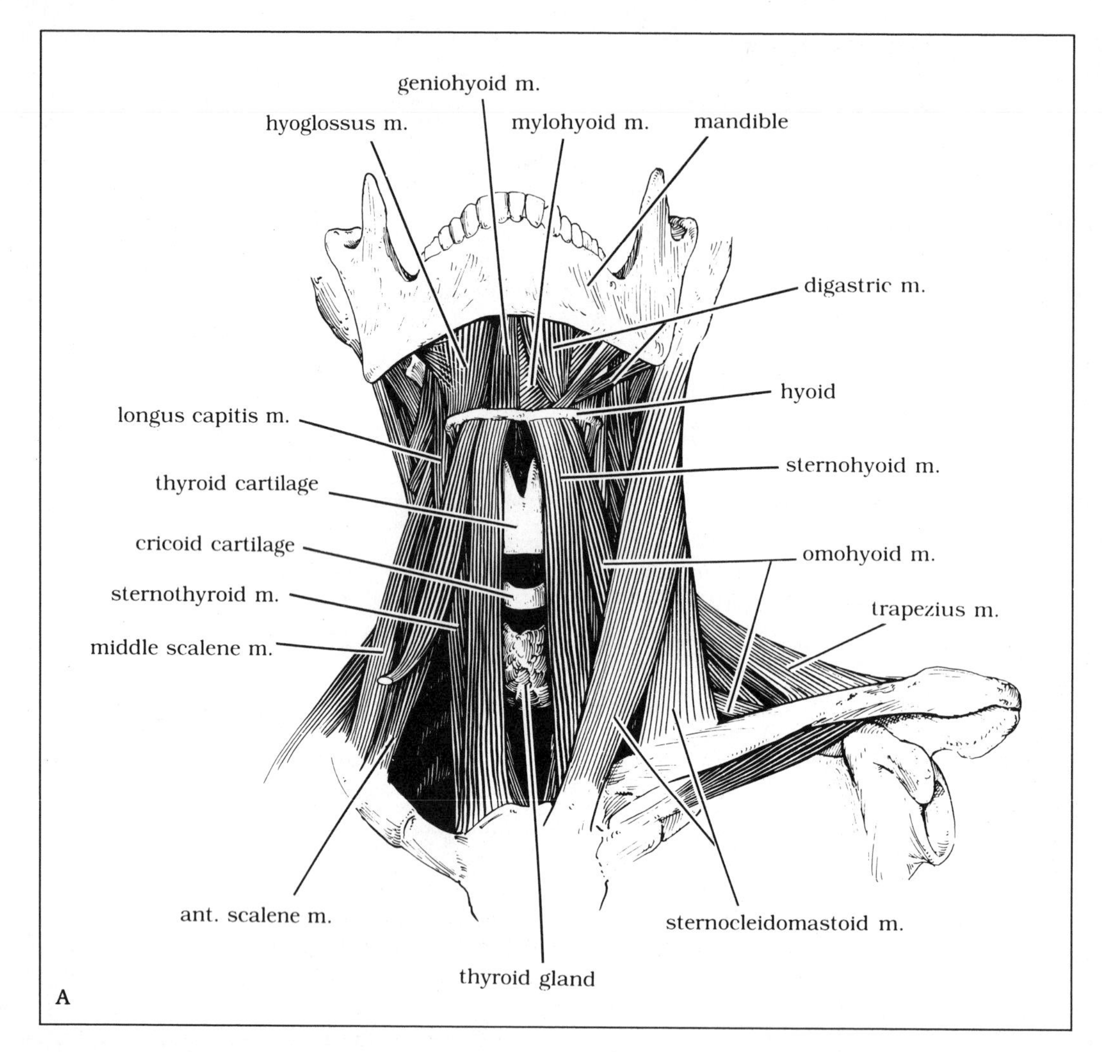

Fig. 8-4. Muscles of the neck. A. Anterior view. B. Lateral view.

rotates the neck to the opposite side. It is innervated through branches of the second to sixth cervical nerves. The **longus capitis** arises in slips from the third to sixth cervical vertebrae. It inserts into the basilar part of the occipital bone. The muscle flexes the head and is innervated by the first to third cervical nerves.

The **rectus capitis anterior** (lying deep to the longus capitis) arises from the atlas and inserts on the basilar part of the occipital bone. It is innervated by the first and second cervical nerves and flexes the head.

The **rectus capitis lateralis** also arises from the atlas. It inserts on the jugular process of the occipital bone. Its action is to bend the head to the same side. Innervation is accomplished through the first and second cervical nerves.

VESSELS

The major vessels of the anterior cervical triangle are contained within the carotid sheath, deep to the sternocleidomastoid and infrahyoid muscles

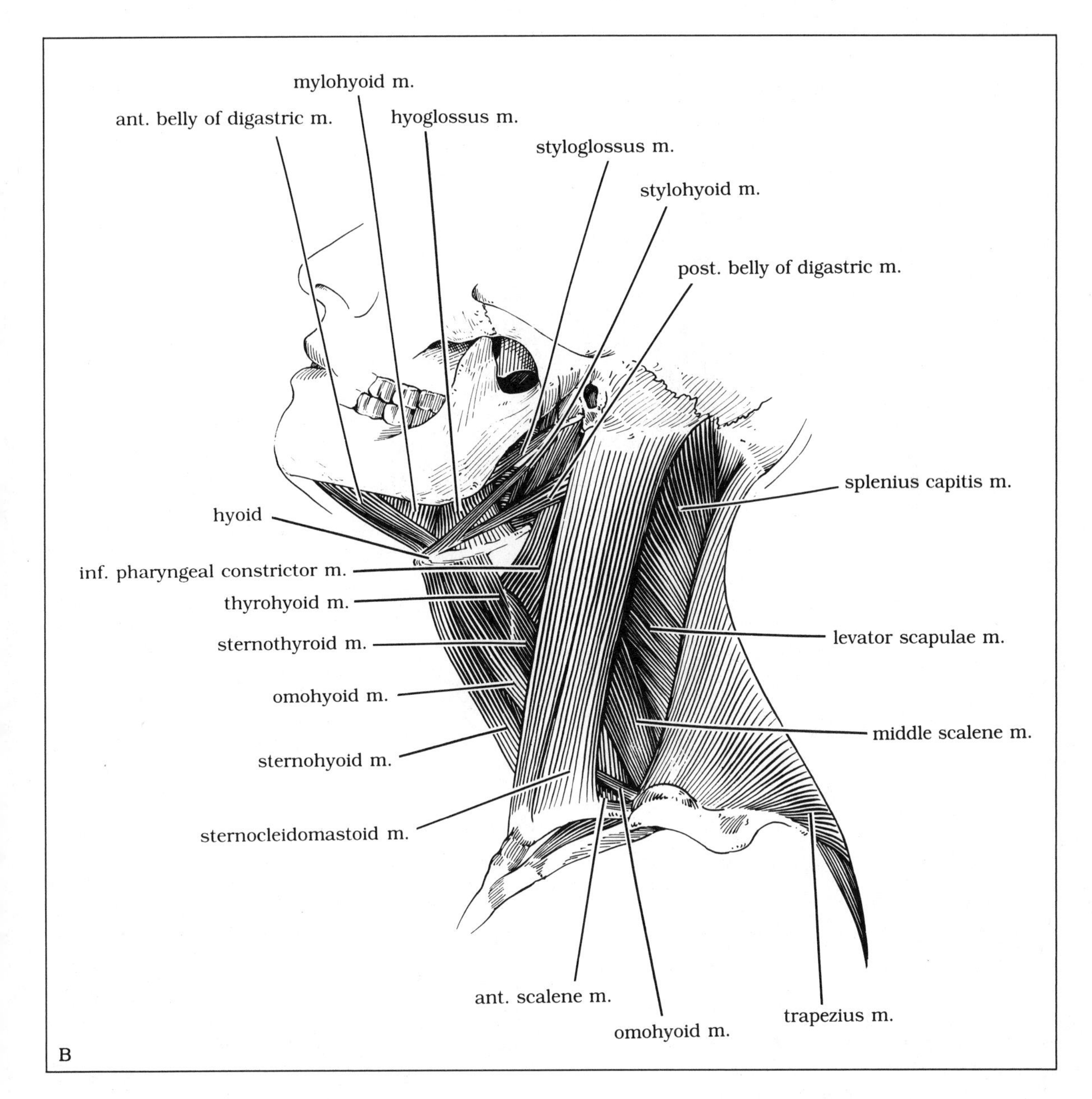

Fig. 8-4 (continued).

(Fig. 8-5). Within the sheath the common carotid artery is located medially, the internal jugular vein laterally, and the vagus nerve posteriorly. Lying superficial to the artery is the superior ramus of the ansa cervicalis.

Common Carotid Arteries

On the right side the common carotid artery arises from a bifurcation of the brachiocephalic artery behind the right sternoclavicular joint. On the left the common carotid artery arises directly from the arch of the aorta; as such it is the longer

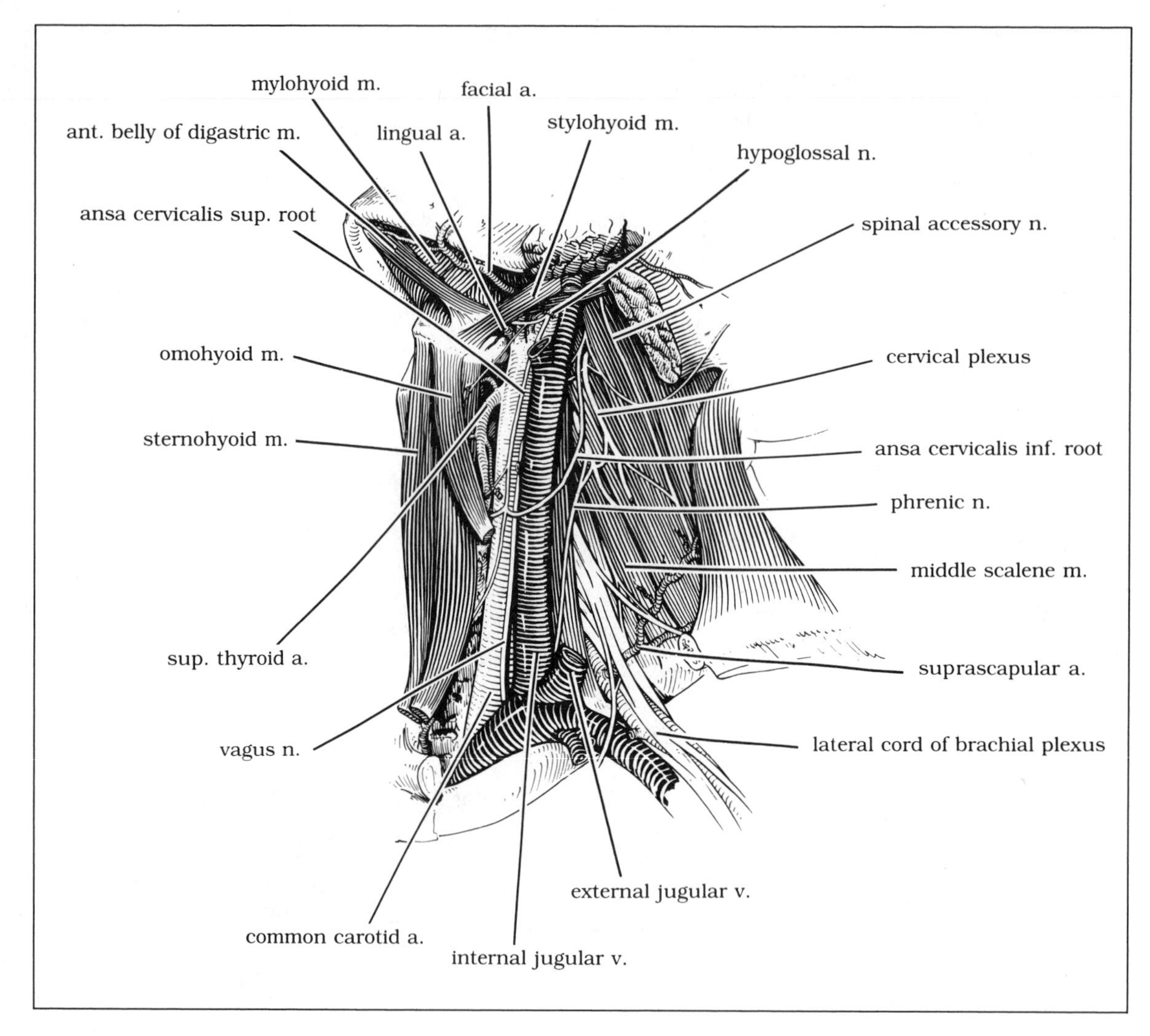

Fig. 8-5. Vessels and nerves of the neck. Lateral view.

artery. Both arteries end at the level of the superior horn of the thyroid cartilage, where they bifurcate into the internal and external carotid arteries. At the bifurcation there is a dilatation of the vessel known as the **carotid sinus.** This sinus is important in monitoring, and regulating, blood pressure through endings of the glossopharyngeal (ninth cranial) nerve.

The common carotid artery has no other branches within the neck besides its two terminal branches. The internal carotid has no branches in the neck; instead it continues within the carotid sheath to enter the cranium through the carotid canal. The external carotid artery, which extends from the superior horn of the thyroid cartilage to the neck of the mandible, provides the entire blood supply to the neck and face through eight branches.

The **superior thyroid artery** (Fig. 8-6; see also Fig. 8-5) is usually the first or second branch of the external carotid. The artery turns downward after leaving the external carotid to reach the thyroid gland, where it splits into anterior and posterior branches that run along the upper border of the gland. (The external branch of the superior laryngeal nerve accompanies the superior thyroid artery.) The artery supplies branches to the infrahyoid, sternocleidomastoid, and cricothyroid muscles, the larynx, the esophagus, and the inferior pharyngeal constrictor.

The **ascending pharyngeal artery** is either the first or the second branch of the external carotid artery. It arises from the medial surface of the artery and ascends along the pharynx. The ascending pharyngeal artery distributes to the prevertebral muscles, the deep cervical lymph nodes, the sympathetic trunk, the superior and middle pharyngeal constrictors, the soft palate, the tympanic cavity, and the dura mater.

The **lingual artery** (see Figs. 8-5 and 8-6) arises from the external carotid at the level of the greater horn of the hyoid. It passes deep to the hyoglossus muscle and the hypoglossal (twelfth cranial) nerve to supply the tongue.

The **facial artery** (see Figs. 8-5 and 8-6) arises at the level of the lingual artery, in common with it, or just above it. It passes upward and medially under the digastric and stylohyoid muscles and the submandibular gland, curves around the inferior margin of the body of the mandible, and reaches the face. In the neck it has four branches: (1) the ascending palatine artery to the pharyngeal wall, soft palate, palatine tonsil, and auditory tube; (2) the tonsillar branch to the pharyngeal tonsils; (3) submandibular branches; and (4) muscular branches to the medial pterygoid, stylohyoid, and mylohyoid muscles and the sublingual gland.

The **occipital artery** arises from the posterior side of the external carotid artery at the lower border of the posterior belly of the digastric. It passes backward to the mastoid process and then pierces the muscles at the back of the neck

Fig. 8-6. Structures of the submental and submandibular triangles.

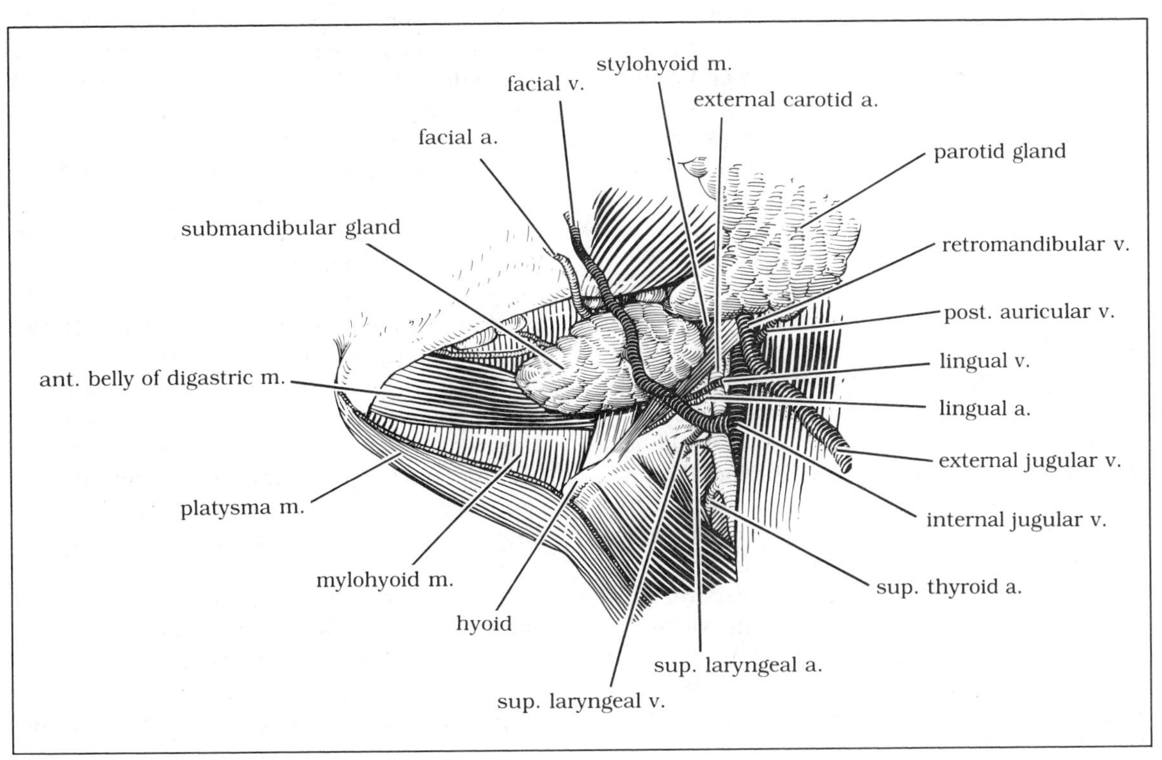

to reach the scalp. The artery supplies the digastric, stylohyoid, sternocleidomastoid, and posterior cervical muscles, the mastoid, the meninges, and the back of the ear.

The **posterior auricular artery** also exits off the posterior side of the external carotid artery at the upper border of the posterior belly of the digastric muscle. It ascends posteriorly behind the ear to reach the scalp, supplying the auricle, parotid gland, and sternocleidomastoid muscle.

The superficial temporal and maxillary arteries are the terminal branches of the external carotid artery. They arise behind the neck of the mandible. A discussion of their distribution is found in Chapter 7.

Internal Jugular Vein

The internal jugular vein collects almost all of the cranial, facial, and cervical blood (see Figs. 8-5 and 8-6). It begins at the jugular foramen, where it connects to the sigmoid sinus, and ends in the brachiocephalic vein. The tributaries in the head and neck are the lingual (from the tongue), facial (from the superficial face), occipital (from the back of the head), and superior and middle thyroid (from the thyroid gland) veins. Each of the above vessels follows the course of its like-named artery. In addition blood is received from the inferior petrosal sinus and a pharyngeal tributary that drains the pharyngeal venous plexus.

INNERVATION

There are two major nerve bundles in the anterior cervical triangle: the vagus nerve and the cervical sympathetic chain. The **vagus** (see Fig. 8-5; see also discussion in Chap. 7) is contained within the carotid sheath from the jugular foramen, where it exits the skull, to the root of the neck, where it enters the thorax. The superior cervical ganglion of the vagus is found at the level of the jugular foramen, the inferior cervical ganglion slightly below it. At this level the vagus receives fibers from the glossopharyngeal, accessory (eleventh cranial), and hypoglossal nerves, the cervical sympathetic trunk, and the ansa cervicalis. Therefore branches of the vagus below this level may also contain fibers from these nerves. Within the neck the vagus sends out branches to the pharynx, larynx, and heart.

The two pharyngeal branches of the vagus nerve arise from the inferior ganglion and pass downward, joining branches to the glossopharyngeal nerve and cervical sympathetic chain to form the pharyngeal plexus. These vagal branches supply the muscles of the pharynx, except the stylopharyngeus and tensor veli palatini, and they send a branch to the **carotid body,** a structure situated near the carotid sinus, which monitors oxygen tension in the blood.

There are two named laryngeal branches of the vagus nerve: the superior and recurrent laryngeal nerves. The **superior laryngeal nerve** descends from the inferior vagal ganglion toward the larynx. It innervates the internal carotid artery. Prior to reaching the larynx the superior laryngeal branches into internal and external laryngeal nerves. The larger internal branch supplies sensory fibers to the mucous membrane of the larynx and also parasympathetic fibers to the glands of the epiglottis, base of the tongue, aryepiglottic fold, and larynx. The external laryngeal branch supplies motor fibers to the inferior constrictor and cricothyroid muscles.

The **recurrent laryngeal nerve** takes a different passage on each side after branching from the vagus. On the right it passes under and behind the subclavian artery, while on the left it passes under the aortic arch and the ligamentum arteriosum. Both nerves then ascend between the esophagus and trachea, giving off cardiac branches to the deep cardiac plexus and tracheal, esophageal, and pharyngeal branches. The nerves terminate as the inferior laryngeal nerves, supplying the intrinsic laryngeal muscles, except the cricothyroid, and the laryngeal mucosa up to the level of the true vocal folds (cords).

There are two vagal cervical cardiac branches. The superior cervical cardiac branch arises at or below the level of the inferior vagal ganglion and passes downward beside the trachea to join the deep cardiac plexus. The inferior vagal cardiac branch arises at the root of the neck and descends to join both the deep and the superficial cardiac plexuses.

Cervical Sympathetic Chain

The cervical sympathetic chain is the ascending continuation of the thoracic sympathetics (Fig. 8-7). The cervical chain receives no preganglionic fibers via white rami from cervical levels; it sends fibers to cervical segmental nerves only via gray rami. The chain lies anterior to the prevertebral muscles and behind the carotid sheath. It may be present either as a solid cord or as a tangle of fibers. Two or four ganglia may be found along the chain.

The **superior cervical sympathetic ganglion** lies at the level of the second or third cervical vertebra and sends fibers to join the first four cervical nerves. In addition branches are sent to the internal carotid artery (ascending into the

Fig. 8-7. Structures of the deep neck. Anterior view.

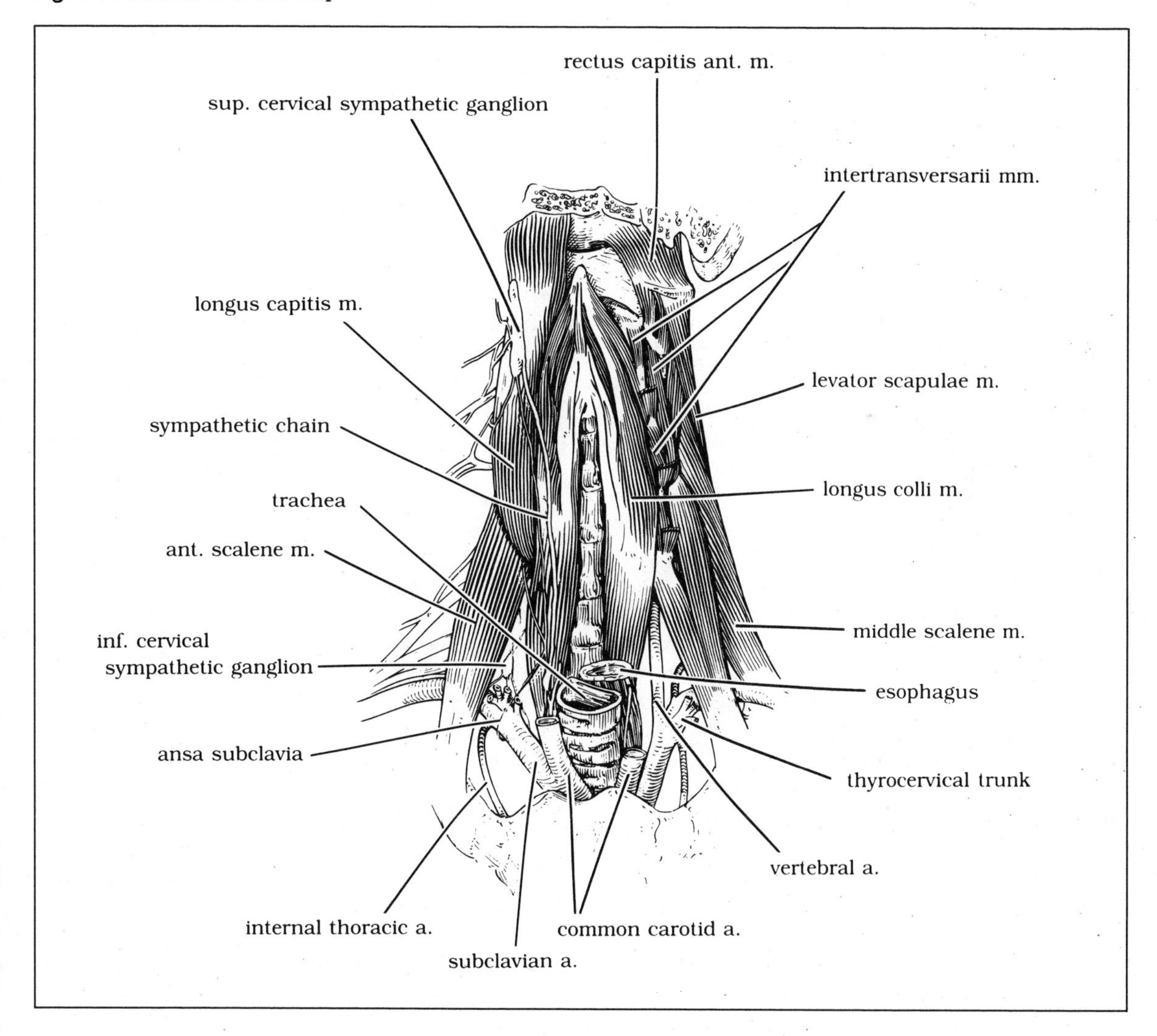

cranium); the glossopharyngeal, vagus, and hypoglossal nerves; and the larynx, pharynx, heart, and plexuses supplying the common and external carotid arteries.

The small and inconstant **middle cervical sympathetic ganglion** is found at the level of the sixth cervical vertebra. It sends branches to the fifth and sixth cervical nerves, the thyroid and parathyroid glands, the cardiac plexus, the subclavian arteries, and the vagal laryngeal nerves. It is connected to the inferior cervical ganglion through both superior and inferior trunks. These trunks loop around the subclavian artery and are known as the ansa subclavia.

The **inferior cervical sympathetic ganglion** is inconstant in size and shape and may be fused with thoracic ganglia, infrequently as low as the level of the fourth thoracic vertebra. When this fusion occurs the resulting structure is known as the cervicothoracic ganglion. It lies at the level of the seventh cervical vertebra or the first rib and supplies the seventh and eighth cervical nerves and whatever thoracic nerves are included as a result of fusion. In addition branches pass to the heart and the subclavian and vertebral arteries.

A fourth ganglion, the **vertebral ganglion,** is sometimes found in front of the vertebral artery. It may contain parts of the middle or inferior ganglia or both, and when present contributes to the formation of the ansa subclavia.

LYMPHATICS

The lymph nodes of the anterior cervical triangle are divided into deep and superficial nodes. The superficial nodes, along with the nodes of the head, drain to the deep cervical lymph nodes. The anterior nodes drain the submental and submandibular triangles and the regions of the anterior and external jugular veins. The deep nodes are divided into two groups: a superior group above the omohyoid, and an inferior group below the omohyoid. The deep nodes drain the nose, sinuses, palate, ear, nasal and oral pharynges, tongue, and tonsils.

In addition two lymphatic trunks are found in the neck. The jugular trunk is formed by the efferent vessels of the deep cervical nodes. It empties into the thoracic duct on the left and the junction of the subclavian and internal jugular veins on the right. The thoracic duct enters at the root of the neck to the left of the esophagus. It arches behind the carotid sheath and then descends to the junction of the left subclavian and left internal jugular veins.

Other deep nodes found in the neck include the infrahyoid, prelaryngeal, pretracheal, and paratracheal groups, all of which drain into the nodes of the anterior cervical triangle. Nodes that lie along the accessory nerve and the transverse cervical vessels drain through the posterior cervical triangle lymphatics.

GLANDS

Two endocrine glands, the thyroid and parathyroid glands, lie within the anterior cervical triangle. The **thyroid gland,** lying in the front of the neck (see Fig. 8-4A), secretes thyroxin, which helps control the body's metabolism. It is contained within both a capsule and a sheath, with the blood vessels lying between them. The sheath is a continuation of the pretracheal layer of cervical fascia.

The thyroid gland consists of two vertical paratracheal limbs that are connected by an isthmus across the upper tracheal rings. Occasionally there is an ascending pyramidal lobe, a remnant of the origin of the gland, whose opening was the foramen cecum at the base of the tongue.

The thyroid gland receives blood via two superior and two inferior thyroid arteries. Venous drainage is via the thyroid plexus into the superior, inferior, or middle thyroid veins. The inferior thyroid veins descend on the trachea to the brachiocephalic veins. If only one vein is present it is the thyroid ima vein. The middle thyroid veins have no corresponding artery.

The gland is innervated by branches of the middle and inferior cervical sympathetic ganglia. Lymphatic drainage is through the superior and inferior deep cervical nodes.

The **parathyroid gland** consists of two to six small glands on the dorsum of the thyroid gland, within its sheath. They secrete parathormone, which regulates blood calcium levels. Their blood supply, innervation, and lymphatic drainage is the same as the thyroid gland's.

Submandibular Triangle

The submandibular triangle is bound by the mandible and the bellies of the digastric muscle (see Figs. 8-3 and 8-4A). The triangle effectively forms the floor of the mouth. Superficially it is covered by the superficial layer of cervical fascia. The vessels of this triangle are discussed with the vessels of the anterior cervical triangle.

MUSCULATURE

The submandibular triangle contains five muscles: the digastric, stylohyoid, mylohyoid, geniohyoid, and hyoglossus. The **digastric muscle** is a two-bellied muscle (Table 8-2; see also Figs. 8-3 and 8-4). The posterior belly arises from the mastoid notch of the temporal bone and descends to the hyoid, where the intermediate muscular tendon is held by a fascial sling to the hyoid bone. The anterior belly arises from the intermediate tendon and ascends to insert into the digastric fossa on the inner side of the lower border of the mandible. The muscle elevates the hyoid, steadies it, and moves it forward and backward. With the hyoid fixed, the digastric aids in opening the jaw. By elevating the hyoid, the larynx and pharynx are raised and food can be moved into the pharynx by raising the floor of the mouth. The muscle is derived from two independent sources and therefore has two separate innervations. The anterior belly is supplied by the nerve to the mylohyoid, a branch of the mandibular division of the trigeminal (fifth cranial) nerve. The posterior belly is supplied by the facial nerve.

The **stylohyoid muscle** (see Fig. 8-4B) lies parallel and medial to the posterior belly of the digastric. It arises from the styloid process of the temporal bone and inserts into the greater horn of the hyoid. The muscle draws the hyoid upward and backward. It is innervated by the facial nerve.

The **mylohyoid muscle** (see Figs. 8-4B and 8-6) forms the floor of the oral cavity. Its origin is the mylohyoid line of the mandible from the symphysis to the level of the last molar, and it inserts into the body of the hyoid and a median fibrous raphe. It is innervated by the nerve to the mylohyoid (mandibular branch of the trigeminal nerve).

The **geniohyoid muscle** (Fig. 8-8) lies deep to

Table 8-2. Musculature of the Submandibular Triangle

Muscle	*Origin*	*Insertion*	*Innervation*	*Action*
Digastric	Mastoid notch of temporal bone	Digastric fossa of mandible	Anterior belly—mandibular nerve; posterior belly—facial nerve	Elevate, steady, or move hyoid
Stylohyoid	Styloid process of temporal bone	Greater horn of hyoid	Facial nerve	Move hyoid up and back
Mylohyoid	Mylohyoid line of mandible	Body of hyoid, median raphe	Nerve to mylohyoid	Elevate floor of oral cavity
Geniohyoid	Mental spine	Body of hyoid	Nerve C1	Draw hyoid anteriorly in swallowing
Hyoglossus	Greater horn of hyoid	Tongue	Hypoglossal nerve	Depress posterolateral sides of tongue

the mylohyoid on both sides of the midline. It arises from the mental spine of the symphysis menti of the mandible and inserts into the body of the hyoid. It is innervated by the first cervical nerve through a branch that travels with the hypoglossal nerve.

The mylohyoid and geniohyoid act together in swallowing. They elevate the tongue and floor of the mouth, move the tongue and hyoid back, and aid in opening the jaw when the hyoid is fixed.

The **hyoglossus muscle** (see Figs. 8-4 and 8-8) is an extrinsic muscle of the tongue. Its origin is the greater horn of the hyoid, and it inserts into the tongue. The hyoglossus is innervated by the hypoglossal nerve.

GLANDS

The glands of the submandibular triangle are salivary glands (see Fig. 8-6). The **submandibular gland** lies below and in front of the angle of the mandible in the most posterior portion of the triangle. It is mostly superficial, but part of it, including the duct, lies deep to the mylohyoid muscle. The gland is grooved by the facial artery, which also supplies it. Venous drainage is via the facial vein, and lymphatic drainage is via the submandibular lymph nodes. Parasympathetic innervation is supplied by the chorda tympani of the facial nerve traveling with the lingual nerve and synapsing in the submandibular ganglion. Sympathetic innervation is from the superior cervical ganglion.

The submandibular duct passes forward and medially, first between the mylohyoid and the hyoglossus and then between the mylohyoid and the genioglossus. The duct opens at the sublingual caruncle at the side of the frenulum of the tongue.

Fig. 8-8. Muscles of the tongue and larynx. Lateral view.

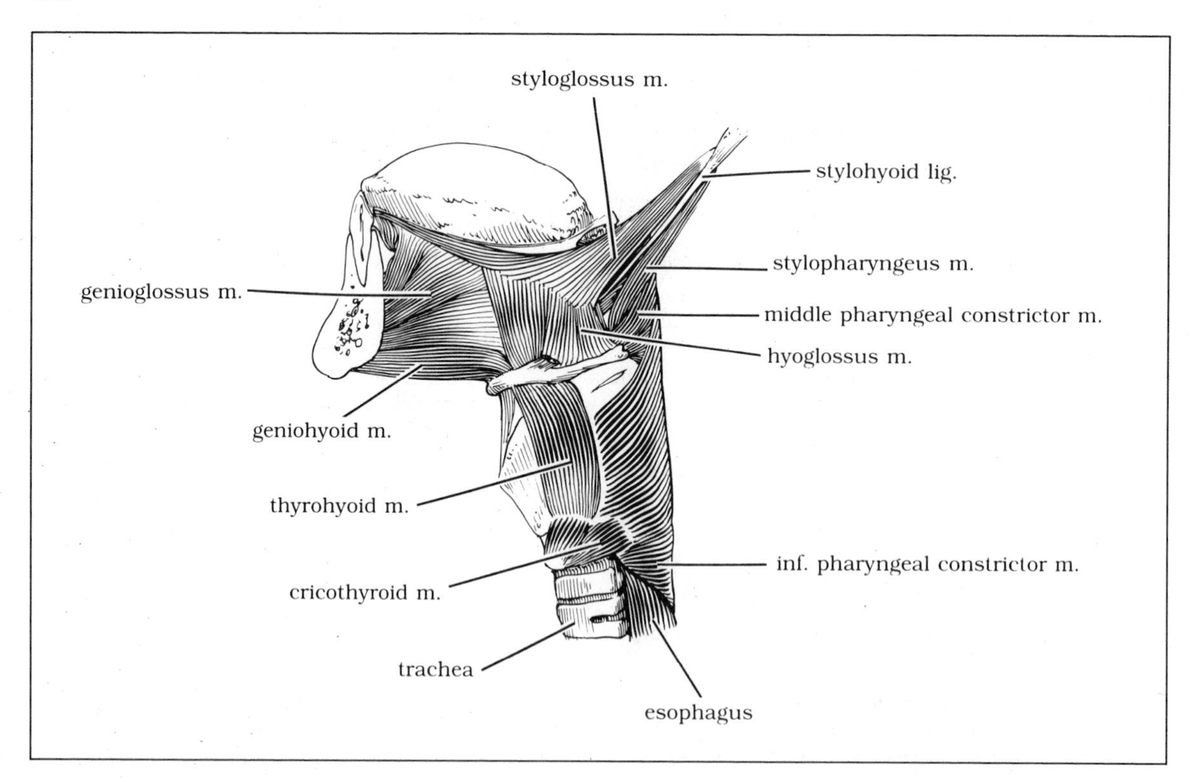

Table 8-3. Musculature of the Posterior Cervical Triangle

Muscle	*Origin*	*Insertion*	*Innervation*	*Action*
Sternocleidomastoid	Clavicle, sternum	Mastoid process	Accessory nerve	Bring head forward and upward
Anterior scalene	Vertebrae C3–C6	Rib 1	Nerves C4–C6	Bend neck forward and laterally
Middle scalene	Vertebrae C2–C7	Rib 1	Nerves C3–C8	Bend neck
Posterior scalene	Vertebrae C4–C6	Rib 2	Nerves C6–C8	Bend neck

The **sublingual salivary gland** lies beneath the floor of the mouth between the mandible and the genioglossus and geniohyoid. The mylohyoid is superficial to it. Its many ducts open directly into the floor of the mouth. Its blood is supplied through the lingual and facial arteries. Venous drainage is through the lingual and facial veins. Its innervation is identical to that of the submandibular gland.

INNERVATION

The **lingual nerve** is a branch of the mandibular division of the trigeminal nerve supplying general sensation to the anterior two-thirds of the tongue. It is accompanied by the chorda tympani. They pass through the submandibular triangle as they spiral around the duct of the submandibular gland to reach the tongue.

The **chorda tympani** is a branch of the facial nerve. It supplies taste to the anterior two-thirds of the tongue and parasympathetic innervation to the submandibular and sublingual glands through synapses in the submandibular ganglion.

The small **submandibular ganglion** appears to hang from the lingual nerve near the posterior border of the mylohyoid muscle. It is the parasympathetic ganglion for the submandibular and sublingual glands.

The **hypoglossal nerve** (see Fig. 8-5) runs through the inferior portion of the submandibular triangle to reach the extrinsic and intrinsic tongue muscles. (The hypoglossal nerve is discussed more fully in Chapter 7.)

Posterior Cervical Triangle

The posterior cervical triangle (see Fig. 8-3) is bounded by the posterior border of the sternocleidomastoid muscle, the trapezius, and the clavicle. Its superficial space contains the cutaneous branches of the cervical plexus.

MUSCULATURE

The **sternocleidomastoid muscle** (Table 8-3; see also Fig. 8-4) arises from a thick clavicular head and a thinner sternal head. The muscle passes upward and backward to insert into the mastoid process and superior nuchal line. Acting independently each muscle serves to turn the head upward and the chin toward the opposite side. Acting together they move the head forward and point the chin up. With the head fixed the muscles act in respiration to raise the clavicle and sternum. The sternocleidomastoid is innervated by the accessory nerve.

The three scalene muscles (see Figs. 8-4A and 8-7) constitute the deep musculature of the posterior cervical triangle. The **anterior scalene muscle** (Fig. 8-9) arises from the third to the sixth cervical vertebrae and inserts onto the first rib. It either elevates the rib or bends the neck forward and laterally, rotating it to the opposite side. The muscle is innervated by branches of the fourth to sixth cervical nerves.

Lying superficial to the anterior scalene muscle the phrenic nerve may be found. Deep to the muscle, between it and the middle scalene, the roots of the brachial plexus emerge.

The **middle scalene muscle** (see Fig. 8-5), the

largest scalene muscle, arises from the second to the seventh cervical vertebrae. It also inserts onto the first rib. The muscle either raises the first rib or bends the neck to the same side. Innervation is accomplished through branches of the third through eighth cervical nerves.

The **posterior scalene muscle** arises from the fourth to the sixth cervical vertebrae. It inserts onto the second rib and either elevates this rib or bends the neck to the same side. It is innervated by the sixth to the eighth cervical nerves.

Through their attachment to the ribs all the scalenes are active in quick breathing.

BLOOD SUPPLY AND LYMPHATICS

The subclavian artery and vein and their branches are the primary vessels of the posterior cervical triangle (see Figs. 8-7 and 8-9). The **subclavian artery** begins in the neck behind the sternoclavicular joint. It ends by becoming the axillary artery as it crosses over the first rib. In its arch through the root of the neck the artery is crossed by the anterior scalene muscle and is thus divided into three parts. The first part gives origin to the vertebral, thyrocervical, and internal thoracic arteries. The costocervical trunk arises from the second part, and the dorsal scapular artery from the third.

The **vertebral artery** ascends medially to enter the costotransverse foramen of the sixth cervical vertebra. It then ascends through these foramina, behind the superior articular process of the atlas, and through the foramen magnum to enter the cranium. Within the cranium the artery branches and then joins its opposite member to form the basilar artery. Before entering the cranium the artery provides branches that accompany the cervical nerves and branches to the vertebrae and deep muscles of the neck.

The **thyrocervical trunk** divides into three branches: the inferior thyroid, suprascapular, and transverse cervical arteries. The **inferior thyroid artery** ascends to the level of the cricoid cartilage before arching downward and medially to reach the thyroid gland. At the top of the arch an ascending cervical branch is given off. This branch ascends parallel to the phrenic nerve. Within the thyroid gland the inferior thyroid artery divides into ascending and inferior branches. The **suprascapular artery** passes downward and laterally to the level of the clavicle. Then it passes parallel to the clavicle to reach the suprascapular notch and over the transverse ligament and through the supraspinous fossa to reach the infraspinous fossa. The **transverse cervical** arterial branch of the thyrocervical trunk passes laterally to the trapezius, which it supplies.

The **internal thoracic artery** descends forward and medially to enter the thorax behind the rib cage. Further discussion of this artery is presented in Chapter 2.

The **costocervical trunk** arises behind the anterior scalene muscle. It passes backward over the dome of the cervical pleura to become the highest intercostal and deep cervical arteries.

The last branch of the subclavian artery, the **dorsal scapular artery,** passes posteriorly through the brachial plexus to reach the levator scapulae. Here it descends to the scapula, supplying the levator, the rhomboids, and the serratus anterior.

The **subclavian vein** follows the course of the subclavian artery. It receives the external jugular vein, the transverse cervical vein, and the suprascapular and dorsal scapular veins. It joins with the internal jugular vein to form the brachiocephalic vein. At this junction, on the left side, the thoracic duct empties into the vessel. On the right side, the right lymphatic duct enters.

INNERVATION

One major nerve and two plexuses are found within the posterior cervical triangle (see Figs. 8-5 and 8-9). The **accessory nerve** exits the cranium through the jugular foramen and descends along with the internal carotid and internal jugular vessels. It then passes laterally, piercing and supplying the sternocleidomastoid, and crosses the posterior cervical triangle to reach the trapezius. (For further discussion of the accessory nerve, see Cranial Nerves in Chapter 7.)

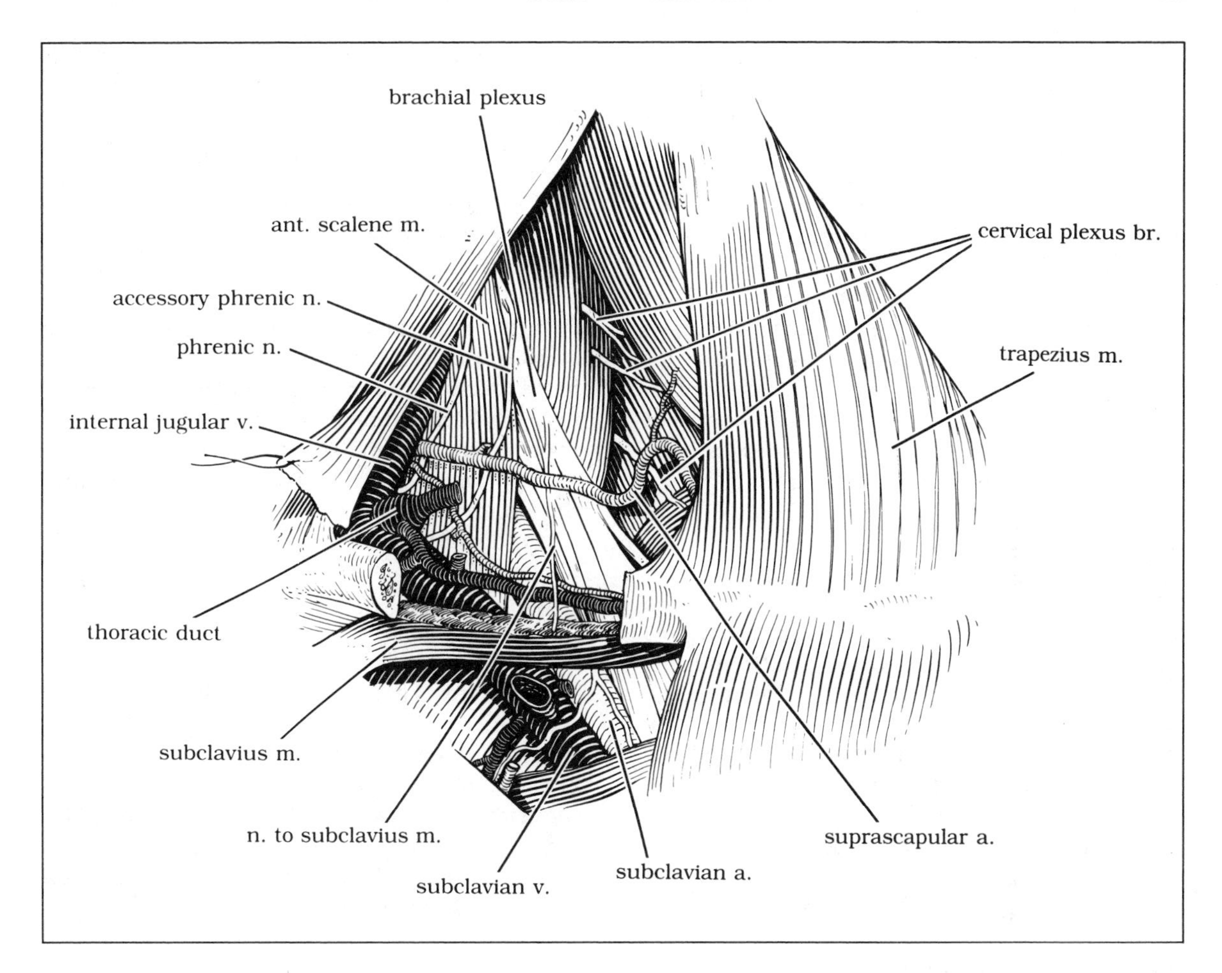

Fig. 8-9. Vessels and nerves at the base of the neck.

The **cervical plexus** is formed from the ventral rami of the first to third cervical nerves (the fourth cervical nerve sometimes contributes fibers also). Sympathetic fibers are supplied by the superior cervical ganglion. The cervical segmental nerves unite to form loops, which in turn give rise to the cervical plexus. The cutaneous branches of this plexus are covered under Superficial Organization. The plexus also supplies much of the musculature associated with the hyoid through the ansa cervicalis.

The superior root of the **ansa cervicalis** is formed by fibers from the first to second cervical nerves that travel briefly with the hypoglossal nerve. Most of the fibers leave the hypoglossal in front of the common carotid artery and loop around to meet the inferior root of the ansa cervicalis, formed by fibers from the second and third cervical nerves, next to the internal jugular vein. The fibers that do not leave the hypoglossal nerve continue on to supply efferent innervation to the thyrohyoid and geniohyoid muscles. The full loop of the ansa cervicalis supplies fibers to the omohyoid, sternohyoid, and sternothyroid muscles.

The cervical nerves also supply branches that accompany the accessory to the trapezius and independent motor innervation to the longus capitis, longus colli, middle scalene, levator scapulae, rectus capitis lateralis, and rectus capitis anterior muscles.

The **phrenic nerve** (and accessory phrenic nerve, when present) arises primarily from the fourth cervical nerve but may have contributions from the third and fifth cervical nerves. The nerve descends over the ventral surface of the anterior scalene muscle to enter the chest. In the chest the phrenic nerve supplies motor innervation to the diaphragm.

The brachial plexus, from the ventral rami of the fifth cervical to the first thoracic nerve, is formed as the roots pass between the anterior and middle scalene muscles in the posterior cervical triangle. Within the neck branches are sent to the longus colli and scalene muscles. The plexus is further discussed in Chapter 5.

Suboccipital Triangle

The suboccipital region (Fig. 8-10), where the head and neck are connected posteriorly, is found deep to the semispinalis capitis muscle. Four muscles are found in this region, all of which are supplied by the dorsal ramus of the first cervical nerve and all of which extend the head and rotate it and the atlas to the same side (Table 8-4).

The **rectus capitis posterior major** arises from the axis and inserts at the level of the inferior nuchal line and below it. The **rectus capitis posterior minor** lies medial to the major. It arises from the atlas, and it also inserts at the level of the inferior nuchal line and below it.

The **obliquus capitis superior** arises from the atlas and inserts above the inferior nuchal line. The **obliquus capitis inferior** arises from the axis and inserts on the atlas.

The suboccipital triangle is filled primarily with fibrofatty tissue. The floor of the triangle is formed by the atlanto-occipital membrane and overlies the vertebral artery. The suboccipital nerve (dorsal primary ramus of the first cervical nerve), which supplies the muscles, pierces the floor to reach them. The greater occipital nerve (dorsal ramus of the second cervical nerve) and the occipital artery cross the suboccipital triangle to reach the scalp.

Cervical Viscera

PHARYNX

The pharynx is a funnel-shaped chamber behind the oral, nasal, and laryngeal cavities, through which it is entered (Figs. 8-11 and 8-12). The pharynx ends at the level of the lower border of the cricoid cartilage by dividing into the anterior larynx and posterior esophagus. The walls of the pharynx are lined by a layer of mucous membrane. Underlying this layer is a fibrous layer,

Table 8-4. Suboccipital Muscles

Muscle	*Origin*	*Insertion*	*Innervation*	*Action*
Rectus capitis posterior major	Spinous process of axis	Inferior nuchal line of occipital bone	Suboccipital nerve (C1)	Extend head, rotate head to same side
Rectus capitis posterior minor	Tubercle on atlas	Medial part of inferior nuchal line of occipital bone	Suboccipital nerve (C1)	Extend head, rotate head to same side
Obliquus capitis inferior	Spinous process of axis	Back of transverse process of atlas	Suboccipital nerve (C1)	Rotate head, turn face to same side
Obliquus capitis superior	Upper surface of transverse process of atlas	Occipital bone between superior and inferior nuchal lines	Suboccipital nerve (C1)	Extend head, bend head laterally

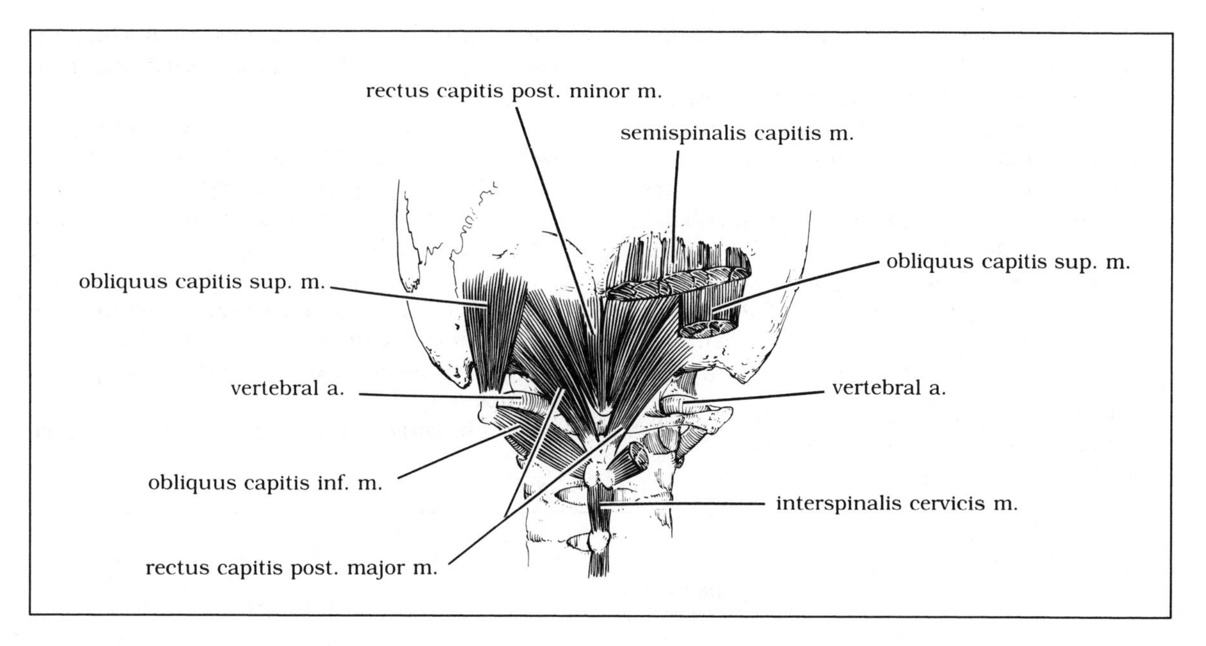

Fig. 8-10. Muscles of the suboccipital triangle.

Fig. 8-11. Musculature of the pharynx.

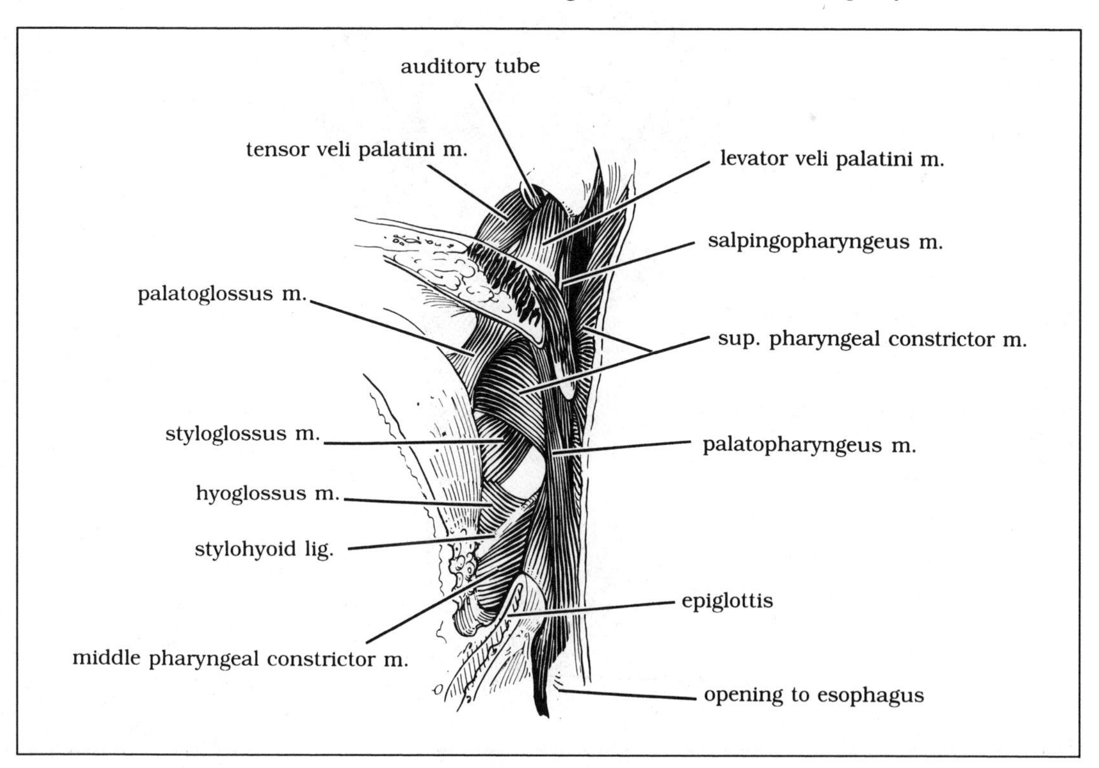

followed by a layer of muscle and then the buccopharyngeal fascia.

The pharynx is divided into three parts, with only the laryngopharynx having an anterior wall. The nasal pharynx, the uppermost region, contains the pharyngeal tonsil situated in the upper region of its posterior wall. The lateral wall contains the salpingopharyngeal fold and the salpingopalatine fold, their underlying muscles, and the levator veli palatini muscle. The auditory tube opens in the lateral wall. The opening is overhung by the torus tubarius, from which the salpingopalatine and salpingopharyngeal folds descend.

The oral pharynx contains the **palatine tonsil** on its lateral wall. The tonsil lies between the palatoglossal and palatopharyngeal folds, which descend from the palate and contain within them their corresponding muscles.

The laryngeal region of the pharynx contains the epiglottis and its associated structures and the openings for the larynx and esophagus. The piriform recess is located posterior to the laryngeal opening.

The pharyngeal wall contains the three pha-

Fig. 8-12. A. The pharynx and larynx. External and posterior view. B. The pharynx. Internal view from behind.

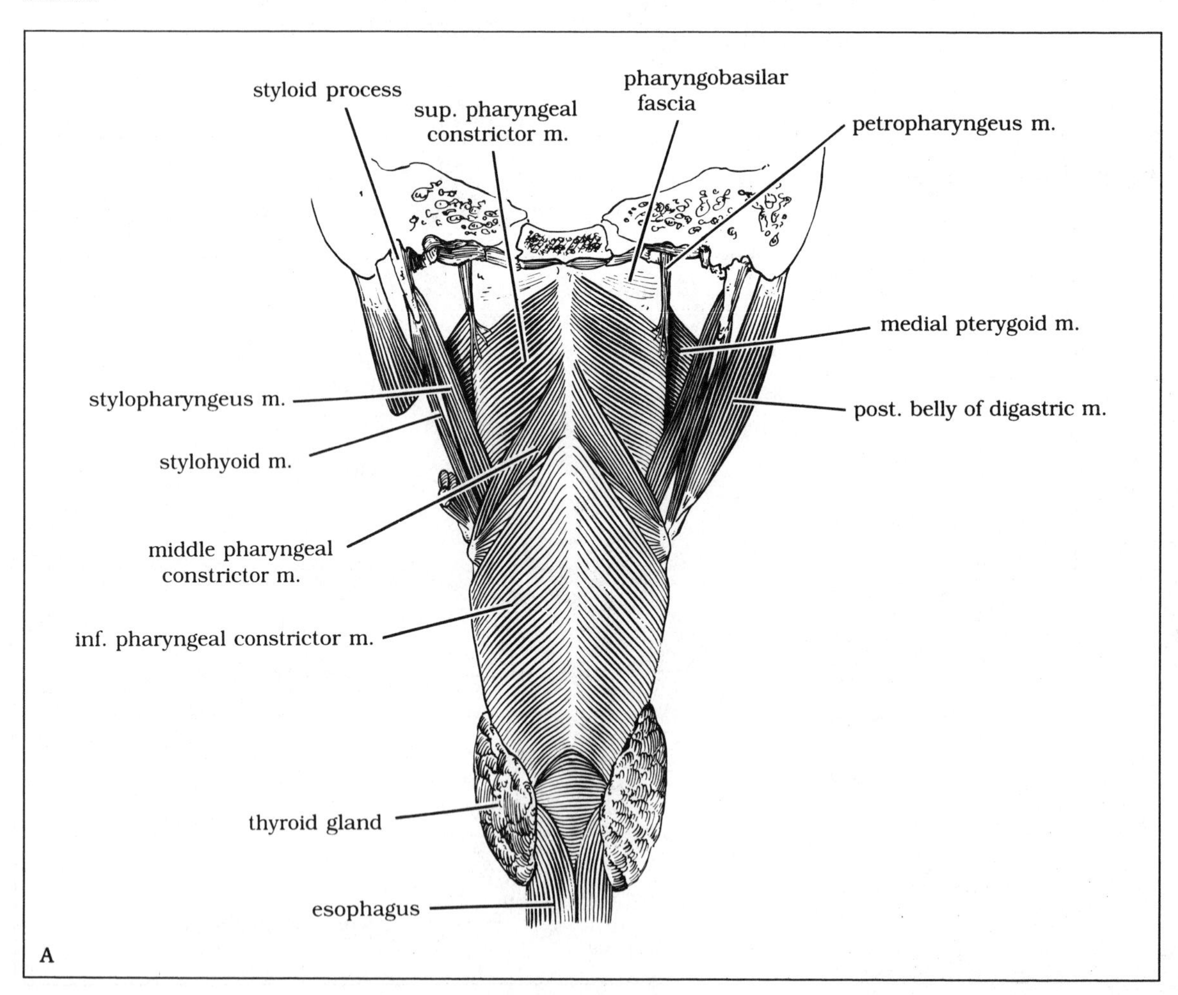

ryngeal constrictors and the stylopharyngeal, palatopharyngeal, and salpingopharyngeal muscles (Table 8-5). The constrictor muscles are paired muscles united in the midline by a median raphe that extends from the pharyngeal tubercle to the esophageal musculature. The **superior pharyngeal constrictor** is the deepest muscle. It arises from the medial pterygoid plate, the hamulus, the pterygomandibular raphe, the mylohyoid line of the mandible, and the root of the tongue. The muscle inserts into the pharyngeal tubercle and the raphe. Superior to the muscle is the pharyngobasilar fascia. The **middle pharyngeal constrictor** arises from the greater and lesser horns of the hyoid and the stylohyoid ligament. The **inferior pharyngeal constrictor,** the most external of the constrictor muscles, arises from the thyroid and cricoid cartilages. Both the middle and inferior constrictors insert into the median raphe.

The **stylopharyngeus** arises from the styloid process. It passes between the superior and middle constrictors to insert into the thyroid cartilage and pharyngeal wall. Through much of its passage it is accompanied by the glossopharyngeal nerve.

The **palatopharyngeus** muscle lies beneath the

Fig. 8-12 (continued).

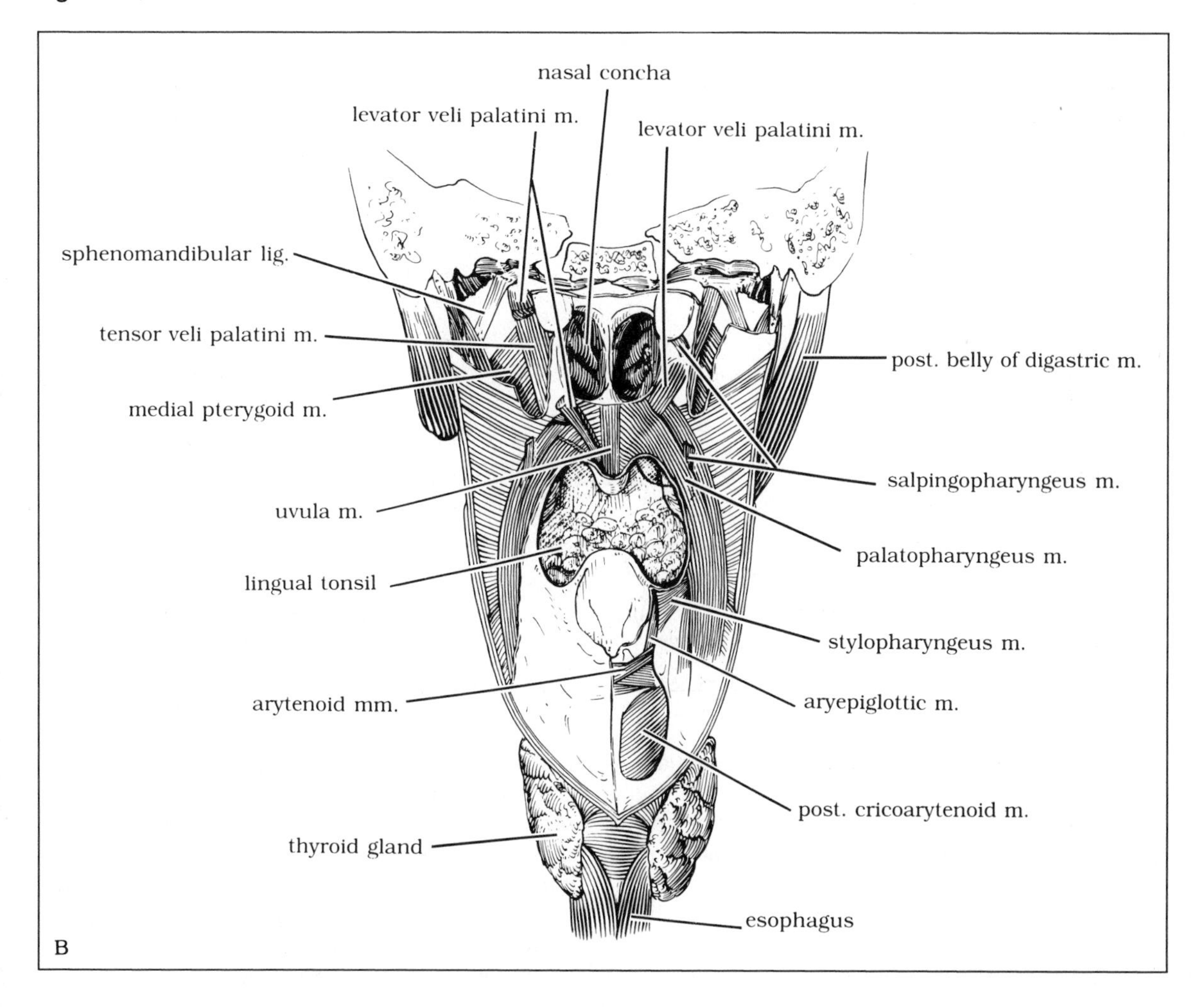

Table 8-5. Musculature of the Cervical Viscera

Muscle	*Origin*	*Insertion*	*Innervation*	*Action*
Pharynx				
Superior pharyngeal constrictor	Medial pterygoid plate, hamulus, pterygomandibular raphe, mylohyoid line, root of tongue	Pharyngeal tubercle, median raphe	Pharyngeal plexus	Swallowing
Middle pharyngeal constrictor	Hyoid, stylohyoid ligament	Median raphe	Pharyngeal plexus	Swallowing
Inferior pharyngeal constrictor	Thyroid and cricoid cartilages	Median raphe	Pharyngeal plexus	Swallowing
Stylopharyngeus	Styloid process	Pharyngeal wall, thyroid cartilage	Glossopharyngeal nerve	Swallowing
Palatopharyngeus	Palate	Pharyngeal wall, thyroid cartilage	Pharyngeal plexus	Swallowing
Salpingopharyngeus	Auditory tube	Pharyngeal wall	Pharyngeal plexus	Swallowing
Larynx				
Cricothyroid	Cricoid cartilage	Thyroid cartilage	Superior laryngeal nerve, external branch	Pull thyroid cartilage down and forward
Posterior cricoarytenoid	Cricoid cartilage	Arytenoid cartilage	Inferior laryngeal nerve	Abduct vocal folds
Lateral cricoarytenoid	Cricoid cartilage	Arytenoid cartilage	Inferior laryngeal nerve	Adduct vocal folds
Transverse arytenoid	Arytenoid cartilage	Arytenoid cartilage	Inferior laryngeal nerve	Adduct vocal folds
Oblique arytenoid	Arytenoid cartilage	Arytenoid cartilage	Inferior laryngeal nerve	Laryngeal sphincter
Thyroarytenoid	Thyroid cartilage	Arytenoid cartilage	Inferior laryngeal nerve	Vestibular sphincter, vocal folds

fold of the same name. It arises from the palate and palatine aponeurosis and inserts, with the stylopharyngeus, into the thyroid cartilage and pharyngeal wall. The **salpingopharyngeus** muscle, underlying the fold of the same name, arises from the auditory tube and inserts into the pharyngeal wall.

All of the pharyngeal muscles, except the stylopharyngeus, are innervated by nerves from the **pharyngeal plexus,** located on the middle constrictor at the level of the greater horn of the hyoid. It is formed by contributions from the vagus, glossopharyngeal, and superior cervical sympathetic nerves. The **glossopharyngeal nerve** descends along the posterior border of the stylopharyngeus muscle before curving laterally to reach the base of the tongue. There it gives off tonsillar branches to the palatine tonsil and branches to the soft palate before distributing fibers within the tongue.

The pharynx is supplied by the ascending pharyngeal and ascending palatine branches of the facial artery, the pharyngeal branch of the maxillary artery, and the muscular branch of the superior thyroid artery. Venous drainage of the pharyngeal plexus is through the pterygoid plexus or the internal jugular vein. Lymph drains to both the retropharyngeal and superior deep cervical groups of lymph nodes.

In the act of swallowing the nasal pharynx is closed through movement of the palate, and the larynx is closed through the action of the laryngeal muscles (see Larynx). The salpingopharyngeus draws the lateral pharyngeal walls upward and inward and then the stylopharyngeus and palatopharyngeus pull up the pharynx. Finally the constrictors contract in descending order to drive food downward.

ESOPHAGUS

The esophagus (see Figs. 8-7 and 8-12) is that portion of the digestive tube that extends from the lower opening of the pharynx through the neck and the chest to end in the abdomen by opening into the stomach. Its walls are muscular and its cylindrical shape averages 2 cm in diameter, being narrowest at its beginning. It arises in the midline in the neck, at the lower border of the cricoid cartilage, but gradually shifts to the left, where it is found at the root of the neck. Laterally it is flanked by the carotid sheath and thyroid gland. The trachea and recurrent laryngeal nerves lie anterior to it, while posteriorly the prevertebral muscles and vertebral column are found.

The esophageal wall consists of an inner circular muscular layer and an outer longitudinal muscular layer. Innervation and blood are supplied by regional nerves and vessels, which in the neck are the cervical sympathetics and recurrent laryngeal nerve and branches of the inferior thyroid artery and vein. Lymph drains to the paratracheal and inferior deep cervical nodes.

LARYNX

The larynx has two functions: to produce sound through controlling expiration and to prevent the passage of food into the airway (Fig. 8-13; see also Figs. 8-2 and 8-8). In form it is composed of nine rigid pieces of cartilage, the muscles that move the cartilage and modify the larynx's shape, and a lining of mucous membrane. The larynx extends from the base of the tongue to the lower border of the cricoid cartilage.

Cartilage

There are nine cartilages in the larynx, three of which are single (thyroid, cricoid, and epiglottic) and three of which are paired (arytenoid, corniculate, and cuneiform). The **thyroid cartilage** is the largest. It forms the laryngeal prominence, or Adam's apple, anteriorly where its two fused laminae meet. In the male they meet at a 90-degree angle, in the female at a 120-degree angle. Anteriorly, at the top of the junction of the laminae is a V-shaped notch, the superior thyroid notch. The thyroid cartilage is suspended from the hyoid bone by the thyrohyoid membrane, which is, in part, composed of the medial and lateral thyrohyoid ligaments. The superior margin (horn) is linked to the greater horn of the hyoid by the lateral thyrohyoid ligament. The inferior margin (horn) articulates with the cricoid cartilage. On the external surface of the thyroid cartilage there is an oblique line that serves as the point of attachment for the sternothyroid, thyrohyoid, and inferior constrictor muscles.

The ring-shaped **cricoid cartilage** has a narrow anterior arch and a wide posterior lamina. The lamina has a vertical midline ridge for the attachment of the esophageal muscle. The posterior cricoarytenoid attaches to either side of this ridge. The cricothyroid and inferior constrictor muscles attach to the arch.

The cricothyroid ligament connects the thyroid, cricoid, and arytenoid cartilages and is part of the fibroelastic lining of the larynx. The anterior part of the ligament is also known as the conus elasticus, a thick, narrow band connecting the thyroid and cricoid cartilages. The lateral portion is thinner and attaches to all three carti-

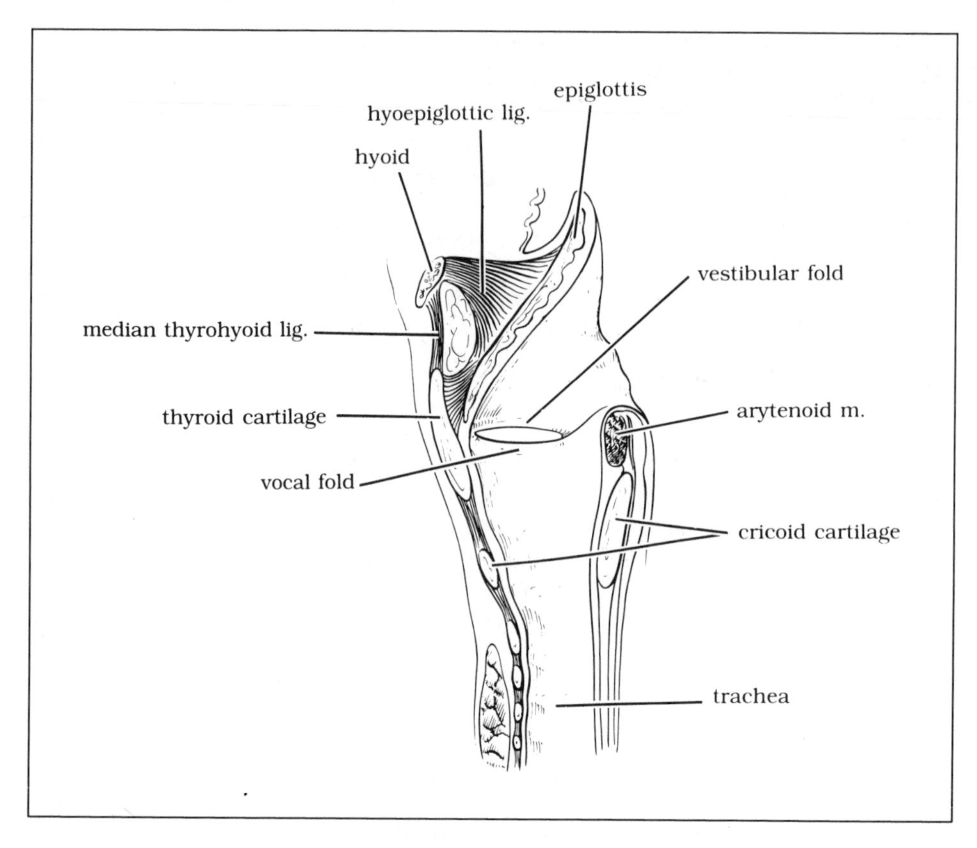

Fig. 8-13. Schematic structures of the larynx.

lages. The vocal ligament is the free lateral edge of the cricoid ligament.

The cricoid cartilage is also connected to the thyroid cartilage by the conus elasticus and the lateral cricoarytenoid muscle. The cricoid is attached to the trachea by the cricotracheal ligament. Posteriorly, on both sides of the midline of the cricoid is an articular surface for the arytenoid cartilages.

The **epiglottis** is found above the trachea. Its superior end is free and is observed behind the root of the tongue and the hyoid bone. It is attached on its anterior surface to the hyoid by the hyoepiglottic ligament. Inferiorly it is attached to the thyroid by the thyroepiglottic ligament. The mucous membrane covering of the epiglottis reflects onto the tongue as one median and two lateral glossoepiglottic folds; the depression between the folds is known as the vallecula epiglottica.

On either side of the midline of the posterior cricoid cartilage an **arytenoid cartilage** may be found. The arytenoid cartilages are contained within the laminae of the thyroid cartilage. They are pyramidal in shape; the anterior apex is known as the vocal process, and the lateral apex is known as the muscular process. The cartilages are attached to the thyroarytenoid muscles. They articulate at their base with the cricoid cartilage, upon which they pivot. As they do so, they either stretch or relax the vocal folds.

The small **corniculate** and **cuneiform cartilages** are found superior to the arytenoids. The corniculates are located on the apex of the arytenoids and extend them. The cuneiforms are located anterior to the corniculates in the quadrangular membrane.

There are two intercartilage joints of importance. The cricothyroid joint is a hinge joint connecting the outside of the cricoid cartilage to the inside of the thyroid cartilage. The joint is contained within a capsule lined with synovial membrane. The cricoarytenoid joint allows the arytenoid to slide over the cricoid. Backward movement tilts the arytenoid up, and forward movement tilts it down. The arytenoid cartilage is also capable of rotary motion.

Interior

The inlet of the larynx is triangular in shape. It is bounded by the epiglottic and arytenoid cartilages and the fold of mucous membrane connecting them, the aryepiglottic fold. The outlet of the larynx is a narrow slit located below the true vocal folds that widens to become the trachea.

The triangular vestibule contains the true folds, ending at the level of the false (vestibular) folds. Between the true and false vocal folds is a depression known as the ventricle. The ventricles may continue upward as the saccule of the larynx lateral to (underneath) the false vocal folds.

The wall of the vestibule contains the quadrangular membrane, a portion of the fibroelastic membrane that lines the larynx deep to the mucous membrane. The quadrangular membrane extends from the level of aryepiglottic folds to the false folds. The vestibular ligaments underlie the false folds and the rima vestibuli is the opening between them.

The true vocal folds extend from the thyroid cartilage anteriorly to the arytenoid cartilage posteriorly. They contain, medially to laterally, the vocal ligaments, vocalis muscle, and thyroarytenoid muscle.

Musculature

There are six major muscles within the larynx: the cricothyroid, posterior cricoarytenoid, lateral cricoarytenoid, transverse arytenoid, oblique arytenoid, and thyroarytenoid muscles (see Table 8-5).

The **cricothyroid muscle** is truly external to the larynx, as it lies between the cricoid and thyroid cartilages. It arises from the anterior and lateral portions of the arch of the cricoid and inserts into the inferior and medial thyroid. The muscle pulls the thyroid cartilage downward and forward, which increases the distance between the thyroid and arytenoid cartilages, increasing the tension of the true vocal folds.

The **posterior cricoarytenoid muscle** is located on the dorsum of the cricoid cartilage. It arises from a depression on either side of the median ridge of the posterior surface of the cricoid and inserts into the muscular (posterior) process of the arytenoid. The posterior cricoarytenoids abduct the vocal folds.

The **lateral cricoarytenoid muscle** arises from the upper border of the cricoid arch. It inserts in front of the muscular process. The muscles adduct the vocal folds.

The single **transverse arytenoid muscle** arises from the lateral border of one arytenoid cartilage and inserts into the lateral border of the other. It also adducts the vocal folds.

The **oblique arytenoid muscles** overlie the transverse muscle. They arise from the muscular processes of the arytenoid cartilage and ascend, crossing, to insert onto the apex of the opposite cartilage. They act together as a laryngeal sphincter.

The **thyroarytenoid muscles** are the most lateral borders of the vocal folds. They arise from the inner surface of the thyroid laminae and insert onto the lateral border of the arytenoids. They lessen the tension in the vocal folds and act as a glottal sphincter. Each thyroarytenoid muscle has two parts. The thyroepiglottic muscle, an upward continuation extending to the quadrangular membrane and the lateral margin of the epiglottis, is a vestibular sphincter. The vocalis muscle comprises the innermost fibers of the thyroarytenoid. It minutely adjusts the vocal folds for speaking.

The laryngeal muscles act together in speaking and swallowing. In speaking the vocal folds are abducted and the rima glottidis is widened by the posterior cricoarytenoids. The folds are adducted and the rima is narrowed by the lateral cricoarytenoids and the transverse arytenoid. Complete closure of the rima is achieved by the oblique arytenoids. Tension in the vocal ligaments is altered by the thyroarytenoid muscle and its vocalis subdivision.

During swallowing the laryngeal inlet is closed and the vestibular walls are brought together by moving the epiglottis backward and medially, tilting the arytenoids forward and medially, and moving the corniculates forward. The muscles involved are the thyroarytenoid, the thyroepiglottic, the aryepiglottic, and the oblique arytenoids.

Blood Supply and Lymphatics

Blood to the larynx is supplied by the superior laryngeal branch of the superior thyroid artery, the inferior laryngeal branch of the inferior thyroid artery, and the cricothyroid branch of the superior thyroid artery. The superior laryngeal artery travels with the internal branch of the superior laryngeal nerve. The inferior laryngeal artery travels with the inferior laryngeal nerve. The laryngeal veins parallel the arterial course.

Lymphatic drainage above the true vocal folds proceeds to the infrahyoid and superior deep cervical nodes. Drainage below the true folds proceeds to the inferior deep cervical nodes via either the prelaryngeal nodes, pretracheal nodes, or paratracheal nodes.

Innervation

Innervation of the larynx is accomplished through either the superior or the inferior laryngeal branches of the vagus. The external branch of the superior laryngeal nerve descends on the oblique line of the thyroid cartilage to reach the cricothyroid muscle. The internal branch pierces the thyrohyoid membrane to supply sensory fibers to the mucous membrane and parasympathetic fibers to the epiglottic glands, base of the tongue, aryepiglottic fold, and interior of the larynx.

The inferior laryngeal nerve, a continuation of the recurrent laryngeal, supplies all the laryngeal muscles except the cricothyroid and provides sensory fibers to the level of the true vocal folds. It ascends under the lower border of the inferior constrictor muscle and divides into anterior and posterior branches. The anterior branch supplies the lateral cricoarytenoid, thyroarytenoid, vocalis, aryepiglottic, and thyroepiglottic muscles. The posterior branch supplies the posterior cricoarytenoids, the transverse arytenoid, and the oblique arytenoids.

TRACHEA

The trachea is the continuation of the respiratory pathway below the larynx (see Figs. 8-2, 8-7, and 8-13). It begins at the lower border of the cricoid cartilage, at the level of the sixth cervical vertebra, and descends to the sternal angle, at the fifth thoracic vertebra. At the level of the sternal angle it divides into left and right bronchi (see Chap. 2). About 6 cm of the trachea is in the neck and about 6 cm is in the chest.

The trachea is composed of 16 to 20 C-shaped cartilaginous rings. Its posterior surface is closed by both fibrous tissue and smooth muscle (the trachealis muscle). The last ring is thick and broad in the middle, with its lower border prolonged downward and backward.

Blood is supplied to the trachea by branches of the inferior thyroid, internal thoracic, and bronchial arteries. Venous drainage proceeds to the inferior thyroid vein. The trachealis receives innervation from the sympathetic chain and the recurrent laryngeal nerve.

National Board Type Questions

Select the one best response for each of the following.

1. Regarding the vertebral artery, all of the following statements are false except
   A. It is the first branch of the subclavian artery.
   B. It is one of the terminal branches of the brachiocephalic artery on the right.
   C. It enters the skull through the jugular foramen.
   D. It is an important source of arterial blood

to structures in the posterior triangle of the neck.
E. Its last branch before entering the skull is the large occipital artery.

2. The muscle primarily responsible for abducting the vocal folds (widening the rima glottidis) is the
A. Cricothyroid
B. Posterior cricoarytenoid
C. Lateral cricoarytenoid
D. Vocalis
E. Oblique arytenoid

3. Which of the following muscles is **not** directly or indirectly involved in depressing the chin?
A. Stylohyoid
B. Omohyoid
C. Thyrohyoid
D. Sternohyoid
E. Sternothyroid

4. All of the following structures are located at the level of the sixth cervical vertebra except the
A. Superior cervical sympathetic ganglion
B. Cricoid cartilage
C. Beginning of the esophagus
D. Lower end of the larynx
E. Isthmus of the thyroid gland

Select the response most closely associated with each numbered item. (The headings may be used once, more than once, or not at all.)

A. Ciliary ganglion
B. Otic ganglion
C. Cervicothoracic ganglion
D. Dorsal root ganglion of C2
E. Geniculate ganglion

5. Cell bodies of postganglionic sympathetic fibers
6. Cell bodies of taste fibers
7. General somatic afferent fibers
8. Supplies postganglionic fibers to the heart

A. Inferior thyroid artery
B. Superior thyroid artery
C. Lingual artery
D. Superior laryngeal artery
E. Transverse cervical artery

9. Lies in close relation to the external branch of the superior laryngeal nerve
10. Reaches the thyroid from a position deep to the common carotid artery
11. Passes between the middle and lower constrictors of the pharynx
12. Is a branch of the superior thyroid artery

Select the response most closely associated with each numbered item.

A. Anterior triangle of the neck
B. Posterior triangle of the neck
C. Both
D. Neither

13. Contains the omohyoid muscle
14. One of its boundaries is the mandible
15. One of its boundaries is the sternocleidomastoid muscle
16. Contains the middle scalene muscle
17. Contains the recurrent laryngeal nerve

For the following, select

A. if only *1, 2, and 3* are correct
B. if only *1 and 3* are correct
C. if only *2 and 4* are correct
D. if only *4* is correct
E. if *all* are correct

18. Which of the following are usual branches of the thyrocervical trunk?
1. Inferior thyroid artery
2. Ascending cervical artery
3. Suprascapular artery
4. Highest (supreme) intercostal artery

19. Operations on the thyroid gland may be complicated by
1. Injury to the recurrent laryngeal nerve
2. Injury to the external branch of the superior laryngeal nerve
3. Injury to the parathyroid glands
4. Mediastinitis due to injury to the esophagus with contamination of the retroesophageal space

20. The contents of the carotid sheath includes the
    1. Internal jugular vein
    2. Phrenic nerve
    3. Vagus nerve
    4. Recurrent laryngeal nerve
21. True statements regarding the larynx include the following:
    1. The posterior cricoarytenoid muscles abduct the vocal folds.
    2. The rima glottidis is widened by the posterior cricoarytenoid muscles.
    3. Sensation to the false vocal folds is supplied by the internal branch of the superior laryngeal nerve.
    4. The recurrent laryngeal nerve supplies all the intrinsic muscles of the larynx except the cricothyroids.
22. The pharyngeal constrictors
    1. Are three in number
    2. Are innervated by the ninth and tenth cranial nerves
    3. Extend from the base of the skull and the pterygomandibular raphe to the esophagus
    4. Are supported by the stylopharyngeus and palatopharyngeus muscles
23. True statement(s) regarding the autonomic ganglia of the head and neck include the following:
    1. Drooping of the eyelid is due to an injury of one of the parasympathetic ganglia.
    2. The superior cervical ganglia has gray rami communicantes.
    3. The inferior cervical ganglion has both gray and white rami communicantes.
    4. The preganglionic sympathetic fibers of the middle cervical ganglion originate in the thoracic region of the spinal cord.
24. The cervical plexus
    1. Provides sensory innervation to the skin over the isthmus of the thyroid gland
    2. Contributes to the formation of the phrenic nerve
    3. Innervates the sternohyoid muscle
    4. Sends white rami to the middle cervical sympathetic ganglion
25. Select the correct statements regarding the arteries of the head and neck.
    1. The external carotid artery is one of the two terminal branches of the common carotid artery.
    2. The pulsation of the facial artery can be palpated over the middle of the body of the mandible.
    3. The vertebral artery is a direct branch of the subclavian artery.
    4. The ophthalmic artery passes through the superior orbital fissure.
26. Various degrees of drooping of the upper eyelid (ptosis) would occur upon section of which of the following:
    1. Oculomotor nerve
    2. Facial nerve
    3. Cervical sympathetic trunk
    4. Ophthalmic division of the trigeminal nerve

Annotated Answers

1. A. The vertebral artery is the first branch of the subclavian, and it is important to remember that, although it sends branches to the deep muscles of the neck, it is not the major artery in the neck.
2. B. Only one muscle acts to abduct the vocal folds.
3. A.
4. A. The sixth cervical vertebra is an important landmark, especially in radiographic views of the neck region. In essence, all of these structures can be "related" to C6 and the cricoid cartilage except the superior cervical ganglia.
5. C., 6. E., 7. D., and 8. C. The cervicothoracic ganglion is another name for the inferior cervical sympathetic ganglion in the neck region that has fused with thoracic ganglia. Remember that the geniculate ganglion of the facial nerve contains cell bodies of the taste fibers (from where on the tongue?) and that general somatic afferent fibers can be found in each spinal nerve, the cell bodies of such fibers being located in the

dorsal root ganglia (see Chap. 7). Finally, be well aware of the pathway by which sympathetic innervation reaches the heart.

9. B.
10. A. The inferior thyroid artery arises from the thyrocervical trunk (which itself arises from what?) and ascends, to pass deep to the common carotid to reach the thyroid gland.
11. D.
12. D.
13. C.
14. A.
15. C. Remember that the sternocleidomastoid muscle divides the anterior and posterior triangles.
16. B.
17. D. The recurrent laryngeal nerve lies deep in the neck between the trachea and esophagus and is not part of the triangles, which serve as surface landmarks.
18. B.
19. A. Injury to the esophagus should not occur because it is deep to the trachea. (See Chapter 2 to understand why one must be aware of infections spreading from the neck region to the thorax via the retropharyngeal space.)
20. B.
21. E. The cricothyroid muscles are innervated by the external branch of the superior laryngeal nerve. All the other laryngeal muscles are innervated by the recurrent laryngeal.
22. E. The pharyngeal muscles are innervated by the *pharyngeal plexus*. This includes input from the vagus and glossopharyngeal nerves. Thus although it is usually said that by definition the glossopharyngeal nerve innervates only one muscle, the stylopharyngeus, it is involved with the constrictors also.
23. C. Remember that preganglionic outflow from the spinal cord (white rami) is limited to the thorax and upper lumbar spinal segments. Review the distribution of the sympathetics in the head so that it is clear why a drooping of the eyelid could be related to sympathetic and not parasympathetic nerve injury (see Question 26).
24. A.
25. A. The facial artery becomes very superficial as it hooks over the body of the mandible and thus can be palpated. Although many things pass through the superior orbital fissure, the ophthalmic artery travels with the optic nerve through the optic canal.
26. B.

9 Back

Objectives

After reading this chapter, you should know the following:

Superficial landmarks of the back

Bony portion of the back, including bones of related regions

Shapes, articulations, and movements of the four types of vertebrae

Structure, curves, ligaments, and movements of the vertebral column

Relationship of spinal nerves to vertebral column and vertebrae

Intrinsic and extrinsic musculature of the back, including muscles involved with the neck and head

Blood supply and innervation (dorsal and ventral rami) of the back

Coverings, tissues, and function of the central nervous system

Proper site for lumbar puncture

Superficial Landmarks

The back includes the vertebral column with its muscles, ligaments, nerves, and blood vessels and the scapulae with the muscles connecting the scapulae to the upper extremities.

The recess down the middle of the back, the **median furrow,** lies over the tips of the spinous processes of the vertebrae. The upper cervical spines are covered by the **nuchal ligament** (ligamentum nuchae) and can be palpated by deep pressure; the lowest cervical spine (vertebra prominens) and all thoracic spines are palpable. The iliac crest and all but the first rib can also be palpated. The scapula with its spine, acromion, and inferior border is apparent.

The following muscles can be visualized: teres major, rhomboid major, trapezius, and latissimus dorsi. The **triangle of auscultation** is formed by the latissimus dorsi, trapezius, and rhomboid major muscles. When the shoulder girdle is moved forward the posterior chest wall, which forms the floor of the triangle, is exposed.

Bones and Ligaments

The bony portion of the back includes portions of two cranial bones—the occipital and temporal—the 33 vertebrae, the medial part of the ribs, the scapulae, and a part of the ilium. Cranial bones are included in the functional considerations of the back because the muscles and ligaments that maintain the vertebral column insert onto these bones.

CRANIAL BONES

Only the outer surface of the occipital bones is involved in steadying the back. As noted in the discussion of the skull (see Chap. 7 and Fig. 7-2), the occipital bone consists of three portions: the basilar, squamous, and lateral portions. Only the squamous portion of the occipital bone is directly related to the back. It has several prominences in the midline: the external occipital protuberance and the inferior and superior nuchal lines. The nuchal ligament attaches to the external occipital protuberance, and the trapezius, occipitalis, splenius capitis, and sternocleidomastoid muscles insert into the superior nuchal line. Only the mastoid portion of the temporal bone is part of the back. The longissimus capitis muscle inserts into this region.

VERTEBRAE

There are 33 individual vertebrae: 7 cervical vertebrae, 12 thoracic vertebrae, 5 lumbar vertebrae, 5 fused sacral vertebrae forming the sacrum, and 4 fused coccygeal vertebrae (Fig. 9-1). A typical vertebra consists of an anterior segment, the body, and a posterior part, the vertebral or neural arch, that encloses the vertebral foramen. The vertebral foramina form the vertebral canal, which houses the spinal cord and its covering, the meninges. The **vertebral arch** consists of a pair of pedicles that extend posteriorly from the body and laminae that extend posteromedially from the pedicles. Seven processes arise from the pedicles and laminae: one spinous, two transverse, and four articular.

The **spinous process** is directed backward and downward from the junction of the laminae; ligaments and muscles attach to this process. The **articular processes,** two superior and two inferior, originate from the junctions of the pedicles and laminae. The direction of the articular surfaces varies with the region. The articular surfaces are covered with hyaline cartilage. **Transverse processes** project on either side at the junction of the laminae and pedicles between the superior and inferior articular processes, these processes also attach muscles and ligaments and are useful in identifying the level of the individual vertebra. The **intervertebral foramina** are formed by the opposition of the deep notch on the inferior edge of each pedicle with the shallow notch on the superior edge of the adjacent pedicle. A spinal nerve and its accompanying vessels are transmitted through each foramen.

The **cervical vertebrae** are the smallest of the vertebrae and can be identified by the presence of a foramen in each transverse process. The first and second cervical vertebrae are unique. The first is called the **atlas** (Fig. 9-2) because it supports the head. It has no body, as its body has become part of the second cervical vertebra (the axis), and it contains no spinous process (Fig. 9-2C). The atlas is ringlike, consisting of thin anterior and posterior arches with two lateral masses. The anterior arch is smaller and contains the anterior tubercle and a facet on its posterior surface for articulation with the axis. The posterior arch contains the posterior tubercle and grooves for the vertebral artery and first cervical nerve. In order to support the head, the lateral masses have evolved into the bulkiest part of this vertebra. These masses have superior facets for articulation with the occipital bone and inferior facets for articulation with the axis. The transverse processes are large, and muscles that assist in rotating the head insert into them. They also contain a large foramen, the transverse foramen, for the vertebral artery and vein.

The second cervical vertebra is named the **axis** (Fig. 9-3) because it forms the pivot upon which the atlas rotates. Its characteristic feature is the **odontoid process** or **dens,** which arises superiorly from the body. The axis articulates in three places with the atlas: the two articular facets of the axis articulate with the two inferior facets of the atlas, and the odontoid process articulates with the anterior arch of the atlas.

The third to sixth cervical vertebrae are similar (Fig. 9-4). They have small bodies, bifid spinous and transverse processes, a transverse foramen, and a triangular vertebral foramen. The articular processes are large and the facets face superiorly and inferiorly. The seventh cervical vertebra differs in that it exhibits a long and prominent spinous process. The seventh cervical vertebra is the first spine that can be palpated in the upper back **(vertebra prominens).**

The **thoracic vertebrae** (Fig. 9-5) are inter-

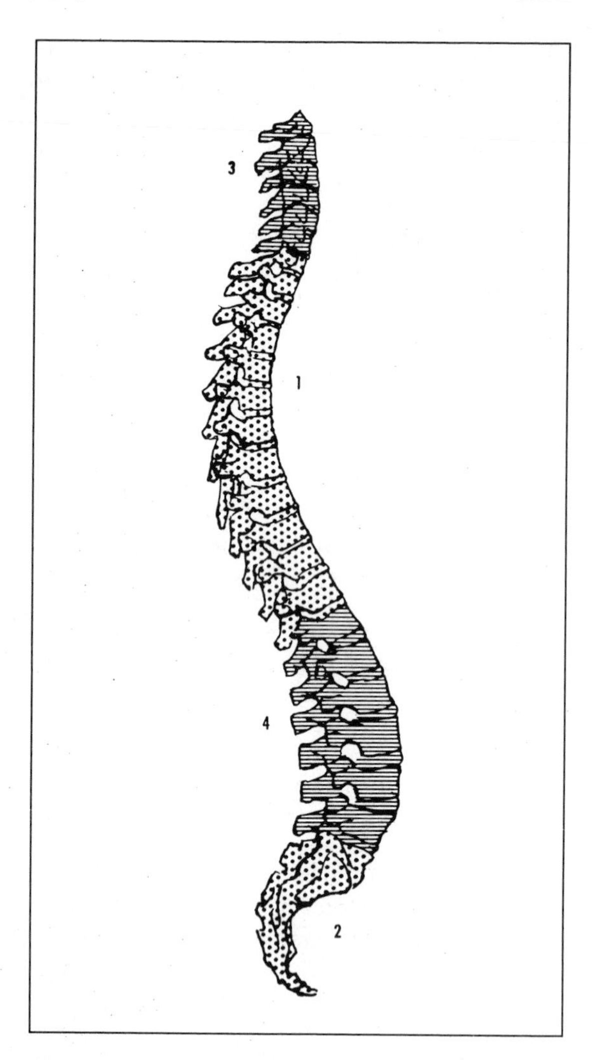

Fig. 9-1. The primary and secondary curves of the vertebral column. The primary curves are the thoracic (1) and sacral (2) curves; the secondary curves are the cervical (3) and lumbar (4) curves. (From R. O'Rahilly, *Gardner, Gray and O'Rahilly Anatomy: A Regional Study of Human Structure* [5th ed.]. Philadelphia: Saunders, 1986. P. 515. With permission.)

mediate in size between those of cervical and lumbar regions and increase in size inferiorly. They can be identified by the presence of facets on the side of the body for articulation with the heads of the ribs, and facets on the transverse processes on all but the eleventh and twelfth vertebrae for articulation with the tubercles of the ribs. Each vertebral body articulates with the rib of the same number until the ninth rib. The bodies, especially in midthoracic regions, are heart shaped, the laminae are flat and broad, and the spinous processes are directed downward and overlap from the fifth to tenth thoracic vertebrae. The transverse processes decrease in size inferiorly until they are replaced by tubercles at the twelfth thoracic vertebra.

The **lumbar vertebrae** (Fig. 9-6) are the largest segments of the moveable parts of the vertebral column. They contain no foramina in the transverse processes and no facets on the sides of the body. The bodies are large, the pedicles short and strong, and the transverse processes long and slender. The spinous processes are thick, broad, and project posteriorly. The superior and inferior articular processes are well marked. The lumbar foramina are triangular and larger than those in the thoracic region, but smaller than those in cervical regions. On the transverse processes, three tubercles are noted, the superior one being the mamillary process. Small accessory processes are also seen.

The **sacrum** (Fig. 9-7) consists of five bones that fuse early in life. This large triangular bone is found at the lower end of the vertebral column and at the upper and back part of the pelvic cavity, where it inserts between the hip bones. Its upper part, the base, articulates with the fifth lumbar vertebra, and its lower part, the apex, connects with the coccyx. The pelvic (anterior) surface is concave and consists of a medial part separated from a lateral part by four foramina for exit of the ventral rami of the sacral nerves. The posterior surface has a midline crest, the middle sacral crest, that is formed by the rudimentary spinous processes of the upper three or four sacral vertebrae. The vertebral canal continues in the sacrum.

The vertebral canal does not extend into the **coccygeal vertebrae** (Fig. 9-8). The first coccygeal vertebra is the largest and resembles the lowest sacral vertebra. The last three are usually smaller.

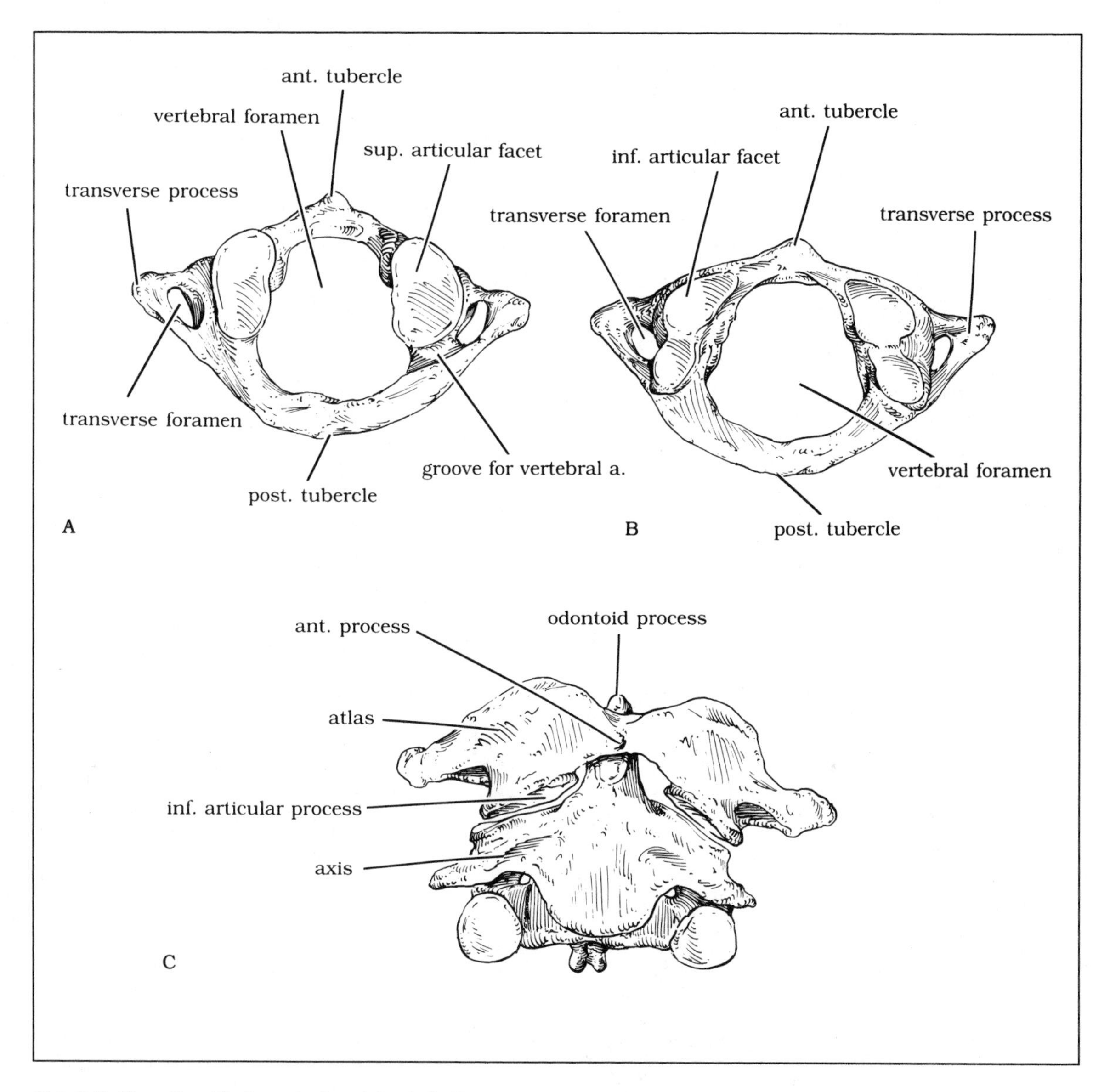

Fig. 9-2. The atlas (first cervical vertebra). A. Superior view. B. Inferior view. C. Anterior view of articulation with the axis.

VERTEBRAL COLUMN

The vertebral column is found in the midline of the back and forms the posterior part of the trunk. The vertebrae increase in size inferiorly. The lateral view is the most striking, as it reveals the curves of the vertebral column (see Fig. 9-1).

The width of the bodies of the vertebrae increases inferiorly, as does the width of the intervertebral spaces. Each intervertebral space is filled by an articular disk. These **articular disks** are fibrocartilaginous, dense in the peripheral anulus fibrosus, and soft in the central nucleus

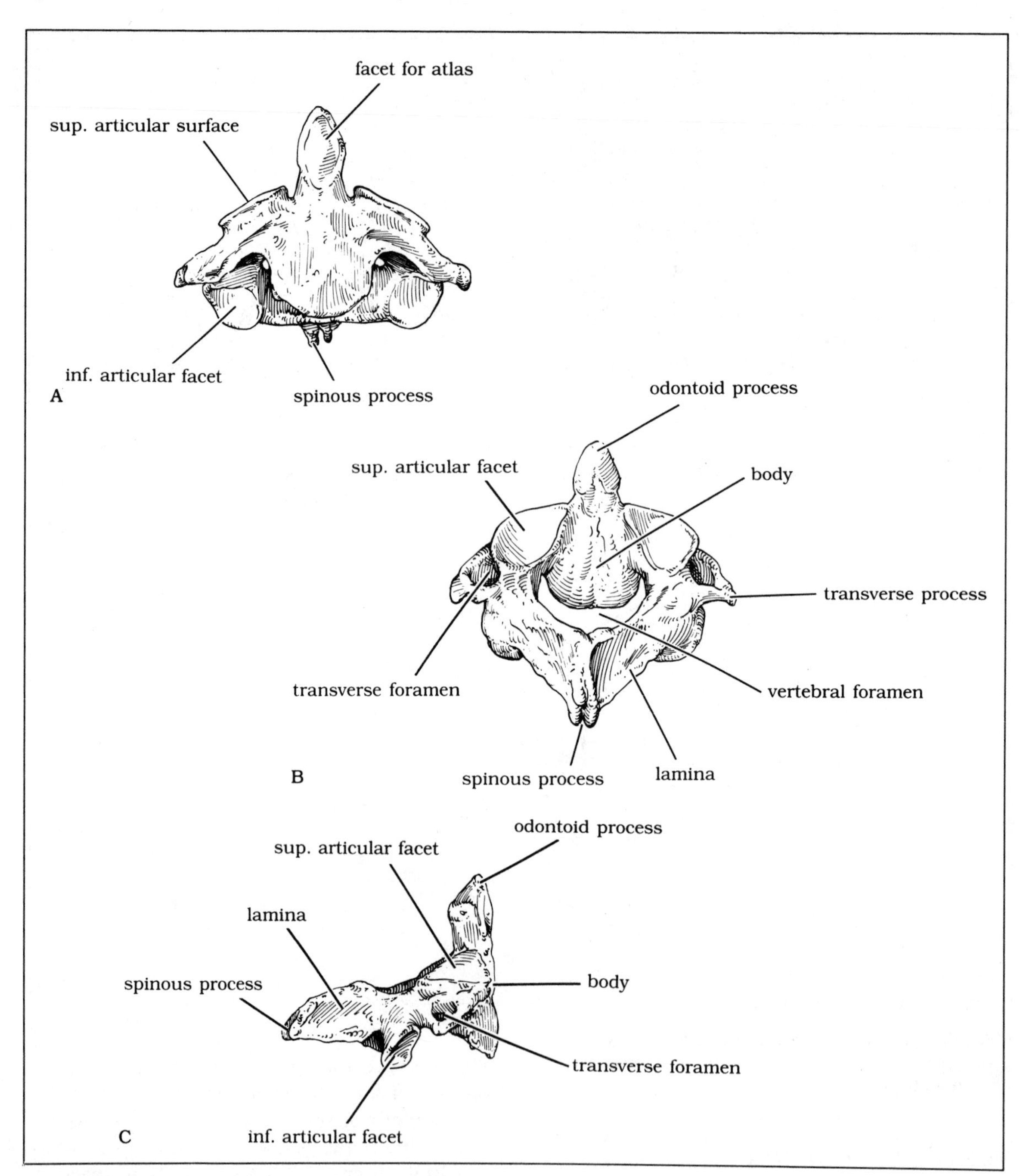

Fig. 9-3. The axis (second cervical vertebra). (Figure 9-2C shows articulation with the atlas.) A. Anterior view. B. Posterior view. C. Lateral view.

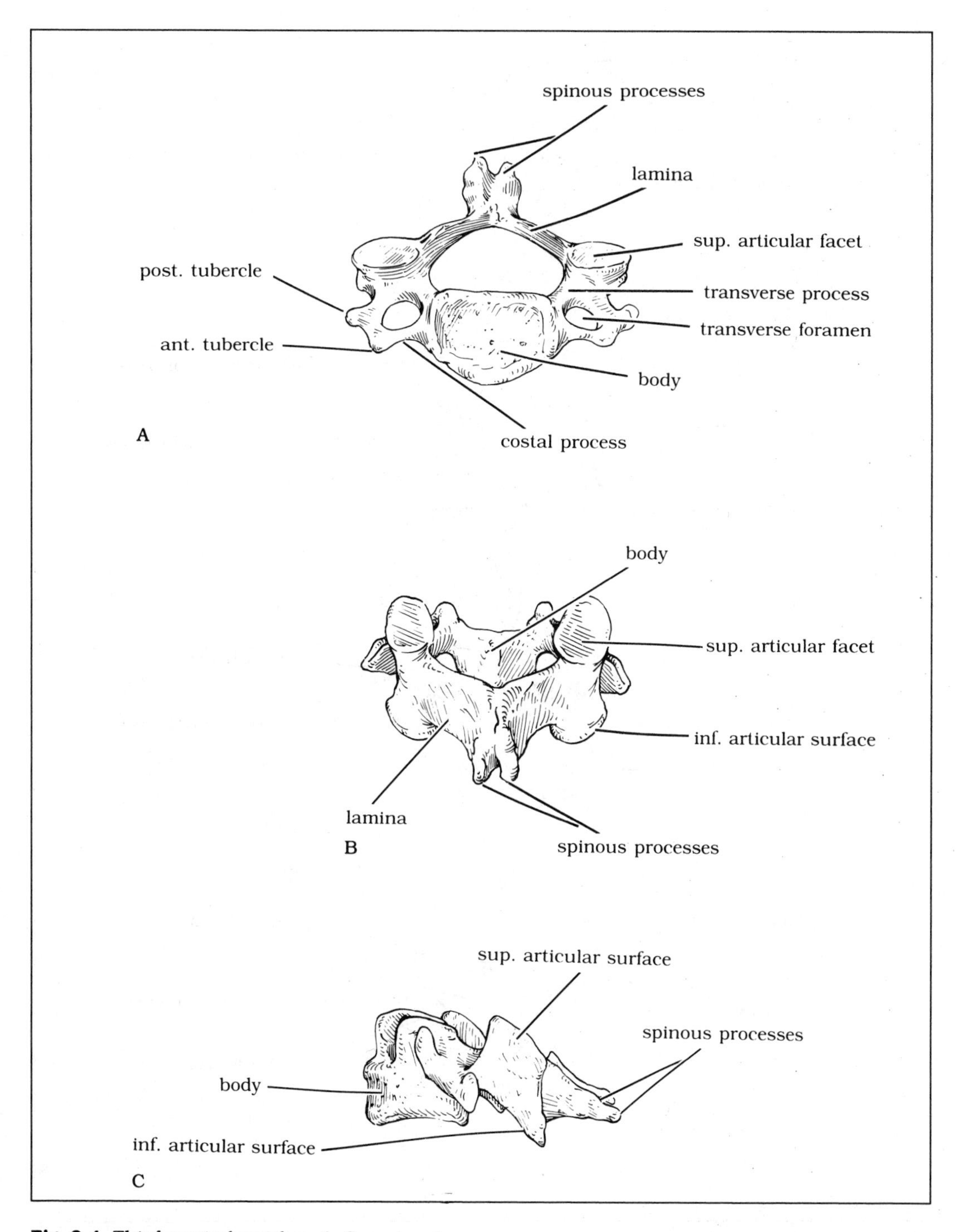

Fig. 9-4. Third cervical vertebra. A. Superior view. B. Posterior view. C. Lateral view.

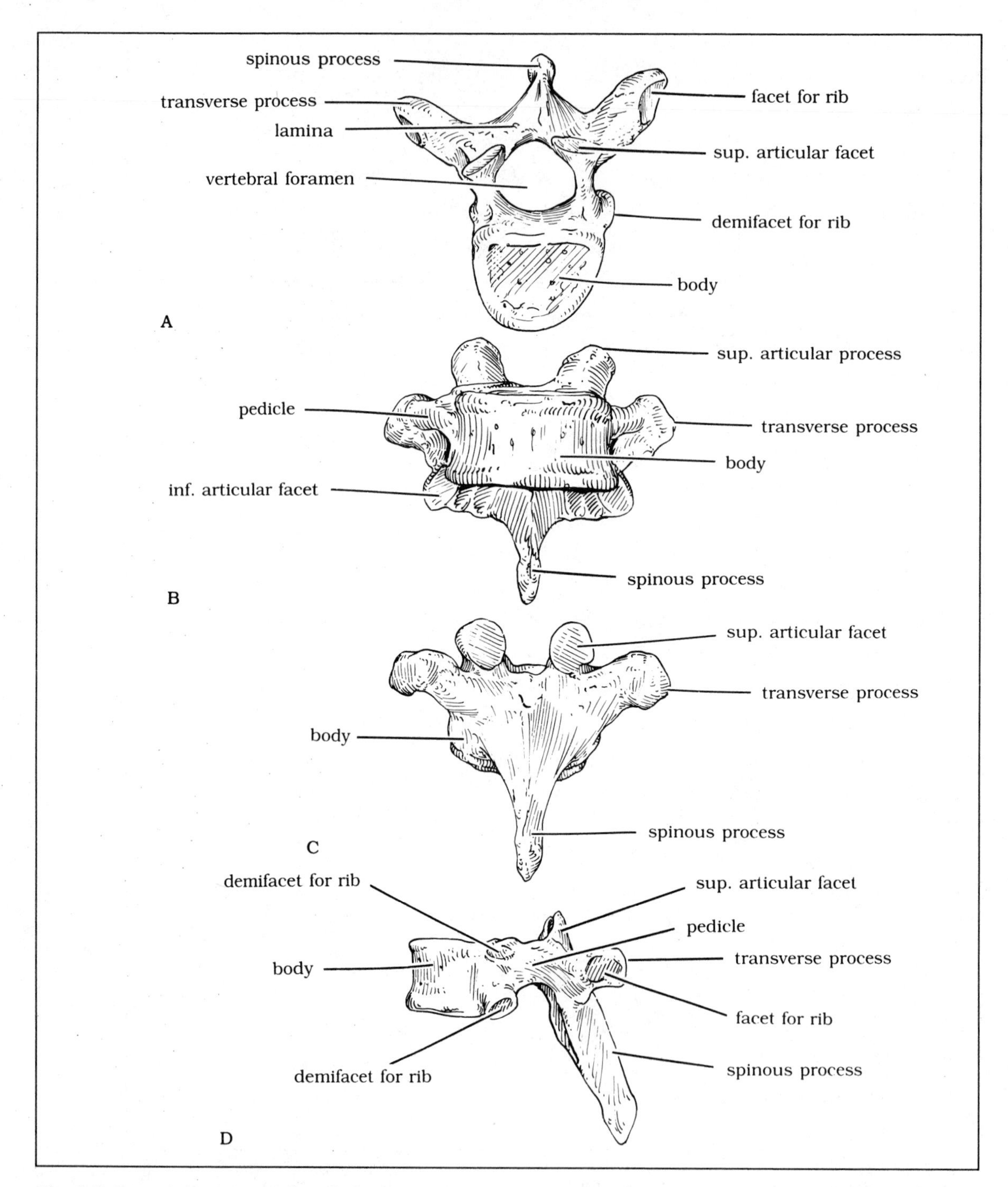

Fig. 9-5. Typical thoracic vertebra. A. Superior view. B. Anterior view. C. Posterior view. D. Lateral view.

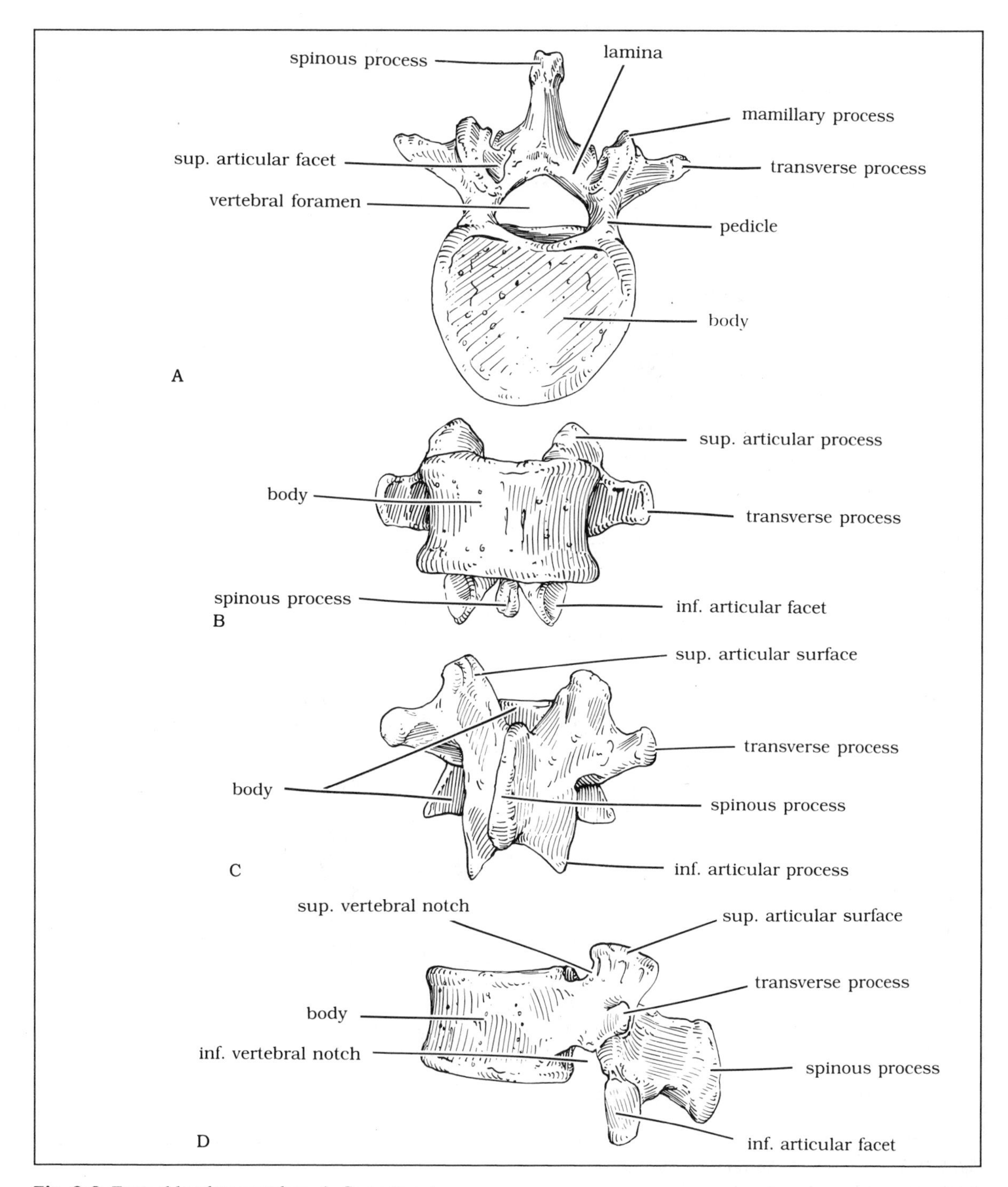

Fig. 9-6. Typical lumbar vertebra. A. Superior view. B. Anterior view. C. Posterior view. D. Lateral view.

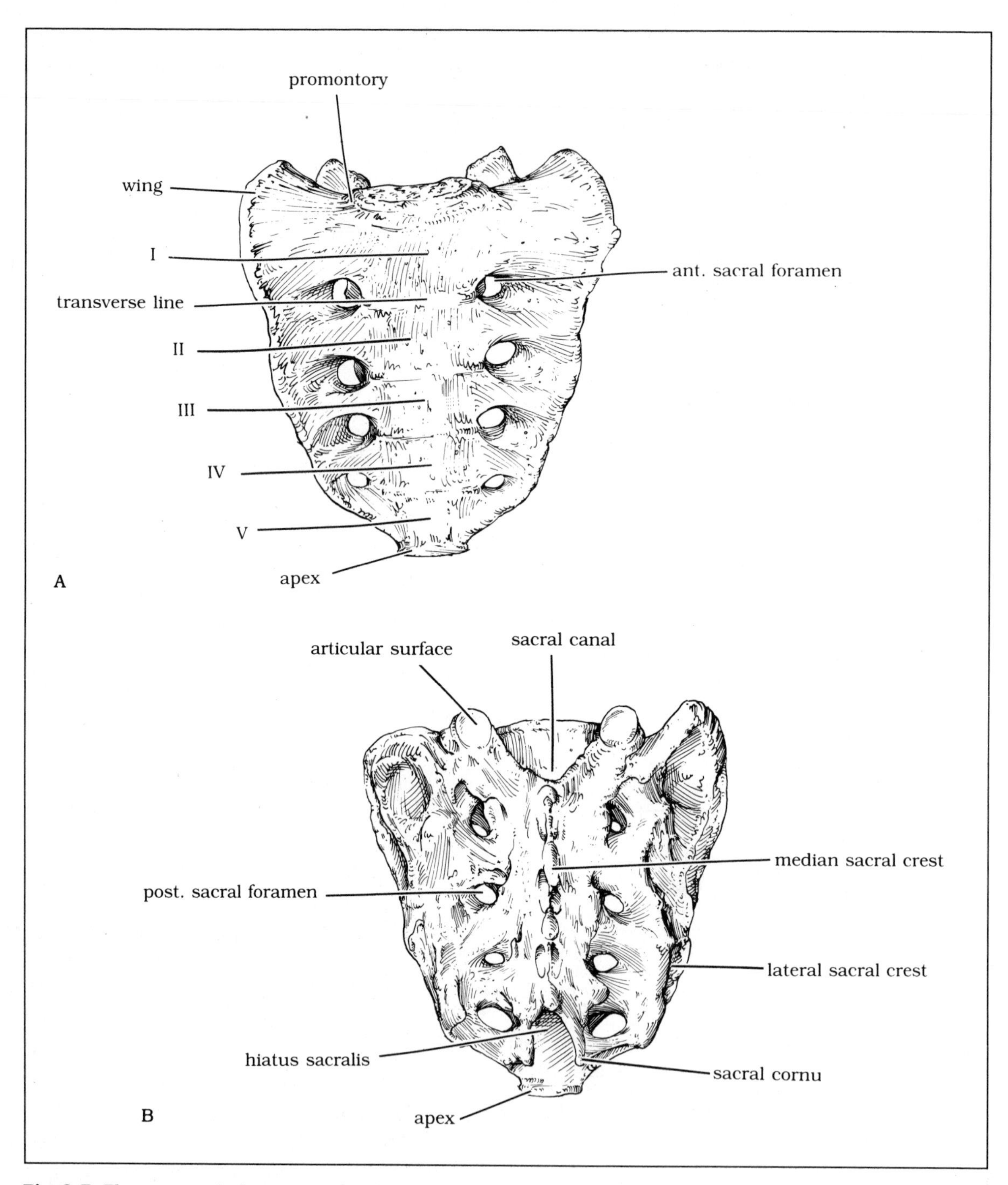

Fig. 9-7. The sacrum. A. Anterior surface (roman numerals indicate fused vertebrae). B. Posterior surface.

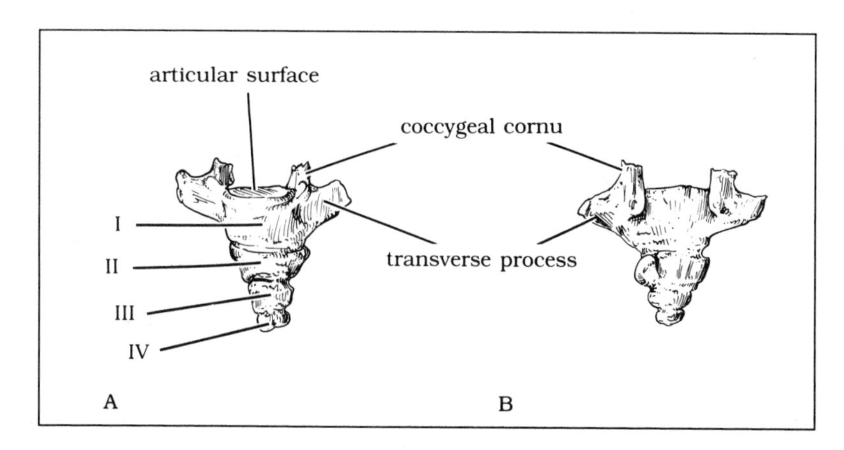

Fig. 9-8. The coccyx. A. Anterior surface (roman numerals indicate fused vertebrae). B. Posterior surface.

pulposus. Herniation of the nucleus pulposus through the fibrous portion produces a herniated disk.

In the midline of the posterior surface are the spinous processes. In the cervical region these spines are short, while in the thoracic region they are long and thin and overlap from the fifth to tenth thoracic vertebrae. In the lumbar levels they are blunt.

The vertebral column has several **curves** that correspond to the different body regions—cervical, thoracic, lumbar, and sacral (see Fig. 9-1). The cervical curve is convex anteriorly. The thoracic curve is concave anteriorly. The lumbar curve is concave posteriorly. In lordosis, the lumbar curvature is anterior rather than posterior. The sacral curve is more marked in the female than in the male and is convex anteriorly. The sacral curve is directed downward and anteriorly. The vertebral column also has a slight lateral curvature, with the convexity directed toward the right. This orientation toward the right may be related to the dominance of the right side in 90 percent of persons; in left-handed persons the orientation is usually to the left.

RIBS

As noted above, the ribs also form a portion of the back. The head and neck of the second to ninth ribs articulate with the vertebra of the same number, and two facets on the head of each rib and a single tubercle on the neck articulate with the corresponding transverse process. Farther laterally at the costal angle, muscles related to the back attach. Further discussion of the ribs may be found in Chapter 2.

CLAVICLE

The clavicle articulates medially with the sternum and laterally with the acromion of the scapula by a syndesmosis.

SCAPULA

The scapula is triangular in shape, attaches to the clavicle laterally, and is supported by its muscles. The entire posterior surface of the scapula, with its spinous process, acromion, and supraspinous and infraspinous fossae, forms a portion of the back. Further discussion of the scapula and clavicular area may be found in Chapter 5.

ILIUM

The ilium, through its articulation with the sacrum and its iliac crests, is the most inferior portion of the bony structure in the back. Further discussion of the ilium may be found in Chapter 4.

LIGAMENTS

Six ligaments hold together the vertebral column:

Supraspinal ligaments, which are the most external, connect the tips of the spinous processes of the vertebrae.
Ligamenta flava interconnect the laminae of adjacent vertebrae.
Interspinal ligaments connect adjacent spinal processes.
The **posterior longitudinal ligament** joins together the posterior surfaces of the vertebral bodies.
The **anterior longitudinal ligament** interconnects the anterior surfaces of the vertebral bodies.
Intertransverse ligaments are small and connect the transverse processes of adjacent vertebrae.

Movement

The articular disk between each two vertebrae allows some bending and movement. Movement is freest in cervical regions, permitting the flexions, extensions, and rotations of the head and neck. The ribs somewhat inhibit movement in the thorax, but there is movement in the lumbar region in all directions. Flexion is limited by the tightening of the posterior longitudinal ligament, while the anterior longitudinal ligament and spines of the vertebrae stop overextension.

Musculature

The extrinsic, or most-superficial, back muscles —trapezius, levator scapulae, latissimus dorsi, rhomboid major, rhomboid minor, and serratus anterior—are related to the upper limb and are discussed in Chapter 5 and shown in Fig. 5-17.

INTRINSIC BACK MUSCLES

The intrinsic or deep muscles of the back extend from the pelvis to the skull and function together as the extensors of the vertebral column. They are found on the posterior surface of the vertebrae and consist of superficial, intermediate, and deep groupings. The superficial bundles are long and straight, and the deep bundles are short and oblique. All intrinsic back muscles are innervated by the dorsal primary rami.

Superficial and Intermediate Groupings

The **splenius** and **erector spinae** muscles constitute the superficial and intermediate groupings of intrinsic back muscles, respectively. These muscles form a longitudinal series from the occipital bone of the skull to the sacrum. The splenius muscle is a large muscle covering the posterior surface of the neck. It consists of two divisions, the splenius capitis and the splenius cervicis, and can be seen in dissection after the trapezius is removed (Fig. 9-9A and Table 9-1).

The erector spinae, or sacrospinalis, consists of three columns of vertically oriented muscles beginning in the sacrum and extending to the skull (Fig. 9-9B and Table 9-2). The medial column, or spinalis muscles, lies along the spines of the vertebrae and consists of the spinalis thoracis, which is hard to identify because it tends to blend with the longissimus thoracis; the spinalis cervicis, which is considered part of the semispinalis; and the spinalis capitis.

The intermediate column, or longissimus muscles, lies along the transverse processes of the vertebrae and forms the bulk of the erector spinae. The column consists of the longissimus thoracis, the largest longissimus muscle; the longissimus cervicis; and the longissimus capitis, which lies medial to the longissimus cervicis and semispinalis.

The lateral column, or iliocostalis muscles, lies in the angle of the ribs and up to the transverse processes of the lower cervical vertebrae and consists of the iliocostalis lumborum, the largest muscle in this column; the iliocostalis thoracis; and the iliocostalis cervicis.

Deep Grouping

The **transversospinalis muscles** are more deeply placed than the erector spinae (Fig. 9-9C and

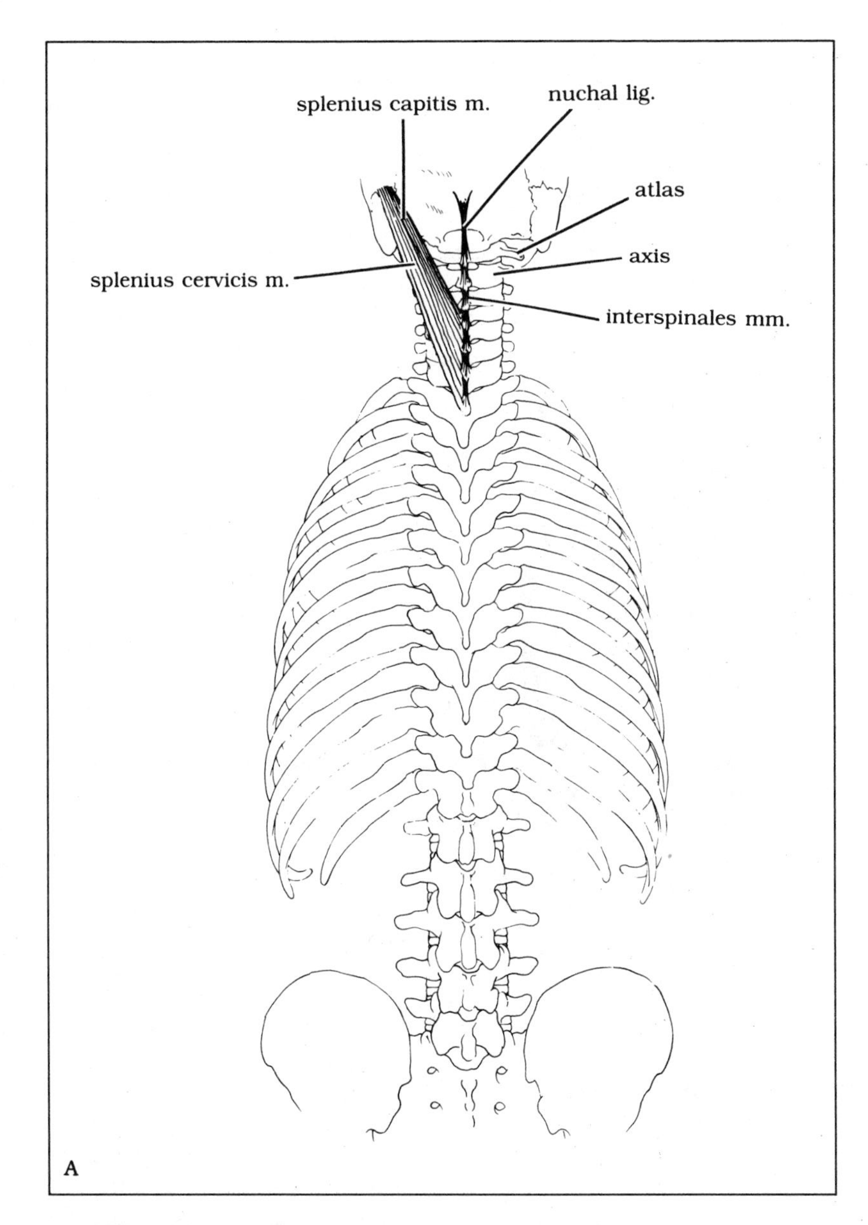

Fig. 9-9. Intrinsic back muscles. A. Splenius and interspinales. B. Erector spinae (longissimus, iliocostalis, and spinalis). C. Transversospinalis.

Table 9-3). They are short, small muscles that run obliquely from transverse processes to spinous processes and consist of five subgroupings arranged superficially to deeply.

The **semispinalis** (thoracis, cervicis, and capitis) form the outermost fibers and are found on the upper half of the vertebral column. The semispinalis capitis, the most superficial muscle of this group, covers the suboccipital triangle and the semispinalis cervicis muscle; it is found beneath the splenius and forms the bulge beside

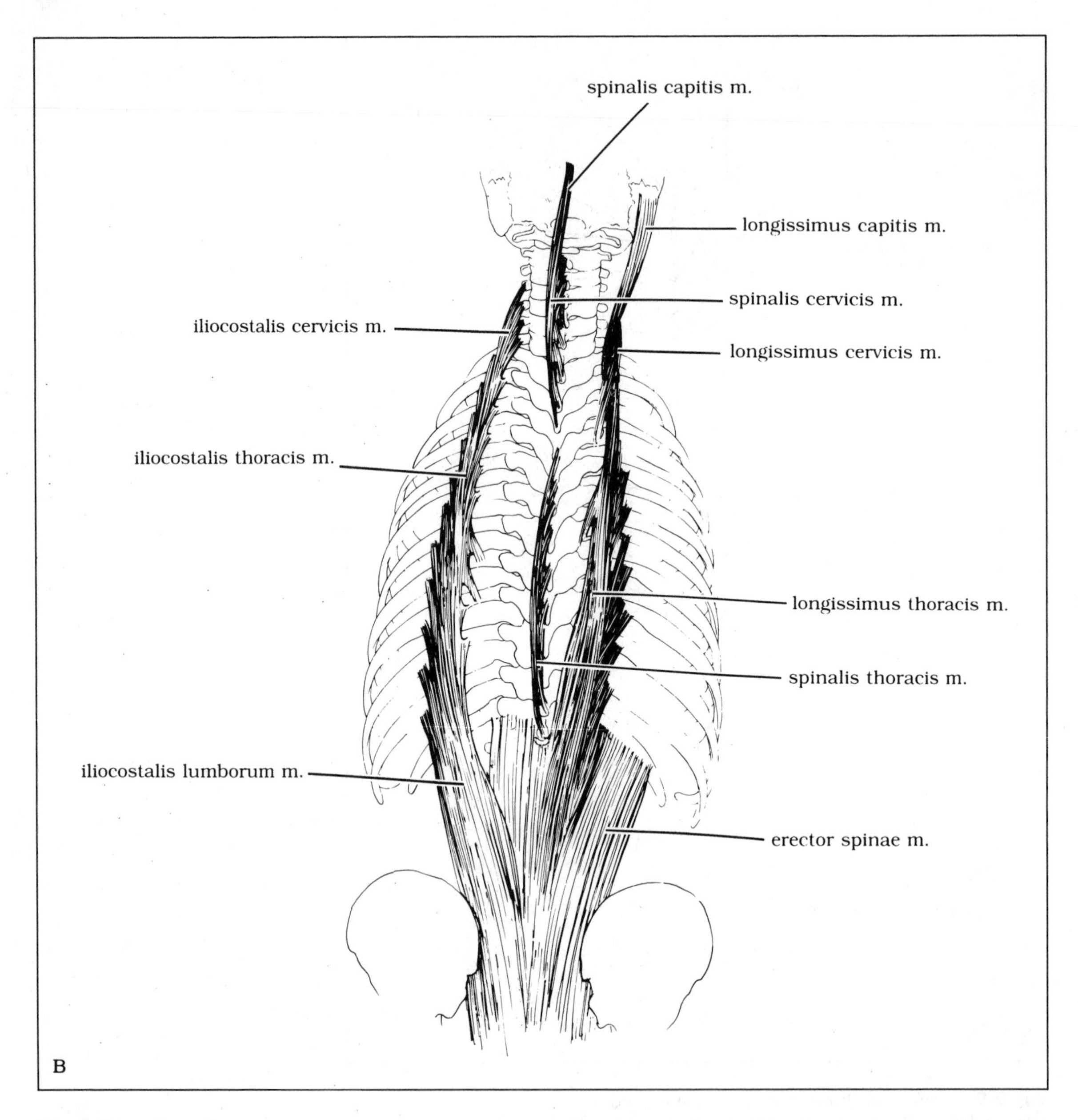

Fig. 9-9 (continued).

the median furrow. The semispinalis thoracis and semispinalis cervicis muscles form a continuous sheath.

The **multifidus** lies deep to the semispinalis in the groove on the spinous processes of the vertebrae from sacrum to axis. The **rotatores** lie under the multifidus and are found deepest in the groove between the spinous and transverse processes along the entire length of the vertebral col-

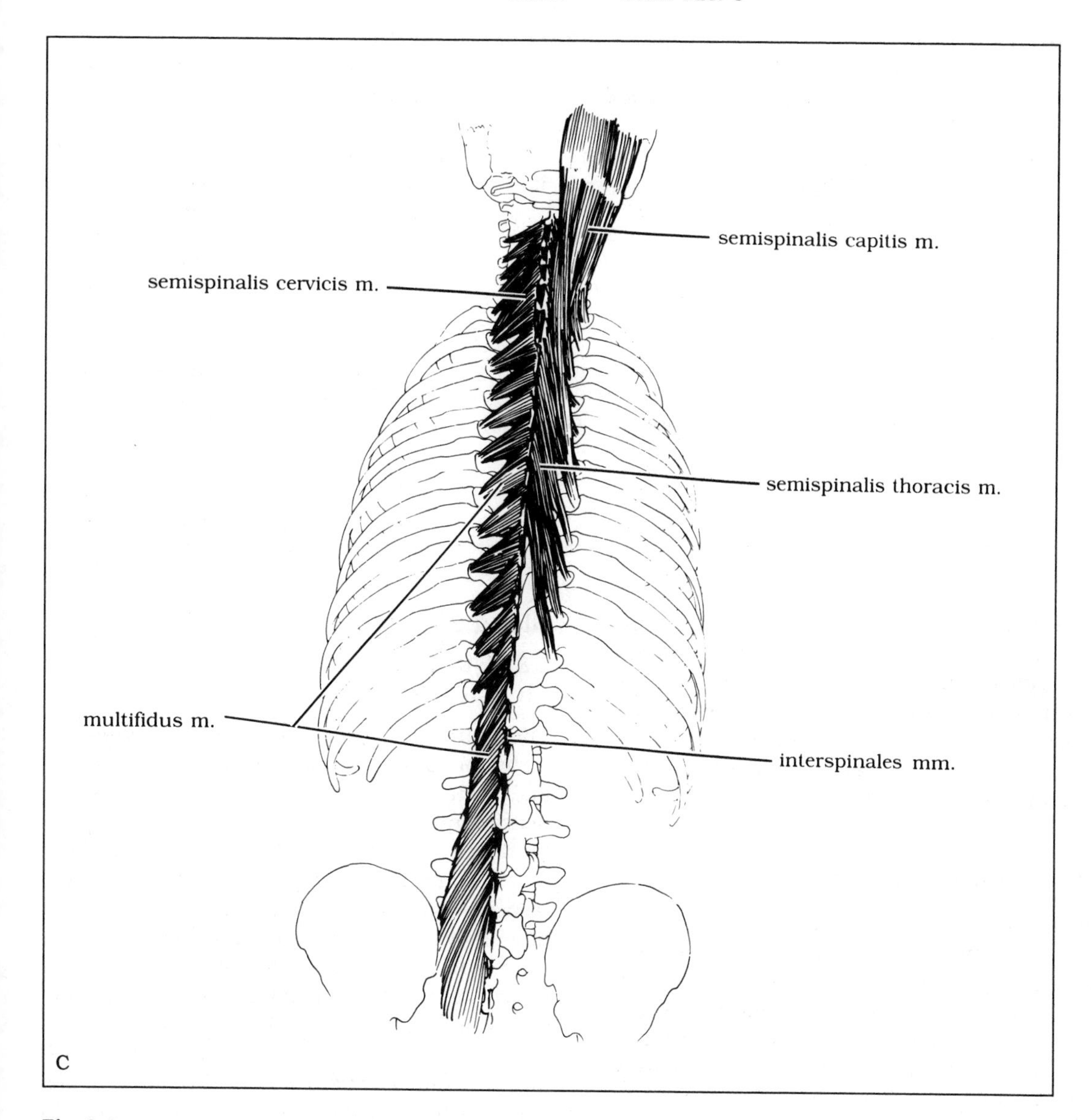

Fig. 9-9 (continued).

umn from sacrum to axis. They are best developed in thoracic levels. The rotatores longi in their oblique course cross one vertebra, while the rotatores breves insert on the next vertebra superiorly.

The **interspinales** are short paired muscles between the spinous processes of continuous vertebrae (see Fig. 9-9A), lying on either side of the interspinal ligament. They are found in cervical, thoracic, and lumbar levels and are best developed in the cervical region.

The **intertransversarii** are small muscles between the transverse processes in cervical levels,

Table 9-1. Superficial Vertebral Musculature of the Neck: The Splenius Muscles

Muscle	*Origin*	*Insertion*	*Innervation*	*Action*
Splenius capitis	Lower half of nuchal ligament	Occipital bone below superior nuchal line, under sternocleidomastoid onto mastoid of temporal bone	Dorsal rami of nerves C4–C8	Draw head and neck backward; turn face to same side; with splenius cervicis, extend head and neck
Splenius cervicis	Spinous processes of vertebrae T3–T6	Posterior tubercles of transverse processes of vertebrae C2 and C3	Dorsal rami of nerves C4–C8	Draw head and neck backward; turn face to same side; with splenius capitis, extend head and neck

where they are best developed; thoracic levels, where they are least developed; and lumbar levels. The intertransversarii are paired, passing between anterior and posterior tubercles of contiguous transverse processes, and are separated from each other by the anterior division of the cervical nerves.

SUBOCCIPITAL MUSCLES

The first cervical, or suboccipital, nerve and the vertebral artery lie in the suboccipital triangle in a deep groove on the surface of the posterior arch of the atlas. The suboccipital muscles—rectus capitis posterior major, rectus capitis posterior minor, obliquus capitis superior, and obliquus capitis inferior—are discussed in Chapter 8 (see Table 8-4).

NUCHAL FASCIA

The nuchal fascia is the deep cervical portion of the fascia covering the upper limb and vertebral column. It covers the splenius and semispinalis muscles and attaches with these muscles onto the skull below the superior nuchal line.

Blood Supply

GENERAL SCHEMA OF SPINAL ARTERIES

The paired posterior spinal arteries enter the vertebral canal via the intervertebral foramina and then divide into three branches: a single neural and two osseous branches. The osseous branches enter the vertebral column and form anterior and posterior branches that penetrate, respectively, the body and laminae of the vertebrae; they anastomose above and below with the arteries of adjacent segments. The neural branch penetrates the dura and divides into anterior and posterior radicular branches, which enter the substance of the spinal cord. The anterior radicular branch usually enters the anterior median fissure and forms the anterior spinal arteries.

The arterial supply to the spinal cord consists of three longitudinal coursing vessels: the paired posterior spinal and the single anterior spinal. The anterior spinal artery supplies the gray matter of the spinal cord and the white matter in the anterior and lateral column of the spinal cord.

The cervical spinal arteries arise from vertebral arteries, and the thoracic spinal arteries arise

Table 9-2. Erector Spinae Muscles

Muscle	*Origin*	*Insertion*	*Innervation*	*Action*
Medial Column				
Spinalis thoracis	Spinous processes of vertebrae T9–L2	Spinous processes of upper thoracic vertebrae	Dorsal division of spinal nerves	Extend vertebral column
Spinalis cervicis (inconstant)	Lower part of nuchal ligament of vertebra C7	Spinous process of axis	Dorsal division of spinal nerves	Extend vertebral column
Spinalis capitis (usually part of semispinalis)	Tips of transverse processes of vertebrae C4–T5	Occipital bone	Dorsal division of spinal nerves	Extend vertebral column
Intermediate Column				
Longissimus thoracis	Accessory and transverse processes of lumbar vertebrae	Tips of transverse processes of all thoracic vertebrae, lower 9 ribs between tubercles and angles	Branches of dorsal rami of thoracic and lumbar spinal nerves	Extend vertebral column and bend it to one side, draw ribs downward
Longissimus cervicis	Transverse processes of vertebrae T4–T5	Posterior tubercles of transverse vertebrae C2–C6	Branches of dorsal rami of cervical and thoracic spinal nerves	Same as above
Longissimus capitis	Transverse and articular processes of vertebrae T1–T4	Posterior margin of mastoid processes of temporal bone, beneath splenius capitis and sternocleidomastoid	Dorsal divisions of nerves C5–C8	Bend head to same side and rotate it to that side, both muscles acting together extend head
Lateral Column				
Iliocostalis lumborum	Middle crest of sacrum; spinous processes of all lumbar vertebrae, T11 and T12; inner lip of iliac crests; spinal ligament	Inferior border of angle of lower 6 or 7 ribs	Dorsal primary division of sacral, lumbar, and thoracic spinal nerves	Extend vertebral column and laterally flex it to one side, draw ribs downward
Iliocostalis thoracis	Upper border of angles of lower 6 ribs	Upper border of angles of upper 6 ribs and transverse process of vertebra C7	Dorsal primary division of thoracic spinal nerves	Extend vertebral column, draw ribs downward
Iliocostalis cervicis (costocervicalis)	Angles of ribs 3–6	Posterior tubercle of transverse processes of vertebrae C4–C6	Dorsal primary division of thoracic and lower cervical nerves	Extend vertebral column and bend it to one side

Table 9-3. Transversospinalis

Muscle	*Origin*	*Insertion*	*Innervation*	*Action*
Semispinalis thoracis	Transverse processes of vertebrae T6–T10	Spinous processes of vertebrae C7–T4	Dorsal primary divisions of nerves C7–T4	Extend vertebral column and rotate it to opposite side
Semispinalis cervicis	Transverse processes of vertebrae T1–T6	Spinous processes of vertebrae C1–C5	Dorsal primary division of cervical spinal nerves	Extend vertebral column and rotate it to opposite side
Multifidus	Dorsal surface of sacrum, mamillary processes of lumbar vertebrae, transverse processes of thoracic vertebrae, articular processes of vertebrae C4–C7	Spinous processes from vertebra C2 to sacrum	Dorsal primary division of sacral, lumbar, thoracic, and cervical spinal nerves	Extend vertebral column and rotate it to opposite side
Rotatores	Transverse process of vertebra C1 to sacrum	Base of spinous process of vertebra above	Dorsal primary branches of all spinal nerves	Extend column and rotate it to opposite side
Interspinales	Spinous processes and interspinal ligaments	Contiguous spinous processes and ligaments	Dorsal primary branches of cervical, thoracic, and lumbar spinal nerves	Extend vertebral column
Intertransversarii anteriores cervicis	Anterior tubercle of cervical and lumbar vertebrae	Adjacent anterior tubercles	Dorsal primary division of lumbar spinal nerves	Bend vertebral column laterally
Intertransversarii posteriores cervicis	Posterior tubercles of cervical and lumbar vertebrae	Adjacent posterior tubercle	Dorsal primary division of cervical and lumbar spinal nerves	Bend vertebral column laterally
Intertransversarii laterales lumborum	Transverse processes of lumbar vertebrae	Adjacent transverse process of inferiorly placed lumbar vertebrae	Dorsal primary division of spinal nerves	Bend vertebral column laterally
Intertransversarii mediales lumborum	Accessory process of lumbar vertebrae	Mamillary processes of inferiorly placed lumbar vertebrae	Dorsal primary division of lumbar spinal nerves	Bend vertebral column laterally

from posterior branches of intercostal arteries. Dorsal branches of lumbar and iliolumbar arteries give rise to the lumbar spinal arteries. Finally, the lateral sacral artery supplies the segmental arteries to the sacral region.

VEINS

The veins of the back form external and internal vertebral venous plexuses that freely anastomose with each other and are devoid of valves, so that blood may flow in either direction. The vertebral venous plexuses also connect the intracranial sinuses with the veins of the neck, thorax, abdomen, and pelvis, meaning that free venous blood flow may occur between the skull and these regions, depending on pressure differences. This is of clinical importance.

The **internal vertebral venous plexus** lies in the vertebral canal and is organized like the arteries, with osseous branches and neural branches draining the bones and spinal cord. The anterior internal plexus, which is lateral to the posterior longitudinal ligament, and the posterior internal plexus are connected by a series of venous rings. The vertebral bodies are drained by **basivertebral veins,** which enter the internal vertebral plexus. The intervertebral spinal veins exit via the intervertebral foramen and connect with the external vertebral plexus.

The **external vertebral plexus** consists of anterior and posterior external plexuses that freely anastomose with each other. The anterior external plexus lies in front of the bodies of the vertebrae and receives venous blood from the basivertebral and intervertebral veins and also from branches within the vertebral bodies. The posterior external plexus is found on the posterior surface of the vertebrae and in the deep dorsal muscles. The external plexus anastomoses with vertebral, occipital, and deep cervical veins. The external branches also drain into the intervertebral veins, which then drain into vertebral, lumbar, or sacral veins.

Innervation

The nerve supply to the back originates from dorsal rami of the spinal nerves and from the meningeal branch of each spinal nerve. The **meningeal branches** arise after the spinal nerves exit the intervertebral foramen. The branch reenters the vertebral canal through the foramen and supplies sensory innervation to the vertebrae and their ligaments and the meninges. Postganglionic sympathetic innervation to the vessels is also provided via meningeal branches. The meningeal branches from the first and third cervical nerves ascend through the foramen magnum and supply the dura on the anterior portion of the posterior cranial fossa.

DORSAL RAMI

The dorsal rami provide motor, sensory, and sympathetic innervation to the back. The dorsal rami, with the exception of the first cervical and the fourth and fifth sacral nerves, divide into lateral and medial rami. Each dorsal ramus anastomoses with nerves from segments above and below. In the cervical region, except for dorsal rami from the first and second cervical nerves, each dorsal ramus passes dorsally, medial to the intertransversarii muscles, and reaches the space between the semispinalis capitis and cervicis.

Cervical Dorsal Rami

The dorsal ramus of the first cervical nerve is called the **suboccipital nerve** and supplies the muscles in the suboccipital triangle and the semispinalis capitis muscle and has a small cutaneous branch. The dorsal rami of the first to fourth cervical nerves are interconnected, forming the posterior cervical plexus.

The dorsal ramus of the second cervical nerve is longer than the ventral ramus. The medial branch is large and is called the **greater occipital nerve.** This nerve ascends between the inferior oblique and semispinalis capitis muscles, pierces the semispinalis capitis and the trapezius muscles, and then becomes subcutaneous. It joins with the occipital artery to distribute to the skin up to the vertex of the skull. The lateral branch is smaller than the medial branch and innervates muscles along the vertebral column.

The medial branch of the dorsal ramus of the third cervical nerve also has a prominent continuation, the **third occipital nerve,** that pierces the trapezius and innervates the skin on the back of the head; the lateral branch supplies the intrinsic muscles in the vertebral column. The dorsal rami of the first and the sixth to eighth cervical nerves have no cutaneous branches to the back; therefore over the back the fifth cervical nerve becomes continuous with the first thoracic nerve.

Thoracic Dorsal Rami

All the thoracic dorsal rami have medial and lateral branches. The medial and lateral cutaneous branches are separated by portions of the longissimus thoracis muscle. The medial branches supply the erector spinae and the muscles, bones, and ligaments of the vertebral column. (The first to third thoracic nerves also have cutaneous branches.) The lateral branches supply the longissimus and iliocostalis and have an extensive cutaneous distribution down to the gluteal region.

Lumbar and Sacral Dorsal Rami

The medial branches of the lumbar and sacral dorsal rami supply the multifidus spinae. The lateral branches of the lumbar dorsal rami supply the skin over the buttock (superior cluneal nerves). The lateral branches of the lower lumbar dorsal rami and those from the first to fourth sacral nerves form the dorsal sacral plexus. From this dorsal sacral plexus the middle cluneal nerves supply the skin over the gluteus maximus.

VENTRAL RAMI

The ventral rami supply the ventral and lateral part of the trunk and limbs and form the major nerve plexuses, including the cervical, brachial, and lumbosacral plexuses (see separate discussions in preceding chapters).

Spinal Cord

The spinal cord lies within the vertebral column and is continuous superiorly with the medulla of the brain stem. Inferiorly, at the lower border of the first lumbar vertebra, it forms the conus medullaris. At birth the lower end of the cord lies at the third lumbar vertebra. During childhood, as the vertebral column elongates, the cord recedes until it reaches the adult position, ending at the disk between the first and second lumbar vertebrae.

MENINGES

The central nervous system, including the spinal cord, is enclosed by three protective membranes collectively called the meninges. The meninges consist of the dura, arachnoid, and pia mater. The **dura mater** is the outer thick layer. The middle layer, the **arachnoid mater,** is a thin connective tissue membrane. The inner layer, the **pia mater,** is attached to the brain and spinal cord.

Dura Mater

When a laminectomy (surgical removal of the vertebral laminae) is performed the dura mater can be seen, and the extradural space, which lies between the vertebrae and the dura, contains blood vessels and fat. The exposed blood vessels are the anterior and the internal vertebral plexuses of veins. At the foramen magnum the dura is continuous with the periosteum of the occipital bone. The dura continues to enclose the cord inferiorly to the second sacral space, where it tapers to a strand of connective tissue, the filum terminale externum, that finally attaches to the inner surface of the coccyx. The dura also extends over the dorsal and ventral roots of the spinal nerves and fuses at the dorsal root ganglion with the epineural sheath on the peripheral nerve. Strands connecting the dura with the posterior longitudinal ligament are also seen. The narrow subdural space between the dura and arachnoid maters is filled wth a lymphlike fluid.

Arachnoid Mater

The arachnoid mater of the spinal cord is continuous with the arachnoid membrane covering the brain. It also sends lateral prolongations to fuse with the coverings of the nerves at the level of the ganglia. The space between the arachnoid and the pia maters is the subarachnoid space, which is filled with the cerebrospinal fluid.

Pia Mater

The pia mater adheres to the spinal cord and is continuous with the pia of the brain. It also continues over the dorsal and ventral rootlets to fuse with the connective tissue coverings of the peripheral nerves at the level of the spinal ganglia. Processes of the pia also extend through the subarachnoid space to attach to the inner surface of the dura between the spinal nerves as the **dentate ligaments,** which are toothlike in appearance. The pia at the lower level of the cord continues as a thin strand of connective tissue, the filum terminale internum, that blends with the filum terminale externum.

Thus, the cord is supported laterally by the dentate ligament and nerve rootlets, inferiorly by the filum terminale, and superiorly by its continuation with the brain stem. The spinal cord is surrounded by the fluid-filled subarachnoid space.

NERVOUS TISSUE

The spinal cord lies within the vertebral canal, continuous superiorly with the medulla of the brain stem. Inferiorly, at the border of the first and second lumbar vertebrae, the lower end of the spinal cord gray and white matter tapers, ending as the **conus medullaris.** Approximately 32 pairs of nerves leave the spinal cord: 8 cervical, 12 thoracic, 5 lumbar, 5 sacral, and 1 to 2 coccygeal (Fig. 9-10). The roots of each segment are formed by many fine strands, and the zone where each set of strands originates marks a spinal cord segment. The spinal cord gray and white matter stops at the first lumbar vertebra,

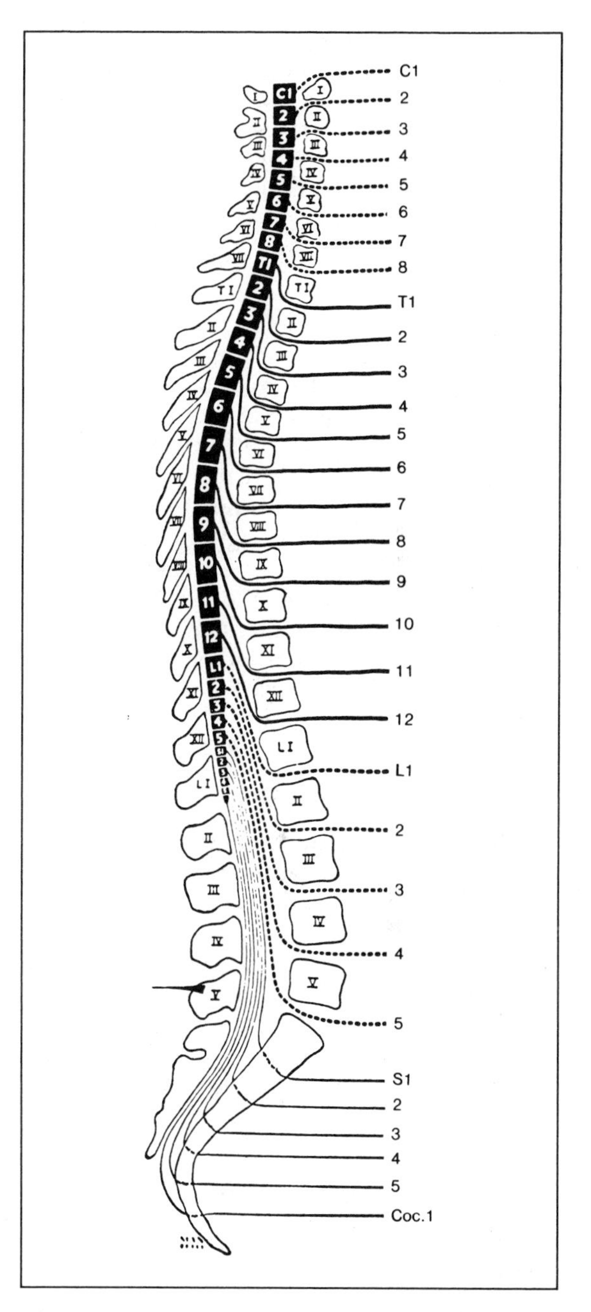

Fig. 9-10. The relation of the segments of the spinal cord and spinal nerves to the vertebral column. (From M. Carpenter and J. Sutin, *Human Neuroanatomy* [8th ed.]. Baltimore: Williams & Wilkins, 1983. With permission.)

but the spinal nerves continue inferiorly to their correct intervertebral foramina. The spinal cord gray matter is centrally located and surrounded by the white matter. The rootlets for lower lumbar, sacral, and coccygeal nerves have an especially long course within the dura and in appearance form a horse's tail, the cauda equina.

Rootlets

The first cervical rootlet exits between the skull and the atlas, the second between the atlas and the axis. Because there are seven cervical vertebrae and eight cervical nerve rootlets, the cervical nerves are numbered according to the vertebrae they exit above, with the eighth cervical rootlet exiting above the first thoracic vertebra. All the other rootlets exit below the vertebra with the same number. The ganglia are also located in the proper intervertebral foramina. An area of skin innervated by a single spinal cord segment is called a dermatome (see Fig. 1-16).

LUMBAR PUNCTURE

At the lower end of the vertebral canal from the second lumbar to the second sacral vertebra only spinal nerves are found in the subarachnoid space. This is the simplest area in which to tap the cerebrospinal fluid with a large needle to determine the existence of hemorrhage, tumors, increased intracranial pressure, or other disease processes. The needle passes through the skin, fascial layers, tendon of the erector spinae, ligaments, dura, and arachnoid.

National Board Type Questions

Select the one best response for each of the following:

1. With regard to the spinal cord and spinal nerves,
   A. The eighth cervical nerve exits between the eighth and ninth vertebrae.
   B. The dorsal roots contain *only* sensory afferent fibers originating in the dermatomes covering the back.
   C. The dorsal and ventral roots join to form the dorsal root ganglion.
   D. The cauda equina is composed of dorsal and ventral roots lying caudal to the conus medullaris.
   E. There are no sympathetic fibers in the thoracic ventral roots.
2. Concerning the spinal cord,
   A. Lumbar puncture can be performed at L1 without fear of damaging the spinal cord.
   B. The denticulate ligaments are extensions of dura mater that stabilize the spinal cord laterally.
   C. The cerebrospinal fluid surrounds the cord, occupying the subdural and subarachnoid spaces.
   D. At birth the lower end of the spinal cord lies at the level of L3.
   E. The extradural (epidural) space is devoid of blood vessels.

Select the response most closely associated with each numbered item.

A. Dorsal primary rami of spinal nerves
B. Ventral primary rami of spinal nerves
C. Both
D. Neither

The following are innervated entirely or chiefly by which nerves?

3. Iliocostalis
4. Latissimus dorsi
5. Trapezius

For the following, select

A. if only *1, 2, and 3* are correct
B. if only *1 and 3* are correct
C. if only *2 and 4* are correct
D. if only *4* is correct
E. if *all* are correct

6. Regarding the anatomy of the spine,
   1. In the cervical region, flexion and extension are freest.

2. In the lumbar region, lateral bending and rotation are prevented.
3. A thoracic vertebra has more articular facets than a cervical or a lumbar vertebra.
4. The anterior longitudinal ligament is usually damaged in a hyperflexion injury.

7. Concerning the exits of cervical nerves,
   1. Nerve C4 exits inferior to vertebra C4.
   2. Nerve C1 exits superior to vertebra C1.
   3. Nerve C7 exits between vertebrae C7 and C8.
   4. Nerve C8 exits inferior to vertebra C7.
8. The spinal dura mater
   1. Is continuous with the cranial dura mater
   2. Is bathed on both sides with cerebrospinal fluid
   3. Blends with the epineurial sheath of spinal nerves
   4. Forms the filum terminale internum
9. Blood supply to the spinal cord
   1. Enters the vertebral canal via the intervertebral foramina
   2. Remains external to the dura
   3. Is derived in the sacral region from the lateral sacral artery
   4. Is not affected if the vertebral arteries are occluded
10. Regarding lumbar puncture,
   1. Cerebrospinal fluid is not found caudal to L2.
   2. The needle will puncture the skin, fascia, dura, and arachnoid.
   3. The posterior longitudinal ligament may block penetration.
   4. The superior border of the iliac crests marks the level of the L4.

Annotated Answers

1. D. By convention the cervical nerves exit cranial to their respective vertebra and there are seven cervical vertebrae (a mammalian characteristic). Remember also that the dorsal roots (as opposed to *rami*) contain sensory fibers from the entire body.
2. D. The spinal cord in the adult ends just caudal to L1, but at birth can be found as low as L3.
3. A.
4. B. Although located "on the back" the latissimus is attached to the upper limb and acts directly on it and therefore is innervated by ventral primary rami.
5. C. The trapezius has a complex innervation. It has motor input from the spinal accessory (eleventh cranial) nerve but also some sensory (proprioceptive) input from cervical nerves (see Chap. 4).
6. B.
7. C.
8. B. The spinal dura is continuous with the cranial dura, is separated from the cerebrospinal fluid by the arachnoid membrane, and blends with the epineurium covering the spinal nerves as they emerge from the intervertebral foramen.
9. B. Arterial supply to the spinal cord enters at each intervertebral foramen and distributes by radicular branches to the substance of the cord. In the sacral region a major source of supply is off the lateral sacral artery.
10. C. A lumbar puncture, which must penetrate the skin, fascia, and arachnoid (do you understand why it *must* do so?) can be safely performed if it is remembered that the iliac crests can be used as markers for the fourth lumbar vertebra.

Index